STEVEN WEINBERG

Selected Papers

STEVEN WEINBERG

Selected Papers

Michael James Duff
Imperial College London, UK

World Scientific
NEW JERSEY • LONDON • SINGAPORE • BEIJING • SHANGHAI • HONG KONG • TAIPEI • CHENNAI • TOKYO

Published by

World Scientific Publishing Co. Pte. Ltd.
5 Toh Tuck Link, Singapore 596224
USA office: 27 Warren Street, Suite 401-402, Hackensack, NJ 07601
UK office: 57 Shelton Street, Covent Garden, London WC2H 9HE

Library of Congress Control Number: 2022941751

British Library Cataloguing-in-Publication Data
A catalogue record for this book is available from the British Library.

First published 2023 (hardcover)
Reprinted 2024 (in paperback edition)
ISBN 978-981-12-9308-5 (pbk)

STEVEN WEINBERG
Selected Papers

ISBN 978-981-126-135-0 (hardcover)
ISBN 978-981-126-136-7 (ebook for institutions)
ISBN 978-981-126-137-4 (ebook for individuals)

For any available supplementary material, please visit
https://www.worldscientific.com/worldscibooks/10.1142/12995#t=suppl

Desk Editor: Joseph Ang

Printed in Singapore

Preface

Steven Weinberg was my scientific hero. So, when KK Phua and Lars Brink invited me to edit a selection of his papers for World Scientific, my first reaction was to welcome it with open arms, and think myself fortunate to have been granted such an honour. But then followed the awesome realization of the heavy responsibility I had taken on. After all, during much of his lifetime Steve was regarded by many as the greatest living scientist. Doing him justice in the choice of papers and commentaries seemed a daunting task. There was no point in dwelling on this for too long, however; I just had to get on with it. So I decided to begin with some personal recollections as in Chapter 1 and to devote Chapter 2 to a brief biography.

Still, the main purpose of this volume is the selection of Steven Weinberg's prominent papers in Chapter 3. Of the 281 papers ordered last to first in the Complete List of Publications, I have selected 37 grouped into six themes: (1) Current Algebra and Effective Lagrangians, (2) The Standard Model, (3) Symmetries, (4) Gravity and Asymptotic Safety, (5) Cosmology and the Multiverse, (6) Popular Articles. This last theme also includes two popular extracts [3.6.1, 3.6.2]. Each of the themes is accompanied by a commentary designed to place the articles in their scientific and historical contexts.

Our aim is to pay tribute to Weinberg's qualities not only as a Nobel-Prize winning physicist, but also as a man, as a teacher, as a communicator of science to the general public and as a friend. We hope we have done him justice.

Michael Duff, Editor

Contents

1
Personal Recollections

One day in 1976 I picked up the phone in my office at King's College London only to hear those five words most designed to instil fear and trembling into the heart of a young postdoc: "Hi, this is Steven Weinberg". Steve had been reading one of my papers and wanted to know whether I had made a sign error. I spent the weekend sweating over my calculations. Anyone who has chased a minus sign will appreciate my discomfort especially when the person waiting at the other end is Steven Weinberg who, for all his qualities, was not famous for suffering fools gladly. Anyway, I stuck to my guns and Steve graciously wrote me a letter (those were the days!) to say he had done the calculations himself and reached the same conclusion. We kept in touch sharing our common interests in the quantum theory of gravity and in 1978–9, while a postdoc with Stanley Deser at Brandeis, I also had desk space at Harvard and talked some more with Steve. See Fig. 1. At the time I was a heavy smoker and was surprised that, even though a non-smoker himself, he did not mind my lighting up in his office. How times have changed.

In 1982 I was pleased to be invited by Steve and his colleague John Wheeler to spend a semester at the University of Texas at Austin. There was a tremendously stimulating atmosphere there and I regard this period as one of the most productive of my career, working with Chris Pope and Bengt Nilsson on attempts to find an eleven-dimensional unification of the fundamental forces. This was made all the more pleasant by the kind hospitality of Steve and his wife Louise (herself a distinguished legal scholar) at their home in Austin. On another occasion, at the 1984 Jerusalem Winter School, Steve, Louise and their daughter Elizabeth invited Christof Wetterich and me to join them for a New Year's Eve celebration at a night club. An unexpected bonus was Steve's lifelike impersonation of a chicken when doing the Chicken Dance (popular in 1984).

The famous Shelter Island Conference of 1947 laid the foundations of Quantum Electrodynamics and much else besides, and in 1983 Nick Khuri organized Shelter Island II [124] to bring together the surviving participants from 1947: Hans Bethe, Dick Feynman, Willis Lamb, Robert Marshak, Abraham Pais, Linus Pauling, Isidor Rabi, Robert Serber, Victor Weisskopf, John Wheeler, and others of that era: Murray Gell-Mann, Tsung-Dao Lee, Yoshio Nishina, Fred Seitz, as well leading figures in 1983: Stephen Adler, Stanley Deser, Stephen Hawking, Roman Jackiw, Toichiro Kinoshita, Yoichiro Nambu, Alexander Polyakov, John Schwarz, Silvan Schweber, Gerard 't Hooft, Steven Weinberg, Bruno Zumino and also some younger physicists making a name for themselves: Michael Duff, Alan Guth, Andre Linde, Peter West and Edward Witten. I think I had Steve to thank for my inclusion in this august list of speakers.

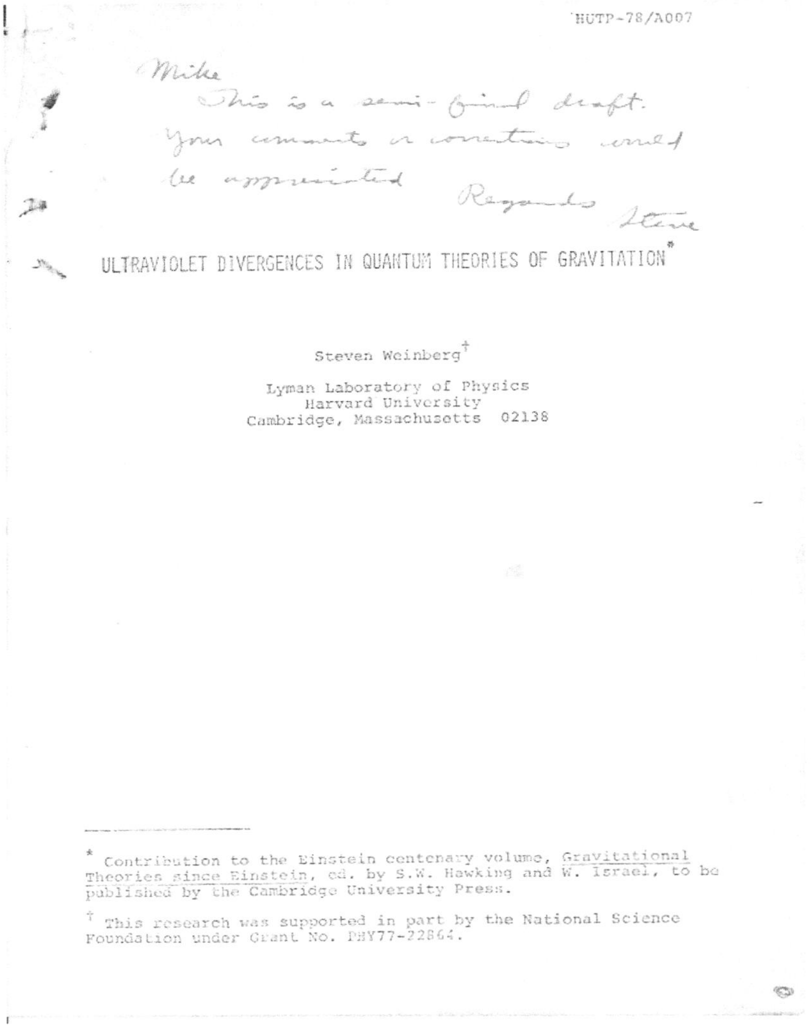

HUTP-78/A007

Mike
This is a semi-final draft.
Your comments or corrections would
be appreciated
Regards Steve

ULTRAVIOLET DIVERGENCES IN QUANTUM THEORIES OF GRAVITATION*

Steven Weinberg†

Lyman Laboratory of Physics
Harvard University
Cambridge, Massachusetts 02138

* Contribution to the Einstein centenary volume, Gravitational Theories since Einstein, ed. by S.W. Hawking and W. Israel, to be published by the Cambridge University Press.

† This research was supported in part by the National Science Foundation under Grant No. PHY77-22864.

Fig. 1. The Einstein Centenary Volume: handwritten request from Steve 1979.

When it was Steve's turn to speak, the light went out on his viewgraph. Hans Bethe, who was sitting in the front row, got up to help, then T. D. Lee, then Dick Feynman, then Linus Pauling, each of them convinced they knew better than the others how to fix the problem. Pretty soon the leading figures of twentieth century physics were posing the question: How many Nobel prizewinners does it take to change a lightbulb? Someone in the audience saw this as a golden photo opportunity and raised their camera but Steve spotted them from the corner of his eye and nimbly stepped aside, thus preserving his dignity when all about him were losing theirs.

I reminded Steve of this episode when giving the vote of thanks for his Imperial College lecture [16] in 2014 celebrating the 80th birthday of Tom Kibble. I remarked that although his physical ability was not what it used to be, his lecture had demonstrated that his mind was as nimble as ever. For example, he reminded us of a 1967 paper by Tom showing that when a gauge group G breaks spontaneously to a subgroup H, the gauge bosons associated with G/H acquire a mass via the Brout-Englert-Higgs-Kibble-Guralnik-Hagen mechanism, while those associated with H remain massless. So when the Standard Model came along this explained not only why the W and the Z are massive but, equally crucial, why the photon is not: "Tom Kibble showed us why light is massless".

From my time in Austin, I remember Steve telling me that his favourite restaurant was Rules, the oldest in London, so my wife Lesley and I took him and Louise to dinner there along with Tom. It was also the only time I got to hail a cab and say "The Ritz" (Steve declined the accommodation offered by Imperial because he and Louise had always wanted to stay at the Ritz and he thought this might be their last opportunity to do so, as indeed it was).

Towering intellect though he was, Steve was not your archetypal absent-minded professor. You would not catch Steven Weinberg wearing odd socks. On the contrary, as the son of a New York taxi-driver, he was always pretty street-smart, an ability that he would bring to bear whether he was answering questions in physics seminars, or testifying before congressional committees. He made heroic but ultimately unsuccessful attempts to convince the House Committee on Science, Space and Technology not to abandon the Superconducting Supercollider (SSC) in Waxahachie, Texas, where 23 km of an 81 km circular tunnel had already been bored and two billion dollars already spent.

He was also one of world's most famous and articulate atheists. I remember once taking him to task when he said [3.5.3]:

Just as Darwin and Wallace explained how the wonderful adaptations of living forms could arise without supernatural intervention, so the string landscape may explain how the constants of nature that we observe can take values suitable for life without being fine-tuned by a benevolent creator.

Suppose the universe were unique, what then? He responded by saying that if there was really only one universe and it was really finely-tuned to support beings like us then maybe the believers had a point! I protested: Does the existence of God really rest on the outcome of some future investigations in theoretical physics? Steve jokingly complained that this was the first time he had been criticised for not being atheist enough!

Fig. 2. Participants of The 2nd Jerusalem Winter School in Theoretical Physics in 1984. Weinberg is in the front row. To Weinberg's right: me and Raoul Bott; to his left: Yuval Ne'eman and Tsvi Piran. Photo courtesy of Tsvi Piran.

2
Brief Biography

Fig. 3. A photograph of Steven Weinberg. Photo courtesy of Louise Weinberg.

1933 Born in New York City

1950 Graduated, Bronx High School

1954 Undergraduate degree at Cornell

1955 Institute for Theoretical Physics, Copenhagen

1957 PhD at Princeton under Sam Treiman on renormalisation of strong and weak interactions

1957–59 Columbia

1959–66 Berkeley, working on Feynman graphs, weak interactions, broken symmetries

1966–69 Loeb Lecturer at Harvard and Visiting Professor at MIT

1969–73 Professor at MIT

1973–82 Higgins Professor of Physics at Harvard

1977 Dannie Heineman Prize for Mathematical Physics

1979 Nobel Prize in Physics

1981 Elected Foreign Member of The Royal Society

1982–2021 Jack S. Josey-Welch Foundation Chair in Science and Regental Professor of Physics, University of Texas at Austin

1991 National Medal of Science

2020 Breakthrough Prize in Fundamental Physics

He met his wife, Louise, while at Cornell and they were married in 1954. Their daughter, Elizabeth, was born in 1963.

With the death of Steven Weinberg in July 2021, the physics community lost its most admired member, its staunchest defender and its most eloquent expositor. Hans Bethe once said that there are two kinds of genius: the "orthodox" and the "magicians". I think Steven Weinberg and my mentor Abdus Salam personify these two species. Salam's sources of inspiration were enigmatic. They had an air of eastern mysticism that left you wondering how to fathom his genius. But there was nothing mystic about Steven Weinberg. His ideas were presented with such devastating logic and clarity they left you feeling "Well I could have done that too! (if only I were smart enough)". He was fond of saying *As is natural for an academic, when I want to learn something, I volunteer to teach a course on the subject.* This resulted in several classic textbooks including *Lectures on Quantum Mechanics*, *Foundations of Modern Physics*, *The Quantum Theory of Fields* and *Cosmology*.

His wife Louise writes:

"I don't mean to flatter myself, but Steven Weinberg's relation to me was rather like Marc Antony's to Cleopatra at the battle of Actium. Not that I am Cleopatra, but he was certainly a Marc Antony. With me in Texas, at an inconvenient distance from Massachusetts, after two years during which he spent too much time in the Atlanta airport, he sought a visiting appointment for a semester during the third year and joined the Physics and Astronomy Departments here at Texas the following year. When he and I taught on the same day he would always take me out to lunch, and then, depositing me at my car parked at the Law School, would drive home behind me all the way, to be sure I got home safely. Sometimes it was hard for me to see how he could give up his happy life in Cambridge. Fortunately there was already a strong group in physics here at U.T., and a strong group of astronomers associated as well with the McDonald Observatory, which with its new hexagonal mirrors and rotation at a seated angle would eventually furnish the prototype for the James Webb. Funds were found for him with which he could form his now celebrated Theory Group. And of course U.T. is a great university. So it wasn't all just Cleopatra. Life in Austin has been enormous fun, and we have made many dear friends in all walks of life. And once more we could work at home together, in our respective home offices, the daily life he loved."

His daughter Elizabeth writes:

"The world knows my father as a great physicist, but few know that he also was a great father. In an era when men typically were hands-off parents, my father took me to school every day, read physics journals at the playground and made me tuna salad. Most importantly, he was always there for me, always interested, always someone I could talk to. A scientist with a great hunger for knowledge, he was a model of learning. He talked with me about poetry, history and philosophy. He loved Shakespeare, could quote all the plays, and also introduced me to science fiction. When we talked, he always asked, "What are you reading?" Although I am not a physicist, I suspect my father's work in physics was not separate from his deep humanism, his appreciation of great art, the majesty of the natural world and the unfolding drama of human history, and also his deep capacity for love."

3
Selected Papers with Commentaries

Fig. 4. A portrait of a younger Weinberg. Photo courtesy of Louise Weinberg.

3.1. Current Algebra and Effective Lagrangians

From my point of view, it started with current algebra.
Steven Weinberg

Joseph Thomson got the Nobel Prize for showing the electron is a particle; his son George Thomson got the Nobel Prize for showing the electron is a wave. Both were correct. A similar Jekyll/Hyde duality appears when comparing Steven Weinberg's two most significant contributions: Effective Lagrangians and Electroweak Unification (for which he shared the Nobel Prize with Glashow and Salam).

Following the success at Shelter Island I in formulating a renormalizable theory of quantum electrodynamics, theorists were able to do precise calculations that agreed with experiment. *Renormalizability* is the requirement that all the ultraviolet divergences may be absorbed into a redefinition of the masses, charges and fields appearing in the classical Lagrangian. What appealed to Weinberg was that renormalization didn't work unless coupling constants were dimensionless, which kept everything simple. The conviction that the weak and strong interactions should also be described by such a simple framework ultimately led to the Standard Model of particle physics, which augments the Glashow-Salam-Weinberg description of the weak and electromagnetic forces with the quantum chromodynamics (QCD) description of the strong force: See Section 3.2. Thus the Standard Model was a triumph for renormalizability:

Dr Jekyll [3.1.7]: *If you had asked me in the mid-1970s about the shape of future fundamental physical theories, I would have guessed that they would take the form of better, more all-embracing, less arbitrary, renormalizable quantum field theories.*

What about Effective Lagrangians? According to Weinberg their origins lie in *Current Algebra*, which focussed on the currents of the weak interactions, the vector and axial vector currents, using their commutation relations and their conservation properties. But Weinberg thought there must be a simpler way to derive useful results such as the Goldberger-Treiman formula for the pion decay amplitude and the Adler-Weisberger sum rule for the axial vector coupling constant. He was able to construct an effective Lagrangian for soft-pion interactions, incorporating chiral $SU(2) \times SU(2)$, such that lowest order perturbation theory precisely reproduces the results of current algebra [3.1.2], in particular for pion scattering lengths [3.1.1]. So we have grouped Current Algebra and Effective Lagrangians together in this section. In the linear approach, such as the linear σ-model, the pions transform according to a linear representation of $SU(2) \times SU(2)$, but in the nonlinear approach, such as the nonlinear σ-model, they transform according to a *nonlinear realization* [3.1.3], an idea that Weinberg attributes [3.1.4] to Julian Schwinger.

Mr Hyde [3.1.7]: *The essential point in using an effective field theory is you're not allowed to make any assumption of simplicity about the Lagrangian. Certainly you're not allowed to assume renormalizability. Such assumptions might be appropriate if you were dealing with a fundamental theory, but not for an effective field theory, where you must include all possible terms that are consistent with the symmetry. The thing that makes this procedure useful is that although the more complicated terms are not excluded because they're non-renormalizable, their effect is suppressed by factors of the ratio of the energy to some fundamental energy scale of the theory. Of course, as you go to higher and higher energies, you have more and more of these suppressed terms that you have to worry about.*

Apart from spontaneous symmetry breaking and quark confinement, the Standard Model is not so very different from Quantum Electrodynamics. It was the same old trick: a renormalizable relativistic quantum field theory where the fields in the Lagrangian were intended to describe elementary particles to arbitrarily high energies. But starting in 1976, and influenced by the work of Kenneth Wilson, Weinberg came to see the Standard Model as a low-energy approximation to a (at present unknown) fundamental theory [3.1.8]. And low energy means energies much less than some extremely high energy scale 10^{15} to 10^{18} GeV. As a low energy approximation we expect non-renormalizable corrections. For example, in contrast with the renormalizable terms they can violate the symmetries of baryon number conservation or lepton number conservation [3.3.5]. Since they are suppressed by inverse powers of the energy scale, however, these symmetries would be approximately conserved. See Section 3.3.

This thinking was based on a "folk theorem" that, assuming we have the right degrees of freedom, if we write down all possible invariant terms in the Lagrangian, and we work to all orders in perturbation theory, the result can't be wrong, because it's just a way of implementing the fundamental principles of Lorentz invariance, chiral symmetry, quantum mechanics, cluster decomposition, and unitarity.

Jekyll/Hyde duality [3.1.9]:

In every order of perturbation theory as you encounter more and more loops, because you are allowing more and more powers of energy, you get more and more infinities, but there are always counterterms available to cancel the infinities. A non-renormalizable theory, like the soft pion theory, is just as renormalizable as a renormalizable theory. You have an infinite number of terms in the Lagrangian, but only a finite number are needed to calculate S-matrix elements to any given order in energy.

More recently, Weinberg used the method of phenomenological Lagrangians [165] to derive the consequences of spontaneously broken chiral symmetry for the forces among pions and nucleons [3.1.5, 3.1.6].

Finally, and perhaps most importantly, we nowadays regard the Einstein-Hilbert Lagrangian as just the low energy approximation to a more fundamental theory of gravity as discussed in Section 3.4.

VOLUME 17, NUMBER 11 PHYSICAL REVIEW LETTERS 12 SEPTEMBER 1966

PION SCATTERING LENGTHS*

Steven Weinberg
Department of Physics, University of California, Berkeley, California
(Received 20 June 1966)

The current commutation relations[1] and partially conserved axial-vector current (PCAC) assumption[2,3] allow the calculation of the matrix elements for emission and absorption of any number of soft pions[4] and, therefore, in particular, determine the scattering length of a pion on any target particle. In this note we give a simple formula for pion scattering on any particle but a pion,[5] and then extend this result to the more difficult case of pion-pion scattering.

Calculations of soft-pion matrix elements may be conveniently performed in three distinct steps:

Step I.—The S matrix is extended off the mass shell, using a pion field defined as proportional to the divergence of the axial-vector current. In our case we define the off-mass-shell invariant pion scattering amplitude $\langle f, qb|M|i,ka\rangle$ by

$$\int d^4x\, d^4y \langle f|T\{\partial_\mu A_b{}^\mu(x), \partial_\nu A_a{}^\nu(y)\}|i\rangle e^{-iqx}e^{iky} \equiv \frac{i(2\pi)^4\delta^4(p_f+q-p_i-k)F_\pi{}^2 m_\pi{}^4}{(q^2+m_\pi{}^2)(k^2+m_\pi{}^2)(2\pi)^3(4E_iE_f)^{1/2}}\langle f, qb|M|i,ka\rangle, \quad (1)$$

where k^μ and q^μ are the initial and final pion momenta, a and b are the initial and final pion isovector indices (running over 1, 2, 3), i and f label the initial and final states of the target particle, $A_a{}^\mu(x)$ is the axial-vector current, and F_π is the pion-decay amplitude, defined by

$$\langle 0|\partial_\nu A_a{}^\nu(0)|\pi_{qb}\rangle \equiv F_\pi m_\pi{}^2 \delta_{ab}(2q^0)^{-\frac{1}{2}}(2\pi)^{-\frac{3}{2}}. \quad (2)$$

VOLUME 17, NUMBER 11 PHYSICAL REVIEW LETTERS 12 SEPTEMBER 1966

Note that Eq. (1) is a definition, not a theorem or an assumption, but that the Lehmann-Symonzik-Zimmerman (LSZ) formalism[6] shows rigorously (and without invoking PCAC) that the S matrix is given in terms of M on the mass shell, by

$$\langle f,\vec{q}b|S|i,\vec{k}a\rangle = \frac{-i(2\pi)^4\delta^4(p_i+k-p_f-q)}{(2\pi)^6(16q^0k^0E_iE_f)^{1/2}} \times[\langle f,qb|M|i,ka\rangle]_{q^2=k^2=-m_\pi^2}. \quad (3)$$

Step II.—The current commutation relations are used to prove an exact theorem about the behavior of the matrix element in the limit of vanishing pion four-momenta. In our case it is convenient to fix $p_i^\mu = p^\mu$, and let q^μ and k^μ go to zero together, so that $p_f^\mu \to p^\mu$. [Since $p_f^2 = -m_\pi^2$, we must require that $p\cdot k = p\cdot q$ to first order, but we do not necessarily take $q^\mu = k^\mu$.] The commutation rules used here are those suggested by the σ model[3,7] and the free-quark model:

$$\delta(x^0-y^0)[A_a^{\ 0}(y),A_b^{\ \mu}(x)] = 2ig_V\epsilon_{abc}V_c^{\ \mu}(x)\delta^4(x-y)+\text{S.T.}, \quad (4)$$

$$\delta(x^0-y^0)[A_b^{\ 0}(x),\partial_\nu A_a^{\ \nu}(y)] = ig_V\sigma_{ab}(x)\delta^4(x-y)+\text{S.T.}, \quad (5)$$

where $V_c^\mu(x)$ is the vector current, $\sigma_{ab}(x)$ is some scalar field which may or may not have something to do with a real 0^+ π-π resonance or enhancement, and "S.T." means possible Schwinger terms. It will be assumed that the Schwinger terms are either c numbers, which do not contribute at all to the connected part of M, or, if operators, involve gradients which kill their contribution to first order in q and k. Our theorem states that, as q^μ and k^ν vanish, the connected part of M approaches

$$\langle f,qb|M|i,ka\rangle \to M_{fb,ia}^{(0)} - 8(g_V/F_\pi)^2(p\cdot q)(T_\pi)_{ba}\cdot(T_t)_{fi} + \text{poles} + O(qq,qk,kk), \quad (6)$$

where $M^{(0)}$ is an unknown constant proportional to $\langle f|\sigma_{ab}(0)|i\rangle$, with $p_f = p_i = p$, and T_π and T_t are the pion and target isospin matrices, with $(T_{\pi c})_{ba} = i\epsilon_{abc}$. The "poles" in Eq. (6) are to be evaluated from the Born terms in gradient coupling theory[4]; for instance, there are no poles in π-π, π-K, or π-Λ scattering, while for π-N scattering the pole terms in Eq. (6) are

$$\text{"poles"} = \left(\frac{g_A}{F_\pi}\right)^2\left(\frac{m_N}{p\cdot q}\right)\bar{u}_f[\gamma_5 \not{q}(-i\not{p}+m_N)\gamma_5\not{k}\tau_b\tau_a + \gamma_5\not{k}(-i\not{p}+m_N)\gamma_5\not{q}\tau_a\tau_b]u_i. \quad (7)$$

The proof follows standard lines. The left-hand side of Eq. (1) is identically equal to

$$\int d^4x\,d^4y\,e^{iky}e^{-iqx} \times\{-\delta(x^0-y^0)\langle f|[A_b^{\ 0}(x),\partial_\nu A_a^{\ \nu}(y)]|i\rangle - iq_\mu\delta(x^0-y^0)\langle f|[A_a^{\ 0}(y),A_b^{\ \mu}(x)]|i\rangle + q_\mu k_\nu\langle f|T\{A_b^{\ \mu}(x),A_a^{\ \nu}(y)\}|i\rangle\}. \quad (8)$$

Using Eqs. (4) and (5) and the known matrix elements of $V_c^\mu(x)$ at zero momentum transfer, the three terms of Eq. (8) yield, respectively, the first three therms of Eq. (6). Note that the first term of Eq. (8) does not produce an additional first-order term in Eq. (6), because Eq. (5) shows that it depends only upon $p\cdot(q-k)$ and $(q-k)^2$, and $p\cdot(q-k)=0$. The pole terms may be identified as the total first-order contribution of the last term in Eq. (8).

Step III.—The exact theorem proved in Step II is used to estimate the matrix element on the mass shell. It is here that we must for the first time invoke PCAC, by which we mean that the M defined in Step I is as smooth a function of q^μ and k^ν as would be expected in a perturbation expansion, based on a Lagrangian field theory in which $\partial_\mu A_a^\mu(x)$ is proportional to the pion field. [The statement, that $\partial_\mu A_a^\mu(x)$ is proportional to the pion field, is by itself empty.] In our present context we will interpret this rather Delphic hypothesis as meaning that, if the pole terms in Eq. (6) are understood to include the poles near $q=k=0$ (such as the 3-3 resonance in π-N scattering) as well as those actually at $q=k=0$ (such as the N pole itself), then the coefficients of

VOLUME 17, NUMBER 11 PHYSICAL REVIEW LETTERS 12 SEPTEMBER 1966

the remaining quadratic terms in Eq. (6) are of order G_π^2/m_i^2, where m_i is some large internal mass, assumed to be of order m_N. Therefore, when the components of q^μ and k^ν in the rest frame of p^μ are of order m_π, the quadratic terms in Eq. (6) are of order $G_\pi^2 m_\pi^2/m_i^2$, while the Goldberger-Treiman relation shows that the linear terms are of order $G_\pi^2 m_\pi m_t/m_N^2$, where m_t is the mass of the target particle. Therefore, if the target mass is much larger than the pion mass, we may get a good approximation to the soft-pion S-matrix element by using Eq. (6) with quadratic and higher terms omitted. We can offer three arguments for also omitting the $M^{(0)}$ term:

(1) The Adler self-consistency argument[8] shows that M must vanish, except for poles, when $q^\mu=0$ and $k^2=-m_\pi^2$. Thus $M^{(0)}$ must be of the same order as the quadratic terms as this point, i.e., of order $G_\pi^2 m_\pi^2/m_i^2$, and is hence negligible. [This argument can be made more explicit by rearranging Eq. (8) to separate the one- and two-pion pole contributions to the last term, as was done in Ref. 4. We then find for M an expression like Eq. (6), but with $M^{(0)}$ multiplied by a factor $(q^2+k^2+m_\pi^2)/m_\pi^2$. The vanishing of this term at $q^\mu=0$, $k^2=-m_\pi^2$ is now automatic, and we see directly that if $M^{(0)}$ were large, it would contribute to M a rapidly varying function of q^μ and k^ν, in contradiction to the spirit of PCAC.][9]

(2) In the σ model,[3] $M^{(0)}$ is of order $G_\pi^2 m_\pi^2/m_\sigma^2$; this may be neglected if $m_\pi/m_t \ll (m_\sigma/m_N)^2$.

(3) The method of Ref. 4 can be used to show that, if $m_\pi=0$ and $\partial_\mu A_a{}^\mu=0$, then M obeys the limiting formula (6), but the $M^{(0)}$ absent. Hence $M^{(0)}$ may be regarded as arising only from the nonvanishing of the internal pion masses.

We still have the poles to consider, but these are generally absent in the s-wave part of the scattering amplitude. (This is true, for example, of the nucleon and 3-3 resonance poles in π-N scattering, the K^* poles in π-K scattering, etc.) Hence the $l=0$ part of the S matrix is given near threshold by just keeping the $p\cdot q$ term in Eq. (6), i.e.,

$$\langle f, qb|S|i, ka\rangle_{l=0} \cong [-i(g_V/F_\pi)^2/2\pi^2](T_\pi)_{ba}(T_t)_{fi} \times(-p\cdot q/m_\pi m_t)\delta^4(p_i+k-p_f-q).$$

The scattering length a_T is defined as $-2i\pi$ times the reduced mass times the coefficient of the δ function in S at threshold, so[5]

$$a_T=-L(1+m_\pi/m_t)^{-1}[T(T+1)-T_t(T_t+1)-2], \quad (9)$$

where T_t is the target isospin, T is the total isospin, and L is a convenient length, given by the Goldberger-Treiman relation as

$$L\equiv\frac{g_V^2 m_\pi}{2\pi F_\pi^2}\simeq\frac{G_\pi^2 m_\pi}{8\pi m_N^2}\left(\frac{g_V}{g_A}\right)^2=0.11 m_\pi^{-1}. \quad (10)$$

The reduced-mass correction $(1+m_\pi/m_t)^{-1}$ in Eq. (9) might well be omitted within the spirit of our approximations, but we keep it because it clearly arises from the definition of a_T.

For π-N scattering Eq. (9) gives

$$a_{1/2}=2L(1+m_\pi/m_N)^{-1}=0.20 m_\pi^{-1},$$
$$a_{3/2}=-L(1+m_\pi/m_N)^{-1}=-0.10 m_\pi^{-1}, \quad (11)$$

results which compare very well with the experimental values[10] $a_{1/2}m_\pi=0.171\pm0.005$ and $a_{3/2}m_\pi=-0.088\pm0.004$. Using the prediction that $a_{1/2}-a_{3/2}=3L(1+m_\pi/m_N)^{-1}$, together with the Goldberger-Miyazawa-Oehme sum rule[11] for $a_{1/2}-a_{3/2}$, we can obtain for (g_A/g_V) a sum rule, which differs from that of Adler and Weisberger[12] only in terms of order m_π^2/m_N^2; however, the sum rule is true only if the odd part of the forward scattering amplitude obeys an unsubtracted dispersion relation, while the derivation of the scattering lengths (11) made no assumption about the high-energy limit of any amplitude. The prediction that $a_{1/2}+2a_{3/2}$ is 0 may be regarded as a threshold version of Adler's self-consistency condition,[8] since we can easily see that if $M^{(0)}$ were non-negligible, $a_{1/2}+2a_{3/2}$ would be proportional to $M^{(0)}$. In whatever form we choose to write the predictions (11), their success probably rules out the presence of any strong low-energy π-π interaction, for our derivation would have failed if M contained a strong singularity in the t channel at a mass near $2m_\pi$.

Equation (9) can also be used to calculate π-K and π-hyperon scattering lengths, but none of these have been measured yet. A few pion-nuclear scattering lengths have been measured[13] and do not compare well with Eq. (9), but this is presumably because 140 MeV is such a high excitation energy for nuclei that we cannot regard a pion at threshold as soft; in particular,

VOLUME 17, NUMBER 11 PHYSICAL REVIEW LETTERS 12 SEPTEMBER 1966

Eq. (9) is real, while in fact pion annihilation makes a_T complex. This point is under further study.

We derived Eq. (9) under the assumption that the target is much heavier than a pion, so Eq. (9) cannot be used for pion-pion scattering. For instance, Eq. (9) gives a nonvanishing scattering length for $T=1$, in contradiction with Bose statistics. In order to calculate the π-π scattering lengths we will have to keep track of $M^{(0)}$ and the qq, qk, and kk terms, because at threshold they are here just as large as the pq term.

First note that crossing symmetry, isospin conservation, and Bose statistics require that the expansion of the off-mass-shell π-π scattering amplitude to second order in momenta takes the form[14]

$$\begin{aligned}\langle ld, qb\,|M|\,pc, ka\rangle &= \delta_{ab}\delta_{cd}[A+B(s+u)+Ct+\cdots]\\ &+\delta_{ad}\delta_{cb}[A+B(s+t)+Cu+\cdots]\\ &+\delta_{ac}\delta_{bd}[A+B(u+t)+Cs+\cdots],\end{aligned} \tag{12}$$

where

$$s\equiv -(p+k)^2,\quad t\equiv -(k-q)^2,\quad u\equiv -(p-q)^2.$$

Also A, B, and C are constant coefficients, and "$+\cdots$" denotes terms of fourth and higher order in the pion four-momenta p, k, l, and q. The crucial point about Eq. (12) is that there is no way that M can contain terms linear in the masses $-p^2$, $-l^2$, $-q^2$, $-k^2$, aside from the terms proportional to s, t, and u.

The PCAC assumption says that if M is defined in analogy with Eq. (1) as proportional to the Fourier transform of the vacuum expectation value of the time-ordered product of four axial-vector divergences, then the quartic and higher order terms in M are small when p^μ, l^μ, k^μ, and q^μ are of order m_π or less. The physical threshold is at $s=4m_\pi^2$, $t=u=0$, so that π-π scattering lengths are

$$a_0 \cong -(1/32\pi m_\pi)[5A+8m_\pi^2B+12m_\pi^2C], \tag{13}$$

$$a_2 \cong -(1/32\pi m_\pi)[2A+8m_\pi^2B]. \tag{14}$$

Equation (6) shows that when $p^\mu = l^\mu$ is on the mass shell and $q^\mu = k^\mu \to 0$, the matrix element approaches

$$\begin{aligned}\langle ld, qb\,|M|\,pc, ka\rangle &\to M_{db,ca}^{(0)} - 8(g_V/F_\pi)^2(p\cdot q)\\ &\times(\delta_{da}\delta_{bc}-\delta_{bd}\delta_{ac}).\end{aligned} \tag{15}$$

(There are no poles here.) In this limit $t=0$, $s\to m_\pi^2-2p\cdot q$, and $u\to m_\pi^2+2p\cdot q$, so comparing Eq. (15) with Eq. (12) gives

$$B-C=4(g_V/F_\pi)^2, \tag{16}$$

$$\begin{aligned}M_{db,ca}^{(0)} &= \delta_{ab}\delta_{cd}[A+2m_\pi^2B]\\ &+(\delta_{ad}\delta_{bc}+\delta_{bd}\delta_{ac})[A+m_\pi^2C+m_\pi^2B].\end{aligned} \tag{17}$$

From Eqs. (13), (14), (16), and (10), we find[15]

$$2a_0-5a_2=6L=0.69m_\pi^{-1}. \tag{18}$$

The Adler self-consistency argument[8] shows that M vanishes when any one of the four pion momenta vanish and the other three are on the mass shell, i.e., $M=0$ when $s=t=u=m_\pi^2$, so

$$A=-m_\pi^2(2B+C). \tag{19}$$

In order to use this result to get another relation for the scattering lengths, it is necessary to add a little new physical information. If we assume that $\partial_\mu A_a^{\ \mu}$ is part of a chiral quadruplet along with an isoscalar field (as in the σ model or the free-quark model), then Eq. (5), and hence $M_{db,ca}^{(0)}$ is proportional to δ_{ba}, so Eq. (17) gives

$$A=-m_\pi^2(B+C). \tag{20}$$

From Eqs. (19) and (20) we have then

$$B=0,\quad A=-m_\pi^2C \tag{21}$$

which, with Eqs. (13) and (14), yields

$$a_0/a_2=-\tfrac{7}{2}. \tag{22}$$

Combining Eqs. (18) and (22), we find

$$a_0=(7/4)L=0.20m_\pi^{-1},\quad a_2=-\tfrac{1}{2}L=-0.06m_\pi^{-1}, \tag{23}$$

and the full low-energy π-π matrix element

VOLUME 17, NUMBER 11 PHYSICAL REVIEW LETTERS 12 SEPTEMBER 1966

is

$$\langle ld, qb|M|pc, ka\rangle = 4(g_V/F_\pi)^2\{[m_\pi^2-t]\delta_{ab}\delta_{cd} + [m_\pi^2-u]\delta_{ad}\delta_{bc} + [m_\pi^2-s]\delta_{ac}\delta_{bd}\}. \quad (24)$$

The striking feature of our result (23) is, of course, that a_0 is predicted to be very much smaller than anyone had thought. It seems appropriate, therefore, to close with some remarks about the theoretical and experimental plausibility of this result:

(1) In the σ model[3] the $\lambda\varphi_\pi^4$ term plus the three one-σ-exchange graphs gives scattering lengths which agree precisely with Eq. (23) in the limit $m_\sigma^2 \gg m_\pi^2$, except that L is given in terms of unrenormalized coupling constants.

(2) If the odd part of the forward π-π scattering amplitude obeys an unsubtracted dispersion relation, then our prediction (18) of $2a_0-5a_2$ may be used to derive the Adler sum rule[16] for π-π scattering total cross sections. (We have already made the corresponding remark for π-N scattering, that our prediction for $a_{1/2}-a_{3/2}$ is essentially equivalent to the Adler-Weisberger relation.) This sum rule seems to require either that a_0 be large, or that there exist a strong π-π resonance. The σ model provides one example where it is a resonance (of arbitrary mass) rather than a large scattering length, that saturates the sum rule.

(3) We have already remarked that the success of our prediction (11) of the π-N scattering lengths (in particular, the prediction $a_{1/2}+2a_{3/2}=0$) would be difficult to understand, if there were any strong low energy π-π interaction. In the same way, the success of a recent calculation[17] of the two K_{e4} form factors provides further experimental evidence that the π-π scattering lengths are quite small.

(4) Experiments on τ and η decay and high-energy "peripheral" two-pion production are ambiguous, since it is not clear whether the effects seen have anything to do with a π-π interaction. Furthermore, the two-pion production, even if peripheral, measures the π-π amplitude with one pion off the mass shell, and the conditions (19) or (20) show that this has a very large effect on M. It would seem desirable to reanalyze these experiments, using formulas (12) or (24) for the off-mass-shell π-π amplitude.

(5) A strong low-energy π-π interaction would have a large effect on the process $\pi+N\to 2\pi+N$ at threshold, where all three pions are soft. A calculation of this process is now under way, in conjunction with Chang.

I am very grateful for valuable discussions with K. Bardakci, D. Caldwell, G. F. Chew, S. Mandelstam, and S. B. Treiman.

---

*Research supported in part by the U. S. Air Force Office of Scientific Research, Office of Aerospace Research, under Grant No. AF-AFOSR-232-66.

[1]M. Gell-Mann, Physics 1, 63 (1964).

[2]Y. Nambu, Phys. Rev. Letters 4, 380 (1960).

[3]M. Gell-Mann and M. Levy, Nuovo Cimento 16, 705 (1960).

[4]S. Weinberg, Phys. Rev. Letters 16, 879 (1966).

[5]While preparing this note I became aware that Y. Tomozawa (to be published) had earlier calculated π-N, π-K, and π-hyperon scattering lengths, with results in agreement with our general formula (9). Since then similar results have also been obtained by B. Hamprecht (to be published), K. Raman and E. C. G. Sudarshan (to be published), and by A. P. Balachandran, M. Gundzik, and F. Nicodemi (to be published). I have nevertheless decided to present the derivation of Eq. (9) here, with a fuller attention to the details of the extrapolation onto the mass shell, partly to make it clear that Eq. (9) applies to the scattering of pions on all elementary particles except pions, and partly to serve as a basis for the π-π calculation, which the above authors do not attempt.

[6]H. Lehmann, K. Symanzik, and W. Zimmerman, Nuovo Cimento 1, 205 (1955).

[7]J. Schwinger, Ann. Phys. (N.Y.) 2, 407 (1957). By using a Jacobi identity and the conservation of $V_c^\mu(x)$, we can show that $\sigma_{ab}(x)$ is always symmetric in a and b, while in the σ model or free-quark model, it is proportional to σ_{ab}. It is only in the case of π-π scattering that we need distinguish different forms of this commutator.

[8]S. L. Adler, Phys. Rev. 137, B1022 (1965); 139, B1638 (1965). Adler uses this argument to give a formula for one of the two functions (A) entering in the symmetric π-N forward scattering amplitude, and then has to use a dispersion relation to compare this result with experiment. It can be seen directly from Adler's formula for A that the full symmetric amplitude vanishes at threshold, A being just cancelled by the pole in the other function B. (Indeed, this could have been seen immediately from the facts that M obviously vanishes as $q\to 0$ gradient coupling theory,[3] and gradient coupling theory satisfies PCAC.) It is clearer to deal with the full amplitude, rather than to divide it into A and B terms, and it is certainly simpler to work at threshold where nucleon and 3-3 resonance poles make no contribution, than to have to use dispersion theory to account for their contribution below threshold. For example, G. F. Chew has reminded me that the relation $f_{1/2}=-2f_{3/2}$, which works beautifully for the forward scattering amplitudes at

VOLUME 17, NUMBER 11 PHYSICAL REVIEW LETTERS 12 SEPTEMBER 1966

threshold, does not work at all well below threshold, presumably because the 3-3 resonance makes a large p-wave contribution except at threshold.

[9]Similar conclusions have been reached by M. L. Goldberger and S. B. Treiman, private communication.

[10]J. Hamilton and W. S. Woolcock, Rev. Mod. Phys. 35, 737 (1963).

[11]M. L. Goldberger, H. Miyazawa, and R. Oehme, Phys. Rev. 99, 986 (1955).

[12]S. L. Adler, Phys. Rev. Letters 14, 1051 (1965); W. I. Weisberger, Phys. Rev. Letters 14, 1047 (1965). In some derivations of the Adler-Weisberger sum rule, a formula is first obtained for the antisymmetric part of the forward pion-nucleon scattering amplitude at $s = m_N^2$ and zero external pion mass; see, e.g., Eq. (49) of Ref. 16. It was originally thought that this formula could only be compared with experiment by using it to derive a sum rule; our point here is that it can be converted into a formula for $a_{1/2} - a_{3/2}$ by direct use of PCAC.

[13]R. Seki and A. H. Cromer, to be published.

[14]The expansion Eq. (12) is certainly no good in the physical region well above threshold, since unitarity requires the presence in M of odd powers of $i(s - 4m_\pi^2)^{1/2}$. However, the scattering lengths we predict are quite small, so it is at least self-consistent to suppose that the unitarity branch point is a weak singularity, which does not prevent our using Eq. (12) up to and somewhat beyond threshold. An analogous expansion is known to work well in the similar case of τ decay. [For rigorous results concerning analyticity in s, t, and u when all external masses are equal, see A. Minguzzi, J. Math. Phys. 7, 679 (1966).] In particular, even where Eq. (12) begins to conflict seriously with unitarity, it would seem quite reasonable to apply it to the real part of the invariant matrix element. These points may in principle be checked by measuring the p-wave scattering length a_1 and the s-wave effective ranges r_0 and r_2, defined by the relation $k^{2l+1}\cot\delta_T \to a_T + \frac{1}{2}k^2 r_T$. In the physical region, Eq. (12) gives the matrix element in terms of just two parameters $A + 4m_\pi^2 B$ and $B - C$; hence, without using any other assumptions, we find that if Eq. (12) holds up to threshold, the scattering lengths must be subject to one relation, which may be written $18m_\pi^2 a_1 = 2a_0 - 5a_2$. If we also believe that Eq. (12) holds for the real part of M somewhat beyond threshold, then we also find that $6m_\pi^2 a_0^2(a_0 + \frac{1}{2}r_0) = a_0 + 10a_2$ and $6m_\pi^2 a_2^2 \times (a_2 + \frac{1}{2}r_2) = 5a_0 - 5a_2$. Similar remarks were made by G. F. Chew and S. Mandelstam, Nuovo Cimento 19, 752 (1960). Note that keeping just the zeroth-order term A in Eq. (12) would yield $a_0/a_2 = \frac{5}{2}$ and $a_1 = 0$, but this would be a very bad approximation because both Eqs. (19) and (20) show that A is of the same order as $m_\pi^2 B$ or $m_\pi^2 C$.

[15]A similar result is derived by F. T. Meire and M. Sugawara (to be published) but without justification of the neglect of mass-extrapolation terms. They then set $a_2 = 0$, and hence get a value of a_0 twice as large as ours.

[16]S. L. Adler, Phys. Rev. 140, B736 (1965).

[17]S. Weinberg, Phys. Rev. Letters 17, 336 (1966).

VOLUME 18, NUMBER 5 PHYSICAL REVIEW LETTERS 30 JANUARY 1967

DYNAMICAL APPROACH TO CURRENT ALGEBRA

Steven Weinberg*
Department of Physics, University of California, Berkeley, California
(Received 12 December 1966)

An effective Lagrangian for soft-pion interactions is constructed such that lowest order perturbation theory precisely reproduces the results of current algebra.

In the last year we have seen the development of a "current-algebra method" for calculating soft-pion matrix elements by direct use of partially conserved axial-vector current (PCAC) and current commutation relations.[1,2] However, despite its successes, this method suffers from some serious deficiencies:

(1) The algebraic effort required goes up rapidly with the number of soft pions.[3] This is of more than academic importance, because we would like to be able to add up the emission rates for arbitrary numbers of soft pions, as we already can do for photons and gravitons.

(2) A related problem is that the current-algebra method yields matrix elements which, although symmetric in the soft-pion labels, are not explicitly so.[4]

(3) In manipulating time-ordered products of currents, we lose sight of the dynamics underlying the results obtained. This is good, in that we are freed from dependence on specific dynamical assumptions, but also bad, in that we are offered no clue on how to go beyond the soft-pion limit.

This note will present a new technique for doing soft-pion calculations, which is guaranteed to give the same results as the current-algebra method, but which also is (1) vastly easier, (2) manifestly in accord with Bose statistics, (3) naturally suggestive of dynamical models which can carry us far beyond the results of current algebra.

First, let us recall the structure of the formulas provided by the current-algebra method. The S-matrix element for emission and absorption of any number of soft pions in a reaction $\alpha \to \beta$ is given as a sum matrix of elements between states β and α of a time-ordered product of a number of vector and axial-vector currents times a factor $(G/2m_N)(g_V/g_A)$ for each soft pion.[5] These currents are to be taken at small momenta; hence their matrix element is evaluated by hooking them on to the external lines of the process $\alpha \to \beta$ (as in the case of inner bremstrahlung), and we pick up an extra factor $(g_V/g_A)_n^{-1}$ for each axial current attached to the nth external line.

The key point in the above is that the soft-pion matrix element is of the lowest possible order in G, except for higher order terms in the factors g_V/g_A and in the matrix element $M_{\beta\alpha}$ for the process without soft pions. Hence we will necessarily get precisely the results of current algebra if we evaluate the soft-pion matrix elements to lowest order in G and then insert the correct values of g_V/g_A and $M_{\beta\alpha}$. To reiterate in more detail, the procedure to be followed is listed below:

(a) Choose any Lagrangian which satisfies PCAC and the proper current commutation relations.

(b) Evaluate the desired soft-pion matrix element to lowest order[6] in G.

(c) Write the result in the form dictated by current algebra, i.e., "trees" of soft pions attached to vector and axial-vector vertices on the external lines of a "core" process $\alpha \to \beta$. We will construct below a Lagrangian which directly yields soft-pion matrix elements of this form, so that this step does not require any further effort.

(d) Supply higher order corrections by multiplying with factors g_V/g_A for each soft pion and g_A/g_{Vn} for each axial-vector vertex on the nth external line, and by using the exact value of the matrix element $M_{\beta\alpha}$ for the process without soft pions. In the simplest cases, like $\pi+N \to \pi+N$ or $2\pi+N$ near threshold,

VOLUME 18, NUMBER 5 PHYSICAL REVIEW LETTERS 30 JANUARY 1967

the transition $\alpha \to \beta$ is a trival one-particle transition and no $M_{\beta\alpha}$ is needed.

The above procedure will be referred to as the chiral-dynamics method. It does not supplant current algebra, since for the present the only justification for the chiral-dynamics method is that its results agree with those of the current-algebra method, but as a computational technique it is far simpler and more transparent.

We will now proceed to the construction of a Lagrangian which in lowest order immediately reproduces the results of the current-algebra method. The time-honored example of a Lagrangian satisfying PCAC and the chiral commutation relations is that of the σ model[7]:

$$\begin{aligned}\mathcal{L} = &-\bar{N}[\gamma^\mu \partial_\mu + m_{N0} - G_0(\sigma + i\vec{\tau}\cdot\vec{\pi}\gamma_5)]N \\ &-\tfrac{1}{2}[\partial_\mu \vec{\pi}\cdot\partial^\mu \vec{\pi} + m_{\pi 0}^2 \vec{\pi}^2] - \tfrac{1}{2}[\partial_\mu \sigma \partial^\mu \sigma + m_{\sigma 0}^2 \sigma^2] \\ &-(m_{\sigma 0}^2 - m_{\pi 0}^2)[(G_0^2/8m_{N0}^2)(\vec{\pi}^2+\sigma^2)^2 - (G_0/2m_{N0})(\vec{\pi}^2+\sigma^2)\sigma].\end{aligned} \quad (1)$$

However, although not incorrect from the standpoint of current algebra, this Lagrangian is not well suited to the purposes of the chiral-dynamics method,[8] because even in lowest order it does not yield matrix elements which are manifestly of the "current-algebra form" required by rule (c) above. For instance, the lowest order σ model yields a low-energy πN scattering matrix equal to the sum of poles for one-nucleon exchange (using pseudoscalar coupling) in the s and u channels, plus one σ exchange in the t channel. It is an exercise in Dirac algebra to show that this result is actually equal to what we expect from current algebra, i.e., a sum of poles for one-nucleon exchange (using pseudovector coupling) in the s and u channels plus an equal-time commutator term which looks like one ρ exchange in the t channel. Rule (d) tells us then to multiply this last term by $(g_V/g_A)^2$, and we emerge with precisely the result[9] of the current-algebra method. A similar calculation is known to work for π-π scattering.[2] More generally, if we wished to use the σ model to calculate soft-pion emission and absorption in a physical process $\alpha \to \beta$ more complicated than a single-particle transition, we should have to add graphs in which pions are emitted from internal as well as external lines of this process, and it would take a major effort to rewrite the result in the current-algebra form of rule (c).

The trouble with the σ model stems from its nonderivative $\bar{N}\gamma_5\vec{\tau}\cdot\vec{\pi}N$ and $\sigma\vec{\pi}^2$ interactions; therefore, let us transform them away. The Lagrangian (1) is invariant (except for the $m_{\pi 0}^2$ terms) under a chiral SU(2)⊗SU(2) group, under which $\vec{\pi}$ and $\sigma - m_{N0}/G_0$ transform as a four-vector. Hence at every point we may define a new nucleon field N' by a chiral transformation

$$N = (1+\vec{\xi}^2)^{-1/2}(1+i\gamma_5\vec{\tau}\cdot\vec{\xi})N', \quad (2)$$

with $\vec{\xi}$ chosen so that

$$\bar{N}[m_{N0} - G_0(\sigma + i\vec{\tau}\cdot\vec{\pi}\gamma_5)]N = \bar{N}'[m_{N0} - G_0\sigma']N', \quad (3)$$

$$m_{N0} - G_0\sigma' \equiv [(m_{N0} - G_0\sigma)^2 + G_0^2\vec{\pi}^2]^{1/2}. \quad (4)$$

The required $\vec{\xi}$ is

$$\vec{\xi} = G_0\vec{\pi}[m_{N0} - G_0\sigma + \{(m_{N0} - G_0\sigma)^2 + G_0^2\vec{\pi}^2\}^{1/2}]^{-1}. \quad (5)$$

Expressions like (4) and (5) are, of course, to be interpreted as power series in G_0.

The transformation (2) replaces the $\vec{\pi}$ dependence of $\mathcal{L}$ with a $\vec{\xi}$ dependence, arising from the presence both of $m_{\pi 0}^2$ terms which break the chiral symmetry, and of derivative terms which are only invariant under constant chiral transformations. We will therefore define a new pion field $\vec{\pi}'$ as a function of $\vec{\xi}$. Different choices of this function will yield different matrix elements off the pion mass shell, but all give the same S matrix, so that we can define $\vec{\pi}'$ almost as we like. One choice would be $\vec{\pi}' \equiv \vec{\pi}$; $\mathcal{L}$ would then reproduce the results of the current-algebra method both on and off the

VOLUME 18, NUMBER 5 PHYSICAL REVIEW LETTERS 30 JANUARY 1967

mass shell, but would be quite complicated. We shall instead make the much more convenient choice

$$\vec{\pi}' \equiv 2m_{N0}\vec{\xi}/G_0. \tag{6}$$

A straightforward calculation shows that the Lagrangian (1) may be written in terms of the new field as

$$\begin{aligned}\mathcal{L} = &-\bar{N}'\left[\gamma^\mu\partial_\mu + m_{N0} - G_0\sigma' + i\gamma^\mu\left(1+\frac{G_0{}^2\vec{\pi}'^2}{4m_{N0}{}^2}\right)^{-1}\left\{\left(\frac{G_0}{2m_{N0}}\right)\gamma_5\vec{\tau}\cdot\partial_\mu\vec{\pi}' + \left(\frac{G_0{}^2}{4m_{N0}{}^2}\right)\vec{\tau}\cdot\vec{\pi}'\times\partial_\mu\vec{\pi}'\right\}\right]N' \\ &-\tfrac{1}{2}[\partial_\mu\sigma'\partial^\mu\sigma' + m_{\sigma 0}{}^2\sigma'^2] - (m_{\sigma 0}{}^2 - m_{\pi 0}{}^2)\left[\left(\frac{G_0{}^2}{8m_{N0}{}^2}\right)\sigma'^4 - \left(\frac{G_0}{2m_{N0}}\right)\sigma'^3\right] \\ &-\frac{1}{2}\left(1-\frac{G_0\sigma'}{m_{N0}}\right)\left[\left(1+\frac{G_0{}^2\vec{\pi}'^2}{4m_{N0}{}^2}\right)^{-2}\partial_\mu\vec{\pi}'\partial^\mu\vec{\pi}' + \left(1+\frac{G_0{}^2\vec{\pi}'^2}{4m_{N0}{}^2}\right)^{-1}m_{\pi 0}{}^2\vec{\pi}'^2\right].\end{aligned} \tag{7}$$

The Lagrangian $\mathcal{L}$ satisfies the requirements of current algebra for all values of $m_{\sigma 0}$; so we are free to take $m_{\sigma 0}$ as large as we like. But inspection of (7) shows that all graphs containing internal σ' lines vanish as $m_{\sigma 0} \to \infty$, and so the σ' field may be dropped everywhere in (7), yielding as our model

$$\begin{aligned}\mathcal{L}' = &-\bar{N}'\left[\gamma^\mu\partial_\mu + m_{N0} + i\gamma^\mu\left(1+\frac{G_0{}^2\vec{\pi}'^2}{4m_{N0}{}^2}\right)^{-1}\left\{\left(\frac{G_0}{2m_{N0}}\right)\gamma_5\vec{\tau}\cdot\partial_\mu\vec{\pi}' + \left(\frac{G_0{}^2}{4m_{N0}{}^2}\right)\vec{\tau}\cdot\vec{\pi}'\times\partial_\mu\vec{\pi}'\right\}\right]N' \\ &-\frac{1}{2}\left(1+\frac{G_0{}^2\vec{\pi}'^2}{4m_{N0}{}^2}\right)^{-2}\partial_\mu\vec{\pi}'\partial^\mu\vec{\pi}' - \frac{1}{2}\left(1+\frac{G_0{}^2\vec{\pi}'^2}{4m_{N0}{}^2}\right)^{-1}m_{\pi 0}{}^2\vec{\pi}'^2.\end{aligned} \tag{8}$$

This last step is only permissible because σ' is a chiral invariant and therefore plays no role in maintaining the chiral invariance of $\mathcal{L}$; in contrast, σ is not a chiral invariant, and graphs generated by (1) which contain internal σ lines do not vanish as $m_{\sigma 0}{}^2 \to \infty$.

Inspection of Eq. (8) shows immediately that its lowest order graphs are automatically of the current-algebra form required by rule (c), i.e., soft pions hang in clusters from vector and axial-vector vertices on the external lines of our process, the derivatives in (8) preventing soft pions from arising from internal lines. We are to use (8) to lowest order in G, not G_0; so G will appear in place of G_0, m_N in place of m_{N0}, etc. Further, rule (d) tells us that each soft pion introduces an extra factor g_V/g_A, and each $\gamma_5\gamma_\mu$ is accompanied with an extra factor g_A/g_V. Thus the prescription of the chiral-dynamics method for calculating soft-pion emission and absorption in a reaction $\alpha \to \beta$ may now be summarized as follows: Calculate the desired matrix element to lowest order in G, treating $M_{\beta\alpha}$ as a known black box of zeroth order, and using for the soft pions the effective Lagrangian[10]

$$\begin{aligned}\mathcal{L}_{\text{eff}} = &-\bar{N}\left[\gamma^\mu\partial_\mu + m_N + i\gamma^\mu\left\{1+\frac{G^2}{4m_N{}^2}\left(\frac{g_V}{g_A}\right)^2\vec{\pi}^2\right\}^{-1}\left\{\left(\frac{G}{2m_N}\right)\gamma_5\vec{\tau}\cdot\partial_\mu\vec{\pi} + \left(\frac{G^2}{4m_N{}^2}\right)\left(\frac{g_V}{g_A}\right)^2\vec{\tau}\cdot\vec{\pi}\times\partial_\mu\vec{\pi}\right\}\right]N \\ &-\frac{1}{2}\left\{1+\frac{G^2}{4m_N{}^2}\left(\frac{g_V}{g_A}\right)^2\vec{\pi}^2\right\}^{-2}\partial_\mu\vec{\pi}\cdot\partial^\mu\vec{\pi} - \frac{1}{2}\left\{1+\frac{G^2}{4m_N{}^2}\left(\frac{g_V}{g_A}\right)^2\vec{\pi}^2\right\}^{-1}m_\pi{}^2\vec{\pi}^2.\end{aligned} \tag{9}$$

For instance, we now get the correct[9] πN scattering lengths by just looking at the $\vec{\pi}\times\partial_\mu\vec{\pi}$ term in

VOLUME 18, NUMBER 5 PHYSICAL REVIEW LETTERS 30 JANUARY 1967

Eq. (9); this is surely the simplest derivation of the Adler-Weisberger relation! Also, by expanding the denominators in the last two terms of (9), we immediately get an effective Lagrangian for pion-pion scattering,

$$\mathcal{L}_{\pi\pi} = \frac{G^2}{8m_N^2}\left(\frac{g_V}{g_A}\right)^2 \vec{\pi}^2[2\partial_\mu\vec{\pi}\cdot\partial^\mu\vec{\pi} + m_\pi^2\vec{\pi}^2]. \qquad (10)$$

This gives the same π-π matrix element (on the mass shell) as does current algebra,[2] a point essential to the consistency of our method since (9) generates peripheral π-π scattering graphs which make important contributions to processes like $\pi + N \to 2\pi + N$.

There is no difficulty in including K mesons and hyperons in our model Lagrangian, or in using it to treat weak and electromagnetic as well as strong interactions. A more intriguing extension of this approach would be to start by adding Yang-Mills vector and axial-vector fields in (1), in such a way that it is only their bare mass, and that of the pion, which break chiral invariance; the $\vec{\pi}\times\partial_\mu\vec{\pi}$ term in the transformed Lagrangian then comes entirely from ρ exchange. This offers hope of reproducing the results of current algebra from a Lagrangian which at the same time is a reasonable phenomenological model of the strong interactions. A paper on this model is in preparation.

Finally, it is remarkable that the Lagrangian (9) is to be used only in lowest order,[11] the effects of loops, etc., being already accounted for by the presence in (9) of the factors g_V/g_A. Just how does this come about? If we knew, we might understand more deeply the dynamical basis of current algebra.

I am very grateful for the hospitality extended to me by the Physics Department of Harvard University. I would also like to thank R. Dashen and F. J. Dyson for interesting discussions.

---

*Presently Morris Loeb Lecturer, Physics Department, Harvard University, Cambridge, Massachusetts.

[1]For a general review of soft-pion calculations see the rapporteur's talk by R. F. Dashen, in Proceedings of the Thirteenth International Conference on High Energy Physics, Berkeley, California, 1966 (unpublished).

[2]The author's previous approach to soft-pion problems is described in Phys. Rev. Letters 17, 616 (1966).

[3]The most thorough statement of the results of current algebra and PCAC for arbitrary numbers of soft pions is that of H. Abarbanel and S. Nussinov, to be published. This article shows how complicated the current-algebra method gets for three or more soft pions.

[4]For instance, see Ref. 3.

[5]Here G is the renormalized pion-nucleon coupling constant, m_N is the nucleon mass, and g_V and g_A are the renormalized vector and axial-vector coupling constants of the nucleon; these factors of course arise from PCAC and the Goldberger-Treiman relation. The currents are defined here not to include the factor g_V, so their commutation relations do not involve any coupling constants.

[6]All unrenormalized pion coupling constants are here to be regarded as power series in the corresponding renormalized coupling constants (of which we ultimately keep only the lowest order terms), and all renormalized pion coupling constants are to be expressed in terms of the pion-nucleon coupling constant G by using Goldberger-Treiman relations.

[7]J. Schwinger, Ann. Phys. (N. Y.) 2, 407 (1957); M. Gell-Mann and M. Lévy, Nuovo Cimento 16, 705 (1960). The label 0 means "unrenormalized."

[8]The σ model is also useless as a phenomenological model of strong interactions; for were we to take it seriously, we should have to give the σ a width $\Gamma_\sigma \cong (3G^2/32\pi)(m_\sigma^3/m_N^2)(g_V/g_A)^2$, which is larger than m_σ if $m_\sigma > 500$ MeV. Such a broad S-wave resonance seems unlikely.

[9]See, e. g., Ref. 2, Eq. (6), or the other references quoted therein.

[10]For processes involving a single soft pion this is just the pseudovector coupling model, a result already known from current algebra; see S. Weinberg, Phys. Rev. Letters 16, 879 (1966). It is of course an old story that there is some sort of equivalence between pseudovector and pseudoscalar coupling, as shown by F. J. Dyson, Phys. Rev. 73, 929 (1948); L. L. Foldy, Phys. Rev. 84, 168 (1951); etc. The complications encountered in these early papers arose because they dealt with theories having no sort of chiral invariance.

[11]In this respect our method is reminiscent of the recently developed "source" approach of J. Schwinger. I wish to thank Professor Schwinger for an interesting discussion on this point.

PHYSICAL REVIEW VOLUME 166, NUMBER 5 25 FEBRUARY 1968

Nonlinear Realizations of Chiral Symmetry*

STEVEN WEINBERG†
Laboratory for Nuclear Science and Department of Physics, Massachusetts Institute of Technology, Cambridge, Massachusetts

(Received 25 September 1967)

We explore possible realizations of chiral symmetry, based on isotopic multiplets of fields whose transformation rules involve only isotopic-spin matrices and the pion field. The transformation rules are unique, up to possible redefinitions of the pion field. Chiral-invariant Lagrangians can be constructed by forming isotopic-spin-conserving functions of a covariant pion derivative, plus other fields and their covariant derivatives. The resulting models are essentially equivalent to those that have been derived by treating chirality as an ordinary linear symmetry broken by the vacuum, except that we do not have to commit ourselves as to the grouping of hadrons into chiral multiplets; as a result, the unrenormalized value of g_A/g_V need not be unity. We classify the possible choices of the chiral-symmetry-breaking term in the Lagrangian according to their chiral transformation properties, and give the values of the pion-pion scattering lengths for each choice. If the symmetry-breaking term has the simplest possible transformation properties, then the scattering lengths are those previously derived from current algebra. An alternative method of constructing chiral-invariant Lagrangians, using ρ mesons to form covariant derivatives, is also presented. In this formalism, ρ dominance is automatic, and the current-algebra result from the ρ-meson coupling constant arises from the independent assumption that ρ mesons couple universally to pions and other particles. Including ρ mesons in the Lagrangian has no effect on the π-π scattering lengths, because chiral invariance requires that we also include direct pion self-couplings which cancel the ρ-exchange diagrams for pion energies near threshold.

I. INTRODUCTION

CURRENT algebra is useful because it allows us to obtain physical predictions from chiral symmetry. We have recently noted[1] that for soft-pion processes the same predictions can also be derived by a different method: Just use the lowest-order graphs generated by any chiral-invariant Lagrangian. The Lagrangian method has since been applied to pion production,[2] η decay,[3] K interactions and decay,[4] and, in various extended versions, to meson mass ratios and decay amplitudes,[5] and to the pion electromagnetic mass difference.[6] Opinions differ[7] as to whether any fundamental significance resides in the Lagrangians that have been used, but there is no doubt that they provide both a convenient method of calculation and a valuable heuristic guide to theorems that can be proved with current algebra.

There are two ways of constructing our chiral-invariant Lagrangians, which mirror two different views of the meaning of chiral symmetry. The first, *conventional* method[8] is to construct $\mathcal{L}$ to be manifestly chiral-invariant, as if chirality were an ordinary linear symmetry like isospin. For example, in the σ model[9] the $\boldsymbol{\pi}$ and σ fields form a four-vector coupled to nucleons in the combination $\sigma+i\boldsymbol{\tau}\cdot\boldsymbol{\pi}\gamma_5$, and the nucleon mass arises from the nonvanishing vacuum expectation value $\langle\sigma\rangle_0=-m_N/G$. In a closely related model[10] the Lagrangian takes the same form, but with σ replaced everywhere with $[(m_N/G)^2-\boldsymbol{\pi}^2]^{1/2}$. Such models suffer from a fundamental disadvantage: They hide the fact that soft pions are emitted in clusters by derivative cou-

* This work is supported in part through funds provided by the Atomic Energy Commission under Contract No. AT(30-1)-2098.

† On leave from the University of California, Berkeley, California.

[1] S. Weinberg, Phys. Rev. Letters **18**, 507 (1967).

[2] L.-N. Chang, Phys. Rev. **162**, 1497 (1967); Ph.D. thesis (unpublished).

[3] W. A. Bardeen, L. S. Brown, B. W. Lee, and H. T. Nieh, Phys. Rev. Letters **18**, 1170 (1967). Precisely the same calculation was done by S. Shei, but not published, because it appeared that the matrix element was too small by a factor m_π^2/m_η^2 to account for the observed decay. Bardeen *et al.* treat the η-π vertex in what seems to me a dubious manner, and thereby escape this difficulty.

[4] B. Zumino (to be published); S. Iwao (to be published).

[5] J. Schwinger, Phys. Letters **24B**, 473 (1967); S. Weinberg, Phys. Rev. Letters **18**, 507 (1967) (see footnote 7); S. Glashow, H. Schnitzer, and S. Weinberg, Phys. Rev. Letters **19**, 139 (1967) [Eq. (13) ff]; M. Lévy (to be published); J. W. Wess and B. Zumino, Phys. Rev. **163**, 1727 (1967); S. Glashow and S. Weinberg (to be published). The decay amplitudes derived using Lagrangian methods by Schwinger [and then in a somewhat more general form by Wess and Zumino] were subsequently rederived using current algebra by H. Schnitzer and S. Weinberg, Phys. Rev. **164**, 1828 (1967).

[6] J. Schwinger (to be published); D. B. Fairlie and K. Yoshida (to be published). The corresponding current-algebra calculation was done by T. Das, G. S. Guralnik, V. S. Mathur, F. E. Low, and J. E. Young, Phys. Rev. Letters **18**, 759 (1967).

[7] In particular, Schwinger has argued that as long as the origin of symmetries remains obscure, the phenomenological Lagrangian provides a suitable arena for their study. [J. Schwinger, Phys. Rev. **152**, 1219 (1966); also Refs. 5 and 6, and private communication.] Others like myself remain uneasy at using a symmetry on the phenomenological level, when it is not clear how *any* fundamental Lagrangian could give rise to the supposed symmetry of phenomena. From this point of view, chirality is in good shape because we have current algebra to underwrite it, but $SU(6)$ remains obscure. Time will tell.

[8] J. Schwinger, Ann. Phys. (N. Y.) **2**, 407 (1957); M. Gell-Mann and M. Lévy, Nuovo Cimento **16**, 705 (1960); F. Gürsey, *ibid.* **16**, 230 (1960); in *Proceedings of the 1960 Rochester Conference on High-Energy Physics* (Interscience Publishers, Inc., New York, (1960), p. 572; Ann. Phys. **12**, 91 (1961); F. Gürsey and B. Zumino, (unpublished); P. Chang and F. Gürsey, Phys. Rev. **164**, 1752 (1967); H. S. Mani, Y. Tomozawa, and Y. P. Yao, Phys. Rev. Letters **18**, 1084 (1967); L. S. Brown, Phys. Rev. **163**, 1802 (1967); and J. A. Cronin, Phys. Rev. **161**, 1483 (1967); and Refs. 1–4.

[9] J. Schwinger and M. Gell-Mann and M. Lévy, Ref. 8.

[10] F. Gürsey and M. Gell-Mann and M. Lévy, Ref. 8.

plings from external lines.[11] For this reason, it proves necessary to perform a chiral field-dependent rotation which eliminates the nonderivative coupling of σ and π and replaces it with a nonlinear derivative coupling of the chiral rotation vector, identified as a new pion field.

In the second, *nonlinear* method,[12] one recognizes from the beginning that chirality is not like other symmetries, because it relates processes involving different numbers of soft pions. Therefore, the Lagrangian is constructed so that it is invariant under chiral transformations expressed, not in terms of isospin matrices and γ_5's, but in terms of isospin matrices and the pion field. In this way we work from the beginning with the nonlinear derivative couplings which emerged only at the end of the first method.

The conventional method has the advantage of expressing the Lagrangian in a manifestly renormalizable form.[9] The nonlinear method has the advantage of putting the Lagrangian in a useful form without our having to work our way through a chiral rotation. One other difference between the two methods is that the second yields *only* those results which can be obtained from current algebra, while the first mixes up these results with others which reflect our prejudices as to how hadrons are grouped into chiral multiplets. For instance, in the σ model one assumes that the nucleon is in a $(\frac{1}{2},0)+(0,\frac{1}{2})$ linear representation of $SU(2)\times SU(2)$, and one finds that the unrenormalized value of the weak-coupling-constant ratio g_A/g_V is unity; in the nonlinear method we never have to ask to what kind of chiral multiplet the nucleon belongs, and the unrenormalized g_A/g_V can be anything we like. It is not clear whether this should be counted as an advantage for the conventional or the nonlinear method.

This article will present a systematic development of the nonlinear approach to chiral invariance. In Sec. II we show that the most general possible nonlinear pion transformation rule is equivalent to

$$[X_a,\pi_b]=-i\lambda^{-1}[\tfrac{1}{2}(1-\lambda^2\boldsymbol{\pi}^2)\delta_{ab}+\lambda^2\pi_a\pi_b], \quad (1.1)$$

where X_a is the chiral generator $(a, b=1, 2, 3)$ and λ is a constant. By "equivalent" we mean that any other possible transformation law can be converted to (1.1) by a suitable redefinition of the pion field.[13] In Sec. III we show that the corresponding transformation for a general field ψ is

$$[\mathbf{X},\psi]=\lambda(\mathbf{t}\times\boldsymbol{\pi})\psi, \quad (1.2)$$

where $\mathbf{t}$ is the isospin matrix for ψ. (We do not limit ourselves to the nucleon field here; ψ could be the field of a K meson, a baryon resonance, etc.) In Sec. IV we show that a chiral-invariant Lagrangian can be constructed simply by forming an arbitrary isospin-invariant Lagrangian out of the ψ, their "covariant derivatives,"

$$D_\mu\psi=\partial_\mu\psi+2i\lambda^2(1+\lambda^2\boldsymbol{\pi}^2)^{-1}\mathbf{t}\cdot(\boldsymbol{\pi}\times\partial_\mu\boldsymbol{\pi})\psi; \quad (1.3)$$

and a pion covariant derivative,

$$D_\mu\boldsymbol{\pi}=(1+\lambda^2\boldsymbol{\pi}^2)^{-1}\partial_\mu\boldsymbol{\pi}. \quad (1.4)$$

By studying the axial-vector current, we find that

$$\lambda=F_{\pi 0}^{-1}, \quad (1.5)$$

where $F_{\pi 0}$ is the unrenormalized value of the pion decay amplitude. We show in Sec. V that the linearly transforming fields of the conventional approach can be constructed from the π and ψ. In Sec. VI we explain what differences may arise among current-algebra results for the pion-pion scattering lengths[14,15] by showing their value is entirely determined by the chiral transformation properties of the symmetry-breaking terms in $\mathcal{L}$ (see Table I). In Sec. VII we discuss the possible role of the ρ meson, and show that the covariant derivative (1.3) can be replaced with

$$\mathfrak{D}_\mu\psi=\partial_\mu\psi-ig_0\mathbf{t}\cdot\boldsymbol{\rho}_\mu\psi, \quad (1.6)$$

provided we use the Yang-Mills Lagrangian for $\boldsymbol{\rho}$, and

TABLE I. Values of the pion-pion scattering lengths (in pion Compton wavelengths) under various assumptions about the chiral transformation properties of the term in the Lagrangian which breaks chiral symmetry. The rows labelled with a value of N represent the cases where this term transforms according to the representation $(N/2,N/2)$, i.e., like a traceless symmetric tensor of rank N. The last two rows represent the possibility that the symmetry-breaking term is simply $-\frac{1}{2}m_\pi^2\boldsymbol{\pi}^2$, with $\boldsymbol{\pi}$ defined to transform according to Eqs. (2.18) or (2.21). (The first, second, and last rows present the results of Refs. 14, 18, and 12, respectively.)

Transformation of symmetry-breaking term	a_0	a_2	a_0/a_2
Four-vector ($N=1$)	0.20	−0.06	−7/2
Tensor ($N=2$)	0.35	0	∞
Tensor ($N=3$)	0.55	0.08	95/14
...	...	...	...
$\boldsymbol{\pi}^2$; see Eq. (2.18)	0.06	−0.11	−1/2
$\boldsymbol{\pi}^2$; see Eq. (2.21)	0.12	−0.09	−3/2

[11] S. Weinberg, Phys. Rev. Letters **16**, 879 (1966).

[12] See particularly, J. Schwinger, Ref. 5.

[13] It perhaps should be emphasized that we are free to use any pseudoscalar isovector object as the pion field; different choices give different matrix elements off the mass shell, but they all give the same S matrix.

[14] The pion-pion scattering lengths were calculated using current algebra by S. Weinberg [Phys. Rev. Letters **17**, 616 (1966)] under the assumption that the symmetry-breaking term in the Lagrangian is the 0 component of a chiral four-vector.

[15] We are not concerned here with the validity of the expansion technique used in Ref. 14. It was clear from the beginning that the low-energy π-π interaction might be so strong that current algebra simply could not be used. However, the consensus of those who have studied this problem is that the expansion technique used in Ref. 14 is at least self-consistent; i.e., it yields scattering lengths small enough so that the unitarity corrections are even smaller. See N. N. Khuri, Phys. Rev. **153**, 1477 (1967); F. J. Meiere, *ibid.* **159**, 1462 (1967); J. Sucher and C. H. Woo, Phys. Rev. Letters **18**, 723 (1967); K. Kang and J. Akiba, Phys. Rev. **164**, 1836 (1967). J. Iliopoulos (to be published). In addition, the successful application of current algebra to a wide variety of multipion processes (discussed at the end of Sec. VI) provides empirical evidence that the scattering lengths *are* small.

provided that for the ρ mass term we use

$$-\tfrac{1}{2}m_\rho^2[\boldsymbol{\rho}_\mu+2g_0^{-1}\lambda^2(1+\lambda^2\boldsymbol{\pi}^2)^{-1}(\boldsymbol{\pi}\times\partial_\mu\boldsymbol{\pi})]^2. \quad (1.7)$$

As a consequence, the contribution of ρ exchange to π-π scattering is effectively cancelled near threshold by a direct $(\boldsymbol{\pi}\times\partial_\mu\boldsymbol{\pi})^2$ interaction. Perhaps, by taking this cancellation into account, it will be possible to construct a model of π-π scattering which applies from threshold up to the ρ mass.

It would be interesting to see what kind of nonlinear realizations are possible for a general symmetry group.

II. TRANSFORMATION OF THE PION FIELD

We will first show that *the nonlinear transformation induced by chiral $SU(2)\times SU(2)$ on the pion field is unique*, up to possible redefinition of the fields.

The operators of chiral $SU(2)\times SU(2)$ will be denoted T_a, X_a, with $a=1, 2, 3$; they satisfy the familiar commutation relations

$$[T_a,T_b]=i\epsilon_{abc}T_c, \quad (2.1)$$

$$[T_a,X_b]=i\epsilon_{abc}X_c, \quad (2.2)$$

$$[X_a,X_b]=i\epsilon_{abc}T_c. \quad (2.3)$$

We will allow the transformation induced by X_a on the pion field π_b to be completely general in form, i.e.,

$$[X_a,\pi_b]=-if_{ab}(\boldsymbol{\pi}), \quad (2.4)$$

where $f_{ab}(\boldsymbol{\pi})$ is as yet an arbitrary function. On the other hand, isotopic spin is an ordinary symmetry, and its action on $\boldsymbol{\pi}$ will be required to take the usual form:

$$[T_a,\pi_b]=i\epsilon_{abc}\pi_c. \quad (2.5)$$

The restrictions imposed by the commutation relations (2.1)–(2.3) on the transformation function $f_{ab}(\boldsymbol{\pi})$ can be most easily determined by using the various Jacobi identities of $\boldsymbol{\pi}$ with pairs of generators. First we use

$$[T_a,[X_b,\pi_c]]=[X_b,[T_a,\pi_c]]+[[T_a,X_b],\pi_c]. \quad (2.6)$$

With Eqs. (2.2)–(2.4), and (2.5), this gives

$$[T_a,f_{bc}(\boldsymbol{\pi})]=i\epsilon_{acd}f_{bd}(\boldsymbol{\pi})+i\epsilon_{abd}f_{dc}(\boldsymbol{\pi}),$$

or in other words,

$$\frac{\partial f_{bc}(\boldsymbol{\pi})}{\partial\pi_d}\epsilon_{ade}\pi_e=f_{bd}(\boldsymbol{\pi})\epsilon_{acd}+f_{dc}(\boldsymbol{\pi})\epsilon_{abd}. \quad (2.7)$$

This just says that $f_{bc}(\pi)$ is an isotopic tensor—not a surprising result.

The other useful Jacobi identity is for $\boldsymbol{\pi}$ with a pair of X's:

$$[X_a,[X_b,\pi_c]]-[X_b,[X_a,\pi_c]]=[[X_a,X_b],\pi_c]. \quad (2.8)$$

With Eqs. (2.3)–(2.5) this gives

$$[X_a,f_{bc}(\boldsymbol{\pi})]-[X_b,f_{ac}(\boldsymbol{\pi})]=-i\epsilon_{abd}\epsilon_{dce}\pi_e,$$

or, more explicitly,

$$\frac{\partial f_{bc}(\boldsymbol{\pi})}{\partial\pi_d}f_{ad}(\boldsymbol{\pi})-\frac{\partial f_{ac}(\boldsymbol{\pi})}{\partial\pi_d}f_{bd}(\boldsymbol{\pi})=\delta_{ac}\pi_b-\delta_{bc}\pi_a. \quad (2.9)$$

Now we solve Eqs. (2.7) and (2.9). As already remarked, Eq. (2.7) is merely the statement that $f_{bc}(\boldsymbol{\pi})$ is an isotopic tensor. Further, $f_{bc}(\boldsymbol{\pi})$ has even parity, so it must be even in $\boldsymbol{\pi}$, and hence takes the form

$$f_{bc}(\boldsymbol{\pi})=\delta_{bc}f(\boldsymbol{\pi}^2)+\pi_b\pi_c g(\boldsymbol{\pi}^2), \quad (2.10)$$

with f and g arbitrary functions of $\boldsymbol{\pi}^2$. Direct calculation gives

$$\frac{\partial f_{bc}(\boldsymbol{\pi})}{\partial\pi_d}f_{ad}(\boldsymbol{\pi})-\frac{\partial f_{ac}(\boldsymbol{\pi})}{\partial\pi_d}f_{bd}(\boldsymbol{\pi})=(fg-2ff'-2\pi^2gf')(\delta_{ac}\pi_b-\delta_{bc}\pi_a),$$

so our other differential equation, Eq. (2.9), imposes *one* further relation on f and g:

$$g(\boldsymbol{\pi}^2)=\frac{1+2f(\boldsymbol{\pi}^2)f'(\boldsymbol{\pi}^2)}{f(\boldsymbol{\pi}^2)-2\boldsymbol{\pi}^2f'(\boldsymbol{\pi}^2)}. \quad (2.11)$$

[A prime denotes differentiation with respect to the argument $\boldsymbol{\pi}^2$.] Our conclusion is that the most general pion transformation law is given by Eqs. (2.4) and (2.5), where $f_{bc}(\boldsymbol{\pi})$ has the form (2.10), with $g(\boldsymbol{\pi}^2)$ specified in terms of $f(\boldsymbol{\pi}^2)$ by Eq. (2.11).

We promised to show that the pion transformation law is essentially unique. Let us consider the effect on the transformation function $f_{ab}(\boldsymbol{\pi})$ of a redefinition $\boldsymbol{\pi}\to\boldsymbol{\pi}^*$ of the pion field.[13] Since the new pion field $\boldsymbol{\pi}^*$ must be an isovector satisfying Eq. (2.5), the most general redefinition is of the form

$$\pi_a^*=\pi_a\Phi(\boldsymbol{\pi}^2). \quad (2.12)$$

We can easily calculate that $\boldsymbol{\pi}^*$ has a chiral transformation law of the same form as $\boldsymbol{\pi}$; i.e.,

$$[X_a,\pi_b^*]=-i\{\delta_{ab}f^*(\boldsymbol{\pi}^{*2})+\pi_a^*\pi_b^*g^*(\boldsymbol{\pi}^{*2})\}, \quad (2.13)$$

where

$$f^*(\boldsymbol{\pi}^{*2})=f(\boldsymbol{\pi}^2)\Phi(\boldsymbol{\pi}^2) \quad (2.14)$$

and

$$g^*(\boldsymbol{\pi}^{*2})=[g(\boldsymbol{\pi}^2)\Phi(\boldsymbol{\pi}^2)+2f(\boldsymbol{\pi}^2)\Phi'(\boldsymbol{\pi}^2)+2\boldsymbol{\pi}^2g(\boldsymbol{\pi}^2)\Phi'(\boldsymbol{\pi}^2)]\Phi^{-2}(\boldsymbol{\pi}^2). \quad (2.15)$$

We see that *by redefining $\boldsymbol{\pi}$ we can make the function f^* anything we like*; Eq. (2.15) then determines the corresponding function g^* in such a way that Eq. (2.11) will still hold.

In particular, we may define the pion field $\boldsymbol{\pi}$ (now dropping superscripts) so that

$$f(\boldsymbol{\pi}^2)=(1/2\lambda)(1-\lambda^2\boldsymbol{\pi}^2), \quad (2.16)$$

with λ an arbitrary constant. Then Eq. (2.11) gives

$$g(\boldsymbol{\pi}^2)=\lambda\,, \tag{2.17}$$

and the pion transformation law is

$$[X_a,\pi_b]=-(i/\lambda)\{\tfrac{1}{2}(1-\lambda^2\boldsymbol{\pi}^2)\delta_{ab}+\lambda^2\pi_a\pi_b\}\,. \tag{2.18}$$

It is easy to show that this is precisely the transformation law (with $\lambda=G/2m_N$) satisfied by the "new pion field" $\boldsymbol{\pi}'\equiv\lambda\boldsymbol{\varepsilon}$ defined in our previous work,[1] and is also the rule adopted by Schwinger.[5] Alternatively, we could also define $\boldsymbol{\pi}$ so that

$$f(\boldsymbol{\pi}^2)=\lambda^{-1}\,, \tag{2.19}$$

in which case Eq. (2.11) would again give

$$g(\boldsymbol{\pi}^2)=\lambda\,, \tag{2.20}$$

and the pion transformation law would be

$$[X_a,\pi_b]=-i\lambda^{-1}\{\delta_{ab}+\lambda^2\pi_a\pi_b\}\,. \tag{2.21}$$

It is to be stressed that (2.18) and (2.21) do not represent inequivalent realizations of $SU(2)\times SU(2)$; these transformation laws, and all other possible pion transformation laws, can be derived from one another by suitable redefinitions of the pion field.

III. TRANSFORMATION OF OTHER FIELDS

Now that we have at hand a $\boldsymbol{\pi}$ field which transforms nonlinearly under chiral $SU(2)\times SU(2)$, what can we do with it? The lesson taught by current algebra is that a symmetry like chirality does not manifest itself in linear γ_5 invariance relations (which for instance would require the vanishing of the nucleon mass), but rather it determines relations between an arbitrary process $\alpha\to\beta$ and the related processes $\alpha\to\beta+n\pi$. The attractive feature of a nonlinear-field transformation law is that it allows us to build this interpretation of chiral symmetry into the Lagrangian from the beginning. Suppose an arbitrary field ψ has a chiral transformation law of the form

$$[X_a,\psi]=v_{ab}(\boldsymbol{\pi})t_b\psi\,, \tag{3.1}$$

where t_b is the Hermitian isospin matrix appropriate to ψ, i.e.,

$$[T_b,\psi]=-t_b\psi\,. \tag{3.2}$$

Then any isospin-invariant function of ψ (not its derivatives) will also be chiral-invariant; for instance $\psi^\dagger\psi$ commutes with T_a and hence with X_a, and chirality will *not* tell us that the mass of ψ vanishes. What it *does* tell us is explored in Sec. IV. In this section we will answer some necessary preliminary questions: Does there exist a function $v_{ab}(\boldsymbol{\pi})$ for which (3.1) and (3.2) are a self-consistent realization of $SU(2)\times SU(2)$? And if so, what is it?

The consistency requirements that must be satisfied by the transformation rules (3.1), (3.2) are embodied in the Jacobi identities analogous to Eqs. (2.6) and (2.8):

$$[T_a,[X_b,\psi]]\equiv[X_b,[T_a,\psi]]+[[T_a,X_b],\psi]\,,$$
$$[X_a,[X_b,\psi]]-[X_b,[X_a,\psi]]\equiv[[X_a,X_b],\psi]\,.$$

The first identity just tells us that $v_{ab}(\boldsymbol{\pi})$ is an isotopic tensor, i.e.,

$$\frac{\partial v_{bc}(\boldsymbol{\pi})}{\partial\pi_d}\epsilon_{ade}\pi_e=v_{bd}(\boldsymbol{\pi})\epsilon_{acd}+v_{dc}(\boldsymbol{\pi})\epsilon_{abd}\,. \tag{3.3}$$

The second identity gives

$$[X_a,v_{be}(\boldsymbol{\pi})]t_e-[X_b,v_{ae}(\boldsymbol{\pi})]t_e +v_{bd}(\boldsymbol{\pi})v_{ac}(\boldsymbol{\pi})[t_d,t_c]=-i\epsilon_{abe}t_e\,.$$

We use the pion transformation law (2.4) and the isospin commutation rule $[t_d,t_c]=i\epsilon_{dce}t_e$ to put this in the form of a differential equation for $v_{ab}(\boldsymbol{\pi})$:

$$\frac{\partial v_{be}(\boldsymbol{\pi})}{\partial\pi_d}f_{ad}(\boldsymbol{\pi})-\frac{\partial v_{ae}(\boldsymbol{\pi})}{\partial\pi_d}f_{bd}(\boldsymbol{\pi}) =-v_{ac}(\boldsymbol{\pi})v_{bd}(\boldsymbol{\pi})\epsilon_{cde}+\epsilon_{abe}\,. \tag{3.4}$$

The function v_{bc} carries negative parity, so it must be odd in $\boldsymbol{\pi}$; with (3.3), this restricts its form to

$$v_{ab}(\boldsymbol{\pi})=\epsilon_{abc}\pi_c v(\boldsymbol{\pi}^2)\,. \tag{3.5}$$

Inserting (3.5) and (2.10) in (3.4), we obtain a nonlinear differential equation for $v(\boldsymbol{\pi}^2)$:

$$\pi_d[\epsilon_{bed}\pi_a-\epsilon_{aed}\pi_b]\times\{v(\boldsymbol{\pi}^2)g(\boldsymbol{\pi}^2)+2v'(\boldsymbol{\pi}^2)[f(\boldsymbol{\pi}^2)+\boldsymbol{\pi}^2g(\boldsymbol{\pi}^2)]\} +2\epsilon_{abe}v(\boldsymbol{\pi}^2)f(\boldsymbol{\pi}^2)=\pi_c\pi_d\epsilon_{bad}v^2(\boldsymbol{\pi}^2)+\epsilon_{abe}\,.$$

It is convenient to rewrite this using the identity

$$\boldsymbol{\pi}^2\epsilon_{abe}\equiv\epsilon_{abc}\pi_c\pi_e+\epsilon_{bec}\pi_c\pi_a+\epsilon_{eac}\pi_c\pi_b\,.$$

We then find

$$\pi_d(\epsilon_{bed}\pi_a-\epsilon_{aed}\pi_b)\times\{v(\boldsymbol{\pi}^2)g(\boldsymbol{\pi}^2)+2v'(\boldsymbol{\pi}^2)[f(\boldsymbol{\pi}^2)+\boldsymbol{\pi}^2g(\boldsymbol{\pi}^2)]-v^2(\boldsymbol{\pi}^2)\} +\epsilon_{abe}[-2v(\boldsymbol{\pi}^2)f(\boldsymbol{\pi}^2)-\boldsymbol{\pi}^2v^2(\boldsymbol{\pi}^2)+1]=0\,. \tag{3.6}$$

For this to be possible, both functions in brackets must vanish; i.e.,

$$vg+2v'(f+\boldsymbol{\pi}^2g)-v^2=0\,, \tag{3.7}$$

$$-2vf-\boldsymbol{\pi}^2v^2+1=0\,. \tag{3.8}$$

The solution of the second equation is

$$v(\boldsymbol{\pi}^2)=-\{f(\boldsymbol{\pi}^2)+[f^2(\boldsymbol{\pi}^2)+\boldsymbol{\pi}^2]^{1/2}\}^{-1}\,. \tag{3.9}$$

We leave it to the reader's pertinacity to show that (3.7) is also satisfied when $v(\boldsymbol{\pi}^2)$ and $g(\boldsymbol{\pi}^2)$ are given by (3.9) and (2.11), respectively. Thus (3.1) is indeed a possible chiral transformation law, with $v_{ab}(\boldsymbol{\pi})$ given *uniquely* by

$$v_{ab}(\boldsymbol{\pi})=\epsilon_{abc}\pi_c\{f(\boldsymbol{\pi}^2)+[f^2(\boldsymbol{\pi}^2)+\boldsymbol{\pi}^2]^{1/2}\}^{-1}\,, \tag{3.10}$$

where $f(\pi^2)$ is the function appearing in the pion transformation equations.

It may be noted that a pion field defined to transform according to Eq. (2.18) will have $v(\pi^2)$ equal to a constant λ, so here (3.1) takes the particularly simple form

$$[\mathbf{X},\psi]=\lambda(\mathbf{t}\times\boldsymbol{\pi})\psi. \tag{3.11}$$

For this reason, this particular choice of the pion transformation rule is somewhat more convenient than other, equivalent choices.

IV. COVARIANT DERIVATIVES

We are now going to see how it is possible to construct a chiral-invariant Lagrangian out of pion fields $\boldsymbol{\pi}$ and other fields ψ which transform according to the rules (2.4) and (3.1). As already remarked, there is no problem in coupling ψ's with each other as long as field derivatives or pion fields do not enter; any isoscalar function of the ψ's alone will be chiral-invariant. [This corresponds to the fact that chiral symmetry tells us nothing about baryon masses, baryon-baryon scattering, etc.] Our task is to learn what to do with the pion fields' derivatives.

First, the pions. It would be possible to treat $\boldsymbol{\pi}$ like any other field ψ if its transformation law had the same form, i.e., if

$$-if_{ab}(\boldsymbol{\pi})=v_{ac}(\boldsymbol{\pi})(t_c)_{bd}\pi_d.$$

But this is not true, so it is not possible to make chiral-invariant interactions by merely coupling $\boldsymbol{\pi}$'s with each other and with ψ's to find form isoscalars. However, we do not run into this difficulty with $\partial_\mu\boldsymbol{\pi}$. That is, we can define a covariant derivative

$$D_\mu\pi_a\equiv d_{ab}(\boldsymbol{\pi})\partial_\mu\pi_b, \tag{4.1}$$

which transforms like an ordinary ψ field; i.e.,

$$[X_b,D_\mu\pi_c]=-iv_{ab}(\boldsymbol{\pi})\epsilon_{bcd}D_\mu\pi_d, \tag{4.2}$$

$$[T_a,D_\mu\pi_c]=-i\epsilon_{acd}D_\mu\pi_d. \tag{4.3}$$

[We are using the appropriate isospin matrix for pions, $(t_b)_{ce}=-i\epsilon_{bce}$.] It is perfectly straightforward, though rather tedious, to show that the function $d_{ab}(\boldsymbol{\pi})$ for which (4.2) and (4.3) are satisfied is uniquely given (up to a multiplicative constant) by

$$d_{ab}(\boldsymbol{\pi})\propto[f^2(\boldsymbol{\pi}^2)+\boldsymbol{\pi}^2]^{-1/2}\delta_{ab} +[f^2(\boldsymbol{\pi}^2)+\boldsymbol{\pi}^2]^{-1}[2f'(\boldsymbol{\pi}^2)-v(\boldsymbol{\pi}^2)]\pi_a\pi_b, \tag{4.4}$$

where $f(\pi^2)$ is the function appearing in the pion transformation equations and $v(\pi^2)$ is the function appearing in the transformation law of other fields;

$$v(\boldsymbol{\pi}^2)=\{f(\boldsymbol{\pi}^2)+[f^2(\boldsymbol{\pi}^2)+\boldsymbol{\pi}^2]^{1/2}\}^{-1}.$$

The covariant derivative is therefore

$$D_\mu\boldsymbol{\pi}\propto[f^2(\boldsymbol{\pi}^2)+\boldsymbol{\pi}^2]^{-1/2}\partial_\mu\boldsymbol{\pi} +[f^2(\boldsymbol{\pi}^2)+\boldsymbol{\pi}^2]^{-1}[f'(\boldsymbol{\pi}^2)-\tfrac{1}{2}v(\boldsymbol{\pi}^2)]\boldsymbol{\pi}\partial_\mu\boldsymbol{\pi}^2. \tag{4.5}$$

By virtue of Eqs. (4.2) and (4.3), any isoscalar function of $D_\mu\boldsymbol{\pi}$ and ψ's will automatically be chiral-invariant. In particular, chirality allows a gradient-coupling pion-nucleon interaction proportional to

$$\bar{N}i\gamma_5\gamma_\mu\boldsymbol{\tau}ND_\mu\boldsymbol{\pi}. \tag{4.6}$$

Also, the free-pion part of the Lagrangian is contained in the self-coupling

$$-\tfrac{1}{2}D_\mu\boldsymbol{\pi}\cdot D^\mu\boldsymbol{\pi}. \tag{4.7}$$

For a pion field defined to transform according to Eq. (2.18) the covariant pion derivative (4.1) is given by

$$D_\mu\boldsymbol{\pi}=(1+\lambda^2\boldsymbol{\pi}^2)^{-1}\partial_\mu\boldsymbol{\pi}, \tag{4.8}$$

so the pion-nucleon interaction is

$$(G_0/2m_N)\bar{N}i\gamma_5\gamma_\mu\boldsymbol{\tau}N(1+\lambda^2\boldsymbol{\pi}^2)^{-1}\partial_\mu\boldsymbol{\pi}, \tag{4.9}$$

and the pion kinematic Lagrangian is contained in

$$-\tfrac{1}{2}(1+\lambda^2\boldsymbol{\pi}^2)^{-2}\partial_\mu\boldsymbol{\pi}\partial^\mu\boldsymbol{\pi}. \tag{4.10}$$

These agree precisely with our previous results,[1] obtained by regarding chirality as a broken linear symmetry. We note once again that chirality forces the interaction of nucleons with single pions to be accompanied by interactions with three pions, five pions, etc., and correspondingly that the kinematic pion term must be accompanied with self-interaction terms representing pion-pion scattering, $2\pi\rightarrow4\pi$, etc.

We will now pause to consider what value should be given to the constant λ. (Had we reserved our full freedom to redefine the pion field, λ would be arbitrary, but in choosing the pion kinematic term to be given by Eq. (4.10) we have committed ourselves to a particular conventional normalization of π, and the value of λ thus has a meaning.) The axial-vector current defined by Noether's theorem is

$$A_a{}^\mu\equiv-2\frac{\partial\mathcal{L}}{\partial(\partial_\mu\pi_b)}f_{ab}(\boldsymbol{\pi})-2i\sum_\psi\frac{\partial\mathcal{L}}{\partial(\partial_\mu\psi)}v_{ab}(\boldsymbol{\pi})t_b\psi. \tag{4.11}$$

Referring back to (4.7) and (2.16), we see that this contains a term $\partial^\mu\pi_a$ with coefficient λ^{-1}. Hence, if $A_a{}^\mu$ is to be identified with the axial-vector current of weak interactions, we must take

$$\lambda=F_{\pi0}{}^{-1}, \tag{4.12}$$

where $F_{\pi0}$ is the unrenormalized value of the usual pion decay amplitude. Note also that if the Lagrangian contains a pion-nucleon interaction (4.9), then $A_a{}^\mu$ will contain a term

$$-(G_0/2m_N\lambda)\bar{N}i\gamma_5\gamma^\mu\tau_aN,$$

so the unrenormalized weak coupling constants satisfy the Goldberger-Treiman relation:

$$(g_A/g_V)_0=G_0/2m_N\lambda=G_0F_{\pi0}/2m_N. \tag{4.13}$$

It is noteworthy that nothing in this formalism forces the unrenormalized ratio $(g_A/g_V)_0$ to be unity.

There remains the problem of incorporating derivatives of the ψ's into a chiral-invariant Lagrangian. We note that $\partial_\mu\psi$ does not transform like ψ; i.e.,

$$[X_a,\partial_\mu\psi]=v_{ab}(\boldsymbol{\pi})t_b\partial_\mu\psi+\frac{\partial v_{ab}(\boldsymbol{\pi})}{\partial\pi_c}t_b\partial_\mu\pi_c\psi. \quad (4.14)$$

Therefore, we will try to construct a covariant derivative

$$D_\mu\psi\equiv\partial_\mu\psi+iM_c(\boldsymbol{\pi})(\partial_\mu\pi_c)\psi, \quad (4.15)$$

such that

$$[X_a,D_\mu\psi]=v_{ab}(\boldsymbol{\pi})t_bD_\mu\psi. \quad (4.16)$$

Here $M_c(\boldsymbol{\pi})$ is a matrix like t_c which acts on the suppressed isospin indices of ψ. Our problem is to find an $M_c(\boldsymbol{\pi})$ for which (4.16) is satisfied; once this is accomplished, we can build a chiral-invariant Lagrangian by coupling $D_\mu\psi$ with ψ and $D_\mu\boldsymbol{\pi}$ in any isospin-invariant way.

Comparing (4.16) with (4.14), we see that the condition to be satisfied by $M_c(\pi)$ is

$$iv_{ab}(\boldsymbol{\pi})[t_b,M_c(\boldsymbol{\pi})]$$
$$=\frac{\partial v_{ab}(\boldsymbol{\pi})}{\partial\pi_c}t_b+\frac{\partial M_c(\boldsymbol{\pi})}{\partial\pi_d}f_{ad}(\boldsymbol{\pi})+M_d(\boldsymbol{\pi})\frac{\partial f_{ad}(\boldsymbol{\pi})}{\partial\pi_c}. \quad (4.17)$$

One particular solution of Eq. (4.17) is

$$M_c(\boldsymbol{\pi})=\epsilon_{cde}t_d\pi_e[f^2(\boldsymbol{\pi}^2)-\boldsymbol{\pi}^2]^{-1/2}v(\boldsymbol{\pi}^2), \quad (4.18)$$

where $f(\boldsymbol{\pi}^2)$ is the function appearing in the pion transformation law; and as before,

$$v(\boldsymbol{\pi}^2)\equiv[f(\boldsymbol{\pi}^2)+(f^2(\boldsymbol{\pi}^2)+\boldsymbol{\pi}^2)^{1/2}]^{-1}.$$

To (4.18) we can add any solution $M_c^{(0)}$ of the homogeneous equation

$$iv_{ab}[t_b,M_c^{(0)}]=\frac{\partial M_c^{(0)}}{\partial\pi_d}f_{ad}+M_d^{(0)}\frac{\partial f_{ad}}{\partial\pi_c};$$

but this just corresponds to introducing the covariant derivative $D_\mu\boldsymbol{\pi}$ of the pion field into interactions, and need not be considered as a separate possibility. Using (4.18) in (4.15), the covariant derivative of a general field ψ is

$$D_\mu\psi=\partial_\mu\psi+iv(\boldsymbol{\pi}^2)[f^2(\boldsymbol{\pi}^2)+\boldsymbol{\pi}^2]^{-1/2}\mathbf{t}\cdot(\boldsymbol{\pi}\times\partial_\mu\boldsymbol{\pi})\psi. \quad (4.19)$$

For a pion field defined to transform according to Eq. (2.18) this is

$$D_\mu\psi=\partial_\mu\psi+2i\lambda^2(1+\lambda^2\boldsymbol{\pi}^2)^{-1}\mathbf{t}\cdot(\boldsymbol{\pi}\times\partial_\mu\boldsymbol{\pi})\psi. \quad (4.20)$$

The kinematic Lagrangian for ψ must be constructed out of $D_\mu\psi$. Using (4.20) with ψ for a Dirac field of arbitrary isospin, this gives the terms

$$-\bar{\psi}\gamma^\mu\partial_\mu\psi-2i\lambda^2(1+\lambda^2\boldsymbol{\pi}^2)^{-1}\ \gamma^\mu\mathbf{t}\psi\cdot(\boldsymbol{\pi}\times\partial_\mu\boldsymbol{\pi}). \quad (4.21)$$

We recognize here the multipion interactions derived earlier from the broken-symmetry approach,[1] and in particular we observe that the term $-2i\lambda^2\bar{\psi}\gamma^\mu\mathbf{t}\psi\cdot(\boldsymbol{\pi}\times\partial_\mu\boldsymbol{\pi})$, with λ given by (4.13), yields the universal pion scattering lengths[16] derived originally from current algebra.

To summarize what we have learned: a Lagrangian will be chiral-invariant if it is isospin-invariant and constructed out of the pion covariant derivative $D_\mu\boldsymbol{\pi}$, any general fields ψ [transforming according to Eq. (3.1)], and their covariant derivatives $D_\mu\psi$.

V. CONVENTIONAL FIELDS

We were originally led to introduce fields with nonlinear transformation rules as a substitute for larger chiral multiplets which transform linearly. We will now show how this process may be reversed. That is, we will use the nonlinearly transforming fields $\boldsymbol{\pi}$, ψ discussed in the last two sections to construct fields, for pions and for other particles of arbitrary isospin, with conventional linear-transformation properties.

Our construction is based on the following:

Lemma: Let x_a, t_a be an arbitrary $N\times N$ matrix representation of the $SU(2)\times SU(2)$ algebra; i.e.,

$$[t_a,t_b]=i\epsilon_{abc}t_c, \quad (5.1)$$

$$[t_a,x_b]=i\epsilon_{abc}x_c, \quad (5.2)$$

$$[x_a,x_b]=i\epsilon_{abc}t_c. \quad (5.3)$$

Then there exists an $N\times N$ matrix function $\Lambda(\pi)$ such that

$$[X_a,\Lambda(\boldsymbol{\pi})]=-x_a\Lambda(\boldsymbol{\pi})-\Lambda(\boldsymbol{\pi})v_{ab}(\boldsymbol{\pi})t_b, \quad (5.4)$$

where $v_{ab}(\boldsymbol{\pi})$ is the function (3.10). (We remind the reader that X_a is not a matrix, but a Hilbert-space operator which does not commute with $\boldsymbol{\pi}$.)

Proof: Using the pion transformation rule (2.4) let us write (5.4) as a Lie differential equation

$$-if_{ab}(\boldsymbol{\pi})\frac{\partial\Lambda(\boldsymbol{\pi})}{\partial\pi_b}=-x_a\Lambda(\boldsymbol{\pi})-\Lambda(\boldsymbol{\pi})v_{ab}(\boldsymbol{\pi})t_b. \quad (5.5)$$

This is soluble if and only if it satisfies an integrability condition:

$$f_{cd}\frac{\partial}{\partial\pi_d}\left(f_{ab}\frac{\partial\Lambda}{\partial\pi_b}\right)-f_{ab}\frac{\partial}{\partial\pi_b}\left(f_{cd}\frac{\partial\Lambda}{\partial\pi_d}\right)$$
$$=\left(f_{cd}\frac{\partial f_{ae}}{\partial\pi_d}-f_{ab}\frac{\partial f_{ce}}{\partial\pi_b}\right)\frac{\partial\Lambda}{\partial\pi_e}. \quad (5.6)$$

It is easy to show that (5.5) does satisfy (5.6) because f_{ab} and v_{ab} satisfy (2.9) and (3.4).

[16] Y. Tomozawa, Nuovo Cimento **46A**, 707 (1966); S. Weinberg, Ref. 14; B. Hamprecht (to be published); K. Raman and E. C. G. Sudarshan, Phys. Rev. **154**, 1499 (1967); A. P. Balachandran, M. Gundzik, and F. Nicodemi, Nuovo Cimento **44**, 1257 (1966).

Now, let us use $\Lambda(\boldsymbol{\pi})$ to construct a conventional chiral quadruplet Π_α of fields representing the pion and a 0^+ meson. Here α is an index running over the values 1, 2, 3, 0, and for t_a and x_a we use the $(\frac{1}{2},\frac{1}{2})$ representation:

$$(t_a)_{bc} = -i\epsilon_{abc}\,, \tag{5.7}$$

$$(t_a)_{b0} = (t_a)_{0b} = (t_a)_{00} = 0\,, \tag{5.8}$$

$$(x_a)_{b0} = -(x_a)_{0b} = i\delta_{ab}\,, \tag{5.9}$$

$$(x_a)_{bc} = (x_a)_{00} = 0\,, \tag{5.10}$$

the indices a, b, c running over 1, 2, 3. We can define a unit four-vector n_α which points in the 0 direction, and construct the four-vector Π_α as

$$\Pi_\alpha = \Lambda_{\alpha\beta}(\boldsymbol{\pi})n_\beta\sigma\,, \tag{5.11}$$

where σ is either a constant or a 0^+ chiral-invariant field. [There is no distinction if the mass of the σ field is sufficiently large.[1] If σ is a constant, then (5.11) really defines only three independent fields Π_a, the fourth being given[10] by $\Pi_0 = (\sigma^2 - \boldsymbol{\Pi}^2)^{1/2}$. If σ is a field,[9] then (5.11) defines four independent fields Π_α in terms of the four fields $\boldsymbol{\pi}$, σ.] The chiral transformation law for Π_α can be determined from (5.4). Since n_β is annihilated by $(t_a)_{\alpha\beta}$, the last term in (5.4) does not contribute here, and we find that

$$[X_a, \Pi_\alpha] = -(x_a)_{\alpha\beta}\Pi_\beta\,, \tag{5.12}$$

so Π_α is indeed a chiral four-vector. This is hardly surprising, for we already knew that a linearly transforming four-vector Π_α can be used to define a nonlinearly transforming triplet $\boldsymbol{\pi}$, and we have shown in Sec. II that such a pion triplet is essentially unique.

Next, consider a field ψ which transforms according to the nonlinear rule (3.1):

$$[X_a, \psi] = v_{ab}(\boldsymbol{\pi})t_b\psi\,.$$

In close analogy with (5.11), define a conventional chiral multiplet Ψ as

$$\Psi = \Lambda(\boldsymbol{\pi})\psi\,. \tag{5.13}$$

From (3.1) and (5.4) we find immediately that

$$[X_a, \Psi] = -x_a\Psi\,, \tag{5.14}$$

so Ψ indeed transforms linearly under chiral $SU(2)\times SU(2)$.

The reader can easily verify that the conventional chiral-invariant Lagrangians constructed from Π_α, $\partial_\mu\Pi_\alpha$, Ψ, and $\partial_\mu\Psi$ may always be re-expressed as functions of $D_\mu\pi$, ψ, and $D_\mu\psi$.

VI. SYMMETRY BREAKING

We will now return to the nonlinear formalism sketched in Secs. II–IV, and use it to discuss the problem of chiral symmetry breaking by the pion mass. There is no doubt that the pion mass does break chirality, for we have seen in Sec. IV that pion fields can only enter the Lagrangian accompanied with at least one derivative. The question is, when we add a term $-\frac{1}{2}m_\pi^2\boldsymbol{\pi}^2$ to the Lagrangian, should we stop there or should we also add terms proportional to $m_\pi^2(\boldsymbol{\pi}^2)^2$, $m_\pi^2(\boldsymbol{\pi}^2)^3$, etc.? We must beware of rejecting such terms on grounds of simplicity alone, for any such hypothesis has meaning only for a particular definition of the pion field. That is, if the only term in the Lagrangian which breaks $SU(2)\times SU(2)$ is the pion-mass term $-\frac{1}{2}m_\pi^2\boldsymbol{\pi}^2$, and we define a new pion field $\boldsymbol{\pi}^*$ so that $\boldsymbol{\pi} = \boldsymbol{\pi}^*(1+\alpha\boldsymbol{\pi}^{*2}+\cdots)$, then the symmetry-breaking term will be expressed in terms of $\boldsymbol{\pi}^*$ as

$$-\tfrac{1}{2}m_\pi^2\boldsymbol{\pi}^{*2}(1+2\alpha\boldsymbol{\pi}^{*2}+\cdots)\,,$$

and the π-π scattering lengths will be different from what they would be if the symmetry-breaking term were $-\frac{1}{2}m_\pi^2\boldsymbol{\pi}^{*2}$. For this reason, chiral symmetry alone can only predict one linear combination $2a_0 - 5a_2$ of the scattering lengths; the ratio a_0/a_2 cannot be determined without a more definite statement of how chirality is broken.[15]

It is natural to characterize the chiral-symmetry-breaking term in the Lagrangian in terms of its chiral transformation properties; in this way we at least avoid making hypotheses which depend on how we define the pion field. Suppose that the symmetry-breaking term is a function $\mathcal{L}_N(\boldsymbol{\pi}^2)$ which transforms according to the $(N/2, N/2)$ representation of $SU(2)\times SU(2)$, or equivalently, using the isomorphism of $SU(2)\times SU(2)$ with $O(4)$, suppose that

$$\mathcal{L}_N = t_{00\cdots0}{}^{(N)}\,, \tag{6.1}$$

where $t_{\alpha\beta\cdots\gamma}{}^{(N)}$ is a traceless symmetric tensor of rank N. The ordinary rules of tensor analysis then give

$$[X_a, \mathcal{L}_N] = -iNt_{a0\cdots0}{}^{(N)}$$

and

$$[X_b,[X_a,\mathcal{L}_N]] = -iN\{-i(N-1)t_{ab0\cdots0}{}^{(N)} + i\delta_{ab}t_{00\cdots0}{}^{(N)}\}\,.$$

But since $t^{(N)}$ is traceless, we have

$$t_{aa0\cdots0}{}^{(N)} = -t_{00\cdots0}{}^{(N)}\,,$$

and so

$$[X_a,[X_a,\mathcal{L}_N]] = N(N+2)\mathcal{L}_N\,. \tag{6.2}$$

We will use (6.2) to distinguish the different possible $\mathcal{L}_N$.

Let us now construct an $\mathcal{L}_N$ which transforms according to Eq. (6.1), i.e., which satisfies (6.2). For convenience, we will adopt a pion field which is defined to transform according to Eq. (2.18); i.e.,

$$[X_a, \pi_b] = -i\lambda^{-1}\{\tfrac{1}{2}(1-\lambda^2\boldsymbol{\pi}^2)\delta_{ab} + \lambda^2\pi_a\pi_b\}\,.$$

We then have

$$[X_a, \boldsymbol{\pi}^2] = -i\lambda^{-1}(1+\lambda^2\boldsymbol{\pi}^2)\pi_a\,,$$

so for an arbitrary function of $\boldsymbol{\pi}^2$

$$[X_a,\mathfrak{F}(\boldsymbol{\pi}^2)]=-i\lambda^{-1}(1+\lambda^2\boldsymbol{\pi}^2)\pi_a\mathfrak{F}'(\boldsymbol{\pi}^2),$$

$$[X_a,[X_a,\mathfrak{F}(\boldsymbol{\pi}^2)]]=-\tfrac{1}{2}\lambda^{-2}(1+\lambda^2\boldsymbol{\pi}^2)(3+\lambda^2\boldsymbol{\pi}^2)\mathfrak{F}'(\boldsymbol{\pi}^2)-\lambda^{-2}(1+\lambda^2\boldsymbol{\pi}^2)^2\boldsymbol{\pi}^2\mathfrak{F}''(\boldsymbol{\pi}^2).$$

Thus Eq. (6.2) just amounts to a second-order differential equation for $\mathfrak{L}(\boldsymbol{\pi}^2)$:

$$(1+\lambda^2\boldsymbol{\pi}^2)^2\boldsymbol{\pi}^2\mathfrak{L}_N''(\boldsymbol{\pi}^2)+\tfrac{1}{2}(1+\lambda^2\boldsymbol{\pi}^2)(3+\lambda^2\boldsymbol{\pi}^2)\mathfrak{L}_N'(\boldsymbol{\pi}^2)+N(N+2)\lambda^2\mathfrak{L}_N(\boldsymbol{\pi}^2)=0. \quad (6.3)$$

Since this differential equation is singular at $\boldsymbol{\pi}^2=0$, its regular solution is unique up to an over-all constant, which we can fix by requiring that the term linear in $\boldsymbol{\pi}^2$ have coefficient $-\frac{1}{2}m_\pi^2$. The solution may then be expressed as a power series in $\lambda^2\boldsymbol{\pi}^2$:

$$\mathfrak{L}_N(\boldsymbol{\pi}^2)=-\tfrac{1}{2}m_\pi^2\{-3[2N(N+2)\lambda^2]^{-1}+\boldsymbol{\pi}^2-\tfrac{1}{5}[N(N+2)+2]\lambda^2(\boldsymbol{\pi}^2)^2+(1/105)[2N^2(N+2)^2+20N(N+2)+27]\times\lambda^4(\boldsymbol{\pi}^2)^3+\cdots\}. \quad (6.4)$$

The constant term is of course without physical significance. The quartic term contributes to the π-π scattering lengths; in conjunction with the kinematic term (4.10), it gives

$$2a_0+a_2=\tfrac{3}{5}L[N(N+2)+2], \quad (6.5)$$

$$2a_0-5a_2=6L, \quad (6.6)$$

where

$$L\equiv\frac{m_\pi\lambda^2}{2\pi}\simeq\frac{G^2m_\pi}{8\pi m_N^2}\left(\frac{g_V}{g_A}\right)^2\simeq 0.115\ m_\pi^{-1}.$$

Higher terms in Eq. (6.4) contribute to more complicated processes, like $2\pi\to 4\pi$, etc. In the previously considered[14] case, $N=1$, where $\mathfrak{L}_N$ and $\partial_\mu A_a{}^\mu$ form a chiral four-vector, it is possible to sum up all these terms by finding a solution of Eq. (6.3) in closed form:

$$\mathfrak{L}_1(\boldsymbol{\pi}^2)=-\tfrac{1}{2}m_\pi^2(1+\lambda^2\boldsymbol{\pi}^2)^{-1}\boldsymbol{\pi}^2+\text{const.}, \quad (6.7)$$

in agreement (what else?) with our previous results.[1]

It is easy to derive the same scattering lengths by the methods of current algebra. Apart from the commutators of current components, what is needed in general in current-algebra calculations[17] is a knowledge of the "σ terms"; e.g.,

$$\sigma_{ab}\equiv[X_a,\partial_\mu A_b{}^\mu],$$

$$\sigma_{abc}\equiv[X_a,\sigma_{bc}],\cdots. \quad (6.8)$$

Since the divergence of the axial current is given by

$$\partial_\mu A_c{}^\mu=-i[X_c,\mathfrak{L}_N],$$

we learn from Eq. (6.2) that

$$\sigma_{abb}=N(N+2)\partial_\mu A_a{}^\mu. \quad (6.9)$$

Also, applying the Jacobi identity and the chiral commutation relations to Eq. (6.8), we find

$$\sigma_{abc}-\sigma_{bac}=\delta_{bc}\partial_\mu A_a{}^\mu-\delta_{ac}\partial_\mu A_b{}^\mu. \quad (6.10)$$

Equations (6.9) and (6.10) provide enough information to compute the π-π scattering matrix with *three* pions off the mass shell.

The numerical values of a_0 and a_2 obtained from Eqs. (6.5) and (6.6) are presented in Table I. The values for $N=1$ are those derived previously by myself,[14] while those for $N=2$ were derived by Meire and Sugawara.[18] The last two rows give the values obtained by assuming that the symmetry-breaking term in $\mathfrak{L}$ is just $-\frac{1}{2}m_\pi^2\boldsymbol{\pi}^2$, with $\boldsymbol{\pi}$ defined to transform according to Eqs. (2.18) or (2.21), respectively. The last row corresponds to Schwinger's model.[5] (Schwinger uses a pion field that transforms according to Eq. (2.18), but gets $a_0/a_2=-\frac{3}{2}$ because he assumes that the pion field *so defined* is proportional to the divergence of the axial-vector current.

Evidently chiral symmetry alone does not suffice to fix a_0 and a_2 separately, without a specific hypothesis as to how the symmetry is broken. Since some such hypothesis is needed, it seems most reasonable to assume that chirality is broken by a term in the Lagrangian whose chiral transformation properties are as simple as possible[19] (rather than by a term which looks like a simple function of some arbitrary defined field). This leads to the choice $N=1$; the symmetry-breaking term $\mathfrak{L}_1$ then forms a chiral four-vector with $\partial_\mu A_a{}^\mu$, and the scattering lengths have the previously derived values $a_0=0.20$, $a_2=-0.06$.

Only experiment can decide whether this is right. Unfortunately, the π-π scattering lengths are not so easy to measure. The high-energy–pion-production experiments which purport to measure π-π scattering can really only do so if the peripheral graph dominates over the other "absorption" graphs. This is true at small momentum transfer if the two pions are produced at a relative velocity at which they can interact strongly (as in ρ production), but if the π-π scattering lengths are as small as we think they are then peripheral production does *not* dominate when the pions are produced at low relative velocities. For similar reasons,

---

[17] See, e.g., H. Abarbanel and S. Nussinov, Ann. Phys. (N. Y.) 42, 467 (1967); also Refs. 2 and 14. It has been suggested by L. S. Kissingler [Phys. Rev. Letters **18**, 861 (1967)] that σ terms might be responsible for the $I=2$ amplitude in $K_{2\pi}$ decay. However, the usual current-algebra calculation neglects "gradient coupling" terms which are probably just as large as the σ terms, and can easily account for the observed rate of $K^+\to\pi^++\pi^0$. In fact, the real puzzle is why $\Delta I=\frac{1}{2}$ works so *well* in the nonleptonic K decays.

[18] F. T. Meiere and M. Sugawara, Phys. Rev. **153**, 1702 (1967). They did not make any explicit assumption about σ terms or the transformation of symmetry-breaking terms in the Lagrangian, but instead arbitrarily set $a_2=0$.

[19] This assumption is analogous to the assumption that $SU(3)$ is broken by an octet term in the Lagrangian, made by M. Gell-Mann [California Institute of Technology Synchrotron Laboratory Report CTSL-20, 1961 (unpublished)], and by S. Okubo [Progr. Theoret. Phys. (Kyoto) **27**, 949 (1962)].

experiments on η decay and τ decay can verify that the π-π scattering lengths are small; but they cannot provide numerical values. The only hope appears to lie in a comparison of data on K_{e4} decay with the Watson-theorem calculations of Cabibbo and Maksymovich,[20] or in a comparison of data on $\pi+N\rightarrow 2\pi+N$ at *low* energy with the current-algebra calculation of Chang.[2] In both cases a tremendous improvement in statistics will be needed before accurate values of the π-π scattering lengths can be obtained. However, the success of current algebra[2,21] in accounting for the existing data on these processes, as well as τ decay[22] and π-N scattering,[16] provides ample evidence that the scattering lengths are small.

VII. ρ MESONS

Our introduction of covariant differentiation in Sec. IV was reminiscent of the Yang-Mills theory of gauge fields.[23] We can make the analogy even closer by introducing ρ mesons to take the place of the direct vector interaction of pion pairs. Consider a general field ψ whose chiral transformation properties are given by Eq. (3.1):

$$[X_a,\psi]=v_{ab}(\boldsymbol{\pi})t_b\psi\,.$$

We may define a covariant derivative

$$\mathfrak{D}_\mu\psi\equiv\partial_\mu\psi-ig_0\mathbf{t}\cdot\boldsymbol{\rho}_\mu\psi\,, \tag{7.1}$$

and require that $\mathfrak{D}_\mu\psi$ transform like ψ; i.e.,

$$[X_a,\mathfrak{D}_\mu\psi]=v_{ab}(\boldsymbol{\pi})t_b\mathfrak{D}_\mu\psi\,. \tag{7.2}$$

This imposes on $\boldsymbol{\rho}$ the transformation law

$$[X_a,\rho_{b\mu}]=-iv_{ad}(\boldsymbol{\pi})\epsilon_{dbc}\rho_{c\mu}-ig_0^{-1}\partial_\mu v_{ab}(\boldsymbol{\pi})\,. \tag{7.3}$$

The last term looks like what we would expect from a gauge transformation of the second kind, and it has similar consequences. In particular, the covariant derivative $\mathfrak{D}_\nu\rho_{b\mu}$ does *not* transform like ψ or $\mathfrak{D}_\mu\psi$, but we can nevertheless define a covariant curl:

$$\mathfrak{F}_{b\mu\nu}\equiv\partial_\mu\rho_{b\nu}-\partial_\nu\rho_{b\mu}+g_0\epsilon_{bcd}\rho_{c\mu}\rho_{d\nu}\,, \tag{7.4}$$

such that

$$[X_a,\mathfrak{F}_{b\mu\nu}]=-iv_{ad}(\pi)\epsilon_{dbc}\mathfrak{F}_{c\mu\nu}\,. \tag{7.5}$$

The kinematic ρ-meson part of the Lagrangian is thus contained in

$$-\tfrac{1}{4}\boldsymbol{\mathfrak{F}}_{\mu\nu}\cdot\boldsymbol{\mathfrak{F}}^{\mu\nu}, \tag{7.6}$$

just as in the Yang-Mills theory.

Of course, we do not want to follow the gauge theory so far that we leave out the ρ-meson mass. The mass term can be introduced without violating chiral invariance if we note that the transformation (7.3) allows the construction of a field

$$\phi_{b\mu}\equiv\rho_{b\mu}+g_0^{-1}v(\boldsymbol{\pi}^2)[f^2(\boldsymbol{\pi}^2)+\boldsymbol{\pi}^2]^{-1/2}\epsilon_{abc}\pi_c\partial_\mu\pi_a \tag{7.7}$$

which transforms like the ψ's; i.e.,

$$[X_a,\phi_{b\mu}]=-iv_{ad}(\boldsymbol{\pi})\epsilon_{dbc}\phi_{c\mu}\,. \tag{7.8}$$

[Note that $\phi_{b\mu}$ has to transform this way because the difference between $\mathfrak{D}_\mu\psi$ and $D_\mu\psi$ is just $-ig\mathbf{t}\cdot\boldsymbol{\phi}_\mu\psi$.] The mass term then is

$$-\tfrac{1}{2}m_\rho^2\boldsymbol{\phi}_\mu^2\,. \tag{7.9}$$

If we define the pion field so that it transforms according to Eq. (2.18), then this is[24]

$$-\tfrac{1}{2}m_\rho^2[\boldsymbol{\rho}_\mu+2g_0^{-1}\lambda^2(1+\lambda^2\boldsymbol{\pi}^2)^{-1}(\boldsymbol{\pi}\times\partial_\mu\boldsymbol{\pi})]^2\,. \tag{7.10}$$

The complete Lagrangian then may be supposed to consist of (7.6) plus (7.9), or (7.10) plus the pion terms (4.7), or (4.10) and (6.4), or (6.7) plus an unknown function of ψ, $\mathfrak{D}_\mu\psi$, and $D_\mu\boldsymbol{\pi}$.

The $\rho\pi\pi$ coupling (for *soft* pions, not necessarily in ρ decay) is given by (7.10) as

$$-2m_\rho^2g_\rho^{-1}\lambda^2\boldsymbol{\rho}_\mu\cdot(\boldsymbol{\pi}\times\partial^\mu\boldsymbol{\pi})\,. \tag{7.11}$$

Thus ρ exchange automatically accounts for the values of the πN scattering lengths given current algebra.[16] It may be surprising that the $\rho\pi\pi$ coupling constant does not automatically come out equal to the $\rho\bar{N}N$ coupling constant g_0; but it should be kept in mind that $\boldsymbol{\rho}_\mu$ is not coupled to $\boldsymbol{\pi}$ like an ordinary gauge field. If we require as a separate condition that the ρ couples universally to pions and nucleons,[25] then we obtain for g_0 the familiar value[26]

$$g_0^2=2m_\rho^2\lambda^2=2m_\rho^2F_{\pi 0}^{-2}\,.$$

We can also now answer a question that has raised some doubts[27] about the validity of the current-algebra calculation[14] of the π-π scattering length: How is it that we are able to get an answer without any explicit reference to the contribution of ρ exchange? We see from (7.11) that at low energy, where the ρ propagator

[20] N. Cabibbo and A. Maksymovich, Phys. Rev. **137**, B438 (1965).

[21] S. Weinberg, Phys. Rev. Letters **17**, 336 (1966).

[22] M. Suzuki, Phys. Rev. **144**, 1154 (1966); Y. Hara and Y. Nambu, Phys. Rev. Letters **16**, 865 (1966); D. K. Elias and J. C. Taylor, Nuovo Cimento **44**, 518 (1966); S. K. Bose and S. N. Biswas, Phys. Rev. Letters **16**, 340 (1966); B. M. K. Nefkens, Phys. Letters **22**, 94 (1966); H. D. I. Abarbanel, Phys. Rev. **153**, 1547 (1967); C. Bouchiat and P. Meyer, Phys. Letters **22**, 198 (1966); L. J. Clavelli, Phys. Rev. **160**, 1384 (1967).

[23] C. N. Yang and R. L. Mills, Phys. Rev. **96**, 191 (1954).

[24] Observe that if we let $m_0\rightarrow 0$ the pion decouples from the ρ meson, and hence from all other hadrons. This corresponds to the remark of Higgs, that Goldstone bosons are not required by a "broken" symmetry like chirality when the theory contains a gauge particle of zero mass; see P. W. Higgs, Phys. Letters **12**, 132 (1964); Phys. Rev. Letters **13**, 508 (1964); Phys. Rev. **145**, 1156 (1966). Also see F. Englert and R. Brout, Phys. Rev. Letters **13**, 321 (1964); G. S. Guralnik, C. R. Hagen, and J. W. B. Hibble, *ibid.* **13**, 585 (1964); J. W. B. Hibble, Phys. Rev. **155**, 1554 (1967). It should be noted that our $\boldsymbol{\rho}_\mu$ is the $\boldsymbol{\rho}'_\mu$ of Wess and Zumino (Ref. 5) while their $\boldsymbol{\rho}_\mu$ is our $\boldsymbol{\phi}_\mu$.

[25] J. J. Sakurai, Ann. Phys. (N. Y.) **11**, 1 (1960).

[26] K. Kawarabayashi and M. Suzuki, Phys. Rev. Letters **16**, 255 (1966); Riazuddin and Fayyazuddin, Phys. Rev. **147**, 1071 (1966); F. J. Gilman and H. J. Schnitzer, *ibid.* **150**, 1362 (1966); J. J. Sakurai, Phys. Rev. Letters **17**, 552 (1966); M. Ademollo, Nuovo Cimento **46**, 156 (1966).

[27] D. F. Greenberg (private communication).

can be taken as m_ρ^{-2}, the three ρ-exchange graphs contribute an effective 4π interaction

$$4m_\rho^2 g_\rho^{-2} \lambda^4 (\boldsymbol{\pi} \times \partial^\mu \boldsymbol{\pi})^2 .$$

But this is exactly cancelled by the $(\boldsymbol{\pi} \times \partial_\mu \boldsymbol{\pi})^2$ term arising directly from (7.10). Thus, if we wish to add ρ-exchange terms to the results of current algebra for low-energy π-π scattering, we must also add compensating terms whose effect is to convert the propagator $(q^2+m_\rho^2)^{-1}$ into $-(q^2/m_\rho^2)(q^2+m_\rho^2)^{-1}$. The total contribution made by ρ exchange plus compensating terms to low-energy π-π scattering is of fourth order in m_π, and hence negligible.

ACKNOWLEDGMENTS

I am very grateful for the hospitality extended to me by the Physics Department of the Massachusetts Institute of Technology. I would also like to thank S. Coleman and J. Schwinger for interesting discussions.

Physica **96A** (1979) 327–340 © *North-Holland Publishing Co.*

PHENOMENOLOGICAL LAGRANGIANS*

STEVEN WEINBERG

Lyman Laboratory of Physics, Harvard University

and

Harvard-Smithsonian Center for Astrophysics, Cambridge, Massachusetts 02138, USA

1. Introduction: A reminiscence

Julian Schwinger's ideas have strongly influenced my understanding of phenomenological Lagrangians since 1966, when I made a visit to Harvard. At that time, I was trying to construct a phenomenological Lagrangian which would allow one to obtain the predictions of current algebra for soft pion matrix elements with less work, and with more insight into possible corrections. It was necessary to arrange that the pion couplings in the Lagrangian would all be derivative interactions, to suppress the incalculable graphs in which soft pions would be emitted from internal lines of a hard-particle process. The mathematical approach I followed[1]) at first was quite clumsy; I started with the old σ-model[2]), in which the pion is in a chiral quartet with a 0+ isoscalar σ; then performed a space-time dependent chiral rotation which transformed $\{\boldsymbol{\pi}, \sigma\}$ everywhere into $\{0, \sigma'\}$ with $\sigma' \equiv (\sigma^2 + \pi)^{1/2}$; and then re-introduced the pion field as the chiral rotation "angle". The Lagrangian obtained in this way had a complicated and unfamiliar non-linear structure, but it did have the desired property of derivative coupling, because any space-time independent part of the rotation "angle" would correspond to a symmetry of the theory, and so would not contribute to the Lagrangian.

Schwinger suggested to me that one might be able to construct a suitable phenomenological Lagrangian directly, by introducing a pion field which from the beginning would have the non-linear transformation property of chiral rotation angles, and then just obeying the dictates of chiral symmetry for such a pion field[3]). Following this suggestion, I worked out a general theory[4]) of non-linear realizations of chiral $SU(2) \times SU(2)$, which was soon after generalized to arbitrary groups in elegant papers of Callan, Coleman, Wess, and Zumino[5]), and has since been applied by many authors[6]). The importance of the approach suggested by Schwinger has been not only that it saves the work

* Research supported in part by the National Science Foundation under Grant No. PHY77-22864.

involved in the transition from an ordinary linear representation like $\{\pi, \sigma\}$ to a non-linear realization, but more important, that it makes clear that the interactions of other hadrons with soft pions does not in any way depend on the chiral transformation properties of whatever fields are associated with these hadrons, but only on their isospin.

In the decade since 1967, Schwinger's ideas have evolved into what he calls "source theory"[7]). I have been pretty much out of touch with this work, mostly because of an involvement with other lines of research, but perhaps also because I found Schwinger's conceptual framework unfamiliar. Recently, several problems have led me to think again about the use of phenomenological Lagrangians, and I find that my ideas have shifted somewhat, to a point of view that seems to me to be now not too different from the point of view of source theory.

To summarize: section 2 presents an argument that phenomenological Lagrangians can be used not only to reproduce the soft pion results of current algebra, but also to *justify* these results, without any use of operator algebra. Section 3 shows how phenomenological Lagrangians can be used to calculate corrections to the leading soft pion results to any desired order in external momenta. In section 4, the renormalization group is used to elucidate the structure of these corrections. Corrections due to the finite mass of the pion are treated in section 5. Section 6 offers speculations about a possible other application of phenomenological Lagrangians.

This article is intended as a review – I doubt that any of the material presented here is entirely new. In particular, although I have not tried here to judge the extent to which the ideas described below overlap those of source theory, I would not be surprised to find that these are points which long ago appeared in Schwinger's work. In that case, I hope that he will take this paper as a little work of translation into the Vulgate, offered as a birthday present to an old friend.

2. Current algebra without current algebra

It is well known that matrix elements for soft-pion interactions can be obtained by "current algebra", that is, by a direct use of the commutation and conservation relations of the vector and axial-vector currents of chiral SU(2) × SU(2). It is also well known that the same matrix elements may also be calculated (usually more easily) from the tree graphs in an SU(2) × SU(2)-invariant phenomenological Lagrangian. However, it has been widely supposed[8]) that the ultimate justification of the results obtained from a phenomenological Lagrangian rests on the foundation of current algebra.

According to this method of derivation, one must first use current algebra to show that soft-pion matrix elements are uniquely determined by the properties of the currents plus certain "smoothness" properties of the matrix elements. One then reflects that any chiral-invariant Lagrangian will have currents with the assumed properties, and that the tree graphs in such a theory will have the assumed smoothness properties. It follows that the matrix elements computed from these tree graphs must automatically reproduce the results of current algebra.

I would like to show in this section that the use of current algebra in the above line of reasoning is actually unnecessary. That is, the phenomenological Lagrangians themselves can be used to justify the calculation of soft-pion matrix elements from the tree graphs, without any use of operator algebra.

This remark is based on a "theorem", which as far as I know has never been proven, but which I cannot imagine could be wrong. The "theorem" says that although individual quantum field theories have of course a good deal of content, quantum field theory itself has no content beyond analyticity, unitarity, cluster decomposition, and symmetry. This can be put more precisely in the context of perturbation theory: if one writes down the most general possible Lagrangian, including *all* terms consistent with assumed symmetry principles, and then calculates matrix elements with this Lagrangian to any given order of perturbation theory, the result will simply be the most general possible S-matrix consistent with analyticity, perturbative unitarity, cluster decomposition and the assumed symmetry principles. As I said, this has not been proved, but any counterexamples would be of great interest, and I do not know of any.

With this "theorem", one can obtain and *justify* the results of current algebra simply by writing down the most general Lagrangian consistent with the assumed symmetry principles, and then deriving low energy theorems by a direct study of the Feynman graphs, without operator algebra. However, in order for this to be a derivation and not merely a mnemonic, it is necessary to include all possible terms in the Lagrangian, and take account of graphs of all orders in perturbation theory.

To illustrate this procedure, let us consider a theory of massless pions, governed by a chiral SU(2) × SU(2) symmetry, for which the pions serve as Goldstone bosons. For simplicity, we will "integrate out" whatever other degrees of freedom may be present – nucleons, ρ mesons, σ mesons, etc. – and consider only the pions. The Lagrangian will be SU(2) × SU(2)-invariant provided it conserves isospin, and is constructed only from a chiral-covariant derivative of the pion field, which by a suitable definition of the pion field may be taken in the form[4])

$$D_\mu \boldsymbol{\pi} = (\partial_\mu \boldsymbol{\pi})/(1 + \boldsymbol{\pi}^2). \qquad (1)$$

The most general such Lagrangian is an infinite series of operators of higher and higher dimensionality[9])

$$\mathscr{L} = -\tfrac{1}{2}g_2 D_\mu \boldsymbol{\pi} \cdot D^\mu \boldsymbol{\pi} - \tfrac{1}{4}g_4^{(1)}(D_\mu \boldsymbol{\pi} \cdot D^\mu \boldsymbol{\pi})^2 \\ -\tfrac{1}{4}g_4^{(2)}(D_\mu \boldsymbol{\pi} \cdot D_\nu \boldsymbol{\pi})(D^\mu \boldsymbol{\pi} \cdot D^\nu \boldsymbol{\pi}) + \cdots, \tag{2}$$

where $g_d^{(n)}$ are constants of dimensionality $[\text{mass}]^{4-d}$. Since the field $\boldsymbol{\pi}$ is dimensionless, d is just equal to the number of derivatives in the interaction. As is well known, the constant g_2 is related to the pion decay amplitude $F_\pi \simeq 190$ MeV by

$$g_2 = F_\pi^2 \tag{3}$$

but it will be more convenient here to treat g_2 in parallel with the other couplings.

According to the "theorem" quoted at the beginning of this section, such a general Lagrangian has no specific dynamical content beyond the general principles of analyticity, unitarity, cluster decomposition, Lorentz invariance, and chirality, so that when it is used to calculate pionic S-matrix elements, it yields the most general matrix elements consistent with these general principles, provided that all terms of all orders in all couplings g_2, $g_4^{(1)}$, $g_4^{(2)}$, etc., are included. One does not need the methods of current algebra for justification here; the Lagrangian (2) is so general that the only conclusions that can be drawn from it are just those which follow from the general principles with which we started.

All this becomes of practical value in the calculation of matrix elements for pions of low energy. Consider the matrix element for a process involving N_e external pion lines, carrying energies proportional to some energy scale E. Such a matrix element will have dimensionality $[\text{mass}]^{D_1}$, where

$$D_1 = 4 - N_e. \tag{4}$$

The coupling constants contributing to a given term in such a matrix element will altogether have dimensionality $[\text{mass}]^{D_2}$, where

$$D_2 = \sum_d N_d(4-d) - 2N_i - N_e. \tag{5}$$

Here N_d is the number of vertices formed from interactions with d derivatives, and N_i is the total number of internal pion lines. (The terms $-2N_i$ and $-N_e$ appear here because the pion field $\boldsymbol{\pi}$ has an unconventional normalization, with a propagator proportional to $1/g_2$ and external line "wave functions" proportional to $1/\sqrt{g_2}$. We could have used a conventionally normalized pion field $\sqrt{g_2}\boldsymbol{\pi}$, in which case the propagators and external lines would not contribute factors involving g_2, but such factors would instead be contributed by the pion fields in the interactions. The final answer is of course the same.) Ultraviolet divergences are to be absorbed into a renormalization of the infinite number of coupling parameters, defined at renormalization points with momenta proportional to some common renormalization energy scale μ. The only quantities with non-vanishing dimensionality are the common energy scale E, the common renormalization scale μ, and the coupling constants themselves, so each term in the matrix element must take the form

$$M = E^D f(E/\mu) \tag{6}$$

with

$$D = D_1 - D_2 = 4 + \sum_d N_d(d-4) + 2N_\mathrm{i}. \tag{7}$$

This can be conveniently re-written by using the well-known formula for the number of loops in a graph

$$N_\mathrm{L} = N_\mathrm{i} - \sum_d N_d + 1. \tag{8}$$

We find then

$$D = 2 + \sum_d N_d(d-2) + 2N_\mathrm{L}. \tag{9}$$

Now suppose that the characteristic pion energy E is very small, and take the renormalization scale μ to be of order E. From (6), we see that the dominant graphs will then be those with the smallest values for the exponent D. According to eq. (9), these are just the tree graphs (i.e., $N_\mathrm{L} = 0$) formed purely from the term in the Lagrangian with the lowest possible number d of derivatives, the $d = 2$ term

$$\mathscr{L}_2 = -\tfrac{1}{2} g_2 D_\mu \boldsymbol{\pi} D^\mu \boldsymbol{\pi}. \tag{10}$$

Thus without using the methods of current algebra, we arrive at the same conclusion, that matrix elements for soft pion processes may be calculated from the effective Lagrangian (10), keeping only tree graphs.

3. Corrections to soft-pion results

The real virtue of the phenomenological Lagrangian approach described in the preceding section is not that it provides an alternative derivation of a known result, but that it allows us in a systematic way to calculate corrections to this result.

For example, suppose that we want to calculate the matrix element for pion–pion scattering with Mandelstam variables s, t, u all of the same order of magnitude, and all very small. As we have seen, eq. (9) tells us that the leading term (which here is of order s) can be calculated using the $d = 2$ term (10) in the tree approximation. This gives the known matrix element[10])

$$M^{(1)}_{\rm abcd} = 4g_2^{-1}[\delta_{\rm ab}\delta_{\rm cd}s + \delta_{\rm ac}\delta_{\rm bd}t + \delta_{\rm ad}\delta_{\rm bc}u]. \tag{11}$$

In the notation used here, a, b, c, d are isovector indices associated with pion lines having four-momenta $P_{\rm A}$, $P_{\rm B}$, $P_{\rm C}$, $P_{\rm D}$ respectively and $s = -(P_{\rm A} + P_{\rm B})^2$, $t = -(P_{\rm A} - P_{\rm C})^2$, $u = -(P_{\rm A} - P_{\rm D})^2$. We here set the pion masses equal to zero, so $s + t + u$ vanishes. Also, the pion–pion scattering matrix element M is normalized so that the S-matrix element is

$$S = {\rm i}(2\pi)^4\delta^4(P_{\rm A} + P_{\rm B} - P_{\rm C} - P_{\rm D})M(2\pi)^{-6}(16E_{\rm A}E_{\rm B}E_{\rm C}E_{\rm D})^{-1/2}.$$

Now, suppose that we want also to calculate the corrections of order $s^2 \sim E^4$. Equation (9) tells us that there will arise from graphs in which there are any number of couplings (10) with $d = 2$, and either *one* vertex with $d = 4$ or *one* loop. A straight-forward calculation gives the order-s^2 corrections to M as:

$$\begin{aligned} M^{(2)}_{\rm abcd} = \frac{\delta_{\rm ab}\delta_{\rm cd}}{g_2^2}\Big[& -\frac{1}{2\pi^2}s^2\ln(-s) - \frac{1}{12\pi^2}(u^2 - s^2 + 3t^2)\ln(-t) \\ & - \frac{1}{12\pi^2}(t^2 - s^2 + 3u^2)\ln(-u) + \frac{1}{\pi^2}(\tfrac{1}{3}s^2 + \tfrac{1}{3}t^2 + \tfrac{1}{3}u^2)\ln\Lambda^2 \\ & - \tfrac{1}{2}g_4^{(1)}s^2 - \tfrac{1}{4}g_4^{(2)}(t^2 + u^2)\Big] + \text{crossed terms}. \end{aligned} \tag{12}$$

Here Λ is an ultraviolet cut-off, and "crossed terms" denotes terms given by the interchanges $b \leftrightarrow c$ and $s \leftrightarrow t$ or $b \leftrightarrow d$ and $s \leftrightarrow u$. The divergence may be eliminated by defining renormalized coupling constants

$$g_4^{(1)}(\mu) \equiv g_4^{(1)} - \frac{2}{3\pi^2}\ln\left(\frac{\Lambda^2}{\mu^2}\right), \tag{13}$$

$$g_4^{(2)}(\mu) \equiv g_4^{(2)} - \frac{4}{3\pi^2}\ln\left(\frac{\Lambda^2}{\mu^2}\right), \tag{14}$$

so that (12) becomes

$$\begin{aligned} M^{(2)}_{\rm abcd} = \frac{\delta_{\rm ab}\delta_{\rm cd}}{g_2^2}\Big[& -\frac{1}{2\pi^2}s^2\ln\left(\frac{-s}{\mu^2}\right) - \frac{1}{12\pi^2}(u^2 - s^2 + 3t^2)\ln\left(\frac{-t}{\mu^2}\right) \\ & - \frac{1}{12\pi^2}(t^2 - s^2 + 3u^2)\ln\left(\frac{-u}{\mu^2}\right) - \tfrac{1}{2}g_4^{(1)}(\mu)s^2 - \tfrac{1}{4}g_4^{(2)}(\mu)(t^2 + u^2)\Big] \\ & + \text{crossed terms}. \end{aligned} \tag{15}$$

It is true that (15) has a polynomial part with unknown coefficients, but it is far from an empty formula: the logarithmic terms have coefficients given by eq. (15) as definite functions of g_2. It is not surprising that chiral symmetry should have consequences of this sort, for the logarithmic branch points arise from intermediate states consisting of soft pion pairs[16]), and the matrix elements for producing and absorbing these soft pions are determined by chiral symmetry. What is noteworthy is that the coefficients of the logarithmic terms can be calculated in detail so easily, by a one-loop calculation using a suitable phenomenological Lagrangian.

4. Application of the renormalization group

Gell-Mann and Low showed long ago[11]) how renormalization group techniques could be used to gain information about the perturbation series for quantum electrodynamics. Without having to calculate Feynman graphs, they were able to show that the photon propagator $\Delta(s)$ is linear in $\log s$ in second order, linear in $\log s$ in fourth order, and quadratic in $\log s$ in sixth order, with the coefficient of $\log^2 s$ determined by the product of the coefficients of $\log s$ in second and fourth order. In much the same way, we can use renormalization group techniques to get detailed information about the perturbation series generated by a non-renormalizable phenomenological Lagrangian.

For illustration, let us consider the matrix element for an arbitrary reaction among soft pions, but now fix all scattering angles and isospin indices, so that the scattering amplitude can be written as function of the center-of-mass energy $E \equiv \sqrt{s}$ alone. According to eq. (9), terms of order E^2 arise from tree graphs involving only g_2; terms of order E^4 arise from one-loop graphs involving only g_2 plus tree graphs linear in $g_4^{(1)}(\mu)$ or $g_4^{(2)}(\mu)$; terms of order E^6 arise from two-loop graphs involving only g_2, plus one-loop graphs linear in $g_4^{(1)}$ or $g_4^{(2)}$, plus tree graphs linear in $g_6^{(n)}$ or the quadratic in the $g_4^{(n)}$; and so on. Hence the matrix element takes the form

$$M(E) = g_2^{1-N_e/2}\Big\{c_2E^2 + g_2^{-1}E^4\Big[F\left(\frac{E}{\mu}\right) + \sum_n c_4^{(n)}g_4^{(n)}(\mu)\Big]$$
$$+ g_2^{-2}E^6\Big[H\left(\frac{E}{\mu}\right) + \sum_n J^{(n)}\left(\frac{E}{\mu}\right)g_4^{(n)}(\mu) + g_2\sum_n c_6^{(n)}g_6^{(n)}(\mu)$$
$$+ \sum_{nm} C_6^{(nm)}g_4^{(n)}(\mu)g_4^{(m)}(\mu)\Big] + \cdots\Big\}. \tag{16}$$

Here the c_i are dimensionless functions of angle and isospin variables. The dimensionless functions F, H, J, etc., arise from loop graphs, and depend only on angle and isospin variables and on the dimensionless ratio E/μ.

The essential idea of the renormalization group method is to exploit the fact that the matrix element must be independent of the arbitrary renormalization scale μ. As applied to eq. (16), this yields the conditions

$$0=-\frac{E}{\mu}F'\left(\frac{E}{\mu}\right)+\sum_n c_4^{(n)}\mu g_4^{(n)\prime}(\mu), \tag{17}$$

$$0=-\frac{E}{\mu}H'\left(\frac{E}{\mu}\right)-\frac{E}{\mu}\sum_n J^{(n)\prime}\left(\frac{E}{\mu}\right)g_4^{(n)}(\mu)$$
$$+\sum_n J^{(n)}\left(\frac{E}{\mu}\right)\mu g_4^{(n)\prime}(\mu)+g_2\sum_n c_6^{(n)}\mu g_6^{(n)\prime}(\mu)+2\sum_{nm} c_6^{(nm)}\mu g_4^{(n)}(\mu)g_4^{(m)\prime}(\mu). \tag{18}$$

Since (17) must hold for arbitrary E and μ, both terms must be constants. For this to be true for all angles and isospins, we must then have

$$\mu g_4^{(n)\prime}(\mu)=b_4^{(n)}, \tag{19}$$

$$\frac{E}{\mu}F'\left(\frac{E}{\mu}\right)=\sum_n c_4^{(n)}b_4^{(n)} \tag{20}$$

with $b_4^{(n)}$ constants that are independent of all angle or isospin variables. Thus the terms of order E^4 can at most contain a single logarithm

$$F\left(\frac{E}{\mu}\right)=f_0+\sum_n b_4^{(n)}c_4^{(n)}\ln\left(\frac{E}{\mu}\right). \tag{21}$$

To determine the $b_4^{(n)}$ we can compare our previous result (13), (14) with the solution of eq. (19)

$$g_4^{(n)}(\mu)=b_4^{(n)}\ln\left(\frac{\mu}{\mu_0}\right) \tag{22}$$

and find

$$b_4^{(1)}=\frac{4}{3\pi^2},\quad b_4^{(2)}=\frac{8}{3\pi^2}. \tag{23}$$

Of course, eq. (21) holds with the same values of $b_4^{(n)}$ for all processes, not just π–π scattering.

Returning now to eq. (18), we can insert (22), and find

$$0=-\frac{E}{\mu}H'\left(\frac{E}{\mu}\right)-\frac{E}{\mu}\sum_n J^{(n)\prime}\left(\frac{E}{\mu}\right)b_4^{(n)}\ln\left(\frac{\mu}{\mu_0}\right)$$
$$+\sum_n J^{(n)}\left(\frac{E}{\mu}\right)b_4^{(n)}+g_2\sum_n c_6^{(n)}\mu g_6^{(n)\prime}(\mu)+2\sum_{nm} c_6^{(nm)}b_4^{(m)}g_4^{(n)}(\mu).$$

Differentiating this with respect to E then yields

$$\frac{E}{\mu}\frac{\partial}{\partial(E/\mu)}\frac{E}{\mu}\frac{\partial}{\partial(E/\mu)}H\left(\frac{E}{\mu}\right)=\sum_n b_4^{(n)}\ln\left(\frac{\mu}{\mu_0}\right)\frac{E}{\mu}\frac{\partial}{\partial(E/\mu)}\frac{E}{\mu}\frac{\partial}{\partial(E/\mu)}J^{(n)}\left(\frac{E}{\mu}\right)$$
$$+\sum_n b_4^{(n)}\frac{E}{\mu}\frac{\partial}{\partial(E/\mu)}J^{(n)}\left(\frac{E}{\mu}\right).$$

Since this must hold for all μ, we can immediately conclude that $J^{(n)}(E/\mu)$ is linear in $\ln E/\mu$ while $H(E/\mu)$ is *quadratic* in $\ln E/\mu$:

$$H(E/\mu)=h_0+h_1\ln E/\mu+h_2\ln^2 E/\mu, \tag{24}$$

$$J^{(n)}(E/\mu)=j_0^{(n)}+j_1^{(n)}\ln(E/\mu) \tag{25}$$

with coefficients of leading logarithms related by

$$h_2=\tfrac{1}{2}\sum_n b_4^{(n)}j_1^{(n)}. \tag{26}$$

It is truly a pleasure to be able to deduce such detailed information about multi-loop graphs, without ever having to calculate any of them.

5. Symmetry breaking

In the real world, chiral symmetry is broken – rather weakly for SU(2) × SU(2), fairly strongly for SU(3) × SU(3). As a result, the "soft π" and "soft K" results of current algebra, which would be precise theorems in the limit of exact chiral symmetry, became somewhat fuzzy, depending for their interpretation on a good deal of unsystematic guesswork about the smoothness of extrapolations off the mass shell. In this section, I wish to show that phenomenological Lagrangians can serve as the basis of an approach to chiral symmetry breaking, which has at least the virtue of being entirely systematic.

Quantum chromodynamics tells us that in a world with only light u and d quark fields, the Lagrangian of the strong interactions takes the form $\mathscr{L}_0+\mathscr{L}_1$, where $\mathscr{L}_0$ is invariant under global SU(2) × SU(2) transformations on the quark fields, and $\mathscr{L}_1$ is in some sense a small perturbation

$$\mathscr{L}_1=m_u\bar{u}u+m_d\bar{d}d. \tag{27}$$

We may write $\mathscr{L}_1$ as the sum of third and fourth components of chiral four-vectors[15])

$$\mathscr{L}_1^{\rm QCD}=V_3+V_4, \tag{28}$$

$$V_3=\tfrac{1}{2}(m_u-m_d)(\bar{u}u-\bar{d}d), \tag{29}$$

$$V_4 = \tfrac{1}{2}(m_u + m_d)(\bar{u}u + \bar{d}d). \tag{30}$$

Thus the S-matrix takes the form of a sum of terms of kth order in $m_u - m_d$ and lth order in $m_u + m_d$, each term having the chiral transformation property of a traceless symmetric tensor of rank $k + l$ with k indices equal to 3 and l indices equal to 4

$$S = \sum_{kl} S^{(kl)},$$

$$S^{(kl)} \propto (m_u - m_d)^k (m_u + m_d)^l. \tag{31}$$

(Strictly speaking, this is true only with a chiral-invariant infrared cut-off.) Now, the most general phenomenological Lagrangian which gives an S-matrix of this form is itself such a sum

$$\mathcal{L}^{\text{EFF}} = \sum_{k,l=0}^{\infty} \mathcal{L}^{(kl)}, \tag{32}$$

$$\mathcal{L}^{(kl)} \propto (m_u - m_d)^k (m_u + m_d)^l. \tag{33}$$

The S-matrix is therefore to be calculated with such a phenomenological Lagrangian, with $\mathcal{L}^{(kl)}$ taken as the most general function of hadronic fields and their derivatives having the chiral transformation property of a component of a traceless symmetric tensor of rank $k + l$ with k 3-indices and l 4-indices.

To see how this works in detail, let us again restrict ourselves to purely pionic processes. As is well known[12]) (and shown below) the square of the pion mass is proportional to $m_u + m_d$, so $S^{(kl)}$ and $\mathcal{L}^{(kl)}$ are of order $m_\pi^{2(k+l)}$. If we calculate a pionic process near threshold, then all of the characteristic energies E are also of order m_π. Hence (9) may be modified to give the number of powers of E and/or m_π contributed by any given graph

$$\bar{D} = 2 + \sum_{d,k,l} N_{dkl}(d - 2 + 2k + 2l) + 2N_{\text{L}}. \tag{34}$$

where N_{dkl} is the number of vertices with d derivatives, k powers of $m_d + m_u$, and l powers of $m_d - m_u$, and N_{L} is again the number of loops. If we regard $E \approx m_\pi$ as a small parameter, then the leading graphs are those with the smallest values of $\bar{D}$.

Now, there is no way to make a chiral scalar out of the pion field alone, with no derivatives, so there is no term in (34) with $d = k = l = 0$. There is a single chiral four-vector that can be formed with no derivatives, with components

$$V_i = (1 + \boldsymbol{\pi}^2)^{-1}\pi_i, \quad V_4 = -\tfrac{1}{2} + \frac{\boldsymbol{\pi}^2}{1 + \boldsymbol{\pi}^2}. \tag{35}$$

The third component is a pseudoscalar, and therefore cannot appear in the strong interaction Lagrangian. However, the fourth component is a scalar, so it yields an interaction with $d = k = 0$, $l = 1$

$$\mathscr{L}^{(01)} = -\tfrac{1}{2}m_\pi^2 g_2(1+\boldsymbol{\pi}^2)^{-1}\boldsymbol{\pi}^2. \tag{36}$$

The coupling constant is fixed here by the condition that the square of the canonically normalized field $\boldsymbol{\pi}' \equiv g_2^{1/2}\boldsymbol{\pi}$ should have coefficient $-m_\pi^2/2$, and a constant term has been discarded. Comparison of (36) with (33) shows as already mentioned that $m_\pi^2 \propto m_u + m_d$.

According to eq. (34), the leading terms in the S-matrix are given by the sum of all tree graphs ($N_L = 0$) constructed from any number of vertices (10) with $k = l = 0$, $d = 2$ and any number of vertices (36) with $k = d = 0$, $l = 1$. For pion–pion scattering, the tree graphs consist of single vertices, formed from the terms in (10) or (36) that are quartic in the pion field. This yields precisely the formulas for the π–π scattering lengths previously derived by operator methods[10]).

As before, phenomenological Lagrangians really come into their own in calculating corrections to the leading terms. Equation (34) shows that the leading corrections arise from one-loop graphs constructed from any number of vertices (10), (36), plus tree graphs constructed from any number of vertices (10), (36) and one vertex with $d + 2k + 2l = 4$. These latter vertices may be formed from functions of the pion field and its derivatives, of three different kinds:

(a) $d = 4$, $k + l = 0$. These are just the chiral scalars with four derivatives appearing in eq. (2).

(b) $d = 2$, $k + l = 1$. This is the chiral four-vector formed by multiplying the four-vector (35) with the scalar (10)

$$U_i = (1+\boldsymbol{\pi}^2)^{-3}\pi_i\partial_\mu\pi_j\partial^\mu\pi_j, \tag{37}$$

$$U_4 = -\tfrac{1}{2}(1+\boldsymbol{\pi}^2)^{-3}(1-\boldsymbol{\pi}^2)\partial_\mu\pi_j\partial^\mu\pi_j. \tag{38}$$

(c) $d = 0$, $k + l = 2$. This is the chiral tensor formed from the direct product of the four-vector (35) with itself

$$T_{ij} = (1+\boldsymbol{\pi}^2)^{-2}\pi_i\pi_j, \tag{39}$$

$$T_{i4} = -\tfrac{1}{2}(1+\boldsymbol{\pi}^2)^{-2}(1-\boldsymbol{\pi}^2)\pi_i, \tag{40}$$

$$T_{44} = \tfrac{1}{4}(1+\boldsymbol{\pi}^2)^{-2}(1-\boldsymbol{\pi}^2)^2. \tag{41}$$

The only operators here of positive parity are the chiral scalars (a), plus the components U_4, T_{ij}, and T_{44}. Hence the new terms in the effective Lagrangian which are needed to calculate the leading corrections to the tree approximation results for soft pion processes are

$$\alpha m_\pi^2(1+\boldsymbol{\pi}^2)^{-3}(1-\boldsymbol{\pi}^2)\partial_\mu\boldsymbol{\pi}\cdot\partial^\mu\boldsymbol{\pi}+\beta m_\pi^4(1+\boldsymbol{\pi}^2)^{-2}\pi_3^2$$
$$+\gamma m_\pi^4(1+\boldsymbol{\pi}^2)^{-2}(1-\boldsymbol{\pi}^2)^2, \tag{42}$$

where α, β, and γ are dimensionless constants of order unity, with

$$\beta/\gamma = 0((m_u-m_d)^2/(m_u+m_d)^2). \tag{43}$$

Current algebra calculations[12]) of the $K^+ - K^0$ mass difference yield the quark mass ratio $m_d/m_u = 1.8$, so the right-hand side of (43) is 0.08. It is interesting that the non-degeneracy of d and u introduces a rather large violation of isotopic spin conservation in the leading corrections to the usual soft pion results.

In calculating terms of a given order in E and/or m_π, we must include graphs with arbitrary numbers of vertices formed from the interactions (10) and (36). For the most part, the number of vertices that can actually occur are limited by the topology of the graphs; for instance for π–π scattering the tree graphs have just a single quartic vertex, the one-loop graphs have two quartic vertices, and so on. However, there is never any limit on the number of times the quadratic part $-\frac{1}{2}m_\pi^2 g_2\pi^2$ of eq. (36) can appear. To put this another and more convenient way, we must calculate all tree and loop graphs using a pion propagator $g_2^{-1}(q^2+m_\pi^2)^{-1}$, and not expand these propagators in powers of m_π. Thus in addition to an over-all power of m_π, the matrix elements we calculate will have singularities in m_π as well as E. These singularities are of course just what is required by perturbative unitarity[13]).

There is not so much need today for refinements in the theory of pion–pion scattering. On the other hand, there is a wide variety of experimentally interesting processes where corrections to soft π or soft K theorems may be important, including $\pi N \to 2\pi N$, $K \to 2\pi$, $K \to 3\pi$, $K \to \pi\mu\nu$, $\eta \to 3\pi$, etc. The approach outlined in this section may serve as a basis for a systematic treatment of all these processes.

6. Phenomenological Lagrangians and QCD

Handy as they are, the phenomenological Lagrangians that we have been using are only phenomenological. This is brought home to us as we calculate graphs to higher and higher order in the pion energy – in each successive order, we encounter more and more unknown parameters. Beneath the phenomenological Lagrangian of the soft pions there must lie a more nearly fundamental quantum field theory of strong interactions, which fixes all the free parameters of our phenomenological Lagrangians.

It now appears increasingly likely that this underlying theory is the renormalizable gauge theory known as quantum chromodynamics. By virtue of its asymptotic freedom, QCD predicts that the strong interactions should become weak at high energies in a certain definite way. The weakness of the interactions allows one to carry out perturbative calculations of certain quantities at high energy, and the results so far are in agreement with experiment.

However, as the energy becomes smaller, the strong interactions in QCD become stronger, and perturbation theory becomes no longer applicable. This of course is just what we want – the richness of the hadron spectrum and the absence of free quarks or gluons show clearly that perturbation theory had better *not* work at all energies. But then how do we do calculations of strong interactions at low energies?

In non-relativistic potential theory, there are well-known methods of solving the problem of calculating scattering amplitudes for potentials that are too strong to allow the use of perturbation theory. One of these methods of solution is the "quasiparticle" approach[14]). In this method, one introduces fictitious elementary particles into the theory, in rough correspondence with the bound states (or more precisely, the eigenvalues of the scattering kernel) of the theory. In order not to change the physics, one must at the same time change the potential. Since the bound states of the original theory are now introduced as elementary particles, the modified potential must not produce them also as bound states. Hence the modified potential is weaker, and can in fact be weak enough to allow the use of perturbation theory.

Following this lead, one might imagine weakening the forces of QCD by introducing some sort of infrared cut-off λ, and preserving the physical content of the theory by introducing the bound states of the theory as fictitious elementary particles. These bound states are just the ordinary hadrons, and they must be described by a chiral-invariant phenomenological Lagrangian. The parameters of the phenomenological Lagrangian would have to be functions of λ, defined by differential equations which guarantee the λ-independence of the S-matrix, with initial condition set by the requirement that the theory goes over to pure QCD in the limit $\lambda \to 0$, where there is no infrared cut-off. The hope would be that at low energy, one could continue the solution of these equations to a value of λ large enough to allow the use of perturbation theory.

It remains to be seen whether this program can be successfully carried through.

References

1) S. Weinberg, Phys. Rev. Lett. **18** (1967) 188.
2) J. Schwinger, Ann. Phys. (N.Y.) **2** (1957) 407.
M. Gell-Mann and M. Levy, Nuovo Cimento **16** (1960) 705.
3) This approach was followed by J. Schwinger, Phys. Lett. **24B** (1967) 473.
4) S. Weinberg, Phys. Rev. **166** (1968) 1568.
5) S. Coleman, J. Wess and B. Zumino, Phys. Rev. **177** (1968) 2239.
C. Callan, S. Coleman, J. Wess and B. Zumino, Phys. Rev. **177** (1968) 2247.
6) For reviews, see B.W. Lee, *Chiral Dynamics* (Gordon and Breach, New York 1972).
S. Weinberg, *Lectures on Elementary Particles and Quantum Field Theory*, S. Deser, M. Grisaru and H. Pendleton, eds. (M.I.T. Press, Cambridge, MA, 1970).
7) J. Schwinger, *Particles, Sources, and Fields* (Addison-Wesley, Reading, MA, 1973).
8) This was my own point of view in ref. 1.
9) Interactions like $(D_\mu D^\mu \boldsymbol{\pi})^2$ are omitted here, because they can be eliminated by a suitable redefinition of the pion field, and hence are not needed in the construction of the most general on-mass-shell matrix elements.
10) S. Weinberg, Phys. Rev. Lett. **17** (1966) 616.
11) M. Gell-Mann and F.E. Low, Phys. Rev. **95** (1954) 1300.
12) See e.g. S. Weinberg, in *A Festschrift for I.I. Rabi* (New York Academy of Sciences, 1977), p. 185, and references quoted therein.
13) The presence and importance of terms in the S-matrix which are not analytic in symmetry breaking parameters like m_u, m_d was pointed out by L.-F. Li and H. Pagels, Phys. Rev. Lett. **26** (1971) 1204; **27** (1971) 1089; Phys. Rev. **D5** (1972) 1509; and P. Langacker and H. Pagels, Phys. Rev. **D8** (1973) 4595.
14) S. Weinberg, Phys. Rev. **130** (1963) 776; **131** (1963) 440.
M. Scadron and S. Weinberg, Phys. Rev. **133** (1964) B1589.
M. Scadron, S. Weinberg and J. Wright, Phys. Rev. **135** (1964) B202.
15) In referring to terms in the Lagrangian as chiral four-vectors or tensors, I am of course making use of the familiar isomorphism of SU(2) × SU(2) with the four-dimensional rotation group.
16) Li and Pagels remarked in ref. 13 that the presence of massless Goldstone bosons in intermediate states makes the matrix element non-analytic in energy-momentum variables. A calculation of the coefficients of the logarithms in pion–pion scattering that uses unitarity instead of the phenomenological Lagrangian has been given by H. Lehmann, Phys. Lett. **41B** (1972) 529; H. Lehmann and H. Trute, Nucl. Phys. **B52** (1973) 280.

Volume 251, number 2 PHYSICS LETTERS B 15 November 1990

Nuclear forces from chiral lagrangians

Steven Weinberg [1]
Theory Group, Department of Physics, University of Texas, Austin, TX 78712, USA

Received 14 August 1990

The method of phenomenological lagrangians is used to derive the consequences of spontaneously broken chiral symmetry for the forces among two or more nucleons.

The forces among nucleons have been studied as much as anything in physics. Much of this work has necessarily been phenomenological: scattering data and deuteron properties are used to determine a two-nucleon interaction, which can then be used as an input to multi-nucleon calculations. As more and more has been learned about the meson spectrum, efforts have been increasingly aimed at calculating the nuclear potential as an expansion in terms of decreasing range arising from the exchange of one or more mesons of various types, but the number of free parameters rises rapidly as more and more meson types are included, especially if one attempts to extend these calculations to forces involving more than two nucleons. This paper applies methods [1] based on the chiral symmetry of quantum chromodynamics to derive an expansion of the potential among any number of low energy nucleons in powers of the nucleon momenta, which is related to but not identical with the expansion in terms of increasing range. It is not clear which expansion will be more useful in dealing with the two-nucleon problem, but the expansion in powers of momenta gives far more specific information about multi-nucleon potentials.

The lagrangian that we shall use in this work will be taken as the most general possible lagrangian involving pions and low-energy nucleons consistent with spontaneously broken chiral symmetry and other known symmetries. It is given by an infinite series of terms with increasing numbers of derivatives and/or nucleon fields, with the dependence of each term on the pion field prescribed by the rules of broken chiral symmetry. Other degrees of freedom, such as heavy vector mesons, Δ's, and antinucleons, are "integrated out": their contribution is buried in the coefficients of the series of terms in the pion–nucleon lagrangian. We shall also integrate out nucleons with momenta greater than some scale Q, which requires that these coefficients in the lagrangian be Q-dependent. Later we will consider how to make a judicious choice of Q; for the moment, it will be enough to specify that Q is substantially less than m_ρ. Any detailed model such as that of Skyrme [2] (also see ref. [3]) that embodies broken chiral symmetry will give results that are consistent with ours, but less general; in particular, we do not specify any particular higher-derivative terms in the lagrangian such as those that are introduced to stabilize skyrmions, but instead we consider all possible terms, with any numbers of derivatives, that are allowed by the symmetries of strong interactions.

Now consider the S-matrix for a scattering process with N incoming and N outgoing nucleons, all with momenta no larger than Q. The non-relativistic nature of the problem makes it appropriate to apply "old-fashioned" time-ordered perturbation theory: there is an energy denominator for every intermediate state, instead of a propagator for every internal particle line. The energy denominators associated with intermediate states involving just N nucleons are small, of order $Q^2/2m_N$, as compared with Q for the

[1] Research supported in part by the Robert A. Welch Foundation and NSF Grant PHY 9009850.

Volume 251, number 2 PHYSICS LETTERS B 15 November 1990

energy denominators associated with intermediate states involving one or more pions, and we know from the existence of shallow nuclear bound states that these small energy denominators cause the perturbation series to diverge at low energies. Therefore we will apply effective lagrangian techniques not to the S-matrix itself, but rather to the "effective potential", defined as the sum of connected old-fashioned diagrams for the S-matrix (or rather the T-matrix, since energy conservation is not imposed) *without* N-nucleon intermediate states. The full S-matrix can be obtained by solving a Lippmann–Schwinger equation (or Schrödinger equation) with this effective potential in place of the interaction hamiltonian, and with *only* N-nucleon intermediate states, each nucleon with momentum less than Q.

Let us then use our lagrangian to calculate the matrix element of the effective potential between N-nucleon states, for which the nucleon momenta are all less than Q but roughly of the same order. Since we are working with a non-renormalizable lagrangian the matrix element will contain ultraviolet divergences of increasing severity as we go to higher and higher order in the perturbation expansion, but these divergences can all be absorbed in a redefinition of the infinite number of constants in the lagrangian. With infinities eliminated in this way, the remaining integrals are effectively cut off at internal momenta of order Q. We wish to expand the effective potential in terms of increasing order in Q and m_π, so let us count powers of these quantities.

An interaction of type i may be characterized by the number d_i of derivatives and m_π factors, the number n_i of nucleon fields, and the number p_i of pion fields. Each derivative contributes one power of Q or m_π (we use the Dirac equation to express time derivatives of nucleon fields in terms of space derivatives) and each meson field factor contributes $-\frac{1}{2}$ powers of Q or m_π, arising from the familiar $1/\sqrt{2E}$ factors that accompany meson emission or absorption. Also, because each intermediate state in the *potential* contains at least one pion, each energy denominator makes a contribution no larger than of order $1/Q$ or $1/m_\pi$. Finally, each loop is accompanied with an integral over a three-momentum, and hence contributes three powers of Q. Putting this together, we see that a graph with V_i vertices of type i, L loops, and D intermediate states, contributes a term with ν powers of Q or m_π, where

$$\nu=\sum_i V_i(d_i-\tfrac{1}{2}p_i)-D+3L\,. \tag{1}$$

The numbers of energy denominators, loops, and vertices of various types as well as the number I of internal lines in a connected graph are related by the following familiar topological identities:

$$D=\sum_i V_i-1\,, \tag{2}$$

$$L=I-\sum_i V_i+1\,, \tag{3}$$

$$2I+2N=\sum_i V_i[p_i+n_i]\,. \tag{4}$$

Eq. (1) can thus be put in the form

$$\nu=2-N+2L+\sum_i V_i(d_i+\tfrac{1}{2}n_i-2)\,. \tag{5}$$

The point of writing the number of powers of Q or m_π in this way is that the coefficient of V_i is non-negative for all interactions allowed by chiral symmetry, because each purely pionic interaction has at least two derivatives, and each interaction involving pions and nucleons has at least one derivative. (Strictly speaking, since chiral symmetry is not exact non-derivative interactions are also present, but these are suppressed by factors of m_π^2, and since we are counting powers of m_π along with powers of Q these symmetry-breaking terms effectively have $d_i \geqslant 2$.) Hence the terms in the potential with the minimum number $2-N$ of powers of Q or m_π are tree graphs ($L=0$) with vertices given by just those terms in the effective hamiltonian that have $d_i=2$, $n_i=0$ or $d_i=1$, $n_i=2$ or $d_i=0$, $n_i=4$, but any number of pion fields. These can be obtained from the terms in the lagrangian with the same restrictions on d_i and n_i:

$$\begin{aligned}\mathscr{L}=&-\tfrac{1}{2}D^{-2}\,\partial_\mu\boldsymbol{\pi}\cdot\partial^\mu\boldsymbol{\pi}-\tfrac{1}{2}D^{-1}m_\pi^2\boldsymbol{\pi}^2\\&-\bar{N}[\not\partial+m_{\rm N}+2{\rm i}D^{-1}F_\pi^{-1}\gamma_5 g_{\rm A}\boldsymbol{t}\cdot\not\partial\boldsymbol{\pi}\\&+2{\rm i}D^{-1}F_\pi^{-2}\boldsymbol{t}\cdot(\boldsymbol{\pi}\times\not\partial\boldsymbol{\pi})]N-(\bar{N}\Gamma_\alpha N)(\bar{N}\Gamma^\alpha N)\,,\end{aligned} \tag{6}$$

where Γ_α and Γ^α are matrices constrained by Lorentz and isospin invariance; $g_{\rm A}\simeq 1.25$ and $F_\pi\simeq 195$ MeV are the usual axial coupling constant and pion decay amplitude; and $D\equiv 1+\boldsymbol{\pi}^2/F_\pi^2$. In the approximation in which we keep only the tree graphs calculated with

Volume 251, number 2 PHYSICS LETTERS B 15 November 1990

this lagrangian, we must also adopt the "static approximation" – that is, we neglect the nucleon kinetic energy in energy denominators, and use zero-momentum Dirac spinors in nucleon matrix elements. The interaction hamiltonian corresponding to the lagrangian (6) then becomes

$$H_{\rm int}=\tfrac{1}{2}(D^2-1)\dot{\boldsymbol{\pi}}^2+\tfrac{1}{2}(D^{-2}-1)\nabla\boldsymbol{\pi}\cdot\nabla\boldsymbol{\pi}$$
$$+\tfrac{1}{2}m_\pi^2(D^{-2}-1)\boldsymbol{\pi}^2+2F_\pi^{-4}[\bar{N}(\boldsymbol{t}\times\boldsymbol{\pi})N]^2$$
$$+\bar{N}[2F_\pi^{-1}g_A D^{-1}\boldsymbol{t}\cdot(\boldsymbol{\sigma}\cdot\nabla\boldsymbol{\pi})+2F_\pi^{-2}D\boldsymbol{t}\cdot(\boldsymbol{\pi}\times\dot{\boldsymbol{\pi}})]N$$
$$+C_S(\bar{N}N)(\bar{N}N)+C_T(\bar{N}\boldsymbol{\sigma}N)(\bar{N}\boldsymbol{\sigma}N)\,, \qquad (7)$$

where C_S and C_T are unknown constants. [Fermi statistics allows the non-derivative four-fermion interactions involving $\boldsymbol{t}$ to be written as linear combinations of the two last terms in eq. (7).

In two-nucleon states the effective potential derived in this way consists of just a conventional one-pion-exchange term, plus a direct two-nucleon interaction produced by the four-fermion terms in (7). Fourier-transforming this momentum-space effective potential gives a local coordinate-space two-nucleon potential:

$$V_{\text{2-nucleon}}=2(C_S+C_T\boldsymbol{\sigma}_1\cdot\boldsymbol{\sigma}_2)\delta^3(\boldsymbol{x}_1-\boldsymbol{x}_2)$$
$$-\left(\frac{2g_A}{F_\pi}\right)^2(\boldsymbol{t}_1\cdot\boldsymbol{t}_2)(\boldsymbol{\sigma}_1\cdot\nabla_1)(\boldsymbol{\sigma}_2\cdot\nabla_2)Y(|\boldsymbol{x}_1-\boldsymbol{x}_2|)$$
$$-(1'\leftrightarrow 2')\,, \qquad (8)$$

where $Y(r)\equiv\exp(-m_\pi r)/4\pi r$ is the usual Yukawa potential. [Throughout it should be understood that these are local potentials, containing a delta function factor like $\delta^3(\boldsymbol{x}_1'-\boldsymbol{x}_1)$ for each nucleon.]

Using eq. (7) in the tree approximation provides just the first term in an expansion in powers of Q/M, where M is some mass scale characteristic of QCD. In particular, the $\boldsymbol{x}$-coordinates in eq. (8) will be smeared out over a volume of order M^{-3} in the full effective potential. Because we have integrated out the ρ and ω vector mesons, which are known to make an important contribution to nucleon–nucleon forces, we may guess that m_ρ can be taken as a representative value for this M.

Now let us consider the effects of a two-nucleon effective potential calculated in this way. It is well known that the one-pion exchange potential itself does a good job of accounting for the higher partial waves in nucleon–nucleon scattering. On the other hand, according to current ideas [3] the dominant attraction in low energy s-wave nucleon–nucleon interactions is of range intermediate between the π and ρ Compton wavelengths, and is due to exchange of s-wave isoscalar pion pairs, with the tensor part of one-pion exchange providing a small correction responsible for the fact that the triplet s-wave two-nucleon state is bound and the single state is not. The intermediate range attraction would have to be represented here by the delta-function terms in (8), with C_S quite large and negative, and C_T much smaller. As already mentioned, constants like these in the effective potential must be given a dependence on the arbitrary momentum scale Q above which nucleons are integrated out, to ensure that final results are Q-independent. The perturbative solution of the Lippmann–Schwinger equation for s-wave two-nucleon states is roughly an expansion in powers of $C_S Q^3/(Q^2/2m_N)$, so apart from the small effects of C_T and pion exchange, we can conclude that C_S scales like $1/Q$. To be specific, in order to have bound or virtual two-nucleon states very near zero binding, it is necessary for the expansion parameter $2m_N C_S Q$ to have a value near $2\pi^2$. (It is quite mysterious here why C_S should have almost precisely the value needed just barely to bind s-wave two-nucleon states, but this is a mystery also in more conventional treatments of nuclear forces.)

Now, it is implicit in the power-counting rules used above that the constants C_S and C_T should be no larger than roughly of order $2\pi^2/m_\rho^2$, so the consistency of our approach seems to require that Q should be greater than or of the order of $m_\rho^2/2m_N$. We also must choose $Q\ll m_\rho$, which makes sense only if we can regard $2m_N$ as much larger than m_ρ. Of course, there are many unknown dimensionless factors that can enter here, so we cannot really tell whether our analysis is actually self-consistent. However, we note that m_N may at least in principle be regarded as an independent parameter which could be arbitrarily large compared with m_ρ, since in QCD with $\mathcal{N}_c$ colors the ratio m_N/m_ρ is proportional to $\mathcal{N}_c$. Even so, with all medium and short-range effects buried in the constants C_S and C_T, it is clear that the terms of lowest order in Q/m_ρ cannot account for the effects usually attributed to a short-ranged repulsive core. For this reason, and also as a check on our approach, it will

Volume 251, number 2 PHYSICS LETTERS B 15 November 1990

be important to calculate the terms in the effective potential of next order in Q/m_ρ, which arise from one-loop graphs calculated with the hamiltonian (7), as well as from tree graphs involving higher-order terms in the hamiltonian.

It is in dealing with multi-nucleon potentials that our approach becomes most useful. The diagrams for the effective potential in three-nucleon states are shown in fig. 1. They give the result

$$\begin{aligned} V_{3\text{-nucleon}} = &-\frac{1}{2}\left(\frac{2g_A}{F_\pi}\right)^4 (\boldsymbol{\sigma}_1\cdot\nabla_1)(\boldsymbol{\sigma}_2\cdot\nabla_1) \\ &\times(\boldsymbol{\sigma}_2\cdot\nabla_3)(\boldsymbol{\sigma}_3\cdot\nabla_3)(\boldsymbol{t}_1\cdot\boldsymbol{t}_2)(\boldsymbol{t}_2\cdot\boldsymbol{t}_3) \\ &\times[\tilde{Y}(|\boldsymbol{x}_1-\boldsymbol{x}_2|)Y(|\boldsymbol{x}_2-\boldsymbol{x}_3|) \\ &+\tilde{Y}(|\boldsymbol{x}_2-\boldsymbol{x}_3|)Y(|\boldsymbol{x}_1-\boldsymbol{x}_2|)] \\ &-\delta^3(\boldsymbol{x}_1-\boldsymbol{x}_2)\left(\frac{2g_A}{F_\pi}\right)^2[C_S+C_T(\boldsymbol{\sigma}_1\cdot\boldsymbol{\sigma}_2)] \\ &\times(\boldsymbol{\sigma}_2\cdot\nabla_2)(\boldsymbol{\sigma}_3\cdot\nabla_3)Y(|\boldsymbol{x}_2-\boldsymbol{x}_3|) \\ &\pm\text{permutations}, \end{aligned} \tag{9}$$

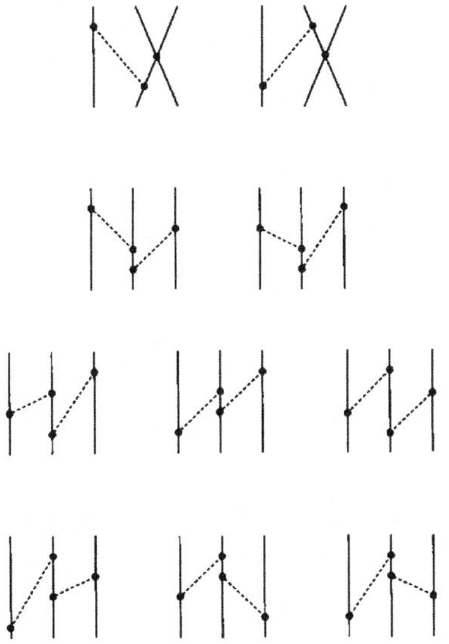

Fig. 1. Diagrams contributing to the three-nucleon effective potential. (Here solid lines are nucleons; dashed lines are pions.) It is necessary to add and subtract terms involving even or odd permutations of initial as well as final nucleons.

where

$$\tilde{Y}(r)\equiv(2\pi)^{-3}\int\frac{\exp(\mathrm{i}\boldsymbol{q}\cdot\boldsymbol{r})\,\mathrm{d}^3r}{(q^2+m_\pi^2)^{3/2}}, \tag{10}$$

and "±permutations" indicate that we must add or subtract terms involving even or odd permutations of initial as well as final nucleons. [Eq. (9) does *not* correspond to any sum of Feynmann diagrams, because we have excluded those pieces of Feynman diagrams that correspond to old-fashioned diagrams with three-nucleon intermediate states.] This does not involve any of the non-linear pion interactions in the hamiltonian (7), because the diagrams in fig. 2 all cancel. Thus in leading order, *the three-nucleon effective potential is built up out of two-nucleon interactions, with no intrinsically three-body forces.* On the other hand, the four-nucleon potential does involve the non-linear pion–nucleon couplings as well as the pion self-interaction, through diagrams like those shown in fig. 3. It will be interesting to see whether

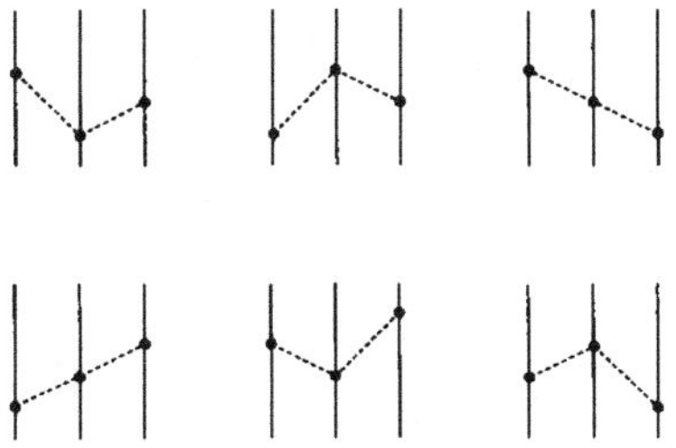

Fig. 2. Diagrams for the three-nucleon effective potential that involve the non-linear pion–nucleon interaction. These diagrams all cancel.

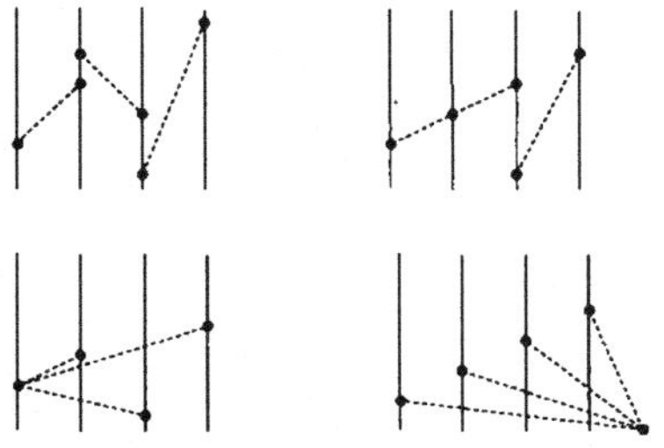

Fig. 3. An assortment of diagrams for the four-nucleon effective potential.

effects of these multi-nucleon effective potentials can be found in nuclear properties.

I am grateful to H. Georgi for his encouragement, and to G.E. Brown and A. Kerman for enlightening conversations about nuclear forces.

References

[1] S. Weinberg, Physica 96A (1979) 327, and references therein.
[2] T.H.R. Skyrme, Proc. R. Soc. A 260 (1961) 127;
E. Witten, Nucl. Phys. B 160 (1979) 57.
[3] For a review, see S.O. Bäckmann, G.E. Brown and J.A. Niskanen, Phys. Rep. 124 (1985) 1.

Nuclear Physics B363 (1991) 3–18
North-Holland

EFFECTIVE CHIRAL LAGRANGIANS FOR NUCLEON–PION INTERACTIONS AND NUCLEAR FORCES

Steven WEINBERG*

Theory Group, Department of Physics, University of Texas, Austin, TX 78712, USA

Received 2 April 1991

The general chiral invariant effective lagrangians is used to study the leading terms in powers of momenta in the S-matrix for a process involving arbitrary numbers of low-momentum pions and nucleons. This work extends and in part corrects an earlier report [S. Weinberg, Phys. Lett. B251 (1990) 288].

1. Introduction

Chiral invariant effective lagrangians were originally introduced as a labor-saving device, allowing a quick and easy derivation of the soft pion theorems that had earlier been deduced by the methods of current algebra. The lagrangians were highly non-linear and non-renormalizable, but this was not a problem because in those early days the lagrangians were only supposed to be used in the tree approximation. Later it was realized that effective chiral lagrangians could be used to calculate soft pion processes to any desired order in the pion energies, including loop as well as tree graphs; all ultraviolet divergences could be absorbed into a redefinition of the coupling constants of the lagrangian, provided one included in the lagrangian all terms consistent with chirality and other symmetry principles.

This widened view of the use of chiral effective lagrangians was first described in the context of purely pionic processes. It would clearly be to our advantage to be able to apply these methods to processes involving low-momentum nucleons as well as pions, where much more experimental data is available, but it takes a little thought to see how to do this. After all, there is no such thing as a soft nucleon.

As we shall see here, the familiar ordering of diagrams by numbers of powers of soft pion momenta can be extended to processes involving a single low-energy nucleon as well as pions. However, a special treatment is needed for problems involving two or more nucleons, including the classic problem of nuclear forces. A

*Research supported in part by the Robert A. Welch Foundation and NSF grant PHY-9009850.

preliminary treatment of such problems has already been presented*. Here we will give a fuller account, including a revised discussion of the case of three or more nucleons.

The picture of nuclear forces that emerges in the leading order in small momenta is quite crude, though not entirely unrealistic. The purpose of this work is not to improve our detailed picture of nuclear forces, which it is hardly likely could be accomplished with these methods, but rather to take a first step toward identifying those aspects of nuclear forces and pion–nucleon interactions that can be derived from the symmetry properties of quantum chromodynamics. An assessment of the success of these methods will have to wait for the evaluation of the corrections of the next few orders in small momenta.

2. Power counting and its problems

We shall deal here with reactions involving arbitrary numbers of low-energy pions and nucleons, all with three-momenta (in the rest frame of any one of the initial nucleons) less than an amount Q, of the order of one or two hundred MeV. To calculate the matrix elements for such reactions, we shall use an effective lagrangian in which we integrate out all other particle types, including heavy mesons and nucleon isobars. We also integrate out those nucleon loops that are connected to the rest of the diagram only by pion lines, burying their contributions in the constants of the effective low-energy pion interactions. The ultraviolet divergences that arise in calculations using this effective theory are absorbed into a renormalization of the parameters of this lagrangian, using renormalization points also characterized by momenta of order Q or less. After renormalization, the effective cut-off is Q, not only on virtual pion four-momenta but also on the four-momenta transferred to or from the remaining internal nucleon lines. Since Q is taken small compared with typical QCD scales such as the nucleon mass, we may treat the nucleons (but not the pions) non-relativistically, with corrections to the non-relativistic limit regarded as additional interactions with extra derivatives, and we may order terms in perturbation theory according to the number ν of powers of Q that they contain. In I this number was calculated using the methods of old-fashioned perturbation theory. It will be instructive to see here how this works in terms of the more familiar Feynman diagrams.

Chiral symmetry dictates that the form of the effective lagrangian for this theory is

$$\begin{aligned}\mathscr{L} = & -\tfrac{1}{2}D^{-2}\partial_\mu\boldsymbol{\pi}\cdot\partial^\mu\boldsymbol{\pi} - \tfrac{1}{2}D^{-1}m_\pi^2\boldsymbol{\pi}^2 \\ & + \bar{N}\left[i\partial_0 - 2D^{-1}F_\pi^{-2}\boldsymbol{t}\cdot(\boldsymbol{\pi}\times\partial_0\boldsymbol{\pi}) - m_{\mathrm{N}} - 2D^{-1}F_\pi^{-1}g_{\mathrm{A}}\boldsymbol{t}\cdot(\vec{\sigma}\cdot\vec{\nabla})\boldsymbol{\pi}\right]N \\ & - \tfrac{1}{2}C_{\mathrm{S}}(\bar{N}N)(\bar{N}N) - \tfrac{1}{2}C_{\mathrm{T}}(\bar{N}\vec{\sigma}N)\cdot(\bar{N}\vec{\sigma}N) + \ldots,\end{aligned} \tag{1}$$

* See ref. [1]. This paper will be referred to as "I" in the text.

where $g_A \simeq 1.25$ and $F_\pi = 190$ MeV are the usual axial coupling constant and pion decay amplitude; $\boldsymbol{t}$ is the isospin matrix; C_S and C_T are unknown constants; $D \equiv 1 + \boldsymbol{\pi}^2/F_\pi^2$; and the dots denote terms with more derivatives or nucleon field factors.

It is necessary to confront an important problem about the structure of this lagrangian. The space-time derivatives of the pion field and the space derivatives of the nucleon fields contribute small factors of order Q to Feynman diagrams, but time derivatives of the nucleon field contribute factors of order m_N, and hence cannot be considered small. On the other hand, by a field redefinition we can replace the chiral invariant time derivative of the nucleon field by its value given by the nucleon field equation

$$\left[i\partial_0 - 2D^{-1}F_\pi^{-2}\boldsymbol{t}\cdot(\boldsymbol{\pi}\times\partial_0\boldsymbol{\pi})\right]N = \left[m_N + 2D^{-1}F_\pi^{-1}g_A\boldsymbol{t}\cdot(\vec{\sigma}\cdot\vec{\nabla})\boldsymbol{\pi} + \ldots\right]N. \quad (2)$$

Hence including a chiral invariant time derivative of the nucleon field in the interaction lagrangian would just change the coefficients of other terms that are allowed by chiral symmetry and hence are already present. For this reason, as mentioned in I we can simply adopt a definition of the fields and of the constants in the lagrangian so that no time derivatives of the nucleon field appear in the lagrangian*.

The propagator of a nucleon with four-momentum $P+q$ (where $P=(0,0,0,m_N)$ is the four-momentum of a nucleon at rest) is given by the lagrangian (1) as** $-(q^0+i\epsilon)^{-1}$. This can be understood as the non-relativistic limit of the usual Dirac propagator,

$$\frac{-i\gamma^\mu(P+q)_\mu + m_N}{(P+q)^2 + m_N^2 - i\epsilon} \simeq \frac{-i\gamma^\mu P_\mu + m_N}{2P\cdot q - i\epsilon} = -\frac{\Lambda}{q^0 + i\epsilon}, \quad (3)$$

where $\Lambda \equiv [-i\gamma^\mu P_\mu + m_N]/2m_N$ is the projection matrix onto positive-energy zero-momentum Dirac wave functions. From eq. (3) we see that each nucleon propagator contributes -1 powers of Q to the S-matrix element. Also, each pion propagator contributes -2 powers of Q, each derivative in any interaction contributes $+1$ powers of Q, and each four-momentum integration contributes $+4$ powers of Q. The total number of powers of Q contributed by a diagram with I_n internal nucleon lines, I_p internal pion lines, V_i vertices of type i, and L loops is then

$$\nu = 4L - I_n - 2I_p + \sum_i V_i d_i, \quad (4)$$

* The similar problem of dealing with derivatives of heavy quark fields has been dealt with by Jenkins and Manohar [2] using a formalism of Georgi [3] (also see ref. [4]), with similar results.

** This is the same propagator as used in heavy quark theories by Eichten and Hill [4].

where d_i is the number of derivatives (or m_π factors) in an interaction of type i. We use the well-known topological relations

$$L = I_p + I_n - \sum_i V_i + 1, \tag{5}$$

$$2I_n + E_n = \sum_i V_i n, \tag{6}$$

where n_i is the number of nucleon fields in an interaction of type i, and E_n is the number of external nucleon lines in the diagram. Eq. (4) may then be rewritten as

$$\nu = 2 - \tfrac{1}{2}E_n + 2L + \sum_i V_i\left[d_i + \tfrac{1}{2}n_i - 2\right]. \tag{7}$$

This is a convenient expression, because chiral symmetry sets a lower bound of zero on the quantity in brackets multiplying V_i; as we see in (1), any chiral invariant interaction must have either no nucleon fields and at least two derivatives (or factors m_π), or two nucleon fields and at least one derivative, or four or more nucleon fields and any number of derivatives. Hence for any given process the leading term will be one with no loops, constructed solely from interactions with*

$$d_i + \tfrac{1}{2}n_i - 2 = 0. \tag{8}$$

The terms shown explicitly in eq. (1) are just those that satisfy this condition.

This power-counting argument provides a justification for the wide-spread use of chiral lagrangians in calculating single-nucleon processes [6], such as pion–nucleon elastic scattering or $\pi + \mathrm{N} \to \pi + \pi + \mathrm{N}$. However, it would be a disaster if this simple ordering of diagrams worked also for processes involving two or more nucleons, because it would rule out the existence of bound nuclear states. Bound states can only arise from a failure of perturbation theory, involving the participation of graphs with arbitrary numbers of loops, while our counting of powers of Q has indicated that the leading terms are those of lowest order, with no loops. Fortunately the supposedly non-leading terms are afflicted with infrared divergences that require a modification in our power counting.

To see the origin of these infrared divergences, consider the simple one-loop graph shown in fig. 1 for nucleon–nucleon scattering at zero kinetic energy. Using the approximation (3) for the nucleon propagator, this gives a matrix element

* Rho [5] has made the nice point that in the interactions of a single slowly varying external electroweak field with nucleons and pions the minimum value of the quantity $d_i + \frac{1}{2}n_i - 2$ is -1 rather than zero, so that the leading terms in these electroweak interactions are to be calculated from graphs involving a single electroweak field interaction with $d_i + \frac{1}{2}n_i - 2 = -1$, *none of which involve more than a single nucleon*, plus any number of interactions among pions and nucleons satisfying eq. (8).

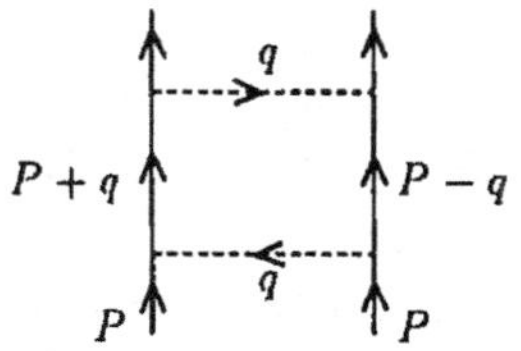

Fig. 1. A graph for nucleon–nucleon scattering that violates the simple power-counting rules. (Solid lines are nucleons; dashed lines are pions.)

proportional to

$$\int \mathrm{d}^4q\,(q^0+i\epsilon)^{-1}(q^0-i\epsilon)^{-1}(q^2+m_\pi^2)^{-2}P(q),$$

where $P(q)$ is a polynomial in the pion four-momentum q. This polynomial includes terms that are non-vanishing in the limit $q^0 \to 0$, so the integral over q^0 has an infrared divergence,

$$\int \mathrm{d}q^0\,(q^0+i\epsilon)^{-1}(q^0-i\epsilon)^{-1}.$$

The contour of integration is pinched between the two poles at $q^0 = \mp i\epsilon$, and so cannot be distorted to avoid these singularities. In contrast, for the crossed ladder graph both poles are on the same side of the integration contour, while in one-nucleon processes there is only one pole, so in these cases there are no infrared divergences.

Of course the infrared divergence in fig. 1 is not real; it only arises because we use the approximation (3) for the nucleon propagators. Including the term q^2 in $(P+q)^2$ in the denominators of the nucleon propagators shifts the poles to $q^0 \simeq \pm[\vec{q}^{\,2}/2m_\mathrm{N} - i\epsilon]$, so that the q^0 integral has the finite value $2m_\mathrm{N} i\pi/\vec{q}^{\,2}$. Equivalently, the infrared divergence forces us to include in the lagrangian the nucleon kinetic energy term,

$$\mathcal{L}_\mathrm{kin} = \bar{N}\nabla^2 N/2m_\mathrm{N}\,. \tag{9}$$

The important point is that although with these corrections the q^0 integral is finite, it is not of the order $|\vec{q}|^{-1}$ called for by our power-counting rules, but is larger by a factor of order $m_\mathrm{N}/|\vec{q}|$. The failure of perturbation theory that is manifested in nuclear binding is to be blamed on such large factors.

Rather than try to keep track of these nearly infrared-divergent graphs, it is much more convenient to switch over to old-fashioned perturbation theory, where the integrals are only over three-momenta, and the problem with our power counting is one of small energy denominators rather than nearly infrared divergent integrals over energies. Intermediate states that contain pions have energy denominators of order Q, while those containing only nucleons have much smaller energy denominators, of order Q^2/m_N. To avoid the small-energy denominators, we define an *effective potential* as the sum of connected old-fashioned perturbation theory graphs for the T-matrix excluding those with pure-nucleon intermediate states. As shown in I, the number ν of powers of Q in each term of perturbation theory for the effective potential is again given by eq. (7). In particular, the leading terms for the effective potential are given by tree graphs (i.e. $L=0$), constructed from the simplest chiral invariant interactions, satisfying eq. (8).

In using old-fashioned perturbation theory we must work with the hamiltonian rather than the lagrangian. The application of the usual rules of canonical quantization to the leading terms in (1) and (9) yields the total hamiltonian,

$$\begin{aligned} H = \int d^3x \Bigg[&\bar{N}\left[m_N - \frac{\nabla^2}{2m_N}\right]N + \tfrac{1}{2}D^2\zeta^2 + \tfrac{1}{2}D^{-2}(\vec{\nabla}\boldsymbol{\pi})\cdot(\vec{\nabla}\boldsymbol{\pi}) + \tfrac{1}{2}D^{-1}m_\pi^2\boldsymbol{\pi}^2 \\ &+ 2DF_\pi^{-2}\bar{N}\mathbf{t}\cdot(\boldsymbol{\pi}\times\boldsymbol{\zeta})N + 2F_\pi^{-1}g_A D^{-1}\bar{N}\mathbf{t}\cdot(\vec{\sigma}\cdot\vec{\nabla})\boldsymbol{\pi}N \\ &+ 2F_\pi^{-4}\left(\bar{N}(\mathbf{t}\times\boldsymbol{\pi})N\right)^2 + \tfrac{1}{2}C_S(\bar{N}N)^2 + \tfrac{1}{2}C_T(\bar{N}\vec{\sigma}N)^2\Bigg], \end{aligned} \tag{10}$$

where $i\bar{N}$ and $\boldsymbol{\zeta}$ are the canonical conjugates to N and $\boldsymbol{\pi}$, respectively. As usual, we separate this into a free-particle term,

$$H_0 = \int d^3x \left[\bar{N}\left[m_N - \frac{\nabla^2}{2m_N}\right]N + \tfrac{1}{2}\zeta^2 + \tfrac{1}{2}(\vec{\nabla}\boldsymbol{\pi})\cdot(\vec{\nabla}\boldsymbol{\pi}) + \tfrac{1}{2}m_\pi^2\boldsymbol{\pi}^2\right], \tag{11}$$

and an interaction $V \equiv H - H_0$. From H_0 we learn that in the interaction picture $\boldsymbol{\zeta} = \dot{\boldsymbol{\pi}}$, and using this in eq. (10) then gives the interaction hamiltonian in the interaction picture,

$$\begin{aligned} H_{\text{int}} = {}& \tfrac{1}{2}(D^2-1)\dot{\boldsymbol{\pi}}^2 + \tfrac{1}{2}(D^{-2}-1)\vec{\nabla}\boldsymbol{\pi}\cdot\vec{\nabla}\boldsymbol{\pi} + \tfrac{1}{2}m_\pi^2(D^{-1}-1)\boldsymbol{\pi}^2 + 2F_\pi^{-4}\left(\bar{N}(\mathbf{t}\times\boldsymbol{\pi})N\right)^2 \\ &+ \bar{N}\left[2F_\pi^{-1}g_A D^{-1}\mathbf{t}\cdot(\vec{\sigma}\cdot\vec{\nabla}\boldsymbol{\pi}) + 2F_\pi^{-2}D\mathbf{t}\cdot(\boldsymbol{\pi}\times\dot{\boldsymbol{\pi}})\right]N \\ &+ \tfrac{1}{2}C_S(\bar{N}N)(\bar{N}N) + \tfrac{1}{2}C_T(\bar{N}\vec{\sigma}N)(\bar{N}\vec{\sigma}N). \end{aligned} \tag{12}$$

As we see, the terms in this interaction hamiltonian also satisfy eq. (8).

3. Two nucleons

In the two-nucleon sector the most general connected graphs consist of ladders of $n = 1,2,3,\ldots$ two-nucleon-irreducible parts connected with $n-1$ pairs of nucleon lines. The sum over n is done by solving a Lippmann–Schwinger equation with the potential taken as the effective potential defined in sect. 2, and with only two-nucleon intermediate states. For two nucleons the effective potential is given simply by the sum of one-pion-exchange and a contact interaction arising from the four-fermion interaction in eq. (12). In coordinate space this gives [1]

$$V_{12} = \left[C_S + C_T \vec{\sigma}_1 \cdot \vec{\sigma}_2\right] \delta^3(\vec{x}_1 - \vec{x}_2)$$
$$- (2g_A/F_\pi)^2 (t_1 \cdot t_2)(\vec{\sigma}_1 \cdot \vec{\nabla}_1)(\vec{\sigma}_2 \cdot \vec{\nabla}_2) Y(|\vec{x}_1 - \vec{x}_2|) - (1' \leftrightarrow 2'), \quad (13)$$

where $Y(r) \equiv e^{-m_\pi r}/4\pi r$ is the usual Yukawa potential.

This yields only a crude simulation of nucleon–nucleon forces. In particular, the $j = T = 0$ pion pair exchange that is believed (for reviews see ref. [7]) to be responsible for the strong medium-range attraction needed for nuclear binding is buried here in the constant C_S; in this leading order it is not possible also to see the effects of the hard repulsive core attributed to the exchange of heavier mesons. Still, (8) is only intended as a lowest-order approximation to the two-nucleon potential, and as such is not so bad.

There is also a deep problem here, noted already in I: despite the small energy denominators contributed by two-nucleon states, it is hard to see why the perturbation series here should diverge at low energies, and in particular why nuclear bound and virtual states should be so shallow. The problem can be seen most easily if we adopt the approximation that the effective potential in coordinate space is here simply

$$V_{12} = C\,\delta^3(\vec{x}_1 - \vec{x}_2). \quad (14)$$

This is true in the $l = 0$ spin singlet partial wave in proton–neutron scattering if the relevant distances are much less than the pion Compton wavelength (which is believed to be a fair approximation.) In this case

$$C \equiv C_S - 3C_T + \frac{g_A^2}{F_\pi^2}. \quad (15)$$

Spin triplet nucleon–nucleon scattering is more complicated because in this case the one-pion-exchange tensor force mixes $l = 0$ and $l = 2$, but (14) may not be too bad a representation of the effective potential at very low energies. If we ignore pion exchange, then the potential in the triplet state takes the form (14) with $C = C_S + C_T$. In general, for an effective potential of the form (14), the Lippmann–Schwinger equation for the $l = 0$ T-matrix in momentum space reads

$$T_{\vec{k}',\vec{k}} = C + \int \frac{d^3k''}{(2\pi)^3} \frac{C T_{\vec{k}'',\vec{k}}}{\frac{\vec{k}^2}{m_N} - \frac{\vec{k}''^2}{m_N} + i\epsilon}, \tag{16}$$

where $\vec{k}$ is the momentum of one of the nucleons in the center-of-mass system. The solution of this sort of equation is well known,

$$T_{\vec{k}',\vec{k}} = \frac{C}{1 - C(2\pi)^{-3} \int d^3k'' \left[\frac{\vec{k}'^2}{m_N} - \frac{\vec{k}''^2}{m_N} + i\epsilon \right]^{-1}}. \tag{17}$$

The integrals here are divergent, but the divergence is removed by renormalization. We may define the renormalized value C_R of C as the value of the T-matrix $T_{\vec{k}',\vec{k}}$ at $\vec{k} = \vec{k}' = 0$. That is

$$C^{-1} = C_R^{-1} - m_N(2\pi)^{-3} \int d^3k'' / \vec{k}''^2. \tag{18}$$

Using eq. (18), we may rewrite eq. (17) as the finite expression

$$T_{\vec{k}',\vec{k}} = \frac{C_R}{1 - m_N C_R (2\pi)^{-3} \int d^3k'' \frac{\vec{k}^2}{\vec{k}''^2 \left[\vec{k}^2 - \vec{k}''^2 + i\epsilon \right]}}$$

$$= \frac{C_R}{1 + i m_N C_R |\vec{k}| / 4\pi}. \tag{19}$$

This is equivalent to the statement that the scattering length is

$$a = m_N C_R / 4\pi \tag{20}$$

and that the effective range is much smaller. Also, if C_R is positive then there is a pole at $\vec{k}^2 = -16\pi^2/m_N^2 C_R^2$, and hence a bound state with binding energy

$$B = \frac{16\pi^2}{m_N^3 C_R^2}\,. \tag{21}$$

The trouble with all this is that our power counting arguments rest on the tacit assumption that any coupling with dimensionality (mass)$^{-d}$ is of the order of some QCD energy scale m_{QCD} to the power $-d$, so that C_R should be roughly of order m_{QCD}^{-2}. If we regard m_N as a typical QCD energy scale, and ignore all factors of 2 and π, then we should expect the nucleon–nucleon scattering lengths be of the order of a typical QCD length scale and the binding energy of the deuteron (and other nuclei) to be of the order of a typical QCD energy scale, which they most certainly are not.

This is a problem for all theories of nuclear forces, not just for the chiral lagrangian approach. One possible way of resolving the problem within the present formalism is to note that the nucleon mass is in a sense an independent parameter, since it is proportional to the number N_c of colors* while other QCD mass scales like m_ρ are not. According to this view, the factors $1/m_N$ in energy denominators like that in eq. (16) enhance the effect of pure-nucleon intermediate states, leading to a divergence of the perturbation series at energies much less than the QCD scale. Be that as it may, because m_N is large the problem of accounting for the large nucleon–nucleon scattering lengths and small deuteron binding energy is not as severe as one might think. Eq. (7) indicates that each loop contributes a factor of Q^2 to the effective potential, but it also contributes a numerical factor of order $1/2\pi^2$, so in our power counting we are really assuming that a coupling of dimensionality (mass)$^{-d}$ is of the order** of $(2\pi^2)^{d/2} m_{QCD}^{-d}$. Hence it is $C_R/2\pi^2$ rather than C_R itself that must be compared with m_{QCD}^{-2}. In fact, eq. (21) indicates that in order to give a deuteron binding energy of 2.2 MeV, $C_R/2\pi^2$ in the triplet state must have the large value (260 MeV)$^{-2}$. The problem is worse in the singlet state; the singlet neutron–proton scattering length $a_s = -24.3 \times 10^{-13}$ cm requires that $-C_R/2\pi^2$ in the singlet state must have a very large value, (110 MeV)$^{-2}$.

* It was shown by Witten [8] that in the limit $N_c \to \infty$ of quantum chromodynamics the nucleon is a soliton, such as a skyrmion, so if the nucleon mass is to be regarded as of a different order of magnitude than the QCD scale then we ought to treat nucleons as solitons. Unfortunately the results obtained for skyrmions depend on the choice of terms introduced into the lagrangian to stabilize the skyrmion. With the simplest choice of such terms, the skyrmion–skyrmion interaction potential in the spin triplet state turns out [9] to be attractive, with a depth at its deepest point of about 25 MeV. (It would be interesting to repeat these calculations for the spin singlet state.) Of course for $N_c \to \infty$ the nucleon kinetic energy is negligible, so this would be the deuteron binding energy. From this point of view, as pointed out to me by Witten, the very small binding energy of the deuteron (and the even smaller energy of the singlet virtual state) would have to be regarded as due to a fortuitous cancellation of the potential and kinetic energies for the physical value $N_c = 3$.

** This is the same sort of counting of powers of 2 and π as used in the "naive dimensional analysis" of Georgi et al. [10].

The same difficulty can be seen in a different way if we define the renormalized value of C as the value of T at a variable renormalization point, $\vec{k}^2 = -\mu^2$. The solution of the renormalization group equation here is obtained by simply substituting $-\mu^2$ for $\vec{k}^2$ in eq. (19),

$$C(\mu) = \frac{C_R}{1 + m_N C_R \mu / 4\pi}. \tag{22}$$

Note that for $\mu \gg 4\pi/m_N C_R$, the dimensionless quantity $C(\mu)\mu m_N$ approaches the fixed point 4π. It is not surprising that the renormalized value of C scales as μ^{-1}. At the renormalization scale μ the delta-function potential (14) is effectively smeared over a radius $b \simeq \mu^{-1}$, so in order to have a volume integral $C(\mu)$, the depth V_0 of the potential must be of order $C(\mu)\mu^3$; the familiar condition $V_0 \sim b^{-2}$ for a short-ranged potential to have a finite effect thus requires that $C(\mu) \propto \mu^{-1}$. What *is* surprising is that for $\mu \gg 4\pi/m_N C_R$, the parameter $C(\mu)\mu m_N$ takes the specific value 4π. Quantum chromodynamics does not seem to provide any reason why this should be the case. (It is like having a square well potential of range b and depth V_0 with $m_N V_0 b^2$ very near the value $\pi^2/4$ that yields a single bound state of zero binding energy.) Of course, the problem only arises if $4\pi/m_N C_R$ is small compared to QCD scales like m_ρ, because otherwise eq. (22) loses its validity before the fixed point is approached. In fact, $4\pi/m_N C_R$ is approximately 46 MeV in the triplet state and 8 MeV in the singlet state, so $C(\mu)\mu m_N$ comes near its fixed point long before eq. (22) becomes invalid.

In any case, we shall from now on adopt the rule that a pure-nucleon intermediate state counts as if it contributed *two* more factors of $1/Q$ than other intermediate states, so that each pure-nucleon intermediate state compensates for one additional loop in the diagrams for the T-matrix.

4. Three nucleons

The leading terms in the three-nucleon effective potential may be identified by the power counting rules described in sect. 2. They are given by the graphs of fig. 2, which yield the result quoted in I,

$$\begin{aligned} V_{123} = &-\frac{1}{2}\left(\frac{2g_A}{F_\pi}\right)^4 (\vec{\sigma}_1 \cdot \vec{\nabla}_1)(\vec{\sigma}_2 \cdot \vec{\nabla}_1)(\vec{\sigma}_2 \cdot \vec{\nabla}_3)(\vec{\sigma}_3 \cdot \vec{\nabla}_3)(t_1 \cdot t_2)(t_2 \cdot t_3) \\ &\times \left[\tilde{Y}(|\vec{x}_1 - \vec{x}_2|)Y(|\vec{x}_2 - \vec{x}_3|) + \tilde{Y}(|\vec{x}_2 - \vec{x}_3|)Y(|\vec{x}_1 - \vec{x}_2|)\right] \\ &- \delta^3(\vec{x}_1 - \vec{x}_2)\left(\frac{2g_A}{F_\pi}\right)^2 \left[C_S + C_T(\vec{\sigma}_1 \cdot \vec{\sigma}_2)\right] \\ &\times (\vec{\sigma}_2 \cdot \vec{\nabla}_2)(\vec{\sigma}_3 \cdot \vec{\nabla}_3) Y(|\vec{x}_2 - \vec{x}_3|) \pm \text{permutations}, \end{aligned} \tag{23}$$

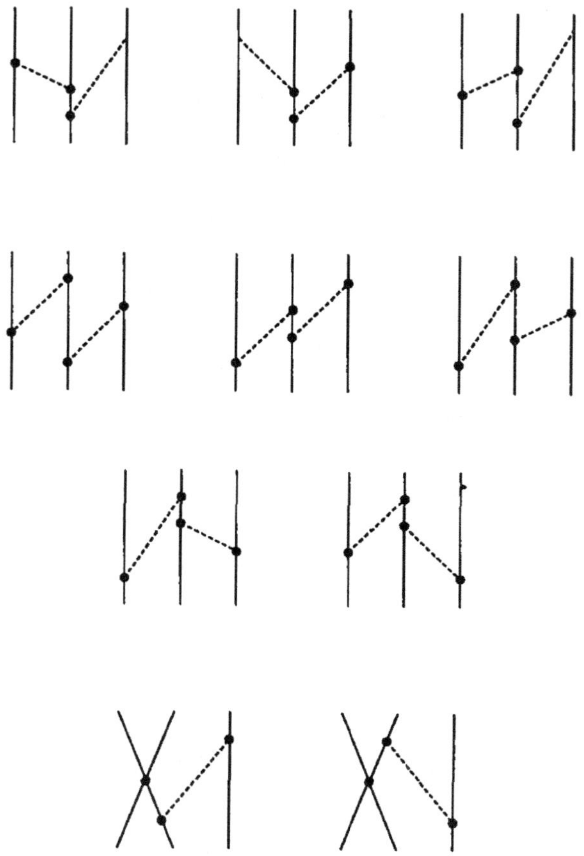

Fig. 2. Graphs for the three-nucleon effective potential. Other graphs involving non-linear pion interactions all cancel.

where

$$\tilde{Y}(r) \equiv (2\pi)^{-3} \int \frac{e^{i\vec{q}\cdot\vec{r}}\, d^3q}{\left(q^2 + m_\pi^2\right)^{3/2}}, \tag{24}$$

and "± permutations" indicates that we must add or subtract terms involving even or odd permutations of initial as well as final nucleons. Nothing new needs to be said here about the three-nucleon effective potential itself, but there is an important complication in the use of the effective potential for three or more nucleons that was not mentioned in I, and needs to be explained here.

It is not true that the complete S-matrix element for interactions among $N \geqslant 3$ nucleons is given by summing ladder graphs with $n = 1, 2, \ldots$ connected N-nucleon irreducible subgraphs (i.e. effective potentials), connected by $(n-1)$ N-nucleon

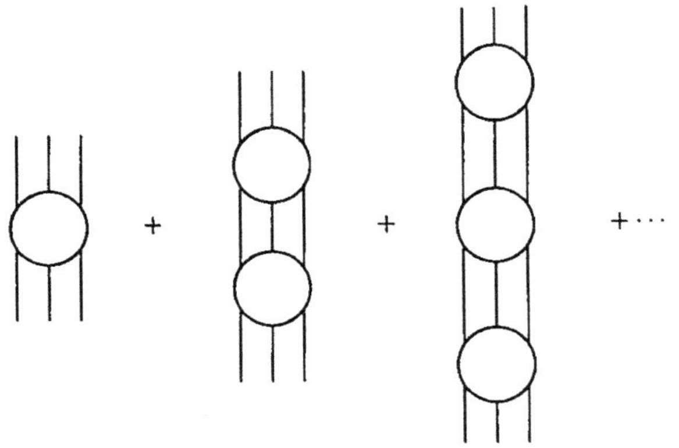

Fig. 3. Sum of graphs for the three-nucleon S-matrix. Circles here denote sums of three-nucleon irreducible subgraphs, shown in fig. 4.

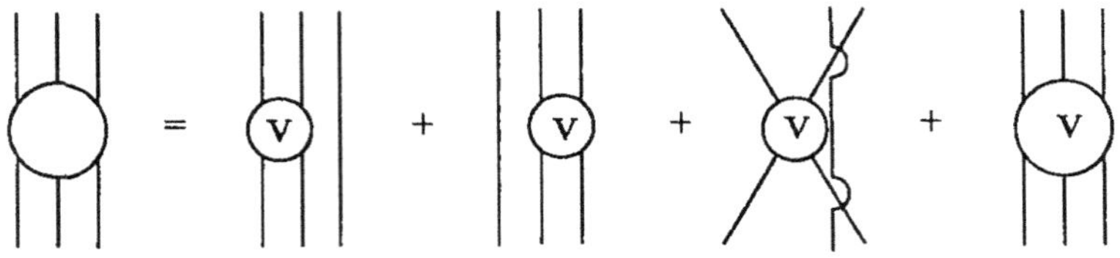

Fig. 4. Sum of graphs for the three-nucleon irreducible subgraphs in fig. 3. Here circles marked "V" denote effective potentials, i.e. connected three-nucleon irreducible graphs.

links. In order to calculate the S-matrix as a series of this sort, shown in fig. 3, we would have to include in the sum of three-nucleon irreducible subgraphs not only the effective potential graphs of fig. 2, but also terms that include disconnected parts*, such as those shown in fig. 4. These disconnected parts in fact make a larger contribution to the three-nucleon irreducible kernel than the three-nucleon effective potential itself. According to eq. (7), the leading tree graphs for the effective two-nucleon and three-nucleon potentials make contributions respectively of order Q^0 and Q^{-1}, but the extra momentum-conservation delta function in the disconnected graphs makes a contribution effectively of order Q^{-3}, so the disconnected graphs are more important by a factor Q^{-2}. For instance, fig. 5 shows two different graphs for the connected three-nucleon S-matrix, each with a *single* three-nucleon intermediate state connecting a pair of three-nucleon irreducible subgraphs. Fig. 5a, in which these two three-nucleon irreducible subgraphs are connected terms appearing in the effective three-nucleon potential, contains two extra loops, and hence makes a contribution to the connected three-nucleon S-matrix which is smaller by a factor Q^4 than fig. 5b, where the two three-nucleon

* I thank S. Coleman for this important remark.

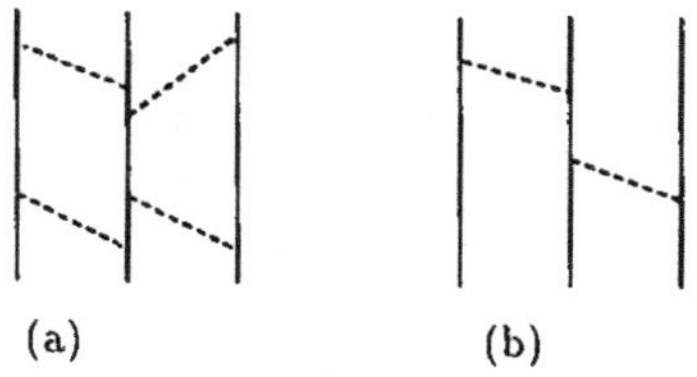

Fig. 5. Two connected graphs for the three-nucleon S-matrix, each with a single three-nucleon intermediate state.

irreducible subgraphs are both disconnected. We conclude that if we sum the series in fig. 3 by rewriting it as a Lippmann–Schwinger integral equation, then the kernel must be calculated using a potential

$$V = V_{12} + V_{13} + V_{23}, \tag{25}$$

with the three-nucleon effective potential (23) entering only in corrections of higher order in Q. This of course is just what has usually been assumed in dealing with the three-nucleon system.

The disconnected terms in the three-nucleon irreducible kernel contain two delta functions, and hence are of order Q^{-6}. Each three-nucleon intermediate state contributes a factor Q^9 from the three-momentum space integrals, so if as discussed above we take the energy denominator for such a state as being of order Q^3, then each such intermediate state contributes a net factor Q^6, just compensating for one of the three-nucleon irreducible kernels. Thus the leading graphs for the three-nucleon S-matrix does indeed consist of a series of terms like those in fig. 3. It is only the presence of the disconnected three-nucleon irreducible terms that makes it necessary to sum the series of fig. 3, or in other words to solve a Lippmann–Schwinger equation*, rather than taking the three-nucleon S-matrix as the effective potential itself.

5. More nucleons

The case of $N \geqslant 4$ nucleons can be dealt with in a similar fashion. The most general N-nucleon irreducible graph consists of c connected parts, to each of which is attached $N_1, N_2, \ldots, N_c$ incoming nucleon lines and an equal number of outgoing nucleon lines, plus d disconnected single lines, with $\Sigma_r N_r + d = N$.

* Because the Lippman–Schwinger integral equation has a disconnected kernel, it cannot be solved by numerical methods that approximate the kernel as a finite matrix. As is well known, for three [11] or more [12] particles it is possible to replace the Lippmann–Schwinger equation by an integral equation with a connected and hence square-integrable kernel, which can be solved by ordinary numerical methods. The leading term in this kernel is calculated using the potential (25), with the connected three-body graphs of fig. 2 providing a higher-order correction.

According to eq. (7), the leading connected N_r-nucleon irreducible tree graphs with N_r incoming and N_r outgoing nucleon lines is of order Q^{2-N_r}, and each extra connected part or single line contributes a three-dimensional delta function to the whole kernel, so the whole graph is of order $Q^{\nu'}$, where

$$\nu' = \sum_{r=1}^{c} (2 - N_r) - 3(c+d) = -c - 2d - N. \tag{26}$$

The leading contributions to the N-nucleon irreducible kernel are hence those with the maximum number $d = N - 2$ of disconnected single lines, plus a single two-nucleon connected piece. Again, the Lippmann–Schwinger equation can be used to calculate the N-nucleon S-matrix, with the potential in the kernel being given by a sum of the $N(N-1)/2$ two-nucleon potentials V_{ij}. In a way it is a pity that the specifically non-linear terms in the pion–nucleon interaction are not manifested except as higher-order corrections.

6. First corrections*

There are several sources of corrections to the leading terms discussed above, but the first corrections with just one extra power of Q are quite limited. Each added loop and each decrease in the number of disconnected pieces of the effective potential contributes two extra powers of Q, so we must calculate the first corrections to the effective potential still using only tree graphs, and still as a sum of two-body terms. However now we must calculate the two-nucleon effective potential including one vertex with $d_i + \frac{1}{2}n_i - 2 = +1$, as well as arbitrary numbers of vertices satisfying eq. (8) that are constructed from the terms in the lagrangian shown explicitly in eq. (1). Parity conservation does not permit any interactions with four nucleon fields and one derivative, and Lorentz invariance does not allow any purely pionic interactions with three spacetime derivatives. This leaves just one-nucleon interactions with $n_i = 2$ and $d_i = 2$. These include the terms

$$D^{-2}\left(\vec{\nabla}\boldsymbol{\pi}\right)^2 \bar{N}N, \tag{27}$$

$$D^{-2}\epsilon_{ijk}\epsilon_{abc}\partial_i\pi_a\,\partial_j\pi_b\,\bar{N}\sigma_k t_c N \tag{28}$$

as well as the interaction terms in the chiral invariant completion of the nucleon kinetic energy term (9):

$$\bar{N}\left[\vec{\nabla} + 2iD^{-1}\boldsymbol{t}\cdot\left(\boldsymbol{\pi}\times\vec{\nabla}\boldsymbol{\pi}\right)\right]^2 N/2m_{\mathrm{N}}. \tag{29}$$

* The work of this section was done in collaboration with C. Ordóñez and U. van Kolck.

In addition there are terms with the chiral symmetry transformation properties of the quark mass terms: one term proportional to $m_u + m_d \propto m_\pi^2$ that transforms as the fourth component of a chiral four-vector*

$$D^{-1}\pi^2\bar{N}N, \tag{30}$$

and another term proportional to $m_u - m_d$ that transforms as the third component of a different chiral four-vector**

$$D^{-1}\pi_3\bar{N}(t\cdot\pi)N. \tag{31}$$

However, apart from any questions of power counting, the only tree graphs that can contribute to the *two*-nucleon effective potential are a direct nucleon–nucleon interaction and one-pion exchange. The interactions (27)–(31) all involve at least two pion fields, so none of them can contribute to the two-nucleon effective potential in the tree approximation. In particular, the absence of any contribution in this order from the isospin-violating term (31) provides some justification for an approximation that seems to be nearly universal among nuclear physicists, that isospin violation arises solely from photon exchange and from the effect of nucleon mass differences in the Lippmann–Schwinger equations (both effects that have not been considered here), but not from effects of quark mass differences on one-pion exchange or other strong interactions among nucleons.

There is another possible term in the lagrangian that appears to contain a term linear in the pion field,

$$D^{-1}\bar{N}(t\cdot\dot{\pi})\,\vec{\sigma}\cdot\left[\vec{\nabla}+2iD^{-1}t\cdot(\pi\times\vec{\nabla}\pi)\right]N,$$

but by integration by parts and using the nucleon field equation this can be written in terms of other chiral invariant terms of the same order, all of which involve at least two powers of the pion field. We conclude that *there are no corrections to the leading terms in the interactions of two or more nucleons that are smaller than the leading terms by just one factor Q*.

There are various corrections of order Q^2, arising from the following sources:

(i) Two-nucleon effective potentials calculated from either (1) a single interaction having four nucleon fields and two derivatives or one quark mass factor, or (2) a one-pion exchange graph with a correction to one of the pion–nucleon vertices involving three derivatives or one derivative and one quark mass factor.

(ii) Two-nucleon effective potentials calculated from one-loop graphs, with all interactions satisfying (8).

* This is the so-called sigma term.

** The presence of this term in pion–nucleon scattering was pointed out in ref. [13].

(iii) Three-nucleon effective potentials calculated from tree graphs with all interactions satisfying (8), as shown in fig. 2.

It will be interesting to see if more of the observed features of nuclear forces emerge in these corrections of order Q^2. Such calculations are being pursued by C. Ordóñez and U. van Kolck.

I am grateful for discussions with many colleagues, especially S. Coleman, J. Negele, C. Ordóñez, J. Polchinski, E.F. Redish, U. van Kolck and E. Witten.

References

[1] S. Weinberg, Phys. Lett. B251 (1990) 288
[2] E. Jenkins and A.V. Manohar, Phys. Lett. B255 (1991) 558
[3] H. Georgi, Phys. Lett. B240 (1990) 447;
N. Isgur and M.B. Wise, Phys. Lett. B232 (1989) 113
[4] E. Eichten and B. Hill, Phys. Lett. B234 (1990) 511
[5] M. Rho, Phys. Rev. Lett. 66 (1991) 1275
[6] J. Gasser, M.E. Saino and A. Svarc, Nucl. Phys. B307 (1988) 779;
J. Gasser, H. Leutwyler, M.P. Locher and M.E. Sainio, Phys. Lett. B213 (1988) 85 and references therein
[7] S.O. Bäckman, G.E. Brown and J.A. Niskanen, Phys. Rep. 124 (1985) 1;
R. Machleidt, K. Hohlinde and Ch. Elster, Phys. Rep. 149 (1987) 1
[8] E. Witten, Nucl. Phys. B160 (1979) 57; B223 (1983) 433;
G.S. Adkins, C.R. Nappi and E. Witten, Nucl. Phys. B228 (1983) 552
[9] E. Braaten and L. Carson, Phys. Rev. Lett. 56 (1986) 1897;
A.J. Schramm, Y. Dothan and L.C. Biedenharn, Phys. Lett. B205 (1988) 151
[10] A. Manohar and H. Georgi, Nucl. Phys. B234 (1984) 189;
H. Georgi and L. Randall, Nucl. Phys. B276 (1986) 241
[11] L.D. Faddeev, Sov. Phys. JETP 12 (1961) 1014; Sov. Phys. Dokl. 6 (1961) 384; 7 (1963) 600;
C.A. Lovelace, *in* Proc. Scottish Universities Summer School, ed. R.G. Moorhouse (Edinburgh, 1963) p. 437
[12] S. Weinberg, Phys. Rev. B133 (1964) 232
[13] S. Weinberg, *in* A Festschrift for I.I. Rabi, Trans. N.Y. Acad. Sci. 38 (1977) 185

Reproduced from Chapter 16 of *Conceptual Foundations of Quantum Field Theory*, Tian Yu Cao (Ed.), Cambridge University Press, 2009. ISBN 9780511470813.

Part Six

16. What is quantum field theory, and what did we think it was?

STEVEN WEINBERG*

Quantum field theory was originally thought to be simply the quantum theory of fields. That is, when quantum mechanics was developed physicists already knew about various classical fields, notably the electromagnetic field, so what else would they do but quantize the electromagnetic field in the same way that they quantized the theory of single particles? In 1926, in one of the very first papers on quantum mechanics,[1] Born, Heisenberg and Jordan presented the quantum theory of the electromagnetic field. For simplicity they left out the polarization of the photon, and took space-time to have one space and one time dimension, but that didn't affect the main results. (Comment from audience.) Yes, they were really doing string theory, so in this sense string theory is earlier than quantum field theory. Born *et al.* gave a formula for the electromagnetic field as a Fourier transform and used the canonical commutation relations to identify the coefficients in this Fourier transform as operators that destroy and create photons, so that when quantized this field theory became a theory of photons. Photons, of course, had been around (though not under that name) since Einstein's work on the photoelectric effect two decades earlier, but this paper showed that photons are an inevitable consequence of quantum mechanics as applied to electromagnetism.

The quantum theory of particles like electrons was being developed at the same time, and made relativistic by Dirac[2] in 1928–1930. For quite a long time many physicists thought that the world consisted of both fields and particles: the electron is a particle, described by a relativistically invariant version of the Schrödinger wave equation, and the electromagnetic field is a field, even though it also behaves like particles. Dirac I think never really changed his mind about this, and I believe that this was Feynman's understanding when he first developed the path integral and worked out his rules for calculating in quantum electrodynamics. When I first learned about the path-integral formalism, it was in terms of electron trajectories (as it is also presented in the book by Feynman and Hibbs[3]). I already thought that wasn't the best way to look at electrons, so this gave me a distaste for the path-integral formalism, which although unreasonable lasted until I learned of 't Hooft's work[4] in 1971. I feel it's all right to mention autobiographical details like that as long as the story shows how the speaker was wrong.

In fact, it was quite soon after the Born–Heisenberg–Jordan paper of 1926 that the idea came along that in fact one could use quantum field theory for everything, not just for electromagnetism. This was the work of many theorists during the period

* Research supported in part by the Robert A. Welch Foundation and NSF grant PHY 9009850. E-mail address: weinberg@utaphy.ph.utexas.edu

1928–1934, including Jordan, Wigner, Heisenberg, Pauli, Weisskopf, Furry, and Oppenheimer. Although this is often talked about as second quantization, I would like to urge that this description should be banned from physics, because a quantum field is not a quantized wave function. Certainly the Maxwell field is not the wave function of the photon, and for reasons that Dirac himself pointed out, the Klein–Gordon fields that we use for pions and Higgs bosons could not be the wave functions of the bosons. In its mature form, the idea of quantum field theory is that quantum fields are the basic ingredients of the universe, and particles are just bundles of energy and momentum of the fields. In a relativistic theory the wave function is a functional of these fields, not a function of particle coordinates. Quantum field theory hence led to a more unified view of nature than the old dualistic interpretation in terms of both fields and particles.

There is an irony in this. (I'll point out several ironies as I go along – this whole subject is filled with delicious ironies.) It is that although the battle is over, and the old dualism that treated photons in an entirely different way from electrons is I think safely dead and will never return, some calculations are actually easier in the old particle framework. When Euler, Heisenberg and Kockel[5] in the mid-1930s calculated the effective action (often called the Euler–Heisenberg action) of a constant external electromagnetic field, they calculated to all orders in the field, although their result is usually presented only to fourth order. This calculation would probably have been impossible with the old fashioned perturbation theory techniques of the time, if they had not done it by first solving the Dirac equation in a constant external electromagnetic field and using those Dirac wave functions to figure out the effective action. These techniques of using particle trajectories rather than field histories in calculations have been revived in recent years. Under the stimulus of string theory, Bern and Kosower,[6] in particular, have developed a useful formalism for doing calculations by following particle world lines rather than by thinking of fields evolving in time. Although this approach was stimulated by string theory, it has been reformulated entirely within the scope of ordinary quantum field theory, and simply represents a more efficient way of doing certain calculations.

One of the key elements in the triumph of quantum field theory was the development of renormalization theory. I'm sure this has been discussed often here, and so I won't dwell on it. The version of renormalization theory that had been developed in the late 1940s remained somewhat in the shade for a long time for two reasons: (1) for the weak interactions it did not seem possible to develop a renormalizable theory, and (2) for the strong interactions it was easy to write down renormalizable theories, but since perturbation theory was inapplicable it did not seem that there was anything that could be done with these theories. Finally all these problems were resolved through the development of the standard model, which was triumphantly verified by experiments during the mid-1970s, and today the weak, electromagnetic and strong interactions are happily all described by a renormalizable quantum field theory. If you had asked me in the mid-1970s about the shape of future fundamental physical theories, I would have guessed that they would take the form of better, more all-embracing, less arbitrary, renormalizable quantum field theories. I gave a talk at the Harvard Science Center at around this time, called 'The renaissance of quantum field theory,' which shows you the mood I was in.

There were two things that especially attracted me to the ideas of renormalization and quantum field theory. One of them was that the requirement that a physical

https://doi.org/10.1017/CBO9780511470813.020 Published online by Cambridge University Press

theory be renormalizable is a precise and rational criterion of simplicity. In a sense, this requirement had been used long before the advent of renormalization theory. When Dirac wrote down the Dirac equation in 1928 he could have added an extra 'Pauli' term[7] which would have given the electron an arbitrary anomalous magnetic moment. Dirac could (and perhaps did) say 'I won't add this term because it's ugly and complicated and there's no need for it.' I think that in physics this approach generally makes good strategies but bad rationales. It's often a good strategy to study simple theories before you study complicated theories because it's easier to see how they work, but the purpose of physics is to find out why nature is the way it is, and simplicity by itself is I think never the answer. But renormalizability was a condition of simplicity which was being imposed for what seemed after Dyson's 1949 papers[8] like a rational reason, and it explained not only why the electron has the magnetic moment it has, but also (together with gauge symmetries) all the detailed features of the standard model of weak, electromagnetic, and strong interactions, aside from some numerical parameters.

The other thing I liked about quantum field theory during this period of tremendous optimism was that it offered a clear answer to the ancient question of what we mean by an elementary particle: it is simply a particle whose field appears in the Lagrangian. It doesn't matter if it's stable, unstable, heavy, light – if it's field appears in the Lagrangian then it's elementary, otherwise it's composite.**

Now my point of view has changed. It has changed partly because of my experience in teaching quantum field theory. When you teach any branch of physics you must motivate the formalism – it isn't any good just to present the formalism and say that it agrees with experiment – you have to explain to the students why this is the way the world is. After all, this is our aim in physics, not just to describe nature, but to explain nature. In the course of teaching quantum field theory, I developed a rationale for it, which very briefly is that it is the only way of satisfying the principles of Lorentz invariance plus quantum mechanics plus one other principle.

Let me run through this argument very rapidly. The first point is to start with Wigner's definition of physical multi-particle states as representations of the inhomogeneous Lorentz group.[9] You then define annihilation and creation operators $a(\vec{p},\sigma,n)$ and $a^\dagger(\vec{p},\sigma,n)$ that act on these states (where $\vec{p}$ is the three-momentum, σ is the spin z-component, and n is a species label). There's no physics in introducing such operators, for it is easy to see that any operator whatever can be expressed as a functional of them. The existence of a Hamiltonian follows from time-translation invariance, and much of physics is described by the S-matrix, which is given by the well known Feynman–Dyson series of integrals over time of time-ordered products of the interaction Hamiltonian $H_{\rm I}(t)$ in the interaction picture:

$$S=\sum_{n=0}^{\infty}\frac{(-i)^n}{n!}\int_{-\infty}^{\infty}dt_1\int_{-\infty}^{\infty}dt_2\cdots\int_{-\infty}^{\infty}dt_n\times T\{H_{\rm I}(t_1)H_{\rm I}(t_2)\cdots H_{\rm I}(t_n)\}. \qquad (1)$$

This should all be familiar. The other principle that has to be added is the cluster decomposition principle, which requires that distant experiments give uncorrelated results.[10] In order to have cluster decomposition, the Hamiltonian is written not

** We should not really give quantum field theory too much credit for clarifying the distinction between elementary and composite particles, because some quantum field theories exhibit the phenomenon of bosonization: at least in two dimensions there are theories of elementary scalars that are equivalent to theories with elementary fermions.

just as any functional of creation and annihilation operators, but as a power series in these operators with coefficients that (aside from a *single* momentum-conservation delta function) are sufficiently smooth functions of the momenta carried by the operators. This condition is satisfied for an interaction Hamiltonian of the form

$$H_1(t) = \int d^3x\, \mathcal{H}(\vec{x}, t), \tag{2}$$

where $\mathcal{H}(x)$ is a power series (usually a polynomial) with terms that are local in annihilation fields, which are Fourier transforms of the annihilation operators:

$$\psi_\ell^{(+)}(x) = \int d^3p \sum_{\sigma,n} e^{ip\cdot x} u_\ell(\vec{p}, \sigma, n) a(\vec{p}, \sigma, n), \tag{3}$$

together of course with their adjoints, the creation fields.

So far this all applies to non-relativistic as well as relativistic theories.[†] Now if you also want Lorentz invariance, then you have to face the fact that the time-ordering in the Feynman–Dyson series (1) for the S-matrix doesn't look very Lorentz invariant. The obvious way to make the S-matrix Lorentz invariant is to take the interaction Hamiltonian density $\mathcal{H}(x)$ to be a scalar, and also to require that these Hamiltonian densities commute at spacelike separations

$$[\mathcal{H}(x), \mathcal{H}(y)] = 0 \quad \text{for spacelike } x - y, \tag{4}$$

in order to exploit the fact that time ordering *is* Lorentz invariant when the separation between space-time points is timelike. In order to satisfy the requirement that the Hamiltonian density commute with itself at spacelike separations, it is constructed out of fields which satisfy the same requirement. These are given by sums of fields that annihilate particles plus fields that create the corresponding antiparticles

$$\psi_\ell(x) = \sum_{\sigma,n} \int d^3p [e^{ip\cdot x} u_\ell(\vec{p}, \sigma, n) a(\vec{p}, \sigma, n) + e^{-ip\cdot x} v_\ell(\vec{p}, \sigma, n) a^\dagger(\vec{p}, \sigma, \bar{n})], \tag{5}$$

where $\bar{n}$ denotes the antiparticle of the particle of species n. For a field ψ_ℓ that transforms according to an irreducible representation of the homogeneous Lorentz group, the form of the coefficients u_ℓ and v_ℓ is completely determined (up to a single over-all constant factor) by the Lorentz transformation properties of the fields and one-particle states, and by the condition that the fields commute at spacelike separations. Thus the whole formalism of fields, particles, and antiparticles seems to be an inevitable consequence of Lorentz invariance, quantum mechanics, and cluster decomposition, without any ancillary assumptions about locality or causality.

This discussion has been extremely sketchy, and is subject to all sorts of qualifications. One of them is that for massless particles, the range of possible theories is slightly larger than I have indicated here. For example, in quantum electrodynamics, in a physical gauge like Coulomb gauge, the Hamiltonian is not of the form (2) – there is an additional term, the Coulomb potential, which is bilocal and serves to cancel a non-covariant term in the propagator. But relativistically invariant quantum theories

[†] By the way, the reason that quantum field theory is useful even in non-relativistic statistical mechanics, where there is often a selection rule that makes the actual creation or annihilation of particles impossible, is that in statistical mechanics you have to impose a cluster decomposition principle, and quantum field theory is the natural way to do so.

https://doi.org/10.1017/CBO9780511470813.020 Published online by Cambridge University Press

always (with some qualifications I'll come to later) do turn out to be quantum field theories, more or less as I have described them here.

One can go further, and ask why we should formulate our quantum field theories in terms of Lagrangians. Well, of course creation and annihilation operators by themselves yield pairs of canonically conjugate variables: from the as and $a^\dagger$s, it is easy to construct qs and ps. The time-dependence of these operators is dictated in terms of the Hamiltonian, the generator of time translations, so the Hamiltonian formalism is trivially always with us. But why the Lagrangian formalism? Why do we enumerate possible theories by giving their Lagrangians rather than by writing down Hamiltonians? I think the reason for this is that it is only in the Lagrangian formalism (or more generally the action formalism) that symmetries imply the existence of Lie algebras of suitable quantum operators, and you need these Lie algebras to make sensible quantum theories. In particular, the S-matrix will be Lorentz invariant if there is a set of 10 sufficiently smooth operators satisfying the commutation relations of the inhomogeneous Lorentz group. It's not trivial to write down a Hamiltonian that will give you a Lorentz invariant S-matrix – it's not so easy to think of the Coulomb potential just on the basis of Lorentz invariance – but if you start with a Lorentz invariant Lagrangian density then because of Noether's theorem the Lorentz invariance of the S-matrix is automatic.

Finally, what is the motivation for the special gauge invariant Lagrangians that we use in the standard model and general relativity? One possible answer is that quantum theories of mass zero, spin one particles violate Lorentz invariance unless the fields are coupled in a gauge invariant way, while quantum theories of mass zero, spin two particles violate Lorentz invariance unless the fields are coupled in a way that satisfies the equivalence principle.

This has been an outline of the way I've been teaching quantum field theory these many years. Recently I've put this all together into a book,[11] now being sold for a negligible price. The bottom line is that quantum mechanics plus Lorentz invariance plus cluster decomposition implies quantum field theory. But there are caveats that have to be attached to this, and I can see David Gross in the front row anxious to take me by the throat over various gaps in what I have said, so I had better list these caveats quickly to save myself.

First of all, the argument I have presented is obviously based on perturbation theory. Second, even in perturbation theory, I haven't stated a clear theorem, much less proved one. As I mentioned there are complications when you have things like mass zero, spin one particles for example; in this case you don't really have a fully Lorentz invariant Hamiltonian density, or even one that is completely local. Because of these complications, I don't know how even to state a general theorem, let alone prove it, even in perturbation theory. But I don't think that these are insuperable obstacles.

A much more serious objection to this not-yet-formulated theorem is that there's already a counter example to it: string theory. When you first learn string theory it seems in an almost miraculous way to give Lorentz invariant, unitary S-matrix elements without being a field theory in the sense that I've been using it. (Of course it is a field theory in a different sense – it's a two-dimensional conformally invariant field theory, but not a quantum field theory in four space-time dimensions.) So before even being formulated precisely, this theorem suffers from at least one counter example.

Another fundamental problem is that the S-matrix isn't everything. Space-time could be radically curved, not just have little ripples on it. Also, at finite temperature there's no S-matrix because particles cannot get out to infinite distances from a collision without bumping into things. Also, it seems quite possible that at very short distances the description of events in four-dimensional flat space-time becomes inappropriate.

Now, all of these caveats really work only against the idea that the final theory of nature is a quantum field theory. They leave open the view, which is in fact the point of view of my book, that although you cannot argue that relativity plus quantum mechanics plus cluster decomposition necessarily leads only to quantum field theory, it is very likely that any quantum theory that at sufficiently low energy and large distances looks Lorentz invariant and satisfies the cluster decomposition principle will also at sufficiently low energy *look* like a quantum field theory. Picking up a phrase from Arthur Wightman, I'll call this a folk theorem. At any rate, this folk theorem is satisfied by string theory and we don't know of any counter examples.

This leads us to the idea of effective field theories. When you use quantum field theory to study low-energy phenomena, then according to the folk theorem you're not really making any assumption that could be wrong, unless of course Lorentz invariance or quantum mechanics or cluster decomposition is wrong, provided you don't say specifically what the Lagrangian is. As long as you let it be the most general possible Lagrangian consistent with the symmetries of the theory, you're simply writing down the most general theory you could possibly write down. This point of view has been used in the last 15 years or so to justify the use of effective field theories, not just in the tree approximation where they had been used for some time earlier, but also including loop diagrams. Effective field theory was first used in this way to calculate processes involving soft π mesons,[12] that is, π mesons with energy less than about $2\pi F_\pi \approx 1200\,\text{MeV}$. The use of effective quantum field theories has been extended more recently to nuclear physics,[13] where although nucleons are not soft they never get far from their mass shell, and for that reason can be also treated by similar methods as the soft pions. Nuclear physicists have adopted this point of view, and I gather that they are happy about using this new language because it allows one to show in a fairly convincing way that what they've been doing all along (using two-body potentials only, including one-pion exchange and a hard core) is the correct first step in a consistent approximation scheme. The effective field theory approach has been applied more recently to superconductivity. Shankar, I believe, in a contribution to this conference is talking about this. The present educated view of the standard model, and of general relativity,[14] is again that these are the leading terms in effective field theories.

The essential point in using an effective field theory is you're not allowed to make any assumption of simplicity about the Lagrangian. Certainly you're not allowed to assume renormalizability. Such assumptions might be appropriate if you were dealing with a fundamental theory, but not for an effective field theory, where you must include all possible terms that are consistent with the symmetry. The thing that makes this procedure useful is that although the more complicated terms are not excluded because they're non-renormalizable, their effect is suppressed by factors of the ratio of the energy to some fundamental energy scale of the theory. Of course, as you go to higher and higher energies, you have more and more of these suppressed terms that you have to worry about.

On this basis, I don't see any reason why anyone today would take Einstein's general theory of relativity seriously as the foundation of a quantum theory of gravitation, if by Einstein's theory is meant the theory with a Lagrangian density given by just the term $\sqrt{g}R/16\pi G$. It seems to me there's no reason in the world to suppose that the Lagrangian does not contain all the higher terms with more factors of the curvature and/or more derivatives, all of which are suppressed by inverse powers of the Planck mass, and of course don't show up at any energy far below the Planck mass, much less in astronomy or particle physics. Why would anyone suppose that these higher terms are absent?

Likewise, since now we know that without new fields there's no way that the renormalizable terms in the standard model could violate baryon conservation or lepton conservation, we now understand in a rational way why baryon number and lepton number are as well conserved as they are, without having to assume that they are exactly conserved.[††] Unless someone has some *a priori* reason for exact baryon and lepton conservation of which I haven't heard, I would bet very strong odds that baryon number and lepton number conservation are in fact violated by suppressed non-renormalizable corrections to the standard model.

These effective field theories are non-renormalizable in the old Dyson power-counting sense. That is, although to achieve a given accuracy at any given energy, you need only take account of a finite number of terms in the action, as you increase the accuracy or the energy you need to include more and more terms, and so you have to know more and more. On the other hand, effective field theories still must be renormalizable theories in what I call the modern sense: the symmetries that govern the action also have to govern the infinities, for otherwise there will be infinities that can't be eliminated by absorbing them into counter terms to the parameters in the action. This requirement is automatically satisfied for unbroken global symmetries, such as Lorentz invariance and isotopic spin invariance and so on. Where it's not trivial is for gauge symmetries. We generally deal with gauge theories by choosing a gauge before quantizing the theory, which of course breaks the gauge invariance, so it's not obvious how gauge invariance constrains the infinities. (There is a symmetry called BRST invariance[15] that survives gauge fixing, but that's non-linearly realized, and non-linearly realized symmetries of the action are not symmetries of the Feynman amplitudes.) This raises a question, whether gauge theories that are not renormalizable in the power-counting sense are renormalizable in the modern sense. The theorem that says that infinities are governed by the same gauge symmetries as the terms in the Lagrangian was originally proved back in the old days by 't Hooft and Veltman[16] and Lee and Zinn-Justin[17] only for theories that are renormalizable in the old power-counting sense, but this theorem has only recently been extended to theories of the Yang–Mills[18] or Einstein type with arbitrary numbers of complicated interactions that are not renormalizable in the power-counting sense.[‡] You'll be reassured to know that these theories are renormalizable in the modern sense, but there's no proof that this will be true of all quantum field theories with local symmetries.

[††] The extra fields required by low-energy supersymmetry may invalidate this argument.

[‡] I refer here to work of myself and Joaquim Gomis,[19] relying on recent theorems about the cohomology of the Batalin–Vilkovisky operator by Barnich, Brandt and Henneaux.[20] Earlier work along these lines but with different motivation was done by Voronov, Tyutin and Lavrov;[21] Anselmi;[22] and Harada, Kugo and Yamawaki.[23]

I promised you a few ironies today. The second one takes me back to the early 1960s when S-matrix theory was very popular at Berkeley and elsewhere. The hope of S-matrix theory was that, by using the principles of unitarity, analyticity, Lorentz invariance and other symmetries, it would be possible to calculate the S-matrix, and you would never have to think about a quantum field. In a way, this hope reflected a kind of positivistic puritanism: we can't measure the field of a pion or a nucleon, so we shouldn't talk about it, while we do measure S-matrix elements, so this is what we should stick to as ingredients of our theories. But more important than any philosophical hang-ups was the fact that quantum field theory didn't seem to be going anywhere in accounting for the strong and weak interactions.

One problem with the S-matrix program was in formulating what is meant by the analyticity of the S-matrix. What precisely are the analytic properties of a multiparticle S-matrix element? I don't think anyone ever knew. I certainly didn't know, so even though I was at Berkeley I never got too enthusiastic about the details of this program, although I thought it was a lovely idea in principle. Eventually the S-matrix program had to retreat, as described by Kaiser in a contribution to this conference, to a sort of mix of field theory and S-matrix theory. Feynman rules were used to find the singularities in the S-matrix, and then they were thrown away, and the analytic structure of the S-matrix with these singularities, together with unitarity and Lorentz invariance, was used to do calculations.

Unfortunately to use these assumptions it was necessary to make uncontrolled approximations, such as the strip approximation, whose mention will bring tears to the eyes of those of us who are old enough to remember it. By the mid-1960s it was clear that S-matrix theory had failed in dealing with the one problem it had tried hardest to solve, that of pion–pion scattering. The strip approximation rested on the assumption that double dispersion relations are dominated by regions of the Mandelstam diagram near the fringes of the physical region, which would only make sense if π–π scattering is strong at low energy, and these calculations predicted that π–π scattering is indeed strong at low energy, which was at least consistent, but it was then discovered that π–π scattering is *not* strong at low energy. Current algebra came along at just that time, and not only was used to predict that low-energy π–π scattering is not strong, but also successfully predicted the values of the π–π scattering lengths.[24] From a practical point of view, this was the greatest defeat of S-matrix theory. The irony here is that the S-matrix philosophy is not that far from the modern philosophy of effective field theories, that what you should do is just write down the most general S-matrix that satisfies basic principles. But the practical way to implement S-matrix theory is to use an effective quantum field theory – instead of deriving analyticity properties from Feynman diagrams, we use the Feynman diagrams themselves. So here's another answer to the question of what quantum field theory is: it is S-matrix theory, made practical.

By the way, I think that the emphasis in S-matrix theory on analyticity as a fundamental principle was misguided, not only because no one could ever state the detailed analyticity properties of general S-matrix elements, but also because Lorentz invariance requires causality (because as I argued earlier otherwise you're not going to get a Lorentz invariant S-matrix), and in quantum field theory causality allows you to derive analyticity properties. So I would include Lorentz invariance, quantum mechanics and cluster decomposition as fundamental principles, but not analyticity.

As I have said, quantum field theories provide an expansion in powers of the energy of a process divided by some characteristic energy; for soft pions this characteristic energy is about a GeV: for superconductivity it's the Debye frequency or temperature; for the standard model it's 10^{15} to 10^{16} GeV; and for gravitation it's about 10^{18} GeV. Any effective field theory loses its predictive power when the energy of the processes in question approaches the characteristic energy. So what happens to the effective field theories of electroweak, strong, and gravitational interactions at energies of order 10^{15}–10^{18} GeV? I know of only two plausible alternatives.

One possibility is that the theory remains a quantum field theory, but one in which the finite or infinite number of renormalized couplings do not run off to infinity with increasing energy, but hit a fixed point of the renormalization group equations. One way that can happen is provided by asymptotic freedom in a renormalizable theory,[25] where the fixed point is at zero coupling, but it's possible to have more general fixed points with infinite numbers of non-zero non-renormalizable couplings. Now, we don't know how to calculate these non-zero fixed points very well, but one thing we know with fair certainty is that the trajectories that run into a fixed point in the ultraviolet limit form a finite-dimensional subspace of the infinite-dimensional space of all coupling constants. (If anyone wants to know how we know that, I'll explain this later.) That means that the condition, that the trajectories hit a fixed point, is just as restrictive in a nice way as renormalizability used to be: it reduces the number of free coupling parameters to a finite number. We don't yet know how to do calculations for fixed points that are not near zero coupling. Some time ago I proposed[26] that these calculations could be done in the theory of gravitation by working in $2+\epsilon$ dimensions and expanding in powers of $\epsilon = 2$, in analogy with the way that Wilson and Fisher[27] had calculated critical exponents by working in $4-\epsilon$ dimensions and expanding in powers of $\epsilon = 1$, but this program doesn't seem to be working very well.

The other possibility, which I have to admit is *a priori* more likely, is that at very high energy we will run into really new physics, not describable in terms of a quantum field theory. I think that by far the most likely possibility is that this will be something like a string theory.

Before I leave the renormalization group, I did want to say another word about it because there's going to be an interesting discussion on this subject here tomorrow morning, and for reasons I've already explained I can't be here. I've read a lot of argument about the Wilson approach[28] vs. the Gell-Mann–Low approach,[29] which seems to me to call for reconciliation. There have been two fundamental insights in the development of the renormalization group. One, due to Gell-Mann and Low, is that logarithms of energy that violate naive scaling and invalidate perturbation theory arise because of the way that renormalized coupling constants are defined, and that these logarithms can be avoided by renormalizing at a sliding energy scale. The second fundamental insight, due to Wilson, is that it's very important in dealing with phenomena at a certain energy scale to integrate out the physics at much higher energy scales. It seems to me these are the same insight, because when you adopt the Gell-Mann–Low approach and define a renormalized coupling at a sliding scale and use renormalization theory to eliminate the infinities rather than an explicit cutoff, you are in effect integrating out the higher-energy degrees of freedom – the integrals converge because after renormalization the integrand begins to fall off rapidly at the energy scale at which the coupling constant is defined. (This is true whether or not the theory is renormalizable in the power-counting sense.) So in other words instead of a sharp cutoff *à la* Wilson, you have a soft cutoff, but it's a cutoff nonetheless and it serves the same purpose of integrating out the short distance degrees of freedom. There are practical differences between the Gell-Mann–Low and Wilson approaches, and there are some problems for which one is better and other problems for which the other is better. In statistical mechanics it isn't important to maintain Lorentz invariance, so you might as well have a cutoff. In quantum field theories, Lorentz invariance is necessary, so it's nice to renormalize *à la* Gell-Mann–Low. On the other hand, in supersymmetry theories there are some non-renormalization theorems that are simpler if you use a Wilsonian cutoff than a Gell-Mann–Low cutoff.[30] These are all practical differences, which we have to take into account, but I don't find any fundamental philosophical difference between these two approaches.

Fisher: Stay tomorrow and you'll hear the error of your ways.

Weinberg: Okay, well maybe, maybe not, you never know. But I can't stay.

On the plane coming here I read a comment by Michael Redhead, in a paper submitted to this conference: 'To subscribe to the new effective field theory programme is to give up on this endeavour' [the endeavor of finding really fundamental laws of nature], 'and retreat to a position that is somehow less intellectually exciting.' It seems to me that this is analogous to saying that to balance your checkbook is to give up dreams of wealth and have a life that is intrinsically less exciting. In a sense that's true, but nevertheless it's still something that you had better do every once in a while. I think that in regarding the standard model and general relativity as effective field theories we're simply balancing our checkbook and realizing that we perhaps didn't know as much as we thought we did, but this is the way the world is and now we're going to go on to the next step and try to find an ultraviolet fixed point, or (much more likely) find entirely new physics. I have said that I thought that this new physics takes the form of string theory, but of course, we don't know if that's the final answer. Nielsen and Oleson[31] showed long ago that relativistic quantum field theories can have string-like solutions. It's conceivable, although I admit not entirely likely, that something like modern string theory arises from a quantum field theory. And that would be the final irony.

References

1. M. Born. W. Heisenberg, and P. Jordan, *Z. Phys.* **35**, 557 (1926).
2. P.A.M. Dirac, *Proc. Roy. Soc.* **A117,** 610 (1928); *ibid.* **A118,** 351 (1928); *ibid.* **A126**, 360 (1930).
3. R.P. Feynman and A.R. Hibbs, *Quantum Mechanics and Path Integrals* (McGraw-Hill, New York, 1965).
4. G. 't Hooft, *Nucl. Phys.* **B35**, 167 (1971).
5. H. Euler and B. Kockel. *Naturwiss.* **23**, 246 (1935); W. Heisenberg and H. Euler, *Z. Phys.* **98,** 714 (1936).
6. Z. Bern and D.A. Kosower, in *International Symposium on Particles, Strings, and Cosmology*, eds. P. Nath and S. Reucroft (World Scientific, Singapore, 1992): 794; *Phys. Rev. Lett.* **66**, 669 (1991).
7. W. Pauli, *Z. Phys.* **37**, 263 (1926); **43**, 601 (1927).
8. F.J. Dyson, *Phys. Rev.* **75**, 486, 1736 (1949).
9. E.P. Wigner, *Ann. Math.* **40**, 149 (1939).
10. The cluster decomposition principle seems to have been first stated explicitly in quantum field theory by E.H. Wichmann and J.H. Crichton, *Phys. Rev.* **132**, 2788 (1963).

https://doi.org/10.1017/CBO9780511470813.020 Published online by Cambridge University Press

11. S. Weinberg, *The Quantum Theory of Fields – Volume I: Foundations* (Cambridge University Press, Cambridge, 1995).
12. S. Weinberg, *Phys. Rev. Lett.* **18**, 188 (1967); *Phys. Rev.* **166**, 1568 (1968); *Physica* **96A**, 327 (1979).
13. S. Weinberg, *Phys. Lett.* **B251**, 288 (1990); *Nucl. Phys.* **B363**, 3 (1991); *Phys. Lett.* **B295**, 114 (1992). C. Ordóñez and U. van Kolck, *Phys. Lett.* **B291**, 459 (1992); C. Ordóñez, L. Ray and U. van Kolck, *Phys. Rev. Lett.* **72**, 1982 (1994); U. van Kolck, *Phys. Rev.* **C49**, 2932 (1994); U. van Kolck, J. Friar and T. Goldman, to appear in *Phys. Lett. B.* This approach to nuclear forces is summarized in C. Ordóñez, L. Ray and U. van Kolck, Texas preprint UTTG-15-95, nucl-th/9511380, submitted to *Phys. Rev. C*; J. Friar, *Few-Body Systems Suppl.* **99**, 1 (1996). For application of these techniques to related nuclear processes, see T.-S. Park, D.-P. Min and M. Rho, *Phys. Rep.* **233**, 341 (1993); Seoul preprint SNUTP 95-043, nucl-th/9505017; S.R. Beane, C.Y. Lee and U. van Kolck, *Phys. Rev.* **C52**, 2915 (1995); T. Cohen, J. Friar, G. Miller and U. van Kolck, Washington preprint DOE/ER/40427-26-N95, nucl-th/9512036.
14. J.F. Donoghue, *Phys. Rev.* **D50**, 3874 (1994).
15. C. Becchi, A. Rouet and R. Stora, *Comm. Math. Phys.* **42**, 127 (1975); in *Renormalization Theory*, eds. G. Velo and A.S. Wightman (Reidel, Dordrecht, 1976); *Ann. Phys.* **98**, 287 (1976); I.V. Tyutin, Lebedev Institute preprint N39 (1975).
16. G. 't Hooft and M. Veltman, *Nucl. Phys.* **B50**, 318 (1972).
17. B.W. Lee and J. Zinn-Justin, *Phys. Rev.* **D5**, 3121, 3137 (1972); *Phys. Rev.* **D7**, 1049 (1972).
18. C.N. Yang and R.L. Mills, *Phys. Rev.* **96**, 191 (1954).
19. J. Gomis and S. Weinberg, *Nucl. Phys.* B **469**, 475–487 (1996).
20. G. Barnich and M. Henneaux, *Phys. Rev. Lett.* **72**, 1588 (1994); G. Barnich, F. Brandt, and M. Henneaux, *Phys. Rev.* **51**, R143 (1995); *Comm. Math. Phys.* **174**, 57, 93 (1995); *Nucl. Phys.* **B455**, 357 (1995).
21. B.L. Voronov and I.V. Tyutin, *Theor. Math. Phys.* **50**, 218 (1982); **52**, 628 (1982); B.L. Voronov, P.M. Lavrov and I.V. Tyutin, *Sov. J. Nucl. Phys.* **36**, 292 (1982); P.M. Lavrov and I.V. Tyutin, *Sov. J. Nucl. Phys.* **41**, 1049 (1985).
22. D. Anselmi, *Class. and Quant. Grav.* **11**, 2181 (1994); **12**, 319 (1995).
23. M. Harada, T. Kugo, and K. Yamawaki, *Prog. Theor. Phys.* **91**, 801 (1994).
24. S. Weinberg, *Phys. Rev. Lett.* **16**, 879 (1966).
25. D.J. Gross and F. Wilczek, *Phys. Rev. Lett.* **30**, 1343 (1973); H.D. Politzer, *Phys. Rev. Lett.* **30**, 1346 (1973).
26. S. Weinberg, in *General Relativity*, eds. S.W. Hawking and W. Israel (Cambridge University Press, Cambridge, 1979): p. 790.
27. K.G. Wilson and M.E. Fisher, *Phys. Rev. Lett.* **28**, 240 (1972); K.G. Wilson, *Phys. Rev. Lett.* **28**, 548 (1972).
28. K.G. Wilson, *Phys. Rev.* **B4**, 3174, 3184 (1971); *Rev. Mod. Phys.* **47**, 773 (1975).
29. M. Gell-Mann and F.E. Low, *Phys. Rev.* **95**, 1300 (1954).
30. V. Novikov, M.A. Shifman, A.I. Vainshtein and V.I. Zakharov, *Nucl. Phys.* **B229**, 381 (1983); M.A. Shifman and A.I. Vainshtein, *Nucl. Phys.* **B277**, 456 (1986); and references quoted therein. See also M.A. Shifman and A.I. Vainshtein, *Nucl. Phys.* **B359**, 571 (1991).
31. H. Nielsen and P. Oleson, *Nucl. Phys.* **B61**, 45 (1973).

UTTG-09-09
TCC-028-09

Effective Field Theory, Past and Future

Steven Weinberg*
Theory Group, Department of Physics, University of Texas
Austin, TX, 78712

Abstract

This is a written version of the opening talk at the 6th International Workshop on Chiral Dynamics, at the University of Bern, Switzerland, July 6, 2009, to be published in the proceedings of the Workshop. In it, I reminisce about the early development of effective field theories of the strong interactions, comment briefly on some other applications of effective field theories, and then take up the idea that the Standard Model and General Relativity are the leading terms in an effective field theory. Finally, I cite recent calculations that suggest that the effective field theory of gravitation and matter is asymptotically safe.

*Electronic address: weinberg@physics.utexas.edu

I have been asked by the organizers of this meeting to "celebrate 30 years" of a paper[1] on effective field theories that I wrote in 1979. I am quoting this request at the outset, because in the first half of this talk I will be reminiscing about my own work on effective field theories, leading up to this 1979 paper. I think it is important to understand how confusing these things seemed back then, and no one knows better than I do how confused I was. But I am sure that many in this audience know more than I do about the applications of effective field theory to the strong interactions since 1979, so I will mention only some early applications to strong interactions and a few applications to other areas of physics. I will then describe how we have come to think that our most fundamental theories, the Standard Model and General Relativity, are the leading terms in an effective field theory. Finally, I will report on recent work of others that lends support to a suggestion that this effective field theory may actually be a fundamental theory, valid at all energies.

It all started with current algebra. As everyone knows, in 1960 Yoichiro Nambu had the idea that the axial vector current of beta decay could be considered to be conserved in the same limit that the pion, the lightest hadron, could be considered massless.[2] This assumption would follow if the axial vector current was associated with a spontaneously broken approximate symmetry, with the pion playing the role of a Goldstone boson.[3] Nambu used this idea to explain the success of the Goldberger-Treiman formula[4] for the pion decay amplitude, and with his collaborators he was able to derive formulas for the rate of emission of a single low energy pion in various collisions.[5] In this work it was not necessary to assume anything about the nature of the broken symmetry – only that there was some approximate symmetry responsible for the approximate conservation of the axial vector current and the approximate masslessness of the pion. But to deal with processes involving more than one pion, it was necessary to use not only the approximate conservation of the current but also the current commutation relations, which of course do depend on the underlying broken symmetry.

[1]S. Weinberg, Physica A96, 327 (1979).

[2]Y. Nambu, Phys. Rev. Lett. 4, 380 (1960).

[3]J. Goldstone, Nuovo Cimento 9, 154 (1961); Y. Nambu and G. Jona-Lasinio, Phys. Rev. 122, 345 (1961); J. Goldstone, A. Salam, and S. Weinberg, Phys. Rev. 127, 965 (1962).

[4]M. L. Goldberger and S. B. Treiman, Phys. Rev. 111, 354 (1956).

[5]Y. Nambu and D. Lurie, Phys Rev. 125, 1429 (1962); Y. Nambu and E. Shrauner, Phys. Rev. 128, 862 (1962).

The technology of using these properties of the currents, in which one does not use any specific Lagrangian for the strong interactions, became known as current algebra.[6] It scored a dramatic success in the derivation of the Adler-Weisberger sum rule[7] for the axial vector beta decay coupling constant g_A, which showed that the current commutation relations are those of $SU(2) \times SU(2)$.

When I started in the mid-1960s to work on current algebra, I had the feeling that, despite the success of the Goldberger-Treiman relation and the Adler-Weisberger sum rule, there was then rather too much emphasis on the role that the axial vector current plays in weak interactions. After all, even if there were no weak interactions, the fact that the strong interactions have an approximate but spontaneously broken $SU(2) \times SU(2)$ symmetry would be a pretty important piece of information about the strong interactions.[8] To demonstrate the point, I was able to use current algebra to derive successful formulas for the pion-pion and pion-nucleon scattering lengths.[9] When combined with a well-known dispersion relation[10] and the Goldberger-Treiman relation, these formulas for the pion–nucleon scattering lengths turned out to imply the Adler-Weisberger sum rule.

In 1966 I turned to the problem of calculating the rate of processes in which arbitrary numbers of low energy massless pions are emitted in the collision of other hadrons. This was not a problem that urgently needed to be solved. I was interested in it because a year earlier I had worked out simple formulas for the rate of emission of arbitrary numbers of soft gravitons or photons in any collision,[11] and I was curious whether anything equally simple could be said about soft pions. Calculating the amplitude for emission of several soft pions by use of the technique of current algebra turned out to be fearsomely complicated; the non-vanishing commutators of the currents associated with the soft pions prevented the derivation of anything as simple as the results for soft photons or gravitons, except in the

[6]The name may be due to Murray Gell-Mann. The current commutation relations were given in M. Gell-Mann, Physics 1, 63 (1964).

[7]S. L. Adler, Phys. Rev. Lett. 14, 1051 (1965); Phys. Rev. 140, B736 (1965); W. I. Weisberger, Phys. Rev. Lett. 14, 1047 (1965).

[8]I emphasized this point in my rapporteur's talk on current algebra at the 1968 "Rochester" conference; see *Proceedings of the 14th International Conference on High-Energy Physics*, p. 253.

[9]S. Weinberg, Phys. Rev. Lett. 17, 616 (1966). The pion-nucleon scattering lengths were calculated independently by Y. Tomozawa, Nuovo Cimento 46A, 707 (1966).

[10]M. L. Goldberger, Y. Miyazawa, and R. Oehme, Phys. Rev. 99, 986 (1955).

[11]S. Weinberg, Phys. Rev. 140, B516 (1965).

special case in which all pions have the same charge.[12]

Then some time late in 1966 I was sitting at the counter of a café in Harvard Square, scribbling on a napkin the amplitudes I had found for emitting two or three soft pions in nucleon collisions, and it suddenly occurred to me that these results looked very much like what would be given by lowest order Feynman diagrams in a quantum field theory in which pion lines are emitted from the external nucleon lines, with a Lagrangian in which the nucleon interacts with one, two, and more pion fields. Why should this be? Remember, this was a time when theorists had pretty well given up the idea of applying specific quantum field theories to the strong interactions, because there was no reason to trust the lowest order of perturbation theory, and no way to sum the perturbation series. What was popular was to exploit tools such as current algebra and dispersion relations that did not rely on assumptions about particular Lagrangians.

The best explanation that I could give then for the field-theoretic appearance of the current algebra results was that these results for the emission of n soft pions in nucleon collisions are of the minimum order, G_π^n, in the pion-nucleon coupling constant G_π, except that one had to use the exact values for the collision amplitudes without soft pion emission, and divide by factors of the axial vector coupling constant $g_A \simeq 1.2$ in appropriate places. Therefore any Lagrangian that satisfied the axioms of current algebra would have to give the same answer as current algebra in lowest order perturbation theory, except that it would have to be a field theory in which soft pions were emitted only from external lines of the diagram for the nucleon collisions, for only then would one know how to put in the correct factors of g_A and the correct nucleon collision amplitude.

The time-honored renormalizable theory of nucleons and pions with conserved currents that satisfied the assumptions of current algebra was the "linear σ-model,"[13] with Lagrangian (in the limit of exact current conservation):

$$\mathcal{L} \;=\; -\frac{1}{2}\left[\partial_\mu \vec{\pi}\cdot\partial^\mu\vec{\pi} + \partial_\mu\sigma\,\partial^\mu\sigma\right]$$

[12]S. Weinberg, Phys. Rev. Lett. 16, 879 (1966).

[13]J. Bernstein, S. Fubini, M. Gell-Mann, and W. Thirring, Nuovo Cimento 17, 757 (1960); M. Gell-Mann and M. Lévy, Nuovo Cimento 16, 705 (1960); K. C. Chou, Soviet Physics JETP 12, 492 (1961). This theory, with the inclusion of a symmetry-breaking term proportional to the σ field, was intended to provide an illustration of a "partially conserved axial current," that is, one whose divergence is proportional to the pion field.

$$-\frac{m^2}{2}\Big(\sigma^2+\vec{\pi}^2\Big)-\frac{\lambda}{4}\Big(\sigma^2+\vec{\pi}^2\Big)^2$$
$$-\bar{N}\gamma^\mu\partial_\mu N - G_\pi\bar{N}\Big(\sigma+2i\gamma_5\vec{\pi}\cdot\vec{t}\Big)N\ , \tag{1}$$

where N, $\vec{\pi}$, and σ are the fields of the nucleon doublet, pion triplet, and a scalar singlet, and $\vec{t}$ is the nucleon isospin matrix (with $\vec{t}^2=3/4$). This Lagrangian has an $SU(2)\times SU(2)$ symmetry (equivalent as far as current commutation relations are concerned to an $SO(4)$ symmetry), that is spontaneously broken for $m^2<0$ by the expectation value of the σ field, given in lowest order by $<\sigma>=F/2\equiv\sqrt{-m^2/\lambda}$, which also gives the nucleon a lowest order mass $2G_\pi F$. But with a Lagrangian of this form soft pions could be emitted from internal as well as external lines of the graphs for the nucleon collision itself, and there would be no way to evaluate the pion emission amplitude without having to sum over the infinite number of graphs for the nucleon collision amplitude.

To get around this obstacle, I used the chiral $SO(4)$ symmetry to rotate the chiral four-vector into the fourth direction

$$\Big(\vec{\pi},\sigma\Big)\mapsto\Big(0,\sigma'\Big)\ ,\qquad \sigma'=\sqrt{\sigma^2+\vec{\pi}^2}\ , \tag{2}$$

with a corresponding chiral transformation $N\mapsto N'$ of the nucleon doublet. The chiral symmetry of the Lagrangian would result in the pion disappearing from the Lagrangian, except that the matrix of the rotation (2) necessarily, like the fields, depends on spacetime position, while the theory is only invariant under spacetime-*independent* chiral rotations. The pion field thus reappears as a parameter in the $SO(4)$ rotation (2), which could conveniently be taken as

$$\vec{\pi}'\equiv F\vec{\pi}/[\sigma+\sigma']\ . \tag{3}$$

But the rotation parameter $\vec{\pi}'$ would not appear in the transformed Lagrangian if it were independent of the spacetime coordinates, so wherever it appears it must be accompanied with at least one derivative. This derivative produces a factor of pion four-momentum in the pion emission amplitude, which would suppress the amplitude for emitting soft pions, if this factor were not compensated by the pole in the nucleon propagator of an external nucleon line to which the pion emission vertex is attached. Thus, with the Lagrangian in this form, pions of small momenta can only be emitted from external lines of a nucleon collision amplitude. This is what I needed.

Since σ' is chiral-invariant, it plays no role in maintaining the chiral invariance of the theory, and could therefore be replaced with its lowest-order expectation value $F/2$. The transformed Lagrangian (now dropping primes) is then

$$\begin{aligned}\mathcal{L} &= -\frac{1}{2}\left[1+\frac{\vec{\pi}^2}{F^2}\right]^{-2} \partial_\mu\vec{\pi}\cdot\partial^\mu\vec{\pi} \\ &\quad -\bar{N}\Bigg[\gamma^\mu\partial_\mu + G_\pi F/2 \\ &\quad + i\gamma^\mu\left[1+\frac{\vec{\pi}^2}{F^2}\right]^{-1}\left[\frac{2}{F}\gamma_5\vec{t}\cdot\partial_\mu\vec{\pi} + \frac{2}{F^2}\vec{t}\cdot(\vec{\pi}\times\partial_\mu\vec{\pi})\right]\Bigg]N\,. \qquad (4)\end{aligned}$$

In order to reproduce the results of current algebra, it is only necessary to identify F as the pion decay amplitude $F_\pi \simeq 184$ MeV, replace the term $G_\pi F/2$ in the nucleon bilinear with the actual nucleon mass m_N (given by the Goldberger–Treiman relation as $G_\pi F_\pi/2g_A$), and replace the pseudovector pion-nucleon coupling $1/F$ with its actual value $G_\pi/2m_N = g_A/F_\pi$. This gives an effective Lagrangian

$$\begin{aligned}\mathcal{L}_{\text{eff}} &= -\frac{1}{2}\left[1+\frac{\vec{\pi}^2}{F_\pi^2}\right]^{-2} \partial_\mu\vec{\pi}\cdot\partial^\mu\vec{\pi} \\ &\quad -\bar{N}\Bigg[\gamma^\mu\partial_\mu + m_N \\ &\quad + i\gamma^\mu\left[1+\frac{\vec{\pi}^2}{F_\pi^2}\right]^{-1}\left[\frac{G_\pi}{m_N}\gamma_5\vec{t}\cdot\partial_\mu\vec{\pi} + \frac{2}{F_\pi^2}\vec{t}\cdot(\vec{\pi}\times\partial_\mu\vec{\pi})\right]\Bigg]N\,. \qquad (5)\end{aligned}$$

To take account of the finite pion mass, the linear sigma model also includes a chiral-symmetry breaking perturbation proportional to σ. Making the chiral rotation (2), replacing σ' with the constant $F/2$, and adjusting the coefficient of this term to give the physical pion mass m_π gives a chiral symmetry breaking term

$$\Delta\mathcal{L}_{\text{eff}} = -\frac{1}{2}\left[1+\frac{\vec{\pi}^2}{F_\pi^2}\right]^{-1} m_\pi^2\,\vec{\pi}^2\,. \qquad (6)$$

Using $\mathcal{L}_{\text{eff}} + \Delta\mathcal{L}_{\text{eff}}$ in lowest order perturbation theory, I found the same results for low-energy pion-pion and pion-nucleon scattering that I had obtained earlier with much greater difficulty by the methods of current algebra.

A few months after this work, Julian Schwinger remarked to me that it should be possible to skip this complicated derivation, forget all about the linear σ-model, and instead infer the structure of the Lagrangian directly from the non-linear chiral transformation properties of the pion field appearing in (5).[14] It was a good idea. I spent the summer of 1967 working out these transformation properties, and what they imply for the structure of the Lagrangian.[15] It turns out that if we require that the pion field has the usual linear transformation under $SO(3)$ isospin rotations (because isospin symmetry is supposed to be not spontaneously broken), then there is a *unique* $SO(4)$ chiral transformation that takes the pion field into a function of itself — unique, that is, up to possible redefinition of the field. For an infinitesimal $SO(4)$ rotation by an angle ϵ in the $a4$ plane (where $a = 1, 2, 3$), the pion field π_b (labelled with a prime in Eq. (3)) changes by an amount

$$\delta_a \pi_b = -i\epsilon F_\pi \left[\frac{1}{2}\left(1 - \frac{\vec{\pi}^2}{F_\pi^2}\right)\delta_{ab} + \frac{\pi_a \pi_b}{F_\pi^2} \right] . \tag{7}$$

Any other field ψ, on which isospin rotations act with a matrix $\vec{t}$, is changed by an infinitesimal chiral rotation in the $a4$ plane by an amount

$$\delta_a \psi = \frac{\epsilon}{F_\pi}\left(\vec{t} \times \vec{\pi}\right)_a \psi . \tag{8}$$

This is just an ordinary, though position-dependent, isospin rotation, so a non-derivative isospin-invariant term in the Lagrangian that does not involve pions, like the nucleon mass term $-m_N \bar{N}N$, is automatically chiral-invariant. The terms in Eq. (5):

$$-\bar{N}\left[\gamma^\mu \partial_\mu + \frac{2i}{F_\pi^2}\gamma^\mu \left[1 + \frac{\vec{\pi}^2}{F_\pi^2}\right]^{-1} \vec{t} \cdot (\vec{\pi} \times \partial_\mu \vec{\pi})\right] N , \tag{9}$$

and

$$-i\frac{G_\pi}{m_N}\left[1 + \frac{\vec{\pi}^2}{F_\pi^2}\right]^{-1} \bar{N}\gamma^\mu \gamma_5 \vec{t} \cdot \partial_\mu \vec{\pi} N , \tag{10}$$

are simply proportional to the most general chiral-invariant nucleon–pion interactions with a single spacetime derivative. The coefficient of the term

---

[14]For Schwinger's own development of this idea, see J. Schwinger, Phys. Lett. 24B, 473 (1967). It is interesting that in deriving the effective field theory of goldstinos in supergravity theories, it is much more transparent to start with a theory with linearly realized supersymmetry and impose constraints analogous to setting $\sigma' = F/2$, than to work from the beginning with supersymmetry realized non-linearly, in analogy to Eq. (7); see Z. Komargodski and N. Seiberg, to be published.

[15]S. Weinberg, Phys. Rev. 166, 1568 (1968).

(9) is fixed by the condition that N should be canonically normalized, while the coefficient of (10) is chosen to agree with the conventional definition of the pion-nucleon coupling G_π, and is not directly constrained by chiral symmetry. The term

$$-\frac{1}{2}\left[1+\frac{\vec{\pi}^2}{F^2}\right]^{-2} \partial_\mu\vec{\pi}\cdot\partial^\mu\vec{\pi} \tag{11}$$

is proportional to the most general chiral invariant quantity involving the pion field and no more than two spacetime derivatives, with a coefficient fixed by the condition that $\vec{\pi}$ should be canonically normalized. The chiral symmetry breaking term (6) is the most general function of the pion field without derivatives that transforms as the fourth component of a chiral four-vector. None of this relies on the methods of current algebra, though one can use the Lagrangian (5) to calculate the Noether current corresponding to chiral transformations, and recover the Goldberger-Treiman relation in the form $g_A = G_\pi F_\pi/2m_N$.

This sort of direct analysis was subsequently extended by Callan, Coleman, Wess, and Zumino to the transformation and interactions of the Goldstone boson fields associated with the spontaneous breakdown of any Lie group G to any subgroup H.[16] Here, too, the transformation of the Goldstone boson fields is unique, up to a redefinition of the fields, and the transformation of other fields under G is uniquely determined by their transformation under the unbroken subgroup H. It is straightforward to work out the rules for using these ingredients to construct effective Lagrangians that are invariant under G as well as H.[17] Once again, the key point is that the invariance of the Lagrangian under G would eliminate all presence of the Goldstone boson field in the Lagrangian if the field were spacetime-independent, so wherever functions of this field appear in the Lagrangian they are always accompanied with at least one spacetime derivative.

[16] S. Coleman, J. Wess, and B. Zumino, Phys. Rev. 177, 2239(1969); C. G. Callan, S. Coleman, J. Wess, and B. Zumino, Phys. Rev. 177, 2247(1969).

[17] There is a complication. In some cases, such as $SU(3)\times SU(3)$ spontaneously broken to $SU(3)$, fermion loops produce G-invariant terms in the action that are not the integrals of G-invariant terms in the Lagrangian density; see J. Wess and B. Zumino, Phys. Lett. 37B, 95 (1971); E. Witten, Nucl. Phys. B223, 422 (1983). The most general such terms in the action, whether or not produced by fermion loops, have been cataloged by E. D'Hoker and S. Weinberg, Phys. Rev. D50, R6050 (1994). It turns out that for $SU(N)\times SU(N)$ spontaneously broken to the diagonal $SU(N)$, there is just one such term for $N\geq 3$, and none for $N=1$ or 2. For $N=3$, this term is the one found by Wess and Zumino.

In the following years, effective Lagrangians with spontaneously broken $SU(2) \times SU(2)$ or $SU(3) \times SU(3)$ symmetry were widely used in lowest-order perturbation theory to make predictions about low energy pion and kaon interactions.[18] But during this period, from the late 1960s to the late 1970s, like many other particle physicists I was chiefly concerned with developing and testing the Standard Model of elementary particles. As it happened, the Standard Model did much to clarify the basis for chiral symmetry. Quantum chromodynamics with N light quarks is automatically invariant under a $SU(N) \times SU(N)$ chiral symmetry,[19] broken in the Lagrangian only by quark masses, and the electroweak theory tells us that the currents of this symmetry (along with the quark number currents) are just those to which the $W^{\pm}$, Z^0, and photon are coupled.

During this whole period, effective field theories appeared as only a device for more easily reproducing the results of current algebra. It was difficult to take them seriously as dynamical theories, because the derivative couplings that made them useful in the lowest order of perturbation theory also made them nonrenormalizable, thus apparently closing off the possibility of using these theories in higher order.

My thinking about this began to change in 1976. I was invited to give a series of lectures at Erice that summer, and took the opportunity to learn the theory of critical phenomena by giving lectures about it.[20] In preparing these lectures, I was struck by Kenneth Wilson's device of "integrating out" short-distance degrees of freedom by introducing a variable ultraviolet cutoff, with the bare couplings given a cutoff dependence that guaranteed that physical quantities are cutoff independent. Even if the underlying theory is renormalizable, once a finite cutoff is introduced it becomes necessary to introduce every possible interaction, renormalizable or not, to keep physics

[18]For reviews, see S. Weinberg, in *Lectures on Elementary Particles and Quantum Field Theory — 1970 Brandeis University Summer Institute in Theoretical Physics, Vol. 1*, ed. S. Deser, M. Grisaru, and H. Pendleton (The M.I.T. Press, Cambridge, MA, 1970); B. W. Lee, *Chiral Dynamics* (Gordon and Breach, New York, 1972).

[19]For a while it was not clear why there was not also a chiral $U(1)$ symmetry, that would also be broken in the Lagrangian only by the quark masses, and would either lead to a parity doubling of observed hadrons, or to a new light pseudoscalar neutral meson, both of which possibilities were experimentally ruled out. It was not until 1976 that 't Hooft pointed out that the effect of triangle anomalies in the presence of instantons produced an intrinsic violation of this unwanted chiral $U(1)$ symmetry; see G. 't Hooft, Phys. Rev. D14, 3432 (1976).

[20]S. Weinberg, "Critical Phenomena for Field Theorists," in *Understanding the Fundamental Constituents of Matter*, ed. A. Zichichi (Plenum Press, New York, 1977).

strictly cutoff independent. From this point of view, it doesn't make much difference whether the underlying theory is renormalizable or not. Indeed, I realized that even without a cutoff, as long as every term allowed by symmetries is included in the Lagrangian, there will always be a counterterm available to absorb every possible ultraviolet divergence by renormalization of the corresponding coupling constant. Non-renormalizable theories, I realized, are just as renormalizable as renormalizable theories.

This opened the door to the consideration of a Lagrangian containing terms like (5) as the basis for a legitimate dynamical theory, not limited to the tree approximation, provided one adds every one of the infinite number of other, higher-derivative, terms allowed by chiral symmetry.[21] But for this to be useful, it is necessary that in some sort of perturbative expansion, only a finite number of terms in the Lagrangian can appear in each order of perturbation theory.

In chiral dynamics, this perturbation theory is provided by an expansion in powers of small momenta and pion masses. At momenta of order m_π, the number ν of factors of momenta or m_π contributed by a diagram with L loops, E_N external nucleon lines, and V_i vertices of type i, for any reaction among pions and/or nucleons, is

$$\nu = \sum_i V_i \left(d_i + \frac{n_i}{2} + m_i - 2 \right) + 2L + 2 - \frac{E_N}{2}\,, \tag{12}$$

where d_i, n_i, and m_i are respectively the numbers of derivatives, factors of nucleon fields, and factors of pion mass (or more precisely, half the number of factors of u and d quark masses) associated with vertices of type i. As a consequence of chiral symmetry, the minimum possible value of $d_i+n_i/2+m_i$ is 2, so the leading diagrams for small momenta are those with $L=0$ and any number of interactions with $d_i+n_i/2+m_i=2$, which are the ones given in Eqs. (5) and (6). To next order in momenta, we may include tree graphs with any number of vertices with $d_i+n_i/2+m_i=2$ and just one vertex with $d_i+n_i/2+m_i=3$ (such as the so-called σ-term). To next order, we include any number of vertices with $d_i+n_i/2+m_i=2$, plus either a single loop, or a single vertex with $d_i+n_i/2+m_i=4$ which provides a counterterm for the infinity in the loop graph, or two vertices with $d_i+n_i/2+m_i=3$. And so on. Thus one can generate a power series in momenta and m_π, in which only a few new constants need to be introduced at each new order. As an

[21] I thought it appropriate to publish this in a festschrift for Julian Schwinger; see footnote 1.

explicit example of this procedure, I calculated the one-loop corrections to pion–pion scattering in the limit of zero pion mass, and of course I found the sort of corrections required to this order by unitarity.[22]

But even if this procedure gives well-defined finite results, how do we know they are true? It would be extraordinarily difficult to justify any calculation involving loop graphs using current algebra. For me in 1979, the answer involved a radical reconsideration of the nature of quantum field theory. From its beginning in the late 1920s, quantum field theory had been regarded as the application of quantum mechanics to fields that are among the fundamental constituents of the universe — first the electromagnetic field, and later the electron field and fields for other known "elementary" particles. In fact, this became a working definition of an elementary particle — it is a particle with its own field. But for years in teaching courses on quantum field theory I had emphasized that the description of nature by quantum field theories is inevitable, at least in theories with a finite number of particle types, once one assumes the principles of relativity and quantum mechanics, plus the cluster decomposition principle, which requires that distant experiments have uncorrelated results. So I began to think that although specific quantum field theories may have a content that goes beyond these general principles, quantum field theory itself does not. I offered this in my 1979 paper as what Arthur Wightman would call a folk theorem: "if one writes down the most general possible Lagrangian, including *all* terms consistent with assumed symmetry principles, and then calculates matrix elements with this Lagrangian to any given order of perturbation theory, the result will simply be the most general possible S-matrix consistent with perturbative unitarity, analyticity, cluster decomposition, and the assumed symmetry properties." So current algebra wasn't needed.

There was an interesting irony in this. I had been at Berkeley from 1959 to 1966, when Geoffrey Chew and his collaborators were elaborating a program for calculating S-matrix elements for strong interaction processes by the use of unitarity, analyticity, and Lorentz invariance, without reference to quantum field theory. I found it an attractive philosophy, because it relied only on a minimum of principles, all well established. Unfortunately, the S-matrix theorists were never able to develop a reliable method of calculation, so I worked instead on other things, including current algebra. Now in 1979

[22]Unitarity corrections to soft-pion results of current algebra had been considered earlier by H. Schnitzer, Phys. Rev. Lett. 24, 1384 (1970); Phys. Rev. D2, 1621 (1970); L.-F. Li and H. Pagels, Phys. Rev. Lett. 26, 1204 (1971); Phys. Rev. D5, 1509 (1972); P. Langacker and H. Pagels, Phys. Rev. D8, 4595 (1973).

I realized that the assumptions of S-matrix theory, supplemented by chiral invariance, were indeed all that are needed at low energy, but the most convenient way of implementing these assumptions in actual calculations was by good old quantum field theory, which the S-matrix theorists had hoped to supplant.

After 1979, effective field theories were applied to strong interactions in work by Gasser, Leutwyler, Meissner, and many other theorists. My own contributions to this work were limited to two areas — isospin violation, and nuclear forces.

At first in the development of chiral dynamics there had been a tacit assumption that isotopic spin symmetry was a better approximate symmetry than chiral $SU(2) \times SU(2)$, and that the Gell-Mann–Ne'eman $SU(3)$ symmetry was a better approximate symmetry than chiral $SU(3) \times SU(3)$. This assumption became untenable with the calculation of quark mass ratios from the measured pseudoscalar meson masses.[23] It turns out that the d quark mass is almost twice the u quark mass, and the s quark mass is very much larger than either. As a consequence of the inequality of d and u quark masses, chiral $SU(2) \times SU(2)$ is broken in the Lagrangian of quantum chromodynamics not only by the fourth component of a chiral four-vector, as in (6), but also by the third component of a different chiral four-vector proportional to $m_u - m_d$ (whose fourth component is a pseudoscalar). There is no function of the pion field alone, without derivatives, with the latter transformation property, which is why pion–pion scattering and the pion masses are described by (6) and the first term in (5) in leading order, with no isospin breaking aside of course from that due to electromagnetism. But there are non-derivative corrections to pion–nucleon interactions,[24] which at momenta of order m_π are suppressed relative to the derivative coupling terms in (5) by just one factor of m_π or momenta:

$$\begin{aligned}\Delta'\mathcal{L}_{\text{eff}} &= -\frac{A}{2}\left(\frac{1-\pi^2/F_\pi^2}{1+\pi^2/F_\pi^2}\right)\bar{N}N \\ &\quad -B\left[\bar{N}t_3N - \frac{2}{F_\pi^2}\left(\frac{\pi_3}{1+\pi^2/F_\pi^2}\right)\bar{N}\vec{t}\cdot\vec{\pi}N\right]\end{aligned}$$

[23] S. Weinberg, contribution to a festschrift for I. I. Rabi, Trans. N. Y. Acad. Sci. 38, 185 (1977).

[24] S. Weinberg, in *Chiral Dynamics: Theory and Experiment — Proceedings of the Workshop Held at MIT, July 1994* (Springer-Verlag, Berlin, 1995). The terms in Eq. (13) that are odd in the pion field are given in Section 19.5 of S. Weinberg, *The Quantum Theory of Fields*, Vol. II (Cambridge University Press, 1996)

$$-\frac{iC}{1+\vec{\pi}^2/F_\pi^2}\bar{N}\gamma_5\vec{\pi}\cdot\vec{t}N$$
$$-\frac{iD\pi_3}{1+\vec{\pi}^2/F_\pi^2}\bar{N}\gamma_5 N\ , \qquad (13)$$

where A and C are proportional to $m_u + m_d$, and B and D are proportional to $m_u - m_d$, with $B \simeq -2.5$ MeV. The A and B terms contribute isospin conserving and violating terms to the so-called σ-term in pion nucleon scattering.

My work on nuclear forces began one day in 1990 while I was lecturing to a graduate class at Texas. I derived Eq. (12) for the class, and showed how the interactions in the leading tree graphs with $d_i + n_i/2 + m_i = 2$ were just those given here in Eqs. (5) and (6). Then, while I was standing at the blackboard, it suddenly occurred to me that there was one other term with $d_i + n_i/2 + m_i = 2$ that I had never previously considered: an interaction with no factors of pion mass and no derivatives (and hence, according to chiral symmetry, no pions), but *four* nucleon fields — that is, a sum of Fermi interactions $(\bar{N}\Gamma N)(\bar{N}\Gamma' N)$, with any matrices Γ and Γ' allowed by Lorentz invariance, parity conservation, and isospin conservation. This is just the sort of "hard core" nucleon–nucleon interaction that nuclear theorists had long known has to be added to the pion-exchange term in theories of nuclear force. But there is a complication — in graphs for nucleon–nucleon scattering at low energy, two-nucleon intermediate states make a large contribution that invalidates the sort of power-counting that justifies the use of the effective Lagrangian (5), (6) in processes involving only pions, or one low-energy nucleon plus pions. So it is necessary to apply the effective Lagrangian, including the terms $(\bar{N}\Gamma N)(\bar{N}\Gamma' N)$ along with the terms (5) and (6), to the two-nucleon irreducible nucleon–nucleon potential, rather than directly to the scattering amplitude.[25] This program was initially carried further by Ordoñez, van Kolck, Friar, and their collaborators,[26] and eventually by several others.

The advent of effective field theories generated changes in point of view and suggested new techniques of calculation that propagated out to numer-

[25]S. Weinberg, Phys. Lett. B251, 288 (1990); Nucl. Phys. B363, 3 (1991); Phys. Lett. B295, 114 (1992).

[26]C. Ordoñez and U. van Kolck, Phys. Lett. B291, 459 (1992); C. Ordoñez. L. Ray, and U. van Kolck, Phys. Rev. Lett. 72, 1982 (1994); U. van Kolck, Phys. Rev. C49, 2932 (1994); U. van Kolck, J. Friar, and T. Goldman, Phys. Lett. B 371, 169 (1996); C. Ordoñez, L. Ray, and U. van Kolck, Phys. Rev. C 53, 2086 (1996); C. J. Friar, Few-Body Systems Suppl. 99, 1 (1996).

ous areas of physics, some quite far removed from particle physics. Notable here is the use of the power-counting arguments of effective field theory to justify the approximations made in the BCS theory of superconductivity.[27] Instead of counting powers of small momenta, one must count powers of the departures of momenta from the Fermi surface. Also, general features of theories of inflation have been clarified by re-casting these theories as effective field theories of the inflaton and gravitational fields.[28]

Perhaps the most important lesson from chiral dynamics was that we should keep an open mind about renormalizability. The renormalizable Standard Model of elementary particles may itself be just the first term in an effective field theory that contains every possible interaction allowed by Lorentz invariance and the $SU(3) \times SU(2) \times U(1)$ gauge symmetry, only with the non-renormalizable terms suppressed by negative powers of some very large mass M, just as the terms in chiral dynamics with more derivatives than in Eq. (5) are suppressed by negative powers of $2\pi F_\pi \approx m_N$. One indication that there is a large mass scale in some theory underlying the Standard Model is the well-known fact that the three (suitably normalized) running gauge couplings of $SU(3) \times SU(2) \times U(1)$ become equal at an energy of the order of 10^{15} GeV (or, if supersymmetry is assumed, 2×10^{16} GeV, with better convergence of the couplings.)

In 1979 papers by Frank Wilczek and Tony Zee[29] and me[30] independently pointed out that, while the renormalizable terms of the Standard Model cannot violate baryon or lepton conservation,[31] this is not true of the higher

[27] G. Benfatto and G. Gallavotti, J. Stat. Phys. 59, 541 (1990); Phys. Rev. 42, 9967 (1990); J. Feldman and E. Trubowitz, Helv. Phys. Acta 63, 157 (1990); 64, 213 (1991); 65, 679 (1992); R. Shankar, Physica A177, 530 (1991); Rev. Mod. Phys. 66, 129 (1993); J. Polchinski, in *Recent Developments in Particle Theory, Proceedings of the 1992 TASI*, eds. J. Harvey and J. Polchinski (World Scientific, Singapore, 1993); S. Weinberg, Nucl. Phys. B413, 567 (1994).

[28] C. Cheung, P. Creminilli, A. L. Fitzpatrick, J. Kaplan, and L. Senatore, J. High Energy Physics 0803, 014 (2008); S. Weinberg, Phys. Rev. D **73**, 123541 (2008).

[29] F. Wilczek and A. Zee, Phys. Rev. Lett. 43, 1571 (1979).

[30] S. Weinberg, Phys. Rev. Lett. 43, 1566 (1979).

[31] This is not true if the effective theory contains fields for the squarks and sleptons of supersymmetry. However, there are no renormalizable baryon or lepton violating terms in "split supersymmetry" theories, in which the squarks and sleptons are superheavy, and only the gauginos and perhaps higgsinos survive to ordinary energies; see N. Arkani-Hamed and S. Dimopoulos, JHEP **0506**, 073 (2005); G. F. Giudice and A. Romanino, Nucl. Phys. B **699**, 65 (2004); N. Arkani-Hamed, S. Dimopoulos, G. F. Giudice, and A. Romanino, Nucl. Phys. B **709**, 3 (2005); A. Delgado and G. F. Giudice, Phys. Lett. B627, 155 (2005).

non-renormalizable terms. In particular, four-fermion terms can generate a proton decay into antileptons, though not into leptons, with an amplitude suppressed on dimensional grounds by a factor M^{-2}. The conservation of baryon and lepton number in observed physical processes thus may be an accident, an artifact of the necessary simplicity of the leading renormalizable $SU(3) \times SU(2) \times U(1)$-invariant interactions. I also noted at the same time that interactions between a pair of lepton doublets and a pair of scalar doublets can generate a neutrino mass, which is suppressed only by a factor M^{-1}, and that therefore with a reasonable estimate of M could produce observable neutrino oscillations. The subsequent confirmation of neutrino oscillations lends support to the view of the Standard Model as an effective field theory, with M somewhere in the neighborhood of 10^{16} GeV.

Of course, these non-renormalizable terms can be (and in fact, had been) generated in various renormalizable grand-unified theories by integrating out the heavy particles in these theories. Some calculations in the resulting theories can be assisted by treating them as effective field theories.[32] But the important point is that the existence of suppressed baryon- and lepton-nonconserving terms, and some of their detailed properties, should be expected on much more general grounds, *even if the underlying theory is not a quantum field theory at all.* Indeed, from the 1980s on, it has been increasingly popular to suppose that the theory underlying the Standard Model as well as general relativity is a string theory.

Which brings me to gravitation. Just as we have learned to live with the fact that there is no renormalizable theory of pion fields that is invariant under the chiral transformation (7), so also we should not despair of applying quantum field theory to gravitation just because there is no renormalizable theory of the metric tensor that is invariant under general coordinate transformations. It increasingly seems apparent that the Einstein–Hilbert Lagrangian $\sqrt{g}R$ is just the least suppressed term in the Lagrangian of an effective field theory containing every possible generally covariant function of the metric and its derivatives. The application of this point of view to long range properties of gravitation has been most thoroughly developed

[32]The effective field theories derived by integrating out heavy particles had been considered by T. Appelquist and J. Carrazone, Phys. Rev. D11, 2856 (1975). In 1980, in a paper titled "Effective Gauge Theories," I used the techniques of effective field theory to evaluate the effects of integrating out the heavy gauge bosons in grand unified theories on the initial conditions for the running of the gauge couplings down to accessible energies: S. Weinberg, Phys. Lett. 91B, 51 (1980).

by John Donoghue and his collaborators.[33] One consequence of viewing the Einstein–Hilbert Lagrangian as one term in an effective field theory is that any theorem based on conventional general relativity, which declares that under certain initial conditions future singularities are inevitable, must be reinterpreted to mean that under these conditions higher terms in the effective action become important.

Of course, there is a problem — the effective theory of gravitation cannot be used at very high energies, say of the order of the Planck mass, no more than chiral dynamics can be used above a momentum of order $2\pi F_\pi \approx 1$ GeV. For purposes of the subsequent discussion, it is useful to express this problem in terms of the Wilsonian renormalization group. The effective action for gravitation takes the form

$$\begin{aligned} I_{\text{eff}} &= -\int d^4x\,\sqrt{-\text{Det}g}\Big[f_0(\Lambda)+f_1(\Lambda)R \\ &\quad +f_{2a}(\Lambda)R^2+f_{2b}(\Lambda)R^{\mu\nu}R_{\mu\nu} \\ &\quad +f_{3a}(\Lambda)R^3+\ldots\Big]\,, \end{aligned} \tag{14}$$

where here Λ is the ultraviolet cutoff, and the $f_n(\Lambda)$ are coupling parameters with a cutoff dependence chosen so that physical quantities are cutoff-independent. We can replace these coupling parameters with dimensionless parameters $g_n(\Lambda)$:

$$\begin{aligned} &g_0 \equiv \Lambda^{-4}f_0\,;\; g_1 \equiv \Lambda^{-2}f_1\,;\; g_{2a} \equiv f_{2a}\,; \\ &\quad g_{2b} \equiv f_{2b}\,;\; g_{3a} \equiv \Lambda^2 f_{3a}\,;\;\ldots\quad . \end{aligned} \tag{15}$$

Because dimensionless, these parameters must satisfy a renormalization group equation of the form

$$\Lambda\frac{d}{d\Lambda}g_n(\Lambda) = \beta_n\Big(g(\Lambda)\Big)\,. \tag{16}$$

In perturbation theory, all but a finite number of the $g_n(\Lambda)$ go to infinity as $\Lambda \to \infty$, which if true would rule out the use of this theory to calculate

[33] J. F. Donoghue, Phys. Rev. D50, 3874 (1884); Phys. Lett. 72, 2996 (1994); lectures presented at the Advanced School on Effective Field Theories (Almunecar, Spain, June 1995), gr-qc/9512024; J. F. Donoghue, B. R. Holstein, B.Garbrecth, and T.Konstandin, Phys. Lett. B529, 132 (2002); N. E. J. Bjerrum-Bohr, J. F. Donoghue, and B. R. Holstein, Phys. Rev. D68, 084005 (2003).

anything at very high energy. There are even examples, like the Landau pole in quantum electrodynamics and the phenomenon of "triviality" in scalar field theories, in which the couplings blow up at a *finite* value of Λ.

It is usually assumed that this explosion of the dimensionless couplings at high energy is irrelevant in the theory of gravitation, just as it is irrelevant in chiral dynamics. In chiral dynamics, it is understood that at energies of order $2\pi F_\pi \approx m_N$, the appropriate degrees of freedom are no longer pion and nucleon fields, but rather quark and gluon fields. In the same way, it is usually assumed that in the quantum theory of gravitation, when Λ reaches some very high energy, of the order of 10^{15} to 10^{18} GeV, the appropriate degrees of freedom are no longer the metric and the Standard Model fields, but something very different, perhaps strings.

But maybe not. It is just possible that the appropriate degrees of freedom at all energies are the metric and matter fields, including those of the Standard Model. The dimensionless couplings can be protected from blowing up if they are attracted to a finite value g_{n*}. This is known as *asymptotic safety.*[34]

Quantum chromodynamics provides an example of asymptotic safety, but one in which the theory at high energies is not only safe from exploding couplings, but also free. In the more general case of asymptotic safety, the high energy limit g_{n*} is finite, but not commonly zero.

For asymptotic safety to be possible, it is necessary that all the beta functions should vanish at g_{n*}:

$$\beta_n(g_*) = 0 \ . \tag{17}$$

It is also necessary that the physical couplings should be on a trajectory that is attracted to g_{n*}. The number of independent parameters in such a theory equals the dimensionality of the surface, known as the *ultraviolet critical surface*, formed by all the trajectories that are attracted to the fixed point. This dimensionality had better be finite, if the theory is to have any predictive power at high energy. For an asymptotically safe theory with a finite-dimensional ultraviolet critical surface, the requirement that couplings lie on this surface plays much the same role as the requirment of renormalizability in quantum chromodynamics — it provides a rational basis for limiting the complexity of the theory.

This dimensionality of the ultraviolet critical surface can be expressed in terms of the behavior of $\beta_n(g)$ for g near the fixed point g_*. Barring

[34] This was first proposed in my 1976 Erice lectures; see footnote 20.

unexpected singularities, in this case we have

$$\beta_n(g) \to \sum_m B_{nm}(g_m - g_{*m}) \ , \quad B_{nm} \equiv \left(\frac{\partial \beta_n(g)}{\partial g_m}\right)_* \ . \tag{18}$$

The solution of Eq. (16) for g near g_* is then

$$g_n(\Lambda) \to g_{n*} + \sum_i u_{in}\,\Lambda^{\lambda_i} \ , \tag{19}$$

where λ_i and u_{in} are the eigenvalues and suitably normalized eigenvectors of B_{nm}:

$$\sum_m B_{nm}\,u_{im} = \lambda_i\,u_{in} \ . \tag{20}$$

Because B_{nm} is real but not symmetric, the eigenvalues are either real, or come in pairs of complex conjugates. The dimensionality of the ultraviolet critical surface is therefore equal to the number of eigenvalues of B_{nm} with negative real part. The condition that the couplings lie on this surface can be regarded as a generalization of the condition that quantum chromodynamics, if it were a fundamental and not merely an effective field theory, would have to involve only renormalizable couplings.

It may seem unlikely that an infinite matrix like B_{nm} should have only a finite number of eigenvalues with negative real part, but in fact examples of this are quite common. As we learned from the Wilson–Fisher theory of critical phenomena, when a substance undergoes a second-order phase transition, its parameters are subject to a renormalization group equation that has a fixed point, with a single infrared-repulsive direction, so that adjustment of a single parameter such as the temperature or the pressure can put the parameters of the theory on an infrared attractive surface of co-dimension one, leading to long-range correlations. The single infrared-repulsive direction is at the same time a unique ultraviolet-attractive direction, so the ultraviolet critical surface in such a theory is a one-dimensional curve. Of course, the parameters of the substance on this curve do not really approach a fixed point at very short distances, because at a distance of the order of the interparticle spacing the effective field theory describing the phase transition breaks down.

What about gravitation? There are indications that here too there is a fixed point, with an ultraviolet critical surface of finite dimensionality. Fixed points have been found (of course with $g_{n*} \neq 0$) using dimensional continuation from $2+\epsilon$ to 4 spacetime dimensions,[35] by a $1/N$ approximation (where N is the number of added matter fields),[36] by lattice methods,[37] and by use of the truncated exact renormalization group equation,[38] initiated in 1998 by Martin Reuter. In the last method, which had earlier been introduced in condensed matter physics[39] and then carried over to particle theory,[40] one derives an exact renormalization group equation for the total vacuum amplitude $\Gamma[g, \Lambda]$ in the presence of a background metric $g_{\mu\nu}$ with an *infrared* cutoff Λ. This is the action to be used in calculations of the true vacuum amplitude in calculations of graphs with an *ultraviolet* cutoff Λ. To have equations that can be solved, it is necessary to truncate these renormalization group equations, writing $\Gamma[g, \Lambda]$ as a sum of just a finite number of terms like those shown explicitly in Eq. (14), and ignoring the fact that the beta function inevitably does not vanish for the couplings of other terms in $\Gamma[g, \Lambda]$ that in the given truncation are assumed to vanish.

Initially only two terms were included in the truncation of $\Gamma[g, \Lambda]$ (a cosmological constant and the Einstein–Hilbert term $\sqrt{g}R$), and a fixed point was found with two eigenvalues λ_i, a pair of complex conjugates with nega-

[35] S. Weinberg, in *General Relativity*, ed. S. W. Hawking and W. Israel (Cambridge University Press, 1979): 700; H. Kawai, Y. Kitazawa, & M. Ninomiya, Nucl. Phys. B 404, 684 (1993); Nucl. Phys. B 467, 313 (1996); T. Aida & Y. Kitazawa, Nucl. Phys. B 401, 427 (1997); M. Niedermaier, Nucl. Phys. B 673, 131 (2003) .

[36] L. Smolin, Nucl. Phys. B208, 439 (1982); R. Percacci, Phys. Rev. D 73, 041501 (2006).

[37] J. Ambjørn, J. Jurkewicz, & R. Loll, Phys. Rev. Lett. 93, 131301 (2004); Phys. Rev. Lett. 95, 171301 (2005); Phys. Rev. D72, 064014 (2005); Phys. Rev. D78, 063544 (2008); and in *Approaches to Quantum Gravity*, ed. D. Oríti (Cambridge University Press).

[38] M. Reuter, Phys. Rev. D 57, 971 (1998); D. Dou & R. Percacci, Class. Quant. Grav. 15, 3449 (1998); W. Souma, Prog. Theor. Phys. 102, 181 (1999); O. Lauscher & M. Reuter, Phys. Rev. D 65, 025013 (2001); Class. Quant. Grav. 19. 483 (2002); M. Reuter & F. Saueressig, Phys Rev. D 65, 065016 (2002); O. Lauscher & M. Reuter, Int. J. Mod. Phys. A 17, 993 (2002); Phys. Rev. D 66, 025026 (2002); M. Reuter and F. Saueressig, Phys Rev. D 66, 125001 (2002); R. Percacci & D. Perini, Phys. Rev. D 67, 081503 (2002); Phys. Rev. D 68, 044018 (2003); D. Perini, Nucl. Phys. Proc. Suppl. C 127, 185 (2004); D. F. Litim, Phys. Rev. Lett. **92**, 201301 (2004); A. Codello & R. Percacci, Phys. Rev. Lett. 97, 221301 (2006); A. Codello, R. Percacci, & C. Rahmede, Int. J. Mod. Phys. A23, 143 (2008); M. Reuter & F. Saueressig, 0708.1317; P. F. Machado and F. Saueressig, Phys. Rev. D77, 124045 (2008); A. Codello, R. Percacci, & C. Rahmede, Ann. Phys. 324, 414 (2009); A. Codello & R. Percacci, 0810.0715; D. F. Litim 0810.3675; H. Gies & M. M. Scherer, 0901.2459; D. Benedetti, P. F. Machado, & F. Saueressig, 0901.2984, 0902.4630; M. Reuter & H. Weyer, 0903.2971.

[39] F. J. Wegner and A. Houghton, Phys. Rev. A8, 401 (1973).

[40] J. Polchinski, Nucl. Phys. B231, 269 (1984); C. Wetterich, Phys. Lett. B 301, 90 (1993).

tive real part. Then a third operator ($R_{\mu\nu}R^{\mu\nu}$ or the equivalent) was added, and a third eigenvalue was found, with λ_i real and negative. This was not encouraging. If each time that new terms were included in the truncation, new eigenvalues appeared with negative real part, then the ultraviolet critical surface would be infinite dimensional, and the theory, though free of couplings that exploded at high energy, would lose all predictive value at high energy.

In just the last few years calculations have been done that allow more optimism. Codello, Percacci, and Rahmede[41] have considered a Lagrangian containing all terms $\sqrt{g}R^n$ with n running from zero to a maximum value n_{max}, and find that the ultraviolet critical surface has dimensionality 3 even when n_{max} exceeds 2, up to the highest value $n_{\text{max}} = 6$ that they considered, for which the space of coupling constants is 7-dimensional. Furthermore, the three eigenvalues they find with negative real part seem to converge as n_{max} increases, as shown in the following table of ultraviolet-attractive eigenvalues:

$n_{\text{max}} = 2$:	$-1.38 \pm 2.32i$	-26.8
$n_{\text{max}} = 3$:	$-2.71 \pm 2.27i$	-2.07
$n_{\text{max}} = 4$:	$-2.86 \pm 2.45i$	-1.55
$n_{\text{max}} = 5$:	$-2.53 \pm 2.69i$	-1.78
$n_{\text{max}} = 6$:	$-2.41 \pm 2.42i$	-1.50

In a subsequent paper[42] they added matter fields, and again found just three ultraviolet-attractive eigenvalues. Further, this year Benedetti, Machado, and Saueressig[43] considered a truncation with a different four terms, terms proportional to $\sqrt{g}R^n$ with $n = 0, 1$ and 2 and also $\sqrt{g}C_{\mu\nu\rho\sigma}C^{\mu\nu\rho\sigma}$ (where $C_{\mu\nu\rho\sigma}$ is the Weyl tensor) and they too find just three ultraviolet-attractive eigenvalues, also when matter is added. If this pattern of eigenvalues continues to hold in future calculations, it will begin to look as if there is a quantum field theory of gravitation that is well-defined at all energies, and that has just three free parameters.

The natural arena for application of these ideas is in the physics of gravitation at small distance scales and high energy — specifically, in the early universe. A start in this direction has been made by Reuter and his collaborators,[44] but much remains to be done.

[41] A. Codello, R. Percacci, & C. Rahmede, Int. J. Mod. Phys. A23, 143 (2008)
[42] A. Codello, R. Percacci, & C. Rahmede, Ann. Phys. 324, 414 (2009)
[43] D. Benedetti, P. F. Machado, & F. Saueressig, 0901.2984, 0902.4630
[44] A. Bonanno and M. Reuter, Phys. Rev. D 65, 043508 (2002); Phys. Lett. B527, 9

I am grateful for correspondence about recent work on asymptotic safety with D. Benedetti, D. Litim, R. Percacci, and M. Reuter, and to G. Colangelo and J. Gasser for inviting me to give this talk. This material is based in part on work supported by the National Science Foundation under Grant NO. PHY-0455649 and with support from The Robert A. Welch Foundation, Grant No. F-0014.

(2002); M. Reuter and F. Saueressig, J. Cosm. and Astropart. Phys. 09, 012 (2005).

Eur. Phys. J. H (2021) 46:6
https://doi.org/10.1140/epjh/s13129-021-00004-x

THE EUROPEAN PHYSICAL JOURNAL H

Regular Article

On the development of effective field theory

Steven Weinberg[a]

Theory Group, Department of Physics, University of Texas, Austin, TX 78712, USA

Received 11 January 2021 / Published online 2 March 2021

Abstract *Editor's note*: One of the most important developments in theoretical particle physics at the end of the 20th century and beginning of the twenty-first century has been the development of effective field theories (EFTs). Pursuing an effective field theory approach is a methodology for constructing theories, where a set of core principles is agreed upon, such as Lorentz symmetry and unitarity, and all possible interactions consistent with them are then compulsory in the theory. The utility of this approach to particle physics (and beyond) is wide ranging and undisputed, as evidenced by the recent formation of the international seminar series *All Things EFT* (Talks in the series can be viewed at https://www.youtube.com/channel/UC1_KF6kdJFoDEcLgpcegwCQ (accessed 21 December 2020).) which brings together each week the worldwide community of EFT practitioners. The text below is a lightly edited version of the talk given by Prof. Weinberg on September 30, 2020, which inaugurated the series. The talk reviews some of the early history of EFTs from the perspective of its pioneer and concludes with a discussion of EFT implications for future discovery.

What is the world made of? This question is perhaps the deepest and earliest in all of science. Greeks were asking this question a hundred years before the time of Socrates. By the time that I became a graduate student an answer had apparently been settled. The world is made not of water, earth, air or fire, but of fields. There is the electromagnetic field that when quantum mechanics is applied to it is manifested in the form of bundles of energy, momentum—particles that are called photons. There is an electron field that similarly when quantized appears as particles called electrons. And there are other fields that we in the late 1950s knew we did not yet know about. The weak and the strong interactions were pretty mysterious. It was clear that there had to be more than just electrons and photons. But we looked forward to a description of nature as consisting fundamentally of fields as the constituents of everything.

The quantum field theory of electrons and photons in the late 1940s had scored a tremendous success. Theorists—Feynman, Schwinger, Tomonaga, Dyson—had figured out after decades of effort how to do calculations preserving not only Lorentz invariance but also the appearance of Lorentz invariance at every stage of the calculation. This allowed them to sort out the infinities in the theory that had been noticed in the early 1930s by Oppenheimer and Waller, and that had been the *bête noire* of theoretical physics throughout the 1930s. They were able to show in the late 1940s that these infinities could all be absorbed into a redefinition, called a renormalization, of the electron mass and charge and the scales of the various fields. And they were able to do calculations of unprecedented precision, which turned out to be verified by experiment: calculations of the Lamb shift and the anomalous magnetic moment of the electron.

More than that, and this particularly appealed to me as a graduate student, renormalization theory didn't always work. It wouldn't work unless the theory had a certain kind of simplicity. Essentially the only coupling constants allowed in the theory, in units where Planck's constant and the speed of light were 1, had to be dimensionless, like the charge of the electron: $e^2/4\pi$ is 1/137. And this provided not only a means of dealing with the infinities but a rationale for the simplicity of the theory. Of course we always like simple theories. But when we have discovered successful simple theories, we always should ask why are they so simple? Renormalizability provided an answer to that question. Of course it was only an answer if we thought that these were really the fundamental theories that described nature at all scales. Otherwise, anything else might intervene to get rid of the infinities.

We hoped that we would see the rest of physics—the mysterious strong and weak nuclear forces—brought into a similar framework. And that is indeed what happened in the following decades in the development of the Standard Model. Once we got past the obscurities produced by spontaneous symmetry breaking in

[a] e-mail: weinberg@physics.utexas.edu (corresponding author)

the weak interactions and color trapping in the strong interactions, the Standard Model was revealed to us as a theory that was really not very different from quantum electrodynamics. We had more gauge fields, not just the electromagnetic field but gluon and W and Z fields. There were more fermions, not just the electron but a whole host of charged leptons and neutrinos and quarks. But the Standard Model seemed to be quantum electrodynamics writ large. One could perhaps have been forgiven for reaching a stage of satisfaction that, although not everything was answered, although there were still outstanding questions, that this was going to be part of nature at the most fundamental level.

Now that has changed. In the decades since the completion of the Standard Model, a new and cooler view has become widespread. The Standard Model, we now see—we being, let me say, me and a lot of other people—as a low-energy approximation to a fundamental theory about which we know very little. And low energy means energies much less than some extremely high energy scale $10^{15} - 10^{18}$ GeV. As a low-energy approximation, we expect corrections to the Standard Model. These corrections are beginning to show up. Some of them have already been found.

This whole point of view goes by the name of effective field theory. It had applications outside elementary particle physics in areas like superconductivity. I am not going to in this talk try to bring the subject up-to-date, including all the applications of effective field theory to hadronic physics, and to areas of physics outside particle physics, like superconductivity. That's going to be done by subsequent lecturers in this series by physicists who played a leading role in the development of effective field theory beyond anything that I knew about it in the early days. They are true experts in the field. I won't dare to try to anticipate what they will say. I'll talk about a subject on which I am undoubtedly the world's expert and that is my own history of how I came to think about these things. I'm a little bit unhappy that I am putting myself too much forward. Other people came to effective field theories through different routes. I'm not going to survey anyone else's intellectual history except my own.

From my point of view, it started with current algebra. The late 1950s and early 1960s were a time of despair about the future—about the practical application of quantum field theory to the strong interactions. Although we could believe that quantum field theory was at the root of things we didn't know how to apply quantum field theory to the strong interactions. Perturbation theory didn't work. Instead, a method was developed called current algebra in which one concentrated on the currents of the weak interactions, the vector and axial vector currents, using their commutation relations, their conservation properties and in particular a suggestion made by Nambu that the divergence of the axial vector current was dominated by one pion states. This current algebra was used in a very clunky way, requiring very detailed calculations that just called out for a simpler approach to derive useful results, including the Goldberger–Treiman formula for the pion decay amplitude and the Adler–Weisberger sum rule for the axial vector coupling constant.

After a while some of us began to think that although these results were important and valuable, perhaps we were giving too much attention to the currents themselves, which of course play a central role in the weak interactions. We ought to concentrate on the symmetry properties of the strong interactions which made all this possible. In particular, the existence of a symmetry which gradually emerged in our thinking, chiral $SU(2) \times SU(2)$, which is just isotopic spin symmetry as applied separately to the left-handed and right-handed parts of what we would now say are quark fields.

This symmetry is a property of the strong interactions which would be important even if there weren't any weak interactions. And it was employed to derive purely strong interaction results, like, for example, the scattering lengths of pions on nucleons and pions on pions and more complicated things like the emission of any number of soft pions in high energy collisions of nucleons or other particles. When this was done using these, as I said, clunky methods of current algebra, looking at the results, they seemed to look like the results of a field theory. You could write down Feynman diagrams just out of the blue which would reproduce the results of current algebra.

And so the question naturally arose, is there a way of avoiding the machinery of current algebra by just writing down a field theory that would automatically produce the same results with much greater ease and perhaps physical clarity? Because after all in using current algebra one had to always wave one's hands and make assumptions about the smoothness of matrix elements, whereas if you could get these results from Feynman diagrams, you could see what the singularity structure of the matrix elements was and make only those smoothness assumptions that were consistent with that.

At the beginning, this was done using a standard theory with the chiral symmetry that we thought was at the bottom of all these results, the linear sigma model, and then redefining the fields in such a way that the results would look like current algebra. The effect of the redefinition was the introduction of a nonlinearly realized chiral symmetry. Eventually, the linear sigma model was scrapped; instead, the procedure was simply to ask, what kind of symmetry transformation for the pion field alone, some transformation into a nonlinear function of the pion field, would have the algebraic properties of chiral symmetry, based on the Lie algebra of $SU(2) \times SU(2)$.

That theory had the property that in lowest order in the coupling constant $1/F_\pi$ the results reproduced the results of current algebra. Why did it? Well, it had to, because a theory having chiral and Lorentz invariance and unitarity and smoothness satisfied the assumptions of current algebra and therefore had to reproduce the same results. For example, current algebra calculations gave a pion–pion scattering matrix element of order $1/F_\pi^2$, where F_π is the pion decay amplitude, so if you use this field theory and just threw away everything except the leading term which is of order $1/F_\pi^2$, this matrix element had to agree with the results of current algebra.

In this way, phenomenological Lagrangians were developed that could be thought of as merely labor-saving devices, which were guaranteed to give the same results as current algebras because they satisfied the same underlying conditions of symmetry and so on, and that could be used in lowest order because current algebra said the result was of lowest order in the $1/F_\pi$ coupling—that is, $1/F_\pi^2$ for $\pi\pi$ scattering and also for pion–nucleon scattering. If you had more pions, you would have more powers of $1/F_\pi$. But the results would always agree with current algebra.

No one took these theories seriously as true quantum field theories at the time. (I am talking about the late 1960s.) No one would have dreamed at this point of using the phenomenological Lagrangian in calculating loop diagrams. What would be the point? We knew that it was the tree approximation that reproduced current algebra. As I said, these phenomenological Lagrangians were simply labor-saving devices.

The late 1960s and 1970s saw many of us engaged in the development of the Standard Model. During this time, I wasn't thinking much about current algebra or phenomenological Lagrangians. The soft-pion theorems had been successful, not only in agreeing with experiment, but also in killing off a competitor of quantum field theory known as *S*-matrix theory. *S*-matrix theory had been the slogan of a school of theoretical physicists headed by Geoff Chew at Berkeley. I had been there at Berkeley in its heyday but had never bought on to it.

Their idea was that field theory is hopeless. It deals with things we will never observe like quantum fields. What we should do is just to study things that are observable, like *S*-matrix elements: apply principles of Lorentz invariance, analyticity and so on, and get results like dispersion relations that we can compare with observation. It was even hoped that stable or unstable composite particles like the ρ meson provide the force that produces these composites, so that using this bootstrap mechanism one could actually do calculations. This never really worked as a calculational scheme, but was extremely attractive philosophically because it made do with very little except the most fundamental assumptions, without introducing things like strongly interacting fields that we really didn't know about. But the chiral symmetry results, the soft-pion results, showed that some of the approximations assumed in using *S*-matrix theory, like strong pion–pion interactions at low energy, just weren't right. Chiral symmetry provided actual calculations of processes like $\pi\pi$ and π-nucleon scattering, which made the ideas of *S*-matrix theory seem unnecessary, attractive as the philosophy was.

S-matrix theory had been largely killed off and chiral symmetry had been put in the books as a success, but we were all involved in applying the ideas of quantum field theory in the weak and the electromagnetic interactions and then the strong interactions, building up the Standard Model, which was beginning to be very successful experimentally. It was a very happy time for the interaction between theory and experiment. During this period of course, I was teaching. It was in the course of teaching that my point of view changed, because in teaching quantum field theory I had to keep confronting the question of the motivation for this theory. Why should these students take seriously the assumptions we were making, in particular the formalism of writing fields in terms of creation and annihilation operators, with their commutation relations. Where did this come from? The standard approach was to take a field theory like Maxwell's theory and quantize it, using the rules of canonical quantization. Lo and behold, you turn the crank, and out come the commutation relations for the operator coefficients of the wave functions in the quantum field.

I found that hard to sell, especially to myself. Why should you apply the canonical formalism to these fields? The answer that the canonical formalism had proved useful in celestial mechanics in the nineteenth century wasn't really very satisfying. In particular, suppose there was a theory that in other ways was successful but couldn't be formulated in terms of canonical quantization, would that bother us? In fact, we have quantum field theories like that. They're not realistic theories. They're theories in six dimensions, or theories we derive by compactifying six dimensional theories. We have quantum field theories that apparently can't be given a Lagrangian formalism—that can't be derived using the canonical formalism. So I looked for some other way of teaching the subject.

I fastened on a point of view that is really not that different from *S*-matrix theory. One starts of course by assuming the rules of quantum mechanics as laid down in the 1920s, together with special relativity and then one makes an additional assumption, the cluster decomposition principle, whose importance was emphasized to me by a colleague at Berkeley, Eyvind Wichmann, while I was there in the 1960s. The cluster decomposition principle says essentially that unless you make special efforts to produce an entangled situation the results of distant experiments are uncorrelated. The results of an experiment at CERN are not affected by the results being obtained by an experiment being done at the same time at Fermilab. The natural way of implementing the cluster decomposition principle is by writing the Hamiltonian as a sum of products of creation and annihilation operators with non-singular coefficients. Indeed, this had been done for many years by condensed matter physicists not because they were interested in quantum field theory as a fundamental principle, but in order to sort out the volume dependence of various thermodynamic quantities. They were managing this by introducing creation and annihilation operators long before I began to teach courses in quantum field theory.

It gradually appeared to me in teaching the subject that although individual quantum field theories like quantum electrodynamics certainly have content, quantum field theory in itself has no content except the principles on which it is based, namely quantum mechanics, Lorentz invariance and the cluster decomposition principle, together with whatever other symmetry principles you may want to invoke, like chiral symmetry and gauge invariance.

This means that if we think we know what the degrees of freedom are—the particles we have to study—like, for example, low-energy pions, that if we write down the most general possible theory involving fields for these particles, including all possible interactions consistent with the symmetries, which in this case are Lorentz invariance and chiral SU(2) × SU(2), if we write down all possible invariant terms in the Lagrangian, and we work to all orders in perturbation theory, the result can't be wrong, because it is just a way of implementing these principles. It is just giving you the most general possible matrix element consistent with Lorentz invariance, chiral symmetry, quantum mechanics, cluster decomposition and unitarity.

Now, this may not sound as if it is a very useful realization. If you tell someone to calculate using the most general possible Lagrangian with an unlimited number of free parameters and calculate to all orders in perturbation theory, they're likely to seek some advice elsewhere on how to spend their time. But it is not that bad, because even if the theory has no small dimensionless couplings, you can use this approach to generate a power series in powers of the energy that you're interested in. For instance, if you're interested in low-energy pions and you don't want to consider energies high enough so that $\pi\pi$ collisions can produce nucleon–antinucleon pairs, you will be dealing with typical energies E well below the nucleon mass. This very general Lagrangian gives you a power series in powers of E. Specifically, aside from a term that depends only on the nature of the process being considered, the number of powers of E arising from a given Feynman diagram is the total number of derivatives acting at all the vertices, plus half the number of nucleon lines connected to all the vertices, plus twice the number of loops.

Now, chiral symmetry dictates that the number of derivatives plus half the number of nucleon fields in each interaction is always equal to or greater than two. The diagrams that give the lowest total number of powers of E are those constructed only from interactions where that number equals two, with no loops. For pion–pion scattering, the leading term comes from a single vertex with just two derivatives. For pion–nucleon scattering, there is a diagram with a single vertex with one derivative to which are connected two nucleon lines. These diagrams are the ones that we had been using since the mid-1960s to reproduce the results of current algebra.

But now the new thing was that you could consider contributions of higher order in energy. If you look for terms that have two additional powers of energy, they could come from diagrams where you have one interaction with the number of derivatives plus half the number of nucleon fields equaling not two but four, plus any number of interactions with this number equal to two, and no loops. Or you could have only interactions with the numbers of derivatives plus half the number of nucleon fields equal to two, that is, just the basic interactions that reproduce current algebra, plus one loop. The infinity in the one loop diagram could be canceled by that one additional vertex that has, say, not two derivatives but four derivatives. In every order of perturbation theory as you encounter more and more loops, because you are allowing more and more powers of energy, you get more and more infinities, but there are always counterterms available to cancel the infinities. A non-renormalizable theory, like the soft-pion theory, is just as renormalizable as a renormalizable theory. You have an infinite number of terms in the Lagrangian, but only a finite number is needed to calculate S-matrix elements to any given order in E.

Similar remarks apply to gravitation, which I think has led to a new perspective on general relativity. Why in the world should anyone take seriously Einstein's original theory, with just the Einstein–Hilbert action in which only two derivatives act on metric fields? Surely that's just the lowest order term in an infinite series of terms with more and more derivatives. In such a theory, loops are made finite by counterterms provided by the higher-order terms in the Lagrangian. This point of view has been actively pursued by Donoghue and his collaborators.

Teaching came to my aid again. At the blackboard one day in 1990, it suddenly occurred to me that one of the kinds of interaction that has the number of derivatives plus half the number of nucleon fields equal to two is the interaction with no derivatives at all and four nucleon fields. It had taken me a decade to realize that four divided by two is two. This sort of interaction is just the kind of hard-core nucleon–nucleon interaction that nuclear physicists had always known would be needed to understand nuclear forces. But now we had a rationale for it. In a calculation of nuclear forces as a power series in energy, the leading terms are just the ones that the nuclear physicists had always been using, pion exchange plus a hard core. This point of view has been explored by Ordoñez and van Kolck and others.

In the theories that I have been discussing, the chiral symmetry theory of soft pions and general relativity, the symmetries don't allow a purely renormalizable interaction. In a theory of gauge fields and quarks and leptons and scalars, you can have a renormalizable Lagrangian. Throwing away everything but renormalizable interactions, you have just the Standard Model, a renormalizable theory of quarks and leptons and gauge fields and a scalar doublet. But now we can look at the Standard Model as one term in a much more general theory, with all kinds of non-renormalizable interactions, which yield corrections if higher order in energy.

In these corrections, the typical energy E must be divided by some characteristic mass scale M. For the chiral theory of soft pions, this mass scale is $M \approx 2\pi F_\pi$, about 1200 MeV. For the theory of weak, strong and electromagnetic interactions, M is probably a much higher scale, perhaps something like the scale of order 10^{15} GeV where Georgi, Quinn and I found the effective gauge couplings of the weak, strong and electromagnetic interactions all come together. Or perhaps it is the characteristic scale of gravitation, the Planck scale

10^{18} GeV . Or perhaps somewhere in that general neighborhood. It is the large scale of M that made it a good strategy in constructing the Standard Model to look for a renormalizable theory .

We now expect that there are corrections to the renormalizable Standard Model of the order of powers of E/M. How in the world are we ever going to find corrections that are suppressed by such incredibly tiny fractions? The one hope is that those corrections can violate symmetries that we once thought were inviolable, but that we now understand are simply accidents, arising from the constraint of renormalizability that we imposed on the Standard Model.

Indeed, quite apart from the development of effective field theory, one of the great things about the Standard Model was that it explained various symmetries that could not be fundamental, because we already knew they were only partial or approximate symmetries. This included flavor conservation, such as strangeness conservation, a symmetry of the strong and electromagnetic interactions that was manifestly violated in the weak interactions. Another example was charge conjugation invariance—likewise a good symmetry of strong and electromagnetic but violated by weak interactions. The same was true of parity, although in this case you have to make special accommodations for non-perturbative effects. All these were accidental symmetries, imposed by the simplicity of the Standard Model necessary for renormalizability plus other symmetries like gauge symmetries and Lorentz invariance that seem truly exact and fundamental. Chiral symmetry itself is such an accidental symmetry, though only approximate. It becomes an accidental exact symmetry of the strong interactions in the limit in which the up and down quark masses are zero, as does isotopic spin symmetry. Since these masses are not zero, but relatively small chiral symmetry is an approximate accidental symmetry.

Now, coming back to effective field theory, there are other symmetries within the Standard Model that are accidental symmetries of the whole renormalizable theory of weak, strong and electromagnetic interactions: In particular, baryon conservation and lepton conservation are respected aside from very small non-perturbative effects (well, very small at least in laboratories, though maybe not so small cosmologically). If baryon and lepton conservations are only accidental properties of the Standard Model, maybe they are not symmetries of nature. In this case, there is no reason why baryon and lepton conservation should be respected by non-renormalizable corrections to the Lagrangian, and so you would expect terms of $\mathcal{O}(E/M)$ or $\mathcal{O}((E/M)^2)$ or higher order as corrections to the Standard Model that violate these symmetries.

Wilczek and Zee and I independently did a catalog of the leading terms of this type. Some of them—those involving baryon number non-conservation—give you corrections of $\mathcal{O}((E/M)^2)$. They have not been yet been discovered experimentally. But there are other terms that produce corrections of $\mathcal{O}(E/M)$ that violate lepton conservation, and they apparently have been discovered, in the form of neutrino masses. I wish I could say that the effective field theory point of view had predicted the neutrino masses. Unfortunately, it must be admitted that neutrino masses were already proposed by Pontecorvo as a solution to the apparent deficit of neutrinos coming from the Sun. So though I can't say that the effective field theory approach had predicted neutrino masses, I do think that it gets a strong boost from the fact that we now have evidence of a correction to the renormalizable part of the Standard Model. But of course there already was a known correction. Gravitation was always there, warning us that renormalizable quantum field theory can't be the whole story.

I expect that sooner or later we will be seeing another departure from the renormalizable Standard Model in the discovery of proton decay, or some other example of baryon non-conservation. In a sense, baryon non-conservation has already been discovered, because we know from the present ratio of baryon number to photon number that in the early universe before the temperature dropped to a few GeV, there was about 1 excess quark over every 10^9 quark–antiquark pairs. This has to be explained by the non-renormalizable corrections to the Standard Model, and indeed, it has been explained, unfortunately not just by one such model but by many different models. We don't know the actual mechanism for producing baryon number in the early universe, but I have no doubt that it will be found.

There are still unnatural elements in the Standard Model. I said that you would expect the leading terms that describe physical reactions to be given by the renormalizable theory—here the Standard Model—with the effects of non-renormalizable terms suppressed by powers of E/M. Those corrections come from interactions that have coupling constants whose dimensions have negative powers of mass, $1/M$, $1/M^2$, and so on. But what about interactions that have positive powers of mass? Why aren't they there at $\mathcal{O}(M^n)$? Well unfortunately we don't have a good explanation.

The cosmological constant is such a term. It has the dimensions of energy per volume, in other words M^4. We don't know why it is as small as it is. There is also the bare mass of the Higgs boson. That's the one term in the Standard Model Lagrangian whose coefficient has the dimensionality of a positive power of mass. We don't know why it is not $\mathcal{O}(10^{15}$ GeV). These are great mysteries that confront us: Why are the terms in our present theory that have the dimensions of positive powers of mass so small compared to the scale that we think is the fundamental scale, somewhere in the neighborhood of 10^{15}–10^{18} GeV ? We don't know.

With the new approach to the Standard Model I think we have to say that this theory in its original form is not what we thought it was. It is not a fundamental theory. But at the same time, I want to stress that the Standard Model will survive in any future textbooks of physics in the same way that Newtonian mechanics has survived as a theory we use all the time applied to the solar system. All of our successful theories survive as approximations to a future theory.

There's a school of philosophy of science associated in particular with the name of Thomas Kuhn that sees the development of science—particularly of physics—as a series of paradigm shifts in which our point of view changes so radically that we can barely understand the theories of earlier times. I don't believe it for a minute. I think our successful theories always survive, as Newtonian mechanics has, and as I'm sure the Standard Model will survive, as approximations.

Now, we have to face the question, approximations to what? We think the Standard Model is a low-energy approximation to a theory whose constituents involve mass scales way beyond what we can approach in the laboratory, scales on the order of 10^{15}, 10^{18} GeV. It may be a field theory. It may be an asymptomatically safe field theory, which avoids coupling constants running off to infinity as the energy increases. Or it seems to me more likely that it is not a field theory at all, that it is something like a string theory. In this case, we will understand the very successful field theories with which we work as effective field theories, embodying the principles of quantum mechanics and symmetry, applied as an approximation valid at low energy where any theory will look like a quantum field theory.

Acknowledgements This work is supported by the National Science Foundation under grant number Phy-1914679 and also with support from the Robert A. Welch Foundation, Grant No. F-0014.

3.2. The Standard Model

Then it suddenly occurred to me that this was a perfectly good sort of theory, but I was applying it to the wrong kind of interaction. The right place to apply these ideas was not to the strong interactions, but to the weak and electromagnetic interactions.
Steven Weinberg

Goldstone's Theorem states that spontaneous breaking of global continuous symmetries inevitably implies the existence of massless particles. The first paper of this section by Goldstone, Salam and Weinberg [3.2.1], written when Weinberg was visiting Imperial College London in 1962, provides the proof. This looked at first like bad news since there were no known candidates for such particles. The resolution of the Goldstone-mode problem came in 1964 from the work of three independent groups: Brout and Englert; Higgs; Guralnik, Hagen and Kibble (BEHGHK). They recognised that Goldstone's theorem could be evaded in a gauge theory: the Goldstone mode can be absorbed into the longitudinal mode of the vector gauge field, rendering it massive. The existence of a residual scalar mode, which was discussed only in Higgs's paper, was not considered important at the time. In 1967, Weinberg thought of looking into the possibility of promoting the $SU(2) \times SU(2)$ chiral symmetry of the strong interactions to a local symmetry and inevitably encountered the BEHGHK effect. The axial vector particle, the $A1$ meson, acquires a mass but, in accordance with Kibble's rule, the vector particle, the ρ meson, associated with the unbroken isospin symmetry, remains massless.

Later that year, while driving across town from Harvard to MIT, he realized that he was working on the wrong problem. Maybe in a theory of weak and electromagnetic interactions (of course with a different spontaneously broken local symmetry group and different matter fields) the massive spin-1 particle would turn out to be not the $A1$ meson, but the W particle that had long been supposed to transmit the weak force. And the massless spin-1 particle would be not the ρ meson, but the photon, associated with unbroken electromagnetic gauge invariance.

A major part of what is now known as the Standard Model, this unification of weak and electromagnetic interactions is based upon the electroweak gauge group $\mathrm{SU}(2)\times\mathrm{U}(1)$, which has four Lie algebra generators. Such a group structure, without spontaneous symmetry breaking, had been anticipated by Glashow and by Salam and Ward. The corresponding four vector gauge fields couple to a Higgs complex scalar field gauge-symmetry doublet with a Lagrangian potential term. The Higgs complex doublet contains four real fields. Three of these belong to the broken-generator coset $(\mathrm{SU}(2) \times \mathrm{U}(1))/\mathrm{U}(1)_{\mathrm{em}}$, where the unbroken electromagnetic subgroup $\mathrm{U}(1)_{\mathrm{em}}$ mixes one combination of SU(2) generators with the U(1) "hypercharge" factor of the original gauge group prior to symmetry breaking. These erstwhile Goldstone fields are absorbed by the Higgs effect into masses for three of the vector gauge fields. The gauge field for the fourth gauge group generator, corresponding to the unbroken $\mathrm{U}(1)_{\mathrm{em}}$ subgroup, remains massless. This is the electromagnetic photon field A_μ. The same conclusions were reached independently by Abdus Salam.

The electroweak Standard Model is one of the most precisely tested and verified achievements of elementary particle physics and Weinberg's paper, *A Model of Leptons*, is now one of the most highly cited papers in all of science [3.2.2]. It got off to a slow start, however, one reason being that there was no proof of renormalizability for such spontaneously broken gauge theories. A major breakthrough was the paper by 't Hooft proving renormalizability, but in the "manifestly renormalizable" gauge where unitarity is not apparent, as opposed to the "manifestly unitary" gauge used by Weinberg where renormalizability is not apparent. Weinberg then showed that his gauge nevertheless gives convergent results [3.2.3] for physical processes consistent with those of 't Hooft, 't Hooft and Veltman and Lee and Zinn-Justin. Moreover, including the quarks, he showed that there was a suppression of strangeness-changing neutral currents by the inclusion of a fourth quark, as pointed out earlier by Glashow, Iliopoulos and Maini in the old current-current theory. See also his paper with Glashow [3.2.5]. The discovery of the J/psi particle by the Ting and Richter groups confirmed the existence of a fourth quark and with the discovery of neutral currents at CERN, the electro-weak part of the story was more-or-less complete. Glashow, Salam and Weinberg were accordingly awarded the Nobel Prize in 1979. (The notoriously cautious Nobel Committee here took a gamble since the Higgs boson had yet to be discovered.)

After the discovery of asymptotic freedom in non-abelian gauge theories of the strong interactions, by Gross and Wilzcek and by Politzer, Weinberg made significant contributions to QCD. He suggested [3.2.4], as did Gross and Wilczek independently, that the reason we do not see quarks and gluons is because they are confined by the growth of the gauge coupling at large distances. Another paper [3.2.6] written with Sterman explained jets in e^+e^- annihilation without assuming the parton model.

Consistency demands that the quarks and leptons appear in families, so the discovery of the tau lepton demanded the existence of the top and bottom quarks. Nature duly obliged. Moreover, three families allowed for (a little) CP violation. The final piece of the Standard Model jigsaw fell into place with the discovery of the Higgs at CERN in 2012.

Many people, and I am one of them, regard the General Theory of Relativity and Standard Model of Particle Physics to be the greatest intellectual achievements of the 20th century. The first was the product of one individual, Albert Einstein, whereas many contributed to the second, but none more than Steven Weinberg.

PHYSICAL REVIEW VOLUME 127, NUMBER 3 AUGUST 1, 1962

Broken Symmetries*

JEFFREY GOLDSTONE
Trinity College, Cambridge University, Cambridge, England

AND

ABDUS SALAM AND STEVEN WEINBERG†
Imperial College of Science and Technology, London, England

(Received March 16, 1962)

Some proofs are presented of Goldstone's conjecture, that if there is continuous symmetry transformation under which the Lagrangian is invariant, then either the vacuum state is also invariant under the transformation, or there must exist spinless particles of zero mass.

I. INTRODUCTION

IN the past few years several authors have developed an idea which might offer hope of understanding the broken symmetrics that seem to be characteristic of elementary particle physics. Perhaps the fundamental Lagrangian is invariant under all symmetries, but the vacuum state[1] is not. It would then be impossible to prove the usual sort of symmetry relations among S-matrix elements, but enough symmetry might remain (perhaps at high energy) to be interesting.

But whenever this idea has been applied to specific models, there has appeared an intractable difficulty. For example, Nambu suggested that the Lagrangian might be invariant under a continuous chirality transformation $\psi \rightarrow \exp(i\boldsymbol{\theta}\cdot\boldsymbol{\tau}\gamma_5)\psi$ even if the fermion physical mass M were nonzero. But then there would be a conserved current J_λ, with matrix element

$$\langle p'|J_\lambda|p\rangle = f(q^2)\bar{u}'\gamma_5[i\gamma_\lambda-(2M/q^2)q_\lambda]u,$$

where $q=p-p'$. The pole at $q^2=0$ can only arise from a spinless particle of mass zero, which almost certainly does not exist. Of course, the pole would not occur if $f(0)=0$, which might be the case if we do not insist on identifying J_λ with the axial vector current of β decay. But Nambu showed that this unwanted massless "pion" also appears as a solution of the approximate Bethe-Salpeter equation.[1]

Goldstone[2] has examined another model, in which the manifestation of "broken" symmetry was the nonzero vacuum expectation value of a boson field. (This was suggested as an explanation of the $\Delta I=\frac{1}{2}$ rule by Salam and Ward.)[3] Here again there appeared a spinless particle of zero mass. Goldstone was led to conjecture that this will always happen whenever a continuous symmetry group leaves the Lagrangian but not the vacuum invariant.

* This research was supported in part by the U. S. Air Force under a contract monitored by the Air Force Office of Scientific Research of the Air Development Command and the Office of Naval Research.

† Alfred P. Sloan Foundation Fellow; Permanent address: University of California, Berkeley, California.

[1] Y. Nambu and G. Jona-Lasinio, Phys. Rev. **122**, 345 (1961); W. Heisenberg, Z. Naturforsch. **14**, 441 (1959).

[2] J. Goldstone, Nuovo cimento **19**, 154 (1961).

[3] A. Salam and J. C. Ward, Phys. Rev. Letters **5**, 512 (1960).

We will present here three proofs of this result. The first uses perturbation theory; the other two are much more general.

II. PERTURBATION THEORY

We will consider a multiplet of n spinless fields ϕ_i which interact among themselves and perhaps also with other fields. The Lagrangian is assumed to be invariant under a set of infinitesimal transformations:

$$\delta^\alpha\phi_i=\epsilon T_{ij}{}^\alpha\phi_j. \tag{1}$$

If the vacuum state were also invariant under these transformations, the vacuum expectation values of the ϕ_i would be subject to a set of linear relations,

$$T_{ij}{}^\alpha\langle\phi_j\rangle_0=0. \tag{2}$$

(Usually, there would be enough such relations to imply that all ϕ_i have zero vacuum expectation value. This is the case in the example to be discussed at the end of this section, where the ϕ_i transform as the defining representation of the orthogonal group, so that the T span the space of all antisymmetric matrices.)

We are going to examine the possibility that the vacuum state is not invariant under these transformations; in particular we will consider the consequences that ensue if

$$T_{ij}{}^\alpha\langle\phi_j\rangle_0\neq0 \tag{3}$$

for some α and some i. It is inconvenient to work with fields with nonzero vacuum expectation value, we we will define

$$\phi_i=\chi_i+\eta_i, \tag{4}$$

where

$$\eta_i\equiv\langle\phi_i\rangle_0,$$

so that χ_i is a quantum field with

$$\langle\chi_i\rangle_0\equiv0. \tag{5}$$

In perturbation theory this means that we should ignore all "tadpole" diagrams with a single external χ line.

The Lagrangian $L(\phi)$, although invariant under (1), will, in general, not be invariant under the "naive" transformations

$$\delta^\alpha\chi_i=\epsilon T_{ij}{}^\alpha\chi_j. \tag{6}$$

Hence, the vanishing of $\langle\chi_i\rangle_0$ provides a nontrivial self-consistency condition which allows us to calculate η_i up to some unavoidable ambiguities. We will now show that the value of η is such that propagators of some of the χ_i have a pole at zero mass.

We begin by defining a function $F(\eta)$ as the sum of all proper connected graphs with no external lines, and with the over-all energy momentum conservation factor $i(2\pi)^4\delta^4(0)$ omitted. Every factor λ_i in each term of $F(\lambda)$ represents a place where we might instead have an external line of type i. This can be seen in general by noting that the interaction Lagrangian density used in calculating these graphs is

$$L'(\chi,\eta)=L(\chi+\eta)-L_0(\chi),$$
$$L_0(\chi)=-\tfrac{1}{2}(\partial_\mu\chi_i)(\partial^\mu\chi_i)-\tfrac{1}{2}m^2\chi_i\chi_i. \tag{7}$$

It follows then that the sum $F^{(N)}$ of all connected proper diagrams with N external lines i, j, $\cdots$ carrying zero energy and momentum is (for $N\neq2$)

$$F_{ij\cdots}{}^{(N)}=(\partial^N/\partial\eta_i\partial\eta_j\cdots)F(\eta). \tag{8}$$

For $N=2$ the mass term in $-L_0$ gives an additional contribution, so

$$F_{ij}{}^{(2)}=(\partial^2/\partial\eta_i\partial\eta_j)F(\eta)+m^2\delta_{ij}. \tag{9}$$

As defined here, $F^{(N)}$ does not include the propagators for its external lines or the over-all factor $i(2\pi)^4\delta(0)$.

It is clear from the definition of $F^{(1)}$ that

$$F_i{}^{(1)}=(\Delta'^{-1}(0))_{ij}\langle\chi_j\rangle_0. \tag{10}$$

Here $\Delta'(p)$ is the complete propagator given by

$$\Delta'_{ij}(p)=\int d^4x\,e^{-ip\cdot x}\langle T\{\chi_i(x),\chi_j(0)\}\rangle_0. \tag{11}$$

Because $\langle\chi_i\rangle_0$ vanishes, the most general improper diagram for $\Delta'(p)$ can be constructed by stringing together proper self-energy parts $\Pi^*(p)$ into a linear chain; hence the inverse of the propagator is given as usual by

$$(\Delta'^{-1}(p))_{ij}=(p^2+m^2)\delta_{ij}-\Pi^*{}_{ij}(p). \tag{12}$$

For zero momentum,

$$\Pi^*{}_{ij}(0)=F_{ij}{}^{(2)}, \tag{13}$$

and so, using (9), we have

$$(\Delta'^{-1})_{ij}=-(\partial^2/\partial\eta_i\partial\eta_j)F(\eta). \tag{14}$$

We are going to prove that (14) has no inverse, so it should be kept in mind that it is $\Delta'^{-1}(0)$ and not $\Delta'(0)$ that is well defined.

To complete the proof, we now must make use of the invariance of $L(\phi)$ under the transformation (1). This has the consequence that it is only the presence of the η terms that breaks invariance under (6), and hence that $F(\eta)$ is invariant under the corresponding transformations

$$\delta^\alpha\eta_i=\epsilon T_{ij}{}^\alpha\eta_j. \tag{15}$$

Thus

$$(\partial F/\partial\eta_i)T_{ij}{}^\alpha\eta_j=0. \tag{16}$$

Differentiating with respect to η, this gives

$$(\partial^2F/\partial\eta_i\partial\eta_j)T_{ik}{}^\alpha\eta_k+(\partial F/\partial\eta_i)T_{ij}{}^\alpha=0. \tag{17}$$

For physically allowed values of η this relation together with (10) and (14), yields at zero momentum

$$(\Delta'^{-1})_{ij}T_{ik}{}^\alpha\eta_k=0. \tag{18}$$

We see that for zero momentum the inverse of the propagator becomes singular and so some elements of the propagator become infinite. This does not prove that there is a pole at zero mass, but we certainly expect the propagator to be infinite at $P^2=0$ only if the theory involves particles of zero mass. The fields with nonvanishing matrix element between the vacuum and states of zero mass are

$$\chi^\alpha \equiv T_{ik}{}^\alpha \eta_k \chi_i. \tag{19}$$

Clearly, none of this trouble would occur if it were not for our assumption (3) that $T_{ik}{}^\alpha \eta_k \neq 0$. It is the broken symmetry, and not merely the nonzero vacuum expectation value η, that necessitates massless bosons.[4]

To see how this all works in a specific example, let us take as the Lagrangian

$$L(\phi) = -\bar{\psi}(\gamma\partial+M)\psi - \tfrac{1}{2}(\partial_\mu\phi_i)(\partial^\mu\phi_i) - \tfrac{1}{2}m^2\phi_i\phi_i - g\bar{\psi}O_i\psi\phi_i - \tfrac{1}{4}\lambda(\phi_i\phi_i)^2. \tag{20}$$

Defining $\phi_i = \chi_i + \eta_i$, this becomes

$$\begin{aligned} L(\phi) = L(\chi) &- m^2(\chi_i\eta_i) - \tfrac{1}{2}m^2(\eta_i\eta_i) - g\bar{\psi}O_i\psi\eta_i \\ &- \lambda(\chi_i\chi_i)(\chi_j\eta_j) - \lambda(\chi_i\eta_i)^2 - \tfrac{1}{2}\lambda(\chi_i\chi_i)(\eta_j\eta_j) \\ &- \lambda(\eta_i\eta_i)(\chi_j\eta_j) - \tfrac{1}{4}\lambda(\eta_i\eta_i)^2. \end{aligned} \tag{21}$$

The low-order contributions to the sum of all proper connected vacuum graphs are

$$\begin{aligned} F(\eta) = F(0) &- \tfrac{1}{2}m^2(\eta_i\eta_i) - \tfrac{1}{4}\lambda(\eta_i\eta_i)^2 + i(2\pi)^{-4}g\eta_i \\ &\times \int d^4p\, \mathrm{Tr}\{O_iS(p)\} + i(2\pi)^{-4}\lambda(\eta_i\eta_i + \tfrac{1}{2}\delta_{ij}\eta_k\eta_k) \\ &\times \int d^4p\, \Delta_{ij}(p) + \frac{i}{2}(2\pi)^{-4}g^2\eta_i\eta_j \int d^4p \\ &\times \mathrm{Tr}\{O_iS(p)O_jS(p)\} - i(2\pi)^{-4}\lambda^2(\eta_i\eta_j + \tfrac{1}{2}\delta_{ij}\eta_k\eta_k) \\ &\times (\eta_a\eta_b + \tfrac{1}{2}\delta_{ab}\eta_c\eta_c) \int d^4p\, \Delta_{ia}(p)\Delta_{jb}(p) \\ &- (2\pi)^{-8}\lambda^2\delta_{ij}\eta_k\delta_{ab}\eta_c \int d^4p\, d^4p' \\ &\times \{\Delta_{ia}(p)\Delta_{jb}(p')\Delta_{kc}(p+p') + \Delta_{ic}(p)\Delta_{jb}(p') \\ &\times \Delta_{ka}(p+p') + \Delta_{ib}(p)\Delta_{jc}(p')\Delta_{ka}(p+p')\}, \end{aligned} \tag{22}$$

where

$$S(p) = [-ip_\mu\gamma^\mu + M]^{-1},$$
$$\Delta_{ij}(p) = \delta_{ij}(p^2+m^2)^{-1}.$$

It can be readily seen that the derivative of $F(\eta)$ with respect to η_i is the sum $F_i{}^{(1)}$ of the proper connected diagrams with one external line, and that the second derivative with respect to η_i and η_j is the sum of the proper connected diagrams with two external lines, except that the $m^2\eta^2$ term in $F(\eta)$ does not contribute to $F^{(2)}$.

We will now assume that $L(\phi)$ is invariant under the group $SO(n)$ of orthogonal transformations on ϕ. This implies that the operators O_i are such that

$$\int d^4p\, \mathrm{Tr}\{O_iS(p)\} = 0, \tag{23}$$

$$\int d^4p\, \mathrm{Tr}\{O_iS(p)O_jS(p)\} = \delta_{ij}I. \tag{24}$$

Thus $F(\eta)$ is a function of $\eta^2 = (\eta_i\eta_i)$,

$$\begin{aligned} F(\eta^2) = F(0) &- \tfrac{1}{2}m^2\eta^2 - \tfrac{1}{4}\lambda\eta^4 \\ &+ i(2\pi)^{-4}\eta^2(1+\tfrac{1}{2}n)\int d^4p(p^2+m^2)^{-1} \\ &+ \tfrac{1}{2}i(2\pi)^{-4}g^2\eta^2 I - i(2\pi)^{-4}\lambda^2\eta^4(2+\tfrac{1}{4}n) \\ &\times \int d^4p(p^2+m^2)^{-1}(p^2+m^2)^{-1} - (2\pi)^{-8}\lambda^2\eta^2(2+n) \\ &\times \int d^4p \int d^4p'(p^2+m^2)^{-1}(p'^2+m^2)^{-1} \\ &\times [(p+p')^2+m^2]^{-1}. \end{aligned} \tag{25}$$

The dependence on η^2 alone implies then that

$$F_i{}^{(1)} = 2\eta_i F', \tag{26}$$

$$F_{ij}{}^{(2)} = (m^2+2F')\delta_{ij} + 4\eta_i\eta_j F''. \tag{27}$$

(A prime denotes differentiation with respect to η^2.) We see that the contribution of the term $-\tfrac{1}{2}m^2\eta^2$ in $F^{(2)}$ is canceled by the term $m^2\delta_{ij}$ in (27) arising from $-L_0$.

We have shown that $F_i{}^{(1)}$ is proportional to $\langle\chi_i\rangle_0$ and so must vanish. This implies that either η is zero or it must satisfy the consistency condition:

$$F'(\eta^2) = 0. \tag{28}$$

Thus if η is not zero, it is determined up to an orthogonal transformation; we certainly could not expect a more unambiguous determination.[5]

[4] It is clear from (19) that the maximum number of zero-mass fields is L, the number of Lie generators. There may in special cases be fewer than L zero-mass fields if not all fields χ^α given by (19) are linearly independent. This happens for example when $T_{ik}{}^\alpha$ correspond to the "tensor" representations of simple Lie groups. For this case $T_{ik}{}^\alpha$ are antisymmetric for all three indices. Therefore $\eta_\alpha\chi^\alpha = \eta_\alpha T_{ik}{}^\alpha\eta_k\chi_i = 0$, and only $(L-1)$ of the fields χ^α are linearly independent. These results are unaltered even if we allow in the theory more than one set of scalar fields ϕ_i with nonzero vacuum expectation values. To take a concrete case, the spurion theory proposed by Salam and Ward (reference 3) to explain the $\Delta I = \frac{1}{2}$ rule rests on assuming $\langle K_1{}^0\rangle \neq 0$. This would mean that the three companion fields to $K_1{}^0$, i.e., K^-, K^+, and $K_2{}^0$, must possess zero masses.

[5] Basic to the entire self-consistency procedure is, of course, the conjecture that $F'(\eta^2)=0$ does possess a root for real η. By considering classical field theories, Goldstone states that a real root would exist provided the bare mass for the ϕ_i fields is pure imaginary. It is interesting to note that if the Lagrangian (20) contains no $-\frac{1}{4}\lambda(\phi_i\phi_i)^2$ term, an application of Lehmann's mass theorem [H. Lehmann, Nuovo cimento **11**, 342 (1945)] shows that Goldstone's condition [(bare mass)$^2<0$] can never be satisfied. If, however, the $-\frac{1}{4}\lambda(\phi_i\phi_i)^2$ term is present in the Lagrangian, Lehmann's theorem gives no indication of the sign of (bare mass)^2.

We have also shown that

$$\begin{aligned}\Delta_{ij}'^{-1}(0)&=m^2\delta_{ij}-F_{ij}^{(2)}\\&=-2F'\delta_{ij}-4\eta_i\eta_jF''.\end{aligned}\tag{29}$$

If $\eta=0$, there is no reason to expect that $F'=0$, so that $(\Delta'^{-1})_{ij}$ is the nonsingular matrix $-2F'\delta_{ij}$. But if $\eta\neq0$, then for physically allowed values of η we must have $F'=0$ so,

$$(\Delta'^{-1}(0))_{ij}=-4\eta_i\eta_jF''.\tag{30}$$

This is certainly a singular matrix. In fact

$$(\Delta'^{-1}(0))_{ij}u_j=0\tag{31}$$

for any u orthogonal to η. All such u can be expressed as

$$u_i=T_{ij}\eta_j,\tag{32}$$

by choosing T_{ij} as an appropriate antisymmetric matrix. We see then that the space of u's is precisely the space indicated by the general considerations above.

III. GENERAL PROOFS

If the Lagrangian is invariant under an n-dimensional set of infinitesimal transformations which transform a general field ϕ_a according to

$$\delta\phi_a=\epsilon T_{ab}{}^{\alpha}\phi_b,\tag{33}$$

then there will exist a set of conserved currents

$$J^{\mu\alpha}=i\frac{\partial L}{\partial(\partial_\mu\phi_a)}T_{ab}{}^{\alpha}\phi_b,\tag{34}$$

$$\partial_\mu J^{\mu\alpha}=0.\tag{35}$$

The usual proof of the conservation equations (35) makes use only of the invariance of the Lagrangian, and hence should not be affected by the noninvariance of the vacuum. Also, from the canonical commutation relations we always expect that

$$[Q^\alpha,\phi_a]=T_{ab}{}^{\alpha}\phi_b,\tag{36}$$

where

$$Q^\alpha=\int d^3x\,J^{0\alpha}(x).\tag{37}$$

We will begin by assuming again that there exists a set of spinless fields ϕ_i transforming according to Eq. (1), i.e.,

$$[Q^\alpha,\phi_i]=T_{ij}{}^{\alpha}\phi_j.\tag{38}$$

The fields ϕ_i need not be "fundamental" here; all our remarks will apply equally well if the ϕ_i are synthetic objects like $\bar{\psi}O_i\psi$.

We shall show that if the vacuum is not annihilated by Q^α, so that

$$T_{ij}{}^{\alpha}\langle\phi_j\rangle\neq0,\tag{39}$$

then the theory must involve massless particles.

The place we will look for zero-mass singularities is in the vacuum expectation value of the commutator of $J^{\mu\alpha}$ with ϕ_i. Using the usual Lehmann-Källén arguments, this can be written

$$\langle[J^{\mu\alpha}(x),\phi_i(y)]\rangle_0=\partial^\mu\int dm^2\,\Delta(x-y,m^2)\rho_i{}^{\alpha}(m^2),\tag{40}$$

where Δ is the usual causal Green's function for mass m and

$$(\Box^2-m^2)\Delta(x-y,m^2)=0,\tag{41}$$

$$(2\pi)^{-3}p^\mu\theta(p^0)\rho_i{}^{\alpha}(-p^2)=-\sum\delta(p-p^n)(0|J^{\mu\alpha}(0)|n)\times\langle n|\phi_i(0)|0\rangle.\tag{42}$$

The current conservation condition (35) together with (40) and (41) implies that

$$m^2\rho_i{}^{\alpha}(m^2)=0,\tag{43}$$

and hence

$$\rho_i{}^{\alpha}(m^2)=N_i{}^{\alpha}\delta(m^2),\tag{44}$$

$$\langle[J^{\mu\alpha}(x),\phi_i(y)]\rangle_0=N_i{}^{\alpha}\partial^\mu D(x-y),\tag{45}$$

where D is Δ for $m=0$. We would normally expect no singularity in $\rho(m^2)$, or in other words, $N_i{}^{\alpha}=0$. (It is well known, for example, that the pion-decay matrix element would vanish if the axial vector current were conserved.) But because of (39) we can show that $N_i{}^{\alpha}\neq0$. For

$$\begin{aligned}N_i{}^{\alpha}&=\int d^3x\,N_i{}^{\alpha}\partial^0D(x)\\&=\langle[Q^\alpha,\phi_i(0)]\rangle_0\\&=T_{ij}{}^{\alpha}\langle\phi_j(0)\rangle_0\neq0.\end{aligned}\tag{46}$$

Thus the sum in (42) must include states of zero mass.

It perhaps does not necessarily follow from the noninvariance of the vacuum that there exists (or can be constructed) a set of spinless fields ϕ_i with $T_{ij}{}^{\alpha}\langle\phi_j\rangle_0\neq0$. We therefore wish to offer a simple nonrigorous argument that if there were no massless particles in the theory, then we would have to conclude that

$$0=|\alpha\rangle\equiv Q^\alpha|0\rangle;\tag{47}$$

for the conservation of current implies that Q^α and hence $|\alpha\rangle$ is invariant under the inhomogeneous Lorentz group. But then

$$\langle\alpha|J^{\alpha\mu}(x)|0\rangle=0,\tag{48}$$

so that

$$\langle\alpha|\alpha\rangle=\int d^3x\,\langle\alpha|J^{\alpha0}(x)|0\rangle=0.\tag{49}$$

Equation (47) follows from (49) and the positive-definiteness assumption.

This "proof" is probably unobjectionable in ordinary theories with no massless particles. But if there are massless particles, the integrals in (37) and (49) become somewhat poorly defined, because then there are states that are not Lorentz invariant but arbitrarily close to Lorentz invariance. If $|\alpha\rangle$ is such a state, the matrix element (48) will be small, but may give a large value to the integral (49).

As an example, let $|\mathbf{p}\rangle$ be a state containing a particle of mass zero, and construct the wave packet

$$|f\rangle = \int d^3p\ |\mathbf{p}|^{-\frac{1}{2}} f(|\mathbf{p}|)\,|\mathbf{p}\rangle. \tag{50}$$

The normalization condition is

$$1 = 4\pi \int |f(E)|^2 E dE, \tag{51}$$

so that if we wish we can choose $f(0) \neq 0$. Lorentz invariance requires that

$$\langle \mathbf{p}| J^{\mu\alpha}(x)|0\rangle = N^\alpha |\mathbf{p}|^{-\frac{1}{2}} p^\mu e^{-ip\cdot x}, \tag{52}$$

where $p^0 = |\mathbf{p}|$.

For particles with mass, current conservation would require that $N^\alpha = 0$, but no such conclusion can be drawn for massless particles. For the wave packet (52), we now have

$$\langle f| J^{\mu\alpha}(x)|0\rangle = N^\alpha \int d^3p (p^\mu/|p|) f^*(|\mathbf{p}|) e^{-ip\cdot x},$$

so that

$$\langle f|\alpha\rangle = \int d^3x\,\langle f|J^{\mu\alpha}(x)|0\rangle$$

$$= (2\pi)^3 N^\alpha f^*(0).$$

If we choose $f(0)$ to be nonzero, then the factor $|\mathbf{p}|^{-\frac{1}{2}}$ in (50) gives $|f\rangle$ a more-or-less Lorentz-invariant component, which has nonzero matrix element with $|\alpha\rangle$.

This becomes a bit more understandable if we ask ourselves what is the meaning of the state

$$(\exp i\epsilon Q^\alpha)\ |0\rangle = |0\rangle + i\epsilon\ |\alpha\rangle.$$

Clearly, this state is degenerate with $|0\rangle$, and is in fact another possible vacuum. It involves an infinitesimal component containing massless bosons of preponderantly low momentum. The role of the massless particles is apparently just to give meaning to the various possible vacua.

IV. PROSPECTS FOR THE UNSYMMETRIC VACUUM

The general proofs of the last section rest entirely on the assumption that there exists a conserved current, and that the integral of its time-component satisfies (38). This follows formally from the invariance of the Lagrangian, but in a *quantum* field theory the noncommutativity of the factors in the current, and the possible nonconvergence of the integral of its time-component, make our arguments essentially nonrigorous.

Therefore, it seems reasonable to defer belief in the necessity of massless bosons in a theory with unsymmetric vacua until such a *bête noire* is found in an actual calculation based on such a theory. We have already shown in Sec. II that the massless bosons do appear when we perform calculations using perturbation theory, provided that the symmetry of the theory is broken only by the choice of the vacuum expectation value η of the boson field.

But this is not the most general possibility. The original work of Nambu[1] indicates that the choice of a fermion mass can also break a symmetry. In this theory the fermion mass is

$$-ip^\mu\gamma_\mu = m_1 + i\gamma_5 m_2,$$

where (m_1, m_2) transform under chirality transformations like the components of a 2-vector. If m_1 and m_2 are not zero they must satisfy a condition of form

$$F(m_1^2 + m_2^2) = 0.$$

Any particular choice of direction for the vector (m_1, m_2) breaks the chirality invariance. (It should be noted that Nambu's choice $m_2 = 0$ is purely arbitrary and not dictated by parity conservation. For a general mass we must simply define the matrix associated with parity transformations to be

$$[(m_1 + im_2\gamma_5)/(m_1^2 + m_2^2)^{\frac{1}{2}}]\beta,$$

rather then just β.)

In Nambu's theory there is no "bare" spinless boson, but it is possible to construct a two-vector

$$\phi_1 = \bar\psi\psi, \quad \phi_2 = i\bar\psi\gamma_5\psi.$$

With Nambu's definition of parity (i.e., $m_2 = 0$) the vacuum expectation value of ϕ_2 but not of ϕ_1 vanishes, so the vector $\langle\boldsymbol{\phi}\rangle_0$ points in the 1-direction. An infinitesimal chirality transformation would rotate $\langle\boldsymbol{\phi}\rangle_0$ towards the 2-axis, so we are led to conjecture that the propagator of ϕ_2 has a zero-mass pole. In fact, just such a pole was found by Nambu in an approximate treatment of the bound-state problem. However, to show that the pole remains at zero mass when more complicated diagrams are considered would require a more thorough understanding of the treatment of bound states in perturbation theory. We are attempting this at present.

In a more complicated situation we could have an invariance broken both by the choice of a vacuum expectation value of a "bare" field and also simultaneously by the choice of a mass. For example, if we specialize the model discussed in Sec. II to the case of chirality invariance, we must take

$$M = 0, \quad O_1 = 1, \quad O_2 = i\gamma_5,$$

so that

$$L = -\bar\psi(\partial\cdot\gamma)\psi - \tfrac{1}{2}(\partial_\mu\phi_i)(\partial^\mu\phi_i) - \tfrac{1}{2}m^2(\phi_i\phi_i)$$
$$- g\bar\psi\psi\phi_1 - ig\bar\psi\gamma_5\psi\phi_2 - \tfrac{1}{4}\lambda(\phi_i\phi_i)^2.$$

In this case our conjecture would be that:

(1) If part of the loss of symmetry is due to the choice of a two-vector $\langle\phi\rangle_0$, then there must appear a zero-mass pole in the part of ϕ perpendicular to this vacuum expectation value.

(2) If part of the loss of symmetry is due to the choice of a nonzero Fermion mass $m_1+im_2\gamma_5$ then there must appear a zero-mass pole in the propagator of

$$\phi'=-m_2\bar{\psi}\psi+m_1\bar{\psi}\gamma_5\psi.$$

[Presumably this is the same pole as for (1). Parity conservation would require (m_1,m_2) to be in the direction of $\langle\phi\rangle_0$.]

(3) If part of the loss of symmetry is due to the choice of a noninvariant boson mass (i.e., if the residue of the pole at mass m in the propagator of ϕ_i and ϕ_j is a matrix which is not just a constant times δ_{ij}), then there must appear a *two-boson* pole at zero mass in the propagator of ϕ^2.

These "conjectures" can be taken as proved if we accept the arguments of Sec. III. We believe that we will also soon be able to prove these conjectures, in general, within the framework of perturbation theory.

If this is so, then there seem only three roads open to an understanding of broken symmetries based on the noninvariance of the vacuum:

(A) The particle interpretation of such theories might be revised (as in the Gupta-Bleuler method) so that the massless particles are not physically present in final states if they are absent in initial states. However, all our attempts in this direction have failed.

(B) The massless particles might really exist. The argument against this based on the Eötvös experiment might not apply if the particles carry quantum numbers, since then the scattering cross section of two macroscopic bodies due to exchange of the massless bosons would be proportional only to the numbers of atoms in each body and not (as for Coulomb forces or gravitation) to the squares of the numbers of atoms. But the couplings of these massless particles would presumably be quite strong, and would have shown up in exotic decay modes.

(C) Goldstone has already remarked that nothing seems to go wrong if it is just discrete symmetries that fail to leave the vacuum invariant. A more appealing possibility is that the "ur symmetry" broken by the vacuum involves an inextricable combination of gauge and space-time transformations.

Note added in proof. Recently, one of us (S. W., Proceedings of the 1962 Geneva Conference on High Energy Nuclear Physics) has developed a method of rewriting any Lagrangian in order to introduce fields for bound as well as "elementary" particles. This allows the proof of Sec. II to be extended to the case where the field with nonvanishing vacuum expectation value is any scalar function of the elementary particle fields, hence completing our argument.

ACKNOWLEDGMENTS

We would like to thank Dr. S. Kamefuchi for some helpful discussions. One of us (S. W.) would like to thank (A. S.) for his hospitality at Imperial College, and we all wish to gratefully acknowledge the hospitality of Professor R. G. Sachs at the 1961 Summer Institute of the University of Wisconsin.

VOLUME 19, NUMBER 21 PHYSICAL REVIEW LETTERS 20 NOVEMBER 1967

A MODEL OF LEPTONS*

Steven Weinberg†
Laboratory for Nuclear Science and Physics Department,
Massachusetts Institute of Technology, Cambridge, Massachusetts
(Received 17 October 1967)

Leptons interact only with photons, and with the intermediate bosons that presumably mediate weak interactions. What could be more natural than to unite[1] these spin-one bosons into a multiplet of gauge fields? Standing in the way of this synthesis are the obvious differences in the masses of the photon and intermediate meson, and in their couplings. We might hope to understand these differences by imagining that the symmetries relating the weak and electromagnetic interactions are exact symmetries of the Lagrangian but are broken by the vacuum. However, this raises the specter of unwanted massless Goldstone bosons.[2] This note will describe a model in which the symmetry between the electromagnetic and weak interactions is spontaneously broken, but in which the Goldstone bosons are avoided by introducing the photon and the intermediate-boson fields as gauge fields.[3] The model may be renormalizable.

We will restrict our attention to symmetry groups that connect the observed electron-type leptons only with each other, i.e., not with muon-type leptons or other unobserved leptons or hadrons. The symmetries then act on a left-handed doublet

$$L \equiv [\tfrac12(1+\gamma_5)]\binom{\nu_e}{e} \tag{1}$$

and on a right-handed singlet

$$R \equiv [\tfrac12(1-\gamma_5)]e. \tag{2}$$

The largest group that leaves invariant the kinematic terms $-\bar{L}\gamma^\mu\partial_\mu L - \bar{R}\gamma^\mu\partial_\mu R$ of the Lagrangian consists of the electronic isospin $\vec{T}$ acting on L, plus the numbers N_L, N_R of left- and right-handed electron-type leptons. As far as we know, two of these symmetries are entirely unbroken: the charge $Q = T_3 - N_R - \frac12 N_L$, and the electron number $N = N_R + N_L$. But the gauge field corresponding to an unbroken symmetry will have zero mass,[4] and there is no massless particle coupled to N,[5] so we must form our gauge group out of the electronic isospin $\vec{T}$ and the electronic hyperchange $Y \equiv N_R + \frac12 N_L$.

Therefore, we shall construct our Lagrangian out of L and R, plus gauge fields $\vec{A}_\mu$ and B_μ coupled to $\vec{T}$ and Y, plus a spin-zero doublet

$$\varphi = \binom{\varphi^0}{\varphi^-} \tag{3}$$

whose vacuum expectation value will break $\vec{T}$ and Y and give the electron its mass. The only renormalizable Lagrangian which is invariant under $\vec{T}$ and Y gauge transformations is

$$\mathcal{L} = -\tfrac14(\partial_\mu\vec{A}_\nu - \partial_\nu\vec{A}_\mu + g\vec{A}_\mu\times\vec{A}_\nu)^2 - \tfrac14(\partial_\mu B_\nu - \partial_\nu B_\mu)^2 - \bar{R}\gamma^\mu(\partial_\mu - ig'B_\mu)R - L\gamma^\mu(\partial_\mu ig\vec{t}\cdot\vec{A}_\mu - i\tfrac12 g'B_\mu)L$$
$$-\tfrac12|\partial_\mu\varphi - ig\vec{A}_\mu\cdot\vec{t}\varphi + i\tfrac12 g'B_\mu\varphi|^2 - G_e(\bar{L}\varphi R + \bar{R}\varphi^\dagger L) - M_1^{\,2}\varphi^\dagger\varphi + h(\varphi^\dagger\varphi)^2. \tag{4}$$

We have chosen the phase of the R field to make G_e real, and can also adjust the phase of the L and Q fields to make the vacuum expectation value $\lambda \equiv \langle\varphi^0\rangle$ real. The "physical" φ fields are then φ^-

VOLUME 19, NUMBER 21 PHYSICAL REVIEW LETTERS 20 NOVEMBER 1967

and

$$\varphi_1 \equiv (\varphi^0 + \varphi^{0\dagger} - 2\lambda)/\sqrt{2} \quad \varphi_2 \equiv (\varphi^0 - \varphi^{0\dagger})/i\sqrt{2}. \quad (5)$$

The condition that φ_1 have zero vacuum expectation value to all orders of perturbation theory tells us that $\lambda^2 \cong M_1^2/2h$, and therefore the field φ_1 has mass M_1 while φ_2 and φ^- have mass zero. But we can easily see that the Goldstone bosons represented by φ_2 and φ^- have no physical coupling. The Lagrangian is gauge invariant, so we can perform a combined isospin and hypercharge gauge transformation which eliminates φ^- and φ_2 everywhere[6] without changing anything else. We will see that G_e is very small, and in any case M_1 might be very large,[7] so the φ_1 couplings will also be disregarded in the following.

The effect of all this is just to replace φ everywhere by its vacuum expectation value

$$\langle\varphi\rangle = \lambda\begin{pmatrix}1\\0\end{pmatrix}. \quad (6)$$

The first four terms in $\mathcal{L}$ remain intact, while the rest of the Lagrangian becomes

$$-\tfrac{1}{8}\lambda^2 g^2[(A_\mu^{\ 1})^2 + (A_\mu^{\ 2})^2] - \tfrac{1}{8}\lambda^2(gA_\mu^{\ 3} + g'B_\mu)^2 - \lambda G_e \bar{e}e. \quad (7)$$

We see immediately that the electron mass is λG_e. The charged spin-1 field is

$$W_\mu \equiv 2^{-1/2}(A_\mu^{\ 1} + iA_\mu^{\ 2}) \quad (8)$$

and has mass

$$M_W = \tfrac{1}{2}\lambda g. \quad (9)$$

The neutral spin-1 fields of definite mass are

$$Z_\mu = (g^2 + g'^2)^{-1/2}(gA_\mu^{\ 3} + g'B_\mu), \quad (10)$$

$$A_\mu = (g^2 + g'^2)^{-1/2}(-g'A_\mu^{\ 3} + gB_\mu). \quad (11)$$

Their masses are

$$M_Z = \tfrac{1}{2}\lambda(g^2 + g'^2)^{1/2}, \quad (12)$$

$$M_A = 0, \quad (13)$$

so A_μ is to be identified as the photon field. The interaction between leptons and spin-1 mesons is

$$\frac{ig}{2\sqrt{2}}\bar{e}\gamma^\mu(1+\gamma_5)\nu W_\mu + \text{H.c.} + \frac{igg'}{(g^2+g'^2)^{1/2}}\bar{e}\gamma^\mu e A_\mu + \frac{i(g^2+g'^2)^{1/2}}{4}\left[\left(\frac{3g'^2 - g^2}{g'^2+g^2}\right)\bar{e}\gamma^\mu e - \bar{e}\gamma^\mu\gamma_5 e + \bar{\nu}\gamma^\mu(1+\gamma_5)\nu\right]Z_\mu. \quad (14)$$

We see that the rationalized electric charge is

$$e = gg'/(g^2 + g'^2)^{1/2} \quad (15)$$

and, assuming that W_μ couples as usual to hadrons and muons, the usual coupling constant of weak interactions is given by

$$G_W/\sqrt{2} = g^2/8M_W^{\ 2} = 1/2\lambda^2. \quad (16)$$

Note that then the e-φ coupling constant is

$$G_e = M_e/\lambda = 2^{1/4}M_e G_W^{\ 1/2} = 2.07\times10^{-6}.$$

The coupling of φ_1 to muons is stronger by a factor M_μ/M_e, but still very weak. Note also that (14) gives g and g' larger than e, so (16) tells us that $M_W > 40$ BeV, while (12) gives $M_Z > M_W$ and $M_Z > 80$ BeV.

The only unequivocal new predictions made by this model have to do with the couplings of the neutral intermediate meson Z_μ. If Z_μ does not couple to hadrons then the best place to look for effects of Z_μ is in electron-neutron scattering. Applying a Fierz transformation to the W-exchange terms, the total effective e-ν interaction is

$$\frac{G_W}{\sqrt{2}}\bar{\nu}\gamma_\mu(1+\gamma_5)\nu\left\{\frac{(3g^2 - g'^2)}{2(g^2+g'^2)}\bar{e}\gamma^\mu e + \tfrac{3}{2}\bar{e}\gamma^\mu\gamma_5 e\right\}.$$

If $g \gg e$ then $g \gg g'$, and this is just the usual e-ν scattering matrix element times an extra factor $\frac{3}{2}$. If $g \simeq e$ then $g \ll g'$, and the vector interaction is multiplied by a factor $-\frac{1}{2}$ rather than $\frac{3}{2}$. Of course our model has too many arbitrary features for these predictions to be

VOLUME 19, NUMBER 21 PHYSICAL REVIEW LETTERS 20 NOVEMBER 1967

taken very seriously, but it is worth keeping in mind that the standard calculation[8] of the electron-neutrino cross section may well be wrong.

Is this model renormalizable? We usually do not expect non-Abelian gauge theories to be renormalizable if the vector-meson mass is not zero, but our Z_μ and W_μ mesons get their mass from the spontaneous breaking of the symmetry, not from a mass term put in at the beginning. Indeed, the model Lagrangian we start from is probably renormalizable, so the question is whether this renormalizability is lost in the reordering of the perturbation theory implied by our redefinition of the fields. And if this model is renormalizable, then what happens when we extend it to include the couplings of $\vec{A}_\mu$ and B_μ to the hadrons?

I am grateful to the Physics Department of MIT for their hospitality, and to K. A. Johnson for a valuable discussion.

---

*This work is supported in part through funds provided by the U. S. Atomic Energy Commission under Contract No. AT(30-1)2098).

†On leave from the University of California, Berkeley, California.

[1]The history of attempts to unify weak and electromagnetic interactions is very long, and will not be reviewed here. Possibly the earliest reference is E. Fermi, Z. Physik 88, 161 (1934). A model similar to ours was discussed by S. Glashow, Nucl. Phys. 22, 579 (1961); the chief difference is that Glashow introduces symmetry-breaking terms into the Lagrangian, and therefore gets less definite predictions.

[2]J. Goldstone, Nuovo Cimento 19, 154 (1961); J. Goldstone, A. Salam, and S. Weinberg, Phys. Rev. 127, 965 (1962).

[3]P. W. Higgs, Phys. Letters 12, 132 (1964), Phys. Rev. Letters 13, 508 (1964), and Phys. Rev. 145, 1156 (1966); F. Englert and R. Brout, Phys. Rev. Letters 13, 321 (1964); G. S. Guralnik, C. R. Hagen, and T. W. B. Kibble, Phys. Rev. Letters 13, 585 (1964).

[4]See particularly T. W. B. Kibble, Phys. Rev. 155, 1554 (1967). A similar phenomenon occurs in the strong interactions; the ρ-meson mass in zeroth-order perturbation theory is just the bare mass, while the A_1 meson picks up an extra contribution from the spontaneous breaking of chiral symmetry. See S. Weinberg, Phys. Rev. Letters 18, 507 (1967), especially footnote 7; J. Schwinger, Phys. Letters 24B, 473 (1967); S. Glashow, H. Schnitzer, and S. Weinberg, Phys. Rev. Letters 19, 139 (1967), Eq. (13) et seq.

[5]T. D. Lee and C. N. Yang, Phys. Rev. 98, 101 (1955).

[6]This is the same sort of transformation as that which eliminates the nonderivative $\vec{\pi}$ couplings in the σ model; see S. Weinberg, Phys. Rev. Letters 18, 188 (1967). The $\vec{\pi}$ reappears with derivative coupling because the strong-interaction Lagrangian is not invariant under chiral gauge transformation.

[7]For a similar argument applied to the σ meson, see Weinberg, Ref. 6.

[8]R. P. Feynman and M. Gell-Mann, Phys. Rev. 109, 193 (1957).

VOLUME 27, NUMBER 24 PHYSICAL REVIEW LETTERS 13 DECEMBER 1971

Physical Processes in a Convergent Theory of the Weak and Electromagnetic Interactions*

Steven Weinberg
Laboratory for Nuclear Science and Department of Physics, Massachusetts Institute of Technology, Cambridge, Massachusetts 02139
(Received 20 October 1971)

A previously proposed theory of leptonic weak and electromagnetic interactions is found to be free of the divergence difficulties present in conventional models. The experimental implications of this theory and its extension to hadrons are briefly discussed.

Several years ago I proposed a unified theory[1] of the weak and electromagnetic interactions of leptons, and suggested that this theory might be renormalizable. This theory is one of a general class of models which may be constructed by a three-step process[2]: (A) First write down a Lagrangian obeying some exact gauge symmetry, in which massless Yang-Mills fields interact with a multiplet of scalar fields[3] and other particle fields. (B) Choose a gauge in which all the scalar field components vanish, except for a few (in our case one) real scalar fields. (C) Allow the gauge group to be spontaneously broken by giving the remaining scalar field a nonvanishing vacuum expectation value. Redefine this field by subtracting a constant λ, so that the "shifted" field φ has zero vacuum expectation value. In the resulting perturbation theory, all vector mesons acquire a mass, except for those (in our case, the photon) associated with unbroken symmetries.

In the proposed theory, this procedure was applied to the gauge group $SU(2)_L \otimes Y$, and resulted in a model involving electrons, electron-type neutrinos, charged intermediate bosons (W_μ), neutral intermediate bosons (Z_μ), photons (A_μ), and massive neutral scalar mesons (φ), with an interaction of the form[1]

$$\begin{aligned}\mathfrak{L}' = {} & \frac{ig}{(g+g'^2)^{1/2}}[gZ^\nu - g'A^\nu][W^\mu(\partial_\mu W_\nu{}^\dagger - \partial_\nu W_\mu{}^\dagger) - W^{\mu\dagger}(\partial_\mu W_\nu - \partial_\nu W_\mu) + \partial^\mu(W_\mu W_\nu{}^\dagger - W_\nu W_\mu{}^\dagger)] \\ & - \frac{g^2}{(g^2+g'^2)} W_\mu W_\nu{}^\dagger (gZ_\rho - g'A_\rho)(gZ_\sigma - g'A_\sigma)(\eta^{\mu\nu}\eta^{\rho\sigma} - \eta^{\mu\rho}\eta^{\nu\sigma}) + \frac{g^2}{2}[|W_\mu W^\mu|^2 - (W_\mu W^{\mu\dagger})^2] \\ & + F(\varphi) - \frac{m_e}{\lambda}\varphi\bar{e}e - \tfrac{1}{8}(\varphi^2 + 2\lambda\varphi)[(g^2+g'^2)Z_\mu Z^\mu + 2g^2 W_\mu W^{\mu\dagger}] \\ & + i(g^2+g'^2)^{-1/2}\bar{e}\gamma^\mu\left[\left(\frac{1-\gamma_5}{2}\right)g'^2 + \frac{1}{2}\left(\frac{1+\gamma_5}{2}\right)(g'^2 - g^2)\right]eZ_\mu + \frac{i}{2}(g^2+g'^2)^{1/2}\bar{\nu}\gamma^\mu\left(\frac{1+\gamma_5}{2}\right)\nu Z_\mu \\ & + \frac{igg'}{(g^2+g'^2)^{1/2}}\bar{e}\gamma^\mu e A_\mu + \frac{ig}{\sqrt{2}}\bar{\nu}\gamma^\mu\left(\frac{1+\gamma_5}{2}\right)eW_\mu{}^\dagger + \frac{ig}{\sqrt{2}}\bar{e}\gamma^\mu\left(\frac{1+\gamma_5}{2}\right)\nu W_\mu. \end{aligned} \tag{1}$$

Here $F(\varphi)$ is a fourth-order polynomial in φ (chosen so that $\langle\varphi\rangle_0 = 0$), and g and g' are independent coupling constants. The electronic charge e, weak coupling constant G, and vector meson masses are given by the formulas

$$e = gg'/(g^2+g'^2)^{1/2}, \quad G/\sqrt{2} = \tfrac{1}{2}\gamma^2, \tag{2}$$

$$m_W = \lambda g/2, \quad m_Z = \lambda(g^2+g'^2)^{1/2}/2. \tag{3}$$

At the time that this theory was proposed, its renormalizability was still a matter of conjecture. It is well known[4] that the Yang-Mills theory with which we start in step A above is indeed renormalizable if quantized in the usual way. However, the shift of the scalar field performed in step C amounts to a rearrangement of the perturbation series, so that the S matrix calculated in perturbation theory corresponds to a representation of the algebra of field operators inequivalent to that with which we started in step A. There is no obvious way to tell that renormalizability is preserved in this shift.

Recently several studies have indicated that various models of this general class actually are renormalizable. By choosing a different gauge in step B, 't Hooft[5] derived effective Lagrangians which appear manifestly renormalizable, but which involve fictitious massless scalar mesons of both positive and negative norm. Subsequently, Lee[6] showed in one case that the renormalization

VOLUME 27, NUMBER 24 PHYSICAL REVIEW LETTERS 13 DECEMBER 1971

program does actually work in this gauge, and that the spurious singularities associated with the fictitious particles all cancel. I would suggest, as an explanation of these results, that, although the shift of fields in step C really does generate an inequivalent representation of the field operators, the choice of gauge in step B does not, so that the S matrix calculated in the manifestly renormalizable gauges of 't Hooft should agree with the S matrix calculated in the "manifestly unitary" gauge used to derive Eq. (1).

If this is correct, then it ought to be possible to carry out calculations of higher-order weak interactions using the interaction (1) directly. This has obvious advantages over the use of Lagrangians of the 't Hooft–Lee form, because (1) involves only physical particles. This paper will explore the results of this model in certain physical processes, both in order to test its renormalizability, and also to gain some insight into its general properties.

First, it is necessary to derive the Feynman rules for this theory. The interaction Hamiltonian here is given by $-\mathcal{L}'$ plus noncovariant terms which are canceled by the noncovariant parts of the propagators of the vector meson fields and their derivatives. However, after this cancelation, there is left over a covariant effective interaction[7]

$$\delta\mathcal{L} = -6i\delta^4(0)\ln(1+\varphi/\lambda).$$

The correct Feynman rules are thus generated by using $-\mathcal{L}' - \delta\mathcal{L}$ as an effective interaction Hamiltonian, keeping only the "naive" covariant parts of the various propagators. No "ghost loops"[4] appear here.

Now let us consider some specific physical processes:

$\nu+\bar{\nu}\to W^+ + W^-$.—This is the reaction used by Gell-Mann, Goldberger, Kroll, and Low[8] to exhibit the difficulties associated with conventional intermediate boson theories. In such theories, the amplitude[9] for production of zero-helicity $W^\pm$ is given in lowest order by

$$f_{\text{GGKL}} = \frac{-iGp^{3/2}\sin\theta e^{-i\varphi}}{2\pi(2E)^{1/2}}\left\{\frac{2E^2[1-(E/p)\cos\theta]-m_W^2}{2E^2[1-(p/E)\cos\theta]-m_W^2+m_e^2}\right\}, \tag{4}$$

where E, p, θ, and φ are the energy, momentum, and scattering angles, respectively, of the W^+ in the center of mass system. For $E\to\infty$ this is dominated by a pure $J=1$ term which grows like E, so that in order to save unitarity it is necessary to introduce a cutoff[8] at energies of order $1/\sqrt{G}$. In the theory proposed here, there is an additional term produced by Z exchange in the s channel:

$$f_Z = \frac{iGp^{3/2}\sin\theta e^{-i\varphi}}{2\pi(2E)^{1/2}}\left[\frac{4E^2+2m_W^2}{4E^2-m_Z^2}\right]. \tag{5}$$

Inspection of (4) and (5) shows that the total scattering amplitude now grows only near threshold and falls off like $1/E$ for $E \gg m_W$. This natural cutoff at $E\sim m_W$ obviates the need for any special unitarity cutoff. To the extent that this is a general phenomenon, we can expect that the perturbation series in G is really an expansion in powers of $Gm_W^2 \sim g^2$, which may be as small as e^2.

$\nu+\nu\to\nu+\nu$.—This is a good process to use as a test of the performance of our theory in loop diagrams, because, as pointed out by Low,[10] the exchange of a pair of W bosons generates a quadratic divergence, related to the failure of unitarity bounds in Eq. (4). In our present theory, there are two additional fourth-order diagrams in which a pair of Z's is exchanged, plus a large number of fourth-order diagrams in which a single Z is exchanged with second-order self-energy or vertex insertions. The former diagrams contain quadratic divergences which cancel among themselves. The latter diagrams contain quartic divergences, which cancel among themselves, plus a large number of quadratically divergent terms. Some of these quadratic divergences can be grouped together as renormalizations of m_Z and the Z-ν coupling constant (and probably cancel) but there remain quadratic divergences in the Z self-energy proportional to $(t-m_Z^2)^2$, and in the Z-ν vertices proportional to $(t-m_Z^2)$. These terms generate an effective quadratically divergent neutrino Fermi interaction, *which turns out to cancel the quadratic divergence found by Low*. (I have not yet checked what happens to the logarithmically divergent terms.)

W-photon interactions.—The first term in Eq. (1) gives the W an "anomalous" magnetic moment, with gyromagnetic ratio $g_W=2$. This is just the value required[11] if the amplitude for Compton scattering on a W behaves well enough at high energies to satisfy a Drell-Hearn sum rule.

$e^+ + e^- \to W^+ + W^-$.—Because the W has an "anomalous" magnetic moment, the electromagnetic pair-production amplitude[9] grows like E as $E \to \infty$. Further, the neutrino t-channel and Z s-channel exchange diagrams do *not* cancel here, so that the weak pair-production amplitude also grows like E. However, *the weak and electromagnetic amplitudes cancel each other as* $E \to \infty$, leaving a scattering amplitude which vanishes like $1/E$ as $E \to \infty$, as required by unitarity bounds. This cooperation between the weak and electromagnetic interactions in solving each other's problems is one of the most satisfying features of this theory.

The weak and electromagnetic interactions of the leptons appear to be in good shape, so let us consider how to incorporate the hadrons. In order to preserve renormalizability, it is necessary to couple Z, W, and A to the currents of an *exact* $SU(2)_L \otimes U(1)$ symmetry of the strong-interaction Lagrangian. This poses a problem, because, apart from any spontaneous symmetry-breaking mechanisms responsible for the baryon masses, it is usually presumed that the nonzero masses of the π and K arise from an *intrinsic* breaking of $SU(2) \otimes SU(2)$ or $SU(3) \otimes SU(3)$. The only way that I can see to save renormalizability is to suppose instead that the π and K masses arise from the same purely spontaneous symmetry-breaking mechanism responsible for the W and Z masses. The problem then is whether it is natural for the strong interactions to conserve parity and isospin.

Leaving aside strange particles, the simplest way to couple the scalar doublet[3] $(\varphi^+,\ \varphi^0 + \lambda)$ of our model to the hadrons is to find some $(\frac{1}{2}, \frac{1}{2})$ $SU(2) \otimes SU(2)$ multiplet $(\sigma, \vec{\pi})$ of hadronic field operators, and write an $SU(2)_L$-invariant interaction,

$$-if\varphi^{+\dagger}(\pi_1 - i\pi_2) + f(\varphi^{0\dagger} + \lambda)(\sigma + i\pi_3) + \text{H.c.}$$

The rest of the strong-interaction Lagrangian is assumed to conserve $SU(2)_R$ as well as $SU(2)_L$, so by an $SU(2)_R$ rotation we can define σ and $\vec{\pi}$ so that f is real. After eliminating φ^+ and $\mathrm{Im}\varphi^0$ in step B, the only remaining symmetry-breaking term is $2f(\lambda + \varphi)\sigma$, which does conserve parity and isospin. Thus the spontaneous symmetry breaking of the weak interactions can act as a seemingly intrinsic symmetry-breaking mechanism for the strong interactions, which in turn is amplified if σ develops a large vacuum expectation value. A particularly attractive aspect of this approach is that the requirement of renormalizability provides the rationale for the conservation or partial conservation of the hadronic weak currents.

The most direct verification of this theory would be the discovery of W's and Z's with the predicted properties. However, the lower limits on m_W and m_Z are, respectively, $\lambda e/2 = 37.3$ GeV and $\lambda e = 74.6$ GeV, so this discovery will take a while.[12] The most accessible effect of the Z's is to change the cross sections for scattering of neutrinos and antineutrinos on electrons. We know nothing about the mass of the scalar meson φ, but its field might contribute to the level shifts in muonic atoms. Higher-order weak interactions produce various "radiative" corrections, including a change of order Gm_μ^2 in the gyromagnetic ratio of the muon.[13] The extension of this theory to strange particles appears to require both strangeness-changing and strangeness-conserving neutral hadronic currents, but the former can be eliminated in an $SU(4) \otimes SU(4)$-invariant model.[14] These matters will be dealt with at greater length in future papers.

I am deeply grateful to Francis Low, both for his indispensable advice and encouragement during the course of this work, and also for discussions over the last several years on the divergence difficulties of the weak interactions.

---

*Work supported in part through funds provided by the the U. S. Atomic Energy Commission under Contract No. AT(30-1)-2098.

[1]S. Weinberg, Phys. Rev. Lett. 19, 1264 (1967).

[2]P. W. Higgs, Phys. Rev. Lett. 12, 132 (1964), and 13, 508 (1964), and Phys. Rev. 145, 1156 (1966); F. Englert and R. Brout, Phys. Rev. Lett. 13, 321 (1964); G. S. Guralnik, C. R. Hagen, and T. W. B. Kibble, Phys. Rev. Lett. 13, 585 (1965); T. W. B. Kibble, Phys. Rev. 155, 1554 (1967). Also see A. Salam, in *Elementary Particle Physics,* edited by N. Svartholm (Almqvist and Wiksells, Stockholm, 1968), p. 367.

[3]There is a mistake in Eq. (3) of Ref. 1. The upper and lower members of the scalar doublet should have charges +1 and 0, not 0 and −1.

[4]R. P. Feynman, Acta Phys. Pol. 24, 697 (1963); B. S. De Witt, Phys. Rev. 162, 1195 (1967); L. D. Fadeev and V. N. Popov, Phys. Lett. 25B, 29 (1967); S. Mandelstam, Phys. Rev. 175, 1580 (1968); E. S. Fradkin and I. V. Tyutin, Phys. Rev. D 2, 2841 (1970); R. Mills, Phys. Rev. D 3, 2969 (1971).

[5]G. 't Hooft, to be published.

[6]B. W. Lee, to be published.

[7]The methods used to derive this result are based on the work of T. D. Lee and C. N. Yang, Phys. Rev. 128, 885 (1962); also see J. Honerkamp and K. Meetz, Phys. Rev. D 3, 1996 (1971); J. Charap, Phys. Rev. D 3, 1998 (1971); I. S. Gerstein, R. Jackiw, B. W. Lee, and

VOLUME 27, NUMBER 24 PHYSICAL REVIEW LETTERS 13 DECEMBER 1971

S. Weinberg, Phys. Rev. D 3, 2486 (1971).

[8]M. Gell-Mann, M. L. Goldberger, N. M. Kroll, and F. E. Low, Phys. Rev. 179, 1518 (1969).

[9]The scattering amplitude f is normalized so that the differential cross section is $|f|$.

[10]F. E. Low, *Lectures on Weak Interactions* (Tata Institute of Fundamental Research, Bombay, 1970), p. 11.

[11]S. Weinberg, in *Lectures on Elementary Particles and Quantum Field Theory*, edited by S. Deser, M. Grisaru, and H. Pendleton (Massachusetts Institute of Technology Press, Cambridge, Mass., 1970), p. 234.

[12]A value of precisely 37.3 GeV has been suggested for the intermediate boson mass by J. Schechter and Y. Ueda, Phys. Rev. D 2, 736 (1970) and more recently by T. D. Lee, Phys. Rev. Lett. D 3, 801 (1971). Lee also gets a gyromagnetic ratio $g_W = 2$. In the present theory, a W mass of 37.3 GeV is only possible if $g = e$, $g' \gg e$, and $m_Z \gg m_W$.

[13]R. Jackiw and S. Weinberg, to be published.

[14]S. L. Glashow, J. Iliopoulos, and L. Maiani, Phys. Rev. D 2, 1285 (1970).

VOLUME 31, NUMBER 7 PHYSICAL REVIEW LETTERS 13 AUGUST 1973

Non-Abelian Gauge Theories of the Strong Interactions*

Steven Weinberg
Lyman Laboratory of Physics, Harvard University, Cambridge, Massachusetts 02138
(Received 30 May 1973)

A class of non-Abelian gauge theories of strong interactions is described, for which parity and strangeness are automatically conserved, and for which the nonconservations of parity and strangeness produced by weak interactions are automatically of order α/m_W^2 rather than of order α. When such theories are "asymptotically free," the order-α weak corrections to natural zeroth-order symmetries may be calculated ignoring all effects of strong interactions. Speculations are offered on a possible theory of quarks.

Recently Gross and Wilczek and Politzer have made the exciting observation that non-Abelian gauge theories can exhibit free-field asymptotic behavior at large Euclidean momenta.[1] However, the physical application of this discovery raises serious problems: (1) Why don't the weak interactions produce parity and stangeness nonconservations of order α? (This problem finds a natural solution when the strong interactions are described by *Abelian* gauge models,[2] but not, to the best of my knowledge, in non-Abelian models of the "Berkeley" type.[3]) (2) Even with asymptotic freedom, when can the strong interactions actually be neglected? (3) Even if asymptotic freedom explains the success of naive quark-model calculations, why don't we see physical quarks? This

VOLUME 31, NUMBER 7 PHYSICAL REVIEW LETTERS 13 AUGUST 1973

note will describe a class of non-Abelian gauge models which provide a complete answer to (1), a partial answer to (2), and a possible answer to (3).

Consider a renormalizable gauge theory of the strong, weak, and electromagnetic interactions with the following properties: (A) The gauge group is a direct product of a "strong" gauge group G_S and a "weak" gauge group G_W. The coupling constants associated with G_S and G_W are of order 1 and e, respectively. The spin-$\frac{1}{2}$ hadrons are non-neutral with respect to both G_S and G_W, but the generators of G_S all commute with the generators of G_W. That is, the quarks would have to form a matrix, with weak and strong interactions producing transitions along columns and rows, respectively, as in the colored-quark[4] or Pati-Salam[5] models. The leptons are neutral under G_S. (B) The strong gauge group G_S is *nonchiral*. (C) There are various weakly coupled spin-0 fields with very large vacuum expectation values (~300 GeV), which break G_W and give the associated vector bosons large masses (~30 GeV) and also produce or contribute to the zeroth-order fermion mass matrix m. However, these spin-0 fields are neutral under G_S, so their vacuum expectation values do not contribute to the mass of the strongly interacting neutral vector bosons.

To avoid massless strongly interacting gauge bosons, it is natural to suppose that G_S is also spontaneously broken. We might assume that this occurs through the vacuum expectation values of various strongly interacting scalar fields, but it appears[1] that the introduction of these fields would prevent a free-field asymptotic behavior. Alternatively, we might assume that there are no strongly interacting spin-0 fields, but that G_S is broken dynamically.[6] However, this still leaves the question of why observed hadron states should belong to simple G_S multiplets, like the color singlets of Ref. 4.

There is one other possible alternative: that G_S is *not* broken, so that the G_S gauge bosons do have zero mass, and the quarks are "color" or "column" degenerate. On the basis of a preliminary analysis, it appears that the infrared divergences in such a theory could make the rate for production of any number of G_S-non-neutral particles in a collision of G_S-neutral particles *vanish*. The world would than consist of composite G_S singlets, like ordinary hadrons, which are not affected by these infrared divergences; the G_S-non-neutral hadrons, such as quarks and massless vector bosons, although present in Feynman diagrams, Wilson expansions, etc., would never appear as physical particles. This conjecture is under further study.

Leaving aside scalar fields, any theory governed by assumptions (A), (B), and (C) will have an effective "zeroth-order" strong-interaction Lagrangian of the form

$$\mathcal{L}_{\text{strong}} = -\bar{\psi}Z_\psi\gamma^\mu D_\mu\psi - \bar{\psi}m\psi - \tfrac{1}{4}(Z_A)_{ab}F_{a\mu\nu}F_b{}^{\mu\nu}, \tag{1}$$

where ψ and $D_\mu\psi$ are the spin-$\frac{1}{2}$ field multiplet and its G_S-gauge-covariant derivative; $F_{a\mu\nu}$ is the G_S-covariant gauge-field-strength term; and Z_ψ, m, and Z_A are G_S-invariant matrices, the first two of which may involve both the γ_5 and 1 Dirac matrices. Because Z_ψ and m commute with the generators of G_S, we can use Schur's lemma to redefine the fermion fields so that Z_ψ becomes unity and m becomes real, diagonal, and γ_5 free, while the generators of G_S remain γ_5 free and commute with m. Once we define the fermion fields in this way, the theory automatically conserves parity, if we assign positive parity to all spin-$\frac{1}{2}$ fields and identify all G_S gauge fields as polar vectors. In addition, any quantum number such as strangeness, charm, etc., which can be expressed in terms of the number of elementary fermions, summing over "color," is also automatically conserved. (The same is true if we introduce strongly interacting spin-0 fields, provided they belong to representations of G_S which prohibit Yukawa couplings to the fermions. All spin-0 fields can then be assigned positive parity and zero strangeness, charm, etc.) Also, it will often happen[7] that, for a wide range of parameters in the original Lagrangian, the spontaneous breakdown of G_W leaves some of the elements of the diagonalized fermion mass matrix equal or zero. This equality will then extend to all colors, and the strong interactions will have a "natural" zeroth-order unitary or chiral symmetry,[7] in addition to the conservation of parity, strangeness, etc.

Now let us consider the effect of second-order weak and electromagnetic interactions which break these symmetries. Just as in the neutral vector gluon model,[2] it is found after canceling gauge-dependent terms that the corrections to a general hadronic S-matrix element fall into three classes: tadpolelike terms, photon terms, and heavy vector boson exchange terms. The tadpolelike terms just amount to a G_S-invariant correc-

VOLUME 31, NUMBER 7 PHYSICAL REVIEW LETTERS 13 AUGUST 1973

tion to the matrix m in (1), and therefore any parity or strangeness nonconservation they introduce can be transformed away as before. The photon terms clearly cannot change parity or strangeness. The remaining terms are of the form

$$\delta S_{FI} = (2\pi)^4 \delta^4(P_F - P_I) \times \int d^4k\, \mathfrak{F}_{\alpha\beta}{}^{FI}(k)(k^2 + \mu'^2)_{\alpha\beta}{}^{-1}, \quad (2)$$

where

$$\mathfrak{F}_{\alpha\beta}{}^{FI}(k) \equiv \tfrac{1}{2} i (2\pi)^{-4} \times \int d^4x \langle F | T(J_{\alpha\mu}(x) J_\beta{}^\mu(y)) | I \rangle e^{ik(x-y)}. \quad (3)$$

Here the $J_{\alpha\mu}$ are the currents of G_W, formed from the spin-$\frac{1}{2}$ hadron fields, and μ' is the G_W-vector boson mass matrix, except that the photon is here given an arbitrary large mass Λ, to compensate for a regulator mass Λ used in evaluating the true photonic term.[2] The only terms in (2) that are of order α rather than α/m_W^2 are those arising from terms in $\mathfrak{F}(k)$ which decrease no faster than k^{-2} as $k \to \infty$. Such terms may be picked out from the Wilson operator-product expansion[8] of the direction-averaged matrix element:

$$\int d\Omega_k \mathfrak{F}_{\alpha\beta}{}^{FI}(k) \underset{k\to\infty}{\longrightarrow} \sum_N \langle F | O_N | I \rangle U_{\alpha\beta}{}^{(N)}(\sqrt{k^2}). \quad (4)$$

The asymptotic behavior of $U_{\alpha\beta}{}^{(N)}(\kappa)$ in perturbation theory is

$$U_{\alpha\beta}{}^{(N)}(\kappa) = O(\kappa^{2-d_N} \times \text{powers of } \ln\kappa), \quad (5)$$

where d_N is the "naive" dimensionality of the operator O_N. Hence the only terms that contribute to δS_{FI} in order α are those with exponent $2 - d_N$ not less than -2. The change in the S matrix is therefore equivalent to a change in the strong-interaction Lagrangian:

$$\delta\mathfrak{L} = \sum_N O_N \int_0^\infty U_{\alpha\beta}{}^{(N)}(\kappa)(\kappa^2 + \mu'^2)_{\alpha\beta}^{-1} \kappa^3 d\kappa, \quad (6)$$

the sum extending only over "renormalizable" operators O_N, with $d_N \leqslant 4$. But these operators are all Lorentz invariant, because we have averaged over momentum-space directions in (4), and they are all G_S-gauge invariant, because the currents $J_{\alpha\mu}$ are neutral under G_S. Hence the order-α correction term (6) must be of precisely the same form as the original zeroth-order Lagrangian (1), and so any order-α violation of parity and strangeness conservation introduced by the weak interactions can be eliminated by redefining the fermion fields as before. The only symmetry breaking remaining after this redefinition will take the form of shifts in the fermion masses, which can break the other natural zeroth-order symmetries[7] mentioned above. On the other hand, the truly weak corrections of order α/m_W^2 cannot be expressed in terms of renormalizable corrections to the Lagrangian, so we expect both parity and strangeness conservation to be violated by such nonleptonic truly weak interactions.

Now, what of the actual calculation of $\delta\mathfrak{L}$? After symmetric integration, the $U^{(N)}$ functions that contribute to (6) all have the naive asymptotic behavior κ^{-2}, so the integral (6) appears logarithmically divergent. However, the trace $U_{\alpha\alpha}$ is G_W invariant, so it cannot contribute to the corrections to natural zeroth-order symmetries, and thus all logarithmic divergences cancel.[2] If we assume that the "free-field" asymptotic behavior sets in at $\kappa < m_W$, then we can evaluate (6) using the asymptotic formula[9]

$$U_{\alpha\beta}{}^{(N)}(\kappa) \to \kappa^{-2} B_{\alpha\beta}{}^{(N)} \exp\{\int_a^k \gamma_N(a') da'/a'\} \quad (7)$$

where $B_{\alpha\beta}{}^{(N)}$ is a constant, independent of the strong-interaction coupling constant; $\gamma_N(a)$ is an anomalous dimension; and the lower limit a depends on the renormalization prescription used to define the operator O_N. In general, γ_N is not zero, so "asymptotic freedom" does *not* say the strong interactions can be ignored in calculating $U^{(N)}$. However, γ_N is in fact zero for the functions that interest us here. By the same arguments as those used in Ref. 2, we can conclude that the only O_N in (6) that matters is the bilinear product $\bar{\psi}_n \psi_m$. This does have a nonzero anomalous dimension $\gamma_{\bar{\psi}\psi}$, but we are interested here in the *nonleading* terms in the corresponding U function of order κ^{-2}, and the presence of a fermion mass factor in these nonleading terms introduces a term $-\gamma_{\bar{\psi}\psi}$ which cancels the previous term. (Details will be published elsewhere.) Using (7) in (6), our result then is that the terms of order α in δS_{FI} are the same as would be produced by a change in the effective strong-interaction Lagrangian

$$\delta\mathfrak{L} = \tfrac{1}{2}\bar{\psi} B_{\alpha\beta} \psi (\ln\mu'^2)_{\alpha\beta},$$

where $(B_{\alpha\beta})_{nm}$ is the B matrix for the operator $\bar{\psi}_n \psi_m$. This is independent of the strong-interaction coupling constants, and therefore must be the same as would be given by a one-loop perturbation calculation. The same is trivially true of the tadpole terms, so, apart from the purely photon terms, *all corrections to the effective Lagrangian are correctly given to order α by neglecting all effects of strong interactions.* Of course, we must not use perturbation theory to

VOLUME 31, NUMBER 7 PHYSICAL REVIEW LETTERS 13 AUGUST 1973

calculate any physical hadron mass or matrix element, but only to calculate $\delta\mathcal{L}$; the result must then be used as an input to current-algebra[10] or parton-model[11] calculations of physical quantities.

I am grateful for valuable discussions with T. Appelquist, C. Callan, S. Coleman, S. Glashow, D. Gross, R. Jackiw, K. Johnson, F. Low, D. Politzer, and H. Schnitzer.

---

*Work supported in part by the National Science Foundation under Grant No. GP-30819X.

[1]D. J. Gross and F. Wilczek, Phys. Rev. Lett. 30, 1343 (1973); H. D. Politzer, *ibid.*, 1346 (1973). Also see G. 't Hooft, to be published.

[2]S. Weinberg, to be published.

[3]I. Bars, M. B. Halpern, and M. Yoshimura, Phys. Rev. Lett. 29, 969 (1972), and to be published. The parity problem may not arise if these are regarded as phenomenological rather than fundamental field theories.

[4]W. A. Bardeen, H. Frtisch, and M. Gell-Mann, to be published. Also see Ref. 1.

[5]J. C. Pati and A. Salam, to be published. The fact that the Pati-Salam theory conserves parity in order α was apparently known to a number of theorists, including S. L. Glashow and also Pati and Salam.

[6]R. Jackiw and K. Johnson, to be published.

[7]S. Weinberg, Phys. Rev. Lett. 29, 338, 1698 (1972); H. Georgi and S. L. Glashow, Phys. Rev. D 6, 2977 (1972), and 8, 2457 (1973).

[8]K. Wilson, unpublished, and Phys. Rev. 179, 1499 (1969).

[9]See Ref. 1. The proof of the renormalization-group equations for the Wilson coefficient functions is given by S. Coleman, in Lectures of the Erice Summer School, Sicily, Italy, 1971 (to be published), and C. Callan, Phys. Rev. D 5, 3202 (1972).

[10]S. L. Glashow and S. Weinberg, Phys. Rev. Lett. 20, 224 (1968); M. Gell-Mann, R. J. Oakes, and B. Renner, Phys. Rev. 175, 2195 (1968); S. L. Glashow, R. Jackiw, and S. S. Shei, Phys. Rev. 187, 1416 (1969); etc.

[11]J. Gunion, to be published.

PHYSICAL REVIEW D VOLUME 15, NUMBER 7 1 APRIL 1977

Natural conservation laws for neutral currents*

Sheldon L. Glashow and Steven Weinberg
Lyman Laboratory of Physics, Harvard University, Cambridge, Massachusetts 02138
(Received 20 August 1976)

We explore the consequences of the assumption that the direct and induced weak neutral currents in an SU(2) ⊗ U(1) gauge theory conserve all quark flavors *naturally*, i.e., for all values of the parameters of the theory. This requires that all quarks of a given charge and helicity must have the same values of weak T_3 and $\vec{T}^2$. If all quarks have charge +2/3 or −1/3 the only acceptable theories are the "standard" and "pure vector" models, or their generalizations to six or more quarks. In addition, there are severe constraints on the couplings of Higgs bosons, which apparently cannot be satisfied in pure vector models. We also consider the possibility that neutral currents conserve strangeness but not charm. A natural seven-quark model of this sort is described. The experimental consequences of charm nonconservation in direct or induced neutral currents are found to be quite dramatic.

I. INTRODUCTION

It has been known for many years that there are no strangeness-changing neutral-current weak interactions, or none with anything like the strength of the familiar charged-current weak interactions. We see this from the slowness of such decays as $K_L^0 \to \mu^+\mu^-$ and $K^+ \to \pi^+\nu\bar{\nu}$, and even more strongly (and independently of the nature of the lepton couplings) from the size of the K_1^0-K_2^0 mass difference. For this reason, until strangeness-conserving neutral-current weak interactions were discovered in 1973,[1] many physicists were inclined to doubt their existence in any form.

The observed suppression of the strangeness-changing neutral currents is so dramatic numerically that we find it hard to believe that it comes about because the parameters of the theory just happen to take certain values. We would prefer instead to believe that the conservation of strangeness by the neutral currents is *natural*, that is, that it follows from the group structure and representation content of the theory, and does not depend on the values taken by the parameters of the theory.

The gauge theory[2] that uses four quark flavors is natural in this sense, but many of the gauge theories proposed recently are not. In some of these theories, the conservation of strangeness by the first-order neutral-current interaction must be arranged by a careful tuning of quark masses and mixing angles. Even after the parameters of the theory are chosen in this way, weak radiative effects can sometimes produce corrections to the quark mass of order α, leading to strangeness-changing neutral currents of order αG_F. In other theories, there are no strangeness-changing neutral currents, but the exchange of pairs of charged intermediate bosons can induce an effective strangeness-changing neutral current of order αG_F. In still other theories, the exchange of Higgs bosons can produce a strangeness-changing neutral current of roughly the same order. In principle these effects may perhaps be eliminated by a retuning of the parameters of the theory (including in the last case the parameters of the Higgs-boson interactions), but we would find a theory much more attractive if the neutral currents conserved strangeness naturally.

In this paper we explore the implications of the condition that the conservation laws obeyed by the neutral current of SU(2) × U(1) gauge theories are obeyed naturally. In Secs. II and III we consider a general SU(2) ⊗ U(1) gauge theory, and demand that the neutral-current interactions conserve all quark flavors naturally: strangeness, charm, and whatever additional flavors may turn out to be necessary. We deduce the necessary and sufficient conditions for this: All quarks of fixed charge and helicity must (1) transform according to the same irreducible representation of weak SU(2), (2) correspond to the same eigenvalue of weak T_3, and (3) receive their contributions in the quark mass matrix from a single source (either from the vacuum expectation value of a single neutral Higgs meson or from a unique gauge-invariant bare mass term). Examples satisfying these very restrictive conditions are discussed. If we limit ourselves to models involving quarks of charges $\frac{2}{3}$ and $-\frac{1}{3}$ exclusively, the only acceptable possibilities with natural flavor conservation to order αG are these: all left-handed quarks in weak doublets and all right-handed quarks in singlets; all quarks of each helicity in doublets. For the latter case (the purely vector model), flavor-changing neutral-current effects due to Higgs exchange cannot naturally be excluded. In Sec. IV we relax our condition, and demand only that the neutral-current interactions conserve strangeness naturally, for this is the only conservation law yet established by experiment. We discuss an amusing seven-quark model satisfying the relaxed condition. In Sec. V we explore the experimental consequences of relaxing the conservation law for charm by neutral-current interactions. These consequences turn out to be quite dramatic. We may soon be in a position to say whether or not the neutral currents do conserve charm, as they do strangeness.

II. ON THE NATURAL ABSENCE OF FLAVOR-CHANGING NEUTRAL-CURRENT INTERACTIONS

For the sake of simplicity and concreteness, we will restrict ourselves throughout this paper to the familiar SU(2) × U(1) gauge theory of weak and electromagnetic interactions, though our remarks have obvious extensions to other gauge theories. Also, we will tacitly assume that the strong interactions are generated by unbroken color SU(3) gauge interactions, but we will suppress color indices everywhere. For the moment, we will put no restriction on the numbers and charges of the quarks in the theory.

The neutral intermediate boson Z will in general couple to a hadronic neutral current of the form

$$J^\mu_z = \bar{q}\gamma^\mu(1+\gamma_5)Y_L q + \bar{q}\gamma^\mu(1-\gamma_5)Y_R q\,. \tag{2.1}$$

Here q is a column vector representing the set of all quark fields, and $Y_{L,R}$ are matrices

$$Y_L = T_{3L} - 2\sin^2\theta Q\,, \tag{2.2}$$

$$Y_R = T_{3R} - 2\sin^2\theta Q\,, \tag{2.3}$$

where θ is the usual mixing angle, T_{3L} and T_{3R} are matrices representing the third component of weak SU(2) on the left- and right-handed quark fields, and Q is the charge matrix. We proceed to impose the condition of natural flavor conservation of neutral-current effects:

Condition I. We demand that the neutral current naturally conserves all quark flavors: strangeness, charm, etc.

In color gauge theories of the strong interactions, each of these quantities is simply the number of quarks of a given flavor; each quark flavor corresponds to a different eigenvalue of the quark mass matrix. Hence, we require that the matrices Y_L and Y_R be diagonal in the basis in which the quark mass matrix is also diagonal.

The mass term in the effective Lagrangian is of the form

$$-\bar{q}(1+\gamma_5)Mq - \bar{q}(1-\gamma_5)M^\dagger q\,, \tag{2.4}$$

where the mass matrix M need not be diagonal nor even Hermitian. M is in general arbitrary, but for the requirement that it conserve charge[3]: $[M, Q] = 0$. We may diagonalize the mass matrix by defining new quark fields

$$q' = \tfrac{1}{2}(1+\gamma_5)U_L q + \tfrac{1}{2}(1-\gamma_5)U_R q\,, \tag{2.5}$$

with U_L and U_R unitary matrices acting on the flavor index of the quarks. In the new basis, the transformed mass matrix,

$$M' = U_L M U_R^{-1}\,, \tag{2.6}$$

is diagonal, and the matrices Y_L and Y_R are replaced by

$$Y'_L = U_L Y_L U_L^{-1}\,,\quad Y'_R = U_R Y_R U_R^{-1}\,. \tag{2.7}$$

Our condition requires that the neutral current conserve quark flavors in this basis, and hence that Y'_L and Y'_R be diagonal.

However, the demand of *natural* flavor conservation of the neutral currents requires that $Y'_{L,R}$ be diagonal whatever the choice of parameters in the theory, in particular, whatever the choice of M. Thus, U_L and U_R must be regarded as *arbitrary* unitary matrices which commute with Q. In order for $Y'_{L,R}$ to be diagonal for all such $U_{L,R}$, it follows that they must act on any set of quarks with the same charge as multiples of the unit matrix; they must in other words be functions of the matrix Q. It follows from Eqs. (2.2) and (2.3) that the same is true of the third component of SU(2):

$$T_{3L} = f_L(Q)\,,\quad T_{3R} = f_R(Q)\,. \tag{2.8}$$

That is, *all quarks with the same charge must have the same value of* T_{3L} *and the same value of* T_{3R}. This condition is clearly necessary and sufficient for the neutral currents (2.1) to conserve all quark flavors.

Condition II. We demand that the effective order αG *neutral-current coupling induced by one-loop radiative corrections naturally conserve all quark flavors.*

Consider the interaction of two fermions (quarks or leptons) without exchange of charge. The problem that concerns us here is whether one-loop diagrams could produce "dangerous" Fermi interactions of order αG_F, such as $(\bar{s}d)(\bar{s}d)$ (ruled out by the K_1^0-K_2^0 mass difference) or $(\bar{s}d)(\bar{\mu}\mu)$ (ruled out by the $K^0\to\mu^+\mu^-$ branching ratio). We are not concerned here with processes that result from the conversion of a virtual photon into a lepton pair, such as $K\to\mu^+\mu^-$.) We are now assuming that Condition I is satisfied, so that the neutral currents themselves conserve strangeness, charm, etc. In consequence, the only diagrams which could produce dangerous four-fermion interactions are those shown in Fig. 1. We will ignore all terms of order $\alpha G_F(m/m_W)^2$,

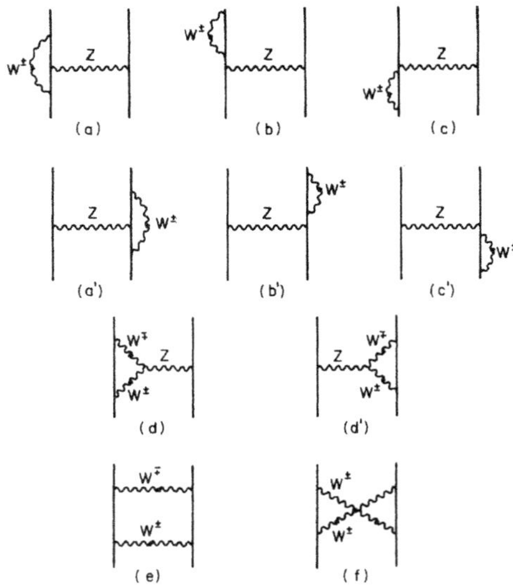

FIG. 1. Diagrams which can potentially produce strangeness-nonconserving (or charm-nonconserving) Fermi interactions of order αG_F. Here straight lines are quarks or leptons, and wavy lines are intermediate vector bosons. Each diagram must be summed over the two directions of flow for the W charges.

where m is a quark or lepton mass or momentum, so the calculations are fairly easy. (In particular, this justifies our neglect of one-loop graphs involving Higgs bosons.)

Diagram (a), together with the fermion-field-renormalization correction induced by diagrams (b) and (c), produces a change in the neutral-current coupling matrix Y in Eq. (2.1) proportional to

$$\Delta Y \propto T_+ Y T_- + T_- Y T_+ - \tfrac{1}{2}\{T_+, T_-\}Y - \tfrac{1}{2}Y\{T_+, T_-\}, \tag{2.9}$$

where $T_\pm$ are the matrices

$$T_\pm = T_1 \pm i T_2 .$$

(A label L or R must be added everywhere, depending on the fermion helicity.) But $T_\pm$ raise or lower the values of Q and T_3, so

$$YT_\pm = T_\pm(Y \pm a), \quad a = 1 - 2\sin^2\theta$$

and therefore (2.8) gives

$$\Delta Y \propto T_+T_-(Y - a) + T_-T_+(Y + a) - \{T_+, T_-\}Y$$
$$= -a[T_+, T_-] = -2aT_3 . \tag{2.10}$$

Under Condition I, this automatically conserves strangeness, charm, etc. The same remarks apply to diagrams (a′), (b′), and (c′).

Diagrams (b), (c), (b′), and (c′) also require a fermion mass renormalization. However, this has no effect, because Condition I ensures that strangeness, charm, etc. are conserved for *all* possible charge-conserving quark mass matrices.

The two diagrams of type (d) together give a contribution proportional to $[T_+, T_-] = 2T_3$, so again Condition I ensures conservation of strangeness, charm, etc. The same applies to diagram (d′).

This leaves diagrams (e) and (f). These diagrams induce an effective Fermi interaction of the form

$$\xi \alpha G_F K_\mu K^\mu , \tag{2.11}$$

where ξ is a dimensionless number of order unity and K is an effective neutral current

$$K^\mu = \bar{q}\gamma^\mu(1+\gamma_5)Z_L q + \bar{q}\gamma^\mu(1-\gamma_5)Z_R q$$
$$+ \text{lepton terms} \tag{2.12}$$

where Z_L and Z_R are the matrices

$$Z_L = 3(\vec{T}_L^{\,2} - T_{3L}^{\,2}) + 10T_{3L} , \tag{2.13}$$

$$Z_R = 3(\vec{T}_R^{\,2} - T_{3R}^{\,2}) - 10T_{3R} . \tag{2.14}$$

As in the derivation of Condition I, we demand that Z_L and Z_R be diagonal in the basis in which the quark masses are diagonal, whatever the quark masses are. This implies that

$$\vec{T}_L^{\,2} = g_L(Q), \quad \vec{T}_R^{\,2} = g_R(Q) .$$

That is, *all quarks with the same charge must have the same value of* $\vec{T}_L^{\,2}$ *and the same value of* $\vec{T}_R^{\,2}$. This condition is clearly both necessary and sufficient for the induced neutral current to conserve all quark flavors, given that Condition I is satisfied.

Let us see how these two conditions work in those SU(2)⊗U(1) theories with quark charges restricted to $\frac{2}{3}$ and $-\frac{1}{3}$. From Condition I we know that quarks of given charge and helicity must either all belong to singlets or all belong to doublets. Because left-handed weak currents are known to exist in nature, we conclude that all left-handed quarks must be in doublets. The right-handed quarks can be all singlets, as in the "standard" unified model,[2] or can be all doublets, as in the pure vector model.[4] Both models would then satisfy Conditions I and II. However, many other popular models do not. In particular, neither Condition I nor Condition II is satisfied by the six-quark model with doublet structure

$$\begin{pmatrix} u \\ d_\theta \end{pmatrix}_L \begin{pmatrix} c \\ s_\theta \end{pmatrix}_L \begin{pmatrix} t \\ b_\theta \end{pmatrix}_L \begin{pmatrix} u \\ b \end{pmatrix}_R .$$

Also, neither Condition I nor Condition II can ever be satisfied in SU(2) × U(1) models in which there are unequal numbers of charge $\frac{2}{3}$ and charge $-\frac{1}{3}$

quarks.

If we allow other values of quark charges, we may construct many other models satisfying our conditions. For example, we may suppose a model involving six quarks with the multiplet structure

$$\begin{pmatrix} u \\ d \end{pmatrix}_L \begin{pmatrix} c \\ s \end{pmatrix}_L (X)_L (Y)_L ,$$

$$\begin{pmatrix} d \\ X \end{pmatrix}_R \begin{pmatrix} s \\ Y \end{pmatrix}_R (u)_R (c)_R ,$$

involving new quarks X and Y which have charge $-\frac{4}{3}$. Both right-handed and left-handed quarks decompose into two doublets and two singlets, but with different charges. All quarks with the same charge and helicity have the same values for $\vec{T}^2$ and T^3, so Conditions I and II are satisfied here.

III. HIGGS-EXCHANGE NEUTRAL CURRENTS

There is a third condition for a natural absence of neutral-current flavor violation, having to do with the system of Higgs mesons. Renormalizable gauge theories of the weak and electromagnetic interactions generally involve physical scalar particles, widely known as Higgs mesons. If a neutral Higgs meson H has off-diagonal interactions such as $d+H \to s$, then its exchange can produce an effective $\Delta S = 2$ Fermi interaction

$$\bar{s} + d \to H \to s + \bar{d} . \tag{3.1}$$

The observed K_1^0-K_2^0 mass difference tells us that any such interaction must have a coupling strength G_H no greater than ~10^{-5} the usual Fermi coupling strength G_F.

How strong do we expect Higgs-exchange effects to be? The interaction of the Higgs field ϕ_i with the quarks will generally be of the form

$$-\bar{q}\Gamma_i(1+\gamma_5)q\phi_i - \bar{q}\Gamma_i^\dagger(1-\gamma_5)q\phi_i^\dagger , \tag{3.2}$$

where Γ_i is a matrix which acts on quark flavor indices. The quark mass matrix is then

$$M = M_0 + \Gamma_i \langle \phi_i \rangle , \tag{3.3}$$

where M_0 is a possible SU(2)⊗U(1)-invariant bare mass, and $\langle\phi_i\rangle$ is the vacuum expectation value of ϕ_i. This shows that the typical value Γ of the coupling of a quark and a Higgs meson is of order

$$\Gamma \approx m_q / \langle \phi \rangle , \tag{3.4}$$

where m_q is a typical quark mass (or mass difference) and $\langle\phi\rangle$ is a typical vacuum expectation value. Also, the Higgs mass is of order

$$m_H \approx \sqrt{f}\,\langle \phi \rangle , \tag{3.5}$$

where f is a typical value of the ϕ^4 coupling constant. Hence the exchange of Higgs bosons produces an effective Fermi interaction with coupling of order

$$G_H \approx \Gamma^2 / m_H^2 \approx m_q^2 / f \langle \phi \rangle^2 . \tag{3.6}$$

If we take $m_q \approx 1$ GeV and $\langle\phi\rangle \approx G_F^{-1/2} = 300$ GeV, then $G_H/G_F \approx 10^{-5}/f$. But as long as the scalar fields couple only weakly to themselves, we must have $f \ll 1$, so $G_H/G_F \gg 10^{-5}$. (Usually one assumes that f is of order α, in which case G_H/G_F is of order 10^{-3}.) We see that an off-diagonal Higgs coupling of the form $d \to H + S$ would be expected to produce much too large a K_1^0-K_2^0 mass difference. Thus we are led to our third condition.

Condition III. We demand that the coupling of each neutral Higgs meson be such as naturally to conserve all quark flavors: strangeness, charm, etc.

Because of our requirement of naturalness, the matrix M_0 must be regarded as an *arbitrary* SU(2)-invariant matrix commuting with Q. Similarly, the matrices Γ_i contain a number of *arbitrary* parameters equal to the number of SU(2)-invariant charge-conserving Yukawa couplings of the Higgs mesons to the quarks. For all such M_0 and for all such Γ_i the couplings of the neutral Higgs mesons must be diagonal in the basis in which M is diagonal.

Suppose that the set of quarks with given charge Q get their mass purely from a single neutral Higgs meson ϕ_Q^0. Then the mass matrix for these quarks of charge Q will be

$$M(Q) = \Gamma_Q \langle \phi_Q^0 \rangle \tag{3.7}$$

and Γ_Q is trivially diagonal in the basis which diagonalizes $M(Q)$. However, if there were more than one neutral Higgs boson contributing to the masses of quarks of a given charge, or if there were both an invariant mass term M_0 and a Higgs contribution, then there would be no reason to expect the couplings of the neutral Higgs bosons to conserve quark flavor. We conclude that Condition III is equivalent to the requirement that *quarks of given charge receive their mass either (1) through the couplings of precisely one neutral Higgs meson or (2) through an SU(2)-invariant mass term, but not by both mechanisms.* This condition might be evaded in special cases, as for instance through the judicious introduction of discrete symmetries. We have not explored these possibilities in detail.

In the standard model, the left-handed quarks are all doublets and the right-handed quarks are all singlets, so the Higgs bosons that couple to the quarks must all be doublets, and there can be no bare mass term M_0. The right-handed charge $+\frac{2}{3}$ quarks must couple to just one neutral Higgs boson

ϕ^0, a member of a doublet (ϕ^0, ϕ^-). Similarly, the right-handed charge $-\frac{1}{3}$ quarks must couple to just one neutral Higgs boson ϕ'^0, a member of a doublet (ϕ'^+, ϕ'^0). These two doublets may or may not be charge conjugates of each other. Condition III is then satisfied.

In the vector model, the left-handed and the right-handed quarks are all doublets, so the Higgs bosons that couple to the quarks must be singlets and/or triplets, and there can also now be an SU(2)-singlet bare mass term μ_0. The experimental values of the quark masses and mixing angles indicates that the quark mass matrix for each charge must have both SU(2)-singlet and SU(2)-triplet parts, so in this model it does not seem possible to eliminate the off-diagonal parts of the neutral-Higgs-boson couplings in a natural way. Condition III is not satisfied.

IV. MODELS WITH CHARM-CHANGING NEUTRAL CURRENTS

The result of the last two sections shows that unless we introduce exotic new quarks with charges of $\frac{5}{3}$ or $-\frac{4}{3}$, the only SU(2)⊗U(1) theories in which neutral-current effects naturally conserve strangeness and charm are the standard model and its generalization to any even number of quarks. It may involve at most two Higgs doublets (ϕ^0, ϕ^-) and $\phi'^+, \phi'^0)$ coupled to the quarks.

The vector model involves a neutral current which conserves charm and strangeness naturally, and this conservation law is respected by induced order-αG neutral-current effects. However, the absence of flavor violation mediated by neutral Higgs mesons cannot be imposed naturally (so far as we can see). Moreover, the vector model appears to be inconsistent with data on elastic[5] and deep-inelastic[6] neutrino-nuclear scattering, which show different neutrino and antineutrino cross sections.

The standard model may naturally satisfy our three conditions, but there are indications that it fails to describe certain data: The reported[7] "high-y anomaly" seems to require the presence of right-handed currents in the charged weak current which involve u and/or d quarks.[8] Of course, the present experimental situation must be regarded as tentative, but it is sufficiently disturbing so as to motivate a search for alternative theories.[9]

The reader will recognize that the three conditions of Secs. II and III result from the requirement that neutral-current effects naturally conserve *all* quark flavors. Yet, the neutral currents are known only to conserve strangeness. Nothing is known experimentally about whether or not they also conserve charm, let alone other proposed quark flavors. We are therefore free to suppose that the neutral currents naturally conserve strangeness and hence conserve all flavors of quarks with charge $-\frac{1}{3}$ but that they may violate the conservation of the other quark flavors, including charm.[10] By following the analysis of Sec. II, we see that the quarks of charge $-\frac{1}{3}$ must all have the same values of T_{3L}, $\vec{T}_L^{\,2}$, T_{3R}, and $\vec{T}_R^{\,2}$, but there is no restriction on the SU(2) transformation properties of quarks with other charges. Similarly, following the reasoning of Sec. III, there can be at most one neutral Higgs boson which couples to the charge $-\frac{1}{3}$ quarks, but any number of Higgs bosons which couple to other quarks.

To see how this can work in practice, consider a seven-quark model, with four quarks of charge $+\frac{2}{3}$ and three quarks of charge $-\frac{1}{3}$. To avoid a premature identification with quarks of definite mass, we will label the charge $+\frac{2}{3}$ quarks as P, C, L, A, and the charge $-\frac{1}{3}$ quarks as G, E, S. In order for strangeness to be conserved naturally, the left-handed G, E, S quarks must all be singlets or doublets, and likewise for the right-handed G, E, S quarks. Since the weak interactions certainly involve some left-handed quarks, the left-handed quarks must be grouped into three doublets and a singlet,

$$\begin{pmatrix} P \\ G \end{pmatrix}_L, \begin{pmatrix} C \\ E \end{pmatrix}_L, \begin{pmatrix} L \\ S \end{pmatrix}_L \text{ doublets, } A_L \text{ singlet.} \quad (4.1)$$

The only remaining alternatives are that the right-handed quarks may be similarly grouped into three doublets and a singlet,

$$\begin{pmatrix} P \\ G \end{pmatrix}_R, \begin{pmatrix} C \\ E \end{pmatrix}_R, \begin{pmatrix} L \\ S \end{pmatrix}_R \text{ doublets, } A_R \text{ singlet,} \quad (4.2)$$

or that they are all singlets. These correspond to modified versions of the standard and the vector models, respectively.

The seven-quark version of the standard model still has the problem of not containing right-handed currents which could account for the high-y anomaly. On the other hand, the seven-quark version of the vector model can now provide a neutral current with an axial-vector part that could produce different ν-N and $\bar{\nu}$-N cross sections, so let us restrict our attention to this case.[11]

We may define the quark states G, E, S to be just the charge $-\frac{1}{3}$ quarks d, s, b of definite mass. The quark states P, C, L, A are then defined by (4.1) and (4.2) in terms of their weak-interaction properties. Each of these is a linear combination of the charge $\frac{2}{3}$ quarks u, c, t, t' of definite mass, with the linear combinations in general different for the left- and right-handed quarks. In particular, if the coefficients of the u_L term in A_L and the u_R term in A_R are different, the weak neutral current will contain an axial-vector part involving u quarks.

The Higgs bosons which couple to quarks in the seven-quark version of the vector model may be SU(2) singlets, doublets, and/or triplets. However, according to the arguments of Sec. III, at most one neutral Higgs boson can couple to the charge $-\frac{1}{3}$ quarks G, E, S. The neutral member of a Higgs doublet (ϕ^0, ϕ^-) does not couple to the charge $-\frac{1}{3}$ quarks at all, so the Higgs fields that couple to the quarks in this theory can in general consist of one doublet (ϕ^0, ϕ^-) and either one singlet $\phi^{0\prime}$ (or a bare quark mass term M_0) or one triplet $(\phi'^+, \phi'^0, \phi'^{-1})$. This appears to be a rich enough set of Higgs fields to produce realistic quark masses and mixing angles, but we have not studied the problem in detail.

V. DO NEUTRAL-CURRENT INTERACTIONS CONSERVE CHARM?

Following the remarks of the preceding section, we now want to consider the experimental consequences of a theory in which the neutral currents themselves, or neutral-current effects induced by two-W or Higgs exchange, do not conserve charm.

We have seen that it is possible to construct an empirically adequate model in which Z^0 is coupled to a current with $\Delta C = 0$ and $\Delta C = \pm 1$ parts. This current can provide an additional mechanism for the production and decay of charmed particles, beyond the expected charged-current mechanism.

Charmed hadrons are ordinarily produced in high-energy neutrino-nucleon collisions by the charged-current couplings of either valence quarks,

$$\nu d \to \mu^- c \,, \tag{5.1}$$

or sea quarks,

$$\nu \to \mu^- c\bar{s} \,, \tag{5.2}$$

$$\bar{\nu} \to \mu^+ \bar{c} s \,. \tag{5.3}$$

If the neutral current is charm-changing, we may anticipate the additional valence processes

$$\nu u \to \nu c \,, \tag{5.4}$$

$$\bar{\nu} u \to \bar{\nu} c \,. \tag{5.5}$$

Events due to (5.4) or (5.5) in which the charmed hadron decays nonleptonically would resemble ordinary neutral-current events, except in that they involve an energy threshold (and a strange final state). Consequently, the existence of a charm-changing neutral current may lead to an observable rise in neutral-current cross sections above charm threshold.

Should a neutrino-produced charmed hadron decay semileptonically, mechanisms (5.1)–(5.3) would yield dilepton events. But (5.4) or (5.5) would yield a single charged lepton of relatively low energy. Events of this kind might be misinterpreted as ordinary charged-current events. However, there would be a very large loss of energy to the two unobserved neutrinos. Kinematically, these events would seem peculiar, and might even seem to lie outside the allowed region of the q^2-ν plot. Conceivably, events induced by (5.5) could account for at least part of the observed "high-y anomaly."

On the other hand, (5.4) and (5.5) probably must be weaker couplings than the conventional charged-current processes

$$\begin{aligned} \nu d &\to \mu^- u \,, \\ \nu u &\to \mu^+ d \,. \end{aligned} \tag{5.6}$$

This is because the charged-current semileptonic decay of a charmed hadron produced by (5.4) could yield a final state μ^+ while (5.5) could yield a final state μ^-. Such events would appear to violate conservation of lepton number, and no such violation has been reported.

Moreover, a charmed hadron produced by the conventional charged-current processes (5.1)–(5.3) could decay semileptonically, via the charm-changing neutral current, producing a charged lepton pair. Overall, this yields a trilepton event. Searches for neutrino-induced trileptons or soft wrong-charge single leptons could provide sensitive limits on the strength of a conjectured charm-changing neutral current. It would be ironic if charmed hadrons were observed to decay into final states including neutral lepton pairs: These are just the analogs of the strange-particle decays that charm was invented to suppress.

Thus far, we have considered possible violations of Condition I which give rise to charm-changing neutral-current effects of order G. If only Condition II or III is violated, these effects will be much smaller. Nevertheless, one or another of the virtual processes

$$u + \bar{c} \to \left\{ \begin{matrix} Z^0 \\ W^+ + W^- \\ \text{Higgs} \end{matrix} \right\} \to \bar{u} + c \tag{5.7}$$

will give rise to an effective $\Delta C = 2$ Fermi coupling. In any of these cases, there will be induced a mass splitting between the CP-even and CP-odd neutral charmed mesons D_1^0 and D_2^0 which is much larger than the D_1^0 or D_2^0 decay rates. Even if the direct decays induced by the charm-changing neutral current are very rare, the effects of the

D_1^0-D_2^0 mass splitting on the decays of neutral D's will be spectacular.

In those theories we are considering, the frequency of D-$\bar{D}$ transitions is large compared to the decay rate. *Thus, D^0 will be expected to decay into channels appropriate to $\bar{D}^0$ as often as not.* If there were no charm-changing neutral-current effects, the Cabibbo-favored decays of D^0 will be into $S=-1$ final states. If there are, D^0 will decay equally into $S=\pm 1$ final states. (The decay of D^0 into an $S=+1$ nonleptonic state is possible in the standard theory, but is doubly Cabibbo-suppressed, by the factor $\tan^4\theta \sim 2\times 10^{-3}$.)

Similar remarks apply to the charged-current semileptonic decays of D^0. Normally, such a decay produces a positively charged lepton. In the presence of charm-changing neutral-current effects, leptons of either sign are produced equally.

Let us consider the consequences of a large D_1^0-D_2^0 mixing on some experiments of current interest. Evidence for the associated production of new hadrons in e^+e^- annihilation has just been reported.[12] If these are interpreted in terms of charm, then the D^0 has been discovered at a mass of 1.865 GeV, in accordance with theoretical anticipation,[13] and has been observed to decay into $K\pi$ and $K\pi\pi\pi$. Two oppositely charmed particles must have been produced in association, possibly together with additional hadrons. One would expect a significant fraction of the weakly decaying charmed particles produced in e^+e^- annihilation to be D^0 and $\bar{D}^0$, perhaps the majority of them.[14] If the Glashow-Iliopoulos-Maiani mechanism *does* apply to charm as well as strangeness, we would expect that the final hadron state would have strangeness zero in the vast majority of cases. On the other hand, if there are charm-changing neutral-current effects, they will lead to complete $D^0\bar{D}^0$ mixing, and we expect $S=\pm 2$ final states to be almost as common as $S=0$ final states. *We are about to learn whether or not neutral currents conserve charm.*

Effects of D_1^0-D_2^0 mixing can also show up in neutrino experiments. Consider the production of charmed particles by high-energy neutrinos off sea quarks, for example

$$\nu p \to p + K^+ + D^0 + \mu^- . \qquad (5.8)$$

If D^0 decayed normally, we would end up with an $S=0$ final state most of the time. If D^0 decayed via $D^0\bar{D}^0$ mixing, $S=2$ would be as common for the final state as $S=0$. A *charm-changing* neut'al-current coupling (either of order G, of or αG, or Higgs-mediated) reveals itself if charged-current $\Delta S=2$, $\Delta Q=2$ events are identified.

Neutrino-induced dimuon events have been detected at Fermilab.[15] About 100 events have been seen involving a final pair of oppositely charmed muons. These may be attributed to the effect of the conventional $\Delta C=\Delta Q$ part of the charged weak current. The incident neutrino has become the final μ^-, while the μ^+ is a decay product of a charmed hadron. Indeed, the observed kinematics favors this interpretation.[16] However, some of the time two μ^- are observed. These events *could* be attributed to the associated production of pairs of charmed hadrons by neutrinos, with one of them decaying semileptonically. This could yield all the $\mu^-\mu^-$ events and some of the $\mu^-\mu^+$ events. They could also be interpreted as "wrong" decays of D^0 which have been produced singly by neutrinos in reactions like (5.4). From the ratio of same-sign to opposite-sign dimuons, one could tell how often, among $\Delta C=1$ charged-current events, a D^0 is produced rather than any other $C=1$ weakly decaying hadron.

Finally, we consider the possibility that there is a significant associated production of charmed hadrons in deep-inelastic muon scattering. Semileptonic decays of the charmed particles would yield dilepton events and, more rarely, trilepton events. Without D_1^0-D_2^0 mixing, one would expect only the trimuon charge signature $\mu^\pm\mu^+\mu^-$, depending on whether the incident beam is μ^+ or μ^-. This is because the $C=1$ hadron normally produces only a positive lepton, and a $C=-1$ hadron normally produces only a negative lepton. If there is D_1^0-D_2^0 mixing, we anticipate as well the charge signatures $\mu^\pm\mu^+\mu^+$ and $\mu^\pm\mu^-\mu^-$.

It is evidently important experimentally to learn whether or not the Glashow-Iliopoulos-Maiani mechanism extends naturally to charm, and we have sketched a few experimental ramifications. It is also important to know if neutral-current effects are charm-conserving for theoretical reasons—as a guide to the development of a complete theory of weak interactions.

Note added in proof. The problems treated here have been discussed recently from a rather different point of view by E. A. Pachos [Phys. Rev. D 15, 1966 (1977)] and by K. Kang and J. E. Kim [Phys. Lett. 64B, 93 (1976)].

ACKNOWLEDGMENT

We wish to thank M. Barnett, A. De Rújula, H. Georgi, and B. W. Lee for helpful discussions.

*Work supported in part by the National Science Foundation under Grant No. MPS75-20427.

[1]F. J. Hasert *et al.*, Phys. Lett. 46, 121 (1973); 46, 138 (1973); A. Benvenuti *et al.*, Phys. Rev. Lett. 32, 800 (1974).

[2]S. Weinberg, Phys. Rev. Lett. 19, 1264 (1967); A. Salam, in *Elementary Particle Physics: Relativistic Groups and Analyticity (Nobel Symposium No. 8)*, edited by N. Svartholm (Almquist and Wiksell, Stockholm, 1968), p. 367; S. L. Glashow, J. Iliopoulos, and L. Maiani, Phys. Rev. D 2, 1285 (1970).

[3]Conceivably, other quantum numbers which, like electric charge, are exactly conserved remain to be discovered; Q stands for these as well.

[4]G. Branco *et al.*, Phys. Rev. D 13, 104 (1976); Report No. COO-223B-84, 1975 (unpublished); A. De Rújula *et al.*, Phys. Rev. D 12, 3589 (1975); H. Fritzsch *et al.*, Phys. Lett. 59B, 256 (1975); S. Pakvasa *et al.*, Phys. Rev. Lett. 35, 703 (1975); F. Wilczek *et al.*, Phys. Rev. D 12, 2768 (1975).

[5]D. Cline *et al.*, Phys. Rev. Lett. 37, 648 (1976).

[6]A. Benvenuti *et al.*, Phys. Rev. Lett. 36, 1478 (1976) and unpublished work; B. Barish *et al.*, unpublished work.

[7]A. Benvenuti *et al.*, Report No. HPWF-76/3 (unpublished).

[8]A. De Rújula *et al.*, Rev. Mod. Phys. 46, 391 (1974).

[9]R. M. Barnett, Phys. Rev. Lett. 36, 1163 (1976); Phys. Rev. D 14, 2990 (1976); E. Derman (unpublished); S. Pakvasa (unpublished); V. Barger and D. V. Nanopoulos Phys. Lett. 63B, 168 (1976); C. Albright, *et al.*, Phys. Rev. D 14, 1780 (1976); F. Gürsey and P. Sikivie, Phys. Rev. Lett. 36, 775 (1976); Y. Achiman *et al.* (unpublished); P. Ramond (unpublished).

[10]The theory and phenomenology of charm-changing neutral currents have been considered before, and some of our results have been discussed in the literature. See R. L. Kingsley *et al.*, Phys. Lett. 61B, 259 (1976); L. B. Okun *et al.*, Lett. Nuovo Cimento 13, 218 (1975).

[11]R. M. Barnett cautions that this model may not be capable of explaining the large observed difference between neutrino and antineutrino neutral-current cross sections (private communication).

[12]G. Goldhaber *et al.*, Phys. Rev. Lett. 37, 255 (1976).

[13]A. De Rújula *et al.*, Phys. Rev. D 12, 147 (1975).

[14]A. De Rújula *et al.*, Phys. Rev. Lett. 37, 398 (1976); K. Lane and E. Eichten, *ibid.* 37, 477 (1976).

[15]A. Benvenuti *et al.*, Phys. Rev. Lett. 34, 419 (1975); B. C. Barish *et al.*, in *La Physique du Neutrino á Haute Énergie*, proceedings of the Colloquium, École Polytechnique, Paris, 1975 (CNRS, Paris, 1975), p. 131.

[16]A. Pais and S. Treiman, Phys. Rev. Lett. 35, 1206 (1975).

VOLUME 39, NUMBER 23 PHYSICAL REVIEW LETTERS 5 DECEMBER 1977

Jets from Quantum Chromodynamics

George Sterman
Institute for Theoretical Physics, State University of New York at Stony Brook, Stony Brook, New York 11790

and

Steven Weinberg
Lyman Laboratory of Physics, Harvard University, Cambridge, Massachusetts 02138
(Received 26 July 1977)

The properties of hadronic jets in e^+e^- annihilation are examined in quantum chromodynamics, without using the assumptions of the parton model. We find that two-jet events dominate the cross section at high energy, and have the experimentally observed angular distribution. Estimates are given for the jet angular radius and its energy dependence. We argue that the detailed results of perturbation theory for production of arbitrary numbers of quarks and gluons can be reinterpreted in quantum chromodynamics as predictions for the production of jets.

The observation[1] of hadronic jets in e^+e^- annihilation provides one of the most striking confirmations of the parton picture.[2] In particular, the distribution of events in the angle θ between the jet axis and the e^+-e^- beam line is observed to be very close to the form $1+\cos^2\theta$ that would be expected for the production of a pair of relativistic charged pointlike particles of spin $\frac{1}{2}$. We shall argue here that the existence, angular distribution, and some aspects of the structure of these jets follow as consequences of the perturbation expansion[3] of quantum chromodynamics[4] (QCD), without assuming the parton picture (in particular, the transverse-momentum cutoff) in advance. Thus, the observed features of jets provide evidence for an underlying asymptotically free gauge field theory with elementary spin-$\frac{1}{2}$ quarks. We also wish here to demonstrate a general approach, which may be applicable to a wide range of high-energy phenomena.

Our procedure is to define a partial cross section for jet production, which in asymptotically free theories like QCD can be calculated perturbatively at high energy. By ordinary dimensional analysis, any sort of total or partial cross section in QCD can be written in the form

$$\sigma = E^{-2} f(m/E, g_E, x), \qquad (1)$$

where E is the energy; x stands for all other dimensionless variables characterizing the final state; m stands for all mass variables; and g_E is the gauge coupling constant, defined at a renormalization point with four-momenta of order E. [We express the cross section in terms of g_E, rather than a coupling g_κ defined at a renormalization point with momenta of arbitrary scale κ, in order to avoid factors of $\ln(E/\kappa)$. Physical quantities are of course independent of the choice of renormalization point.] Even in asymptotically free theories, where $g_E \to 0$ as $E \to \infty$, it is generally not possible to calculate the cross section perturbatively for large E, because the cross section will exhibit singularities for $m/E \to 0$. It is of course for this reason that asymptotic freedom has as a rule been used directly to justify perturbative calculations of Green's functions and Wilson coefficient functions, rather than cross sections themselves.

However, by performing various sums over states, it is possible to define a wide range of cross sections which are free of $m \to 0$ singularities. To learn what they are, we observe that quantum field theories of massless particles have always been found (in the absence of superrenormalizable couplings) to be physically sensible, i.e., that any cross section which would actually be measurable in such a massless theory is free of infrared divergences in each order of perturbation theory.[5] Hence in the real world with $m \neq 0$, any sort of partial cross section which would be measurable for $m=0$ is expected to be free of singularities in m as $m \to 0$, and can therefore be calculated perturbatively[3] in QCD for $E \to \infty$.

For instance, the cross section for production of a definite number of particles does have singularities for $m \to 0$, because for $m=0$ we could not expect to be able to tell the difference between one particle or several particles moving in the same direction. At the opposite extreme, the total cross section for $e^+e^- \to$ hadrons would clearly be measurable even for zero quark mass, and hence must be free of singularities in m (to

VOLUME 39, NUMBER 23 PHYSICAL REVIEW LETTERS 5 DECEMBER 1977

lowest order in α) for $m\to 0$. Indeed, although the original application of asymptotic freedom to this process was by way of the vacuum-polarization Green's function at Euclidean momentum,[6] it is easier to justify the use of QCD perturbation theory[3] here directly, by working with the cross section itself.

To study jets, we consider the partial cross section $\sigma(E, \theta, \Omega, \epsilon, \delta)$ for e^+e^- hadron production events, in which all but a fraction $\epsilon \ll 1$ of the total e^+e^- energy E is emitted within some pair of oppositely directed cones of half-angle $\delta \ll 1$, lying within two fixed cones of solid angle Ω (with $\pi\delta^2 \ll \Omega \ll 1$) at an angle θ to the e^+e^- beam line. We expect this to be measurable for $m=0$, because the only quarks or gluons which are likely to be diffracted or radiated away from a calorimeter at θ have very long wavelength, and so carry negligible energy. Thus σ should be free of mass singularities for $m\to 0$, and calculable by a perturbation expression in g_E for $E\to\infty$.

We have calculated $\sigma(E, \theta, \Omega, \epsilon, \delta)$ to order g_E^2. It proved algebraically convenient to set the quark masses equal to zero from the beginning, but to use a finite gluon mass $\mu \ll \epsilon E$ as an infrared cutoff in intermediate stages of the calculation. To order g_E^2, σ receives contributions from three distinct kinds of final state[7]: (a) One jet may consist of a quark or antiquark plus a hard (energy $\geqslant \epsilon E$) gluon, the other jet of just an antiquark or quark; (b) there may be a quark in one jet, an antiquark in the other, and a soft (energy $\leqslant \epsilon E$) gluon which may or may not be in one of the jets; (c) there may be just a quark and antiquark, one in each of the jets. Working to order g_E^2, we evaluate the contributions of (a) and (b) using only tree graphs, while for (c) we include the tree graph and its interference with one-loop graphs. The respective contributions to σ are then

$$\sigma_a = (d\sigma/d\Omega)_0\, \Omega(g_E^2/3\pi^2)[-3\ln(E\delta/\mu) - 2\ln^2 2\epsilon - 4\ln(E\delta/\mu)\ln(2\epsilon) + \tfrac{17}{4} - \pi^2/3], \tag{2}$$

$$\sigma_b = (d\sigma/d\Omega)_0\, \Omega(g_E^2/3\pi^2)[2\ln^2(2\epsilon E/\mu) - \pi^2/6], \tag{3}$$

$$\sigma_c = (d\sigma/d\Omega)_0\, \Omega\{1 + (g_E^2/3\pi^2)[-2\ln^2(E/\mu) + 3\ln(E/\mu) - \tfrac{7}{4} + \pi^2/6]\}, \tag{4}$$

where $(d\sigma/d\Omega)_0$ is the cross section for $e^+e^-\to q\bar{q}$ in Born approximation:

$$\left(\frac{d\sigma}{d\Omega}\right)_0 = \frac{\alpha^2}{4E^2}(1+\cos^2\theta)\sum_{\text{flavors}} 3Q^2. \tag{5}$$

As expected, each separate contribution is singular for $\mu\to 0$, but cancellations[8] occur in the sum, and the final result is free of mass singularities:

$$\sigma(E, \theta, \Omega, \epsilon, \delta) = (d\sigma/d\Omega)_0\, \Omega[1 - (g_E^2/3\pi^2)(3\ln\delta + 4\ln\delta\ln 2\epsilon + \pi^2/3 - \tfrac{5}{2})]. \tag{6}$$

This formula immediately demonstrates the dominance of two-jet final states at very high energy where $g_E^2/3\pi^2$ is small. By summing Eq. (6) over a set of cones of solid angle Ω that fill the 4π steradians around the e^+e^- collision, and comparing the result with the QCD expression[6] $(1+g_E^2/4\pi^2)\sigma_0$ for the total cross section, we see that the fraction of all events which have all but a fraction ϵ of their energy in some pair of opposite cones of half-angle δ is

$$f = 1 - (g_E^2/3\pi^2)(3\ln\delta + 4\ln\delta\ln 2\epsilon + \pi^2/3 - \tfrac{7}{4}). \tag{7}$$

If $g_E^2/3\pi^2 \ll 1$, then even if we take ϵ and δ to be moderately small, the two-jet probability f will be close to unity. To be quantitative, suppose we define a jet angular radius $\delta(E)$, by requiring that 70% of all events have at least 80% of their energy emitted within two cones of half-angle $\delta(E)$. Setting $f=0.7$ and $\epsilon=0.2$ in Eq. (7), and using the asymptotic QCD formula[9] $g_E^2 = 24\pi^2/25\ln(E/\Lambda)$ with $\Lambda = 500$ MeV, we find that $\delta(E)$ is about 13° at the energy $E = 7.4$ GeV of current experiments,[1] and decreases as $E^{-0.25}$ at higher energies. In contrast, with a fixed transverse-momentum cutoff $P_\perp$, we would expect a jet angular radius $\varphi(E)$ which would decrease much faster, like $1/E$ or $(\ln E)/E$. At relatively low energy $\varphi(E)$ will be greater than $\delta(E)$, so that our calculation of the jet radius is probably invalidated by the nonperturbative effects[3] associated with $P_\perp$. However, at sufficiently high energy $\delta(E)$ becomes greater than $\varphi(E)$, and perturbation theory becomes valid for angular radii down to $\delta(E)$. The angle $\delta(E)$ then defines the outermost angular distance from the jet axis at which any appreciable hadron energy is to be found. Even at such high energies,

VOLUME 39, NUMBER 23 PHYSICAL REVIEW LETTERS 5 DECEMBER 1977

it is possible that the fixed-transverse-momentum jet of the parton model will survive deep within the cone of half-angle $\delta(E)$, beyond the reach of perturbative methods, but the angle $\psi(E)$ becomes so small for sufficiently high energy that the jet angular distribution is in any case constrained to have the $1+\cos^2\theta$ form of QCD perturbation theory.

Our definition of two-jet events has an obvious generalization to arbitrary numbers of jets. To order g_E^2, the fraction f of hadronic e^+e^- events that (by definition) are not of the two-jet type consists entirely of three-jet events.[10] However, in order to determine the angular radius of the jets in three-jet events, it would be necessary to carry our calculations to order g_E^4, where four-jet events are beginning to enter. Continuing in this way, it should be possible to test the detailed predictions of QCD for production of any numbers of quarks and gluons, but always reinterpreting these particles in terms of jets.[11]

We can also conclude from Eq. (6) that the two-jet events have just the same $1+\cos^2\theta$ angular distribution as in the Born approximation for $e^+e^- \to q\bar{q}$. This result is expected to persist for massless quarks to all orders in g_E^2, because for $e^+e^- \to q\bar{q}$ the conservation of current and chirality limit the matrix element $\langle q\bar{q}|J^\mu|0\rangle$ to just a γ^μ term, while the effect of adding more gluons or quarks to these jets is merely to convert $\ln(E/\mu)$ factors to factors of $\ln\epsilon$ or $\ln\delta$. The dominant corrections to this angular dependence come from the finite quark mass and from the ambiguity between three-jet and two-jet events; the former gives an angular distribution $1+a\cos^2\theta$, with $1-a=4\langle m^2\rangle/E^2$.

It might be thought that the partial jet cross section $\sigma(E, \theta, \Omega, \epsilon, \delta)$ should be measurable for massless theories, and hence free of mass singularities in the limit of zero mass, even if we specify the charge in each jet. If this were the case (and if there are no failures of perturbation theory[3] in QCD when jet charges are measured) then our calculation would not account for real jets with integer total charge, since to order g_E^2 it is only possible to produce jets of third-integral charge.[12] However, direct calculation to order g_E^4 shows that the cross section for final states with a definite value for the charge emitted in a given solid angle will have singularities in the limit that the quark mass vanishes. As far as we can tell, the reason that cross sections for the emission of massless particles with a definite total charge into a definite solid angle cannot be measured is that any attempt to stop these particles would result in the emission of soft charged massless particles in all directions.[13] Fortunately, as long as we define jets in terms of energy but not charge, we must sum over final states in which soft quarks are emitted in arbitrary directions, and the mass singularities are expected to cancel.

The methods of this paper can be applied to any field theory, not just QCD. However only in an asymptotically free field theory like QCD can we deduce the simple behavior which seems to be observed experimentally: a total cross section dominated at high energy by two-jet events, with an angular distribution characteristic of the lowest-order production of elementary particles.

This work grew out of extensive discussions of one of us (G.S.) at the University of Illinois with Shau-Jin Chang and Jeremiah Sullivan, and out of the stimulus provided to the other (S.W.) from a seminar given at Stanford Linear Accelerator Center by Nathan Weiss. In addition, we would like to thank James Bjorken, Howard Georgi, Gail Hanson, Tom Kinoshita, Benjamin Lee, T. D. Lee, Michael Nauenberg, David Politzer, Helen Quinn, John Stack, Roberto Suaya, Frank Wilczek, and Edward Witten for helpful comments. One of us (S.W.) wishes also to thank the Physics Department of Stanford University for their kind hospitality. This work was supported in part by the National Science Foundation Grants No. PHY-76-15328 and No. PHY 75-20427.

---

[1]G. Hanson *et al.*, Phys. Rev. Lett. 35, 1609 (1975); R. F. Schwitters, in *Proceedings of the International Symposium on Leptons and Photon Interactions at High Energy, Stanford, California, 1975*, edited by W. T. Kirk (Stanford Linear Accelerator Center, Stanford, Calif., 1975), p. 5; G. Hanson, SLAC Report No. SLAC-PUB-1814, September 1976 (unpublished).

[2]For early theoretical predictions of jets in parton models, see S. D. Drell, D. J. Levy, and T. M. Yan, Phys. Rev. 187, 2159 (1969), and Phys. Rev. D 1, 1617 (1970); N. Cabibbo, G. Parisi, and M. Testa, Lett. Nuovo Cimento 4, 35 (1970); J. D. Bjorken and S. D. Brodsky, Phys. Rev. D 1, 1416 (1970); R. P. Feynman, *Photon-Hadron Interactions* (Benjamin, New York, 1972), p. 166.

[3]We will not attempt to deal here with the nonperturbative effects which in QCD presumably account for the trapping of quarks and gluons. Instead, we adopt the rule of thumb, that when colored particles are not

VOLUME 39, NUMBER 23 PHYSICAL REVIEW LETTERS 5 DECEMBER 1977

explicitly isolated, these effects become negligible at sufficiently high energy. [This is the case for instance if these effects behave like $\exp(-\text{const.}/g_E^2)$, where g_E is the gauge coupling defined at a renormalization point with momenta of order E.] We can offer no proof of this assumption, and we cannot predict in advance at what energy the nonperturbative effects become negligible, but we note that a rule of this sort has had to be assumed in every application of QCD to physical problems, including the calculation of the e^+e^- total cross section itself.

[4]H. Fritzsch, M. Gell-Mann, and H. Leutwyler, Phys. Lett. 47B, 365 (1973); D. J. Gross and F. Wilczek, Phys. Rev. D 8, 3633 (1973); S. Weinberg, Phys. Rev. Lett. 31, 494 (1973).

[5]We know of no rigorous proof of this principle in the form stated here. It is supported by arguments of T. Kinoshita, J. Math. Phys. (N.Y.) 3, 650 (1962), especially Appendix A; and T. D. Lee and M. Nauenberg, Phys. Rev. 133, B1549 (1964). Recent work indicates that no special problems arise in non-Abelian gauge theories: see Y.-P. Yao, Phys. Rev. Lett. 36, 653 (1976); T. Appelquist, J. Carazzone, H. Kluberg-Stern, and M. Roth, Phys. Rev. Lett. 36, 768 (1976); L. Tyburski, Phys. Rev. Lett. 37, 319 (1976); E. C. Poggio and H. R. Quinn, Phys. Rev. D 14, 578 (1976); G. Sterman, Phys. Rev. D 14, 2123 (1976); F. G. Krausz, Phys. Lett. 66B, 251 (1977). We hope to prove the cancellation of mass singularities described here to all orders in a future publication.

[6]T. Appelquist and H. Georgi, Phys. Rev. D 8, 4000 (1973); A. Zee, Phys. Rev. D 8, 4038 (1973). For a discussion closer in spirit to that of the present paper, see T. Appelquist and H. D. Politzer, Phys. Rev. D 12, 1404 (1975).

[7]We are dropping terms here which vanish for $\mu \to 0$, and in the remaining expression we drop finite terms of order ϵ or δ or Ω. In consequence, there is no contribution to order g_E^2 from final states with a soft quark or antiquark outside the jets, or with both quark and antiquark in the same jet.

[8]To the order studied here, this cancellation can be derived from the theorem of Lee and Nauenberg, Ref. 5 (see especially Appendix D and Sec. II, remark 3), using the fact that the only state which is degenerate with a specific physical quark-antiquark state and which can be produced from it by a single action of the interaction Hamiltonian consists of a quark, an antiquark, and a soft or collinear gluon.

[9]D. J. Gross and F. Wilczek, Phys. Rev. Lett. 30, 1343 (1973); H. D. Politzer, Phys. Rev. Lett. 30, 1346 (1973).

[10]A perturbative analysis of three-jet events in order g_E^2 is given by J. Ellis, M. K. Gaillard, and G. G. Ross, Nucl. Phys. B111, 253 (1976).

[11]It has been widely conjectured that QCD predictions for the production of quarks and gluons can be taken seriously at sufficiently high energy if reinterpreted in terms of jets. For instance, this is the guiding assumption of Ellis *et al.*, Ref. 10. Our aim here is to show that this hypothesis can actually be *derived* in QCD, by using the absence of mass singularities in suitably defined jet cross sections. The suggestion that the QCD result for the total two-jet probability can be derived in this way was first made in 1975 by one of us (G.S.) in an unpublished University of Illinois preprint. (This paper explicitly exhibited the cancellation of mass singularities to order g_E^2 in the jet probability, but did not give the correct results for the finite part.) Later, the other author (S.W.) independently suggested that QCD results for jet total probabilities can be justified in this manner, and extended this reasoning to jet distributions as well. The present paper is intended to incorporate this earlier unpublished work of both authors, but goes beyond it in various respects, including the calculations leading to Eqs. (2)–(7).

[12]For discussions of this point in the context of parton models, see R. P. Feynman, in *Neutrino '72-Proceedings*, edited by A. Frenkel and G. Marx (OMDK-Technoinform, Budapest, 1972), Vol. II, p. 75; G. R. Farrar and J. L. Rosner, Phys. Rev. D 7, 2747 (1973); R. N. Cahn and E. W. Colglazier, Phys. Rev. D 9, 2658 (1974); S. J. Brodsky and N. Weiss, SLAC Report No. SLAC-PUB-1926 (to be published).

[13]This was independently suggested to one of us (S.W.) by E. Witten.

3.3. Symmetries

One thing on which we could all agree was the importance of symmetries.
Steven Weinberg

Throughout his career Weinberg wrestled with the power of symmetry. Indeed, the philosophy of Effective Lagrangians — to include every term compatible with your chosen symmetry — was that if you had the right symmetry you had the right theory (assuming it to be a quantum field theory).

This section brings together papers with a diversity of themes but with symmetry as a vital common ingredient. The first [3.3.1] poses the question of whether the broken symmetries of elementary particle physics would be restored at high temperatures, in the same way as the rotational invariance of a ferromagnet is restored. Extending earlier results by Kirzhnits and Linde, Weinberg considered gauge theories and calculated the critical temperatures. Continuing the analogy with ferromagnets, he asks whether the universe could consist of domains, in which symmetries are broken differently.

The second [3.3.2] with Georgi and Quinn notes that the energy at which the three $SU(3) \times SU(2) \times U(1)$ coupling constants meet in grand unified theories such as $SU(5)$ can be as large as 10^{17} GeV, almost the Planck mass. This disparity between the unification scale and the weak scale became known as the *Gauge Hierarchy Problem* and remains unsolved to this day.

The third [3.3.3] considers the possibility, as did Susskind and others, that vector bosons acquire their mass not through the vacuum expectation values of scalar fields but dynamically, the Goldstone bosons being bound states of fermions (though Weinberg is at pains to point out its limitations). The discovery of the Higgs boson would appear to settle the issue but this paper is certainly of historical interest and, given the twists and turns of ideas in theoretical physics, may one day be of physical interest too. It provided the inspiration for a paper of mine in 1975 discussing how one might give a mass to the graviton through a dynamical breaking of general covariance. This involves a spin one Goldstone bound state similar to the one that imparts a mass to the graviton in the 2001 Karch-Randall brane-world.

As shown by 't Hooft, instanton effects allow for CP violation by the strong interactions, parameterized by an angle θ. Yet experiment constrains θ to be close to zero. A solution to this strong CP problem was proposed by Peccei and Quinn which involves a global chiral symmetry $U(1)_{PQ}$. Weinberg [3.3.4] and, independently, Wilczek noted that the spontaneous breaking of $U(1)_{PQ}$ leads to a light (pseudo-)Goldstone pseudoscalar boson, called the axion. It has other attractive features which include being a dark-matter candidate, but has not yet been detected.

As we saw in Section 3.1, nonrenormalizable terms in the effective action approach to the Standard Model can violate the symmetries of baryon number conservation or lepton number conservation [3.3.5], as was also shown independently by Wilczek and Zee. Since they are suppressed by inverse powers of the energy scale, however, these symmetries would be approximately conserved.

No discussion of symmetry would be complete without supersymmetry. Hopes that the gauge hierarchy problem might be cured by supersymmetry were dealt a blow by the failure to detect superpartners at the Large Hadron Collider in Geneva. It is perhaps worth mentioning, however, that supersymmetry was not "invented" to solve the gauge hierarchy problem as some journalists would have us believe (it predates it) nor does this spell the death of supersymmetry since this was not why the majority of super-enthusiasts[1] were enamoured of supersymmetry in the first place: global supersymmetry is the square root of special relativity. Moreover, as stressed by Deser and Teitelboim, local supersymmetry is the square root of general relativity and so provides a very natural framework in which to reconcile gravity with the other forces.

Weinberg wrote influential papers both from the point of view that supersymmetry is broken at energies accessible to accelerators, for example [3.3.6] and [3.3.7] with Farrar, and from the point of view that it is broken at much higher energies, for example [3.3.8] with Hall and Lykken.

[1]Who tend to publish in arXiv hep-th rather than arXiv hep-ph.

PHYSICAL REVIEW D VOLUME 9, NUMBER 12 15 JUNE 1974

Gauge and global symmetries at high temperature*

Steven Weinberg
Lyman Laboratory of Physics, Harvard University, Cambridge, Massachusetts 02138
(Received 19 February 1974)

It is shown how finite-temperature effects in a renormalizable quantum field theory can restore a symmetry which is broken at zero temperature. In general, for both gauge symmetries and ordinary symmetries, such effects occur only through a temperature-dependent change in the effective bare mass of the scalar bosons. The change in the boson bare mass is calculated for general field theories, and the results are used to derive the critical temperatures for a few special cases, including gauge and nongauge theories. In one case, it is found that a symmetry which is unbroken at low temperature can be broken by raising the temperature above a critical value. An appendix presents a general operator formalism for dealing with higher-order effects, and it is observed that the one-loop diagrams of field theory simply represent the contribution of zero-point energies to the free energy density. The cosmological implications of this work are briefly discussed.

I. INTRODUCTION

The idea of broken symmetry was originally brought into elementary-particle physics on the basis of experience with many-body systems.[1] Just as a piece of iron, although described by a rotationally invariant Hamiltonian, may spontaneously develop a magnetic moment pointing in any given direction, so also a quantum field theory may imply physical states and *S* matrix elements which do not exhibit the symmetries of the Lagrangian.

It is natural then to ask whether the broken symmetries of elementary-particle physics would be restored by heating the system to a sufficiently high temperature, in the same way as the rotational invariance of a ferromagnet is restored by raising its temperature. A recent paper by Kirzhnits and Linde[2] suggests that this is indeed the case. However, although their title refers to a gauge theory, their analysis deals only with ordinary theories with broken global symmetries. Also, they estimate but do not actually calculate the critical temperature at which a broken symmetry is restored.

The purpose of this article is to extend the analysis of Kirzhnits and Linde to gauge theories,[3] and to show how to calculate the critical temperature for general renormalizable field theories, with either gauge or global symmetries. Our results completely confirm the more qualitative conclusions of Kirzhnits and Linde.[2]

The diagrammatic formalism[4] used here is described in Sec. II. Any finite-temperature Green's function is given by a sum of Feynman diagrams, just as in field theory, except that energy integrals are replaced with sums over a discrete imaginary energy. The justification of the use of this formalism in gauge theories is discussed briefly.

Section III lays the general foundation for calculations of the critical temperature. Our work here is based on the observation that a symmetry which is broken or unbroken in the lowest order of perturbation theory will remain broken or unbroken to all orders, unless there is some circumstance which invalidates the perturbation expansion. We assume that the theory is characterized by a small dimensionless coupling constant $e \ll 1$, so it might be thought that the symmetries of the theory are simply determined by the minima of the scalar-field polynomial $P(\phi)$ in the Lagrangian, and therefore could not be affected by an increase in temperature. However, at very high temperatures, powers of the temperature θ can compensate for powers of e, leading to a breakdown of the perturbation expansion. The leading effect of this sort arises from the $e^2\theta^2$ terms which accompany single-loop quadratic divergences. Since the theory is renormalizable, and we work in a renormalizable gauge, all such terms can be absorbed into a redefinition of the mass terms in $P(\phi)$. Once this is done, the validity of the perturbation expansion is restored. In a general renormalizable theory involving scalar fields ϕ_i, the change in the effective polynomial is calculated here as

$$\Delta P(\phi) = \tfrac{1}{48}\theta^2\{f_{ijkk} + 6(\theta_\alpha\theta_\alpha)_{ij} + \mathrm{Tr}[\gamma_4\Gamma_i\gamma_4\Gamma_j]\}\phi_i\phi_j \;, \qquad (1.1)$$

where f_{ijkl} is the coefficient (of order e^2) of the quartic term in $P(\phi)$, θ_α is the matrix (of order e) representing the αth generator of the gauge group on the scalar fields, and Γ_i is the Yukawa coupling matrix (of order e) for the scalar-spinor interaction. (This notation is the same as used in Refs. 7 and 25, and is reviewed here in Sec. II.) In general a critical temperature is reached at values of θ for which the symmetries of the minimum of $P(\phi) + \Delta P(\phi)$ are gained or lost; usually this occurs when one of the eigenvalues of the bare mass matrix in $P + \Delta P$ vanishes.

This general formalism is used in Sec. IV to calculate critical temperatures in three special cases. The first case is a scalar field theory with an O(n) global symmetry group; it is found that the spontaneous symmetry breakdown encountered at low temperature disappears at a finite temperature θ_c, given by

$$\theta_c = \left(\frac{6}{n+2}\right)^{1/2}\left(\frac{M(0)}{e}\right) , \qquad (1.2)$$

where the quadrilinear self-coupling is taken as $\frac{1}{4}e^2(\phi_i\phi_i)^2$, and $M(0)$ is the mass of the single non-Goldstone boson at zero temperature. The second case is a gauge theory with a local O(n) symmetry group; it is found that there is again a critical temperature θ_c above which the gauge symmetry is restored, now given by

$$\theta_c = [\tfrac{1}{8}(n+2)e^2 + \tfrac{1}{2}(n-1)e'^2]^{-1/2}M(0) , \qquad (1.3)$$

where e' is the gauge coupling constant and e^2 is again the quadrilinear coupling constant. [In gauge theories of the weak and electromagnetic interactions,[5] we typically have $M(0)/e$ of the order of $G_F^{-1/2}$, so these examples indicate that θ_c will be of the order of 300 GeV.] The third case is a scalar field theory with a global O(n)$\times$O(n) symmetry; it is found that for certain ranges of the parameters in the theory it is possible for one of the O(n)'s to be broken at low temperatures and restored at high temperatures, while the other O(n) is unbroken at low temperatures and broken at high temperatures. This has the appearance of a violation of the second law of thermodynamics, but this is not the case: In fact, certain crystals, such as the ferroelectric known as Rochelle or Seignette salt, also have a smaller invariance group above some critical temperature than below it.[6]

Section V compares these results with those that would be found by calculation in a non-renormalizable "unitarity" gauge.[7] In general, in this gauge the introduction of an effective polynomial would not eliminate the θ^2 terms which accompany quadratic divergences, and therefore would not restore perturbation theory at high temperature. In certain simple cases the θ^2 terms which accompany tadpole graphs are eliminated by introduction of an effective polynomial, but the critical temperatures deduced in this way disagree with those calculated in renormalizable gauges, and are argued to be physically irrelevant.

The problem of calculating the critical temperature more accurately and of determining the nature of the phase transition is discussed briefly in Sec. VI. The difficulty here is that as we approach the critical temperature we encounter infrared divergences which invalidate perturbation theory, even after introducing an effective polynomial. It is estimated that the true critical temperature differs from the critical temperature calculated in Sec. IV by an amount at most of order $e^2\theta_c$.

Section VII deals with the question of the observability of phase transitions in gauge theories. It is concluded that spontaneous symmetry breaking can be detected by measurement of Green's functions for gauge-invariant operators carrying zero energies and moderate momenta. Also, although the pressure and energy and entropy densities are continuous at the critical temperature, the specific heat per unit volume has a discontinuity of order $e^2\theta_c^{\ 3}$.

An appendix deals with the problem of defining and calculating a "potential" whose minimum will be at the precise thermodynamic mean value of the scalar field. The potential is defined, using operator rather than diagram methods, as the free energy per unit volume, and it is observed that the corresponding potential calculated earlier for field theories at zero temperature[8] simply represents the contribution of zero-point energies to the free energy. The calculations go through smoothly for scalar-field theories, with the same results as found in Sec. III and IV. However, for gauge theories this operator formalism requires canonical quantization in the unitarity gauge, and in consequence divergences appear in the potential which cannot be eliminated by renormalization of the scalar-field polynomial. Suggestions are offered for further progress along these lines.

This paper is mainly concerned with the study of the phase transition itself, but the existence of this phase transition has wider implications. These are not discussed in the body of this paper, but a word about them may be in order here.

One implication is philosophical. It has been suggested[9] that all the complicated properties of a theory that are usually derived from an assumed broken gauge symmetry may also be derived from the requirements either of perturbative unitarity or of renormalizability. If this is so, then perhaps the gauge symmetry is in some sense a fiction, not representing any truly fundamental invariance principle. It is not clear to me whether this is a question of words or of substance. However, if a gauge symmetry becomes unbroken for sufficiently high temperature, then it is difficult to doubt its reality.

Another implication is cosmological. In "big-bang" cosmologies the critical temperature was presumably reached at some time in the past (unless the richness of hadron states imposes some upper limit on the temperature).[10] In earlier epochs the weak interactions would have produced long-range forces similar to Coulomb forces, with the difference that while the universe appears to be electrically neutral, it may not be neutral with respect to the conserved quantities to which the intermediate vector bosons couple. Such long-range forces would have had profound effects on the evolution of the universe; in particular, as noted by Kirzhnits and Linde,[2] the universe could not have been isotropic and homogeneous if permeated by these lines of force. Long-range vector fields would also play an important role in determining the nature of the initial singularity[11] (if any). Finally, the analogy with ferromagnetism suggests a strange possibility that may occur as the universe cools below the critical temperature.[12] Field theorists are used to the idea that whenever a continuous or discrete symmetry is broken by the appearance of a nonvanishing vacuum expectation value $\langle\phi_i\rangle$ of a scalar-field multiplet, it can be broken in a variety of ways, represented by the different directions of $\langle\phi_i\rangle$. Usually we regard these different directions as entirely equivalent, and ignore the multiplicity of broken-symmetry solutions. However, when a ferromagnet cools below its critical temperature, it does not acquire a single magnetization in some arbitrary direction; rather it breaks up into domains, each with its own direction of magnetization. Does the universe consist of domains, in which symmetries are broken in equivalent but different directions? If so, what happens when a particle or an observer travels from one domain to another?

For reasons of simplicity, it is assumed in this paper that we are interested in states of thermodynamic equilibrium in which all conserved quantum numbers have mean value zero, so that all chemical potentials vanish. (For this reason, the phase transition found here is quite unrelated to the superfluid transition in liquid helium.) It would not be at all difficult to include a chemical potential μ for an absolutely conserved quantity like baryon number; in this case the baryonic part of the term $\mathrm{Tr}[\gamma_4\Gamma_i\gamma_4\Gamma_j]$ in Eq. (1.1) would simply be multiplied with a factor

$$\frac{6}{\pi^2}\int_0^\infty\left(\frac{1}{e^{x-\mu}+1}+\frac{1}{e^{x+\mu}+1}\right)x\,dx\,,$$

with no change in any other results. This is an increasing function of the absolute value of the chemical potential μ, so the presence of a net baryon number would lower the critical temperature. However, μ appears to have a very small cosmological value,[13] of order 10^{-9}, in which case such effects would be quite negligible.

A much more interesting and challenging problem is presented by the possibility of a nonvanishing net mean value for some quantum number carried by bosons as well as fermions, which is exactly conserved only above the critical temperature. In this case we would have to consider not only the effects of a chemical potential but also the possibility of a true superfluid condensate at sufficiently high densities. Work on this problem is being continued.

II. GENERAL FORMALISM

We will consider a general renormalizable quantum field theory, which can be either a simple scalar theory, or a theory of scalar and spinor fields, or a full-fledged gauge theory, with or without spinor fields. For a simple scalar-field theory we would take the Lagrangian to be

$$\mathcal{L} = -\tfrac{1}{2}\partial_\mu\phi_i\partial^\mu\phi_i - P(\phi)\,, \tag{2.1}$$

where ϕ_i is a set of Hermitian spin-zero fields and $P(\phi)$ is a quartic polynomial. In this case we will assume $\mathcal{L}$ and $P(\phi)$ to be invariant under a group of global transformations with generators θ_α:

$$\frac{\partial P(\phi)}{\partial\phi_i}(\theta_\alpha)_{ij}\phi_j = 0\,, \tag{2.2}$$

$$\theta_\alpha^\dagger = \theta_\alpha\,. \tag{2.3}$$

For a gauge theory, we would take the Lagrangian as

$$\mathcal{L} = -\tfrac{1}{2}(D_\mu\phi)_i(D^\mu\phi)_i - \tfrac{1}{4}F_{\alpha\mu\nu}F_\alpha^{\mu\nu} - \bar\psi\gamma^\mu D_\mu\psi - \bar\psi m_0\psi - P(\phi) - \bar\psi\Gamma_i\psi\phi_i, \tag{2.4}$$

where ϕ_i is a set of Hermitian spin-zero fields, $(D_\mu\phi)_i$ is their gauge-covariant derivative, $A_{\alpha\mu}$ is a set of gauge fields, $F_{\alpha\mu\nu}$ is their gauge-covariant curl, ψ_n is a set of spin-$\frac{1}{2}$ fields, $(D_\mu\psi)_n$ is their gauge-covariant derivative, m_0 is a gauge-invariant bare mass matrix, Γ_i is a gauge-covariant Yukawa coupling matrix, and $P(\phi)$ is a gauge-invariant quartic polynomial. This notation is explained more fully in Refs. 7 and 25; for our present purposes it will suffice to note that if θ_a are the Hermitian matrices representing the gauge generators on the scalar multiplet, then

$$(D_\mu\phi)_i \equiv \partial_\mu\phi_i - (\theta_\alpha)_{ij}\phi_j A_{\alpha\mu}\,, \tag{2.5}$$

and Eq. (2.2) now furnishes the necessary gauge-invariance condition on $P(\phi)$. Almost all of our discussion will apply equally well to theories described by the Lagrangian (2.1) or (2.4) or anything in between.

We shall need to impose some sort of weak-coupling condition in order to justify the use of perturbation theory. For the sake of both simplicity and physical relevance, it will be assumed that the orders of magnitude of the various parameters in the Lagrangian are characterized by a mass parameter $\mathfrak{M}$ and a *small* dimensionless coupling parameter $e \ll 1$, with

$$\begin{aligned}
&\text{coefficient of quartic term in } P(\phi) \approx e^2\,,\\
&\text{coefficient of cubic term in } P(\phi) \approx e\mathfrak{M}\,,\\
&\text{coefficient of quadratic term in } P(\phi) \approx \mathfrak{M}^2\,,\\
&\text{gauge couplings } (\theta_\alpha) \approx e\,,\\
&\text{Yukawa couplings } (\Gamma_i) \approx e\,,\\
&\text{Fermion bare mass } (m_0) \approx \mathfrak{M}\,.
\end{aligned} \tag{2.6}$$

[For simplicity we are assuming that the Lagrangian involves no gauge-invariant scalar fields, so there are no linear terms in $P(\phi)$. Of course, we do not rule out the possibility that some of the parameters in the theory may be anomalously small; in particular the symmetries of the theory may require m_0, Γ_i, and/or the cubic term in $P(\phi)$ to vanish.] With this form of the weak-coupling assumption, the expansion of any given S-matrix element at zero temperature in powers of e^2 is the same as an expansion in the number of loops appearing in Feynman diagrams.

It will further be assumed that the symmetries of the Lagrangian are spontaneously broken at zero temperature. This symmetry breakdown is manifested in the appearance of a nonvanishing lowest-order vacuum expectation value λ_i of the scalar fields ϕ_i, given by

$$\frac{\partial P(\phi)}{\partial\phi_i} = 0 \text{ at } \phi = \lambda\,. \tag{2.7}$$

The criterion for spontaneous symmetry breaking in lowest order is

$$(\theta_\alpha)_{ij}\lambda_j \neq 0\,. \tag{2.8}$$

It follows from the weak-coupling assumptions (2.6) that λ is of order

$$\lambda \approx \mathfrak{M}/e\,. \tag{2.9}$$

At this point, the reader may wonder how raising the temperature can possibly restore a broken symmetry. The effects of a finite temperature appear only through diagrams of higher order in e^2, so it appears that at all temperatures the leading term in the mean value of ϕ_i will be the temperature-independent term λ_i given by (2.7). The answer, to be discussed in the next section, is that the symmetries of the theory can only be affected by a finite temperature when the temperature is so high that powers of temperature can compensate for powers of e. However, before we can discuss such matters, we need to review the formalism for perturbative calculations at general finite temperature.

In general, the physical quantities with which we will be concerned here are the partition function

$$Q \equiv \mathrm{Tr}[e^{-H/\theta}] \tag{2.10}$$

and its variational derivatives with respect to external perturbations, the temperature Green's functions[4]

$$Q\langle T_\tau\{A(\vec{x}_1,\tau_1)B(\vec{x}_2,\tau_2)\cdots\}\rangle \equiv \mathrm{Tr}[T_\tau\{A(\vec{x}_1,\tau_1)B(\vec{x}_2,\tau_2)\cdots\}e^{-H/\theta}]\ , \tag{2.11}$$

where H is the Hamiltonian, θ is the temperature (times Boltzmann's constant), $A(\vec{x},\tau)$ is an operator defined in terms of the Schrödinger-representation operators $A(\vec{x})$ by

$$A(\vec{x},\tau)\equiv e^{H\tau}A(\vec{x})e^{-H\tau}\ , \tag{2.12}$$

and T_τ denotes ordering according to the values of τ, with τ values decreasing from left to right, and with an extra minus sign for odd permutations of fermion operators. The perturbation in Q caused by the addition of terms proportional to the operators $A(\vec{x}_1)$, $B(\vec{x}_2)$, etc. is an integral involving the temperature Green's functions (2.11) at τ values in the range

$$0\leq\tau\leq 1/\theta\ . \tag{2.13}$$

It is therefore convenient to express these Green's functions as a Fourier integral over momenta and a Fourier *sum* over discrete energies. However, the Green's functions satisfy a periodicity property of having the same (opposite) values when any one of the τ's for a boson (fermion) operator has the values 0 and $1/\theta$; for instance

$$\begin{aligned}\langle T_\tau\{A(\vec{x}_1,1/\theta)B(\vec{x}_2,\tau_2)\cdots\}\rangle &= Q^{-1}\mathrm{Tr}[A(\vec{x}_1,1/\theta)T_\tau\{B(\vec{x}_2,\tau_2)\cdots\}e^{-H/\theta}]\\ &= Q^{-1}\mathrm{Tr}[T_\tau\{B(\vec{x}_2,\tau_2)\cdots\}e^{-H/\theta}A(\vec{x}_1,1/\theta)]\\ &= Q^{-1}\mathrm{Tr}[T_\tau\{B(\vec{x}_2,\tau_2)\cdots\}A(\vec{x}_1,0)e^{-H/\theta}]\\ &= \pm\langle T_\tau\{A(\vec{x}_1,0)B(\vec{x}_2,\tau_2)\cdots\}\rangle\end{aligned} \tag{2.14}$$

with a $+(-)$ sign when A is a boson (fermion) operator. The same periodicity property also applies to arbitrary τ derivatives of the Green's function, and therefore requires that the Fourier sums contain only even or only odd Fourier components.[14] We can therefore write

$$\langle T_\tau\{A(\vec{x}_1,\tau_1)B(\vec{x}_2,\tau_2)\cdots\}\rangle=\int d^3p_1d^3p_2\cdots\sum_{\omega_1}\sum_{\omega_2}\cdots G(\vec{p}_1,\omega_1,\vec{p}_2,\omega_2,\cdots)\times\exp[i\vec{p}_1\cdot\vec{x}_1-i\omega_1\tau_1+i\vec{p}_2\cdot\vec{x}_2-i\omega_2\tau_2+\cdots]\ , \tag{2.15}$$

where

$$\omega=\pi\theta\times\begin{cases}\text{even integer (bosons)}\\ \text{odd integer (fermions)}\ .\end{cases} \tag{2.16}$$

(These integers can, of course, be positive or negative.)

There is a well-known diagrammatic procedure[4] for calculating the G's: Simply draw all Feynman graphs (dropping vacuum fluctuations) with one external line for each operator $A, B, \ldots$, and evaluate as usual in field theory, except that every internal energy p^0 is replaced with a quantity $i\omega$ satisfying the "quantization" conditions (2.16), and all energy integrals are replaced with ω sums:

$$p^0\to i\omega\ ,$$
$$\int d^4p\to 2i\pi\theta\int d^3p\sum_\infty\ , \tag{2.17}$$
$$\delta^4(p-p')\to(2i\pi\theta)^{-1}\delta_{\omega\omega'}\delta^3(\vec{p}-\vec{p}')\ .$$

The same procedure gives $\ln Q$ as the sum of connected diagrams with no external lines.

In what follows this diagrammatic procedure will be used to calculate Green's functions of gauge-invariant operators using the renormalizable "ξ gauge" of Fujikawa, Lee, and Sanda.[15] This use of a "nonunitarity" gauge may be justified by a three-step argument:

(a) First quantize the theory in the unitarity gauge,[7] and use the Hamiltonian in this gauge to derive finite-temperature Feynman rules as indicated above.

(b) In the same manner as in field theory,[7] show that these Feynman rules are equivalent to the Feynman rules for a ξ gauge with $\xi=0$.

(c) Either directly or by functional methods, show that the results obtained for the partition function or for the Green's functions of gauge-invariant generators are ξ-independent, and therefore correctly given by renormalizable ξ gauges[15] with $\xi\neq0$.

The last two steps go through just as in field theory,[16] because none of the algebraic manipulations depend on whether we integrate over real energies or sum over complex energies. The same result can also be obtained by a more direct functional approach.[17]

One important advantage of our use of a renormalizable perturbative formalism is that we can check that the counterterms which remove divergences in S matrix elements at zero temperature also remove the divergences in finite-temperature Green's functions. For the milder divergences this can be seen from the classic formula[18]

$$h\sum_{n=-N}^{+N}f(nh)-\int_{-(N+1/2)h}^{(N+1/2)h}f(\omega)d\omega=\frac{-h^3}{24}\sum_{n=-N}^{+N}f''(\xi_nh)\ , \tag{2.18}$$

where $f(\omega)$ is an arbitrary twice-differentiable function, h is an arbitrary interval (in our case taken as $2\pi\theta$), and ξ_n is for each n some definite point in the range

$$n-\tfrac{1}{2}\leq\xi_n\leq n+\tfrac{1}{2}\ .$$

Even if the sum and the integral on the left-hand side diverge for $N\to\infty$, their difference is finite in this limit, as long as the divergence is mild enough so that the right-hand side converges. This will in particular be the case for the linear and logarithmic divergences encountered in physical theories, for which $f(\omega)$ behaves like 1 or $1/\omega$ times powers of $\ln\omega$ as $|\omega|\to\infty$, so that $f''(\omega)$ behaves like $1/\omega^2$ or $1/\omega^3$ times powers of $\ln\omega$. In these cases we can pass to the limit $N\to\infty$ in (2.18), and we see that the divergences in the temperature-*dependent* sum on the left are removed by whatever temperature-*independent* subtraction renders the integral convergent. A similar result is obtained for the quadratic divergences in the next section.

III. CALCULATION OF THE EFFECTIVE POLYNOMIAL

We now begin our calculation of the temperature at which a broken symmetry is restored. As already mentioned in the last section, this can only happen in a weak-coupling theory at a temperature so high that powers of the temperature can compensate for powers of the coupling. The number of factors of e in a given graph is simply given by the number of loops, increasing by two units for each additional loop. Hence we must ask how many powers of θ are contributed by each loop.

Consider a single loop, with superficial divergence D, determined by counting powers of momenta as usual, including $+4$ for each loop. We can rescale all internal momenta as well as energies by a factor θ, so that the whole loop takes the form

$$\theta^D I(p_{\text{ext}}/\theta,\ \omega_{\text{ext}}/\theta,\ m_{\text{int}}/\theta), \tag{3.1}$$

where p_{ext} and ω_{ext} represent the various external momenta and energies, and m_{int} represents the various internal masses. Thus for $\theta\to\infty$, the loop behaves like θ^D, unless there are infrared divergences when the arguments of the function I vanish. If $D<1$ there are such infrared-divergences, but they occur only for a finite number of terms in the energy sum, in which two or more of the internal lines of the loop represent a boson carrying zero energy. Such terms are convergent three-dimensional integrals, and therefore can increase no faster than θ as $\theta\to\infty$ because aside from the factor θ in (2.17), the integrands are decreasing functions of θ. (Note that it *is* possible to get a factor θ even from a convergent loop with $D<0$, in particular from the term in which all internal boson energies vanish.) On the other hand, for $D>1$, there are no infrared divergences in $I(0,0,0)$, so the loop simply contributes a factor θ^D. The leading terms for large θ therefore come from those loops with $D>1$ which are as divergent as possible.

Now, aside from an uninteresting quartic divergence in $\ln Q$, the worst divergences in any renormalizable field theory are quadratic. We therefore conclude that *the leading terms for e small and θ large are those in which all loops beyond the lowest order are quadratically divergent.* The convergent part of such a loop contributes a factor $\theta^2 e^2$, so we can anticipate that the critical temperature is reached when θ^2e^2 is of order $\mathfrak{M}^2$, i.e.,

$$\theta_c\approx\mathfrak{M}/e\ . \tag{3.2}$$

At temperatures of this order of magnitude, the contribution of other loops is suppressed either by a factor $e^2\theta/\mathfrak{M}\approx e$ or by a factor e^2.

Further, we know that in any renormalizable field theory, including renormalizable gauge theories, all quadratic divergences can be eliminated by a renormalization of the quadratic term in the polynomial $P(\phi)$. [We are assuming here that there are no gauge-invariant scalar fields in the Lagrangian, in which case the quadratic terms in $P(\phi)$ are the only terms in the Lagrangian with the correct dimensionality needed to cancel quadratic divergences.] We therefore expect that *at finite temperature, all leading terms contributed by multiloop graphs, which survive when $e\ll1$ with $\theta\approx\mathfrak{M}/e$, as well as all quadratic divergences, are canceled by a redefinition of the polynomial part of the Lagrangian,*

$$P_{\text{eff}}(\phi)=P(\phi)+\tfrac{1}{2}Q_{ij}(\theta)\phi_i\phi_j\ , \tag{3.3}$$

and the compensating introduction of a counterterm in the interaction

$$\delta\mathcal{L}'=\tfrac{1}{2}Q_{ij}(\theta)\phi_i\phi_j\ , \tag{3.4}$$

where Q_{ij} is some gauge-invariant quadratically divergent matrix. Since the nonleading terms are suppressed by factors $e^2\theta/\mathfrak{M}\approx e$ or e^2 for each loop, we conclude that any *Green's function is given to a lowest approximation for $e\ll1$ and $\theta\lesssim\mathfrak{M}/e$ by just the lowest-order graphs, but calculated using $P_{\text{eff}}(\phi)$ in place of $P(\phi)$.* In particular, we must define the perturbation expansion by using a shifted field

$$\phi_i'=\phi_i-\lambda_i\ , \tag{3.5}$$

with λ_i a minimum of the new polynomial

$$\left.\frac{\partial P_{\mathrm{eff}}(\phi)}{\partial\phi_i}\right|_{\phi=\lambda}=0\,. \tag{3.6}$$

Thus the presence or absence of spontaneous symmetry breaking at any given temperature can be determined by an examination of the minimum of $P_{\mathrm{eff}}(\phi)$.

In order to calculate Q_{ij}, we note that the only graphs that contain quadratic divergences in any renormalizable theory are the tadpole T_i and the boson self-energy Π_{ij}. After we perform the shift (3.5), the interaction term (3.4) provides counterterms for both of these:

$$\delta\mathcal{L}'=\tfrac{1}{2}Q_{ij}\phi_i'\phi_j'+Q_{ij}\lambda_i\phi_j'+\tfrac{1}{2}Q_{ij}\lambda_i\lambda_j\,. \tag{3.7}$$

Hence we can determine Q_{ij} by requiring that the divergences and θ^2e^2 terms in Π_{ij} and T_i are canceled by the divergences and θ^2e^2 terms in (3.7). (To the order in e that concerns us here, we do not need to worry about other divergences.)

The one-loop tadpole graphs were calculated in a renormalizable ξ gauge at zero temperature in Ref. 25 (see Fig. 1). The result was

$$\begin{aligned}T_i^{(\theta=0)}=&-\tfrac{1}{2}\int d^4k f_{ijk}(k^2+M^2)^{-1}{}_{jk}\\&+\int d^4k\,\mathrm{Tr}[\Gamma_i(i\gamma_\lambda k^\lambda+m)^{-1}]\\&-3(\theta_\beta\theta_\alpha\lambda)_i\int d^4k(k^2+\mu^2)^{-1}{}_{\alpha\beta}\\&+\tfrac{1}{2}(M^2\theta_\alpha\theta_\beta\lambda)_i\\&\times\int d^4k(k^2)^{-1}(\xi k^2+\mu^2)^{-1}{}_{\alpha\beta}\,.\end{aligned} \tag{3.8}$$

(A $-i\epsilon$ term is understood in all denominators.) Here f_{ijk} is the trilinear coupling

$$f_{ijk}=\left.\frac{\partial^3P(\phi)}{\partial\phi_i\partial\phi_j\partial\phi_k}\right|_{\phi=\lambda}, \tag{3.9}$$

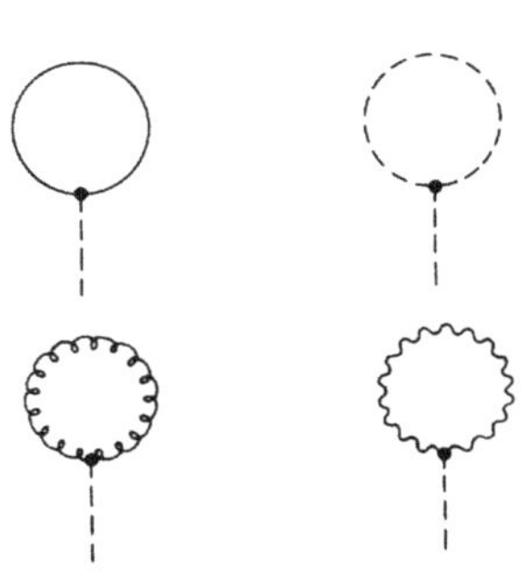

FIG. 1. Feynman graphs for the tadpole T. (Here dashed lines refer to scalar fields, solid lines refer to spinor fields, wavy lines refer to gauge fields, and looped lines refer to Faddeev-Popov "ghost" fields.)

and M, m, and μ are the lowest-order scalar, spinor, and vector mass matrices:

$$M^2{}_{ij}=\left.\frac{\partial^2P(\phi)}{\partial\phi_i\partial\phi_j}\right|_{\phi=\lambda}, \tag{3.10}$$

$$m=m_0+\Gamma_i\lambda_i\,, \tag{3.11}$$

$$\mu^2{}_{\alpha\beta}=\lambda_i\lambda_j(\theta_\alpha\theta_\beta)_{ij}\,. \tag{3.12}$$

Also recall that θ_α is the Hermitian matrix (of order e) which appears in the gauge-covariant derivative of the scalar fields

$$(D_\mu\phi)_i\equiv\partial_\mu\phi_i-(\theta_\alpha)_{ij}\phi_jA_{\alpha\mu}\,. \tag{3.13}$$

We can easily extract the quadratically divergent part, and note that for $\xi\neq0$ it is ξ-independent:

$$\begin{aligned}T_i^\infty=&\{-\tfrac{1}{2}f_{ikk}+\mathrm{Tr}[\Gamma_i\gamma_4m\gamma_4]-3(\theta_\alpha\theta_\alpha\lambda)_i\}\\&\times\int d^4k(k^2)^{-1}\,.\end{aligned} \tag{3.14}$$

[Note that Γ_i may contain terms proportional to γ_5, which anticommute with both $(\gamma_\lambda k^\lambda)^{-1}$ and γ_4.] Under our assumption that the Lagrangian contains no gauge-invariant scalar fields, the first two terms are purely of first order in λ:

$$f_{ikk}=f_{ijkk}\lambda_j\,, \tag{3.15}$$

$$\mathrm{Tr}[\Gamma_i\gamma_4m\gamma_4]=\mathrm{Tr}[\Gamma_i\gamma_4\Gamma_j\gamma_4]\lambda_j\,, \tag{3.16}$$

where f_{ijkl} is the coefficient of the quadrilinear term in $P(\phi)$

$$f_{ijkl}\equiv\frac{\partial^4P(\phi)}{\partial\phi_i\partial\phi_j\partial\phi_k\partial\phi_l}\,. \tag{3.17}$$

Therefore, T_i^∞ is also of first order in λ:

$$\begin{aligned}T_i^\infty=&\{-\tfrac{1}{2}f_{ijkk}+\mathrm{Tr}[\Gamma_i\gamma_4\Gamma_j\gamma_4]-3(\theta_\alpha\theta_\alpha)_{ij}\}\\&\times\lambda_j\int d^4k(k^2)^{-1}\,.\end{aligned} \tag{3.18}$$

In accordance with the finite-temperature Feynman rules discussed in the last section, the leading terms in the tadpole are obtained by replacing the energy integral with a sum over the discrete energies (2.16):

$$\begin{aligned}T_i=&-i(2\pi)^4[\tfrac{1}{2}f_{ijkk}+3(\theta_\alpha\theta_\alpha)_{ij}]\lambda_jI_B(\theta)\\&+i(2\pi)^4\mathrm{Tr}[\Gamma_i\gamma_4\Gamma_j\gamma_4]\lambda_jI_F(\theta)\,.\end{aligned} \tag{3.19}$$

where

$$I_B(\theta)\equiv(2\pi)^{-4}(2\pi\theta)\sum_{n=-\infty}^{\infty}\int d^3k[\vec{k}^2+4n^2\pi^2\theta^2]^{-1}\,, \tag{3.20}$$

$$I_F(\theta)\equiv(2\pi)^{-4}(2\pi\theta)\sum_{n=-\infty}^{\infty}\int d^3k[\vec{k}^2+(2n+1)^2\pi^2\theta^2]^{-1}\,. \tag{3.21}$$

The counterterm (3.7) supplies an additional contribution

$$\delta T_i = i(2\pi)^4 Q_{ij}\lambda_j \,, \tag{3.22}$$

so in order to cancel the leading terms in the one-loop tadpole we must choose Q_{ij} as the matrix

$$Q_{ij}(\theta) = [\tfrac{1}{2}f_{ijkk} + 3(\theta_\alpha\theta_\alpha)_{ij}]I_B(\theta) - \mathrm{Tr}[\Gamma_i\gamma_4\Gamma_j\gamma_4]I_F(\theta)\,. \tag{3.23}$$

Before discussing the calculation of I_B and I_F, let us check that the counterterms in (3.7) now also cancel the leading terms in the scalar-boson self-energy. The one-loop self-energy graphs at zero boson momentum and zero temperature were also calculated in Ref. 25 (see Fig. 2). The result was

$$\begin{aligned}
\Pi_{ij}^{(\theta=0)}(0) = {} & \frac{-3i}{2(2\pi)^4}(\{\theta_\alpha,\theta_\beta\}\lambda)_i(\theta_\gamma,\theta_\delta\}\lambda)_i \int d^4k(k^2+\mu^2)^{-1}{}_{\alpha\gamma}(k^2+\mu^2)^{-1}{}_{\beta\delta} \\
& - \frac{i}{2(2\pi)^4} f_{ikl}f_{jpq}\int d^4k(k^2+M^2)^{-1}{}_{kp}(k^2+M^2)^{-1}{}_{lq} \\
& - \frac{i}{(2\pi)^4}\int d^4k(k^2)^{-1}(\xi k^2+\mu^2)^{-1}{}_{\alpha\beta}[M^2\theta_\alpha(k^2+M^2)^{-1}\theta_\beta M^2 + M^2\theta_\alpha(k^2+M^2)^{-1}M^2\theta_\beta \\
& \qquad + \theta_\alpha M^2(k^2+M^2)^{-1}\theta_\beta M^2]_{ij} \\
& - \frac{i}{(2\pi)^4}\int d^4k\,\mathrm{Tr}[\Gamma_i(i\gamma_\lambda k^\lambda+m)^{-1}\Gamma_j(i\gamma_\lambda k^\lambda+m)^{-1}] \\
& - \frac{i}{2(2\pi)^4}(M^2\theta_\gamma\theta_\alpha\lambda)_i(M^2\theta_\beta\theta_\delta\lambda)_j\int d^4k(k^2)^{-1}(\xi k^2+\mu^2)^{-1}{}_{\alpha\beta}(\xi k^2+\mu^2)^{-1}{}_{\gamma\delta} \\
& + \frac{3i}{2(2\pi)^4}(\{\theta_\beta,\theta_\alpha\})_{ij}\int d^4k(k^2+\mu^2)^{-1}{}_{\alpha\beta} + \frac{i}{2(2\pi)^4}f_{ijkl}\int d^4k(k^2+M^2)^{-1}{}_{kl} \\
& - \frac{i}{2(2\pi)^4}f_{ijk}M^{-2}{}_{kl}f_{lpq}\int d^4k(k^2+M^2)^{-1}{}_{pq} + \frac{i}{(2\pi)^4}f_{ijk}M^{-2}{}_{kl}\int d^4k\,\mathrm{Tr}[\Gamma_l(i\gamma_\lambda k^\lambda+m)^{-1}] \\
& - \frac{3i}{2(2\pi)^4}f_{ijk}M^{-2}{}_{kl}(\{\theta_\alpha,\theta_\beta\}\lambda)_l\int d^4k(k^2+\mu^2)^{-1}{}_{\alpha\beta}\,.
\end{aligned}$$

The quadratically divergent part for $\xi\neq 0$ is ξ-independent:

$$\begin{aligned}
\Pi_{ij}^{\infty} = \frac{i}{(2\pi)^4}\{ & -\mathrm{Tr}[\Gamma_i\gamma_4\Gamma_j\gamma_4] + 3(\theta_\alpha\theta_\alpha)_{ij} + \tfrac{1}{2}f_{ijkk} \\
& - \tfrac{1}{2}f_{ijk}M^{-2}{}_{kl}f_{lpp} + f_{ijk}M^{-2}{}_{kl}\mathrm{Tr}[\Gamma_l\gamma_4 m\gamma_4] - 3f_{ijk}M^{-2}{}_{kl}(\theta_\alpha\theta_\alpha\lambda)_l\}\int d^4k(k^2)^{-1}\,.
\end{aligned}$$

Replacing energy integrals by energy sums and using (3.15), (3.16), and (3.23), the leading term at finite temperature may be written

$$\Pi_{ij} = -Q_{ij} + f_{ijk}M^{-2}{}_{kl}Q_{lm}\lambda_m\,.$$

The first term here is immediately canceled by the first term in (3.7), while the second is canceled by the tadpole (3.22) produced by the second term in (3.7).

Returning now to the functions (3.20) and (3.21), we note that the sums may be turned back into integrals,

$$I_B(\theta) = \frac{-i}{2(2\pi)^4}\int d^3k\oint_C d\omega(\vec{k}^2+\omega^2)^{-1}\cot\left(\frac{\omega}{2\theta}\right)\,,$$

$$I_F(\theta) = \frac{+i}{2(2\pi)^4}\int d^3k\oint_C d\omega(\vec{k}^2+\omega^2)^{-1}\tan\left(\frac{\omega}{2\theta}\right)\,.$$

The contour C runs from $+\infty$ to $-\infty$ just above the real axis and then back from $-\infty$ to $+\infty$ just below the real axis. By closing the two halves of this contour with large semicircles in the upper and

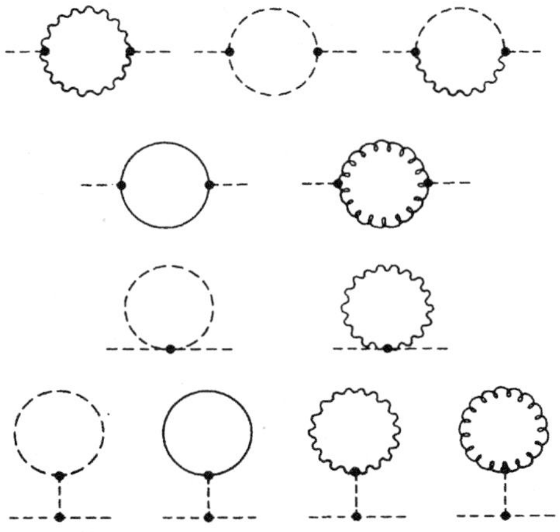

FIG. 2. Feynman graphs for the scalar self-energy Π. (Conventions same as in Fig. 1.)

lower half-planes, we pick up the poles at $\omega = \pm |\vec{k}|$,

$$I_B(\theta) = (2\pi)^{-3} \int \frac{d^3k}{2|\vec{k}|} \coth \frac{|\vec{k}|}{2\theta} ,$$

$$I_F(\theta) = (2\pi)^{-3} \int \frac{d^3k}{2|\vec{k}|} \tanh \frac{|\vec{k}|}{2\theta} .$$

These integrals are of course divergent, but their divergences can be separated out by extracting their values at $\theta = 0$:

$$I_B(\theta) = I_B(0) + \tfrac{1}{12}\theta^2 , \qquad I_F(\theta) = I_F(0) - \tfrac{1}{24}\theta^2 . \tag{3.24}$$

This is important because it shows that the same infinite counterterm which removes divergences at zero temperature continues to remove them at all temperatures; in fact, we may recognize the divergences here as just the same ones we encounter in field theory:

$$I_B(0) = I_F(0) = (2\pi)^{-3} \int \frac{d^3k}{2|\vec{k}|} = \frac{-i}{(2\pi)^4} \int \frac{d^4k}{k^2 - i\epsilon} .$$

Using (3.24) in (3.23), we have finally

$$Q_{ij}(\theta) = Q_{ij}(0) + \tfrac{1}{24}\theta^2 \{ f_{ijkk} + 6(\theta_\alpha\theta_\alpha)_{ij} + \mathrm{Tr}[\Gamma_i\gamma_4\Gamma_j\gamma_4] \} . \tag{3.25}$$

[We note that f_{ijkl} is of order e^2, while θ_α and Γ_i are of order e, so $Q - Q(0)$ is of order $e^2\theta^2$, as anticipated.] The term $Q_{ij}(0)$ in (3.25) just serves to provide a temperature-independent quadratically divergent renormalization of the mass parameters in the polynomial, so we may write (3.3) as

$$P_{\mathrm{eff}}(\phi) = P_{\mathrm{ren}}(\phi) + \tfrac{1}{48}\theta^2 \{ f_{ijkk} + 6(\theta_\alpha\theta_\alpha)_{ij} + \mathrm{Tr}[\Gamma_i\gamma_4\Gamma_j\gamma_4] \} \phi_i\phi_j , \tag{3.26}$$

where $P_{\mathrm{ren}}(\phi)$ is just the original polynomial $P(\phi)$, but with masses replaced by renormalized values. This formula will be used in the next section to determine the critical temperature.

Even though this has so far been a one-loop calculation, it is actually valid to lowest order in e but to all orders in $e\theta$. We could insert another loop in the single-loop diagram used to calculate the tadpoles or scalar self-energies, and if this new loop were quadratically divergent it would contribute a non-negligible factor $e^2\theta^2$, but the *old* loop would then not be quadratically divergent, and therefore would be suppressed at least by a factor e (see Fig. 3). More generally, we expect multiloop as well as single-loop diagrams

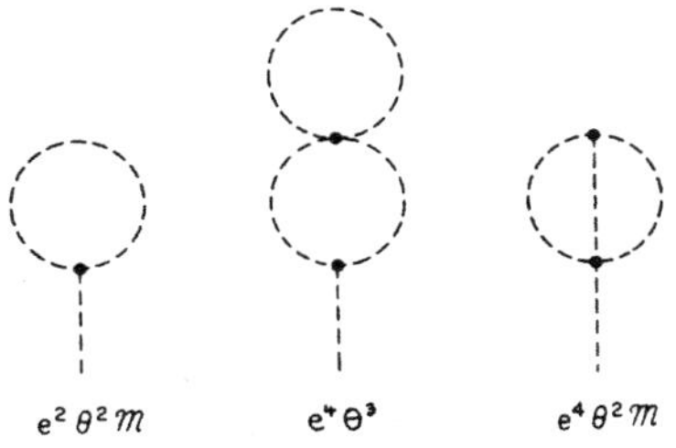

FIG. 3. One- and two-loop graphs for the tadpole T in a scalar field theory. The order of magnitude of the various contributions is indicated below each graph.

for T_i to involve only a *single* factor of θ^2, so it is only the one-loop diagram that survives when $e \ll 1$ and $\theta \approx \mathfrak{M}/e$.

IV. THE CRITICAL TEMPERATURES

We have learned in the last section that the leading effect of multiloop graphs at temperatures θ of order $\mathfrak{M}/e$ is to change the polynomial $P(\phi)$ in the Lagrangian to the effective polynomial given by Eq. (3.26). The symmetry group of the Green's functions at a given temperature consists simply of that subgroup of the invariance group of the Lagrangian which leaves invariant the point λ_i at which $P_{\mathrm{eff}}(\phi)$ has its minimum. We can therefore locate the various critical temperatures of the theory by asking at what temperature the symmetries of λ_i are gained or lost.

In particular, we note that if the temperature-dependent part of $P_{\mathrm{eff}}(\phi)$ is a positive-definite function of ϕ, then at sufficiently high temperatures the minimum of $P_{\mathrm{eff}}(\phi)$ must be at $\phi = 0$. This is because the quartic part of $P_{\mathrm{eff}}(\phi)$ is in any case positive-definite (otherwise the energy would be unbounded below for large ϕ), while for sufficiently large θ the total quadratic term in $P_{\mathrm{eff}}(\phi)$ will also be positive-definite (and large enough to overwhelm any cubic term that is not overwhelmed by the quartic terms). Thus we conclude that if the θ^2 term in P_{eff} is positive-definite then there is always a highest critical temperature, above which the Green's functions exhibit the full symmetry group of the theory.

The $\theta_\alpha\theta_\alpha$ and $\Gamma_i\Gamma_i$ in (3.26) are indeed positive matrices, because the θ and Γ matrices satisfy Hermiticity conditions,

$$\theta_\alpha^\dagger = \theta_\alpha, \quad \Gamma_i^\dagger = \gamma_4\Gamma_i\gamma_4 .$$

In fact, these terms are positive-*definite*, unless there are no gauge couplings or Yukawa couplings in the theory at all. However, the f term in (3.26), while usually positive-definite, is not always so.[19] We shall take a look at two examples where symmetry is restored at high temperature, and one example where it is not.

Example 1: Global O(n) with one n-vector

Let us consider a nongauge theory, invariant under a global group O(n), involving a single n-vector multiplet of scalar fields ϕ_i. The polynomial P will be of the form

$$P(\phi)=\tfrac{1}{2}\mathfrak{M}_0^2\phi_i\phi_i+\tfrac{1}{4}e^2(\phi_i\phi_i)^2\ ,$$

where $\mathfrak{M}_0^2$ and e^2 are real quantities, with $e^2>0$ but $\mathfrak{M}_0^2$ of arbitrary sign. The quadrilinear coupling coefficient here is

$$f_{ijkl}=2e^2(\delta_{ij}\delta_{kl}+\delta_{ik}\delta_{jl}+\delta_{il}\delta_{jk})\ ,$$

so (3.25) gives

$$P_{\text{eff}}(\phi)=\tfrac{1}{2}\mathfrak{M}^2(\theta)\phi_i\phi_i+\tfrac{1}{4}e^2(\phi_i\phi_i)^2\ ,$$

where

$$\mathfrak{M}^2(\theta)=\mathfrak{M}^2(0)+\tfrac{1}{12}(n+2)e^2\theta^2\ ,$$

and $\mathfrak{M}^2(0)$ is $\mathfrak{M}_0^2$ plus renormalization counter-terms (see Fig. 4). If $\mathfrak{M}^2(0)$ is negative then for sufficiently low temperatures $\mathfrak{M}^2(\theta)$ will also be negative, and $P_{\text{eff}}(\phi)$ will have an O(n)-noninvariant minimum at $\phi_i=\lambda_i$, with

$$e^2\lambda_i\lambda_i=-\mathfrak{M}^2(\theta)>0\ .$$

The full O(n) symmetry is therefore restored at a temperature θ_c such that

$$\mathfrak{M}^2(\theta_c)=0\ ,$$

or[20]

$$\theta_c=\left(\frac{12}{n+2}\right)^{1/2}\frac{|\mathfrak{M}(0)|}{e}\ .$$

In order to express this in terms of observables,

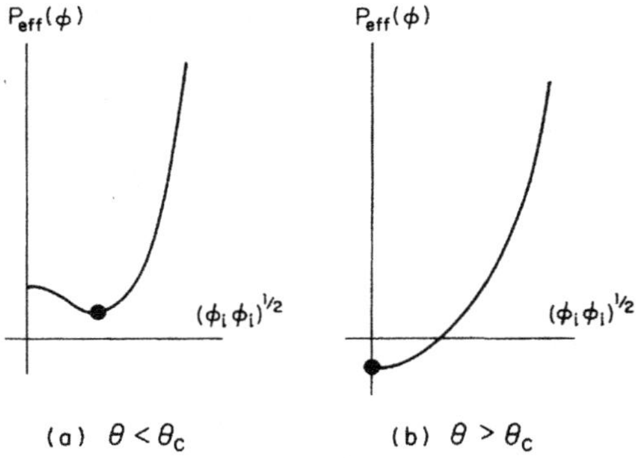

FIG. 4. Schematic representation of the effective polynomial in Examples 1 and 2 of Sec. IV, below and above the critical temperature. The dark dot indicates the state of thermal equilibrium.

we may note that the physical zeroth-order mass matrix of the scalar fields is

$$\begin{aligned}M^2{}_{ij}(\theta)&=\left.\frac{\partial^2P_{\text{eff}}(\phi)}{\partial\phi_i\,\partial\phi_j}\right|_{\phi=\lambda}\\&=\mathfrak{M}^2(\theta)\delta_{ij}+e^2(\delta_{ij}\lambda_k\lambda_k+2\lambda_i\lambda_j)\\&=2e^2\lambda_i\lambda_j\ ,\end{aligned}$$

so for $\theta<\theta_C$ the excitation spectrum consists of $n-1$ Goldstone bosons of zero mass and one boson of mass

$$M^2(\theta)=2e^2\lambda_i\lambda_i=-2\mathfrak{M}^2(\theta)\ .$$

We can therefore rewrite the critical temperature in terms of the single nonzero boson mass at zero temperature:

$$\theta_c=\left(\frac{6}{n+2}\right)^{1/2}\left(\frac{M(0)}{e}\right)\ .$$

Above the critical temperature the excitation spectrum consists of n degenerate bosons with mass $\mathfrak{M}(\theta)$.

Example 2: Local O(n) with one n-vector

Next, let us consider a *gauge* theory based on the group O(n), again with a single n-vector multiplet of scalar fields. The generators of the gauge group may be represented by matrices,

$$(\theta_{kl})_{ij}=ie'(\delta_{ki}\delta_{lj}-\delta_{kj}\delta_{li})\ ,\quad 1\le k<l\le n$$

with a prime on the gauge coupling constant to distinguish it from the boson self-coupling constant e. The Casimir operator here is

$$\sum_{k<l}(\theta_{kl}\theta_{kl})_{ij}=(n-1)e'^2\delta_{ij}\ .$$

We take the polynomial in the Lagrangian again of the form

$$P(\phi)=\tfrac{1}{2}\mathfrak{M}_0^2\phi_i\phi_i+\tfrac{1}{4}e^2(\phi_i\phi_i)^2\ .$$

Equation (3.26) now gives, as in case (a),

$$P_{\text{eff}}(\phi)=\tfrac{1}{2}\mathfrak{M}^2(\theta)\phi_i\phi_i+\tfrac{1}{4}e^2(\phi_i\phi_i)^2\ ,$$

where

$$\mathfrak{M}^2(\theta)=\mathfrak{M}^2(0)+\tfrac{1}{12}(n+2)e^2\theta^2+\tfrac{1}{4}(n-1)e'^2\theta^2\ .$$

For $\mathfrak{M}^2(0)<0$ there is again a critical temperature θ_c, determined by the condition that $\mathfrak{M}^2(\theta_c)$ should vanish[20]:

$$\theta_c=[\tfrac{1}{12}(n+2)e^2+\tfrac{1}{4}(n-1)e'^2]^{-1/2}|\mathfrak{M}(0)|\ .$$

We see here an example of the general phenomenon, that adding gauge fields lowers the critical temperature. For $\theta<\theta_c$, this theory has an excitation spectrum consisting of one scalar boson of mass

$$M^2(\theta) = -2\mathfrak{M}^2(\theta)$$

(the remaining $n-1$ massless scalars are now unphysical) plus $n-1$ vector bosons with mass given by Eq. (3.12):

$$\mu^2(\theta) = e'^2\lambda_i(\theta)\lambda_i(\theta) = \left(\frac{e'}{e}\right)|\mathfrak{M}^2(\theta)|$$

plus $(n-1)(n-2)/2$ vector bosons of zero mass, corresponding to the unbroken $O(n-1)$ subgroup. For $\theta > \theta_c$ the theory has n scalar bosons of mass $\mathfrak{M}(\theta)$ plus $n(n-1)/2$ vector bosons of zero mass. At θ_c there is evidently a transmutation of the zero-helicity states of $n-1$ vector bosons into $n-1$ scalar bosons, with the mass of all these bosons vanishing at θ_c to make the transmutation possible (see Fig. 5).

Example 3: Global $O(n) \times O(n)$ with two n-vectors

As an example with a very different behavior, let us consider an $O(n)\times O(n)$-invariant theory with two independent scalar multiplets χ_A and η_a transforming according to the representations $(n, 1)$ and $(1, n)$. The polynomial in the Lagrangian must take the form

$$P(\chi, \eta) = \tfrac{1}{2}\mathfrak{M}_\chi{}^2\chi_A\chi_A + \tfrac{1}{2}\mathfrak{M}_\eta{}^2\eta_a\eta_a + \tfrac{1}{4}e_{\chi\chi}{}^2(\chi_A\chi_A)^2 - \tfrac{1}{2}e_{\chi\eta}{}^2(\chi_A\chi_A)(\eta_a\eta_a) + \tfrac{1}{4}e_{\eta\eta}{}^2(\eta_a\eta_a)^2 ,$$

with parameters subject to the positivity constraints

$$e_{\chi\chi}{}^2 > 0, \quad e_{\eta\eta}{}^2 > 0, \quad e_{\chi\eta}{}^2 < |e_{\chi\chi}e_{\eta\eta}| .$$

(We use capital indices $A, B, \ldots$ for the χ's and lower-case indices $a, b, \ldots$ for the η's but all these indices run from 1 to n.) The nonvanishing elements of the quadrilinear coupling coefficient f_{ijkl} are now

$$f_{ABCD} = 2e_{\chi\chi}{}^2(\delta_{AB}\delta_{CD} + \delta_{AC}\delta_{BD} + \delta_{AD}\delta_{BC}) ,$$

$$f_{ABab} = -2e_{\chi\eta}{}^2\delta_{AB}\delta_{ab} ,$$

$$f_{abcd} = 2e_{\eta\eta}{}^2(\delta_{ab}\delta_{cd} + \delta_{ac}\delta_{bd} + \delta_{ad}\delta_{bc}) ,$$

together with other elements obtained by permutation of indices. The effective polynomial (3.26) therefore has the form

$$P(\chi, \eta) = \tfrac{1}{2}\mathfrak{M}_\chi{}^2(\theta)\chi_A\chi_A + \tfrac{1}{2}\mathfrak{M}_\eta{}^2(\theta)\eta_a\eta_a + \tfrac{1}{4}e_{\chi\chi}{}^2(\chi_A\chi_A)^2 - \tfrac{1}{2}e_{\chi\eta}{}^2(\chi_A\chi_A)(\eta_a\eta_a) + \tfrac{1}{4}e_{\eta\eta}{}^2(\eta_a\eta_a)^2 ,$$

where

$$\mathfrak{M}_\chi{}^2(\theta) = \mathfrak{M}_\chi{}^2(0) + \tfrac{1}{12}\theta^2[(n+2)e_{\chi\chi}{}^2 - ne_{\chi\eta}{}^2] ,$$

$$\mathfrak{M}_\eta{}^2(\theta) = \mathfrak{M}_\eta{}^2(0) + \tfrac{1}{12}\theta^2[(n+2)e_{\eta\eta}{}^2 - ne_{\chi\eta}{}^2] .$$

For $e_{\chi\eta}{}^2$ positive, which as we shall see is the interesting case, there are four possible phases (see Fig. 6):

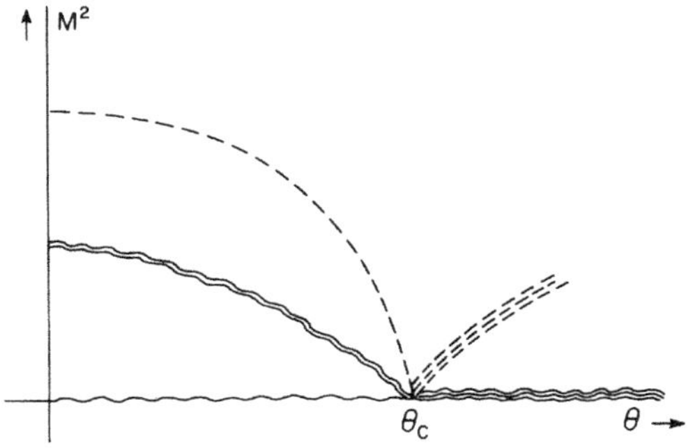

FIG. 5. Schematic representation of the excitation spectrum as a function of temperature for the gauge field theory discussed in Example 2 of Sec. IV. The gauge group here is taken as O(3). Wavy lines indicate particles of spin 1; dashed lines particles of spin 0. Note the continuity in the total numbers of helicity states at the critical temperature.

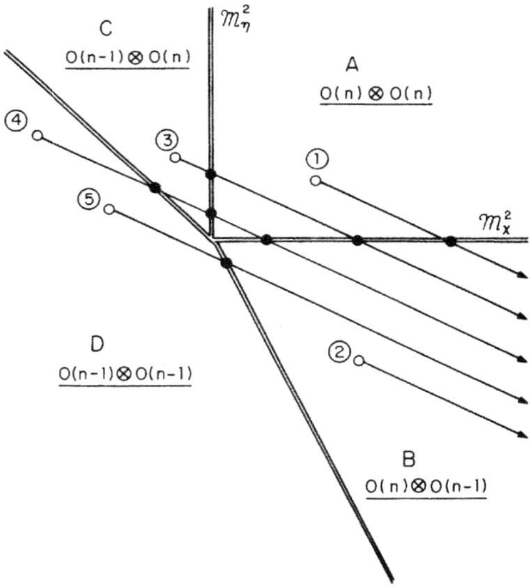

FIG. 6. Phase diagram for the theory described in Example 3 of Sec. IV. Phase boundaries are indicated by double lines. The values of $\mathfrak{M}_\chi{}^2$ and $\mathfrak{M}_\eta{}^2$ at zero temperature are indicated by open circles; the arrows indicate the behavior of $\mathfrak{M}_\chi{}^2$ and $\mathfrak{M}_\eta{}^2$ for large temperature. Critical temperatures are indicated by dark circles. The numbers in circles refer to the cases listed in the text.

(A) $\mathfrak{M}_\chi^2(\theta)>0, \quad \mathfrak{M}_\eta^2(\theta)>0$.

In this case the only minimum of $P_{\rm eff}(\chi,\eta)$ is at $\chi_A=\eta_a=0$, so the symmetry $O(n)\times O(n)$ is unbroken.

(B) $\mathfrak{M}_\chi^2(\theta)>0$,

$$0>\mathfrak{M}_\eta^2(\theta)>-(e_{\eta\eta}^2/e_{\chi\eta}^2)\mathfrak{M}_\chi^2(\theta)\,.$$

In this case the only minimum of $P_{\rm eff}(\chi,\eta)$ is at $\chi_A=0$, $\eta_a\neq 0$, so the symmetry is broken down to $O(n)\times O(n-1)$.

(C) $\mathfrak{M}_\eta^2(\theta)>0$,

$$0>\mathfrak{M}_\chi^2(\theta)>-(e_{\chi\chi}^2/e_{\chi\eta}^2)\mathfrak{M}_\eta^2(\theta)\,.$$

In this case the only minimum of $P_{\rm eff}(\chi,\eta)$ is at $\eta_a=0$, $\chi_A\neq 0$, so the symmetry is broken down to $O(n-1)\times O(n)$.

(D) For all other values of $\mathfrak{M}_\chi^2(\theta)$ and $\mathfrak{M}_\eta^2(\theta)$ the *deepest* minimum is at $\chi_A\neq 0$, $\eta_a\neq 0$, so the symmetry is broken down to $O(n-1)\times O(n-1)$.

The phase for $\theta\to\infty$ is always (A), (B), or (C), depending on the relative values of the coupling constants. If for example we choose

$$(n+2)e_{\chi\chi}^2>ne_{\chi\eta}^2>(n+2)e_{\eta\eta}^2$$

[which is consistent with the positivity requirements on $P(\chi,\eta)$ for large fields] then for $\theta\to\infty$ the system is necessarily in phase (B). We see that with this choice of coupling constants, *the symmetry is necessarily broken down to $O(n)\times O(n-1)$ at high temperature*. (The same is true if we introduce gauge fields, providing that the gauge coupling constant is sufficiently small compared with $|e_{\chi\eta}|$.) The critical points encountered at lower temperature depend on the signs and relative magnitude of $\mathfrak{M}_\chi^2(\theta)$ and $\mathfrak{M}_\eta^2(\theta)$ at zero temperature. We may distinguish the following cases (see Fig. 6):

(1) $\mathfrak{M}_\chi^2(0)>0, \quad \mathfrak{M}_\eta^2(0)>0$.

There is a single critical temperature, at which the phase changes from type (A) at low temperature to type (B) at high temperature.

(2) $\mathfrak{M}_\chi^2(0)>0$,

$$0>\mathfrak{M}_\eta^2(0)>-(e_{\eta\eta}^2/e_{\chi\eta}^2)\mathfrak{M}_\chi^2(0)\,.$$

There are no critical temperatures; the phase is of type (B) at all temperatures.

(3) $\mathfrak{M}_\eta^2(0)>0$,

$$0>\mathfrak{M}_\chi^2(0)>-(e_{\chi\chi}^2/e_{\chi\eta}^2)\mathfrak{M}_\eta^2(0)\,.$$

There are *two* critical temperatures, at which the phase changes from type (C) at low temperature to type (A) at medium temperature to type (B) at high temperature.

(4) $\mathfrak{M}_\eta^2(0)>0$,

$$-(e_{\chi\chi}^2/e_{\chi\eta}^2)\mathfrak{M}_\eta^2(0)>\mathfrak{M}_\chi^2(0) > -\left(\frac{(n+2)e_{\chi\chi}^2-ne_{\chi\eta}^2}{ne_{\chi\eta}^2-(n+2)e_{\eta\eta}^2}\right) \times \mathfrak{M}_\eta^2(0)\,.$$

There are *three* critical temperatures, at which the phase changes from type (D) at low temperature, to type (C) and then to type (A) at medium temperature, and finally to type (B) at high temperature.

(5) For other values of $\mathfrak{M}_\chi^2(0)$ and $\mathfrak{M}_\eta^2(0)$, there is a single critical temperature, at which the phase changes from type (D) at low temperature to type (B) at high temperature.

The existence of an (A)→(B) critical point, at which the symmetry shifts with increasing temperature from the group $O(n)\times O(n)$ to a *smaller* group $O(n)\times O(n-1)$, runs counter to most of our experience with macroscopic systems. For instance, heating a superconductor restores gauge invariance; heating a ferromagnet restores rotational invariance; heating a crystal restores translational invariance. However, the example of Rochelle salt[6] reassures us that there is nothing impossible about a loss of symmetry with increasing temperature.

V. COMPARISON WITH UNITARITY GAUGE CALCULATIONS

It is instructive to compare the results we have obtained by calculating in the renormalizable ξ gauges with the corresponding results that would be obtained in the unitarity gauge, for which $\xi=0$. In this case, we would have had no reason to expect that the counterterm (3.7) would cancel all the $e^2\theta^2$ terms. In particular, the quadratically divergent part of the tadpole (3.8) would have contained an additional term,

$$T_{iU}^\infty=T_i^\infty+\tfrac{1}{2}(\mu^{-2})_{\alpha\beta}(M^2\theta_\alpha\theta_\beta\lambda)_i\int d^4k(k^2)^{-1}. \tag{5.1}$$

(A subscript U denotes the use of the unitarity gauge; quantities without this subscript are calculated in the renormalizable gauges with $\xi\neq 0$.) This implies a new temperature-dependent term,

$$T_{iU}=T_i+\tfrac{1}{2}i(2\pi)^4 I_B(\theta)(\mu^{-2})_{\alpha\beta}(M^2\theta_\alpha\theta_\beta\lambda)_i\,. \tag{5.2}$$

This cannot in general be canceled by a counterterm of the form (3.7), because

$$\frac{\partial T_{iU}}{\partial\lambda_j}\neq\frac{\partial T_{jU}}{\partial\lambda_i}\,.$$

The counterterm (3.7) does in general cancel the quadratic divergences and $e^2\theta^2$ terms even for $\xi=0$ in a general *gauge invariant* Green's function, but this cancellation occurs because for $\xi=0$ there are quadratic divergences and $e^2\theta^2$ terms contributed by a wide variety of diagrams besides tadpoles and scalar self-energies. Thus it is not possible to determine the matrix Q_{ij} in (3.7) by inspection of tadpole graphs evaluated in the unitarity gauge.

The inadequacy of the unitarity gauge for our purposes may be obscured by the fact that in certain specially simple gauge theories[21] T_{iU} does take the form of a λ gradient. These theories are characterized by the condition that the scalar fields belong to a representation of the gauge group which is "transitive on the sphere," i.e., for which any direction in the representation space of the scalar fields may be rotated into any other direction by a gauge transformation. In this case $P_{\rm eff}(\phi)$ must be of the same form as in an O(n)-invariant theory,

$$P_{{\rm eff},U}(\phi)=\tfrac{1}{2}\mathfrak{M}_U{}^2(\theta)\phi_i\phi_i+\tfrac{1}{4}e^2(\phi_i\phi_i)^2, \qquad (5.3)$$

so we can take over the results of examples 1 and 2 of the last section. In particular, we again have

$$M^2{}_{ij}=2e^2\lambda_i\lambda_j,$$

so (5.2) gives

$$T_{iU}=T_i+ie^2(2\pi)^4I_B(\theta)\lambda_i\,. \qquad (5.4)$$

The new term here can be canceled by a new term in the matrix Q_{ij} in (3.7):

$$Q_{ijU}(\theta)=Q_{ij}(\theta)-e^2I_B(\theta)\delta_{ij}\,. \qquad (5.5)$$

This new term in $Q_{ij}(\theta)$ leads to a decrease in the temperature-dependent mass term in $P_{\rm eff}$, and hence to an increase in the critical temperature. For instance, the mass term calculated in example 2 of the last section using the renormalizable gauges was

$$\mathfrak{M}^2(\theta)=\mathfrak{M}^2(0)+\tfrac{1}{12}(n+2)e^2\theta^2+\tfrac{1}{4}(n-1)e'^2\theta^2, \qquad (5.6)$$

so the new term in (5.5) changes this to

$$\mathfrak{M}_U{}^2(\theta)=\mathfrak{M}_U{}^2(0)+\tfrac{1}{12}(n+1)e^2\theta^2+\tfrac{1}{4}(n-1)e'^2\theta^2, \qquad (5.7)$$

and the critical temperature for $\mathfrak{M}_U{}^2(0)<0$ is[20]

$$\theta_c=|\mathfrak{M}_U(0)|\left[\tfrac{1}{12}(n+1)e^2+\tfrac{1}{4}(n-1)e'^2\right]^{-1/2}. \qquad (5.8)$$

Since unitarity-gauge calculations give different results from renormalizable-gauge calculations, which should we believe? The answer is provided by our discussion in Sec. III: It is only in the renormalizable gauge that a change in the scalar-field polynomial restores the validity of perturbation theory at high temperatures, and therefore only in the renormalizable gauges can we use the effective polynomial to study the pattern of symmetry breaking. When the scalar field representation is transitive on a sphere, the introduction of an effective polynomial does eliminate the $e^2\theta^2$ terms found in the unitarity gauge in tadpole graphs, but there are plenty of other quadratic divergences and $e^2\theta^2$ terms in this gauge which are not thereby eliminated, and there is no reason to regard the effective polynomial as being of any special importance.

VI. HIGHER-ORDER EFFECTS

As frequently emphasized, these calculations of the effective polynomial are valid to all orders in $e^2\theta^2$ but only to lowest order in e^2. Suppose we wish to locate the critical temperature more exactly, or try to determine the precise nature of the phase transition. How would we go about it?

The introduction of the counterterm (3.7) had the purpose of restoring the validity of perturbation theory for $e^2\ll 1$ at temperatures of order $\mathfrak{M}/e$. As long as perturbation theory is valid, we can infer the symmetry properties of the theory by a study of its lowest-order terms, i.e., of $P_{\rm eff}(\phi)$. However, even with the counterterm (3.7) working to cancel the $e^2\theta^2$ terms, the perturbation theory still breaks down when θ is very near θ_c, because in this case one or more of the mass terms in $P_{\rm eff}(\phi)$ becomes very small, and so powers of e can be canceled by factors which become infrared-divergent for $\theta=\theta_c$.

At first sight, this problem appears very similar to that studied by Coleman and E. Weinberg.[8] Instead of adjusting the temperature so that the mass term in an effective polynomial vanishes, they considered a relativistic quantum field theory at zero temperature, with the bare scalar masses adjusted so that the renormalized scalar masses vanish. In order to decide whether the symmetries of the theory were spontaneously broken, they searched for the minima of an effective potential, the first term of which is just $P(\phi)$, using renormalization-group methods to sum up the logarithms associated with infrared divergences.

The difference here is that the infrared divergences are profoundly affected by a finite temperature.[22] According to the Feynman rules discussed in Sec. II, the only denominators which can ever vanish are those in boson propagators carrying zero energy. Since the energy is a discrete variable, the degree of infrared divergence must be determined counting *three* rather than four powers of momentum for each loop.

Hence the infrared divergences are those expected in a superrenormalizable rather than an ordinary renormalizable theory, and the appropriate methods needed here are those of Wilson[23] rather than those of Coleman and Weinberg.[8] The difference between our problem and that studied by Wilson is that we do not have a cutoff; ultraviolet divergences are eliminated by the renormalization procedure.

Without pursuing this problem too far, we can at least estimate how close the actual critical temperature, at which a broken symmetry is restored, is to the approximate critical temperature θ_c calculated in Sec. IV. For temperatures well below or well above θ_c (say, $|\theta-\theta_c|>\frac{1}{2}\theta_c$) perturbation theory is presumably valid [with the counterterm (3.7) canceling the $e^2\theta^2$ terms], so the symmetry of the theory must be just that of the lowest-order terms. On the other hand, when $\theta-\theta_c$ is small, the mass terms $\mathfrak{M}(\theta)$ in $P_{\text{eff}}(\phi)$ vanish, with

$$\mathfrak{M}(\theta)\approx[e^2(\theta^2-\theta_c^2)]^{1/2}\approx e\theta_c^{1/2}(\theta-\theta_c)^{1/2}.$$

Our order-of-magnitude analysis at the beginning of Sec. III is then valid only if we replace the characteristic mass $\mathfrak{M}$, which is roughly $\mathfrak{M}(0)$, with the much smaller quantity $\mathfrak{M}(\theta)$. In particular, if we consider a Green's function all of whose external lines are zero-energy bosons with momenta of order $\mathfrak{M}(\theta)$, and allow as internal lines only zero-energy bosons with masses of order $\mathfrak{M}(\theta)$, then each loop contributes a factor e^2, a factor θ_c from Eq. (2.17), and, since the dimensionality of all graphs must be the same, also a factor $\mathfrak{M}^{-1}(\theta)$. The condition for the validity of perturbation theory is therefore that

$$e^2\left(\frac{\theta_c}{\mathfrak{M}(\theta)}\right)\ll 1,$$

or in other words

$$|\theta-\theta_c|\gg e^2\theta_c.$$

The true critical temperature is therefore expected to lie somewhere in the range

$$|\theta-\theta_c|\lesssim e^2\theta_c.$$

However, to locate it more precisely in this range, or to determine whether the phase transition is of first or second order,[24] we would need to carry out a renormalization-group analysis beyond the scope of the present article.

VII. OBSERVABILITY OF THE PHASE TRANSITION IN GAUGE THEORIES

In gauge theories, the mean value of the field ϕ_i is not an observable, because it is not gauge-invariant. Indeed, we saw in Sec. V that even the quadratically divergent and $e^2\theta^2$ terms in $\langle\phi_i\rangle$ are different in the unitarity gauge and the renormalizable ξ gauges, though they are the same in all the renormalizable ξ gauges. A suspicion may therefore arise as to the reality of the phase transition, which seems to occur when the invariance group of the lowest-order term λ_i in $\langle\phi_i\rangle$ suddenly expands or contracts. Could physical measurements really reveal a different symmetry group just above the critical temperature than just below it?

To some extent, this question already arises in gauge field theories at zero temperature. There, also, one cannot rely on the properties of Green's functions of gauge-noninvariant operators to tell us which symmetries are broken and which are not. However, at zero temperature these Green's functions have poles whose residues, the S-matrix elements, are gauge-invariant, and can be used to diagnose the symmetries of the theory. At finite temperature there is no such thing as a single collision (particles interact with the thermal background on their way into and out of any encounter), so there are no S-matrix elements as such, and this approach fails us. In particular, we cannot *directly* measure the particle masses discussed in Sec. IV.

We can, however, presumably measure the partition function in the presence of various gauge-invariant perturbing operators of the form $\phi_i\phi_i$, $F_{\alpha\mu\nu}F_\alpha{}^{\mu\nu}$, $\bar{\psi}\psi$, etc., and from these measurements we can infer values for the temperature Green's functions (2.11) for these operators. It is not immediately obvious, though, that we can use such Green's functions to learn about λ values and masses, because the gauge-invariant operators are necessarily at least bilinear, and so the lowest-order graphs for their Green's functions are not trees but loops, and we cannot adjust the four-momenta carried by internal lines of such loops to arbitrary values. In particular, if the typical energy or momentum carried by the internal lines of the loop is of order θ, then the insertion of a quadratically divergent subgraph of order $e^2\theta^2$ in an internal line of the loop produces two more powers of θ in the denominator, so the over-all effect is to introduce a factor e^2. From this standpoint, it does not seem that the "failure" of perturbation theory, which forced us to introduce the effective polynomial in Sec. III, is real at all.

The answer is that we must consider not the Green's functions of arbitrary gauge-invariant operators, but the Green's functions of gauge-invariant operators carrying moderate momenta ($P\approx\mathfrak{M}$ rather than $P\approx\theta$) and *zero* energy. As long as there are enough operators so that the over-all dimensionality D of the Green's function is less than 1, the leading lowest-order terms for $\theta \gg \mathfrak{M}$ will be the graphs in which all internal lines have zero energy and moderate momenta. (As discussed in Sec. III, the temperature dependence of such terms in entirely contained in the single multiplicative factor θ; all other terms are then suppressed by factors of order $\theta/\mathfrak{M}$.) These leading graphs will be sensitive functions of the masses of their internal lines, and can be used to study the symmetries of the theory. In particular, if we did not introduce our effective polynomial, then the insertion of quadratically divergent subgraphs in the zero-energy internal lines of such leading graphs would introduce factors of $e^2\theta^2/\mathfrak{M}^2$ rather than e^2, and hence would lead to a breakdown of perturbation theory for $\theta \approx \mathfrak{M}/e$. The masses and λ values which can be inferred from a study of zero-energy gauge-invariant Green's functions are therefore the ones derived from our temperature-dependent effective polynomial, and these are the ones which show the phase transition.

Of course, all we can ever measure in this way are invariant quantities, such as $\lambda_i\lambda_i$, $\mathrm{Tr}[M^2]$, $\mathrm{Tr}[\mu^2]$, etc. However, these can easily be used to tell whether the symmetries of the theory are broken. For instance, if a symmetry requires that λ_i or μ vanish, or that M^2 be proportional to the unit matrix, and our measurements reveal that $\lambda_i\lambda_i$ or $\mathrm{Tr}[\mu^2]$ is not zero, or that $\mathrm{Tr}[M^4]$ is not equal to $(\mathrm{Tr}[M^2])^2$ then we know the symmetry is broken, although of course we never find out in which direction the symmetry breaking occurs.

It is also possible to infer the existence of a phase transition by studying the partition function itself. The lowest-order graph is the "no-loop" term

$$[\ln Q]_{\text{no loop}} = -\frac{\Omega}{\theta} P_{\text{eff}}(\lambda), \tag{7.1}$$

with Ω the volume of the system, and P_{eff} evaluated at its minimum. The one-loop corrections involve λ-dependent terms which are canceled by the last term in Eq. (3.7), plus λ-independent terms of order θ^4 and $\mathfrak{M}^2\theta^2$. The latter terms are quite large, and in a sense represent a breakdown of perturbation theory which has so far been ignored because it occurs only in $\ln Q$ rather than in the Green's functions. [These terms are calculated for scalar field theories in the Appendix, and included there in P_{eff}; see Eq. (A36).] However, Eq. (7.1) correctly represents the leading terms in the part of $\ln Q$ which is λ-dependent and hence nonanalytic at the critical temperature. We will therefore rewrite Eq. (7.1) as

$$\{\ln Q\}_{\text{NA}} = -\frac{\Omega}{\theta} P_{\text{eff}}(\lambda), \tag{7.2}$$

with the subscript NA denoting the nonanalytic part. Familiar thermodynamic arguments then give the nonanalytic part of the pressure

$$\{P\}_{\text{NA}} = -P_{\text{eff}}(\lambda), \tag{7.3}$$

the entropy density

$$\{s\}_{\text{NA}} = -\frac{\partial P_{\text{eff}}(\lambda)}{\partial\theta}, \tag{7.4}$$

and the energy density

$$\{u\}_{\text{NA}} = P_{\text{eff}}(\lambda) - \theta\frac{\partial P_{\text{eff}}(\lambda)}{\partial\theta}. \tag{7.5}$$

To see how this works in practice, let us return to example 2 of Sec. IV. The effective polynomial was

$$P_{\text{eff}}(\phi) = \tfrac{1}{2}\mathfrak{M}^2(\theta)\phi_i\phi_i + \tfrac{1}{4}e^2(\phi_i\phi_i)^2,$$

where

$$\mathfrak{M}^2(\theta) = \mathfrak{M}^2(0) + \tfrac{1}{12}(n+2)e^2\theta^2 + \tfrac{1}{4}(n-1)e'^2\theta^2.$$

For $\theta < \theta_c$, the value of the polynomial at its minimum is

$$\begin{aligned} P_{\text{eff}}(\lambda) &= -\frac{1}{4e^2}\mathfrak{M}^4(\theta) \\ &= -\frac{1}{4e^2}\left[\tfrac{1}{12}(n+2)e^2 + \tfrac{1}{4}(n-1)e'^2\right]^2(\theta^2-\theta_c^2)^2. \end{aligned}$$

On the other hand, for $\theta > \theta_c$, the minimum of the polynomial is at $\phi_i = 0$, where

$$P_{\text{eff}}(0) = 0.$$

Thus the pressure and the energy and entropy densities are continuous at $\theta = \theta_c$, but their derivatives are not. In particular, the specific heat shows a discontinuity:

$$\begin{aligned} \Delta C_V &\equiv \left(\frac{\partial u}{\partial\theta}\right)_{\theta=\theta_c+\epsilon} - \left(\frac{\partial u}{\partial\theta}\right)_{\theta=\theta_c-\epsilon} \\ &= \theta_c\left[\frac{\partial^2 P_{\text{eff}}(\lambda)}{\partial\theta^2}\right]_{\theta=\theta_c-\epsilon} \\ &= -\frac{\theta_c^3}{e^2}\left[\tfrac{1}{12}(n+2)e^2 + \tfrac{1}{4}(n-1)e'^2\right]^2. \end{aligned}$$

This certainly reveals the presence of a phase transition, though we need to go beyond thermodynamics to determine what symmetries are restored in this transition.

APPENDIX: OPERATOR APPROACH TO THE POTENTIAL

In order to take account of the higher-order effects discussed in Sec. VI, we need to generalize the effective polynomial by defining a potential,[8] the symmetries of whose minima will truly determine the symmetries of the theory. In this appendix, I will describe an operator approach to this problem, which incidentally leads to an interesting interpretation of the radiative corrections in field theories at zero temperature in terms of the zero-point energies of the various degrees of freedom. Unfortunately, it will be seen that this operator approach breaks down badly in gauge theories.

We will consider a system with a large but finite volume Ω, and define the spatial average of the Schrödinger representation operator $\phi(\vec{x})$ as

$$\bar{\phi}_i \equiv \frac{1}{\Omega}\int_\Omega d^3x\, \phi_i(\vec{x}). \tag{A1}$$

In order to allow us to vary the thermodynamic mean value of $\bar{\phi}_i$, we include in the Hamiltonian a perturbation $\Omega J_i \bar{\phi}_i$, with J_i a variable c-number "current." The mean value of $\bar{\phi}_i$ is then

$$\eta_i(J) \equiv \langle \bar{\phi}_i \rangle_J = \frac{\mathrm{Tr}[\bar{\phi}_i \exp\{-(1/\theta)(H+\Omega J_i \bar{\phi}_i)\}]}{\mathrm{Tr}[\exp\{-(1/\theta)(H+\Omega J_i\bar{\phi}_i)\}]}\,. \tag{A2}$$

This may be expressed in terms of the Helmholtz free energy per unit volume

$$A(J) \equiv -\frac{\theta}{\Omega}\ln \mathrm{Tr}[\exp\{-(1/\theta)(H+\Omega J_i\bar{\phi}_i)\}] \tag{A3}$$

as

$$\eta_i(J) = \frac{\partial A(J)}{\partial J_i}\,. \tag{A4}$$

(Of course, in addition to its dependence on J_i, the free energy also depends on θ and Ω and on any other parameters appearing in H.) Our potential is defined as a function of η rather than J, using a Legendre transformation to introduce the analog of the Gibbs free energy:

$$V(\eta) \equiv A(J) - J_i \eta_i\,. \tag{A5}$$

Using (A4), we may now define J as a function of η by the condition

$$\frac{\partial V(\eta)}{\partial \eta_i} = -J_i(\eta). \tag{A6}$$

In particular, the possible mean values of $\bar{\phi}_i$ when the current vanishes are given by the points where $V(\eta)$ is stationary,

$$\frac{\partial V(\eta)}{\partial \eta_i} = 0 \text{ if } J_i = 0, \tag{A7}$$

so $V(\eta)$ is a suitable potential for our purposes. In fact, the actual mean value of $\bar{\phi}_i$ for $J=0$ must be at a point where $V(\eta)$ is not only stationary but also a local *minimum*, because we can easily show that

$$\frac{\partial^2 V(\eta)}{\partial\eta_i \partial\eta_j} = \frac{\theta}{\Omega}\Delta^{-1}{}_{ij}(\eta),$$

with Δ the positive matrix

$$\Delta_{ij}(\eta) \equiv \langle(\bar{\phi}_i - \eta_i)(\bar{\phi}_j - \eta_j)\rangle_J\,.$$

The function $V(\eta)$ must be defined by analytic continuation at η values where $\partial^2 V/\partial\eta_i\partial\eta_j$ is not positive, because no current J can produce such η values in a state of thermal equilibrium.

So far, this has been quite general. Let us now consider the simple scalar field theory described by the Lagrangian (2.1). The Hamiltonian here is

$$H = \int d^3x\,[\tfrac{1}{2}\pi_i \pi_i + \tfrac{1}{2}\vec{\nabla}\phi_i \cdot \vec{\nabla}\phi_i + P(\phi)], \tag{A8}$$

where π_i is the canonical conjugate to ϕ_i. As in Sec. II, we assume the parameters in $P(\phi)$ to be characterized by a typical mass $\mathfrak{M}$ and a typical dimensionless coupling $e \ll 1$, in the sense of Eq. (2.6). We will construct a perturbative expansion for $V(\eta)$ in powers of e, for temperatures θ ranging from 0 to order $\mathfrak{M}/e$.

First, in order to cancel the divergences and the attendant temperature-dependent terms in $V(\eta)$, we must again introduce a polynomial counterterm, writing

$$H = \int d^3x\,[\tfrac{1}{2}\pi_i \pi_i + \tfrac{1}{2}\vec{\nabla}\phi_i \cdot \vec{\nabla}\phi_i + P_{\text{eff}}(\phi) - \Delta P(\phi)], \tag{A9}$$

where

$$P_{\text{eff}}(\phi) = P(\phi) + \Delta P(\phi), \tag{A10}$$

with ΔP a quartic temperature-dependent polynomial to be constructed as we go along. To the order we will be studying here, it will be sufficient to treat $\Delta P(\phi)$ as a quantity of zeroth order in e for $\phi \approx \eta$; in higher order we would need to include other renormalization counterterms, including higher terms in ΔP.

Next, since we are interested in values of the variable η_i of order $\mathfrak{M}/e$ [compare Eq. (2.9)] we must shift ϕ_i, defining a new field ϕ_i' by

$$\phi_i \equiv \phi_i' + \eta_i\,. \tag{A11}$$

The Hamiltonian may now be written as a sum of terms $H^{(n)}$ of order n in e:

$$H = H^{(-2)} + H^{(-1)} + H^{(0)} + H^{(1)} + H^{(2)}, \tag{A12}$$

where

$$H^{(-2)} = \Omega P_{\text{eff}}(\eta), \tag{A13}$$

$$H^{(-1)} = \frac{\partial P_{\text{eff}}(\eta)}{\partial \eta_i}\int_\Omega d^3x\, \phi_i', \tag{A14}$$

$$H^{(0)} = \int_\Omega d^3x \left[\tfrac{1}{2}\pi_i\pi_i + \tfrac{1}{2}\vec{\nabla}\phi_i'\cdot\vec{\nabla}\phi_i' + \frac{1}{2}\frac{\partial^2 P_{\rm eff}(\eta)}{\partial\eta_i\partial\eta_j}\phi_i'\phi_j'\right] - \Omega\Delta P(\eta), \tag{A15}$$

$$H^{(1)} = \frac{1}{6}\frac{\partial^3 P_{\rm eff}(\eta)}{\partial\eta_i\partial\eta_j\partial\eta_k}\int_\Omega d^3x\,\phi_i'\phi_j'\phi_k' - \frac{\partial\Delta P(\eta)}{\partial\eta_i}\int_\Omega d^3x\,\phi_j', \tag{A16}$$

$$H^{(2)} = \frac{1}{24}\frac{\partial^4 P_{\rm eff}(\eta)}{\partial\eta_i\partial\eta_j\partial\eta_k\partial\eta_l}\int_\Omega d^3x\,\phi_i'\phi_j'\phi_k'\phi_l' - \frac{\partial^2\Delta P(\eta)}{\partial\eta_i\partial\eta_j}\int_\Omega d^3x\,\phi_i'\phi_j'. \tag{A17}$$

We shall attempt to calculate $V(\eta)$ as a sum of terms $V^{(n)}(\eta)$ of order n in e:

$$V(\eta) = \sum_n V^{(n)}(\eta), \tag{A18}$$

using the formula

$$V(\eta) = -\frac{\theta}{\Omega}\ln\mathrm{Tr}\left[\exp\left\{-\frac{1}{\theta}\left(H + J_i\int_\Omega d^3x\,\phi_i'\right)\right\}\right] \tag{A19}$$

[see Eqs. (A5) and (A3)]. The current here is given by Eq. (A6), so it may also be expanded as a sum of terms $J_i^{(n)}$ of order n in e:

$$J_i(\eta) = \sum_n J_i^{(n)}(\eta), \tag{A20}$$

with

$$J_i^{(n)}(\eta) = \frac{\partial V^{(n-1)}(\eta)}{\partial\eta_i}. \tag{A21}$$

(Recall that η is of order $\mathfrak{M}/e$.) Our calculation will therefore be recursive: Given $V^{(n)}(\eta)$, we use (A21) to calculate $J_i^{(n+1)}(\eta)$, and then insert the result back in Eq. (A19) to determine $V^{(n+1)}(\eta)$.

To start this recursive calculation, we tentatively assume that $J^{(n)}$ vanishes for $n \leqslant -2$, so (A19) gives the leading term in $V(\eta)$ as

$$V^{(-2)}(\eta) = -\frac{\theta}{\Omega}\ln\mathrm{Tr}\left[\exp\left(\frac{-H^{(-2)}}{\theta}\right)\right] = P_{\rm eff}(\eta). \tag{A22}$$

This can be used as in Sec. IV to study the symmetries of the theory to lowest order in e, but we do not yet know $Q_{ij}(\theta)$.

To go to the next order, we use Eq. (A21) to determine

$$J_i^{(-1)}(\eta) = -\frac{\partial P_{\rm eff}(\eta)}{\partial\eta_i}. \tag{A23}$$

Thus in first order the current term (A23) cancels the Hamiltonian (A14) in (A19), and therefore

$$V^{(-1)}(\eta) = 0. \tag{A24}$$

In the next order, Eq. (A21) gives

$$J_i^{(0)}(\eta) = 0, \tag{A25}$$

so from Eq. (A19)

$$V^{(0)}(\eta) = -\frac{\theta}{\Omega}\ln\mathrm{Tr}\left[\exp\left(\frac{-H^{(0)}}{\theta}\right)\right]. \tag{A26}$$

Aside from the c-number term in (A15), this is nothing but the free energy per unit volume of a noninteracting mixture of ideal Bose gases. It may therefore be written as a sum,

$$V^{(0)}(\eta) = -\Delta P(\eta) + \mathrm{Tr}[G(M^2(\eta))], \tag{A27}$$

where $G(M^2)$ is the free energy per unit volume of an ideal Bose gas with mass M,

$$G(M^2) = -\frac{\theta}{\Omega}\ln\prod_{\vec{k}}\sum_{N=0}^{\infty}\exp[-(N+\tfrac{1}{2})(\vec{k}^2+M^2)^{1/2}/\theta], \tag{A28}$$

and $M^2{}_{ij}(\eta)$ is the mass matrix in (A15)

$$M^2{}_{ij}(\eta) \equiv \frac{\partial^2 P_{\rm eff}(\eta)}{\partial\eta_i\partial\eta_j}. \tag{A29}$$

[Note that we are keeping the zero-point energy in the exponential in Eq. (A28), a point that will be of some importance later on.] Passing to the limit of infinite volume and performing the sum over N in (A28), we find

$$G(M^2) = \frac{1}{(2\pi)^3}\int d^3k\left(\tfrac{1}{2}(\vec{k}^2+M^2)^{1/2} + \theta\ln\{1-\exp[-(1/\theta)(\vec{k}^2+M^2)^{1/2}]\}\right). \tag{A30}$$

Before carrying out our renormalization, it is useful to compare our results so far with those of relativistic field theory. At zero temperature, Eqs. (A22), (A24), (A27), and (A30) give

$$V(\eta) \simeq P(\eta) + \frac{1}{2(2\pi)^3}\mathrm{Tr}\left[\int d^3k\,(\vec{k}^2+M^2(\eta))^{1/2}\right].$$

If $P(\eta)$ has a minimum at $\eta=\lambda$, then the true mean value of $\overline{\phi}_i$ for $J=0$ is determined by the condition

$$0 = \left.\frac{\partial V(\eta)}{\partial\eta_i}\right|_{\eta=\langle\overline{\Phi}\rangle} \simeq M^2{}_{ij}[\langle\overline{\phi}_j\rangle - \lambda_j] + \frac{1}{4(2\pi)^3}f_{ijk}\int d^3k\,(\vec{k}^2+M^2)^{-1/2}{}_{jk},$$

where

$$M^2_{ij}=\frac{\partial^2 P(\lambda)}{\partial\lambda_i\partial\lambda_j}\,,\quad f_{ijk}=\frac{\partial^3 P(\lambda)}{\partial\lambda_i\partial\lambda_j\partial\lambda_k}\,.$$

Therefore,

$$\langle\overline{\phi}_i\rangle=\lambda_i-\tfrac{1}{4}(2\pi)^{-3}M^{-2}{}_{il}f_{ljk}\int d^3k(\vec{k}^2+M^2)^{-1/2}{}_{jk}\,.$$

On the other hand, Eq. (3.8) shows that the mean value of the scalar field calculated by Feynman-diagram methods is

$$\begin{aligned}\langle\overline{\phi}_i\rangle&=\lambda_i-i(2\pi)^{-4}M^{-2}{}_{il}T_l\\&=\lambda_i+\tfrac{1}{2}i(2\pi)^{-3}M^{-2}{}_{il}f_{ljk}\\&\quad\times\int d^4k(k^2+M^2-i\epsilon)^{-1}{}_{jk}\,.\end{aligned}$$

By performing the k^0 integration, we easily see that this agrees with the result obtained by operator methods above. Thus *the one-loop diagrams simply represent the contribution of zero-point energies to the total free energy.*

Returning now to the main line of our calculation, we note that the first term of the free-energy function (A30) may be written

$$\frac{1}{2(2\pi)^3}\int d^3k(\vec{k}^2+M^2)^{1/2}=G_\infty(M^2)+\frac{1}{64\pi^2}M^4\ln M^2\,,\tag{A31}$$

where $G_\infty(M^2)$ is a quadratic polynomial in M^2 with divergent coefficients. Then $\mathrm{Tr}G_\infty(M^2(\phi))$ is a quartic polynomial in ϕ which satisfies all the symmetry requirements imposed on $P(\phi)$, so we can adjust the parameters in $P(\phi)$ to cancel this term, leaving over a finite-temperature-independent renormalized polynomial

$$P_{\rm ren}(\phi)\equiv P(\phi)+\mathrm{Tr}[G_\infty(M^2(\phi))]\,.\tag{A32}$$

For "moderate" temperatures, say $\theta\lesssim\mathfrak{M}$, the only counterterm we need is just the term $\mathrm{Tr}[G_\infty]$ in (A32), and we can take $P_{\rm ren}$ as our effective polynomial. Equations (A22), (A24), (A27), and (A30)–(A32) then give the potential as

$$V(\eta)\simeq P_{\rm ren}(\eta)+\mathrm{Tr}[G_1(M^2(\eta))]\,,\tag{A33}$$

with the function G_1 defined by

$$\begin{aligned}G_1(M^2)&\equiv G(M^2)-G_\infty(M^2)\\&=\frac{1}{64\pi^2}M^4\ln M^2\\&\quad+\frac{\theta}{(2\pi)^3}\int d^3k\ln\{1-\exp[-(1/\theta)(\vec{k}^2+M^2)^{1/2}]\}.\end{aligned}\tag{A34}$$

For $\theta\approx\mathfrak{M}$ the correction term $\mathrm{Tr}[G_1]$ is of order $\mathfrak{M}^4$, while $P_{\rm ren}(\eta)$ is of order $\mathfrak{M}^4/e^2$, so it is the temperature-independent polynomial $P_{\rm ren}$ that governs the pattern of broken symmetries.

On the other hand, for "high" temperatures, say $\theta\approx\mathfrak{M}/e$, the free energy (A30) becomes

$$G(M^2)\simeq G_\infty(M^2)-\tfrac{1}{90}\pi^2\theta^4+\tfrac{1}{24}\theta^2M^2\qquad(\text{for }\theta\gg M).\tag{A35}$$

All these terms are at least as large as $P_{\rm ren}(\eta)$, so they all must be included in the effective polynomial, which now becomes

$$P_{\rm eff}(\phi)=P_{\rm ren}(\phi)-\tfrac{1}{90}B\pi^2\theta^4+\tfrac{1}{24}\theta^2\,\mathrm{Tr}[M^2(\phi)]\,,\tag{A36}$$

where B is the number of boson fields. By expanding in powers of ϕ, we find

$$\mathrm{Tr}[M^2(\phi)]=\tfrac{1}{2}f_{ijkk}\phi_i\phi_j+\text{constant},\tag{A37}$$

so Eq. (A36) is the same as our previous result (3.26), except for a constant (the Stefan-Boltzmann term), which of course has no effect on the location of the minima of $P_{\rm eff}$. We might try to improve this calculation by including the terms of order M^4 in (A36), but this improvement would be illusory; the "second-order" term $V^{(2)}(\eta)$ in the potential includes η-dependent contributions of order $e^2\theta^2\mathfrak{M}^2\approx\mathfrak{M}^4$, which are just as large as the corrections to Eq. (A36). In any case, even if we took these effects into account and correctly calculated all terms in V of order $\mathfrak{M}^4$, we still would not be able to calculate the behavior of the minimum of the potential near the critical temperatures, where one of the eigenvalues of $M^2(\eta)$ vanishes.

There is no problem in including fermions in this sort of calculation. Rather than go into such inessential complications here, let us turn immediately to the more challenging problem of a gauge theory described by the Lagrangian (2.4), leaving fermions aside for simplicity.

We need first to construct a Hamiltonian for this theory. To the best of my knowledge, this has so far been possible only in the "unitarity" gauges, defined by the condition that ϕ should have no components along the directions $\theta_\alpha\lambda$

$$\phi_i(\theta_\alpha\lambda)_i=0,\tag{A38}$$

where λ is some fixed vector. It is usual to choose λ to be a vector at which $P(\phi)$ is stationary,

$$\left.\frac{\partial P(\phi)}{\partial\phi_i}\right|_{\phi=\lambda}=0,$$

because then the directions $(\theta_\alpha\lambda)$ define the eigenvectors of the mass matrix with eigenvalue zero,

$$\left.\frac{\partial^2 P(\phi)}{\partial\phi_i\partial\phi_j}\right|_{\phi=\lambda}(\theta_\alpha\lambda)_j=0,$$

and the condition (A38) just amounts to the exclusion of Goldstone bosons from the theory. However, we shall leave λ arbitrary here.

The dynamical variables in the gauge (A38) are the spatial components $\vec{A}_\alpha$ of the gauge fields and the scalar field components

$$\phi_a = n_{ai}\phi_i \,, \tag{A39}$$

where n_{ai} forms a complete orthonormal set of vectors orthogonal to all $\theta_\alpha\lambda$:

$$n_{ai}n_{bi} = \delta_{ab}\,, \quad n_{ai}(\theta_\alpha\lambda)_i = 0\,. \tag{A40}$$

The Hamiltonian in this gauge is given by the unpleasant-looking formula[7]

$$\begin{aligned} H = \int d^3x\,\{&\tfrac{1}{2}\omega^{-1}{}_{\alpha\beta}(\phi)[\vec{\nabla}\cdot\vec{P}_\alpha - C_{\alpha\gamma\delta}\vec{P}_\delta\cdot\vec{A}_\gamma + i(\theta_\alpha)_{ab}\pi_a\phi_b][\vec{\nabla}\cdot\vec{P}_\beta - C_{\beta\epsilon\zeta}\vec{P}_\zeta\cdot\vec{A}_\epsilon + i(\theta_\beta)_{cd}\pi_c\phi_d] \\ &+\tfrac{1}{2}\vec{P}_\alpha\cdot\vec{P}_\alpha + \tfrac{1}{2}\pi_a\pi_a + \tfrac{1}{2}(\vec{\nabla}\times\vec{A}_\alpha - C_{\alpha\beta\gamma}\vec{A}_\beta\times\vec{A}_\gamma)\cdot(\vec{\nabla}\times\vec{A}_\alpha - C_{\alpha\delta\epsilon}\vec{A}_\delta\times\vec{A}_\epsilon) \\ &+\tfrac{1}{2}\vec{\nabla}\phi_a\cdot\vec{\nabla}\phi_a + i\vec{A}_\alpha\cdot\vec{\nabla}\phi_a(\theta_\alpha)_{ab}\phi_b + \tfrac{1}{2}\vec{A}_\alpha\cdot\vec{A}_\beta[\omega_{\alpha\beta}(\phi) - (\theta_\alpha)_{ac}(\theta_\beta)_{bc}\phi_a\phi_b] + P(\phi)\}\,, \end{aligned} \tag{A41}$$

where $\vec{P}_\alpha$ and π_a are the "momenta" canonically conjugate to A_α and ϕ_α, and

$$(\theta_\alpha)_{ab} \equiv n_{ai}n_{bj}(\theta_\alpha)_{ij}\,, \tag{A42}$$

$$\omega_{\beta\delta}(\phi) \equiv (\theta_\beta\phi)_i(\theta_\alpha\lambda)_i(\nu^{-2})_{\alpha\gamma}(\theta_\gamma\lambda)_j(\theta_\delta\phi)_j\,, \tag{A43}$$

$$\nu^{-2}{}_{\alpha\beta} \equiv -(\theta_\alpha\lambda)_i(\theta_\beta\lambda)_i\,. \tag{A44}$$

At this point we face a problem. If we identify λ with the argument η of the potential V, then changes in λ will change V not only directly, through changes in the current $J(\lambda)$, but *also through changes in the choice of gauge.* The λ dependence of the Hamiltonian would invalidate the general formalism used here; indeed we already know[25] that the vacuum expectation value of ϕ_i at zero temperature in the one-loop approximation is not given by the minimum of *any* potential depending on the single variable λ. To avoid this problem, we fix λ_i and our choice of gauge once and for all,[26] and use as the independent variables in the potential the thermodynamic mean values η_a of the fields ϕ_a in this gauge.

With this understanding, our general formalism becomes applicable again. We define a shifted scalar field

$$\phi'_a = \phi_a - \eta_a\,, \tag{A45}$$

with η_a of order $\mathfrak{M}/e$. The Hamiltonian can then be written as a sum of terms $H^{(n)}$ of order e^n:

$$H = H^{(-2)} + H^{(-1)} + H^{(0)} + H^{(1)} + H^{(2)}, \tag{A46}$$

where, up to zeroth order,

$$H^{(-2)} = \Omega P_{\text{eff}}(\eta), \tag{A47}$$

$$H^{(-1)} = \frac{\partial P_{\text{eff}}(\eta)}{\partial\eta_i}\int d^3x\,\phi'_i\,, \tag{A48}$$

$$\begin{aligned} H^{(0)} = \int d^3x\,\{&\tfrac{1}{2}\omega^{-1}{}_{\alpha\beta}(\eta)[\vec{\nabla}\cdot\vec{P}_\alpha + i(\theta_\alpha)_{ab}\pi_a\eta_b][\vec{\nabla}\cdot\vec{P}_\beta + i(\theta_\beta)_{cd}\pi_c\eta_d] + \tfrac{1}{2}\vec{P}_\alpha\cdot\vec{P}_\alpha + \tfrac{1}{2}\pi_a\pi_a + \tfrac{1}{2}(\vec{\nabla}\times\vec{A}_\alpha)\cdot(\vec{\nabla}\times\vec{A}_\alpha) \\ &+\tfrac{1}{2}\vec{\nabla}\phi'_a\cdot\vec{\nabla}\phi'_a + i\vec{A}_\alpha\cdot\vec{\nabla}\phi'_a(\theta_\alpha)_{ab}\eta_b + \tfrac{1}{2}\mu^2{}_{\alpha\beta}(\eta)\vec{A}_\alpha\cdot\vec{A}_\beta + \tfrac{1}{2}M^2{}_{ab}(\eta)\phi'_a\phi'_b\} - \Omega\Delta P(\eta). \end{aligned} \tag{A49}$$

with

$$P_{\text{eff}}(\phi) = P(\phi) + \Delta P(\phi), \tag{A50}$$

$$\mu^2{}_{\alpha\beta}(\eta) \equiv \omega_{\alpha\beta}(\eta) - (\theta_\alpha)_{ac}(\theta_\beta)_{bc}\eta_a\eta_b\,, \tag{A51}$$

$$M^2{}_{ab}(\eta) = \frac{\partial^2 P_{\text{eff}}(\eta)}{\partial\eta_a\partial\eta_b}\,. \tag{A52}$$

[In evaluating $\omega(\eta)$ and $P(\eta)$ and its derivatives, we set η_i equal to $n_{ai}\eta_i$.] For future reference, we note that since the n_{ai} span the space orthogonal to the $\theta_\alpha\lambda$, they satisfy the sum rule

$$n_{ai}n_{aj} = \delta_{ij} + (\theta_\alpha\lambda)_i\,\nu^{-2}{}_{\alpha\beta}(\theta_\beta\lambda)_j$$

[see (A44)]. Contracting this with $(\theta_\gamma\eta)_i(\theta_\delta\eta)_j$, we find

$$(\theta_\alpha)_{ab}(\theta_\delta)_{ac}\eta_b\eta_c = (\theta_\gamma\eta)_i(\theta_\delta\eta)_i + \omega_{\gamma\delta}(\eta),$$

and therefore

$$\mu^2{}_{\alpha\beta}(\eta) = -(\theta_\alpha\eta)_i(\theta_\beta\eta)_i\,.$$

Following precisely the same reasoning as for the scalar field theory discussed above, we find here for the potential up to zeroth order in e:

$$V(\eta) \simeq P_{\text{eff}}(\eta) - (\theta/\Omega)\ln \mathrm{Tr}[\exp(-H^{(0)}/\theta)]\,. \tag{A53}$$

The calculation of the second term is now not entirely straightforward, because $H^{(0)}$ is not in the familiar form of a free-particle Hamiltonian. In order to bring it to this form, we must first perform a canonical transformation to a new set of canonical variables $\vec{a}_\alpha, \Phi_a$ and their conjugates $\vec{p}_\alpha, \Pi_a$:

$$\vec{a}_\alpha = \vec{A}_\alpha + i(\mu^{-2}(\eta))_{\alpha\beta}(\theta_\beta)_{ab}\eta_b\vec{\nabla}\phi'_a\,, \tag{A54}$$

$$\vec{p}_\alpha = \vec{P}_\alpha, \tag{A55}$$

$$\Phi_a = (S^{-1}(\eta))_{ab}\phi_b', \tag{A56}$$

$$\Pi_a = (S(\eta))_{ab}[\pi_b + i(\mu^{-2}(\eta))_{\alpha\beta}(\theta_\beta)_{bc}\eta_c \vec{\nabla}\cdot\vec{P}_\alpha], \tag{A57}$$

where $S(\eta)$ is any matrix such that

$$S_{ba}(\eta)S_{ca}(\eta) = \delta_{bc} - (\theta_\alpha)_{ba}\eta_a(\theta_\beta)_{cd}\eta_d\omega^{-1}{}_{\alpha\beta}(\eta). \tag{A58}$$

It is straightforward, though tedious, to check that this *is* a canonical transformation, and that it brings $H^{(0)}$ to the diagonalized form:

$$\begin{aligned} H^{(0)} = \int d^3x\,[&\tfrac{1}{2}\mu^{-2}{}_{\alpha\beta}(\eta)(\vec{\nabla}\cdot\vec{p}_\alpha)(\vec{\nabla}\cdot\vec{p}_\beta) + \tfrac{1}{2}\vec{p}_\alpha\cdot\vec{p}_\beta \\ &+\tfrac{1}{2}\Pi_a\Pi_a + \tfrac{1}{2}(\vec{\nabla}\times\vec{a}_\alpha)\cdot(\vec{\nabla}\times\vec{a}_\alpha) \\ &+\tfrac{1}{2}\vec{\nabla}\Phi_a\cdot\vec{\nabla}\Phi_a + \tfrac{1}{2}\mu^2{}_{\alpha\beta}(\eta)\vec{a}_\alpha\cdot\vec{a}_\beta \\ &+\tfrac{1}{2}\tilde{M}^2{}_{ab}(\eta)\Phi_a\Phi_b] - \Omega\Delta P(\eta), \end{aligned} \tag{A59}$$

where

$$\tilde{M}^2{}_{cd}(\eta) \equiv \frac{\partial^2 P_{\text{eff}}(\eta)}{\partial\eta_a\partial\eta_b}(S(\eta))_{ac}(S(\eta))_{bd}. \tag{A60}$$

We can now immediately write down the free energy:

$$\begin{aligned} V(\eta) \simeq P_{\text{eff}}(\eta) + 3\,\text{Tr}[G(\mu^2(\eta))] + \text{Tr}[G(\tilde{M}^2(\eta))] \\ -\Delta P(\eta), \end{aligned} \tag{A61}$$

where G is the function (A30).

At this point our calculation breaks down. The divergent part of the function $G(M^2)$ is a quadratic polynomial in M^2, so if $\tilde{M}^2(\eta)$ and $\mu^2(\eta)$ are quadratic polynomials in η the divergent parts of $\text{Tr}[G(\mu^2(\eta))]$ and $\text{Tr}[G(\tilde{M}^2(\eta))]$ will be quartic polynomials in η, and hence can be removed by the counterterm ΔP. However, although $\mu^2(\eta)$ is a quadratic polynomial in η, $\tilde{M}^2(\eta)$ is not [because of the matrices $S(\eta)$ in (A60)], and *therefore the infinite part of* $\text{Tr}[G(\tilde{M}^2(\eta))]$ *is not a quartic polynomial in* η *and cannot be removed by renormalization.*

The reason for this difficulty is not hard to find. In general, the only reason that we would have to believe that renormalization should work in a calculation based on the unitarity gauge is that the results must be equivalent to those obtained in one of the renormalizable gauges, such as the ξ gauges used here in Secs. II–IV. However, in defining the potential, we have perturbed the Hamiltonian by a term linear in the scalar fields in the unitarity gauge, and these scalar fields are nonpolynomial functions of the scalar fields of the renormalizable gauge. [For instance, in an O(2) gauge theory with one 2-vector scalar field multiplet, the single U-gauge scalar field ϕ is related to the two R-gauge scalar fields ϕ_1, ϕ_2 by $\phi = (\phi_1{}^2 + \phi_2{}^2)^{1/2}$.] From the point of view of the R gauge, the perturbed Hamiltonian corresponds to a manifestly nonrenormalizable interaction, so of course renormalization theory does not work, whether we use the R gauge or the U gauge. [Actually the divergent part of $\text{Tr}G(\tilde{M}^2(\eta))$ is a quartic polynomial in η in the special case where ϕ_i furnishes a representation transitive on the sphere, because in this case the matrix S_{ba} is the unit matrix. However, as discussed in Sec. V, even though we can calculate an effective potential in unitarity gauge in such simple theories, it is not particularly useful to do so.]

This analysis suggests two possible directions for construction of a suitable potential in gauge theories:

(a) Instead of perturbing the Hamiltonian by a linear function of the U-gauge scalar fields, we could use a quadratic or higher-order function which can be written as a gauge-invariant polynomial function of the R-gauge scalar fields. A preliminary analysis indicates that this would cure the nonrenormalizability we have found here, but the formalism needs further development.

(b) We could give up the operator methods altogether, and return to a diagrammatic R-gauge analysis along the lines of Secs. II–V. The difficulty here is in defining a suitable potential; it is known[25] that even at zero temperature, the tadpoles and other boson Green's functions with zero external four-momenta are not given by derivatives of any potential in any of the ξ gauges except the Landau ($\xi = \infty$) gauge. The solution would be to use either the Landau gauge or one of the ξ gauges defined by a fixed vector λ_i different from the argument η_i of the potential,[26] as in this appendix. The potential defined in this way would be gauge-dependent, because the perturbation $J_i\bar{\phi}_i$ is a gauge-dependent operator. However, the values of the potential at its local minima or maxima are gauge-independent, because J_i vanishes at these stationary points. Hence if there is a minimum of $V(\eta)$ with a value lower than $V(0)$ in one gauge, then there will be such a minimum in any gauge (although its position will generally be different), and the symmetries will definitely be broken.[27]

At any rate, the definition of a suitable potential is only a first step toward a solution of the real problem, the summation to all orders of the infrared divergences at the critical temperature.

ACKNOWLEDGMENTS

I am grateful for frequent valuable discussions throughout the course of this work with C. Bernard, S. Coleman, R. Jackiw, and P. Martin. I also wish to thank E. Brezin, A. Duncan, B. Harrington, J. R. Schrieffer, B. Vul, and K. Wilson for interesting conversations on this and related matters.

---

*Work supported in part by the National Science Foundation under Grant No. GP40397X.

[1]Y. Nambu and G. Jona-Lasinio, Phys. Rev. 122, 345 (1961). Also see W. Heisenberg, Z. Naturforsch. 14, 441 (1959).

[2]D. A. Kirzhnits and A. D. Linde, Phys. Lett. 42B, 471 (1972). Also see D. A. Kirzhnits, Zh. Eksp. Teor. Fiz. Pis'ma Red 15, 745 (1972) [JETP Lett. 15, 529 (1972)].

[3]The critical temperatures in various field theories have also been calculated, using rather different methods, by L. Dolan and R. Jackiw, this issue, Phys. Rev. D 9, 3320 (1974).

[4]For a general introduction and references to the original literature, see A. L. Fetter and J. D. Walecka, *Quantum Theory of Many-Particle Systems* (McGraw-Hill, New York, 1971), Chap. 7.

[5]S. Weinberg, Phys. Rev. Lett. 19, 1264 (1967). Also see A. Salam, in *Elementary Particle Theory: Relativistic Groups and Analyticity (Nobel Symposium No. 8)*, edited by N. Svartholm (Almqvist and Wiksells, Stockholm, 1968), p. 367.

[6]See, e.g., F. Jona and G. Shirane, *Ferroelectric Crystals* (Pergamon, Oxford, 1962), p. 280. Rochelle salt has a lower Curie point at −18 °C, below which the crystal is orthorhombic, and above which it is monoclinic.

[7]S. Weinberg, Phys. Rev. D 7, 1068 (1973).

[8]Potentials have been used for the study of broken symmetries in quantum field theories at zero temperature by J. Goldstone, A. Salam, and S. Weinberg [Phys. Rev. 127, 965 (1962)], G. Jona-Lasinio [Nuovo Cimento 34, 1790 (1964)], S. Coleman and E. Weinberg [Phys. Rev. D 7, 1888 (1973)], S. Weinberg (Ref. 25); J. M. Cornwall and R. E. Norton [Phys. Rev. D 8, 3338 (1973)], and others. Functional methods for the construction of potentials in gauge theories at zero temperature have recently been developed by R. Jackiw [Phys. Rev. D 9, 1686 (1974)] and have been applied to the problem of gauge symmetry breaking by L. Dolan and R. Jackiw [Phys. Rev. D 9, 2904 (1974)]. For a review of some of these developments, see S. Coleman, Lectures given at the 1973 International Summer School of Physics "Ettore Majorana" (unpublished).

[9]J. M. Cornwall, D. N. Levin, and G. Tiktopoulos, Phys. Rev. Lett. 30, 1268 (1973); 31, 572(E) (1973); C. H. Llewellyn Smith, Phys. Lett. 46B, 233 (1973); S. Joglekan, SUNY-Stony Brook report (unpublished); J. Sucher and C. H. Woo, Phys. Rev. D 8, 2721 (1973). Also see S. Weinberg, Phys. Rev. Lett. 27, 1688 (1971); A. I. Vainshtein and I. B. Khriplovich. Yad. Fiz. 13, 198 (1971) [Sov. J. Nucl. Phys. 13, 111 (1971)]; J. Bell Nucl. Phys. B60, 427 (1973).

[10]R. Hagedorn, Astron. Astrophys. 5, 184 (1970); K. Huang and S. Weinberg, Phys. Rev. Lett. 25, 895 (1971); S. Frautschi, Phys. Rev. D 3, 2821 (1971); etc.

[11]See, e.g., V. A. Belinskii, E. M. Lifschitz, and I. M. Khalatnikov, Usp. Fiz. Nauk. 102, 463 (1970) [Sov. Phys.–Usp. 13, 745 (1971)]; Zh. Eksp. Teor. Fiz. 62, 1606 (1972) [Sov. Phys.—JETP 35, 838 (1972)]; V. A. Belinskii and I. M. Khalatnikov, Zh. Eksp. Teor. Fiz. 63, 1121 (1972) [Sov. Phys.—JETP 36, 591 (1973).]

[12]These remarks are based on a conversation with J. R. Schrieffer.

[13]For a general review, see S. Weinberg, *Gravitation and Cosmology: Principles and Applications of the General Theory of Relativity* (Wiley, New York, 1972), Chap. XV, Secs. V and VI.

[14]A. A. Abrikosov, L. P. Gorkov, and I. E. Dzvaloskinskii, Zh. Eksp. Teor. Fiz. 36, 900 (1959) [Sov. Phys.—JETP 9, 636 (1959)]; E. S. Fradkin, Zh. Eksp. Teor. Fiz. 36, 1286 (1959) [Sov. Phys.—JETP 9, 912 (1959)]; P. C. Martin and J. Schwinger, Phys. Rev. 115, 1342 (1959).

[15]K. Fujikawa, B. W. Lee, and A. I. Sanda, Phys. Rev. D 6, 2923 (1972). Also see Y.-P. Yao, Phys. Rev. D 7, 1647 (1973); G. 't Hooft and M. Veltman, Nucl. Phys. B50, 318 (1972); S. Weinberg, Ref. 25, Appendix A.

[16]This has been checked in detail by C. Bernard (private communication).

[17]C. Bernard, this issue, Phys. Rev. D 9, 3312 (1974); L. Dolan and R. Jackiw, Ref. 3.

[18]G. H. Hardy, *A Course of Pure Mathematics* (Cambridge Univ. Press, Cambridge, 1949), p. 330, Ex. 3.

[19]An example of a positive-definite quartic form $f_{ijkl} \times \phi_i\phi_j\phi_k\phi_l$ for which $f_{ijkk}\phi_i\phi_j$ is not a positive quadratic form was originally suggested to me by S. Coleman (private communication). Example 3 of Sec. IV is based on the quartic form suggested by Coleman.

[20]This result was also found by L. Dolan and R. Jackiw, Ref. 3, for the case $n = 2$.

[21]These are the simple theories considered in Sec. VII of Ref. 7.

[22]The effect of a background of black-body radiation on the infrared divergences in the quantum theory of photons or gravitons was considered by S. Weinberg, in *Contemporary Physics* (International Atomic Energy Agency, Vienna, 1969), Vol. I, p. 560.

[23]For a review, see K. G. Wilson and J. Kogut (unpublished); S.-K. Ma, Rev. Mod. Phys. 45, 589 (1973).

[24]The examples in Sec. IV exhibit at least an approximate second-order transition, in the sense that any discontinuity in the "order parameter" λ can only be due to effects of higher order in e. However, it remains an open possibility that the phase transition may be *weakly* of first order, i.e., that λ exhibits a small discontinuity at the true critical temperature. It should be noted that the presence of weakly coupled gauge fields in superconductors and certain liquid

crystals leads to phase transitions that are weakly of first order, as recently shown by B. I. Halperin, T. C. Lubensky, and S.-K. Ma, Phys. Rev. Lett. 32, 292 (1974).

[25]S. Weinberg, Phys. Rev. D 7, 2887 (1973).

[26]This was suggested to me by R. Jackiw (private communication).

[27]This remark is due to C. Bernard (private communication).

VOLUME 33, NUMBER 7 PHYSICAL REVIEW LETTERS 12 AUGUST 1974

Hierarchy of Interactions in Unified Gauge Theories*

H. Georgi,† H. R. Quinn, and S. Weinberg
Lyman Laboratory of Physics, Harvard University, Cambridge, Massachusetts 02138
(Received 15 May 1974)

We present a general formalism for calculating the renormalization effects which make strong interactions strong in simple gauge theories of strong, electromagnetic, and weak interactions. In an SU(5) model the superheavy gauge bosons arising in the spontaneous breakdown to observed interactions have mass perhaps as large as 10^{17} GeV, almost the Planck mass. Mixing-angle predictions are substantially modified.

The scaling observed in deep inelastic electron scattering suggests that what are usually called the strong interactions are not so strong at high energies. Asymptotically free gauge theories of the strong interactions[1] provide a possible explanation: The gluon coupling constant $g(\mu)$ (defined as the value of a three-gluon or gluon-fermion-fermion vertex with momenta characterized by a mass μ) is small when μ is several GeV or larger, but becomes large when μ is small, through the piling up of the logarithms encountered in perturbation theory. In one recent calculation[2] a fit was found for a gauge coupling [in a color SU(3) model][3] with $g^2(\mu)/4\pi \simeq 0.1$ when $\mu \simeq 2$ GeV.

If $g(\mu)$ is small when μ is large, then perhaps the strong gauge coupling at some large fundamental mass is of the same order as the couplings in gauge theories of the weak and electromagnetic interactions.[4] Georgi and Glashow[5] have recently gone one step farther, and proposed a model based on the *simple* gauge group SU(5), in which there naturally appears only one free gauge coupling. In their model, SU(5) suffers a spontaneous breakdown to the gauge subgroups SU(3) and SU(2)⊗U(1), which are associated respectively with the strong[3] and the weak and electromagnetic[6] interactions. In order to suppress unobserved interactions, Georgi and Glashow made the necessary assumption[7] that some vector bosons are superheavy.

We find the notion of a simple gauge group uniting strong, weak, and electromagnetic interactions extraordinarily attractive. However, as emphasized by Georgi and Glashow, the success of any such scheme hinges on an understanding of the effects which produce the obvious disparity in strength between the strong and the weak and electromagnetic interactions at ordinary energies. We therefore wish to present in this paper a general formalism for the calculation of such effects. This will lead us to an estimate of the mass of the superheavy gauge bosons. Where a specific model of the gauge groups of the observed interactions is needed as an example, we shall assume that the strong and the weak and electromagnetic interactions are described by color SU(3)[3] and by SU(2)⊗U(1), respectively, and where a specific example of a unifying simple gauge group is needed, we shall use SU(5).

If we neglect all renormalization effects, the embedding of the gauge groups G_i of the observed interactions in a larger simple group G imposes a relation among their coupling constants. We

VOLUME 33, NUMBER 7 PHYSICAL REVIEW LETTERS 12 AUGUST 1974

normalize the generators T_α of G so that in any representation D of G we have

$$\mathrm{Tr}(T_\alpha T_\beta) = N_D \delta_{\alpha\beta}, \quad (1)$$

where N_D may depend on the representation but not on α and β. We use the same normalization conventions for the gauge groups of the observed interactions. Then invariance under G implies that the coupling constants g_G, g_3, g_2, and g_1 associated with the group G and the subgroups SU(3), SU(2), and U(1), respectively, are equal. The usual SU(2) and U(1) coupling constants[6] may be identified as

$$g = g_2, \quad g' = g_1/C, \quad (2)$$

where C is a constant entering the relation between the charge Q and the SU(2) and U(1) generators $\vec{T}$ and T_0, normalized according to Eq. (1):

$$Q = T_3 - CT_0. \quad (3)$$

The weak mixing angle[6] is then given by

$$\sin^2\theta = e^2/g^2 = g'^2/(g^2 + g'^2) = (1 + C^2)^{-1}. \quad (4)$$

In any representation of G, reducible or irreducible,

$$\mathrm{Tr}(Q^2) = (1 + C^2)\mathrm{Tr}(T_3^{\,2}). \quad (5)$$

If we take our representation to consist of the left-handed states of three quartets of colored quarks, three antiquark quartets, and ν_e, ν_μ, e^-, e^+, μ^-, μ^+, then there are eight SU(2) doublets, and so $\mathrm{Tr}(T_3^{\,2}) = 4$, while $\mathrm{Tr}(Q^2) = \frac{32}{3}$, so that

$$C^2 = \tfrac{5}{3}, \quad \sin^2\theta = \tfrac{3}{8}. \quad (6)$$

This is the case for the SU(5) model.[5] We shall leave C arbitrary in what follows, and will find that the choice of the simple unifying group G enters the calculation only through the single parameter C.

Now let us see how to take renormalization effects into account. The gauge couplings are functions of the momentum scale μ, and the above relations among gauge couplings really only apply when μ is much larger than the superheavy boson masses, where the breaking of G may be neglected. However, the observed values of the gauge couplings refer to much smaller values of μ, of the order of the W and Z masses, or even smaller. The problem is to bridge the gap between superlarge values of μ, where G imposes relations among the gauge couplings, and ordinary values of μ, where the gauge couplings are observed.

In order to accomplish this, we make use of the theorem[8] that all matrix elements involving only "ordinary" external particles with momenta and masses much less than all superheavy masses may be calculated in an effective renormalizable field theory, which is just the original field theory with all superheavy particles omitted, but with coupling constants that may depend on the superheavy masses. All other effects of the superheavy particles are suppressed by factors of an ordinary mass divided by a superheavy mass.

When μ is large compared with all ordinary masses but small compared with all superheavy masses, the μ dependence of the couplings is governed by a renormalization-group equation,[9]

$$\mu \frac{d}{d\mu} g_i(\mu) = \beta_i(g_i(\mu)), \quad (7)$$

with β_i calculated in the effective field theory based on the "observed" gauge group G_0. If all $g(\mu)$ are small, then β_i depends only on g_i, with[10]

$$\beta_i(g(\mu)) \simeq b_i g_i^{\,3}(\mu) \text{ for } |g_i| \ll 1, \quad (8)$$

so that

$$g_i^{-2}(\mu) \simeq \text{const} - 2b_i \ln\mu. \quad (9)$$

The integration constants are determined by the underlying simple group G. Specifically, if we suppose that all superheavy gauge bosons have masses of the order of some typical superheavy mass M, and if we take μ to be of the order of but somewhat smaller than M, then the $g_i(\mu)$ may be calculated perturbatively in the simple gauge theory based on G, and as long as $g_i(M)$ is sufficiently small, each gauge coupling will be essentially given by its group-theoretic value g_G neglecting all renormalizations. Thus Eq. (9) gives

$$g_i^{-2}(\mu) = g_G^{-2}(M) + 2b_i \ln(M/\mu) \quad (10)$$

for $\mu \leq M$.

The gauge coupling constants g_{i_0} observed in present experiments are essentially given by the values of the $g_i(\mu)$ when μ is some "ordinary" mass m, of the order of 10 GeV. Since all these couplings are small and therefore slowly varying in the range of interest, our result is not particularly sensitive to the value of the "ordinary" mass at which we choose to study the couplings.

Let us now specifically assume that the "observed" gauge group is SU(3)⊗SU(2)⊗U(1), but for the moment leave open the choice of the group G. Choosing convenient linear combinations of

(10), we have

$$\frac{C^2}{g_{1o}^{\ 2}}+\frac{1}{g_{2o}^{\ 2}}-\frac{(1+C^2)}{g_{3o}^{\ 2}}=\frac{1}{e^2}-\frac{(1+C^2)}{g_{3o}^{\ 2}}$$

$$=2\ln\left(\frac{M}{m}\right)[b_1C^2+b_2-b_3(1+C^2)], \qquad (11)$$

$$C^2(g_{1o}^{\ -2}-g_{2o}^{\ -2})=e^{-2}[1-(1+C^2)\sin^2\theta]$$

$$=2C^2(b_1-b_2)\ln(M/m). \qquad (12)$$

To calculate the b's, we note that any multiplet of particles forming a representation of G does not contribute at all to the b's if *all* particles are superheavy, while it contributes equally to all b's if *no* particles in the multiplet are superheavy, and therefore in either case has no effect in Eqs. (11) and (12). We shall assume that the only multiplet which contains *both* ordinary and superheavy particles is the gauge multiplet itself, in which case[1]

$$b_2=-\tfrac{22}{3}(4\pi)^{-2}+b_1, \quad b_3=-11(4\pi)^{-2}+b_1; \qquad (13)$$

so (11) and (12) give

$$\ln\left(\frac{M}{m}\right)=\frac{3(4\pi)^2}{22(1+3C^2)}\left(\frac{1}{e^2}-\frac{(1+C^2)}{g_{3o}^{\ 2}}\right), \qquad (14)$$

$$\sin^2\theta=(1+3C^2)^{-1}(1+2C^2e^2/g_{3o}^{\ 2}). \qquad (15)$$

For $C^2=\frac{5}{3}$ [the SU(5) value], $m=10$ GeV, and reasonable values of $g_{3o}^{\ 2}/4\pi$, we obtain the results displayed in the following table:

$g_{3o}^{\ 2}/4\pi$	M (GeV)	$\sin^2\theta$
0.5	2×10^{17}	0.175
0.2	2×10^{16}	0.187
0.1	5×10^{14}	0.207
0.05	2×10^{11}	0.248

It is intriguing that we are led to contemplate elementary particle masses as high as 2×10^{17} GeV, of about the same order of magnitude as the Planck mass, $G^{-1/2}=1.2206\times10^{19}$ GeV. Perhaps gravitation has something to do with the superstrong spontaneous symmetry breaking, or perhaps the spontaneous breakdown of the simple gauge group has something to do with setting the scale of the gravitational interaction.

Equation (15) predicts lower values for $\sin\theta$ than does Eq. (4). While the available data favor the higher value, they are rather preliminary, and the strong constraints on $\sin^2\theta$ follow only if the Z mass is assumed to satisfy the relation $M_W^{\ 2}=M_Z^{\ 2}\cos^2\theta_W$, which depends on the Higgs structure of the model.[11] If this relation is abandoned, $\sin^2\theta_W$ must be inferred from the ratio of the ratio of neutral- to charged-current events seen in neutrino scattering to that seen in antineutrino scattering.[11]

In the SU(5) example, it is not necessarily true that all effects of order m/M are negligible, because some of the superheavy vector bosons mediate proton decay into lepton plus pions. Since the proton is otherwise stable, such very small effects may be observable. Calculation of the proton lifetime involves details of the strong interactions at small momenta, but we can give an order-of-magnitude estimate on dimensional grounds. The lifetime must be proportional to M^4 and so it must approximately equal $M^4/m_p^{\ 5}$. Taking $M=5\times10^{15}$ GeV, for example, gives a proton lifetime of about 6×10^{31} yr. The present experimental lower limit is 10^{30} yr.[12] The observation of proton decay with a lifetime of this order of magnitude would be a startling confirmation of the ideas discussed here.

Before concluding, we emphasize again the approximation which went into the derivation of Eqs. (14) and (15). We have idealized the two transition regions: the region in momentum scale around M where the three coupling constants are merging into one, and the region from m into the timelike domain where we actually measure e^2. The corrections to (14) and (15) due to changes of the coupling constants in these regions can be calculated using perturbation theory in the relevant coupling constants. The corrections for the second region are electromagnetic and therefore small (g_{3o} can in principle be measured directly in the spacelike region through the observation of logarithmic violations of scaling in electroproduction). The corrections from the first region will be small if $g_G(M)$ is small. To calculate $g_G(M)$ we need to know the fermion and scalar-meson content of the theory. For the SU(5) model with $M=5\times10^{15}$ GeV (see the table), $g_G^{\ 2}(M)/4\pi=(48\pm1)^{-1}$.

We have also assumed that the lowest-order form for β_i, Eq. (8), is valid down to $\mu=m$. Next-order corrections to β have been calculated by Belavin and Migdal.[13] We use their results and find the ratio of the correction to the lowest-order value in the SU(5) theory to be about $0.6g^2/4\pi$. Such corrections are obviously only relevant for $g_3(\mu)$, and even for $g_{3o}^{\ 2}/4\pi=0.5$, the largest value used in the table, the correction is only 30%.

Finally we want to emphasize what seems to us to be the most disturbing feature of the class of models discussed here, that is, the existence of

VOLUME 33, NUMBER 7 PHYSICAL REVIEW LETTERS 12 AUGUST 1974

two stages of spontaneous symmetry breaking characterized by radically different mass scales. In the context of the conventional Higgs mechanism, we can find no natural explanation of the enormous ratio of superheavy mass to ordinary mass. We have nothing quantitative to say about this mystery, but the following speculation seems attractive to us. Suppose that only superstrong breaking takes place via the Higgs mechanism. There is only one mass scale in the theory and it is superheavy. All of the scalar mesons are either superheavy or Goldstone bosons (note that this obviates the difficulties associated with superlarge trilinear couplings among ordinary-mass scalars). Well below the superheavy mass scale the theory is an effective SU(3)⊗SU(2)⊗U(1) theory containing only gauge fields and fermions. The next stage of symmetry breaking is dynamical and hence nonperturbative. The mass scale associated with this stage is the mass at which $g_3(\mu)$ gets large enough that nonperturbative effects become important.

We are grateful for discussions with T. Appelquist, S. Coleman, S. L. Glashow, and H. D. Politzer.

---

*Work supported in part by the National Science Foundation under Grant No. GP40397X.

†Junior Fellow, Harvard University Society of Fellows.

[1]H. D. Politzer, Phys. Rev. Lett. 30, 1346 (1973); D. J. Gross and F. Wilczek, Phys. Rev. Lett. 30, 1343 (1973).

[2]H. D. Politzer, to be published. Of course, given the available data, any such estimate is necessarily very crude.

[3]W. Bardeen, H. Fritsch, and M. Gell-Mann, in *Scale and Conformal Invariance in Hadron Physics*, edited by R. Gatto (Wiley, New York, 1973), p. 139. Also see R. H. Dalitz, in *High Energy Physics*, edited by C. DeWitt and M. Jacob (Gordon and Breach, New York, 1966), p. 287.

[4]S. Weinberg, J. Phys. (Paris), Colloq. 34, C1-45 (1973), and to be published.

[5]H. Georgi and S. L. Glashow, Phys. Rev. Lett. 32, 438 (1974).

[6]S. Weinberg, Phys. Rev. Lett. 19, 1264 (1967); A. Salam, in *Elementary Particle Physics*, edited by N. Svartholm (Almquist and Wiksels, Stockholm, 1968), p. 367.

[7]S. Weinberg, Phys. Rev. D 5, 1962 (1972).

[8]T. Appelquist and J. Carrazone, to be published. A different proof of the same result for graphs including only a single superheavy line is given by S. Weinberg, Phys. Rev. D 8, 605, 4482 (1973). This theorem and hence our entire discussion could be invalidated if there were superlarge trilinear couplings among scalar fields. The survival of any light scalars already requires that certain scalar self-couplings in the symmetric theory were made extremely small. It appears that the problem of large trilinear couplings can always be avoided by some device. We will discuss later in this paper a possible point of view which avoids this unattractive situation altogether.

[9]M. Gell-Mann and F. E. Low, Phys. Rev. 95, 1300 (1954); C. G. Callan, Phys. Rev. D 2, 1541 (1970); K. Symanzik, Commun. Math. Phys. 18, 227 (1970).

[10]It was H. D. Politzer who pointed out to us that perturbation theory could be used down to the energy where the strong interactions begin to be much stronger than the weak and electromagnetic interactions, so that it is not necessary here to worry about the region where the strong interactions are really strong.

[11]A. De Rújula, H. Georgi, S. L. Glashow, and H. Quinn, to be published.

[12]F. Reines and M. F. Crouch, Phys. Rev. Lett. 32, 493 (1974).

[13]A. A. Belavin and A. A. Migdal, to be published. We learned recently that the same calculation has been performed by W. E. Caswell [Phys. Rev. Lett. 33, 244 (1974)]. His result differs slightly from that given by Belavin and Migdal, but the difference is insignificant as far as our estimate is concerned.

PHYSICAL REVIEW D VOLUME 13, NUMBER 4 15 FEBRUARY 1976

Implications of dynamical symmetry breaking

Steven Weinberg*
Lyman Laboratory of Physics, Harvard University, Cambridge, Massachusetts
(Received 8 September 1975)

An analysis is presented of the physical implications of theories in which the masses of the intermediate vector bosons arise from a dynamical symmetry breaking. In the absence of elementary spin-zero fields or bare fermion masses, such theories are necessarily invariant to zeroth order in the weak and electromagnetic gauge interactions under a global $U(N)\otimes U(N)$ symmetry, where N is the number of fermion types, not counting color. This symmetry is broken both intrinsically by the weak and electromagnetic interactions and spontaneously by dynamical effects of the strong interactions. An effective Lagrangian is constructed which allows the calculation of leading terms in matrix elements at low energy; this effective Lagrangian is used to analyze the relative direction of the intrinsic and spontaneous symmetry breakdown and to construct a unitarity gauge. Spontaneously broken symmetries which belong to the gauge group of the weak and electromagnetic interactions correspond to fictitious Goldstone bosons which are removed by the Higgs mechanism. Spontaneously broken symmetries of the weak and electromagnetic interactions which are not members of the gauge group correspond to true Goldstone bosons of zero mass; their presence makes it difficult to construct realistic models of this sort. Spontaneously broken elements of $U(N)\otimes U(N)$ which are not symmetries of the weak and electromagnetic interactions correspond to pseudo-Goldstone bosons, with mass comparable to that of the intermediate vector bosons and weak couplings at ordinary energies. Quark masses in these theories are typically less than 300 GeV by factors of order α. These theories require the existence of "extra-strong" gauge interactions which are not felt at energies below 300 GeV.

I. INTRODUCTION

When unified gauge theories of the weak and electromagnetic interactions were first proposed, it was assumed[1] that the spontaneous symmetry breakdown responsible for the intermediate-vector-boson masses is due to the vacuum expectation values of a set of spin-zero fields. For a variety of reasons, the attention of theorists has since been increasingly drawn to the possibility that this symmetry breaking is of a purely dynamical nature.[2] That is, it is supposed that there may be no elementary spin-zero fields in the Lagrangian, and that the Goldstone bosons associated with the spontaneous symmetry breakdown are bound states.

Almost all the effort that has been put into analyses of dynamical symmetry breaking has been directed to the difficult mathematical problem, of whether and how this phenomenon can occur in a variety of field-theoretic models. In this article I would like to address quite a different question: Assuming that dynamical symmetry breaking is a mathematical possibility in gauge field theories, what are the consequences for the real world?

Why should we believe that the masses of the intermediate vector bosons arise from dynamical symmetry breaking? The absence of *strongly* interacting elementary spin-zero fields is indicated by a number of requirements: asymptotic freedom,[3] electroproduction sum rules,[4] and the naturalness of order-α parity and strangeness conservation.[5] On the other hand, the absence of *weakly* interacting elementary spin-zero fields is much less certain. Apart from simplicity, the best reason for this assumption comes from the requirement for a natural hierarchy of gauge symmetry breaking.[6] In order to put together the observed weak and electromagnetic interactions into a simple gauge group, it is necessary to suppose[7] that in the spontaneous breakdown of this simple group to the nonsimple gauge group of the observed interactions, vector-boson masses are generated that are much larger than the masses expected for the W and Z; this conclusion is even stronger if we try to include the strong interactions as well.[8] This superstrong symmetry breakdown may well be due to the vacuum expectation values of elementary spin-zero fields. However, at ordinary energies, far below the superheavy vector-boson masses, physics is described by an effective field theory[9] involving those fermions and vector bosons that did not get masses from the superstrong spontaneous symmetry breakdown, but no spin-zero fields. Likewise, the gauge group of this effective field theory consists of a direct product of those simple and U(1) subgroups of the simple gauge group that were not broken at the superstrong level. The only way that the non-superheavy fermions and vector bosons can then acquire masses is from a dynamical breakdown of this remaining gauge group. Futhermore, the mass scale determined by the dynamical symmetry breakdown is expected to be of the order of magnitude of the renormalization point at which the largest of the gauge couplings of the effective field theory reaches a value of order unity; this mass scale is in general enormously different from the mass scale of the superstrong symmetry breakdown.

For the purposes of this article, we will not need to commit ourselves to the general picture described above. However, our assumptions are those inspired by this picture: We assume that weak, strong, and electromagnetic interactions are described by a gauge field theory (perhaps an "effective" field theory[9]) involving fermions and gauge fields but no elementary spin-zero fields or bare fermion masses, and we suppose that the vector-boson and fermion masses arise from a dynamical breakdown of this gauge group. These assumptions are spelled out more precisely in Sec. II.

In further support of these assumptions, it should be mentioned that it is the absence of bare fermion masses that makes it natural for a spontaneous dynamical symmetry breakdown to occur. Any spontaneous symmetry breaking requires the appearance of massless Goldstone bosons,[10] whether or not they are eventually eliminated by the Higgs mechanism.[11] For dynamical symmetry breaking, these Goldstone bosons would have to be bound states. However, we would normally expect that any bound state at zero mass would move away from zero mass if we changed the strength of the binding interactions, in which case dynamical symmetry breaking could only occur for a discrete set of coupling strengths,[12] and could not be considered "natural." In the theories considered here, with zero bare fermion mass, there is only one mass scale, defined by the renormalization point at which the gauge couplings are specified; thus a small change in the gauge coupling constant corresponds to a general change of mass scale,[13] and cannot shift a massless bound state away from zero mass.

The consequences of our assumptions turn out to be quite striking. Before turning on the weak and electromagnetic interactions, the strong interactions are necessarily invariant, not only under the strong gauge group, but also, as shown in Sec. III, under a global $U(N)\otimes U(N)$ group, where N is the number of fermion types, not counting color or possible other strong gauge indices. This global group is broken in two different ways, described in Sec. IV. It is *spontaneously* broken down to some subgroup H by dynamical effects of the strong interactions. It is also *intrinsically* broken to that subgroup S_W of $U(N)\otimes U(N)$ which leaves the weak and electromagnetic interactions invariant. And of course the gauge group of the weak and electromagnetic interactions is a subgroup G_W of S_W.

It is this double breakdown of $U(N)\otimes U(N)$, shown symbolically in Fig. 1, that will occupy most of our attention in this paper. Indeed, aside from the final section, the bulk of this paper can be regarded as a mathematical analysis of general theories in which there is both a strong spontaneous symmetry breaking and a weak intrinsic symmetry breaking induced by gauge interactions. The property that is specific to theories without spinless fields is that the over-all global group is $U(N)\otimes U(N)$, but most of our discussion would apply to any other global group.

Our analysis is complicated by three factors:

(1) We cannot use perturbation theory to describe the strong interactions responsible for the spontaneous symmetry breaking. This problem is evaded here by restricting ourselves to processes at relatively "low" energies, not greater than the expected masses of the intermediate vector bosons, or roughly $e\times 300$ GeV. It is shown in Sec.

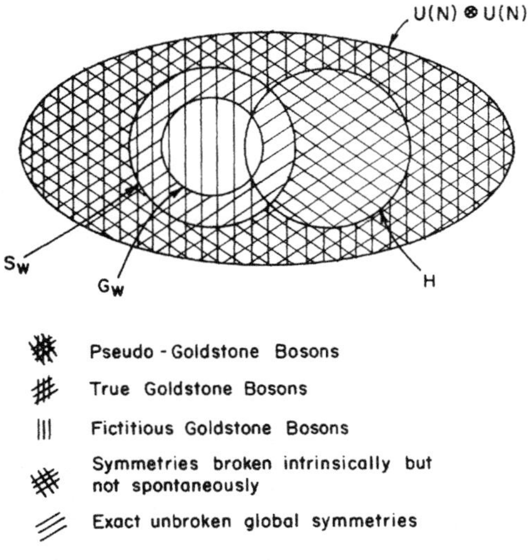

FIG. 1. Schematic representation of the various subgroups of $U(N)\otimes U(N)$. Cross hatchings indicate the various ways that global or local symmetries are broken; the unhatched lens represents the unbroken exact local symmetries, such as electromagnetic gauge invariance. As discussed in the text, H is the subgroup of $U(N)\otimes U(N)$ which is not spontaneously broken; S_W is the global symmetry group of the weak and electromagnetic interactions; and G_W is the weak and electromagnetic gauge group.

V that at such energies the terms of leading order in e in the matrix element for any process are given by calculating tree graphs, using an effective Lagrangian[14] of reasonably simple structure. The effective Lagrangian involves those fermions and strong gauge vector bosons that did not acquire masses from the spontaneous dynamical symmetry breaking (which we identify as the usual quarks and gluons) plus the Goldstone bosons that accompany the spontaneous symmetry breakdown, and the gauge bosons of the weak and electromagnetic interactions.

(2) In defining a particular theory, it is not enough to specify the group structure of the nonspontaneously broken subgroup H and the gauge subgroup G_W; we must also say how these subgroups line up with each other[15] within the overall group $U(N)\otimes U(N)$. In Sec. VI it is shown that the alignment of these subgroups is determined by the condition that the Goldstone bosons must not have tadpoles; otherwise perturbation theory breaks down.

(3) There is a Goldstone boson for every independent broken symmetry[10] in $U(N)\otimes U(N)$, but those Goldstone bosons that correspond to generators of the gauge group G_W are "fictitious" Goldstone bosons, which are eliminated by the Higgs mechanism.[11] In Sec. VII we show how in general to define a unitarity gauge[16] in which these fictitious Goldstone bosons are absent. The masses of the intermediate vector bosons can be determined (to leading order in e) by inspecting the effective Lagrangian in the unitarity gauge. The other Goldstone bosons which are not eliminated by the Higgs phenomenon are studied in Sec. VIII. This class consists of "true" Goldstone bosons of zero mass, corresponding to broken symmetries in S_W but not G_W, and "pseudo"-Goldstone bosons[17] with mass of order $e\times 300$ GeV, corresponding to broken symmetries of $U(N)\otimes U(N)$ which are neither in S_W nor G_W (see Fig. 1).

Different aspects of this analysis have been discussed before, but not to the best of my knowledge all together. Thus, effective Lagrangians for both broken global[14] and broken gauge symmetries[18] are an old story, but not for the case where the broken-symmetry group consists of a group of approximate global symmetries with an exact gauge subgroup. Also, the problem of subgroup alignment mentioned in item (2) above has been studied in the presence of strong interactions,[15] but with a nongauge perturbation, and also with a gauge perturbation,[19] but in the absence of strong interactions. Finally, previous attempts at a general definition of the unitarity gauge[16] and the pseudo-Goldstone bosons[17] dealt only with a spontaneous symmetry breakdown produced by vacuum expectation values of elementary spin-zero fields.

In Sec. IX we take up an unrealistic example which is designed to show how this analysis can be applied to specific theories. As usual in models with spontaneously broken symmetries, we can obtain quite detailed information about the interaction of soft Goldstone bosons with quarks and vector bosons, despite the presence of strong interactions. Surprisingly, one can also solve the subgroup alignment problem explicitly. There are just two possible ways that the subgroups can line up, corresponding to the possible signs of a single unknown parameter. In one case there are two massive vector bosons, one "photon," one true Goldstone boson, two pseudo-Goldstone bosons, and a finite quark mass of second order in e; in the other case there are three massive vector bosons, no photons, no true Goldstone bosons, two pseudo-Goldstone bosons, and an exact symmetry which keeps the quark mass zero to all orders in e. Evidently the subgroup alignment is crucial in determining the physical content of theories with a given group structure. It is striking that for both alignments the theory contains unwelcome massless particles: in one case a true Goldstone boson, in the other a massless quark. This is a common problem in theories with dynamical symmetry breaking.

The last section offers a series of remarks about the application of the formalism developed in this article to models of the real world.

This article is not intended as an argument that the masses of the intermediate vector bosons actually do arise from dynamical symmetry breaking. Indeed, some of the difficulties of constructing realistic models based on dynamical symmetry breaking are emphasized in Sec. X. However, it would be wise at least to keep in mind that the experiments designed to find intermediate vector bosons may discover pseudo-Goldstone bosons as well.

II. GENERAL ASSUMPTIONS

The theories to be discussed in this paper are governed by the following general assumptions:

(a) The Lagrangian is locally invariant under a gauge group

$$G_S\otimes G_W. \tag{2.1}$$

Here G_S describes the strong interactions, and has gauge couplings roughly of order unity; G_W describes the weak and electromagnetic interactions, and has gauge couplings roughly of order e. [Both G_S and G_W may themselves be direct products of simple and/or U(1) gauge groups.] As discussed in Sec. X, it is likely that G_S is larger than the usual color SU(3) gauge group.

(b) In addition to the vector gauge fields required under (a) there is a set of fermion fields $\psi_{nm}(x)$. Here n is an N-valued row or "flavor" index labeling fermion type $\mathcal{P}$, $\mathcal{N}$, λ, $\mathcal{P}'$, etc., on which G_W acts, and m is a column or "color" index, on which G_S acts. (The status of the leptons is considered briefly in Sec. X.)

(c) The Lagrangian contains no fermion mass terms and no elementary spin-zero fields.

(d) The theory is renormalizable.

These assumptions are made here because they seem to be required[6] in simple gauge theories with hierarchies of symmetry breaking, of the type described in the Introduction. However, for the purpose of this paper it will not be necessary to suppose that the strong, weak, and electromagnetic interactions arise from a superstrongly broken simple gauge group; it will only be assumed that physics at "ordinary" energies (say, up to a few thousand GeV) is governed by assumptions (a)–(d).

Under these assumptions, the Lagrangian must take the form[20]

$$\mathcal{L} = -\bar{\psi}\gamma^\mu \mathfrak{D}_\mu \psi - \tfrac{1}{4}\mathcal{F}_{\alpha\mu\nu}\mathcal{F}_\alpha^{\mu\nu} - \tfrac{1}{4}G_{\sigma\mu\nu}G_\sigma^{\mu\nu}, \tag{2.2}$$

where $\mathfrak{D}_\mu\psi$ is the gauge-covariant derivative of the fermion field

$$(\mathfrak{D}_\mu \psi)_{nm} = \partial_\mu \psi_{nm} - i\sum_{n'\alpha}(w_\alpha)_{nn'}\psi_{n'm}W_{\alpha\mu} - i\sum_{m'\sigma}(s_\sigma)_{mm'}\psi_{nm'}S_{\sigma\mu}, \tag{2.3}$$

with $W_{\alpha\nu}$ and $S_{\sigma\mu}$ the G_W and G_S gauge fields, and w_α and s_σ the matrices representing the corresponding group generators. (The gauge coupling constants are included as factors in w_α and s_σ.) Also, $F_{\alpha\mu\nu}$ and $G_{\sigma\mu\nu}$ are the usual covariant curls of $W_{\alpha\mu}$ and $S_{\sigma\mu}$, respectively.

III. U(N)⊗U(N) SYMMETRY

The most striking consequence of the general assumptions outlined in the last section is the existence of an "accidental" approximate global symmetry of the Lagrangian. In the limit $e \to 0$ the Lagrangian is automatically invariant not only under the *local* G_S transformations on the fermion column (i.e., color) indices, but also under a group $U(N)\otimes U(N)$ of *global* transformations on the N-valued fermion row (i.e., $\mathcal{P}$, $\mathcal{N}$, λ, $\mathcal{P}'$, . . .) index. That is, for each of the $2N^2$ independent Hermitian matrices λ_A (with Dirac matrix factor of 1 or γ_5) there is a vector or axial-vector current

$$J_A^\mu = -i\sum_{mn'n}\bar{\psi}_{n'm}\gamma^\mu(\lambda_A)_{n'n}\psi_{nm}. \tag{3.1}$$

Apart from triangle anomalies (about which more will be said later) these are all conserved:

$$\partial_\mu J_A^\mu = 0 \quad (\text{for } e=0). \tag{3.2}$$

In what follows it will be convenient to normalize the λ_A and hence the J_A so that

$$\mathrm{Tr}(\lambda_A\lambda_B) = 8\delta_{AB}. \tag{3.3}$$

(An unusual extra factor of 4 appears here because the trace includes a trace on Dirac indices; this is necessary because half the λ_A are proportional to the Dirac matrix γ_5.)

It should be emphasized that the $U(N)\otimes U(N)$ symmetry arises only because of our assumptions that the Lagrangian contains no fermion mass terms $m\bar{\psi}\psi$, no scalar field couplings $\phi\bar{\psi}\psi$, and no nonrenormalizable interactions, such as a Fermi interaction $\bar{\psi}\psi\bar{\psi}\psi$. Any one of these terms might in general destroy the $U(N)\otimes U(N)$ symmetry. On the other hand, once we make these assumptions, the $U(N)\otimes U(N)$ symmetry is inescapable—the fermion fields must enter the Lagrangian only in the form $\bar{\psi}\gamma^\mu\mathfrak{D}_\mu\psi$, and in the limit $e\to 0$ the covariant derivative $\mathfrak{D}_\mu$ contains only matrices which commute with all λ_A.

IV. SYMMETRY BREAKING

The $U(N)\otimes U(N)$ symmetry is in general broken by the weak and electromagnetic interactions. This intrinsic symmetry breaking can be quantitatively described by writing the generators w_α of the weak and electromagnetic gauge group as linear combinations of the $U(N)\otimes U(N)$ operators λ_A:

$$w_\alpha = \sum_A e_{\alpha A}\lambda_A. \tag{4.1}$$

In accordance with our previous assumptions, the coefficients $e_{\alpha A}$ are all of order e. Emission and absorption of virtual W bosons will produce order-e^2 perturbations which are not expected to be $U(N)\otimes U(N)$ invariant.

We shall assume that in addition to this intrinsic symmetry breaking, even in the limit $e\to 0$, there is a *spontaneous* breakdown of the symmetry group $G_S\otimes U(N)\otimes U(N)$ of the strong interactions, caused by strong forces among fermions, antifermions, and G_S gauge bosons. As usual in any spontaneous symmetry breaking, it is perfectly natural for some subgroup U to be left unbroken. For the sake of simplicity and definiteness, we shall assume that U does not mix the strong gauge group G_S with the accidental global symmetry group $U(N)\otimes U(N)$; that is, the unbroken symmetry group for $e\to 0$ is a direct product

$$U = H_S \otimes H, \tag{4.2}$$

where H_S is local and a subgroup of G_S, while H is global and a subgroup of $U(N)\otimes U(N)$. Most of the considerations below would also apply to the more general case where the unbroken subgroup is not of the form (4.2), but our discussion would have to be made considerably more elaborate to deal with this case.

For any independent generator of G_S which is not a generator of the unbroken gauge subgroup H_S, the corresponding strongly interacting vector boson gets a mass from the Higgs phenomenon. Also, depending on the nature of the unbroken global subgroup H, some of the fermions will acquire a mass from the spontaneous breakdown of U(N) ⊗ U(N) to H. In the limit $e\to 0$ this theory contains no very large or very small dimensionless parameters, so we would expect all these masses to be of the same order of magnitude, say M. The only dimensional parameter in the theory for $e\to 0$ is the scale characterizing the renormalization point of the G_S gauge couplings, so we would also expect M to be determined by the condition that the largest G_S gauge coupling reaches a critical value of order unity at a renormalization point characterized by momenta of order M.

We will see below that M is likely to be quite large, of order 300 GeV. The physics of strong interactions at lower energies $E \ll M$ can therefore be described in terms of those fermions and G_S gauge bosons which do *not* pick up masses of order M from the spontaneous symmetry breakdown, and hence remain massless in the limit $e\to 0$. These may be identified as the ordinary quarks and gluons, respectively. (Of course, we do not at this point rule out the possibility that H_S is just G_S, so that *all* of the G_S gauge bosons remain massless.) As we shall see, turning on the weak and electromagnetic interactions will give the quarks masses of order e^2M, while the gluons will remain massless.

In addition to quarks and gluons, this theory necessarily contains one other class of hadrons with masses which vanish for $e\to 0$, the Goldstone bosons. For every linearly independent generator of U(N) ⊗ U(N) which is not a generator of the unbroken subgroup H, there must appear a Goldstone boson Π_a. Since U(N) ⊗ U(N) is not a gauge group, there is no Higgs phenomena which can eliminate these Goldstone bosons in the limit $e\to 0$.

The coupling of the ath Goldstone boson to the Ath U(N) ⊗ U(N) current is described by a parameter F_{aA}, defined by

$$\langle 0|J_A^\mu(0)|\Pi_a\rangle = F_{aA}P_\Pi^\mu(2\pi)^{-3/2}(2E_\Pi)^{-1/2}. \tag{4.3}$$

These F_{aA} have the dimensions of a mass, and will play a role here like that played by the parameter F_π of current algebra. We expect that all F_{aA} are of the order of the mass M introduced earlier,

$$F_{aA}\approx M,$$

because in the limit $e\to 0$ this is the only mass in the theory.

It will be very convenient to adapt the basis for U(N) ⊗ U(N) to the pattern of symmetry breaking. We may define the generators of the unbroken subgroup H as linear combinations of the $2N^2$ generators λ_A of U(N) ⊗ U(N),

$$t_i=\sum_A C_{iA}\lambda_A, \tag{4.4}$$

with the C_{iA} chosen as orthonormal vectors so that

$$\sum_A C_{iA}C_{jA}=\delta_{ij}, \tag{4.5}$$

$$\mathrm{Tr}(t_it_j)=8\delta_{ij}. \tag{4.6}$$

The unbroken-symmetry currents have no couplings to the Goldstone bosons, so that

$$\sum_A C_{iA}F_{aA}=0. \tag{4.7}$$

Further, by a suitable unitary transformation we can always choose the Π_a states so as to diagonalize the positive Hermitian matrix

$$\sum_A F_{aA}F_{bA}.$$

If the element with $b=a$ is denoted $F_a{}^2$, we have then

$$F_{aA}=F_aB_{aA}, \tag{4.8}$$

with

$$\sum_A B_{aA}B_{bA}=\delta_{ab}, \tag{4.9}$$

and also

$$\sum_A B_{aA}C_{iA}=0. \tag{4.10}$$

There is one Goldstone boson for each independent broken symmetry, so the B's and C's form a complete orthonormal set of vectors

$$\sum_a B_{aA}B_{aB}+\sum_i C_{iA}C_{iB}=\delta_{AB}. \tag{4.11}$$

Correspondingly, we can define a set of broken symmetry generators

$$x_a\equiv\sum_A B_{aA}\lambda_A, \tag{4.12}$$

with

$$\mathrm{Tr}(t_ix_a)=0, \tag{4.13}$$

$$\mathrm{Tr}(x_ax_b)=8\delta_{ab}. \tag{4.14}$$

The generators t_i and x_a span the algebra of U(N) ⊗ U(N).

V. EFFECTIVE LAGRANGIAN

We now want to consider the special phenomena which arise because U(N)⊗U(N) is simultaneously broken both intrinsically by the weak and electromagnetic interactions and also spontaneously by dynamical effects of the strong interactions. This is not a mere matter of expanding in powers of e, because the theory has infrared singularities which, at momenta p much less than the characteristic mass M, introduce factors M/p which can compensate for factors e.

In order to explore this problem, let us consider the leading pole singularities produced by soft virtual quarks, Goldstone bosons, *and* G_W vector bosons in a general Green's function with external quark, gluon, Goldstone boson, and/or G_W vector bosons carrying momenta of order $p \ll M$.[21] These singularities can be calculated from the sum of all tree graphs for this Green's function, constructed from an effective Lagrangian[14] involving quark, G_W vector boson, gluon, and Goldstone boson fields. [For the moment, we are ignoring the effects of loops containing *hard* virtual G_W vector bosons. These produce U(N)⊗U(N)-breaking corrections of order e^2 in the effective Lagrangian, and will be considered at the end of this section.]

The effective Lagrangian here takes the general form

$$\mathcal{L} = -\tfrac{1}{4} F_{\alpha\mu\nu} F_{\alpha}^{\mu\nu} + \mathcal{L}_1 . \qquad (5.1)$$

The term $\mathcal{L}_1$ is subject to three conditions:

(1) In the limit $e \to 0$ the W dependence drops out, and $\mathcal{L}_1$ becomes invariant under U(N)⊗U(N), with fields transforming according to one of the usual nonlinear realizations of U(N)⊗U(N), in which the unbroken subgroup H is realized algebraically. It will be convenient to define the fields so that Π transforms like the so-called exponential parametrization[22] of the cosets in U(N)⊗U(N)/H. That is, a general U(N)⊗U(N) transformation, which is represented on the fermion fields in the original Lagrangian by a matrix g, induces on the fields in the effective Lagrangian the nonlinear transformations[22]

$$\Pi_a \to \Pi_a'(\Pi, g), \qquad (5.2)$$

$$q \to \exp\Big(i \sum_i \mu_i(\Pi, g) t_i\Big) q, \qquad (5.3)$$

where q is the quark field multiplet (with components corresponding to those fermion fields that do not acquire masses of order M from the spontaneous symmetry breaking), and Π' and μ are functions defined by the relation

$$g \exp\Big(i \sum_a \Pi_a x_a / F_a\Big) = \exp\Big(i \sum_a \Pi_a'(\Pi, g) x_a / F_a\Big) \times \exp\Big(i \sum_i \mu_i(\Pi, g) t_i\Big). \qquad (5.4)$$

The gluon fields are of course invariant under these transformations.

(2) The currents to which the products $W_{\alpha\mu} W_{\beta\nu} \cdots$ of G_W gauge fields (or their derivatives) couple in $\mathcal{L}_1$ take the form

$$\sum_{A,B,\ldots} e_{\alpha A} e_{\beta B} \cdots T_{AB\cdots}, \qquad (5.5)$$

where $T_{AB\cdots}$ is a quantity, formed out of quark, gluon, and Goldstone boson fields and their derivatives, which transforms like a U(N)⊗U(N) tensor (i.e., like $\lambda_A \lambda_B \cdots$) when Π and q undergo the transformations (5.2) and (5.3).

(3) $\mathcal{L}_1$ is locally as well as globally invariant under the unbroken subgroup H_S of the strong gauge group and under the weak and electromagnetic gauge group G_W.

It is shown in Appendix A that these conditions require $\mathcal{L}_1$ to be constructed from just the following ingredients:

(i) The quark fields q.

(ii) Their covariant derivatives (for notation, see below):

$$D_\mu q \equiv \partial_\mu q - i \sum_{ai} t_i q E_{ai}(\Pi) \partial_\mu \Pi_a F_a^{-1} - i \sum_{BA\alpha i} t_i q \Lambda_{AB}(\Pi) e_{\alpha A} W_{\alpha\mu} C_{iB} + \text{gluon terms}. \qquad (5.6)$$

(iii) Covariant derivatives of the Goldstone boson fields:

$$D_\mu \Pi_a \equiv F_a \Big[\sum_b D_{ba}(\Pi) \partial_\mu \Pi_b F_b^{-1} + \sum_{A\alpha B} \Lambda_{AB}(\Pi) e_{\alpha A} W_{\alpha\mu} B_{aB}\Big]. \qquad (5.7)$$

(iv) A covariant curl of the W field:

$$\mathcal{F}_{A\mu\nu} = \sum_{B\alpha} \Lambda_{BA}(\Pi) e_{\alpha A} F_{\alpha\mu\nu}. \qquad (5.8)$$

(v) The usual covariant curl of the gluon field.

Here $F_{\alpha\mu\nu}$ is the usual Yang-Mills G_W-covariant curl,[20] and the functions D, E, and Λ are defined by the formulas

$$S^{-1}(\Pi) \frac{\partial}{\partial \Pi_a} S(\Pi) = i F_a^{-1} \Big[\sum_b D_{ab}(\Pi) x_b + \sum_i E_{ai}(\Pi) t_i\Big], \qquad (5.9)$$

$$S^{-1}(\Pi)\lambda_A S(\Pi) = \sum_B \Lambda_{AB}(\Pi)\lambda_B, \tag{5.10}$$

with

$$S(\Pi) \equiv \exp\left(i\sum_b \Pi_b x_b/F_b\right). \tag{5.11}$$

With our normalization convention (3.3), the Λ matrices are orthogonal,

$$\Lambda^T(\Pi) = \Lambda^{-1}(\Pi). \tag{5.12}$$

[At this point, the reader may wish to be reminded that indices A, B, etc. label all generators of $U(N)\otimes U(N)$; i, j, etc. label the unbroken generators; and a, b, etc. label the broken generators.]

In addition to these limitations on the ingredients in $\mathcal{L}_1$, the conditions (1)–(3) also require that $\mathcal{L}_1$ must be invariant under formal global H transformations, with $D_\mu\Pi$, $D_\mu q$, and $\mathfrak{F}_{A\mu\nu}$ transforming according to whatever (linear) representations of H they happen to contain.

At this point the effective Lagrangian we have derived still has an extremely complicated structure, involving unlimited numbers of q, $D_\mu q$, $D_\mu\Pi$, $G_{\sigma\mu\nu}$, and $\mathfrak{F}_{A\mu\nu}$ functions. Indeed, if we were to take this Lagrangian seriously as a basis for higher-order calculations, we would have to keep all these interactions in order to provide counterterms for the infinite number of primitive divergents that would arise. However, the structure of the effective Lagrangian can be very much simplified if we use it only to determine the matrix elements to lowest order in e and p/M.

Ordinary dimensional analysis leads us to expect that an interaction v appearing in the effective Lagrangian with n_{vq} quark fields, n_{vs} gluon fields, $n_{v\Pi}$ Goldstone boson fields, n_{vw} G_W-gauge boson fields, and n_{vd} derivatives, will have a coupling constant of order

$$g_v \approx e^{n_{vw}} M^{n_{vm}}, \tag{5.13}$$

where M is the characteristic mass introduced in the last section (of order F_a) and

$$n_{vm} = 4 - \tfrac{3}{2}n_{vq} - n_{vs} - n_{v\Pi} - n_{vd}. \tag{5.14}$$

If a Green's function has external lines with momenta of order

$$p \approx eM, \tag{5.15}$$

then each power of M counts like a factor $1/e$. The total number of factors of e *or* p/M is

$$N_e = \sum_v N_v (n_{vw} - n_{vm}), \tag{5.16}$$

where N_v is the number of vertices of type v. However, a tree graph with E_q external quark lines, E_Π external Goldstone boson lines, and E_W external G_W-gauge bosons will obey the topological relation

$$\sum_v (n_{vq} + n_{v\Pi} + n_{vw} - 2)N_v = E_q + E_\Pi + E_W - 2. \tag{5.17}$$

(Gluon terms on both sides cancel, because we exclude *internal* gluon lines.[21]) Therefore, the total number of factors of e or p/M is

$$N_e = E_q + E_\Pi + E_W - 2 + \sum_v N_v \Delta_v, \tag{5.18}$$

where

$$\Delta_v \equiv n_{vw} + \tfrac{1}{2}n_{vq} + n_{vd} + n_{vs} - 2. \tag{5.19}$$

Thus for any given set of external lines, the terms of lowest order in e or p/M will be given by graphs composed of vertices with the smallest possible values of Δ_v.

In fact, the smallest values of Δ_v for any allowed interaction is $\Delta_v = 0$. Keeping only terms with $\Delta_v = 0$, and normalizing fields appropriately, gives an effective Lagrangian of the general form

$$\mathcal{L}_1 = -\bar{q}\gamma^\mu D_\mu q - \tfrac{1}{2}\sum_a D_\mu\Pi_a D^\mu\Pi_a - \sum_a \bar{q}\gamma^\mu\Gamma_a q D_\mu\Pi_a + \text{Fermi interactions}. \tag{5.20}$$

Here Γ_a is a constant matrix, proportional to γ_5 and/or 1, and of order $1/M$, which has the same H-transformation properties as Π_a; also, the "Fermi interactions" are $\bar{q}q\bar{q}q$ terms of order $1/M^2$ which are H invariant, but may involve any of the 16 Dirac covariants. Note that $D_\mu q$ and $D_\mu\Pi$ have just the right dependence on quark, gluon, G_W vector boson, and Goldstone boson fields, so that it is possible to construct an effective Lagrangian obeying all necessary symmetry conditions with only $\Delta_v = 0$ terms. (To the extent that a graph with loops is dominated by states with energy $E \ll M$, it can also be calculated with this effective Lagrangian.)

There is, in fact, one other term which in effect has $\Delta_v = 0$ and therefore should be added to the effective Lagrangian (5.20). The emission and reabsorption of a hard G_W vector boson produces an effective interaction of second order in e. For any such term, the Δ_v in Eq. (5.19) should be increased by two units, so that it becomes

$$\Delta_v = n_{vw} + \tfrac{1}{2}n_{vq} + n_{vd} + n_{vs}. \tag{5.21}$$

Thus we get a term with $\Delta_v = 0$ of second order in e if it is constructed *solely* from Goldstone boson fields, with no G_W gauge fields, quark fields, gluon fields, or derivatives. (See Fig. 2.)

This term will take the form

$$\mathcal{L}_2(\Pi) = -\sum_{\alpha AB} e_{\alpha A} e_{\alpha B} J_{AB}(\Pi), \tag{5.22}$$

where J_{AB} is the time-ordered product of two $U(N)\otimes U(N)$ currents J^{μ}_{A} and J^{ν}_{B}, integrated over momenta (including a $1/k^2$ weight factor from the W^{μ}_{α} propagator) and contracted over space-time indices. This two-point function is a $U(N)\otimes U(N)$ tensor, i.e., it transforms like $\lambda_A\lambda_B$. As shown in Appendix B, the most general form of such a tensor is

$$J_{AB}(\Pi) = -\sum_{CD}\Lambda_{AC}(\Pi)\Lambda_{BD}(\Pi)I_{CD}, \tag{5.23}$$

with I a Π-independent quantity of order M^4 which behaves as a tensor under the unbroken subgroup H. Thus, the $O(e^2)$ term with $\Delta_\nu=0$ is

$$\mathcal{L}_2(\Pi) = -\sum_{\alpha ABCD} e_{\alpha A}e_{\alpha B}\Lambda_{AC}(\Pi)\Lambda_{BD}(\Pi)I_{CD}. \tag{5.24}$$

This contains both a Π mass term (with some Goldstone boson masses of order eM) and a nonderivative Π self-interaction, like that in pion-pion scattering.

Incidentally, the integral in J_{AB} is expected to receive its major contribution from momenta of order M, because in these theories there is no reason why the integral should start to converge at any lower momentum. This is why we do not include the effects of G_W vector-boson masses here; these masses will turn out to be of order eM, so their effect is a higher-order correction.[23]

So far, we have seen that the leading $(\Delta_\nu=0)$ part of the effective Lagrangian takes the form

$$\mathcal{L}_{\text{eff}} = -\tfrac{1}{4}F_{\alpha\mu\nu}F^{\mu\nu}_{\alpha} + \mathcal{L}_1 + \mathcal{L}_2. \tag{5.25}$$

We could go on and describe the structure of higher terms with $\Delta_\nu=1$, $\Delta_\nu=2$, etc. However, we will content ourselves with describing one term of particular interest, associated with the quark masses.

The quarks are *defined* as the fermions that do not pick up a mass of order M from the spontaneous breakdown of $U(N)\otimes U(N)$ to H, so there is no quark mass term $\bar{q}q$ with $\Delta_\nu=-1$ in the effective Lagrangian. However, emission and absorption of a G_W vector boson can produce a term of order e^2 with $\Delta_\nu=+1$. Following the same reasoning as for $\mathcal{L}_2$, this term must take the form (see Fig. 3)

$$\mathcal{L}_m = -\sum_{\alpha AB}\bar{q}N_{AB}(\Pi)qe_{\alpha A}e_{\alpha B}, \tag{5.26}$$

where $\bar{q}Nq$ transforms as a tensor under $U(N)\otimes U(N)$. As shown in Appendix C, the most general form of N is

$$N_{AB}(\Pi) = \sum_{CD}\Lambda_{AC}(\Pi)\Lambda_{BD}(\Pi)Q_{CD}, \tag{5.27}$$

where Q_{CD} is a Π-independent matrix of order M which behaves as a tensor under the unbroken subgroup H. Thus the quark mass term is

$$\mathcal{L}_m = -\sum_{\alpha ABCD}\bar{q}Q_{CD}q\Lambda_{AC}(\Pi)\Lambda_{BD}(\Pi)e_{\alpha A}e_{\alpha B}. \tag{5.28}$$

The quark mass matrix is then

$$\mathfrak{M} = \sum_{\alpha} Q_{AB}e_{\alpha A}e_{\alpha B}, \tag{5.29}$$

so quark masses are expected to be of order e^2M. In addition, there are multilinear interactions of Goldstone bosons with quarks, including a $\Pi\bar{q}q$ coupling of order $e^2M/F\sim e^2$.

VI. ALIGNMENT OF SUBGROUPS

There is one further step that must be taken before the effective Lagrangian can be used for actual calculations. The symmetry $U(N)\otimes U(N)$ is supposed to be spontaneously broken to some subgroup H by dynamical effects of the strong interactions. However, although we can presume that the structure of H is determined by dynamical considerations, the strong interactions alone do not determine *which* subgroup of $U(N)\otimes U(N)$ with this structure is left unbroken. Given any solution of the strong-interaction dynamics with a subgroup H left invariant, we can find another solution in which the subgroup left invariant is

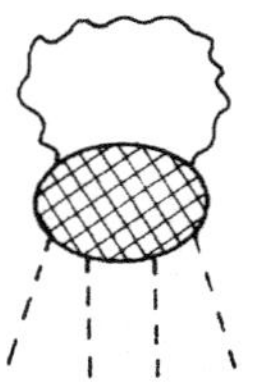

FIG. 2. Diagrams which contribute to the term $\mathcal{L}_2$ in the effective Lagrangian. Wavy lines are intermediate vector bosons and dashed lines are Goldstone bosons.

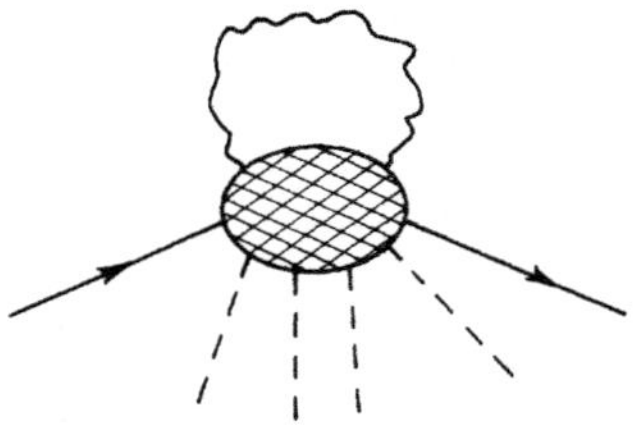

FIG. 3. Diagrams which contribute to the term $\mathcal{L}_m$ in the effective Lagrangian. Wavy lines are intermediate vector bosons, dashed lines are Goldstone bosons, and straight lines are quarks.

$$H(g) = gHg^{-1}, \tag{6.1}$$

where g is any element of $U(N) \otimes U(N)$.

Normally we do not concern ourselves with this sort of ambiguity, for these different solutions are usually physically equivalent. However, in our case, the theory contains a perturbation, the weak and electromagnetic interactions, which also break $U(N) \otimes U(N)$ down to some fixed subgroup S_W. (Of course, S_W contains the gauge group G_W of the weak and electromagnetic interactions, but as shown in Sec. X, in general it is larger.) Thus the different solutions corresponding to the different unbroken subgroups (6.1) are physically inequivalent, and we must decide which is the correct one.

This problem was encountered in a different context some time ago by Dashen,[15] who gave a general solution. It is necessary to construct a potential, given to lowest order as

$$V(g) = \langle 0, g | \mathcal{H}' | 0, g \rangle, \tag{6.2}$$

where $\mathcal{H}'$ is the symmetry-breaking perturbation (in our case an operator of second order in the weak and electromagnetic interactions) and $|0,g\rangle$ is the vacuum corresponding to the solution that has unbroken subgroup $H(g)$:

$$H(g)|0,g\rangle = |0,g\rangle. \tag{6.3}$$

The g that defines the "correct" solution in the presence of the perturbation $\mathcal{H}'$ is defined by the condition that $V(g)$ be a *minimum*.

For our purposes, it will be much more convenient to keep the solution of the strong- interaction symmetry breaking fixed, and instead vary the way that the weak and electromagnetic gauge group is inserted into the larger $U(N) \otimes U(N)$ global group. That is, we now fix the vacuum and the unbroken subgroup H, and instead let the gauge group be

$$G_W(g) = g^{-1} G_W g, \tag{6.4}$$

where g runs over all elements of $U(N) \otimes U(N)$. This has the advantage that we can fix the choice of the generators t_i and x_a from the beginning; the whole effect of varying the gauge group is that the "charges" $e_{\alpha A}$ are replaced with

$$e_{\alpha A}(g) = \sum_B R_{BA}(g) e_{\alpha B}, \tag{6.5}$$

where $R(g)$ is the regular representation of $U(N) \otimes U(N)$.

Using $e_{\alpha A}(g)$ in place of $e_{\alpha A}$ in Eq. (5.24), we see that the $O(e^2)$ term in the leading part of the effective Lagrangian is

$$\mathcal{L}_2(\Pi, g) = -\sum_{\alpha ABCD} e_{\alpha A}(g) e_{\alpha B}(g) \Lambda_{AC}(\Pi) \Lambda_{BD}(\Pi) I_{CD}. \tag{6.6}$$

The potential (6.2) is just given by the vacuum-fluctuation part of $\mathcal{L}_2$:

$$V(g) = -\mathcal{L}_2(0, g) \tag{6.7}$$

or more explicitly

$$V(g) = -\sum_{\alpha AB} e_{\alpha A}(g) e_{\alpha B}(g) I_{AB}. \tag{6.8}$$

Another interpretation can now be put on the condition that $V(g)$ be a minimum. Using Eqs. (6.5), (B2), (B6), and (5.2) in (6.6), we have

$$\mathcal{L}_2(\Pi, g_1 g) = \mathcal{L}_2(\Pi'(\Pi, g_1), g), \tag{6.9}$$

where g_1 and g are arbitrary elements of $U(N) \otimes U(N)$, and Π' is the image of Π under g_1. Thus the variation of $\mathcal{L}_2(\Pi, g)$ with respect to g may be determined from its variation with respect to Π. In particular, for $\Pi = 0$ and g_1 infinitesimal, Eq. (6.7) and (6.9) give

$$V\left(\left(1 + i\sum_a \frac{\epsilon_a x_a}{F_a}\right) g\right) = -\mathcal{L}_2(\epsilon, g), \tag{6.10}$$

so the condition that $V(g)$ be stationary with respect to g is equivalent to the condition that

$$\left.\frac{\partial \mathcal{L}_2(\Pi, g)}{\partial \Pi_a}\right|_{\Pi=0} = 0. \tag{6.11}$$

Graphically, this says that tadpole graphs, in which a single Goldstone boson disappears into the vacuum, necessarily vanish. The rationale for this condition is that otherwise perturbation theory in e would break down; the dominator of the propagator of a Goldstone boson at zero four-momentum is at most of order e^2 (see below) so a tadpole produced by second-order effects of the weak and electromagnetic interactions would be of zeroth order in e.[24]

Not only must we choose the gauge group $G_W(g)$ so that $V(g)$ is stationary; we must choose it so that $V(g)$ is at least a local *minimum*. The reason is again to be found in Eq. (6.9); the condition that $V(g)$ be a minimum is equivalent to the condition that $\Pi = 0$ be a minimum of $-\mathcal{L}_2(\Pi, g)$, and this in turn ensures the positivity of the mass matrix

$$m^2_{ab}(g) = -\left.\frac{\partial^2}{\partial \Pi_a \partial \Pi_b} \mathcal{L}_2(\Pi, g)\right|_{\Pi=0}. \tag{6.12}$$

From now on we will assume that g has been chosen from the beginning so that (6.11) is satisfied and (6.12) is positive. We can therefore drop the explicit argument g everywhere.

In order to put these conditions in a more useful form, we note that

$$\Lambda(\Pi) = \exp\left(i\sum_a X_a \Pi_a / F_a\right), \tag{6.13}$$

where $(X_a)_{AB}$ is the matrix representing the generator x_a in the adjoint representation of U(N)⊗U(N). Hence (5.24) gives

$$\mathcal{L}_2(\Pi) = -\operatorname{Tr}\left[E \exp\left(i\sum_a \frac{X_a\Pi_a}{F_a}\right) I \exp\left(-i\sum_a \frac{X_a\Pi_a}{F_a}\right)\right], \tag{6.14}$$

where E is the matrix

$$E_{AB} \equiv \sum_\alpha e_{\alpha A} e_{\alpha B}. \tag{6.15}$$

Condition (6.11) therefore reads

$$\operatorname{Tr}\{E[X_a, I]\} = 0, \tag{6.16}$$

and (6.12) gives the mass matrix as

$$m^2{}_{ab} = -\frac{1}{2F_aF_b}\operatorname{Tr}\{E[X_a,[X_b,I]] + E[X_b,[X_a,I]]\}. \tag{6.17}$$

Since I_{AB} is H invariant, we can generalize (6.16) to

$$\operatorname{Tr}\{E[\Lambda_A, I]\} = 0 \tag{6.18}$$

where Λ_A is the adjoint representation of an arbitrary generator λ_A of U(N)⊗U(N). Also, Eq. (6.17) can be simplified because the two terms on the right are equal; the Jacobi identity gives their difference as

$$\operatorname{Tr}\{E[X_a,[X_b,I]]\} - \operatorname{Tr}\{E[X_b,[X_a,I]]\} = \operatorname{Tr}\{E[[X_a,X_b,],I]\},$$

and this vanishes according to (6.18). Thus, (6.17) may be written

$$m^2{}_{ab} = m^2{}_{ba} = -\frac{1}{F_aF_b}\operatorname{Tr}\{E[X_a,[X_b,I]]\}. \tag{6.19}$$

VII. UNITARITY GAUGE AND VECTOR BOSON MASSES

The effective Lagrangian derived in Sec. V is still locally invariant under the gauge group G_W of the weak and electromagnetic interactions. We are therefore free to adopt a "unitarity gauge," in which the particle content of the theory is explicitly displayed.[16]

Suppose for a moment that the gauge group G_W were a sufficiently large subgroup of U(N)⊗U(N), so that its generators w_α, together with the generators t_i of the unbroken subgroup H, would completely span the algebra of U(N)⊗U(N). (This includes the case usually discussed, where G_W is the whole of the original symmetry group.) Then any element of U(N)⊗U(N) could be written as a product of an element of G_W times an element of H. This would in particular be true of the element $\exp(i\Sigma_a x_a\Pi_a/F_a)$, and therefore we could write

$$\exp\left(i\sum_a x_a\Pi_a/F_a\right) = \exp\left(-i\sum_\alpha \Theta_\alpha(\Pi)w_\alpha\right) \times \exp\left(i\sum_i \mu_i(\Pi)t_i\right)$$

for some real parameters $\theta_\alpha(\Pi)$ and $\mu_i(\Pi)$. But by comparing this with Eq. (5.4), we see that the gauge transformation $\exp(i\Sigma_\alpha\theta_\alpha w_\alpha)$ would carry the Goldstone boson field Π_a into $\Pi'_a = 0$. Thus, in this case there would be a choice of gauge which eliminates all Goldstone bosons.

In the general case, we do *not* expect the generators of G_W and H to span the algebra of U(N)⊗U(N). Note that if G_W is too large, then the weak and electromagnetic interactions will not break the symmetries of H sufficiently; we would then find that any quarks which do not get masses of order M from the spontaneous breakdown of U(N)⊗U(N) to H will remain massless to all orders in e (see Sec. X).

However, as shown in Appendix D, we can always write a general element λ of the algebra of U(N)⊗U(N) in the form

$$\lambda = -\sum_\alpha \theta_\alpha w_\alpha + \sum_a \phi_a x_a + \sum_i \mu_i t_i, \tag{7.1}$$

with ϕ_a constrained by the condition that

$$0 = \operatorname{Tr}\left(\theta_\alpha \sum_a \phi_a F_a{}^2 x_a\right) = \sum_{aA} e_{\alpha A} B_{aA} F_a{}^2 \phi_a. \tag{7.2}$$

(The reason for adopting this particular constraint, and in particular for inserting the factor $F_a{}^2$, will be made clear below.) Hence, since every element of U(N)⊗U(N) that is infinitesimally close to the identity may be expressed in the form

$$\exp\left(-i\sum_\alpha \theta_\alpha w_\alpha\right)\exp\left(i\sum_a \phi_a x_a\right)\exp\left(i\sum_i \mu_i t_i\right)$$

[with ϕ_a satisfying (7.2)], it follows that every element of U(N)⊗U(N) in at least some finite region around the identity may be written in this form. In particular we may write

$$\exp\left(i\sum_a x_a\Pi_a/F_a\right) = \exp\left(-i\sum_\alpha \theta_\alpha(\Pi)w_\alpha\right) \times \exp\left(i\sum_a \phi_a(\Pi)x_a\right) \times \exp\left(i\sum_i \mu_i(\Pi)t_i\right). \tag{7.3}$$

But Eq. (5.4) then tells us that the gauge transformation $\exp[+i\Sigma_a\theta_\alpha(\Pi)w_\alpha]$ carries the Goldstone boson field Π_a into

$$\Pi'_a/F_a = \phi_a(\Pi). \tag{7.4}$$

That is, we can adopt a gauge in which, according to Eq. (7.2),

$$\sum_{aA} e_{\alpha A} B_{aA} F_a \Pi_a' = 0 \tag{7.5}$$

or equivalently

$$\mathrm{Tr}\left(w_\alpha \sum_a x_a F_a \Pi_a'\right) = 0. \tag{7.6}$$

This is the unitarity gauge.

The unitarity gauge as we have defined it has the crucial property of eliminating the zeroth-order mixing between the Goldstone bosons and the G_W vector bosons. From Eqs. (5.7) and (5.15), we see that the part of the effective Lagrangian that is quadratic in gauge and/or Goldstone fields is

$$-\tfrac{1}{2}\sum_a (D_\mu \Pi_a')_{\mathrm{lin}} (D^\mu \Pi_a')_{\mathrm{lin}}, \tag{7.7}$$

with $(D_\mu \Pi')_{\mathrm{lin}}$ the linear part of the covariant derivative

$$(D_\mu \Pi_a')_{\mathrm{lin}} = \partial_\mu \Pi_a' + F_a \sum_{A\alpha} e_{\alpha A} W_{\alpha\mu}' B_{aA}. \tag{7.8}$$

Hence Eq. (7.5) has the effect of insuring that the Π-W cross terms drop out in the quadratic part of the effective Lagrangian. It was to bring this about that we inserted the factor F_a^2 in Eq. (7.2), and it is this feature of the unitarity gauge that justifies the statement that it correctly displays the particle content of the theory.

The same result can be obtained by a simple generalization of the method of Jackiw and Johnson and Cornwall and Norton.[2] In a general gauge there are "black boxes," connecting a single G_W-gauge boson with a single Goldstone boson line. If we sum up the pole singularities produced by a linear chain of Π and W lines, we find that the only poles in the sum which correspond to particles of zero spin and zero mass are those in channels described by precisely the condition (7.5). It is also easy to see that the unitarity gauge as usually defined in theories with elementary scalar fields does satisfy Eq. (7.5).

Now that we have eliminated the Π-W cross terms, the mass of the G_W vector bosons may be read off from the effective Lagrangian. Equation (7.7) contains a term quadratic in W

$$-\tfrac{1}{2}\sum_{\alpha\beta} \mu^2{}_{\alpha\beta}\, W_{\alpha\mu}' W_\beta'^{\mu}. \tag{7.9}$$

with a vector-boson mass

$$\mu^2{}_{\alpha\beta} = \sum_{aAB} F_a{}^2 B_{aA} B_{aB} e_{\alpha A} e_{\alpha B} \tag{7.10}$$

or equivalently

$$\mu^2{}_{\alpha\beta} = \tfrac{1}{64}\sum_a F_a{}^2 \mathrm{Tr}(x_a w_\alpha)\mathrm{Tr}(x_a w_\beta). \tag{7.11}$$

It is immediately apparent from (7.11) that $\mu^2{}_{\alpha\beta}$ vanishes if either w_α or w_β is the generator of a symmetry that is not spontaneously broken. Also, the μ's that are not zero are of order eF, and since the Fermi coupling constant G_F must be of order e^2/μ^2, we can conclude that

$$M \approx F \approx G_F{}^{-1/2} \approx 300 \text{ GeV}$$

as previously indicated.

VIII. CLASSIFICATION OF THE GOLDSTONE BOSONS

The other side of the Higgs phenomenon, complementary to the appearance of vector-boson masses, is the disappearance of Goldstone bosons.[11] We will now consider the nature and the mass spectrum of the Goldstone bosons that are *not* eliminated by the Higgs phenomenon.

It is useful to begin by studying the eigenvectors and eigenvalues of the formal Goldstone boson mass matrix $m^2{}_{ab}$, before transformation to the unitarity gauge. First, note that there is an eigenvector of $m^2{}_{ab}$ with eigenvalue zero for every independent broken gauge generator. Every generator w_α of G_W may be written in the form

$$w_\alpha = \sum_a \frac{x_a u_{a\alpha}}{F_a} + h_\alpha, \tag{8.1}$$

where h_α is a linear combination of the generators t_i of the unbroken symmetry subgroup H. But then Eq. (6.19) gives

$$\sum_b m^2{}_{ab} u_{b\alpha} = -\frac{1}{F_a}\mathrm{Tr}\{E[(W_\alpha - H_\alpha), [X_a, I]]\},$$

where W_α and H_α are the matrices representing w_α and h_α in the regular representation of $U(N) \otimes U(N)$. The W term can be rewritten in terms of $[W, E]$, which vanishes because the sum $\sum_\alpha e_{\alpha A} e_{\alpha B}$ is G_W invariant. The H term can be rewritten in terms of the double commutator $[X_a, [H_\alpha, I]]$, which vanishes because I is invariant under H, plus the double commutator $[[H_\alpha, X_\alpha], I]$, which gives no contribution because of the "subgroup-alignment" condition (6.18). Thus, we see that $u_{a\alpha}$ is our eigenvector

$$\sum_b m^2{}_{ab} u_{b\alpha} = 0. \tag{8.2}$$

These will be called the *fictitious Goldstone bosons*, because as we shall see, it is just these that are eliminated by the unitarity gauge condition.

The number of independent fictitious Goldstone bosons is evidently equal to the dimensionality of G_W minus the dimensionality of that subgroup H_W of G_W which is unbroken by the spontaneous symmetry breakdown of $U(N)\otimes U(N)$ to H. Of course, for every generator of H_W there will be a vector boson whose mass remains zero to all orders in e, so the number of fictitious Goldstone bosons equals the number of *massive* vector bosons. (In the real world, H_W would presumably consist solely of electromagnetic gauge transformations.)

The argument that led from (8.1) to (8.2) did not actually depend on the fact that W_α is a generator of G_W, but only on the fact that it is a symmetry of the weak and electromagnetic interactions, so that W_α commutes with E. But in general there will be broken symmetries of the weak interactions that are not themselves generators of G_W. Exactly the same reasoning tells us that for each of these there will be another eigenvector of $m^2{}_{ab}$ with eigenvalue zero. These will be called the *true Goldstone bosons*, because they remain massless to all orders in e but, as we shall see, they are not eliminated by the Higgs phenomenon. The number of true and fictitious Goldstone bosons is equal to the dimensionality of the complete global symmetry group S_W of the weak and electromagnetic interactions, minus the dimensionality of that subgroup of S_W which is left unbroken by the spontaneous breakdown of $U(N)\otimes U(N)$ to H. Of course, the occurrence of true Goldstone bosons would present grave difficulties for any theory that has pretensions of providing a realistic model of the actual world. These problems are further discussed in Sec. X.

Finally, there will in general be eigenvectors u_a of $m^2{}_{ab}$ for which the quantity $\sum_a u_a x_a/F_a$ *cannot* be expressed as a linear combination of generators of S_W and generators of H. There is no reason in this case why the eigenvalue should vanish, so we expect a mass m of order

$$m^2 \approx e^2 I/F^2 \approx e^2 M^2 \tag{8.3}$$

about the same as for the vector bosons. These are called the *pseudo-Goldstone bosons*,[17] because they are not the Goldstone bosons of any true symmetry of the whole theory, but only of an accidental approximate symmetry which appears exact in the limit $e \to 0$. As we shall see, the pseudo-Goldstone bosons, like the true Goldstone bosons, are not eliminated by the Higgs phenomenon. The total number of all Goldstone bosons, fictitious, true and pseudo, is simply equal to the dimensionality $2N^2$ of $U(N)\otimes U(N)$ minus the dimensionality of the unbroken subgroup H.

By use of the familiar Schmidt orthogonalization technique, we can choose an orthonormal set of eigenvectors of $m^2{}_{ab}$

$$\sum_b m^2{}_{ab} u_b^n = m_n{}^2 u_a^n , \tag{8.4}$$

$$\sum_a u_a^n u_a^m = \delta_{nm} , \tag{8.5}$$

with each u^n representing either a fictitious, true, or pseudo-Goldstone bosons. That is, we first choose an orthonormal set of u_a^n vectors corresponding to fictitious Goldstone bosons, for which $\sum u_a^n x_a/F_a$ is a linear combination of H and G_W generators; then add an orthonormal set corresponding to true Goldstone bosons, for which $\sum u_a^n x_a/F_a$ is a linear combination of H and S_W but not H and G_W generators; and finally add an orthonormal set corresponding to pseudo-Goldstone bosons, for which $\sum u_a^n x_a/F_a$ is not a linear combination of H and S_W generators. With a set of orthonormal vectors u^n constructed in this way, we can define a corresponding set of Goldstone boson fields

$$\Pi^n \equiv \sum_a u_a^n \Pi_a , \tag{8.6}$$

each of definite mass and type (fictitious, true, or pseudo). Further, since the u^n form a complete set, we can also write

$$\Pi_a = \sum_n u_a^n \Pi^n . \tag{8.7}$$

In particular, we have

$$\sum_a \partial_\mu \Pi^a \partial^\mu \Pi^a = \sum_n \partial_\mu \Pi^n \partial^\mu \Pi^n ,$$

so the Π^n are canonically normalized.

Now let us impose the condition (7.6) that defines the unitarity gauge. A vector u^n which corresponds to a true or a pseudo-Goldstone boson will be orthogonal to all u_a^n corresponding to the fictitious Goldstone bosons, and hence also to the vectors $u_{a\alpha}$ defined by Eq. (8.1). But it follows then that

$$\mathrm{Tr}\Big(w_\alpha \sum_a x_a u_a^n F_a\Big) = \mathrm{Tr}\Big(\sum_b \frac{x_b u_{b\alpha}}{F_b} \sum_a x_a u_a^n F_a\Big)$$

$$= 8\sum_a u_{a\alpha} u_a^n = 0 .$$

Hence the unitarity gauge condition (7.6) imposes no constraint on the fields Π^n representing true or pseudo-Goldstone bosons. On the other hand, a u_a^n which corresponds to a fictitious Goldstone boson allows the decomposition

$$\sum_n \frac{u_a^n x_a}{F_a} = w^n + h^n ,$$

where w^n and h^n are generators of G_W and H, respectively. It follows that

$$\mathrm{Tr}\Big(w^n \sum_a x_a u_a^m F_a\Big) = \mathrm{Tr}\Big(\sum_b \frac{x_b u_b^n}{F_b} \sum_a x_a u_a^m F_a\Big)$$

$$= 8\sum_a u_a^n u_a^m = 8\delta_{nm} .$$

Thus (7.5) requires that

$$0=\mathrm{Tr}\left(w^n\sum_a x_a\Pi_a' F_a\right)=8\Pi^{n\prime}.$$

We conclude that *the whole effect of the condition of unitarity gauge is to eliminate the* Π^n *corresponding to fictitious Goldstone bosons,* leaving the masses and fields of the true and the pseudo-Goldstone bosons unchanged.

IX. AN EXAMPLE

We shall now descend from the generality of the previous discussion to the consideration of one specific example. It probably is unnecessary to remark that this model is totally unrealistic as a theory of real particles or interactions; it is presented solely for the purposes of illustration.

Our model contains two color multiplets of fermions, called q and h. (Color indices are dropped everywhere.) In the limit $e\to 0$, the strong interactions are necessarily invariant under a group $U(2)\otimes U(2)$ of global transformations

$$\begin{pmatrix} q\\ h\end{pmatrix}\to U\begin{pmatrix} q\\ h\end{pmatrix}, \tag{9.1}$$

with U a unitary matrix (commuting with color), involving both the Dirac matrices 1 and γ_5. The generators of this algebra are defined as

$$\begin{aligned}\vec{\lambda}_L&\equiv\frac{1}{\sqrt2}(1+\gamma_5)\vec{\tau},\quad \lambda_{0L}\equiv\frac{1}{\sqrt2}(1+\gamma_5),\\ \vec{\lambda}_R&\equiv\frac{1}{\sqrt2}(1-\gamma_5)\vec{\tau},\quad \lambda_{0R}\equiv\frac{1}{\sqrt2}(1-\gamma_5),\end{aligned}\tag{9.2}$$

with $\vec{\tau}$ the usual 2×2 Pauli matrices.

We assume that the $U(2)\otimes U(2)$ symmetry is dynamically broken down to the largest subgroup which will allow *one* of the two fermion multiplets to acquire a mass:

$$H=U(1)\otimes U(1)\otimes U(1). \tag{9.3}$$

(The color gauge group G_S is assumed to remain unbroken.) We can always define the fermion fields so that it is q that remains massless; the labels "q" and "h" thus stand for "quark" and "heavy fermion." With this definition, the generators of H are

$$\begin{aligned} t_L&\equiv\frac{1}{\sqrt2}(\lambda_{0L}+\lambda_{3L})=\tfrac12(1+\gamma_5)(1+\tau_3),\\ t_R&\equiv\frac{1}{\sqrt2}(\lambda_{0R}+\lambda_{3R})=\tfrac12(1-\gamma_5)(1+\tau_3),\\ t_0&\equiv\tfrac12(\lambda_{0L}+\lambda_{0R}-\lambda_{3L}-\lambda_{3R})=\frac{1}{\sqrt2}(1-\tau_3).\end{aligned}\tag{9.4}$$

We can complete an orthonormal basis for $U(N)\otimes U(N)$ with the five additional generators

$$\begin{aligned} x_{1L}&\equiv\lambda_{1L}=\frac{1}{\sqrt2}(1+\gamma_5)\tau_1,\\ x_{2L}&\equiv\lambda_{2L}=\frac{1}{\sqrt2}(1+\gamma_5)\tau_2,\\ x_{1R}&\equiv\lambda_{1R}=\frac{1}{\sqrt2}(1-\gamma_5)\tau_1,\\ x_{2R}&\equiv\lambda_{2R}=\frac{1}{\sqrt2}(1-\gamma_5)\tau_2,\\ x_0&=\tfrac12(\lambda_{0L}-\lambda_{0R}-\lambda_{3L}+\lambda_{3R})=\frac{1}{\sqrt2}\gamma_5(1-\tau_3).\end{aligned}\tag{9.5}$$

With respect to the $O(2)\otimes O(2)$ group generated by t_L and t_R, these generators transform according to the representations

$\{x_{1L},x_{2L}\}$ (2, 1),
$\{x_{1R},x_{2R}\}$ (1, 2),
x_0 (1, 1).

The Goldstone bosons Π_a may be chosen to belong to corresponding $O(2)\otimes O(2)$ multiplets. With this definition, we automatically have

$$\langle 0|J_A^\mu|\Pi_a\rangle\propto B_{aA}F_a, \tag{9.6}$$

where B_{aA} are the coefficients in (9.5),

$$x_a=\sum_A B_{aA}F_a\lambda_A,$$

and the F's are equal within irreducible H multiplets,

$$F_{1L}=F_{2L}\equiv F_L,\quad F_{1R}=F_{2R}\equiv F_R. \tag{9.7}$$

We expect that the F's are of the same order of magnitude as the heavy quark mass,

$$F_L\approx F_R\approx F_0\approx M_h. \tag{9.8}$$

It is straightforward to calculate the covariant derivatives (5.6) and (5.7) for $e=0$ as power series in the Π_a. For instance, the effective Lagrangian contains a bilinear interaction of Goldstone bosons with quarks

$$-\frac{1}{F_L^2}\bar q\gamma^\mu(1+\gamma_5)q(\Pi_{1L}\partial_\mu\Pi_{2L}-\Pi_{2L}\partial_\mu\Pi_{1L})-\frac{1}{F_R^2}\bar q\gamma^\mu(1-\gamma_5)q(\Pi_{1R}\partial_\mu\Pi_{2R}-\Pi_{2R}\partial_\mu\Pi_{1R}) \tag{9.9}$$

and a trilinear self-interaction of Goldstone bosons

$$-\frac{1}{\sqrt2 F_0}\left(1-\frac{F_0^2}{F_L^2}\right)\partial_\mu\Pi_0(\Pi_{1L}\partial^\mu\Pi_{2L}-\Pi_{2L}\partial^\mu\Pi_{1L})+\frac{1}{\sqrt2 F_0}\left(1-\frac{F_0^2}{F_R^2}\right)\partial_\mu\Pi_0(\Pi_{1R}\partial^\mu\Pi_{2R}-\Pi_{2R}\partial^\mu\Pi_{1R}). \tag{9.10}$$

The vector and axial-vector bilinear covariants that can be formed from the q field do not share the same H transformation properties as any of the Π_a, so there is no $\bar{q}qD\Pi$ term in the leading part of the effective Lagrangian. There is, however, a Fermi interaction, which after Fierz transformations may be put in the form

$$-G_{LL}[\bar{q}\gamma^{\mu}(1+\gamma_5)q][\bar{q}\gamma_{\mu}(1+\gamma_5)q]-G_{LR}[\bar{q}\gamma^{\mu}(1+\gamma_5)q][\bar{q}\gamma_{\mu}(1-\gamma_5)q]-G_{RR}[\bar{q}\gamma_{\mu}(1-\gamma_5)q][\bar{q}\gamma_{\mu}(1-\gamma_5)q] \; . \quad (9.11)$$

We expect the constants G_{LL}, G_{LR}, and G_{RR} to be of order $1/M_h^2$.

Now let us turn on the "weak" interactions. We will assume that the weak gauge group is

$$G_W = \mathrm{SU}(2) \quad (9.12)$$

with both left- and right-handed fermion fields $[(1\pm\gamma_5)q, (1\pm\gamma_5)h]$ transforming as G_W doublets. However, we do not immediately know *which* SU(2) subgroup of U(2)⊗U(2) generates the weak interactions. In general, the generators of SU(2) might be any matrices of the form

$$w_\alpha = e\sum_\beta [(g_L)_{\alpha\beta}\lambda_{L\beta}+(g_R)_{\alpha\beta}\lambda_{R\beta}], \quad (9.13)$$

where e is the SU(2) gauge coupling constant; α and β run over the values 1, 2, 3; $\vec{\lambda}_L$ and $\vec{\lambda}_R$ are the matrices (9.2); and g_L and g_R are *unknown* 3×3 orthogonal matrices. Nor is it arbitrary which g matrices we choose; the definition of the fermions has been fixed (up to an H transformation) by our convention that the spontaneous breakdown of U(2)⊗U(2) to H gives a mass to h, not q.

In order to settle this question, we must examine the "potential" term in the effective Lagrangian. In general, this has the form (6.8):

$$V(g) = -\sum_{\alpha AB} e_{\alpha A}(g)e_{\alpha B}(g)I_{AB}, \quad (9.14)$$

where I_{AB} is some unknown H invariant of order M_h^4, and $e_{\alpha A}(g)$ are the coefficients which give the G_W generators as linear combinations of the U(2)⊗U(2) generators

$$e_{\alpha,\beta L}(g) = e(g_L)_{\alpha\beta}, \qquad e_{\alpha,\beta R}(g) = e(g_R)_{\alpha\beta}. \quad (9.15)$$

Thus V here takes the form

$$V(g) = -e^2\sum_{\alpha\beta\gamma}[(g_L)_{\alpha\beta}(g_L)_{\alpha\gamma}I_{L\beta,L\gamma} + 2(g_L)_{\alpha\beta}(g_R)_{\alpha\gamma}I_{L\beta,R\gamma} + (g_R)_{\alpha\beta}(g_R)_{\alpha\gamma}I_{R\beta,R\gamma}].$$

But the g matrices are orthogonal, so this immediately simplifies to

$$V(g) = -2e^2\sum_{\beta\gamma}(g_L^{-1}g_R)_{\beta\gamma}I_{L\beta,R\gamma} + \text{constant}.$$

In addition, the H invariance of I_{AB} requires that it be invariant under *independent* rotations around the 3 axis on either the $L\alpha$ and/or $R\alpha$ indices; thus, in particular,

$$I_{L\beta,R\gamma} = In_\beta n_\gamma, \quad (9.16)$$

where $\vec{n}$ is a unit vector pointing in the 3-direction

$$\vec{n} = (0, 0, 1),$$

and I is some unknown constant of order M_h^4. The potential has now become simply

$$V(g) = -2e^2 I(n, g_L^{-1}g_R n). \quad (9.17)$$

This is to be minimized over the whole range of orthogonal matrices g_L, g_R. The location of such a minimum is quite obvious:

$$\text{(A) For } I>0, \quad g_L^{-1}g_R n = +n. \quad (9.18a)$$

$$\text{(B) For } I<0, \quad g_L^{-1}g_R n = -n. \quad (9.18b)$$

This does not, of course, entirely determine g_L and g_R; given any solution, we can find another of the form

$$g_L' = g_1 g_L g_2, \quad g_R' = g_1 g_R g_3,$$

where g_1 is an arbitrary orthogonal matrix, and g_2 and g_3 are orthogonal matrices representing arbitrary independent rotations around the 3 axis. But g_1 represents a redefinition of the weak gauge couplings by an SU(2) transformation belonging to the gauge group G_W, while g_2 and g_3 represent a redefinition of the fermion fields by a transformation belonging to that subgroup H of U(2)⊗U(2) which is not spontaneously broken. Clearly, there is no way that this remaining ambiguity in the g's could ever be resolved, nor is there any reason why we would wish to do so. Thus, we can freely choose *any* orthogonal g_L and g_R matrices which satisfy the condition for a minimum, Eq. (9.18).

We will now need to consider the two cases separately.

A. $I>0$

Here it is convenient to choose g_L and g_R as unit matrices

$$g_L = g_R = 1. \quad (9.19)$$

The generators of the weak gauge group are then given by (9.13) as

$$\vec{w} = e(\vec{\lambda}_L + \vec{\lambda}_R) = e\sqrt{2}\,\vec{\tau}, \quad (9.20)$$

with $\vec{\tau}$ the usual 2×2 Pauli matrices. There is a single 2×2 matrix which is a generator of *both* a G_W gauge transformation and an unbroken H transformation

$$e\tau_3 = \frac{1}{\sqrt{2}} w_3 = \frac{e}{2}(t_L + t_R) - \frac{e}{\sqrt{2}} t_0 . \quad (9.21)$$

This corresponds to a "photon," which keeps zero mass despite the spontaneous symmetry breaking. The gauge bosons corresponding to the other two generators of G_W, w_1, and w_2, acquire a mass by the Higgs mechanism, given by (7.10) is

$$\mu_1{}^2 = \mu_2{}^2 = e^2(F_L{}^2 + F_R{}^2). \quad (9.22)$$

These massive gauge bosons have w_3 charges $+1$ and -1. Also, q has w_3 charge $+1$, while the Goldstone bosons $\Pi_{1L} \pm i\Pi_{2L}$, $\Pi_{1R} \pm i\Pi_{2R}$, and Π_0 have w_3 charges ± 1, ± 1, and 0, respectively.

In order to classify the Goldstone bosons, we note first that there are two independent linear combinations of the x_a that can be expressed as linear combinations of generators of H and G_W:

$$e(x_{1L} + x_{1R}) = w_1 , \quad (9.23)$$

$$e(x_{2L} + x_{2R}) = w_2 , \quad (9.24)$$

and there is also one other linear combination of the x_a that can be expressed as a linear combination of a generator of H and the generator of the exact global symmetry $\psi \to \exp(i\gamma_5\epsilon)\psi$ of the whole Lagrangian

$$x_0 = \sqrt{2}\,\gamma_5 1 - \frac{1}{\sqrt{2}}(t_L + t_R). \quad (9.25)$$

In accordance with the conclusions of Sec. VIII, we must therefore expect that this theory has two fictitious Goldstone bosons, one true Goldstone boson, and $5 - 2 - 1 = 2$ pseudo-Goldstone bosons. Their fields are of the form

$$\Pi^n = \sum_a u_a^n \Pi_a ,$$

with u^n a set of *orthonormal* vectors subject to certain conditions: For the fictitious Goldstone bosons $\sum_a u_a^n x_a/F_a$ must be a linear combination of (9.23) and (9.24), so the fields are

$$\Pi^{\text{I}} \equiv (F_L{}^2 + F_R{}^2)^{-1/2}(F_L \Pi_{1L} + F_R \Pi_{1R}), \quad (9.26)$$

$$\Pi^{\text{II}} \equiv (F_L{}^2 + F_R{}^2)^{-1/2}(F_L \Pi_{2L} + F_R \Pi_{2R}). \quad (9.27)$$

For the true Goldstone boson, $\sum_a u_a^n x_a/F_a$ must be proportional to (9.25), so the field is simply

$$\Pi^0 \equiv \Pi_0 . \quad (9.28)$$

For the pseudo-Goldstone bosons the u_a^n need only be orthogonal to all the others, so the fields are

$$\Pi^1 \equiv (F_L{}^2 + F_R{}^2)^{-1/2}(-F_R \Pi_{1L} + F_L \Pi_{1R}), \quad (9.29)$$

$$\Pi^2 \equiv (F_L{}^2 + F_R{}^2)^{-1/2}(-F_R \Pi_{2L} + F_L \Pi_{2R}). \quad (9.30)$$

The masses of Π^{I}, Π^{II}, and Π^0 are zero, while second-order weak (and "electromagnetic") effects will give Π^1 and Π^2 masses that are equal (because w_3 invariance is unbroken) and of order eM_h. These masses are proportional to the constant I in Eq. (9.16), but I is unknown, so the calculation is not worthwhile. The effect of the transformation to unitarity gauge is just to eliminate the fields of the fictitious Goldstone bosons:

$$\Pi^{\text{I}'} = \Pi^{\text{II}'} = 0 .$$

The five "old" fields Π_a' may then be expressed in terms of the three "new" fields $\Pi^{n'}$ as

$$\Pi_{1L}' = -\frac{F_R}{(F_L{}^2 + F_R{}^2)^{1/2}} \Pi^{1'},$$

$$\Pi_{1R}' = \frac{F_L}{(F_L{}^2 + F_R{}^2)^{1/2}} \Pi^{1'},$$

$$\Pi_{2L}' = -\frac{F_R}{(F_L{}^2 + F_R{}^2)^{1/2}} \Pi^{2'},$$

$$\Pi_{2R}' = \frac{F_L}{(F_L{}^2 + F_R{}^2)^{1/2}} \Pi^{2'},$$

and, of course

$$\Pi_0' = \Pi^{0'} .$$

In particular, the bilinear interaction (9.9) of the nonfictitious Goldstone bosons with the quarks is (now dropping primes)

$$(F_L{}^2 + F_R{}^2)^{-1/2}\left[-\frac{F_R}{F_L{}^2}\bar{q}\gamma^\mu(1+\gamma_5)q - \frac{F_L}{F_R{}^2}\bar{q}\gamma^\mu(1-\gamma_5)q\right](\Pi^1\partial_\mu\Pi^2 - \Pi^2\partial_\mu\Pi^1), \quad (9.31)$$

and their trilinear self-interaction (9.10) is

$$2^{-1/2}(F_L{}^2 + F_R{}^2)^{-1/2}F_0{}^{-1}\left[F_R\left(1 - \frac{F_0{}^2}{F_L{}^2}\right) + F_L\left(1 - \frac{F_0{}^2}{F_R{}^2}\right)\right]\partial_\mu\Pi^0(\Pi^1\partial^\mu\Pi^2 - \Pi^2\partial^\mu\Pi^1). \quad (9.32)$$

The quark mass matrix is given to order e^2M_h by Eq. (5.29), which here becomes

$$M_q = e^2 \sum_\alpha (Q_{\alpha L,\alpha L} + 2Q_{\alpha L,\alpha R} + Q_{\alpha R,\alpha R}),$$

where Q_{AB} is some constant matrix of order M_h which transforms as a tensor under H, in the sense of Eq. (C4). (Again, we let α and β run over the values 1, 2, 3.) It is straightforward to show that the most general such tensor has

$$Q_{\alpha L,\beta L} = Q_{\alpha R,\beta R} = 0,$$

and also (with a suitable choice of relative phase for the left- and right-handed quark fields)

$$Q_{\alpha L,\beta R} = Q[\tfrac{1}{2}(1+\gamma_5)n^+_\alpha n^-_\beta + \tfrac{1}{2}(1-\gamma_5)n^-_\alpha n^+_\beta],$$

where $Q \approx M_h$ is some unknown constant, and

$$\vec{n}^\pm \equiv \frac{1}{\sqrt{2}}(1, \pm i, 0).$$

Thus, the quark here does acquire a mass of order e^2M_h

$$M_q = 2e^2Q. \tag{9.33}$$

Also, using Eq. (5.28), the Yukawa coupling here is

$$\frac{4iM_q}{\sqrt{2}F_0}\bar{q}\gamma_5 q\Pi^0 \tag{9.34}$$

as required by a Goldberger-Treiman relation.

It happens that in this model either the massive vector bosons or the pseudo-Goldstone bosons are absolutely stable. This is just because they have w_3 charges ± 1; the only lighter states into which they could decay consist of "photons," true Goldstone bosons, and quark-antiquark pairs, all of which are w_3 neutral.

B. $I<0$

Here it is convenient to choose g_L and g_R as equal and opposite rotations of 90° about the 1 axis. The generators of the weak gauge group are then

$$w_1 = e(\lambda_{1L} + \lambda_{1R}) = e\sqrt{2}\,\tau_1,$$

$$w_2 = e(\lambda_{3L} - \lambda_{3R}) = e\sqrt{2}\gamma_5\tau_3,$$

$$w_3 = e(-\lambda_{2L} + \lambda_{2R}) = -e\sqrt{2}\gamma_5\tau_2,$$

with $\vec{\tau}$ again the 2×2 Pauli matrices. No linear combination of these generators is a generator of the unbroken subgroup H, so there is no "photon" here—every vector boson gets a mass. From (7.11), we find that the masses are

$$\mu_1^2 = \mu_3^2 = e^2(F_L^2 + F_R^2),$$

$$\mu_2^2 = F_0^2.$$

In order to classify the Goldstone bosons, we note first that there are three independent linear combinations of the x_a that can be expressed as linear combinations of the generators of H and G_W:

$$x_{1L} + x_{1R} = w_1/e, \tag{9.35}$$

$$x_0 = \frac{1}{\sqrt{2}}(t_L - t_R) - w_2/e, \tag{9.36}$$

$$x_{2L} - x_{2R} = -w_3/e, \tag{9.37}$$

while there is no other linear combination of the x_a that can be expressed as a linear combination of a generator of H and the generator of any exact global symmetry of the whole Lagrangian. According to Sec. VIII, we must now expect that this theory has three fictitious Goldstone bosons, no true Goldstone bosons, and $5-3=2$ pseudo-Goldstone bosons. Their fields are of the form

$$\Pi^n = \sum_a u^n_a \Pi_a, \tag{9.38}$$

with u^n_a a set of orthonormal vectors subject to certain conditions: For the fictitious Goldstone bosons $\sum_a u^n_a x_a/F_a$ must be a linear combination of (9.35)–(9.37), so the fields are

$$\Pi^{\mathrm{I}} = (F_L^2 + F_R^2)^{-1/2}(F_L\Pi_{1L} + F_R\Pi_{1R}), \tag{9.39}$$

$$\Pi^0 = \Pi_0, \tag{9.40}$$

$$\Pi^{\mathrm{II}} = (F_L^2 + F_R^2)^{-1/2}(F_L\Pi_{2L} - F_R\Pi_{2R}). \tag{9.41}$$

For the pseudo-Goldstone bosons the u^n must simply be orthogonal to the others, so the fields are

$$\Pi^1 = (F_L^2 + F_R^2)^{-1/2}(-F_R\Pi_{1L} + F_L\Pi_{1R}), \tag{9.42}$$

$$\Pi^2 = (F_L^2 + F_R^2)^{-1/2}(F_R\Pi_{2L} + F_L\Pi_{2R}). \tag{9.43}$$

The masses of Π^{I}, Π^{II}, and Π^0 are zero, while second-order weak effects give Π^1 and Π^2 masses proportional to I and of order eM_h. These latter masses are equal, because the whole Lagrangian has an exact global symmetry which is not spontaneously broken, generated by

$$\lambda_{0L} - \lambda_{0R} + \lambda_{3L} - \lambda_{3R} = t_L - t_R = \lambda_{0L} - \lambda_{0R} + w_2/e, \tag{9.44}$$

and this rotates Π^1 and Π^2 into each other.

The effect of the transformation to unitarity gauge is to eliminate the fields of the fictitious Goldstone bosons

$$\Pi^{\mathrm{I}\prime} = \Pi^{0\prime} = \Pi^{\mathrm{II}\prime} = 0.$$

The five fields Π'_a may be expressed in terms of the two remaining Π^n, fields, as

$$\Pi'_{1L} = -(F_L{}^2+F_R{}^2)^{-1/2}F_R\Pi^{1\prime},$$
$$\Pi'_{1R} = (F_L{}^2+F_R{}^2)^{-1/2}F_L\Pi^{1\prime},$$
$$\Pi'_{2L} = +(F_L{}^2+F_R{}^2)^{-1/2}F_R\Pi^{2\prime},$$
$$\Pi'_{2R} = (F_L{}^2+F_R{}^2)^{-1/2}F_L\Pi^{2\prime},$$
$$\Pi'_0 = 0.$$

In particular, the bilinear interaction (9.9) of the nonfictitious Goldstone bosons with the quarks is (now dropping primes)

$$\left[\frac{F_R}{F_L{}^2(F_L{}^2+F_R{}^2)^{1/2}}\,\bar{q}\gamma^\mu(1+\gamma_5)q-\frac{F_L}{F_R{}^2(F_L{}^2+F_R{}^2)^{1/2}}\,\bar{q}\gamma^\mu(1-\gamma_5)q\right](\Pi^1\partial_\mu\Pi^2-\Pi^2\partial_\mu\Pi^1). \tag{9.45}$$

The trilinear coupling (9.10) now vanishes. Also, as a consequence of the exact unbroken symmetry generated by (9.44), the quark mass remains zero to all orders in e. The Yukawa $\Pi\bar{q}q$ coupling vanishes, as required by a Goldberger-Treiman relation, even though it is not forbidden by the symmetry generated by (9.44).

The moral of this analysis is twofold. First, a theory with a given group-theoretic character and a given field content can have enormously different physical consequences depending on how the subgroups G_W and H line up with each other. The differences between the two cases found in this section are summarized in Table I. In addition, although these theories do not have the predictive power of a theory in which the spontaneous symmetry breaking is due to vacuum expectation values of weakly coupled scalar fields, the predictive power of theories with dynamical symmetry breaking is by no means negligible.

X. IMPLICATIONS

The foregoing analysis has been chiefly concerned with mathematical formalism rather than physical applications. We close with some remarks about the implications of this analysis for real particles and interactions.

The weak interactions in this class of theories arise both from the exchange of intermediate vector bosons and also from a direct Fermi interaction in the effective Lagrangian. Both are of the same order of magnitude; the direct Fermi interaction has a coupling constant of order M^{-2} (where M is the scale associated with the dynamical symmetry breaking), while the exchange of vector bosons of mass $\mu\approx eM$ produces an effective Fermi coupling of order $e^2/\mu^2\approx M^{-2}$. Either way, we are led to the estimate that $M\approx$ 300 GeV.

The two kinds of weak interactions can of course be distinguished by their energy dependence at energies of order eM or greater. They may also be distinguished even at lower energies by their symmetry properties; the direct Fermi interaction is invariant under the unbroken subgroup H of U(N) ⊗U(N), while the vector-boson exchange interaction is not.

Some of the fermions of these theories may get masses of order M from the dynamical symmetry breaking. However, the ordinary quarks $\mathcal{P}, \mathcal{N}, \lambda$, $\mathcal{P}'$ (etc.?) can hardly be this heavy,[25] so we must suppose that the unbroken subgroup H must be large enough to prevent the appearance of masses of order M. The masses of the ordinary quarks would then have to arise from higher-order corrections, which would presumably give them values of order e^2M. This is a gratifying result, for it offers at least a qualitative explanation of the mysterious fact that the ratio of the mass scale of the hadrons (say, 1 GeV) to that of the Fermi interaction (300 GeV) is roughly of order α.

Of course, in order to produce quark masses of order e^2M, the unbroken symmetry group H must not be too large. Specifically, there must be no chiral symmetries in H which are also symmetries of the weak and electromagnetic interactions. The breakdown of U(N) ⊗ U(N) to H may be signaled by the appearance of fermion masses of order M, but

TABLE I. Summary of the properties of the model discussed in Section IX, in two cases corresponding to the two different possible minima of the potential $V(g)$.

	A	B
Massive vector bosons	2 degenerate	3 (2 degenerate)
Massless vector bosons	1	0
Unbroken global symmetries (including fermion conservation)	1	2
True Goldstone bosons	1	0
Pseudo-Goldstone bosons	2 degenerate	2 degenerate
Quark mass	$\approx e^2M_h$	0

it is also possible that H forbids all fermion masses of order M while allowing other nonchiral interactions, such as scalar, tensor, or pseudoscalar Fermi interactions of order M^{-2}.

In all cases that I have examined, H must be sufficiently small so that there are some broken symmetries, not in H, that are also not in the weak and electromagnetic gauge group G_W. In particular, in a theory with N' heavy fermions and $N-N'$ ordinary quarks, the broken symmetry generators include all Hermitian matrices, chiral or nonchiral, which connect ordinary quark and heavy fermion fields. If these were all generators of the weak and electromagnetic gauge group, their multiple commutators would be also gauge generators. But these commutators span the algebra of $SU(N) \otimes SU(N)$. This is not possible because then all the unbroken chiral symmetries in $SU(N) \otimes SU(N)$, which keep the ordinary quarks from getting masses of order M, would also be symmetries of the weak and electromagnetic interactions, so that the ordinary quarks could not get masses of any order in e. Also, the weak and electromagnetic gauge group cannot include $SU(N) \otimes SU(N)$; triangle anomalies would make such a theory nonrenormalizable.[26]

For every broken symmetry which is not a symmetry of the weak and electromagnetic interactions, there is a pseudo-Goldstone boson that is not eliminated by the Higgs phenomenon. These particles have masses of order $e \times 300$ GeV, and do not interact strongly at ordinary energies, but the charged pseudo-Goldstone bosons could of course be pair-produced by the electromagnetic interactions. Thus, it will be important to distinguish carefully between pseudo-Goldstone bosons and intermediate vector bosons when colliding beams reach energies adequate to produce such particles in pairs.

It makes a great difference in the description of the decay modes and interactions of pseudo-Goldstone bosons whether they can interact with ordinary hadrons and leptons singly, or only in pairs. The broken symmetry generator corresponding to a given pseudo-Goldstone boson might have no matrix elements between quark or lepton states, but only between states of which one is a heavy (≈ 300 GeV) fermion. The pseudo-Goldstone bosons would still interact in pairs with ordinary hadrons, as in Eqs. (9.31) and (9.45), but they could only decay into each other. On the other hand, the quarks might not be entirely neutral under the various broken symmetry generators in $U(N) \otimes U(N)$, in which case some of the pseudo-Goldstone bosons would be able to decay into ordinary hadrons. In the absence of a candidate for a realistic model, it is not worth pursuing these various possibilities in great detail.

In addition to pseudo-Goldstone bosons, such theories will usually have true Goldstone bosons of zero mass which also are not eliminated by the Higgs mechanism. This is because there are always some broken global symmetries of the weak and electromagnetic interactions which are not themselves elements of the weak and electromagnetic gauge group. For instance, one such global symmetry is the U(1) chiral transformation which multiplies all fermion fields with a common factor $\exp(i\gamma_5\theta)$; this must not be a member of the weak and electromagnetic gauge group, because if it were then triangle anomalies would make the theory nonrenormalizable, and it must be spontaneously broken, because otherwise none of the fermions in the theory could pick up any mass. If the generator x of any such broken exact global symmetry could be written as a sum of a gauge symmetry generator w and a nonspontaneously broken symmetry generator h, then the corresponding Goldstone boson would be eliminated by the Higgs mechanism; however, in this case the theory would have an extra exact nonspontaneously broken symmetry (apart from fermion conservation) generated by $x - w = h$. This is what happens in case B of the model discussed in Sec. IX; the extra symmetry there keeps the quark massless to all orders in e. If we do not want to allow such extra exact symmetries then the generators of broken nongauge symmetries of the weak and electromagnetic interactions must not be linear combinations of gauge and unbroken symmetry generators, and the corresponding Goldstone bosons cannot be eliminated by the Higgs mechanism.

In the particular case of the chiral U(1) symmetry mentioned above, it is possible that the massless true Goldstone boson, although not eliminated by the Higgs mechanism, would nevertheless not be observable as a free particle. There is a triangle anomaly connecting one $\gamma_\mu\gamma_5$ vertex to two colored gluons; this anomaly forces us to include gluon terms in the conserved chiral current, which make it not gluon-gauge invariant. Since the corresponding true Goldstone boson cannot be proved to appear as a pole in any gluon-gauge-invariant operator, there is at least a chance that it is a trapped particle, like the unobserved ninth pseudoscalar meson[27] with mass $< \sqrt{3}\, m_\pi$.

In a variety of models there are also true Goldstone bosons which *could* be observed as free particles. For instance, in the familiar four-quark version[28] of the $SU(2) \otimes U(1)$ model there is (in the absence of elementary spin-zero fields) an exact global symmetry of the weak and electromagnetic as well as the strong interactions, of the form

$$\mathfrak{N}_R \to \cos\phi\, \mathfrak{N}_R + \sin\phi\, \lambda_R,$$
$$\lambda_R \to -\sin\phi\, \mathfrak{N}_R + \cos\phi\, \lambda_R,$$
$$\mathcal{P}_R, \mathcal{P}'_R, \mathcal{P}_L, \mathfrak{N}_L, \mathcal{P}'_L, \lambda_L \text{ invariant.}$$

[As usual a subscript L or R denotes multiplication with $(1+\gamma_5)$ or $(1-\gamma_5)$, respectively.] This must be spontaneously broken, for otherwise λ and $\mathfrak{N}$ quarks could not have any mass, and for the same reason it cannot be expressed as a sum of a gauge and an unbroken generator. Also, in this case the symmetry current is both conserved and gluon-gauge-invariant. Thus, a four-quark SU(2) ⊗ U(1) model with purely dynamical symmetry breaking will have an untrapped massless true Goldstone boson.

I do not know whether present experiments rule out the possibility of electrically neutral and weakly interacting spin-zero bosons of zero mass. However, if we assume (as seems reasonable) that such particles do not in fact exist, then their absence puts a strong constraint on theories of dynamical symmetry breaking. In particular, it is probably necessary to have weak interactions that connect the ordinary quarks with heavy (~ 300-GeV) fermions, not only as a means of giving masses of order $e^2 \times 300$ GeV to the ordinary quarks, but also to avoid the unwanted anomaly-free global symmetries of the weak interactions.

We now come to one of the most puzzling and unsatisfactory features of dynamical symmetry breaking. In the currently popular gauge theories of strong interactions, the strong gauge coupling constant is fairly small at a renormalization point of order 2–3 GeV, and decreases further with increasing energy.[3] How then can the strong interaction produce a spontaneous symmetry breaking characterized by parameters F_a of order 300 GeV? Indeed, we believe that the strong interactions do induce a spontaneous symmetry breakdown, with the pion octet playing the role of Goldstone bosons, but the parameter F_π is 190 MeV, not 300 GeV.

Another difficulty arises when we try to include the leptons. If it is the ordinary strong interactions that produce the dynamical symmetry breaking discussed in this article, then can the color-neutral leptons get a mass in any order of e?

One way to approach these problems is to suppose that in addition to the color SU(3) associated with the observed strong interactions, there is another gauge group whose generators commute with color SU(3), associated with a new class of "extra-strong" interactions, which act on leptons as well as other fermions. If the gauge coupling constant of the extra-strong interactions reaches a value of order unity at a renormalization point of scale 300 GeV, then the extra-strong interactions could produce the dynamical symmetry breaking discussed in this article. Also, we would not observe direct effects of the extra-strong interactions at ordinary energies, provided that this dynamical symmetry breaking left no subgroup of the extra-strong gauge group unbroken, so that all vector bosons of the extra strong interactions got masses of order 300 GeV.

At first sight this possibility seems quite natural in the framework of the unified simple gauge theories of weak electromagnetic, and strong interactions discussed in the Introduction.[c] The spontaneous superstrong breakdown of the original simple gauge group can leave any number of subgroups unbroken, and some of these may have gauge couplings which grow faster with decreasing renormalization-point energy than the coupling constant of the ordinary strong interactions. However, this naturalness disappears on closer examination. Within the realm of validity of perturbation theory, the gauge couplings g_i of the various simple subgroups of the original unified simple group are given by

$$g_i^2(\mu) = \frac{g^2(\bar{M})}{1 + 2b_i g^2(\bar{M}) \ln(\bar{M}/\mu)},$$

where μ is the scale of a variable renormalization point; $\bar{M}$ is the superlarge mass at which all the $g_i(\mu)$ become equal; and b_i is the coefficient of g_i^3 in the Gell-Mann–Low function $\beta_i(g)$. Suppose we identify the onset of strong coupling for any simple subgroup as the point μ_i at which $g_i(\mu)$ reaches some definite value, of order unity, but taken sufficiently small so that perturbation theory is still valid. Then the ratio of the μ's for two subgroups will be given by

$$\mu_j/\mu_i = (\mu_i/\bar{M})^{b_i/b_j - 1}.$$

But $\bar{M}$ is likely to be enormous,[6] perhaps as large as 10^{19} GeV. Thus, unless b_i and b_j are unreasonably close, the onset of strong interaction will differ by many orders of magnitude for different simple subgroups. From this point of view, it is hard to understand how the onset of strong coupling for the ordinary strong interactions (a few hundred MeV) and the extra-strong interactions (300 GeV) could be so close.

Another possibility is that the color SU(3) gauge group is a subgroup of a larger gauge group which acts on leptons as well as on other fermions, and whose coupling constant reaches a value of order unity at a renormalization point of order 300 GeV. This could produce a dynamical symmetry breakdown of the larger group to color SU(3). In the effective field theory[9] which describes physics below 300 GeV, there could be a color SU(3) gauge symmetry, but since perturbation theory breaks down at 300 GeV, the strong gauge coupling in this effective theory would have no simple relation to the strong gauge coupling above 300 GeV, and might well be somewhat smaller. It would rise very slowly with decreasing renormalization-point energy, and even if it started just under 300 GeV at a value only a little less than its value above 300 GeV, it would not regain this value until much smaller renormalization scales were reached.

In either case, it is not the ordinary color SU(3) gluons that could produce the dynamical symmetry breaking which gives masses to the intermediate vector bosons. These gluons presumably *do* produce the dynamical breakdown of the previously unbroken subgroup H to SU(3) or SU(4) at energies of order $F_\pi = 190$ MeV, with the pion octet as Goldstone bosons, and with the quark masses of order $e^2 \times 300$ GeV furnishing the intrinsic H breaking which gives masses to the pion octet.

In closing, it is interesting to compare the conclusions of this article with the results obtained in theories with elementary spin-zero fields. In order to give the weak interactions the right strength, the vacuum expectation values of some of these fields must be of order 300 GeV. However, we can still distinguish between three kinds of theory:

I. It may be that the spin-zero fields have weak [say, $O(e^2)$] couplings to themselves and to the fermions. This is the case originally considered,[1] and it is the sort of theory with by far the greatest predictive power. The quark and lepton masses in such theories could arise directly from vacuum expectation values of the spin-zero fields, so there would be no need for heavy fermions. A characteristic feature of these theories is the appearance of Higgs scalars with masses that are less than 300 GeV by a factor of order $\sqrt{f}$, where f is the coupling constant of the quartic interaction.

II. It may be that the spin-zero fields have weak couplings to fermions, but strong interactions to themselves. In this case much of the gauge-theory phenomenology would survive, but it would be impossible to relate the vector-boson mass ratio to mixing angles, or to say anything at all about the existence of Higgs scalars. Again, there would be no need for heavy fermions.

III. It may be that the spin-zero fields have strong couplings both to fermions and to themselves. In general, such a theory would have very little predictive power; we would not even be able to say that weak processes like β decay arise from exchange of single vector bosons rather than from complicated higher-order effects.

Viewed in this way, gauge theories with dynamical symmetry breaking seem hardly distinguishable from theories of type III. The one significant difference, which gives theories of the type discussed in this article much greater predictive power than theories of type III, is the occurrence of a natural accidental symmetry, $U(N)\otimes U(N)$.

APPENDIX A: STRUCTURE OF COVARIANT DERIVATIVES

First, we note that under the requirements (1) and (2) of Sec. V, the effective Lagrangian $\mathcal{L}_1$ can be made formally invariant under $U(N)\otimes U(N)$ by introducing the G_W gauge fields

$$W_{A\mu} = \sum_\alpha e_{\alpha A} W_{\alpha\mu}, \tag{A1}$$

and imagining that these fields transform under $U(N)\otimes U(N)$ like λ_A. That is, we give $W_{A\mu}$ the formal transformation rule

$$W_{A\mu} \to W'_{A\mu} = \sum_B R_{AB}(g) W_{B\mu}, \tag{A2}$$

where g is an arbitrary element of $U(N)\otimes U(N)$, and $R_{AB}(g)$ is the corresponding orthogonal matrix in the regular representation of $U(N)\otimes U(N)$:

$$g^{-1}\lambda_A g = \sum_B R_{AB}(g)\lambda_B. \tag{A3}$$

[This is just like the well-known trick of introducing a fictitious octet of "photons" in order to study the SU(3) properties of electromagnetic corrections.]

Equation (A2) is a linear transformation rule, while $\mathcal{L}_1$ also contains quark and Goldstone boson fields which transform according to the nonlinear rules (5.2) and (5.3). It is therefore convenient, in order to make the whole of $\mathcal{L}_1$ invariant under $U(n)\otimes U(N)$, to replace $W_{A\mu}$ with a field which belongs to the same sort of nonlinear realization of $U(N)\otimes U(N)$ as does q:

$$\tilde{W}_{A\mu} \equiv \sum_B \Lambda_{BA}(\Pi) W_{B\mu} = \sum_{B,\alpha} \Lambda_{BA}(\Pi) e_{\alpha B} W_{\alpha\mu}, \tag{A4}$$

where Λ_{AB} is an orthogonal matrix defined by

$$\Lambda_{AB}(\Pi) = R_{AB}\left(\exp\left(i\sum_a \frac{\Pi_a x_a}{F_a}\right)\right). \tag{A5}$$

It is straightforward, using the transformation rules (A2) and (5.2)–(5.4), to check that $\tilde{W}_{A\mu}$ undergoes the $U(N)\otimes U(N)$ transformation

$$\tilde{W}_{A\mu} \to \tilde{W}'_{A\mu} = \sum_B R_{AB}\left(\exp\left(i\sum_i \mu_i(\Pi, g) t_i\right)\right)\tilde{W}_{B\mu}. \tag{A6}$$

The field $\tilde{W}_{A\mu}$ thus behaves under $U(N)\otimes U(N)$ transformations just like q, except of course that it belongs to a different linear representation of the unbroken subgroup H.

The Lagrangian $\mathcal{L}_1$ will thus be globally $U(N)$

$\otimes U(N)$ invariant if it is algebraically invariant under the unbroken subgroup H and if it is composed of just the following ingredients: quark fields q, their $U(N)\otimes U(N)$ covariant derivatives

$$\partial_\mu q - i\sum_{i,a} t_i q E_{ai}(\Pi)\partial_\mu \Pi, \tag{A7}$$

together with Goldstone boson fields Π_a and their covariant derivatives

$$D_\mu \Pi_a \equiv F_a \sum_b D_{ba}(\Pi)\partial_\mu \Pi_b F_b^{-1}, \tag{A8}$$

and $\tilde{W}_{A\mu}$ fields and their covariant derivatives

$$\partial_\mu \tilde{W}_{A\nu} - i\sum_{i,B}(t_i)_{AB}\tilde{W}_{B\nu}E_{ai}(\Pi)\partial_\mu \Pi_a,$$

plus gluon fields and their derivatives. [The functions D_{ba} and E_{ai} are defined by Eq. (5.9).] Aside from terms which are separately $U(N)\otimes U(N)$ invariant and hence may be dropped, the covariant derivative of the gauge field may also be written

$$D_\mu \tilde{W}_{A\nu} = \sum_B \Lambda_{BA}(\Pi)\partial_\mu W_{B\nu}. \tag{A9}$$

We now must impose requirement (3) of Sec. V, that the Lagrangian be locally as well as globally invariant under G_W and H_S. For a space-time dependent $U(N)\otimes U(N)$ transformation $g(x)$, the gauge field $W_{\alpha\mu}$ transforms according to the rule

$$w_\alpha W'_{\alpha\mu} = g(w_\alpha W_{\alpha\mu})g^{-1} - (\partial_\mu g)g^{-1}. \tag{A10}$$

Also, derivatives of g appear in the G_W transformation rules for the quantities (A7)–(A9). By using the derivative of Eq. (5.4) to evaluate these g derivatives, we can easily see that in order to cancel them, we must add gauge field terms to (A7) and (A8), so that those derivatives become G_W-covariant quantities[18]:

$$\partial_\mu q - i\sum_{i,a} t_i q E_{ai}(\Pi)\partial_\mu \Pi_a - i\sum_{A,i} t_i q C_{iA}\tilde{W}_{A\mu}, \tag{A11}$$

$$\sum_b D_{ba}(\Pi)\partial_\mu \Pi_b + F_a \sum_a C_{aA}\tilde{W}_{A\mu}. \tag{A12}$$

Note that these quantities are still formally locally $U(N)\otimes U(N)$ covariant, as well as globally G_W covariant.

Also, G_W invariance requires that derivatives of the G_W gauge field only appear in the Yang-Mills curl[20]

$$F_{\alpha\mu\nu} \equiv \partial_\mu W_{\alpha\nu} - \partial_\nu W_{\alpha\mu} - \sum_{\beta\gamma} C_{\alpha\beta\gamma}W_{\beta\mu}W_{\gamma\nu}, \tag{A13}$$

where $C_{\alpha\beta\gamma}$ are the structure constants of G_W. It is elementary to show that

$$\sum_{B\alpha}\Lambda_{BA}(\Pi)e_{\alpha B}F_{\alpha\mu\nu} = D_\mu \tilde{W}_{A\nu} - D_\nu \tilde{W}_{A\mu} - \sum_{BC} f_{ABC}\tilde{W}_{B\mu}\tilde{W}_{C\nu}, \tag{A14}$$

where f_{ABC} are the $U(N)\otimes U(N)$ structure constants. Thus, this quantity is both locally G_W covariant and (formally) globally $U(N)\otimes U(N)$ covariant.

Finally, we must impose invariance under the unbroken strong gauge group H_S. According to our assumptions, it is only q that transforms nontrivially under H_S, so we must simply add a gluon term in (A11). This, together with (A12) and (A14), comprise the three sorts of fully covariant derivatives allowed in the effective Lagrangian $\mathcal{L}_1$.

APPENDIX B: STRUCTURE OF $\mathcal{L}_2$

The quantity $J_{AB}(\Pi)$ in Eq. (5.10) must be an $U(N)\otimes U(N)$ tensor in the sense that for any element g of $U(N)\otimes U(N)$, we have

$$J_{AB}(\Pi') = \sum_{CD} R_{AC}(g)R_{BD}(g)J_{CD}(\Pi), \tag{B1}$$

with Π' and R defined by Eqs. (5.4) and (A3). But it follows from (A4) and (A5) that

$$R(g)\Lambda(\Pi) = \Lambda(\Pi')R\left(\exp\left(i\sum_i \mu_i t_i\right)\right), \tag{B2}$$

so contracting Eq. (B1) on the left-hand side with $\Lambda^{-1}(\Pi')$ gives

$$I_{AB}(\Pi') = \sum_{CD} R_{AC}\left(\exp\left(i\sum_i \mu_i t_i\right)\right) \times R_{BD}\left(\exp\left(i\sum_i \mu_i t_i\right)\right) I_{CD}(\Pi), \tag{B3}$$

where

$$I_{AB}(\Pi) \equiv \sum_{CD}\Lambda^{-1}_{AC}(\Pi)\Lambda^{-1}_{BD}(\Pi)J_{CD}(\Pi). \tag{B4}$$

If we now choose $\Pi_a = 0$ and

$$g = \exp\left(i\sum_a x_a \pi_a / F_a\right),$$

we find from (5.4) that

$$\Pi'_a = \pi_a, \quad \mu'_i = 0,$$

so Eq. (B3) reads here

$$I_{AB}(\pi) = I_{AB}(0) \equiv I_{AB}. \tag{B5}$$

Hence I_{AB} is a constant. Equation (B3) then says that it is a constant tensor under H. That is,

$$I_{AB} = \sum_{CD} R_{AC}(h)R_{BD}(h)I_{CD} \tag{B6}$$

for arbitrary elements $h \in H$.

APPENDIX C: STRUCTURE OF $\mathcal{L}_m$

The quantity $N_{AB}(\Pi)$ in Eq. (5.26) is an U(N) ⊗U(N) tensor, in the sense that for an arbitrary element g of U(N)⊗U(N), we have

$$\exp\Big(-i\sum_i t_i\mu_i(\Pi,g)\Big)N_{AB}(\Pi')\exp\Big(i\sum_i t_i\mu_i(\Pi,g)\Big)=\sum_{CD}R_{AC}(g)R_{BD}(g)N_{CD}(\Pi), \tag{C1}$$

with μ_i, Π', and R defined by Eqs. (5.4) and (A3). Using Eq. (B2) and contracting Eq. (C1) on the left with $\Lambda^{-1}(\Pi')$ gives

$$\exp\Big(-i\sum_i t_i\mu_i\Big)Q_{AB}(\Pi')\exp\Big(i\sum_i t_i\mu_i\Big)=\sum_{CD}R_{AC}\Big(\exp\Big(i\sum_i t_i\mu_i\Big)\Big)R_{BD}\Big(\exp\Big(i\sum_i t_i\mu_i\Big)\Big)Q_{CD}(\Pi'). \tag{C2}$$

If we choose $\Pi_a=0$ and

$$g=\exp\Big(i\sum_a x_a\pi_a/F_a\Big),$$

Then Eq. (5.4) gives

$$\Pi'_a=\pi_a,\quad \mu_i=0,$$

so Eq. (C2) in this case just tells us that $Q(\pi)$ is π independent,

$$Q_{AB}(\pi)=Q_{AB}(0)\equiv Q_{AB}. \tag{C3}$$

Equation (C2) may thus be written

$$h^{-1}Q_{AB}h=\sum_{CD}R_{AC}(h)R_{BD}(h)Q_{CD} \tag{C4}$$

for arbitrary elements $h\in H$.

APPENDIX D: CONSTRUCTION OF UNITARITY GAUGE

Recall first that the x_a and t_i are defined to span the algebra of U(N)⊗U(N), so that any element λ of this algebra may be written

$$\lambda=\sum_i \mu_i{}^0 t_i+\sum_a \phi_a{}^0 x_a,$$

with $\mu_i{}^0$ and $\phi_a{}^0$ so far unconstrained. We then define

$$\phi_a(\theta)\equiv\phi_a{}^0+\sum_{\alpha A}\theta_\alpha e_{\alpha A}B_{aA},$$

and choose θ to minimize the positive continuous function

$$\sum_a F_a{}^2\phi_a{}^2(\theta).$$

Denoting the value of $\phi_a(\theta)$ at this minimum as ϕ_a, we then have

$$0=2\sum_a F_a{}^2\phi_a\sum_A e_{\alpha A}B_{aA},$$

so ϕ_a satisfies the constraint (7.2). Also, we can rewrite λ as

$$\lambda=\sum_i \mu_i{}^0 t_i+\sum_a\phi_a x_a-\sum_{a\alpha A}\theta_\alpha e_{\alpha A}B_{aA}x_a$$

$$=\sum_i\mu_i{}^0 t_i+\sum_a\phi_a x_a-\sum_\alpha\theta_\alpha w_\alpha+\sum_{i\alpha A}\theta_\alpha e_{\alpha A}C_{iA}t_i,$$

so that λ may be decomposed as in Eq. (7.1), with

$$\mu_i=\mu_i{}^0+\sum_{\alpha A}\theta_\alpha e_{\alpha A}C_{iA}.$$

ACKNOWLEDGMENTS

I am grateful for frequent valuable discussions on this subject with colleagues at Harvard and M.I.T. I also thank D. J. Gross and H. Pagels for enlightening conversations.

*Work supported in part by the National Science Foundation under grant No. NSF-GP-13547X.

[1] S. Weinberg, Phys. Rev. Lett. 19, 1264 (1967); A. Salam, in *Elementary Particle Physics*, edited by N. Svartholm (Almqvist and Wiksells, Stockholm, 1968), p. 367.

[2] Dynamical mechanisms for spontaneous symmetry breaking were first discussed by Y. Nambu and G. Jona-Lasinio, Phys. Rev. 122, 345 (1961); J. Schwinger, *ibid.* 125, 397 (1962); 128, 2425 (1962). The subject was revived in the context of modern gauge theories by R. Jackiw and K. Johnson, Phys. Rev. D 8, 2386 (1973); J. M. Cornwall and R. E. Norton, *ibid.* 8, 3338 (1973). More recent references include D. J. Gross and A. Neveu, *ibid.* 10, 3235 (1974); E. J. Eichten and F. L. Feinberg, *ibid.* 10, 3254 (1974); J. M. Cornwall, R. Jackiw, and E. Tomboulis, *ibid.* 10, 2438 (1975); E. C. Poggio, E. Tomboulis, and S-H. H. Tye, *ibid.* 11, 2839 (1975); K. Lane, *ibid.* 10, 1353 (1974); 10, 2605 (1974); S. Weinberg, in *Proceedings of the XVII International Conference on High Energy Physics, London, 1974*, edited by J. R. Smith (Rutherford Laboratory, Chilton, Didcot, Berkshire, England, 1974), p. III–59.

[3]D. J. Gross and F. Wilczek, Phys. Rev. Lett. 30, 1343 (1973); H. D. Politzer, *ibid.* 30, 1346 (1973). The problem of maintaining asymptotic freedom in theories with strongly interacting spin-zero fields has been discussed by D. J. Gross and F. J. Wilczek, Phys. Rev. D 8, 3633 (1973); T. P. Cheng, E. Eichten, and L. F. Li, *ibid.* 9, 2259 (1974).

[4]C. Callan and D. Gross, Phys. Rev. Lett. 22, 156 (1969).

[5]S. Weinberg, Phys. Rev. Lett. 31, 494 (1973); Phys. Rev. D 8, 4482 (1973).

[6]H. Georgi, H. Quinn, and S. Weinberg, Phys. Rev. Lett. 33, 451 (1974).

[7]S. Weinberg, Phys. Rev. D 5, 1962 (1972).

[8]See Ref. 6. Specific models which unify the weak, electromagnetic, and strong interactions have been proposed by J. C. Pati and A. Salam, Phys. Rev. D 8, 1240 (1973); H. Georgi and S. L. Glashow, Phys. Rev. Lett. 32, 438 (1974); H. Fritzsch and P. Minkowski (unpublished).

[9]At this point we use the term "effective field theory" in the sense of T. Appelquist and J. Carrazone, Phys. Rev. D 11, 2856 (1975).

[10]J. Goldstone, Nuovo Cimento 19, 154 (1961); J. Goldstone, A. Salam, and S. Weinberg, Phys. Rev. 127, 965 (1965). Also see Ref. 2.

[11]P. W. Higgs, Phys. Lett. 12, 132 (1965); Phys. Rev. Lett. 13, 508 (1964); Phys. Rev. 145, 1156 (1966); F. Englert and R. Brout, Phys. Rev. Lett. 13, 321 (1964); G. S. Guralnik, C. R. Hagen, and T. W. B. Kibble, *ibid.* 13, 585 (1964); T. W. B. Kibble, Phys. Rev. 155, 1554 (1967).

[12]For instance, in the model of Nambu and Jona-Lasinio, Ref. 2, the dynamical symmetry breakdown can occur for only one value of the Fermi coupling constant.

[13]This is the "dimensional transmutation" of S. Coleman and E. Weinberg, Phys. Rev. D 7, 1888 (1973).

[14]The use of effective Lagrangians, to reproduce the low-energy theorems associated with a spontaneously broken symmetry, was initiated by S. Weinberg, Phys. Rev. Lett. 18, 188 (1967). The formalism was further developed by J. Schwinger, Phys. Lett. 24B, 473 (1967); S. Weinberg, Phys. Rev. 166, 1568 (1968); S. Coleman, J. Wess, and B. Zumino, *ibid.* 177, 2239 (1968); C. G. Callan, S. Coleman, J. Wess, and B. Zumino, *ibid.* 177, 2247 (1968). An explicit construction of an effective Lagrangian involving composite Goldstone boson fields has been given by S. Coleman (unpublished).

[15]R. F. Dashen, Phys. Rev. D 3, 1879 (1971). For a recent application, see H. Pagels, *ibid.* 11, 1213 (1975).

[16]See Ref. 11. For a more general definition of the unitarity gauge, see S. Weinberg, Phys. Rev. D 7, 1068 (1973).

[17]Pseudo-Goldstone bosons are the Goldstone bosons associated with the spontaneous breakdown of any "accidental" symmetry of the $e=0$ terms in the Lagrangian. They were originally discussed in theories with elementary spin-zero fields; see S. Weinberg, Phys. Rev. Lett. 29, 1698 (1972); Phys. Rev. D 7, 2887 (1973). For a generalization, see H. Georgi and A. Pais, Phys. Rev. D 12, 508 (1975).

[18]See Callan, Coleman, Wess, and Zumino, Ref. 14, Sec. IV. Also see Jackiw and Johnson, Ref. 2, Sec. III.

[19]S. Coleman and E. Weinberg, Ref. 13; S. Weinberg, Ref. 17.

[20]C. N. Yang and R. L. Mills, Phys. Rev. 96, 191 (1954); R. Utiyama, *ibid.* 101, 1597 (1956); M. Gell-Mann and S. Glashow, Ann. Phys. (N.Y.) 15, 437 (1961).

[21]Note that we exclude pole terms produced by internal gluons here. These can be included by stitching together the black boxes calculated from this effective Lagrangian with soft gluon lines.

[22]See Coleman, Wess, and Zumino, Ref. 14.

[23]This may be contrasted with the calculation described in Ref. 17. There, the integrals began to converge for virtual momenta of the order of the intermediate-vector-meson mass, and the pseudo-Goldstone boson mass was of order e^2M, not eM.

[24]This is essentially the same argument that forces us to diagonalize the matrix elements of a perturbation between degenerate states before beginning to construct a perturbation series. Also, compare S. Weinberg, Phys. Rev. Lett. 31, 494 (1973), Sec. III.

[25]It is not so clear what meaning should be attached to quark masses if quarks are not observable as free particles. One interpretation is that the quark masses are the inputs to current-algebra calculations, as used by M. Gell-Mann, R. J. Oakes, and B. Renner, Phys. Rev. 175, 2195 (1968); S. L. Glashow and S. Weinberg, Phys. Rev. Lett. 20, 224 (1968). The success of these calculations indicates that the quark masses are small compared with some scale characteristic of the strong interactions.

[26]I thank H. Georgi for this remark.

[27]J. Kogut and L. Susskind, Phys. Rev. D 10, 3468 (1974); K. Lane, *ibid.* 10, 1353 (1974); 10, 2605 (1974); S. Weinberg, Ref. 2; J. Kogut and L. Susskind, Phys. Rev. D 11, 3594 (1975); S. Weinberg, *ibid.*, 11, 3583 (1975).

[28]S. L. Glashow, J. Iliopoulos, and L. Maiani, Phys. Rev. D 2, 1285 (1970); also see S. Weinberg, *ibid.* 5, 1412 (1972).

VOLUME 40, NUMBER 4 PHYSICAL REVIEW LETTERS 23 JANUARY 1978

A New Light Boson?

Steven Weinberg
Lyman Laboratory of Physics, Harvard University, Cambridge, Massachusetts 02138
(Received 6 December 1977)

It is pointed out that a global U(1) symmetry, that has been introduced in order to preserve the parity and time-reversal invariance of strong interactions despite the effects of instantons, would lead to a neutral pseudoscalar boson, the "axion," with mass roughly of order 100 keV to 1 MeV. Experimental implications are discussed.

One of the attractive features of quantum chromodynamics[1] (QCD) is that it offers an explanation of why C, P, T, and all quark flavors are conserved by strong interactions, and by order-α effects of weak interactions.[2] However, the discovery of quantum effects[3] associated with the "instanton" solution of QCD has raised a puzzle with regard to P and T conservation. Because of Adler-Bell-Jackiw anomalies, the chiral transformation which is needed in QCD to bring the quark-mass matrix to a real, diagonal, γ_5-free form will in general change the phase angle θ associated[3] with instanton effects, leaving $\bar{\theta} \equiv \theta + \arg\det m$ invariant. [Here m is the coefficient of $\frac{1}{2}(1+\gamma_5)$ in a decomposition of the quark-mass matrix into $\frac{1}{2}(1\pm\gamma_5)$.] The condition for P and T conservation is that $\theta = 0$ when the quark fields are defined so that m is real, or more generally, that $\bar{\theta} = 0$. But θ is a free parameter, and in QCD there is no reason why it should take the value $-\arg\det m$. Furthermore, even if we simply demanded that the strong interactions in isolation conserve P and T, so that $\bar{\theta} = 0$, there would still be a danger that the weak interactions would introduce P- and T-nonconserving phases of order $10^{-3}\alpha$ in m, leading to an unacceptable neutron electric dipole moment, of order 10^{-18} $e \cdot$cm.

An attractive resolution of this problem has been proposed by Peccei and Quinn.[5] They note that the quark-mass matrix is a function $m(\langle\varphi\rangle)$ of the vacuum expectation values of a set of weakly coupled scalar fields φ_i. Although θ is arbitrary, $\langle\varphi\rangle$ is not; it is determined by the minimization of a potential $V(\varphi)$ which depends on θ. Peccei and Quinn assume that the Lagrangian has a global U(1) chiral symmetry [which I will call $U(1)_{PQ}$], under which $\det m(\varphi)$ changes by a phase. The phase of $\det m(\varphi)$ at the minimum of $V(\varphi)$ is then undetermined in any finite order of perturbation theory, and is fixed only by instanton effects which break the $U(1)_{PQ}$ symmetry. However, the potential will then depend on $\bar{\theta}$, but not separately on θ and $\arg\det m$, so that it is not a miracle if the phase of $\det m(\varphi)$ at the minimum of $V(\varphi)$ happens to have the P- and T-conserving value $-\theta$. Peccei and Quinn[5] show in a number of examples that this is just what happens.

Now, the $U(1)_{PQ}$ symmetry of the Lagrangian is intrinsically broken by instantons, and so at first sight one might not expect that it would have any further physical consequences. Certainly it does not lead to the strongly interacting isoscalar pseudoscalar meson below $\sqrt{3}m_\pi$,[6] that was the bugbear of the old U(1) problem. However, the scalar fields φ do not know about instantons, except through a semiweak ($\propto G_F^{1/2}$) coupling to quarks. Hence the spontaneous breakdown of the chiral $U(1)_{PQ}$ symmetry associated with the appearance of nonzero vacuum expectation values $\langle\varphi\rangle$ leads[7] to a very light pseudoscalar pseudo-Goldstone boson,[8] the "axion," with m_a^2 proportional to the Fermi coupling G_F.

For insight in to the properties of the axion, it is useful to examine how they appear in the simplest realistic model that admits a $U(1)_{PQ}$ symmetry. We assume an $SU(2)\otimes U(1)$ gauge group, with quarks in $N/2$ left-handed doublets and N right-handed singlets, and just two scalar doublets $\{\varphi_i^+, \varphi_i^0\}$, carrying $U(1)_{PQ}$ quantum numbers such that φ_1 (φ_2) couples right-handed quarks of charge $-\frac{1}{3}$ $(+\frac{2}{3})$ to left-handed quarks. By writing the Yukawa interaction in terms of quark fields of definite mass, we easily see that the interaction of neutral scalar fields with quarks is[9]

$$\mathcal{L}_N = -[m_d\bar{d}_R d_L + m_s\bar{s}_R s_L + m_b\bar{b}_R b_L + \cdots]\varphi_1^{0*}\langle\varphi_1^0\rangle^{*-1} - [m_u\bar{u}_R u_L + m_c\bar{c}_R c_L + m_t\bar{t}_R t_L + \cdots]\varphi_2^0\langle\varphi_2^0\rangle^{-1} + \text{H.c.}, \quad (1)$$

where L and R indicate multiplication with $\frac{1}{2}(1\pm\gamma_5)$. The part of $\mathcal{L}_N$ involving the light quarks u, d, and s may be treated as a perturbation $\mathcal{L}_{uds}$, while terms in $\mathcal{L}_N$ involving c, t, b,... must be included in the

VOLUME 40, NUMBER 4 PHYSICAL REVIEW LETTERS 23 JANUARY 1978

unperturbed QCD Lagrangian $\mathcal{L}_0$.

Putting together SU(2)⊗ U(1) and $U(1)_{PQ}$, we see that in the limit $\mathcal{L}_{uds}\to 0$ the Lagrangian would be invariant under five independent phase transformations on $\bar{u}_R u_L$, $\bar{d}_R d_L$, $\bar{s}_R s_L$, φ_1, and φ_2, with the latter two transformations supplemented with suitable phase transformations on heavy-quark bilinears $\bar{c}_R c_L$, $\bar{b}_R b_L$, ... However, because of instanton effects, we will only have a true symmetry if we eliminate anomalies by supplementing each of these phase transformations with a suitable phase transformation on $\bar{u}_R u_L$, $\bar{d}_R d_L$, and $\bar{s}_R s_L$. There are then *four* massless neutral pseudoscalar Goldstone bosons for $\mathcal{L}_{uds}=0$, which can be taken as π^0 and η^0 plus the two bosons associated with the phase transformation on φ_1^0 and φ_2^0.

The perturbation $\mathcal{L}_{uds}$ produces a 4×4 squared-mass matrix for the four Goldstone bosons, which may be calculated by the usual methods of current algebra.[10] After diagonalization, we find a π^0 and η^0 with essentially the usual masses; a strictly massless boson that is removed by the SU(2)⊗ U(1) Higgs mechanism; and the axion, with mass[11]

$$m_a \simeq \frac{N m_\pi F_\pi}{2(m_u+m_d)^{1/2}}\left[\frac{m_u m_d m_s}{m_u m_d + m_d m_s + m_s m_u}\right]^{1/2}\frac{2^{1/4}G_F^{1/2}}{\sin 2\alpha} = (23\ \text{keV})\times N/\sin 2\alpha . \tag{2}$$

Here m_u, m_d, and m_s are the quark masses appearing in the Lagrangian, with ratios[10] $m_s/m_d=20$, $m_d/m_u=1.8$; $F_\pi \simeq 190$ MeV; N is the number of quark flavors; and α is an unknown angle defined by the relations $|\langle\varphi_1^0\rangle| = 2^{-1/4}G_F^{-1/2}\sin\alpha$ and $|\langle\varphi_2^0\rangle| = 2^{-1/4}G_F^{-1/2}\cos\alpha$.

Axion emission or absorption can take place through a mixing of a^0 with π^0 or η^0, with an amplitude of form $\xi_\pi A_\pi + \xi_\eta A_\eta$, where $A_{\pi,\eta}$ are the amplitudes for emission and absorption of a massless π^0 or η^0, and $\xi_{\pi,\eta}$ are the components of the physical axion along the bare π^0 and η^0, given[11] for $N=4$ and $m_s \gg m_{d,u}$ by

$$\xi_\pi = \xi\left[\left(\frac{3m_d - m_u}{m_d+m_u}\right)\tan\alpha - \left(\frac{3m_u - m_d}{m_u+m_d}\right)\cot\alpha\right], \tag{3}$$

$$\xi_\eta = \xi[3^{1/2}\tan\alpha + 3^{-1/2}\cot\alpha], \tag{4}$$

$$\xi = \tfrac{1}{4}\, 2^{1/4}G_F^{1/2}F_\pi = 1.9\times 10^{-4}. \tag{5}$$

These mixing effects should donimate for processes involving only u, d, and s quarks, because other terms are suppressed by factors m_u or m_d or m_s. [Using (3)–(5) together with the Goldberger-Treiman relation, we see that the effect of the π^0 and η^0 poles is to convert the "current algebra" masses in Eq. (1) into constituent quark masses.] We do not know α, and so ξ_π could have any value; where numerical estimates are needed we will take $\xi_\pi \simeq \xi$. On the other hand, ξ_η has a lower bound of 2ξ, but the $\eta\bar{N}N$ coupling is considerably weaker than the $\pi^0\bar{N}N$ coupling, so that π^0-a^0 mixing should dominate in most processes. There is also a direct coupling of a^0 to heavy quarks, of the form

$$\mathcal{L}_{aq} = i2^{1/4}G_F^{1/2}a^0[m_c\bar{c}\gamma_5 c\tan\alpha + m_b\bar{b}\gamma_5 b\cot\alpha + \cdots]. \tag{6}$$

Finally, with only two doublets, either φ_1 or φ_2 would have to couple to leptons, giving the axion a coupling

$$\mathcal{L}_{al} = +i2^{1/4}G_F^{1/2}a^0[m_e\bar{e}\gamma_5 e + m_\mu\bar{\mu}\gamma_5\mu \cdots][\tan\alpha \text{ or } \cot\alpha]. \tag{7}$$

If the axion has a mass below $2m_e$, it will decay chiefly by the processes $a^0\to 2\gamma$, with a rate of order $(4N/3)^2(m_a/m_\pi)^3\xi^2$ times $\Gamma(\pi^0\to 2\gamma)$, or $\approx (10^4\ \text{MeV}^{-3}\ \text{sec}^{-3})m_a^3$. For $m_a > 2m_e$, we also have $a^0\to e^+e^-$, with a rate of order $2^{1/2}G_F m_e^2 P_e/4\pi$, or $(3\times 10^8\ \text{MeV}^{-1}\ \text{sec}^{-1})m_a$ for $m_a \gg m_e$.

Would the axion have been seen in existing experiments?[12] One can think of several possibilities:

(1) Axion exchange would introduce a term in the gyromagnetic ratio of the muon of order $G_F m_\mu^2/\pi^2 \approx 10^{-8}$, comparable to the uncertainty in present calculations[13] of g_μ.

(2) Axion exchange would produce spin-spin interactions in atoms and molecules, but even for $m_a=0$ these are weaker than corresponding magnetic interactions by factors 10^{-8} in H atoms, 10^{-6} in muonic hydrogen, 3×10^{-9} in muonium, and 10^{-5} for the pp interaction in H_2 molecules, and thus

VOLUME 40, NUMBER 4 PHYSICAL REVIEW LETTERS 23 JANUARY 1978

well below current theoretical or experimental uncertainties.[14]

(3) The absence of a spike at the upper end of the pion spectrum in searches for $K^+ \to \pi^+\nu\bar{\nu}$ gives an upper limit[15] of 1.2×10^{-6} on the ratio $(K^+ \to \pi^+a^0)/(K^+ \to \pi^+\pi^0)$. This is safely larger than the ratio $\xi^2 \sim 4\times10^{-8}$ expected if $K^+ \to \pi^+a^0$ proceeds through π^0-a^0 mixing. However, the axion can also be emitted through η^0-a^0 mixing, and $K^+ \to \pi^+\eta$ is not like $K^+ \to \pi^+\pi^0$, suppressed by the $\Delta I = \frac{1}{2}$ rule, so that the ratio $(K^+ \to \pi^+a^0)/(K^+ \to \pi^+\pi^0)$ might be expected to be 2–3 orders of magnitude larger than ξ^2. On the other hand, $K^+ \to \pi^+\eta$ cannot occur through octet terms in the effective weak Hamiltonian, and so axion emission by a^0-η^0 mixing may also be somewhat suppressed.

(4) In accelerator neutrino experiments there is generally about one π^0 produced for each ν_μ from π^+ decay, and so the a^0/ν_μ ratio should be of order $a^0/\pi^0 \approx \xi^2$. The cross section of high-energy axions on nucleons is expected to be of order $\xi^2\sigma_{\pi N}$, so that the ratio of a^0 to ν_μ events should be of order $\xi^4\sigma_{\pi N}/\sigma_{\nu N} \approx (3\times10^{-3}\ \mathrm{GeV})/E_\nu$. In several "beam dump" experiments[16] the number of neutrinos (with $E_\nu \approx 1$ GeV) was reduced by 2–3 orders of magnitude, and so the number of axion events should have been comparable to the number of neutrino events. It is not clear to me whether the extra events would have been noticed.

(5) Nuclear reactors are expected to emit axions at a rate of order $(v_{\mathrm{nuc}}/c)^2\xi^2G_{\pi N}^2/4\pi\alpha \approx 10^{-6}$ axion per prompt γ. There is also about one $\bar{\nu}_e$ per prompt γ, so that the axion flux in reactor neutrino experiments[17] should be about 10^{-6} the $\bar{\nu}_e$ flux, or $2\times10^7\, a/\mathrm{cm}^2\cdot\mathrm{sec}$. These axions can produce electron recoils by the reaction $a^0e^- \to \gamma e^-$ or $a^0 \to 2\gamma$ followed by Compton scattering but very few of these events would be mistaken for elastic $\bar{\nu}_e e^-$ scattering, because the extra photons would produce veto pulses in the scintillator of NaI annulus. However, about one-fifth of the axions would have energies above 1.5 MeV, and thus would contribute to the measured background of NaI pulses if they decayed anywhere within the 10^5-cm^3 shielded volume, or if they were absorbed in the 300 kg of NaI. The axion absorption coefficient in NaI is of the order of $2^{1/2}G_F m_e^2/4\pi\alpha$ times the photon absorption coefficient, or about $10^{-12}\mathrm{cm}^2/\mathrm{g}$, so that the axion absorption rate should be of order 10^5/day. For $m_a \gtrsim 100$ keV, axion decay also produces over 10^6 pulses/day. Both rates are much faster than the measured background[17] of -160 ± 260 pulses/day. Further, about one-tenth of the axions would be above threshold for the reaction $a^0+d \to p+n$, with cross section of order $[4\xi^2G_{\pi N}^2/4\pi\alpha(2.79+1.91)^2]\times\sigma_{M1}(\gamma+d \to p+n)$, or 5×10^{-33} cm^2. Thus with 178 kg of D_2O and an efficiency of 0.043, there should have been about 4×10^5 neutron counts/day, as compared with a measured reactor-associated rate[17] of (-2.9 ± 7.2)/day.

We see that there are already several experiments which provide evidence against the existence of axions. However, our estimates of axion production and detection rates are highly uncertain, and in particular refer to a specific model with just two scalar doublets, involving the unknown angle α. Perhaps judgment should be reserved.

The reactor evidence against axions would disappear if α took a value for which $|\xi_\pi| \ll \xi$, or if axions decayed or were absorbed so rapidly that very few reached the detector. A search for monochromatic photons from the decay $J/\psi \to a^0\gamma$ may provide a good way to look for axions,[18] which does not depend on how they couple to light quarks or leptons, or how they are absorbed or decay. We expect $\Gamma(J/\psi \to a^0\gamma)/\Gamma(J/\psi \to e^+e^-)$ to be of order $m_{J/\psi}^2\,\xi^2\,\Gamma(\omega^0 \to \pi^0\gamma)/m_\omega^2\Gamma(\omega^0 \to e^+e^-) \approx 6\times10^{-4}$.

If axions are found not to exist, it will show that there is no $U(1)_{PQ}$ symmetry, and an alternative explanation for P and T invariance will have to be found. One possibility is that one of the quark masses may be zero, so that θ can be taken to have any value we like. However, the quark masses produce a $K^0 - K^+$ mass difference[10] $(m_\pi^2/2m_K)(m_d - m_u)/(m_d+m_u)$, which for $m_u=0$ or $m_d=0$ is ±18 MeV. Electromagnetic effects are expected[19] to produce an additional contribution of only about -1 MeV, and, although this value is subject to large uncertainties,[20] it seems highly unlikely that electromagnetism could shift the K-mass difference to the observed value of $+4$ MeV.

I have benefitted from conversations with a great many colleagues, including C. Baltay, J. Bjorken, S. Coleman, G. Feinberg, H. Georgi, S. Glashow, M. Goldhaber, R. Hildebrand, V. Hughes, J. D. Jackson, L. Lederman, A. Mann, D. V. Nanopoulos, H. Pagels, R. Peccei, M. Peskin, H. Quinn, N. Ramsey, F. Reines, A. Salam, J. P. Schiffer, M. Schwartz, R. Shrock, K. Strauch, L. Sulak, S. Ting, F. Wilczek, and R. Wilson.

This research was supported in part by the National Science Foundation under Grant No. PHY 75-20427.

[1]H. Fritzsch, M. Gell-Mann, and H. Leutwyler, Phys. Lett. 47B, 365 (1972); D. J. Gross and F. Wilczek, Phys. Rev. D 8, 3497 (1973); S. Weinberg, Phys. Rev. Lett. 31, 494 (1973).

[2]S. Weinberg, Ref. 1, and Phys. Rev. D 8, 4482 (1973).

[3]G. 't Hooft, Phys. Rev. Lett. 37, 8 (1976), and Phys. Rev. D 14, 3432 (1976); R. Jackiw and C. Rebbi, Phys. Rev. Lett. 37, 172 (1976); C. G. Callan, R. F. Dashen, and D. J. Gross, Phys. Lett. 63B, 334 (1976).

[4]A. A. Belavin, A. M. Polyakov, A. S. Schwartz, and Yu. S. Tyuplin, Phys. Lett. 59B, 85 (1975).

[5]R. D. Peccei and H. R. Quinn, Phys. Rev. Lett. 38, 1440 (1977), and Phys. Rev. D 16, 1791 (1977).

[6]S. Weinberg, Phys. Rev. D 11, 3583 (1975).

[7]This was independently noted by F. Wilczek, Phys. Rev. Lett., to be published. I am grateful to Dr. Wilczek for informing me of his work prior to publication.

[8]S. Weinberg, Phys. Rev. Lett. 29, 1698 (1972).

[9]See, e.g., S. Weinberg, Phys. Rev. Lett. 17, 657 (1976), Eq. (5).

[10]S. Weinberg, in *A Festschrift for I. I. Rabi*, edited by Lloyd Motz (New York Academy of Sciences, New York, 1977), and references quoted therein.

[11]This result was derived independently for four quark flavors by myself and by M. Peskin (private communication). The generalization to arbitrary N is due to Peskin. Details will be published elsewhere. {With one scalar doublet, there is a lower bound of 6.1 GeV (for $\sin^2\theta = 0.27$) on the Higgs boson mass; see A. Linde, Pis'ma Zh. Eksp. Teor. Fiz. 23, 73 (1976) [JETP Lett. 23, 64 (1976)]; S. Weinberg, Phys. Rev. Lett. 36, 294 (1976). This lower bound does not apply here, because $U(1)_{PQ}$ requires at least two scalar doublets.}

[12]Empirical lower bounds on Higgs boson masses have been discussed by J. Ellis, M. K. Gaillard, and D. V. Nanopoulos, Nucl. Phys. B106, 292 (1976), and references cited therein. However, these bounds refer specifically to 0^+ particles, not to a 0^- particle like the axion.

[13]For a review, see J. Calmet, S. Narison, M. Perrottet, and E. de Rafael, Rev. Mod. Phys. 49, 21 (1977).

[14]For a review, see, e.g., B. E. Lautrup, A. Peterman, and E. de Rafael, Phys. Rep. 3C, 193 (1972). For the pp interaction in H_2, see R. F. Code and N. F. Ramsey, Phys. Rev. A 4, 1945 (1971).

[15]J. Klems, R. Hildebrand, and R. Steining, Phys. Rev. D 4, 66 (1971); R. Hildebrand, private communication.

[16]R. Burns *et al.*, Phys. Rev. Lett. 15, 830 (1965); unpublished results quoted by L. M. Lederman, in *Old and New Problems of Elementary Particles*, edited by G. Puppi (Academic, London, 1968); A. F. Rothenberg, SLAC Report No. 147, 1972 (unpublished); Gargamelle Collaboration, 1972 (unpublished); D. S. Baranov *et al.*, to be published; L. Lederman, C. Baltay, and M. Schwartz, private communications. I am grateful to L. Lederman for first drawing my attention to the beam-dump experiments.

[17]F. Reines, H. S. Gurr, and H. W. Sobel, Phys. Rev. Lett. 37, 315 (1976); H. S. Gurr, F. Reines, and H. W. Sobel, Phys. Rev. Lett. 33, 179 (1974). I am grateful to J. P. Schiffer and to C. Baltay and G. Feinberg for pointing out to me that the first reference quotes a useful upper limit on all NaI pulses, and to G. Feinberg and M. Goldhaber for alerting me to the stringent limit on $a^0 d \to pn$ events provided by the second reference. The results obtained here for this reaction rate are in essential agreement with the rate calculated by Feinberg.

[18]The suggestion of using $\Upsilon \to H^0\gamma$ decay as a source of Higgs bosons is due to F. Wilczek, Phys. Rev. Lett. 39, 1304 (1977). The result here for J/ψ decay is consistent with Wilczek's result for Υ decay, scaled down by the ratio of squared masses. Also see Ref. 7.

[19]R. Dashen, Phys. Rev. 183, 1245 (1964).

[20]P. Langacker and H. Pagels, Phys. Rev. D 8, 4620 (1973).

VOLUME 43, NUMBER 21 PHYSICAL REVIEW LETTERS 19 NOVEMBER 1979

Baryon- and Lepton-Nonconserving Processes

Steven Weinberg
Lyman Laboratory of Physics, Harvard University, Cambridge, Massachusetts 02138, and Harvard-Smithsonian Center for Astrophysics, Cambridge, Massachusetts 02138
(Received 13 August 1979)

A number of properties of possible baryon- and lepton-nonconserving processes are shown to follow under very general assumptions. Attention is drawn to the importance of measuring μ^+ polarizations and $\bar{\nu}_e/e^+$ ratios in nucleon decay as a means of discriminating among specific models.

Of the supposedly exact conservation laws of physics, two are especially questionable: the conservation of baryon number and lepton number. As far as we know, there is no necessity for an *a priori* principle of baryon and lepton conservation. As we shall see, even without such a principle, the fact that the weak, electromagnetic, and strong interactions of ordinary quarks and leptons conserve baryon and lepton number can be understood as simply a consequence of the $SU(2)\otimes U(1)$ and $SU(3)$ gauge symmetries. Also, in contrast with the conservation of charge, color, and energy and momentum, the conservation of baryon number and lepton number are almost certainly not unbroken local symmetries.[1] Not only is baryon conservation unnecessary as a fundamental principle, the apparent excess of baryons over antibaryons in our universe provides a positive clue that some sort of physical processes have actually violated baryon-number conservation.[2] Violations of baryon and lepton conservation are likely to occur in grand unified theories that combine the gauge theory of weak and electromagnetic interactions with that of strong interactions and have leptons and quarks in the same gauge multiplets, and such violations have been found in various of these models.[3]

The purpose of this paper is to point out those features of baryon- or lepton-nonconserving processes that are to be expected on very general grounds. Other features will be indicated that may be used to discriminate among specific models.

No grand unified model or other specific gauge model of baryon- and lepton-nonconserving processes will be adopted here. Instead, it will simply be assumed that these processes are mediated by some unspecified "superheavy" particles, with a characteristic mass M above, say, 10^{14} GeV. Such large masses are indicated by the experimental lower bound[4] on the proton lifetime, and are also required in order that these parti-

VOLUME 43, NUMBER 21 PHYSICAL REVIEW LETTERS 19 NOVEMBER 1979

cles should decay sufficiently early in the history of the universe to yield an appreciable baryon number.[2] Large masses are also required by general ideas of grand unification[5]: The strong coupling g_S of quantum chromodynamics (QCD) decreases so slowly with increasing energy that we must go up to very high energies before g_S becomes comparable with the weak and electromagnetic couplings g, g'.

In addition, it will be assumed here that the only particles with masses much less than the superheavy mass scale M are the "ordinary" particles of the sort with which we are familiar: left-handed lepton and quark doublets l_{iaL} and $q_{i\alpha aL}$; right-handed lepton and quark singlets l_{aR}, $u_{\alpha aR}$, and $d_{\alpha aR}$; and color-neutral bosons $W^{\pm}$, Z^0, γ, gluons, and Higgs scalars. [Here $\alpha = 1, 2, 3$ is an SU(3) index; $i = 1, 2$ is an SU(2) index; and $a = 1, 2, 3, \ldots$ is a "generation" index, distinguishing $e, \mu, \tau, \ldots$; $u, c, t, \ldots$; $d, s, b, \ldots$.] As it stands, this is a fairly restrictive assumption, but there are many other types of particles whose presence at ordinary mass levels would not affect our conclusions. Of course, one can also give examples of possible exotic particle types, whose presence at mass levels below the superheavy mass scale M could invalidate the general rules derived below; any observed departures from these rules would then provide valuable data on the nature of such exotic particles. Most of our results do not depend in any way on the nature of the superheavy particles of mass M, but we shall also see what consequences follow from the further assumption that baryon instability is due to exchange of a single superheavy vector or scalar boson.

Physical processes which occur at ordinary energies, including proton decay, can be described in terms of an SU(3) ⊗ SU(2) ⊗ U(1)-invariant effective field theory, which is obtained by integrating out all the superheavy degrees of freedom. The effective theory involves only the "ordinary" particles whose mass is much less than the characteristic superheavy mass scale M. The ordinary bosons which appear in the effective theory all have vanishing baryon number and lepton number, so that purely bosonic terms in the effective Lagrangian conserve baryon number and lepton number trivially. Also, for interactions involving a pair of ordinary fermions and any number of derivatives and ordinary bosons, SU(3) of QCD immediately implies baryon conservation.[6] Hence the terms in the effective Lagrangian which violate baryon conservation must involve at least four fermion fields.

These operators have dimensionality (mass)^d, with $d \geq 6$. But the only mass scale entering in the determination of the effective Lagrangian is the characteristic mass M of the superheavy particles,[7] and so the effective coupling constants associated with these operators must on dimensional grounds be roughly of order M^{4-d}. With M as large as assumed here, the only baryon-nonconserving interactions of practical interest are those given by the operators with $d = 6$. These have just four fermion fields and no derivatives or boson fields.

It is straightforward to enumerate all possible operators of this type which are SU(3) ⊗ SU(2) ⊗ U(1) invariant and do not conserve baryon number:

$$O^{(1)}_{abcd} = (\bar{d}^C_{\alpha aR} u_{\beta bR})(\bar{q}^C_{i\gamma cL} l_{jdL})\epsilon_{\alpha\beta\gamma}\epsilon_{ij}, \tag{1}$$

$$O^{(2)}_{abcd} = (\bar{q}^C_{i\alpha aL} q_{j\beta bL})(\bar{u}^C_{\gamma cR} l_{dR})\epsilon_{\alpha\beta\gamma}\epsilon_{ij}, \tag{2}$$

$$O^{(3)}_{abcd} = (\bar{q}^C_{i\alpha aL} q_{j\beta bL})(\bar{q}^C_{k\gamma cL} l_{ldL})\epsilon_{\alpha\beta\gamma}\epsilon_{ij}\epsilon_{kl}, \tag{3}$$

$$O^{(4)}_{abcd} = (\bar{q}^C_{i\alpha aL} q_{j\beta bL})(\bar{q}^C_{k\gamma cL} l_{ldL})\epsilon_{\alpha\beta\gamma} \times (\vec{\tau}\epsilon)_{ij} \cdot (\vec{\tau}\epsilon)_{kl}, \tag{4}$$

$$O^{(5)}_{abcd} = (\bar{d}^C_{\alpha aR} u_{\beta bR})(\bar{u}^C_{\gamma cR} l_{dR})\epsilon_{\alpha\beta\gamma}, \tag{5}$$

$$O^{(6)}_{abcd} = (\bar{u}^C_{\alpha aR} u_{\beta bR})(\bar{d}^C_{\gamma cR} l_{dR})\epsilon_{\alpha\beta\gamma}. \tag{6}$$

Here α, β, and γ are SU(3) indices; i, j, k, and l are SU(2) indices; a, b, c, and d are generation indices; l_{iaL} and $q_{i\alpha aL}$ are generic left-handed lepton and quark SU(2) doublets; l_{aR}, $u_{\alpha aR}$, and $d_{\alpha aR}$, are generic right-handed charged lepton and quark SU(2) singlets; C denotes the Lorentz-invariant complex conjugate; and ϵ_{ij} and $\epsilon_{\alpha\beta\gamma}$ are the totally antisymmetric SU(2) and SU(3) tensors with $\epsilon_{12} \equiv \epsilon_{123} \equiv +1$. Fierz transformations have been used to put the various Fermi interactions in the form of Eqs. (1)–(6), and in particular, to eliminate all vector and tensor Dirac matrices.

Inspection of Eqs. (1)–(6) leads immediately to a number of general rules which govern baryon-nonconserving interactions:

(A) $\Delta L = \Delta B$.—All interactions (1)–(6) conserve the difference of the baryon number B and lepton number L. Hence nucleons can only decay into antileptons, not leptons. The conservation of $B - L$ has already been noted[8] as a general consequence of SU(3) ⊗ SU(2) ⊗ U(1) invariance in the couplings of arbitrary superheavy scalar or vector bosons to pairs of ordinary fermions, but this argument leaves open the possibility that conservation of $B - L$ could be violated in baryon decay by graphs of higher order in α, involving

VOLUME 43, NUMBER 21 PHYSICAL REVIEW LETTERS 19 NOVEMBER 1979

$(B-L)$-nonconserving violating couplings of superheavy bosons to each other, or to superheavy fermions. The fact that (1)–(6) conserve $B-L$ implies that $(B-L)$-nonconserving processes like $n\to e^-\pi^+$ are suppressed relative to $(B-L)$-conserving processes like $n\to e^+\pi^-$ or $p\to e^+\pi^0$ by factors of order $m_W/M \lesssim 10^{-12}$, and not just by powers of α.

(B) $\Delta S/\Delta B \leq 0$.—The $\Delta B = \Delta L = -1$ operators (1)–(6) can contain 0, 1, or 2 fields which destroy s quarks, but no fields which create s quarks, so processes like $p\to \bar{K}^0 l^+$ or $n\to \bar{K}^- l^+$ with $\Delta S = \Delta B$ are forbidden.[9]

(C) $\Delta I = 1/2$.—Interactions with $\Delta S = 0$ and $\Delta B = -1$ are obtained by replacing the generic quark fields of charge $\frac{2}{3}$ and $-\frac{1}{3}$ in Eqs. (1)–(6) with u and d, respectively. Explicitly,[10]

$$O^{(1)}_{\Delta S=0} = (\bar{d}^C_{\alpha R} u_{\beta R})(\bar{u}^C_{\gamma L} l_L^- - \bar{d}^C_{\gamma L}\nu_L)\epsilon_{\alpha\beta\gamma}, \tag{7}$$

$$O^{(2)}_{\Delta S=0} = -2(\bar{d}^C_{\alpha L} u_{\beta L})(\bar{u}^C_{\gamma R} l_R^-)\epsilon_{\alpha\beta\gamma}, \tag{8}$$

$$O^{(3)}_{\Delta S=0} = -2(\bar{d}^C_{\alpha L} u_{\beta L})(\bar{u}^C_{\gamma L} l_L^- - \bar{d}^C_{\gamma L}\nu_L)\epsilon_{\alpha\beta\gamma}, \tag{9}$$

$$O^{(4)}_{\Delta S=0} = O^{(6)}_{\Delta S=0} = 0, \tag{10}$$

$$O^{(5)}_{\Delta S=0} = (\bar{d}^C_{\alpha R} u_{\beta R})(\bar{u}^C_{\gamma R} l_R^-)\epsilon_{\alpha\beta\gamma}. \tag{11}$$

(Here l^- is any of e, μ, τ; and ν is the corresponding neutrino.) The antisymmetric part of $\bar{d}^C_\alpha u_\beta$ is an isoscalar, so the operators multiplying the l_L^- and ν_L fields form isotropic doublets, and the operators multiplying the l_R^- field are the top members of other isotopic doublets. This leads directly to a number of simple relations among rates,[11] such as

$$\Gamma(p\to l_R^+\pi^0) = \tfrac{1}{2}\Gamma(n\to l_R^+\pi^-) = \tfrac{1}{2}\Gamma(p\to\bar{\nu}\pi^+) = \Gamma(n\to\bar{\nu}\pi^0) \tag{12}$$

and

$$\Gamma(p\to l_L^+\pi^0) = \tfrac{1}{2}\Gamma(n\to l_L^+\pi^-). \tag{13}$$

There are also relations among inclusive rates, such as

$$\Gamma(p\to l_R^+X) = \Gamma(n\to\bar{\nu}X),\quad \Gamma(n\to l_R^+X) = \Gamma(p\to\bar{\nu}X). \tag{14}$$

For experiments in which the charged lepton helicities are not measured, the relations (12)–(14) must be combined to give

$$\Gamma(p\to l^+\pi^0) = \tfrac{1}{2}\Gamma(n\to l^+\pi^-) \geq \tfrac{1}{2}\Gamma(p\to\bar{\nu}\pi^+); \tag{15}$$

$$\Gamma(p\to l^+X) \geq \Gamma(n\to\bar{\nu}X),\quad \Gamma(n\to l^+X) \geq \Gamma(p\to\bar{\nu}X). \tag{16}$$

(Relations connecting $\bar{\nu}$ and l^+ processes are only valid in the limit of relativistic l^+ velocities.)

We will now consider one further assumption. To lowest order in α, baryon nonconservation would presumably be due to exchange of a single vector or scalar superheavy boson. It has been shown[8] that in general there are just five kinds of superheavy vector or scalar bosons that can have $SU(3)\otimes SU(2)\otimes U(1)$-invariant baryon-nonconserving interactions to a pair of ordinary fermions: These are SU(3)-triplet, SU(2)-doublet vector bosons X_V, X_V' of charges $(\frac{4}{3}, \frac{1}{3})$ and $(\frac{2}{3}, -\frac{1}{3})$; SU(3)-triplet, SU(2)-singlet scalar bosons X_S, X_S' of charges $-\frac{1}{3}, -\frac{4}{3}$; and SU(3)-triplet, SU(2)-triplet scalar bosons X_S'' of charge $(\frac{2}{3}, -\frac{1}{3}, -\frac{4}{3})$; plus their antiparticles. It is straightforward[12] to check that X_S exchange can contribute to interactions of form $O^{(1)}$, $O^{(2)}$, $O^{(3)}$, and $O^{(5)}$; X_S' and X_S'' can contribute only to $O^{(6)}$ and $O^{(4)}$, respectively; while X_V exchange can contribute only to $O^{(1)}$ and $O^{(2)}$; and X_V' exchange only to $O^{(1)}$. If we assume that baryon nonconservation is due to exchange of any sort of *vector* boson, then only $O^{(1)}$ and $O^{(2)}$ enter, and we can write the effective Lagrangian for $\Delta S=0$ and $\Delta S = -\Delta B$ baryon-nonconserving processes as

$$\mathcal{L} = [g_1(\bar{d}^C_{\alpha R}u_{\beta R})(\bar{u}^C_{\gamma L}l_L^- - \bar{d}^C_{\gamma L}\nu_L) + g_2(\bar{d}^C_{\alpha L}u_{\beta L})(\bar{u}^C_{\gamma R}l_R^-) + g_1'(\bar{s}^C_{\alpha R}\bar{u}_{\beta R})(\bar{u}^C_{\gamma L}l_L^- - \bar{d}^C_{\gamma L}\nu_L) + g_1''(\bar{d}^C_{\alpha R}u_{\beta R})(\bar{s}^C_{\gamma L}\nu_L) + g_2'(\bar{s}^C_{\alpha L}u_{\beta L})(\bar{u}^C_{\gamma R}l_R^-) + \text{H.c.}]\epsilon_{\alpha\beta\gamma}. \tag{17}$$

This leads to an immediate result for charged-lepton emission.

(D) *Universal polarization.*— The hadronic operator associated with l_L^- in $\Delta S=0$ processes is just g_1/g_2 times the parity transform of the hadronic operator associated with l_R^-, and similarly for $\Delta S=1$ processes. It follows that for relativistic charged leptons, the lepton polarizations take constant values, which depend only on whether the decay mode has $\Delta S=0$ or $\Delta S=-\Delta B$:

$$P_{\Delta S=0} = \frac{|g_1|^2 - |g_2|^2}{|g_1|^2 + |g_2|^2},\quad P_{\Delta S=-\Delta B} = \frac{|g_1'|^2 - |g_2'|^2}{|g_1'|^2 + |g_2'|^2}, \tag{18}$$

VOLUME 43, NUMBER 21 PHYSICAL REVIEW LETTERS 19 NOVEMBER 1979

where

$$P \equiv \frac{\Gamma(N \to l_R^{\,+}H) - \Gamma(N \to l_L^{\,+}H)}{\Gamma(N \to l_R^{\,+}H) + \Gamma(N \to l_L^{\,+}H)} \tag{19}$$

with H any specific hadronic channel.

An experimental check of the results (A)–(D) would be useful as a test of the general assumptions described above. In particular, verification of $\Delta I = \frac{1}{2}$ relations like (12)–(16) would provide a good indication that baryon nonconservation is due to virtual particles so heavy that $SU(2) \otimes U(1)$ is effectively unbroken in their interactions. Also, verification of the universality of the charged-lepton polarizations would indicate an absence of significant scalar-boson exchange; the inclusion of operators $O^{(3)}$, $O^{(5)}$ which could be produced by X_S exchange would lead to lepton polarizations which depend on the relative values of matrix elements of $(\bar{d}^C_{\alpha R} u_{\beta R}) u_{\gamma L} \epsilon_{\alpha\beta\gamma}$ and $(\bar{d}^C_{\alpha L} u_{\beta L}) u_{\gamma L} \epsilon_{\alpha\beta\gamma}$, and hence which depend on the details of the decay mode. But a check of (A)–(D) cannot be used to verify any specific gauge model of baryon decay.

Within the context described here, different models of baryon nonconservation can be distinguished only through measurements of the five parameters $g_1, g_2, g_1', g_1'', g_2'$ for each lepton type. But as already mentioned above, as far as charged leptons are concerned, the operators multiplying g_1 and g_2 in Eq. (17) are simply space inversions of each other, and likewise for g_1' and g_2', and so if parity-odd operators are not measured, the only quantities that can be determined in observations of charged lepton modes are the overall coupling scales $|g_1|^2 + |g_2|^2$ and $|g_1'|^2 + |g_2'|^2$. Under our general assumptions, for $\Delta S = 0$ and $\Delta S = \Delta B$ baryon-nonconserving charged-lepton processes, as long as no pseudoscalars are measured, *all models must give the same results for the relative rates of different decay modes*, and can differ only in the total $\Delta S = 0$ and $\Delta S = \Delta B$ decay rates for each lepton type.

This conclusion serves to emphasize the importance of measuring charged lepton polarizations or $\bar{\nu}/l^+$ ratios in nucleon decay. Different models will give quite different polarizations: For instance, if X_V' exchange is dominant then $g_2 = g_2' = 0$, so that $P = +1$ in both $\Delta S = 0$ and $\Delta S = -\Delta B$ processes, while if X_V exchange is significant then P will depend on ΔS and on the details of the model.

Fortunately, if baryon nonconservation is discovered, it should be feasible to determine the lepton polarization, perhaps in a second round of experiments. The μ^+ polarization can be determined from the direction of positrons from stopped muons, using the same detection system that is used to detect positrons from nucleon decay. The e^+ polarization would probably have to be determined indirectly, using the $\Delta I = \frac{1}{2}$ relations (12) or (14). Once these polarizations are measured, Eq. (18) can be used to discriminate among models, with no need to worry about complications due to strong interactions.

The sort of analysis used here in treating baryon nonconservation can also be applied to lepton nonconservation. A great difference is that there is a possible lepton-nonconserving term in the effective Lagrangian with dimensionality $d = 5$:

$$f_{abmn} \bar{l}^C_{iaL} l_{jbL} \varphi_k^{(m)} \varphi_l^{(n)} \epsilon_{ik} \epsilon_{jl} + f_{abmn}' \bar{l}^C_{iaL} l_{jbL} \varphi_k^{(m)} \varphi_l^{(n)} \epsilon_{ij} \epsilon_{kl}\,, \tag{20}$$

where $\varphi^{(m)}$ are one or more scalar doublets. We expect f and f' to be roughly of order $1/M$; one-loop graphs would give values of order α^2/M.[13] The interaction (20) would produce a neutrino mass $m_\nu \simeq G_F^{-1} f$, or roughly 10^{-5} to 10^{-1} eV. This is well below any existing laboratory or cosmological limits, but there is no reason why this neutrino-mass matrix should be diagonal, and masses of this order might perhaps be observable in neutrino oscillation experiments.

I am very grateful for valuable conversations with H. Georgi, M. Machacek, D. V. Nanopoulos, and L. Sulak.

This research was supported in part by the National Science Foundation under Grant No. PHY77-22864.

Note added.—After this paper was submitted for publication, I received a preprint from F. Wilczek and A. Zee [following Letter, Phys. Rev. Lett. 43, 1571 (1979)], which reaches similar conclusions about baryon-nonconserving processes.

[1] A massless boson coupled to baryon or lepton number would introduce discrepancies in the Eötvös experiment unless its couplings were incredibly weak; see T. D. Lee and C. N. Yang, Phys. Rev. 98, 101 (1955).

[2] M. Yoshimura, Phys. Rev. Lett. 41, 281 (1978), and 42, 746(E) (1979), and Tohoku University Reports No. TU/79/192 and No. TU/79/193 (to be published); L. Dimopoulos and L. Susskind, Phys. Rev. D 18, 4500 (1978), and Phys. Lett. 81B, 416 (1979); A. Yu. Ignatiev, N. V. Krasnikov, V. A. Kuzmin, and A. V. Tavhelidze, Phys. Lett. 76B, 436 (1978); B. Toussaint, S. B. Treiman, F. Wilczek, and A. Zee, Phys. Rev. D 19, 1036 (1979); J. Ellis, M. K. Gaillard, and D. V. Nanopoulos,

VOLUME 43, NUMBER 21 PHYSICAL REVIEW LETTERS 19 NOVEMBER 1979

Phys. Lett. 80B, 360 (1979), and 82B, 464(E) (1979); S. Weinberg, Phys. Rev. Lett. 42, 850 (1979); N. J. Papastamatiou and L. Parker, Phys. Rev. D 19, 2283 (1979); D. V. Nanopoulos and S. Weinberg, Harvard University Report No. HUTP-79/A023 (to be published); S. Barr, G. Segrè, and H. A. Weldon, to be published. For early suggestions along this line, see S. Weinberg, in *Lectures on Particles and Fields*, edited by S. Deser and K. Ford (Prentice-Hall, Englewood Cliffs, N. J., 1964), p. 482; A. D. Sakharov, Zh. Eksp. Teor. Fiz., Pis'ma 5, 32 (1967) [JETP Lett. 5, 24 (1967)].

[3]This was first noted by J. C. Pati and A. Salam, Phys. Rev. D 8, 1240 (1973), and 10, 275 (1974). Other leading grand unified models which violate baryon conservation include those based on SU(5) [H. Georgi and S. L. Glashow, Phys. Rev. Lett. 32, 438 (1974)] on SO(10) [H. Georgi, in *Particles and Fields—1974*, edited by C. E. Carlson, AIP Conference Proceedings No. 23 (American Institute of Physics, New York, 1975); H. Fritzsch and P. Minkowski, Ann. Phys. (N.Y.) 93, 193 (1975); H. Georgi and D. V. Nanopoulos, Phys. Lett. 82B, 392 (1979), and Harvard University Report No. HUTP-79/A001, 1979 (to be published)]; on E_6, E_7 [F. Gürsey, P. Ramond, and P. Sikivie, Phys. Lett. B60, 177 (1975); F. Gürsey and P. Sikivie, Phys. Rev. Lett. 36, 775 (1976); P. Ramond, Nucl. Phys. B110, 214 (1976)]; etc.

[4]F. Reines and M. F. Crouch, Phys. Rev. Lett. 32, 493 (1974); J. Learned, F. Reines, and A. Soni, Phys. Rev. Lett. 43, 907, 1626(E) (1979).

[5]H. Georgi, H. R. Quinn, and S. Weinberg, Phys. Rev. Lett. 33, 451 (1974). This paper showed that for a broad range of grand unified gauge models, the mass scale M is of order 10^{15} to 10^{16} GeV; the proton lifetime for typical M is very roughly of order 10^{32} yr; and $\sin^2\theta$ is close to 0.2. [This class of theories includes essentially all gauge models in which a simple grand gauge group is spontaneously broken in a single step at M to $SU(3)\otimes SU(2)\otimes U(1)$, and in which the fermions form generations of the same sort as for observed quarks and leptons.] These estimates have been improved by more detailed studies: A. Buras, J. Ellis, M. K. Gaillard, and D. V. Nanopoulos, Nucl. Phys. B135, 66 (1978); D. Ross, Nucl. Phys. B140, 1 (1978); T. J. Goldman and D. A. Ross, Phys. Lett. 84B, 208 (1979); W. Marciano, Rockefeller University Report No. COO-2232B-173, 1979 (to be published); N. P. Chang, A. Das, and J. Perez-Mercader, to be published; C. Jarlskog and F. J. Yndurain, to be published; M. Machacek, Harvard University Report No. HUTP-79/A021, 1979 (to be published).

[6]This was not the case for the Pati-Salam model of Ref. 3, because SU(3) was assumed there to be spontaneously broken.

[7]Strictly speaking, this is correct if the couplings in the effective Lagrangian are defined at renormalization scales of order M. At ordinary energies E there are $\alpha \ln E/M$ renormalization effects (some calculated by Buras *et al.*, Ref. 5) but these corrections are at most of order unity, and do not affect our conclusions. Our approach is related to that of T. Appelquist and J. Carazzone, Phys. Rev. 11, 2856 (1975).

[8]Weinberg, Ref. 2, note (1). [The scalars X_S', X_S'' make no contribution to nucleon decay, and were previously omitted.]

[9]This conclusion was reached on essentially the same grounds by Machacek, Ref. 5.

[10]I understand that the general isospin properties of the effective interaction were worked out in a similar way in unpublished work by H. Georgi.

[11]Essentially the same relations have already been found by Machacek, Ref. 5. However, her derivation was in the context of specific models, and it was not clear which of these results would be more generally valid.

[12]General formulas for the possible interactions of X_S, X_V, and X_V' with quarks and leptons were given by Nanopoulos and Weinberg, Ref. 2.

[13]In the O(10) model, a term of form (20) is produced by a tree graph, with $f \approx G_F m_t^2/M$ [Georgi and Nanopoulos, Ref. 3]. Of course, f would vanish if $B-L$ were exactly conserved.

PHYSICAL REVIEW D VOLUME 26, NUMBER 1 1 JULY 1982

Supersymmetry at ordinary energies. Masses and conservation laws

Steven Weinberg
Lyman Laboratory of Physics, Harvard University, Cambridge, Massachusetts 02138
and Department of Physics, University of Texas, Austin, Texas 78712
(Received 30 September 1981)

An assessment is made of the general problems encountered in formulating a realistic supersymmetric theory in which the spontaneous breakdown of supersymmetry occurs at ordinary energies accessible to accelerators. As a starting point, three problems are identified in SU(3)×SU(2)×U(1) supersymmetric models with only quark and lepton chiral superfields: the up quarks get no masses, baryon and lepton (B and L) conservation are violated by renormalizable and hence unsuppressed interactions, and the scalar counterparts of the quarks and leptons are too light. An interesting SU(3)×SU(2)×U(1) model of Dimopoulos and Georgi that avoids these problems is considered; it is found that this model contains B- and L-nonconserving effective interactions of dimensionality 5 that lead to proton decay at too rapid a rate. To guarantee natural B and L conservation in effective interactions of dimensionality 4 and 5, it is suggested that the gauge group that describes physics at ordinary energies contains a factor, such as another U(1), in addition to SU(3)×SU(2)×U(1). Such theories do not contain dimension-5 L-nonconserving interactions which could produce an observable neutrino mass, but they do allow dimension-6 B- and L-nonconserving interactions that would lead to proton decay at an observable rate. Supersymmetry is found to constrain the matrix elements for proton decay in a phenomenologically interesting way. A general explanation is given of how such theories naturally avoid the problem of light scalars, as found by Fayet. The formalism is used to derive general approximate mass relations for the scalar superpartners of the quarks and leptons. The problem of anomalies in the new U(1) current is considered, and one attractive scheme for avoiding them is offered, in which the anomalies cancel for precisely three generations of quarks and leptons.

I. INTRODUCTION

We know that if nature at a fundamental level really obeys supersymmetry,[1] then the supersymmetry must be spontaneously broken. However, we do not know whether the vacuum expectation values involved in this breakdown are of an "ordinary" scale, say of order 300 GeV, like those involved in the breakdown of the electroweak gauge symmetry, or whether they are much larger, perhaps as high as the Planck mass. One reason to suspect that supersymmetry is broken only at ordinary energies arises from the hierarchy problem[2]: if supersymmetry is unbroken at higher energies, then it can protect some scalar fields from getting enormous masses in the spontaneous breakdown of whatever symmetry connects strong and electroweak interactions; these scalars would then survive to provide a second stage of symmetry breaking, in which the electroweak gauge symmetry and supersymmetry are both spontaneously broken at ordinary energies. At any rate, the hypothesis that supersymmetry is unbroken above some ordinary energy scale of order 300 GeV is worth careful attention, because it has direct experimental implications at the energies that will soon be accessible to accelerators. Models which are supersymmetric down to ordinary energies have already been developed and their consequences studied, most notably by Fayet.[3]

The purpose of this paper is to take a fresh look at the implications of supersymmetry at ordinary energies, and especially to apply to supersymmetric theories certain developments that have occurred in the last few years, particularly in our point of view regarding baryon and lepton conservation.

One of the successes of the SU(3)×SU(2)×U(1) gauge theory of strong and electroweak interactions was that it explained the experimentally observed conservation laws for baryon and lepton number (B and L) without needing to invoke separate global conservation laws. The most general renormalizable interaction that one can write down involving just ordinary quarks, leptons, Higgs doublets, and

SU(3)×SU(2)×U(1) gauge fields is forced by SU(3)×SU(2)×U(1) gauge invariance to conserve B and L. However, supersymmetric theories necessarily contain additional particles, including scalar superpartners of the quarks and leptons, and in general these could participate in renormalizable B- or L-nonconserving interactions. Of course, one could simply impose B or L conservation laws as global symmetries of supersymmetric models. This does not seem to me to be the most fruitful approach. Global symmetries, like strangeness, isospin, etc., increasingly appear to us as incidental consequences of gauge symmetries and renormalizability, with no status as *a priori* constraints on a fundamental level. It would be preferable to hold on to the feature of SU(3)×SU(2)×U(1) gauge theories, that B and L conservation follow automatically from gauge symmetries and renormalizability, when the particle content of the theories is extended to incorporate supersymmetry. As we shall see, this requirement proves to be a useful guide in constructing a satisfactory supersymmetric model of physics at ordinary energies.

The approach advocated here will turn out to be indispensable if B and L are found to be not exactly conserved. If the renormalizable interactions of particles of ordinary masses ($\lesssim 300$ GeV) automatically conserve B and L, then any observable B- and L-nonconserving interactions among these particles would have to be due to the effects of virtual superheavy (10^{15} GeV?) particles with different gauge quantum numbers. Such effects would show up in the effective Lagrangian which describes physics at ordinary energies as nonrenormalizable effective interactions of dimensionality $d>4$, with coupling constants suppressed by $d-4$ powers of the superheavy mass. Now, just as SU(2)×U(1) can be used[4] directly to study the structure of the B- and L-nonconserving nonrenormalizable effective interactions, because even though SU(2)×U(1) is spontaneously broken at ordinary energies it is not broken at the superheavy masses of the particles whose exchange violates B and L conservation, in the same way if supersymmetry is unbroken at these superhigh energies, then it can be used to constrain the effective interactions with $d>4$ that are responsible for B- and L-nonconserving interactions of ordinary particles. Supersymmetry, like SU(2)×U(1), may become manifest in the selection rules for proton decay.

The general rules for constructing renormalizable and nonrenormalizable supersymmetric effective interactions are very simple. Suppose a theory involves a set of left-handed chiral scalar superfields, generically called S, together with their right-handed adjoints S^*. Leaving aside the possibility of extra derivatives and gauge couplings, the most general supersymmetric Lagrangian will be of the form

$$\mathscr{L}=f(S)_F+f(S)_F^*+d(S^*,S)_D\ , \tag{1}$$

where f and d are arbitrary functions, and as usual F and D denote the terms in these functions proportional to $\theta_L\theta_L$ or $\theta_L\theta_L\theta_R\theta_R$, respectively, with θ_L and θ_R the left- and right-handed parts of the fermionic superfield coordinate θ. Note that we do not include F terms of functions of both S^* and S, because these would not be supersymmetric, and we do not include D terms of functions of S or S^* alone, because these would be total derivatives. It is easy to extend this Lagrangian so that it has a local gauge symmetry: in the same way as is familiar for renormalizable theories,[1] we add a Yang-Mills-type term for the real vector superfield $V\equiv\sum_\alpha g_\alpha t_\alpha V_\alpha$ and replace S^* in the D term with S^*e^{2V}. However, the inclusion of gauge couplings (or extra derivatives) in an effective interaction would generally increase its dimensionality without introducing new possibilities for B or L nonconservation. The expression for the Lagrangian (1) in terms of ordinary component fields is given in Appendix A.

Now, the dimensionality of a scalar superfield is $+1$ (in powers of mass), and the dimensionality of the F term or D term of any function is equal to the dimensionality of the function plus 1 or 2, respectively. Thus the terms in the effective Lagrangian (1) of various dimensionalities have the structure

$$\begin{aligned}
d&=2:\ (S)_F\ ,\\
d&=3:\ (SS)_F\ ,\\
d&=4:\ (S^*S)_D\ ,\quad (SSS)_F\ ,\\
d&=5:\ (S^*SS)_D\ ,\quad (SSSS)_F\ ,\\
d&=6:\ (S^*SSS)_D\ ,\quad (S^*S^*SS)_D\ ,\quad (SSSSS)_F
\end{aligned} \tag{2}$$

and so on, plus the Hermitian conjugates. Of course, the S superfields have a number of indices which must be suitably contracted in order to maintain invariance under all gauge symmetries. In this way, we can construct all possible terms in the effective Lagrangian which can occur up to some definite degree of suppression by superheavy masses.

In order to implement this program, it is necessary at least to know what are the gauge symmetries and particle spectrum which appear in a supersymmetric model at ordinary energies, and this opens up the whole range of phenomenological problems faced by supersymmetric theories. For orientation, in Sec. II, we inspect the properties of a minimal supersymmetric model, with only a $SU(3)\times SU(2)\times U(1)$ gauge symmetry and only chiral scalar superfields corresponding to the known quarks and leptons. This turns out to have severe problems: no up-quark masses, renormalizable (and hence unsuppressed) B- and L-nonconserving interactions, and unobserved light scalars. Then in Sec. III, we explore a model proposed by Dimopoulos and Georgi,[5] in which renormalizable B- and L-nonconserving interactions are ruled out by a discrete reflection symmetry; Higgs superfields are introduced to give masses to all quarks and leptons; and light scalars are avoided by supposing that supersymmetry is explicitly but softly broken by terms in the Lagrangian with $d=2$ and $d=3$. We find that although proton decay is suppressed in this model, it is not suppressed enough: there are B- and L-nonconserving terms with $d=5$ in the effective Lagrangian, and these lead to processes like $p\to\mu^+K^0$ with a proton lifetime of order 10^{28} yr. To avoid such catastrophes, we consider in Sec. IV the addition of an extra gauge symmetry to the invariance group of the model. It is found that all $d=5$ terms are forbidden, including not only the B- and L-nonconserving terms which gave trouble in the Dimopoulos-Georgi model, but also any L-nonconserving terms which could give an observable neutrino mass. However, there remain allowed $d=6$ terms, which could produce a "normal" $(B-L)$-conserving proton decay at an observable rate. Supersymmetry and gauge symmetries constrain the matrix elements for this decay, with interesting phenomenological consequences.

The extra gauge symmetry which is introduced in order automatically to suppress B and L nonconservation also has important implications for particle masses. These are described in Sec. V, and as an illustration a special case of the models described by Fayet is analyzed in detail in Appendix B. It is shown that supersymmetric theories can quite naturally account for the fact that we observe quarks and leptons with masses much less than those of their scalar superpartners, without having to break supersymmetry explicitly in the Lagrangian. Supersymmetry is broken only spontaneously here, the scale of this breaking and of $SU(2)\times U(1)$ breaking being set by the coefficients of Fayet-Iliopoulos terms[6] $(V)_D$. As shown by Witten,[7] it is natural for these coefficients not to get large values from the spontaneous breakdown of a semisimple group at superlarge energies, or from perturbative corrections, so this part of the hierarchy problem is solved in such theories. There remains the vexing question of why the $(V)_D$ terms are present at all, and with such small coefficients relative to the superheavy masses. One can hope that this will be explained by nonperturbative effects[7] yielding $(V)_D$ terms proportional to $\exp(-\text{constant}/g^2)$, but such questions will not be addressed in this paper.

Section VI describes one way of canceling the Adler-Bell-Jackiw (ABJ) anomalies[8] in a model with precisely three generations of quarks and leptons. A future paper[9] will apply to "R invariance" the same sort of analysis used here for B and L, but with special attention to the effects of ABJ anomalies and instantons, and will attempt to summarize the problems that still face the formulation of a realistic supersymmetric theory.

II. VARIETIES OF TROUBLE: A MINIMAL MODEL

To appreciate the difficulties encountered in formulating a realistic model that is supersymmetric down to ordinary energies, consider first an $SU(3)\times SU(2)\times U(1)$ gauge theory containing just those chiral scalar superfields whose spin-$\frac{1}{2}$ components are the ordinary leptons and quarks. We will denote these superfields by capital letters: the spin-$\frac{1}{2}$ component of the left chiral scalar superfield $Q_L=(U_L,D_L)$ is the left-handed quark doublet $q_L=(u_L,d_L)$; the spin-$\frac{1}{2}$ components of the left chiral superfields U_R^*,D_R^* are the antiparticles u_R^*,d_R^* of the right-handed singlet quarks; and similarly the superfields L_L and E_R^* have as spin-$\frac{1}{2}$ components the lepton doublet $l_L^-=(\nu_L,e_L^-)$ and singlet e_R^*. The scalar components of these superfields are labeled by capital script letters:

$$\mathcal{Q}_L=(\mathcal{U}_L,\mathcal{D}_L)\ ,\ \mathcal{U}_R^*\ ,\ \mathcal{D}_R^*\ ,\ \mathcal{L}_L=(\mathcal{N}_L,\mathcal{E}_L)\ ,\ \mathcal{E}_R^*\ .$$

Here and wherever not explicitly otherwise indicated, the symbols for particles in the lowest generation are used to represent any particle with the same quantum numbers; thus e stands for e, μ, or

TABLE I. Summary of notation for superfields, with their spin-$\frac{1}{2}$ and spin-0 components and $SU(3)\times SU(2)\times \tilde{U}(1)$ quantum numbers. Note that where not otherwise indicated, the letters u,d,ν,e refer to quarks and leptons of any generation with the indicated quantum numbers.

Left-chiral scalar superfield	Spin-$\frac{1}{2}$ component	Spin-0 component	SU(3)	SU(2)	Y
$Q_L=(U_L,D_L)$	$q_L=(u_L,d_L)$	$(\mathscr{Q}=(\mathscr{U}_L,\mathscr{D}_L)$	3	2	$-\frac{1}{6}$
U_R^*	u_R^*	$\mathscr{U}_R^*$	$\bar{3}$	1	$\frac{2}{3}$
D_R^*	d_R^*	$\mathscr{D}_R^*$	$\bar{3}$	1	$-\frac{1}{3}$
$L_L=(N_L,E_L)$	$l_L=(\nu_L,e_L)$	$\mathscr{L}_L=(\mathscr{N}_L,\mathscr{E}_L)$	1	2	$\frac{1}{2}$
E_R^*	e_R^*	$\mathscr{E}_R^*$	1	1	-1
H_L	h_L	$\mathscr{H}_L=(\mathscr{Y}_L^0,\mathscr{U}_L^-)$	1	2	$\frac{1}{2}$
H_L'	h_L'	$\mathscr{H}_L'=(\mathscr{H}_L^{+\prime},\mathscr{U}_L^{0\prime})$	1	2	$-\frac{1}{2}$

τ, and so on. For convenience this notation is summarized in Table I.

Apart from kinematic terms and gauge couplings, the most general renormalizable supersymmetric $SU(3)\times SU(2)\times U(1)$-invariant interaction among the quark and lepton superfields is a linear combination of the trilinear F terms

$$(L_L L_L E_R^*)_F\ ,\quad (L_L Q_L D_R^*)_F\ ,\quad (D_R^* D_R^* U_R^*)_F\ , \tag{3}$$

with SU(3) and SU(2) indices contracted in an obvious way. There are three conspicuous things wrong with such a theory.

(1) Although a vacuum expectation value of the neutral scalar field $\mathscr{N}_L$ of L_L will break $SU(2)\times U(1)$ in the usual way and give mass to the charged leptons and charge $-\frac{1}{3}$ quarks, there is no neutral scalar field here whose vacuum expectation value can give mass to the charge $+\frac{2}{3}$ quarks.

(2) There is no way of extending the definition of baryon and lepton number (B and L) to the scalar fields so that B and L are conserved. In particular, exchange of the $\mathscr{D}_R$ scalar between the last two interactions in (3) can produce the proton decay process $q_L d_R u_R \to \bar{l}_L$ at a catastrophic rate. (However, $B-L$ can be still defined as a conserved R symmetry.[10])

(3) Such theories contain unobserved light scalars. This is shown most clearly by a theorem of Dimopoulos and Georgi[5] which states that in any supersymmetric theory with gauge group $SU(3)\times SU(2)\times U(1)$ and in which quark and gluon superfields are the only colored fields, the spontaneous breakdown of supersymmetry and $SU(2)\times U(1)$ must leave at least one scalar lighter than the lightest d or u quark.

Different attempts at realistic supersymmetry models can be conveniently characterized by the features that are put in to avoid these three problems.

III. SU(3)×SU(2)×U(1) SUPERSYMMETRIC MODELS

Let us first consider what must be done to surmount the problems discussed in the last section, if we do not wish to expand the gauge group at ordinary energies beyond $SU(3)\times SU(2)\times U(1)$. In this case the B- and L-nonconserving renormalizable interaction $(D_R^* D_R^* U_R^*)_F$ must be prohibited by some sort of global symmetry. This could of course be B or L conservation itself, but it is also possible to forbid B- and L-nonconserving interactions of dimensionality 4 with weaker symmetries, which would not require complete conservation of B and L. It is also necessary to add at least one Higgs superfield to give mass to the charge $\frac{2}{3}$ quarks, and if the global symmetry which rules out the B- and L-nonconserving term $(D_R^* D_R^* U_R^*)_F$ also rules out the other terms in (3), we have to add a second Higgs superfield to give mass to the charged leptons and charge $-\frac{1}{3}$ quarks as well. [The second Higgs superfield also serves to cancel an $SU(2)\times U(1)$ ABJ anomaly introduced by the first spin-$\frac{1}{2}$—Higgs—fermion doublet.] Finally, there is still the problem of the light scalars. The theorem of Dimopoulos and Georgi[5] makes it clear that in this class of models, the unacceptably low values of some scalar masses can be avoided only by supposing that supersymmetry is explicitly broken, but perhaps only "softly," by terms in the Lagrangian of dimensionality $d<4$.

Such a model has been developed by Dimopoulos and Georgi[5] (DG). In their work, the effective Lagrangian that governs physics at ordinary energies arises from an underlying grand unified model, but it can be described here in terms of the effective Lagrangian itself.

In the DG model, the problem of B and L nonconservation is dealt with by imposing a discrete symmetry: invariance under a change of sign of all quark and lepton superfields. Such a symmetry immediately rules out all the interactions (3), B and L conserving as well as nonconserving ones.

To restore the possibility of quark and lepton masses, DG add a pair of color-singlet electroweak-doublet left-chiral Higgs superfields H_L and H_L' with weak hypercharge ($Y\equiv T_3-Q$) equal to $+\frac{1}{2}$ and $-\frac{1}{2}$, which are even under the above discrete symmetry. Apart from kinematic terms and gauge couplings, the complete set of renormalizable supersymmetric interactions which are invariant under $SU(3)\times SU(2)\times U(1)$ and the discrete symmetry consists of just the F terms

$$\begin{aligned}&(H_L H_L')_F\ ,\quad (H_L L_L E_R^*)_F\ ,\\&(H_L Q_L D_R^*)_F\ ,\quad (H_L' Q_L U_R^*)_F\end{aligned}\tag{4}$$

with obvious index contractions. Up as well as down quarks can now get masses, and B and L are automatically conserved, with H_L and H_L' assigned vanishing B and L values.

The remaining problem found in Sec. II was (3), the problem of light scalars, To obviate this, DG suppose that only interactions of dimension 4 are supersymmetric, and that supersymmetry is explicitly (though softly) broken by terms of dimension 2 and 3. With this assumption, particle masses can be given values that do not conflict with observations.

Now, what about the suppressed nonrenormalizable terms in the effective Lagrangian? The least suppressed terms are those of dimensionality 5: either trilinear D terms or quadrilinear F terms. It is easy to see that the most general $SU(3)\times SU(2)\times U(1)$-invariant $d=5$ terms, which are also invariant under the DG discrete symmetry, have the form

$$\begin{aligned}&(L_L E_R^* H_L'^{\,*})_D\ ,\quad (Q_L D_R^* H_L'^{\,*})_D\ ,\quad (Q_L U_R^* H_L^*)_D\ ,\\&(L_L L_L H_L' H_L')_F\ ,\ (Q_L Q_L U_R^* D_R^*)_F\ ,\ (Q_L U_R^* L_L E_R^*)_F\ ,\\&(Q_L Q_L Q_L L_L)_F\ ,\quad (U_R^* U_R^* D_R^* E_R^*)_F\ ,\end{aligned}\tag{5}$$

again with SU(3) and SU(2) indices contracted in the obvious way. (For notation, see Sec. II or Table I.) The coupling constants of these effective interactions are in general expected to be of order

$$G_5\approx f^2/M\ ,\tag{6}$$

where f is a typical superheavy-particle coupling constant (Higgs or gauge), and M is a typical superheavy mass.

Most of the interactions (5) are innocuous, conserving both B and L. The $(L_L L_L H_L' H_L')_F$ term provides the sort of $d=5$ L-nonconserving interaction[11] $\nu_L\nu_L\mathscr{H}_L'^{0}\mathscr{H}_L'^{0}$ which would give the neutrino a small but possibly observable neutrino mass of order $f^2G_F{}^{-1}/M$. The dangerous terms are the last two, which violate both B and L. These are $d=5$ interactions, so the B- and L-violating matrix elements here are suppressed by only one power of the superheavy mass M, in contrast with the usual case[4] of theories without scalar superpartners of quarks and lepton, where the B- and L-violating effective interactions had $d=6$ and were suppressed by two powers of M. The DG model thus runs the risk of predicting much too fast a rate of proton decay.

We still must ask whether the B- and L-nonconserving interactions $(Q_L Q_L Q_L L_L)_F$ and $(U_R^* D_R^* D_R^* E_R^*)_F$ actually appear in the effective Lagrangian with couplings of order (6), and if so, whether these interactions really produce proton decay at an unacceptable rate.

The first question can of course only be answered in the context of a specific theory of the superheavy degrees of freedom. In the grand unified form of the DG model, there are superheavy color-triplet superfields, where exchange actually does produce the effective interactions $(Q_L Q_L Q_L L_L)_F$ and $(U_R^* D_R^* D_R^* E_R^*)_F$ in tree approximation, with coupling of order (6), where f is a typical Higgs-particle coupling.[12]

As to the effect of these interactions, note that they include the two-fermion—two-scalar terms (see Appendix A)

$$\begin{aligned}&q_L q_L \mathscr{Q}_L \mathscr{L}_L\ ,\quad q_L l_L \mathscr{Q}_L \mathscr{Q}_L\ ,\\&u_R^* u_R^* \mathscr{D}_R^* \mathscr{E}_R\ ,\quad u_R^* e_R^* \mathscr{U}_R^* \mathscr{D}_R^*\ ,\\&e_R^* d_R^* \mathscr{U}_R^* \mathscr{U}_R^*\ ,\quad d_R^* u_R^* \mathscr{U}_R^* \mathscr{E}_R^*\ .\end{aligned}\tag{7}$$

(Recall that capital script letters denote scalars.) Each of the heavy-scalar pairs $\mathscr{Q}_L\mathscr{L}_L$, $\mathscr{Q}_L\mathscr{Q}_L$, $\mathscr{U}_R\mathscr{D}_R$, and $\mathscr{U}_R\mathscr{E}_R$ can be created from a fermion pair u_Re_R, u_Rd_R, q_Lq_L, and q_Ll_L, respectively, by emitting a pair of fermion partners h_L, h_L' of the Higgs bosons, which then annihilate through the Majorana mass of these Higgs fermions. This yields the proton decay effective interactions $q_Lq_Lu_Re_R$ and $q_Ll_Lu_Rd_R$. However, the proton decay rate produced in this way is enormously suppressed by the four powers of small Higgs-particle couplings in the matrix element, and is probably too small to be observed. On the other hand, the heavy boson pairs in (7) can also be produced from the corresponding light fermion pairs q_Ll_L, q_Lq_L, u_Rd_R, u_Re_R by emitting a pair of the fermion superpartners of the gauge bosons Z^0 or $W^\pm$, which then annihilate through their Majorana mass term (which DG explicitly included in their model as a soft supersymmetry breaking). This yields the proton-decay effective interactions $q_Lq_Lq_Ll_L$ and $u_R^*u_R^*d_R^*e_R^*$, with coefficients of order

$$\frac{1}{8\pi^2}\frac{f^2}{M}\frac{e^2}{m_W} . \tag{8}$$

[The factor $1/8\pi^2$ is put in because this is a one-loop graph; the factor of f^2/M is the coupling constant (6) of the $d=5$ effective interaction; the factor e^2 arises from the emission and absorption of the fermion superpartners of the $W^\pm$ and Z; and the factor $1/m_W$ arises from the integral on the assumption that m_W is the characteristic mass of the scalar superpartners of the quarks and leptons and the fermion superpartners of the $W^\pm$ and Z^0.] The proton decay rate is then roughly of order

$$\Gamma \approx m_p{}^5 \left| \frac{1}{8\pi^2}\frac{f^2}{M}\frac{e^2}{m_W} \right|^2 . \tag{9}$$

For the process $p \to \mu^+ K^0$, the coupling f^2 is a product of Higgs-particle couplings,

$$f^2 \approx G_F m_d m_s \simeq 1.3\times 10^{-8} . \tag{10}$$

For M of order 10^{15} GeV, the proton lifetime would be of order 10^{24} yr. Even for $M=10^{17}$ GeV (the appropriate value for the grand unified version of the DG model), the proton lifetime is only 10^{28} yr, too short by $2-3$ orders of magnitude.

This difficulty could be avoided, if we do not insist on embedding the DG model in a grand unified theory, by adding additional global symmetries. This is not very attractive—even the modest discrete symmetry of the DG model was not entirely appealing. It seems better to attempt a different solution.

IV. SU(3)×SU(2)×U(1)×$\widetilde{G}$ THEORIES: B AND L CONSERVATION

We will now consider a different way of avoiding the problem of insufficient suppression of B and L violations encountered in Secs. II and III. Instead of imposing global symmetries, we will suppose that the gauge group which survives down to ordinary energies is not just SU(3)×SU(2)×U(1), but contains an additional factor $\widetilde{G}$. Depending on the transformation properties of the quark and lepton superfields under $\widetilde{G}$, it is then quite plausible that B and L nonconservation could be ruled out in $d=4$ and $d=5$ interactions, and yet be not altogether forbidden.

To take one example, suppose that $\widetilde{G}=\widetilde{\text{U}}(1)$, where $\widetilde{\text{U}}(1)$ is another U(1) group, commuting with the original SU(3)×SU(2)×U(1). Suppose also that all of the left-chiral quark and lepton superfields $Q_L, U_R^*, D_R^*, L_L, E_R^*$ have values of the $\widetilde{\text{U}}(1)$ quantum number $\widetilde{Y}$ with the same sign. Then clearly *all* F terms that involve only quarks and lepton superfields are forbidden, including the B- and L-nonconserving effective interactions with $d=4$ and $d=5$ in (3) and (5), respectively.

To allow us to catalog the varieties of possible B and L nonconservation which remain, let us suppose for definiteness that the $\widetilde{\text{U}}(1)$ quantum number $\widetilde{Y}$ has equal values, say $+1$, for all left-chiral quark and lepton superfields. In order to provide for quark and lepton masses, we must suppose, as in Sec. III, that in addition there are SU(3)-singlet SU(2)-doublet Higgs superfields H_L and H_L', with ordinary weak hypercharges $Y=+\frac{1}{2}$ and $Y=-\frac{1}{2}$, respectively, and now also with the new $\widetilde{\text{U}}(1)$ quantum number $\widetilde{Y}=-2$. (As discussed in Sec. VI and Appendix B, other superfields will also have to be added in order to cancel anomalies and to obtain suitable symmetry-breaking solutions.) A complete list (apart, as always, from kinematic terms, gauge couplings, and extra derivatives) of SU(3)×SU(2)×U(1)×$\widetilde{\text{U}}(1)$-invariant supersymmetric interactions involving quark, lepton, and Higgs superfields up to dimensionality $d=6$ is

$$d=4:\ (L_LE_R^*H_L)_F\ ,\ \ (Q_LD_R^*H_L)_F\ ,\ \ (Q_LU_R^*H_L')_F\ , \tag{11}$$

$$d=5:\ \text{None}\ ,$$

$$d=6:\ (Q_L Q_L U_R E_R)_D\ ,\ (Q_L U_R D_R L_L)_D\ ,$$

$$\text{etc.}\ , \tag{12}$$

where etc. denotes a large number of allowed terms of form $(S^*SS^*S)_D$ that conserve both B and L.

Some remarks are now in order about the implications of this list.

(1) The baryon-number-violating effective interactions of lowest dimensionality have $d=6$, and are therefore adequately suppressed, by two powers of superheavy masses. According to the rules given in Appendix A, these superfield interactions contain the usual four-fermion operators[4]

$$q_L q_L u_R e_R\ ,\ q_L u_R d_R l_L\ , \tag{13}$$

which produce proton decay directly.

(2) Not all of the four-fermion proton-decay interactions which would be allowed by SU(3) $\times$SU(2)$\times$U(1) are produced in this way; we do not obtain the other two interactions[4]

$$q_L q_L q_L l_L\ ,\ u_R u_R d_R e_R\ . \tag{14}$$

Thus whatever the underlying mechanism of proton decay, it would be expected to have matrix elements of the *LLRR* form which would be expected from the exchange of vector bosons, and not of the *LLLL* or *RRRR* forms which could only be produced by scalar-boson exchange. This has well-known phenomenological consequences,[4] including model-independent (but strong-interaction-dependent) predictions for ratios of all decay rates for $\Delta S=0$ modes, and universal lepton polarizations. Although this result (that *LLRR* terms may occur while *LLLL* and *RRRR* terms are forbidden) can be derived here from the $\widetilde{\text{U}}(1)$ symmetry, it is actually quite general, and follows directly from the supersymmetric nature of the effective interaction, whatever the structure of $\widetilde{G}$, as shown by the last term of Eq. (A7).

(3) There are no $d=5$ effective interactions which could produce an observable neutrino mass. To generate such interactions, one would need Higgs superfields with $\widetilde{Y}=-1$ as well as $\widetilde{Y}=-2$.

(4) The $\widetilde{\text{U}}(1)$ symmetry rules out Higgs mass terms $(H_L H_L')_F$ which would otherwise be allowed by SU(3)$\times$SU(2) $\times$U(1) and supersymmetry. Thus supersymmetry together with SU(3)$\times$SU(2)$\times$U(1)$\times\widetilde{\text{U}}(1)$ is serving its hoped for role of prohibiting scalar mass terms.

V. SU(3)$\times$SU(2)$\times$U(1)$\times G$ THEORIES: MASSES

The extra gauge symmetry $\widetilde{G}$, which was introduced in the previous section in order to preserve the natural suppression of B and L nonconservation, has another attractive feature: it helps in understanding how a reasonable spectrum of masses can arise in a supersymmetric theory. An extra gauge symmetry [with $\widetilde{G}=\text{U}(1)$] was introduced for this purpose by Fayet.[3] I will come back at the end of this section to the relation between his approach and that taken here.

First, what is the problem? It is often said that the chief difficulty faced in constructing models which are supersymmetric down to ordinary energies is to understand why the scalar superpartners of the quarks and leptons do not occur at masses low enough for them to have been seen. In my opinion this somewhat misstates the real problem. After all, there is a natural scale of masses to be expected in a gauge theory, the scale of the masses of the gauge bosons associated with broken symmetries, such as the $W^\pm$ at 80 GeV. Clearly any number of scalar counterparts of quarks and leptons could be lurking at energies of order 80–100 GeV, and we would not yet know it.[13] The real problem is one that has bedeviled gauge theories of electroweak intereactions from the start—not why are the scalars so heavy, but why are at least some quarks and leptons so light?

I have no new answer to this problem, but at least we can try to apply a not very satisfying old answer in a supersymmetric context: The observed quarks and leptons are much lighter than the W or Z because they have very weak Yukawa couplings to the Higgs bosons. In a supersymmetric model, this would mean that the $d=4$ interactions of the quark, lepton, and Higgs superfields in (11) all have very small coupling constants. We do not know why this should be the case, but if by assuming these couplings to be small we can understand qualitatively why the quarks and leptons are much lighter than the $W^\pm$ while the scalars are not, then we will at least be no worse off than without supersymmetry.

To simplify matters, let us go all the way, and imagine that the coupling constants of renormalizable $(S^3)_F$-type interactions involving quark and lepton superfields [like those in (11)] all vanish. We will however leave it an open possibility that the Higgs superfields may have $(S^3)_F$-type interactions with other superfields, denoted X_L. [It is implicit in these assumptions that the quark and lepton superfields form representations of the SU(3)×SU(2)×U(1)×$\tilde{U}(1)$ gauge group which are separate from those of the Higgs and X superfields.] Our aim then is to see if the broken-symmetry solutions of such a theory with $\langle \mathscr{H}^0 \rangle \neq 0$, $\langle \mathscr{H}^{0\prime} \rangle \neq 0$ will have massless quarks and leptons and massive scalars. If so, then we will not have to worry about whether the scalars are heavy enough; they would have to have masses of the order of the only remaining mass scale in the theory, that of the $W^{\pm}$ and Z^0.

The potential of the scalar fields in this sort of theory will have the general form[14]

$$V(\phi,\sigma)=\sum_n \left| \frac{\partial f(\phi)}{\partial \phi_n} \right|^2 + \tfrac{1}{2}\sum_\alpha (g_\alpha \phi^\dagger t_\alpha^\phi \phi + g_\alpha \sigma^\dagger t_\alpha^\sigma \sigma + \xi_\alpha)^2 , \tag{15}$$

where σ stands for all the scalar superpartners of the quarks and leptons, ϕ stands for the scalar components of all other superfields, including the scalar Higgs fields $\mathscr{H}_L, \mathscr{H}'_L$ and the scalar components of whatever other superfields X_L may have $(S^3)_F$-type interactions with them, $f(H_L, H'_L, \ldots)$ is the trilinear function whose F component describes any such interactions, t_α^ϕ and t_α^σ are the matrices representing the αth gauge generator on the ϕ and σ fields, g_α is the corresponding coupling constant, and ξ_α is the coefficient of the term[6] $(V_\alpha)_D$ which may appear in the Lagrangian for U(1) gauge superfields.

This potential can usefully be rewritten in the form

$$V(\phi,\sigma)=V(\phi,0)+\sum_\alpha g_\alpha D_\alpha(\phi)\sigma^\dagger t_\alpha^\sigma \sigma + \tfrac{1}{2}\sum_\alpha g_\alpha^2 (\sigma^\dagger t_\alpha^\sigma \sigma)^2 , \tag{16}$$

where

$$D_\alpha(\phi) \equiv g_\alpha \phi^\dagger t_\alpha^\phi \phi + \xi_\alpha . \tag{17}$$

It is immediately apparent that if we find a value ϕ_0 of ϕ which minimizes the potential $V(\phi,0)$ with $\sigma=0$, then the point $\phi=0$, $\sigma=0$ will be a local minimum of the full potential if and only if the matrix

$$\mathscr{M}^2 \equiv \sum_\alpha g_\alpha D_\alpha(\phi_0) t_\alpha^\sigma \tag{18}$$

is positive definite. In practice if $\mathscr{M}^2$ is positive definite, this is usually (but not always; see the Added Notes) also an absolute minimum, because the only way to get a lower value of $V(\phi,\sigma)$ would be to go to ϕ's for which some of the eigenvalues of $\sum_\alpha g_\alpha D_\alpha(\phi) t_\alpha^\sigma$ have changed sign, and these will be far from the point $\phi=\phi_0$ which minimizes $V(\phi,0)$. On the other hand, if $\mathscr{M}^2$ is not positive, then the true minimum of $V(\phi,\sigma)$ definitely has $\sigma \neq 0$.

The importance here of finding a minimum with $\sigma=0$ is that this is a sufficient condition to have vanishing quark and lepton masses. (Of course, we want the scalar counterparts of the quarks and charged leptons to have vanishing vacuum expectation values also in order to preserve color and charge conservation.) In general, the mass matrix of the left-handed fermions in a renormalizable supersymmetric theory is given in tree approximation by[15]

$$m_{nm}=\left[\frac{\partial^2 f(\mathscr{S})}{\partial \mathscr{S}_n \partial \mathscr{S}_m} \right]_0 , \tag{19}$$

$$m_{\alpha n}=m_{n\alpha}=-\sqrt{2} g_\alpha (\mathscr{S}_0^* t_\alpha)_n , \tag{20}$$

$$m_{\alpha\beta}=0 , \tag{21}$$

where n and α label the left-handed fermion fields in left-chiral scalar superfields and real gauge vector superfields, respectively, $\mathscr{S}_n$ is the scalar component (σ or ϕ) of the left-chiral scalar superfield S_n, $(t_\alpha)_{nm}$ is the matrix representing the αth generator of the gauge group on the S_n, $f(S)$ is the function whose F term appears in the renormalizable part of the interaction (1), and the subscript zero means that these quantities are evaluated at the minimum of the potential $V(\mathscr{S})$. For the scalar superpartners σ of the quarks and leptons, (19) vanishes under the assumption that σ does not appear in $f(\mathscr{S})$, and (20) vanishes because we are considering a minimum of $V(\mathscr{S})$ with $\sigma_0=0$. Hence all quarks and leptons have zero mass. Of course in the real world $f(\phi,\sigma)$ would be assumed to depend weakly on σ, and the quarks and leptons would not be massless but only relatively light.

We see that the positivity of $\mathscr{M}^2$ will lead to the result we would like, that quarks and leptons are very light, so let us suppose $\mathscr{M}^2$ *is* positive definite. What about the scalars? Equation (16) shows that their mass-squared matrix is just $\mathscr{M}^2$, which we have just now assumed is positive definite. There is no reason to expect that any of the positive-definite eigenvalues of (18) would be much smaller than the natural scale $g\phi_0 \approx m_W$, so all scalar counterparts of the quarks and leptons are expected to be too heavy to have been observed yet.

We can now see why it was necessary to introduce the extra gauge group $\tilde{G}$. The matrices t_α^σ which represent the generators of SU(3)×SU(2)×U(1) on the known quarks and leptons all have zero trace, so if there were no extra gauge generator $\tilde{Y}$, then (18) would give $\mathrm{Tr}\mathcal{M}^2=0$, and hence $\mathcal{M}^2$ could not be positive definite. (This argument is closely related to that of Dimopoulos and Georgi.[5]) Of course, to be of use the generators $\tilde{t}_\alpha^\sigma$ which represents $\tilde{G}$ on the quarks and leptons must not themselves be traceless. [However, if the $\mathrm{SU}(3)\times\mathrm{SU}(2)\times\mathrm{U}(1)\times\tilde{G}$ theory comes from a semisimple grand unified theory, $\mathrm{Tr}\tilde{t}_\alpha^\phi+\mathrm{Tr}\tilde{t}_\alpha^\sigma$ would have to vanish.] We see that $\tilde{G}$ must contain at least one $\tilde{\mathrm{U}}(1)$ factor, whose generator $\tilde{Y}$ has nonvanishing trace on the known quarks and leptons. We are almost inevitably led again to the conclusion that $\tilde{G}$ is just $\tilde{\mathrm{U}}(1)$, with quarks and leptons all having $\tilde{Y}$ values of the same sign, as assumed in Sec. IV to enforce a natural suppression of B and L nonconservation.

The general picture outlined here has interesting immediate consequences for the scalar-particle masses, which became exact in the limit of vanishing quark and lepton masses and Yukawa couplings. We note that all members of a given SU(2) multiplet have the same behavior under SU(3), U(1), and $\tilde{G}$, as (18) tells us that the mass splitting in any doublet has a common value

$$M^2(\mathscr{U}_L)-M^2(\mathscr{D}_L)=M^2(\mathscr{N}_L)-M^2(\mathscr{E}_L)\equiv\Delta M^2\ , \tag{22}$$

independent of the generations of the corresponding quarks and leptons. We can also set an upper bound on this splitting. Since SU(2) has $\xi=0$, Eqs. (17) and (18) give

$$\Delta M^2=g^2(\phi^\dagger t_3\phi)_0 \tag{23}$$

with g and t_3 now specifically denoting the SU(2) coupling and generator. This may be compared with the usual formula for the $W^\pm$ mass

$$m_W{}^2=g^2[\phi^\dagger(t_1{}^2+t_2{}^2)\phi]\ . \tag{24}$$

We see that

$$\Delta M^2/m_W{}^2=\langle t_3\rangle/\langle t_1{}^2+t_2{}^2\rangle\ , \tag{25}$$

the averages being weighted with the SU(2)-nonsinglet ϕ-field vacuum expectation values. If only electroweak singlets and doublets have vacuum expectation values (so that m_Z and m_W have the usual ratio), then $|\langle t_3\rangle|\le\frac{1}{2}$, while $\langle t_1{}^2+t_2{}^2\rangle=\frac{1}{2}$, so (25) gives

$$|\Delta M^2|\le m_W{}^2\ . \tag{26}$$

Many more mass relations arise if we make simple assumptions about the extra gauge group $\tilde{G}$. For instance, suppose that $\tilde{G}$ is a U(1) group, and that its generator $\tilde{Y}$ has equal values for all left-handed quark and lepton superfields, say $\tilde{Y}=1$. Then (18) gives the masses of all scalar superpartners of the quarks and leptons in terms of just three unknowns, the values of $g_\alpha D_\alpha(\phi_0)$ for the generators T_3, Y, and $\tilde{Y}$. The scalars corresponding to a quark or lepton of a given $\mathrm{SU}(3)\times\mathrm{SU}(2)\times\mathrm{U}(1)$ type will thus have a mass which is equal for all generations. Also, these generation-independent masses will be subject to the relations

$$M^2(\mathscr{U}_L)+M^2(\mathscr{D}_L)=\tfrac{1}{3}M^2(\mathscr{U}_R^*)+\tfrac{5}{3}M^2(\mathscr{D}_R^*)\ , \tag{27}$$

$$M^2(\mathscr{N}_L)+M^2(\mathscr{E}_L)=\tfrac{5}{3}M^2(\mathscr{U}_R^*)+\tfrac{1}{3}M^2(\mathscr{D}_R^*)\ , \tag{28}$$

$$M^2(\mathscr{E}_R^*)=-\tfrac{2}{3}M^2(\mathscr{U}_R^*)+\tfrac{5}{3}M^2(\mathscr{D}_R^*)\ . \tag{29}$$

If $\tilde{Y}$ varies from generation to generation but is the same within each generation, then the scalar masses will no longer be generation-independent, but all mass splittings within each generation will be the same for all generations, and subject to (27)–(29). Of course, all these results are only approximate, with corrections of order $(m_{\mathrm{quark}}/m_W)^2$.

Fayet[3] has derived Eq. (18) as a formula for the difference between scalar and fermion squared masses, without using the approximation of weak Yukawa couplings. From this, it is straightforward to derive an improved version of Eqs. (22)–(29), and also the Dimopoulos-Georgi theorem.[5] However, as recognized by Fayet, although his derivation applies when there is an unbroken R symmetry, and in some other cases, in general there may appear additional terms in (18). The assumption of weak F-term couplings of the quark and lepton superfields was made here in order to avoid having to invoke R invariance or the details of specific models in deriving (18), and also because we need some such assumption to explain why the quarks and leptons are so light. By using color conservation, Dimopoulos and Georgi were able to obtain their result on the existence of light scalar quarks, without having to make any of the above assumptions.

Our conclusion is that there is no difficulty in understanding why quarks and leptons are so much lighter than scalar and vector bosons, provided the model without quark and lepton fields has a minimum of the potential at which the matrix (18) is positive definite. Whether or not this is the case is a question that must be checked in specific models. As an "existence proof," a semirealistic model which satisfies this positivity condition is given in Appendix B.

VI. ANOMALIES

The problems raised by the minimal model of Sec. II have been satisfactorily avoided in the $SU(3)\times SU(2)\times U(1)\times\widetilde{G}$ models discussed in Secs. IV and V and Appendix B. These models have a variety of other phenomenological problems, including new neutral currents and massless fermionic superpartners of gluons. However, before tinkering with the models to avoid these other problems, it will be more useful first to address the outstanding problem of mathematical consistency raised by the introduction of the extra gauge group, the problem of Adler-Bell-Jackiw (ABJ) anomalies.[8] After seeing what new superfields need to be added to cancel these anomalies, we will be better able (in a future paper[9]) to consider what phenomenological problems may remain.

Let us assume for definiteness that the gauge group is $SU(3)\times SU(2)\times U(1)\times\widetilde{U}(1)$, and that among the left-chiral scalar superfields there are N_g "generations" $(Q_L, U_R^*, D_R^*, L_L, E_R^*)$ of quark and lepton superfields with $\widetilde{Y}=+1$ and N_h pairs of Higgs doublets H_L, H_L' with $Y=\pm\frac{1}{2}$ and $\widetilde{Y}=-2$. Then the introduction of the extra $\widetilde{U}(1)$ gauge group produces the ABJ anomalies

$$SU(3)^2\widetilde{U}(1):\quad \mathrm{Tr}({T_{SU(3)}}^2\widetilde{Y})=2N_g\,, \tag{30}$$

$$SU(2)^2\widetilde{U}(1):\quad \mathrm{Tr}({T_{SU(2)}}^2\widetilde{Y})=2N_g-2N_h\,, \tag{31}$$

$$U(1)^2\widetilde{U}(1):\quad \mathrm{Tr}(Y^2\widetilde{Y})=\tfrac{10}{3}N_g-2N_h\,, \tag{32}$$

$$\widetilde{U}(1)^3:\quad \mathrm{Tr}(\widetilde{Y}^3)=15N_g-32N_h\,. \tag{33}$$

We note in particular that colored fields with $\widetilde{Y}$ negative must be added to cancel the $SU(3)^2\widetilde{U}(1)$ anomaly. This raises the possibility that the problem of B and L nonconservation which was solved by the introduction of $\widetilde{U}(1)$ may reappear.

As an example of the sort of trouble we can get into in adding new colored superfields, suppose we try to cancel anomalies by embedding $SU(3)\times SU(2)\times U(1)\times\widetilde{U}(1)$ in a larger group whose representations are known to be anomaly free, E_6.[16] This group is chosen because E_6 has an $SO(10)\times\widetilde{U}(1)$ subgroup, and therefore leads to generations of quarks and leptons all with the same value of $\widetilde{Y}$. Specifically, the $\underline{27}$ of E_6 consists of one generation of quarks and leptons with $\widetilde{Y}=+1$, including an extra $SU(3)\times SU(2)\times U(1)$-neutral neutrino N_R^* with $\widetilde{Y}=+1$, plus a pair of Higgs doublets H_L, H_L' with $\widetilde{Y}=-2$, plus a pair of color-triplet (antitriplet) SU(2)-singlets with $Y=+\frac{1}{3}$ $(-\frac{1}{3})$ and $\widetilde{Y}=-2$, plus an $SU(3)\times SU(2)\times U(1)$-neutral singlet with $\widetilde{Y}=+4$. All anomalies automatically cancel. It is striking that we find here all the superfields of the models we have been considering, including the $\widetilde{Y}=+4$ singlet discussed in Appendix B. Unfortunately, we find other superfields as well. The extra singlet neutrino ν_R^* cannot be superheavy [since $\widetilde{U}(1)$ has to survive down to ordinary energies], so the neutrinos would be expected to get large Dirac masses from their Higgs couplings. Even worse, the color-antitriplet superfield with $Y=-\frac{1}{3}$ and $\widetilde{Y}=-2$ would be expected to have renormalizable F-term interactions with both L_LQ_L and $D_R^*U_R^*$, leading to unsuppressed proton decay at a disastrous rate, just as in the minimal model of Sec. II. One other unattractive feature of this set of superfields is that to maintain the cancellations of anomalies, we must add new Higgs etc. fields for each new generation of quarks and leptons.

Inspection of (30)–(33) suggests a different approach to the cancellation of anomalies. Note that for $N_g=3$ generations, the trace (30) has the value $+6$. This is neatly canceled by a *single* color octet O_L of chiral superfields with $\widetilde{Y}=-2$. To avoid reintroducing an $SU(3)^2U(1)$ anomaly, we must take O_L to be neutral under U(1) as well as SU(2); in other words, O_L is a member of the adjoint representation of $SU(3)\times SU(2)\times U(1)$. Suppose we try adding another "adjoint superfield" with $\widetilde{Y}=-2$, a *single* SU(2) triplet T_L which is neutral under SU(3) and U(1). No new anomalies are produced, and T_L contributes -4 to the trace (31), which cancels the $SU(2)^2\widetilde{U}(1)$ anomaly for a single Higgs pair and three generations of quarks and leptons. Doubtless these cancellations could be understood by embedding $SU(3)\times SU(2)\times U(1)\times\widetilde{U}(1)$ in some large group with anomaly-free representations containing just three generations of nonsuperheavy quarks and leptons.

Not all anomalies have been canceled. With $N_g=3$ and $N_h=1$, the trace (32) has the value $+8$, so new charged $SU(3)\times SU(2)$-singlet superfields must be added with $\widetilde{Y}$ negative to cancel the $U(1)^2\widetilde{U}(1)$ anomaly. One possibility which would not reintroduce a $U(1)\widetilde{U}(1)^2$ or $U(1)^3$ anomaly is to add two pairs of additional singlet superfields J_L, J_L' with charges ± 1 and $\widetilde{Y}=-2$. [At any rate

TABLE II. An anomaly-free set of left-chiral scalar superfields for an SU(3)×SU(2)×U(1)×$\tilde{\mathrm{U}}$(1)-invariant supersymmetric theory.

	SU(3)	SU(2)	Y	$\tilde{Y}$	
Q_L	3	2	$-\frac{1}{6}$	1	three generations
U_R^*	$\bar{3}$	1	$\frac{2}{3}$	1	three generations
D_R^*	$\bar{3}$	1	$-\frac{1}{3}$	1	three generations
L_L	1	2	$\frac{1}{2}$	1	three generations
E_R^*	1	1	-1	1	three generations
H_L	1	2	$\frac{1}{2}$	-2	
H_L'	1	2	$-\frac{1}{2}$	-2	
O_L	8	1	0	-2	
T_L	1	3	0	-2	
J_L	1	1	1	-2	two each
J_L'	1	1	-1	-2	two each
X_L	1	1	0	?	several

it is encouraging that we do not have to add color-neutral particles with noninteger charges, which as seen from (32) is another special feature of three generations.] The only remaining anomaly is $\tilde{\mathrm{U}}(1)^3$, but this can be canceled in any number of ways by adding SU(3)×SU(2)×U(1)-neutral fields with various values of $\tilde{Y}$, one of which could be the X_L of Appendix B.

We now must check that we have not reintroduced the possibility of renormalizable B- and L-violating interactions. To be specific, suppose that the only left-chiral scalar superfields are those listed in Table II. Then the only trilinear F terms which involve a pair of the left-chiral quark and/or lepton superfields are those listed in Eq. (11), plus the new interaction $(L_L L_L J_L')_F$. Also, as long as we do not introduce new X_L superfields with $\tilde{Y}_L=+1$, there are no trilinear F terms which involve 1 or 3 of the quark and lepton superfields. Hence B and L are automatically conserved by all renormalizable SU(3)×SU(2)×U(1)×$\tilde{\mathrm{U}}$(1)-invariant supersymmetric interactions, with B and L assigned conventional values for quark and lepton superfields; zero values for H_L, H_L', O_L, T_L, and X_L; and $B=0$ and $L=+1$ (-1) for J_L (J_L'). Also, there is no trouble with $d=5$ effective interactions; as long as the X_L are limited to certain $\tilde{Y}$ values (including 1, -2, 4, 5, etc.) no $d=5$ terms are allowed, B and L conserving or not.

The phenomenological viability of a theory with this set of superfields now depends on the pattern of scalar field vacuum expectation values. In particular, $\mathscr{H}_L^0$ and $\mathscr{H}_L'^0$ must both get nonvanishing vacuum expectation values to give masses to all quarks and leptons, the charged scalar components of H_L, H_L', T_L, J_L, and J_L' must have vanishing vacuum expectation values to preserve charge conservation, $\mathscr{O}_L$ must have vanishing vacuum expectation value to preserve color invariance, and the neutral scalar component of T_L must have zero or small vacuum expectation value to preserve the usual result for the Z/W mass ratio. Finally, some of the $\mathscr{X}_L$ must have nonvanishing vacuum expectation values. Otherwise the neutral U(1) gauge boson $\tilde{Z}$ would get its mass solely from the same $\mathscr{H}_L^0$ and $\mathscr{H}_L'^0$ vacuum expectation values as Z^0, and the effective low-energy neutral-current couplings would be related by

$$\frac{\tilde{g}^2}{m^2(\tilde{Z}^0)}=\frac{1}{16}\frac{g^2+g'^2}{m^2(Z^0)}\;. \tag{34}$$

(This is for $\langle\mathscr{H}^0\rangle=\langle\mathscr{H}'^0\rangle$, for $\langle\mathscr{H}^0\rangle\neq\langle\mathscr{H}'^0\rangle$ the Z^0 and $\tilde{Z}^0$ are mixed.) Since quarks and leptons have larger values of $\tilde{Y}$ than of the SU(2)×U(1) quantum numbers, the neutral-current coupling (34) is probably inconsistent with experiment,[17] and so we must suppose that $\tilde{Z}^0$ gets part of its mass from other neutral scalar fields $\mathscr{X}_L^0$ which carry nonvanishing values of $\tilde{Y}$. Unfortunately, the question of whether or not the vacuum expectation values have all these necessary properties cannot be settled until we decide what sorts of X_L superfields to include in the theory.

The particular set of superfields that has been introduced in this section of course represents just one solution to the problem of canceling anomalies without reintroducing an unsuppressed violation of B and L conservation. It has the attractive feature that it works only for three generations of quarks and leptons. In addition, this solution has one other significant feature, of canceling anomalies in R symmetries,[9] which will be discussed in a future paper.

Added notes

(1) After this article was submitted for publication, I received a paper by N. Sakai and T. Yanagida [Munich Report No. MPI-PAE/PThSS/81 (unpublished)] which deals with the problem of baryon and lepton nonconservation in supersymmetric theories, using an approach similar to that presented here.

(2) M. Claudson and M. Wise have pointed out to me that the supersymmetry-breaking solution of the model presented in Appendix B is actually destabilized by the introduction of leptons: With two generations of leptons it is possible to find a deeper potential minimum at which charge is broken and supersymmetry is unbroken. They also note that this new supersymmetry-preserving solution could be avoided if there were an SU(2) triplet as well as an SU(2)-singlet field with $\widetilde{Y}=+4$. One can go further, and show that if all left-chiral superfields have $\widetilde{Y}=+1$, -2, and $+4$, and if there are enough superfields with $\widetilde{Y}=+4$ to allow all pairs of $\widetilde{Y}=-2$ superfields to have separate couplings with them, and if $\widetilde{\xi}/\widetilde{g}>0$, then supersymmetry must be spontaneously broken. (This is because a supersymmetry-preserving solution would have to have $[\partial f(\mathscr{S})/\partial \mathscr{S}_a]_0$ for all scalar fields, and in particular for all $\widetilde{Y}=4$ fields, which would imply that all $\widetilde{Y}=-2$ fields vanish; and it would also have to have $D_\alpha=0$ for all gauge fields, and in particular for the $\widetilde{\mathrm{U}}(1)$ field, which would only be possible if some fields with $\widetilde{Y}$ negative do not vanish.) Once one eliminates the possibility of a supersymmetry-preserving solution, the heights of the potential at the various local minima will generally be functions of the parameters of the theory, so that the sort of solution found in Appendix B will be the deepest minimum for at least a finite range of parameters. Unfortunately, as Claudson and Wise point out, it is very difficult to construct a theory with enough $\widetilde{Y}=+4$ fields to avoid a supersymmetry-conserving solution without also introducing a $\widetilde{Y}^3$ ABJ anomaly.

ACKNOWLEDGMENTS

I am very grateful to Glennys Farrar, Daniel Freedman, Howard Georgi, Paul Ginsparg, Marc Grisaru, Mark Wise, and Edward Witten for their helpful comments.

This research was supported in part by the National Science Foundation under Grant No. PHY77-22864.

APPENDIX A: SUPERSYMMETRIC LAGRANGIANS

This appendix will present expressions for the Lagrangian in terms of ordinary fields, corresponding to the general Lagrangian for left-chiral scalar superfields $S_n(x,\theta)$ discussed in Sec. I:

$$\mathscr{L}=[f(S)]_F+[f(S)]_F^*+[d(S,S^*)]_D \ . \tag{A1}$$

First, let us establish our notation. Since we want our results to appear at the end in a conventional Dirac formalism, the superfield coordinate θ in $S(x,\theta)$ will be taken as a Majorana four-component spinor, and all spinors and associated matrices will be four-dimensional. We define the component fields $\mathscr{S}_n$ (scalar), s_n (spinor), and $\mathscr{M}_n$ (auxiliary scalar) of a left-chiral scalar superfield S_n by the expansion

$$S_n=\mathscr{S}_n-i\sqrt{2}(\theta_L^T\epsilon s_{nL})-i(\theta_L^T\epsilon\theta_L)\mathscr{M}_n+\tfrac{1}{2}(\theta^T\epsilon\gamma^\mu\theta)\partial_\mu\mathscr{S}_n+\frac{i}{\sqrt{2}}(\theta_L^T\epsilon\theta_L)(\theta_R^T\epsilon\gamma^\mu\partial_\mu s_{nL})-\tfrac{1}{8}(\theta^T\epsilon\theta)^2\Box\mathscr{S}_n \ . \tag{A2}$$

Our metric is $+++-$, and our Dirac matrices can be taken (in supermatrix notation) as

$$\vec{\gamma}=\begin{bmatrix} 0 & -i\vec{\sigma} \\ i\vec{\sigma} & 0 \end{bmatrix}, \quad \gamma^0=\begin{bmatrix} 0 & -i \\ -i & 0 \end{bmatrix}$$

with $\vec{\sigma}$ the 2×2 Pauli matrices. Also, γ_5 and ϵ are two diagonal supermatrices

$$\gamma_5=\begin{bmatrix}1 & 0\\ 0 & -1\end{bmatrix},\quad \epsilon=\begin{bmatrix}i\sigma_2 & 0\\ 0 & i\sigma_2\end{bmatrix}.$$

A subscript L or R denotes multiplication with $\frac{1}{2}(1+\gamma_5)$ or $\frac{1}{2}(1-\gamma_5)$, respectively. Finally, θ and s are Majorana spinors, in the sense that

$$s_n=\epsilon\gamma_5\beta s_n^*,\quad \theta=\epsilon\gamma_5\beta\theta^*,$$

where $\beta=i\gamma^0$. It is useful to note that $\bar{s}\equiv s^\dagger\beta=s^T\epsilon\gamma_5$.

The F term of any function of left-chiral superfields $S_n(x,\theta)$ is defined as the coefficient of $\theta_L^T\epsilon\theta_L$ in the expansion of the function in powers of θ_L and θ_R. Using the expansion (A2), it is straightforward to show that the first term in the general Lagrangian (A1) has the form

$$[f(S)]_F=\tfrac{1}{2}(s_{Ln}^T\epsilon s_{Lm})\left[\frac{\partial^2 f(\mathscr{S})}{\partial\mathscr{S}_n\partial\mathscr{S}_m}\right]-i\mathscr{M}_n\frac{\partial f(\mathscr{S})}{\partial\mathscr{S}_n}. \tag{A3}$$

The D term of any real function of left-handed chiral superfields $S_n(x,\theta)$ and their adjoints $S_n^*(x,\theta)$ is defined as the coefficient of $-\frac{1}{2}(\theta^T\epsilon\theta)^2$ in the expansion of the function in powers of θ_L and θ_R. By a still straightforward, though somewhat tedious, calculation, this gives the last term in (A1):

$$\begin{aligned}[d(S,S^*)]_D={}&\frac{1}{4}\frac{\partial d}{\partial\mathscr{S}_n}\Box\mathscr{S}_n+\frac{1}{4}\frac{\partial d}{\partial\mathscr{S}_n^*}\Box\mathscr{S}_n^*-\tfrac{1}{2}\partial_\mu\mathscr{S}_n\partial^\mu\mathscr{S}_m^*\frac{\partial^2 d}{\partial\mathscr{S}_n\partial\mathscr{S}_m^*}\\
&+\tfrac{1}{4}\partial_\mu\mathscr{S}_n\partial^\mu\mathscr{S}_m\frac{\partial^2 d}{\partial\mathscr{S}_n\partial\mathscr{S}_m}+\tfrac{1}{4}\partial_\mu\mathscr{S}_n^*\partial^\mu\mathscr{S}_m^*\frac{\partial^2 d}{\partial\mathscr{S}_n^*\partial\mathscr{S}_n^*}\\
&-\tfrac{1}{2}(s_n^T\epsilon\gamma_5\gamma^\mu\partial_\mu s_m)\frac{\partial^2 d}{\partial\mathscr{S}_n\partial\mathscr{S}_m^*}+\tfrac{1}{2}(s_{Ln}^T\epsilon\gamma^\mu s_{Rm})\left[\frac{\partial^3 d}{\partial\mathscr{S}_n\partial\mathscr{S}_m^*\partial\mathscr{S}_l}\partial_\mu\mathscr{S}_l-\frac{\partial^3 d}{\partial\mathscr{S}_n\partial\mathscr{S}_m^*\partial\mathscr{S}_l^*}\partial_\mu\mathscr{S}_l^*\right]\\
&-\tfrac{1}{4}(s_{Ln}^T\epsilon s_{Lm})(s_{Rl}^T\epsilon s_{Rk})\frac{\partial^4 d}{\partial\mathscr{S}_n\partial\mathscr{S}_m\partial\mathscr{S}_l^*\partial\mathscr{S}_k^*}+\mathscr{M}_n\mathscr{M}_m^*\frac{\partial^2 d}{\partial\mathscr{S}_n\partial\mathscr{S}_m^*}\\
&-\frac{i}{2}\mathscr{M}_n^*(s_{Lm}^T\epsilon s_{Ll})\frac{\partial^3 d}{\partial\mathscr{S}_n^*\partial\mathscr{S}_m\partial\mathscr{S}_l}+\frac{i}{2}\mathscr{M}_n(s_{Rm}^T\epsilon s_{Rl})\frac{\partial^3 d}{\partial\mathscr{S}_n\partial\mathscr{S}_m^*\partial\mathscr{S}_l^*}.\end{aligned} \tag{A4}$$

Note that although d and f are here not constrained to be polynomials, the Lagrangian (A1) is still quadratic in the auxiliary fields $\mathscr{M}_n$ and $\mathscr{M}_n^*$. Hence by requiring that the action be stationary with respect to these fields, we can obtain the closed-form result for $\mathscr{M}$:

$$\mathscr{M}_n=J_{nm}{}^{-1}\left[-\frac{i}{2}(s_{Lk}^T\epsilon s_{Ll})\frac{\partial^3 d}{\partial\mathscr{S}_m^*\partial\mathscr{S}_k\partial\mathscr{S}_l}-i\left[\frac{\partial f}{\partial\mathscr{S}_m}\right]^*\right], \tag{A5}$$

where J is the ubiquitous matrix

$$J_{nm}=\frac{\partial^2 d}{\partial\mathscr{S}_n\partial\mathscr{S}_m^*}. \tag{A6}$$

Inserting this in (A3) and (A4) and discarding terms that vanish on integrations, we find the Lagrangian

$$\mathscr{L}=-J_{nm}\partial_\mu\mathscr{S}_n\partial^\mu\mathscr{S}_m^*-J_{nm}{}^{-1}\left[\frac{\partial f}{\partial\mathscr{S}_n}\right]\left[\frac{\partial f}{\partial\mathscr{S}_m}\right]^*-\tfrac{1}{2}J_{nm}(s_n^T\epsilon\gamma_5\gamma^\mu\partial_\mu s_m)$$

$$+\tfrac{1}{2}(s_{Ln}^T\epsilon s_{Lm})\left[\frac{\partial^2 f}{\partial\mathscr{S}_n\partial\mathscr{S}_m}-J_{kl}{}^{-1}\frac{\partial f}{\partial\mathscr{S}_k}\frac{\partial^3 d}{\partial\mathscr{S}_l^*\partial\mathscr{S}_n\partial\mathscr{S}_m}\right]$$

$$+\tfrac{1}{2}(s_{Rn}^T\epsilon s_{Rm})\left[\left[\frac{\partial^2 f}{\partial\mathscr{S}_n\partial\mathscr{S}_m}\right]^*-J_{lk}{}^{-1}\left[\frac{\partial f}{\partial\mathscr{S}_k}\right]^*\left[\frac{\partial^3 d}{\partial\mathscr{S}_l\partial\mathscr{S}_n^*\partial\mathscr{S}_m^*}\right]\right]$$

$$+\tfrac{1}{2}(s_{nL}^T\epsilon\gamma^\mu s_{mR})\left[\frac{\partial^3 d}{\partial\mathscr{S}_n\partial\mathscr{S}_m^*\partial\mathscr{S}_l}\partial_\mu\mathscr{S}_l-\frac{\partial^3 d}{\partial\mathscr{S}_n\partial\mathscr{S}_m^*\partial\mathscr{S}_l^*}\partial_\mu\mathscr{S}_l^*\right]$$

$$-\tfrac{1}{4}(s_{nL}^T\epsilon s_{mL})(s_{lR}^T\epsilon s_{kR})\left[\frac{\partial^4 d}{\partial\mathscr{S}_n\partial\mathscr{S}_m\partial\mathscr{S}_l^*\partial\mathscr{S}_k^*}+J_{qp}{}^{-1}\frac{\partial^3 d}{\partial\mathscr{S}_p^*\partial\mathscr{S}_n\partial\mathscr{S}_m}\frac{\partial^3 d}{\partial\mathscr{S}_q\partial\mathscr{S}_l^*\partial\mathscr{S}_k^*}\right]. \tag{A7}$$

(This can be easily written in a more familiar notation by using the relation $s^T\epsilon=\bar{s}\gamma_5$.) This result is already well known[14] for the special case of a quadratic D term and arbitrary F term, where J is a constant matrix.

It is interesting that even though we have allowed $f(S)$ and $d(S,S^*)$ to contain terms of arbitrarily high order in the superfields, the Lagrangian has turned out to involve only terms quadratic and quartic in the fermion fields. Furthermore, the quartic fermion interactions are only of the *LLRR* type.

It is not difficult to extend these results to gauge theories, or to Lagrangians with extra derivatives. It may be that these results will prove useful not only for the analysis of effective B- and L-nonconserving interactions, but also for the study of radiative corrections in renormalizable supersymmetric theories, because any term that does not appear in the most general supersymmetric Lagrangian cannot be produced by radiative corrections.

APPENDIX B: AN ILLUSTRATIVE MODEL

One necessary condition that must be met in order to apply the results of Sec. V to a specific model is that when the quark and lepton superfields are omitted the model must have an absolute potential minimum at which the matrix $\Sigma_\alpha g_\alpha D_\alpha t_\alpha$ is positive definite. For a proof that this is possible we may refer to a specific version of a class of models developed by Fayet.[3] To the best of my knowledge, Fayet has not yet presented a complete analysis of the symmetry-breaking solutions of his models, so I will go into some detail here.

The superfield content of this model is limited to just the left-chiral scalar quark, lepton, and Higgs superfields and $SU(3)\times SU(2)\times U(1)\times\widetilde{U}(1)$ gauge vector superfields described in Secs. II–V, plus one additional $SU(3)\times SU(2)\times U(1)$-neutral left-chiral superfield X_L. We assume that all quark and lepton superfields have $\widetilde{U}(1)$ quantum number $\widetilde{Y}=+1$, the Higgs superfields H_L, H_L' have $\widetilde{Y}=-2$, and the neutral superfield X_L has $\widetilde{Y}=+4$. With these quantum numbers, the most general renormalization interaction among the chiral scalar superfields has the form

$$g_U(Q_L U_R^* H_L')_F+g_D(Q_L D_R^* H_L)_F+g_E(L_L E_R^* H_L)+g_H(H_L H_L' X_L)_F+\text{H.c.}$$

In accordance with the discussion in Sec. V, we are interested in the case where g_U, g_D, and g_E are much smaller than the gauge couplings, and as an approximation we can begin by dropping these terms altogether. With all quark and lepton scalars set equal to zero, the potential takes the form

$$V=g_H{}^2\left[\,|\mathcal{H}_L^T\epsilon\mathcal{H}'_L|^2+(\mathcal{H}_L^\dagger\mathcal{H}_L)|\mathcal{L}_L|^2+(\mathcal{H}'^\dagger_L\mathcal{H}'_L)|\mathcal{L}_L|^2\right]+\tfrac{1}{2}\left[g(\mathcal{H}_L^\dagger\vec{\mathrm{t}}\mathcal{H}_L)+g(\mathcal{H}'^\dagger_L\vec{\mathrm{t}}\mathcal{H}'_L)\right]^2$$
$$+\tfrac{1}{2}\left[\frac{g'}{2}(\mathcal{H}_L^\dagger\mathcal{H}_L)-\frac{g'}{2}(\mathcal{H}'^\dagger_L\mathcal{H}'_L)+\xi\right]^2+\tfrac{1}{2}\left[-2\tilde{g}(\mathcal{H}_L^\dagger\mathcal{H}_L)-2\tilde{g}(\mathcal{H}'^\dagger_L\mathcal{H}'_L)+4\tilde{g}|\mathcal{L}_L|^2+\tilde{\xi}\right]^2, \tag{B1}$$

where ξ and $\tilde{\xi}$ are the coefficients of the term $[V_a]_D$ in the Lagrangian for U(1) and $\tilde{\mathrm{U}}(1)$, respectively. [We are now using a somewhat less schematic notation for SU(2) indices: $\mathcal{H}_L$ and $\mathcal{H}'_L$ are two-component columns, ϵ is the totally antisymmetric matrix with $\epsilon_{12}=+1$, and $\vec{\mathrm{t}}$ is the matrix of electroweak isospin.]

It is now straightforward to show the following.

(a) At the minimum of the potential, $\mathcal{H}_L^\dagger\vec{\mathrm{t}}\mathcal{H}_L$ and $\mathcal{H}'^\dagger_L\vec{\mathrm{t}}\mathcal{H}'_L$ will be parallel or antiparallel according as $2g_H{}^2$ is greater or less than g^2. We will adopt the assumption that

$$2g_H{}^2<g^2\,, \tag{B2}$$

because then with these vectors antiparallel we can [by an SU(2) transformation] bring the Higgs doublets at the potential minimum to the form

$$\mathcal{H}_L=\begin{bmatrix}\mathcal{H}_L^0\\0\end{bmatrix},\quad \mathcal{H}'_L=\begin{bmatrix}0\\\mathcal{H}'^0_L\end{bmatrix}, \tag{B3}$$

corresponding to conservation of electric charge.

(b) There is no stationary point of the potential with $\mathcal{H}_L^0$, $\mathcal{H}'^0_L$, and $\mathcal{L}_L$ all nonzero.

(c) The minimum of the potential lies at a point with $\mathcal{H}_L^0\neq0$, $\mathcal{H}'^0_L\neq0$, and $\mathcal{L}_L=0$ if and only if

$$\frac{|\xi g'|}{g^2+g'^2-2g_H{}^2}<\frac{2\tilde{\xi}\tilde{g}}{8\tilde{g}^2+g_H{}^2}\,. \tag{B4}$$

Equation (B2) ensures that the denominator on the left-hand side is positive so (B4) requires in particular that $\tilde{\xi}$ should have the same sign as $\tilde{g}$. We assume that (B4) is satisfied because we want $\langle\mathcal{H}_L^0\rangle$ and $\langle\mathcal{H}'^0_L\rangle$ to have nonvanishing values so that all quarks and leptons can get small masses when the Yukawa couplings are turned on.

(d) Under the conditions (B2) and (B4), the Higgs fields at the minimum of the potential have the values

$$\left.\begin{matrix}|\mathcal{H}_L^0|^2\\|\mathcal{H}'^0_L|^2\end{matrix}\right\}=\frac{2\tilde{\xi}\tilde{g}}{8\tilde{g}^2+g_H{}^2}\pm\frac{\xi g'}{g^2+g'^2-2g_H{}^2}\,. \tag{B5}$$

With this information, we can now compute the mass-squared matrix (18) of the scalar superpartners of the quarks and leptons. Our result is

$$\mathcal{M}^2=\frac{-\xi g'}{g^2+g'^2-2g_H{}^2}\left[g^2t_3-(g^2-2g_H{}^2)y\right]+\frac{g_H{}^2\tilde{\xi}\tilde{g}}{g_H{}^2+8\tilde{g}^2}\tilde{y}\,, \tag{B6}$$

where t_3, y, and $\tilde{y}$ are the SU(2), U(1), and $\tilde{\mathrm{U}}(1)$ generators for the corresponding quarks and leptons. The matrix $\tilde{y}$ is positive definite (in the special case considered here, it is the unit matrix) and the conditions (B2) and (B4) set no upper bound on $\tilde{\xi}$, so as long as g_H and $\tilde{g}$ are nonzero we can clearly make $\mathcal{M}^2$ positive by making $\tilde{\xi}$ sufficiently large.

Incidentally, (B6) shows why the field $\mathcal{L}_L$ has to be added to the model, even though its value vanishes at the minimum of the potential. Without $\mathcal{L}_L$ we would in effect have $g_H=0$, and although the conditions (B2) and (B4) could still be satisfied, the matrix $\mathcal{M}^2$ could not be positive definite.

For comparison, we also note the vector-boson mass-matrix elements

$$m_W{}^2=\tfrac{1}{2}g^2(|\mathcal{H}_L^0|^2+|\mathcal{H}_L^{0\prime}|^2)=\frac{2\tilde{\xi}\tilde{g}g^2}{8\tilde{g}^2+g_H{}^2}\,,$$

$$m_Z{}^2=(g^2+g'^2)m_W{}^2/g^2\,,$$

$$m_{\tilde{Z}}{}^2=16\tilde{g}^2m_W{}^2/g^2\,,$$

$$m_{Z\tilde{Z}}{}^2=g\tilde{g}(|\mathcal{H}_L^0|^2-|\mathcal{H}_L^{0\prime}|^2)=\frac{2\xi g'g\tilde{g}}{g^2+g'^2-2g_H{}^2}\,.$$

It is interesting that it is $\tilde{\xi}$ alone that sets the scale of the W mass and the overall scale of the Z and $\tilde{Z}$ masses, while ξ only determines the Z-$\tilde{Z}$ mixing.

[1]J. Wess and B. Zumino, Nucl. Phys. B70, 39 (1974); Phys. Lett. 49B, 52 (1974); Nucl. Phys. B78, 1 (1974). The superfield formalism used in the present paper is due to A. Salam and J. Strathdee, Nucl. Phys. B76, 477 (1974); B80, 499 (1974). For a general review with references to the original literature, see P. Fayet and S. Ferrara, Phys. Rep. 32, 249 (1977).

[2]E. Gildener and S. Weinberg, Phys. Rev. D 13, 3333 (1976), especially Sec. II; S. Weinberg, Phys. Lett. 82B, 387 (1979).

[3]P. Fayet, Phys. Lett. 69B, 489 (1977); 70B, 461 (1977); in *New Frontiers in High Energy Physics*, edited by A. Perlmutter and L. F. Scott (Plenum, New York, 1978), p. 413; in *Unification of the Fundamental Particle Interactions*, proceedings of the Europhysics Study Conference, Erice, 1980, edited by S. Ferrara, J. Ellis, and P. van Nieuwenhuizen (Plenum, New York, 1980), p. 587; Ecole Normale Supérieure Report No. LPTENS 81/9, 1981 (unpublished). Also see M. Sohnius, Nucl. Phys. B122, 291 (1977).

[4]S. Weinberg, Phys. Rev. Lett. 43, 1566 (1979); F. Wilczek and A. Zee, *ibid.* 43, 1571 (1979).

[5]S. Dimopoulos and H. Georgi, Nucl. Phys. B193, 150 (1981). A similar model has been proposed by N. Sakai, Tohoku University Report No. YU/81/225 (unpublished).

[6]P. Fayet and J. Illiopoulos, Phys. Lett. 31B, 461 (1974).

[7]E. Witten, Princeton report, 1980 (unpublished). This paper did much to instigate the present work. On the vanishing of the renormalization of $(V)_D$ terms, also see W. Fischler, H. P. Nilles, J. Polchinski, S. Raby, and L. Susskind, Phys. Rev. Lett. 47, 757 (1981). In a more recent work [Phys. Lett. 105B, 267 (1981)], Witten proposes an intriguing new view of gauge hierarchies, in which supersymmetry is again supposed to be spontaneously broken at ordinary energies. Models have also been proposed in which supersymmetry is spontaneously broken at ordinary energies through the binding of spin-$\frac{1}{2}$ Goldstone fermions by extra-strong forces; see S. Dimopoulos and S. Raby, Santa Barbara report, 1981 (unpublished) and M. Dine, W. Fischler, and M. Srednicki, Princeton report, 1981 (unpublished). Such models will not be considered here, but some of our considerations would also apply for such dynamical symmetry breaking.

[8]S. L. Adler, Phys. Rev. 177, 2426 (1969); J. S. Bell and R. Jackiw, Nuovo Cimento 60A, 47 (1969).

[9]G. Farrar and S. Weinberg (in preparation).

[10]R symmetries were introduced by A. Salam and J. Strathdee, Nucl. Phys. B87, 85 (1975); P. Fayet, *ibid.* B90, 104 (1975), and will be discussed in detail in the next paper of this series.

[11]S. Weinberg, Ref. 4.

[12]It might be thought that nonrenormalizable F terms cannot be produced by integrating out superheavy degrees of freedom if they were not present to begin with, according to the theorem that radiative correcting can only produce D terms, not F terms. [On this theorem, see M. T. Grisaru, W. Siegel, and M. Rocek, Nucl. Phys. B159, 429 (1979); Witten, Ref. 7.] However, it should be kept in mind that this theorem only limits the terms in an effective Lagrangian that can be produced by loop graphs, not tree graphs. It is the tree graphs in the grand unified form of the DG model that produce the baryon-nonconserving $d=5$ effective interactions. I would like to take this opportunity to thank Howard Georgi for discussions of these points in general, for his calculation of baryon-nonconserving tree graphs in the DG model, and also for his remark that the spin-$\frac{1}{2}$-Higgs-fermion exchange contribution to proton decay is so suppressed by the smallness of the Higgs couplings that it probably could not be observed.

[13]The phenomenology of the scalar superpartners of quarks and leptons is discussed by G. Farrar and P. Fayet, Phys. Lett. 89B, 191 (1980); C. R. Nappi, Institute for Advanced Study report, 1981 (unpublished); also see Ref. 3.

[14]For the general form of the potential in supersymmetric gauge theories, see, e.g., S. Ferrara, L. Girardello, and F. Palumbo, Phys. Rev. D 20, 403 (1979) and Refs. 5 and 15.

[15]B. deWit and D. Z. Freedman, Phys. Rev. D 12, 2286 (1975). In this reference a factor $\frac{1}{4}$ appears in place in the $\frac{1}{2}$ in Eq. (15) because the authors use an unconventional normalization for complex scalar fields in which the kinematic term is $-\frac{1}{2}\partial_\mu\phi^\dagger\partial^\mu\phi$.

[16]The fact that E_6 provides a rationale for Higgs superfields was pointed out to me by both Howard Georgi and Paul Ginsparg, and has also been noted by S. Dimopoulos and F. Wilczek, Santa Barbara report, 1981 (unpublished).

[17]The phenomenology of a $\widetilde{Z}^0$ boson has been considered by P. Fayet, Phys. Lett. 96B, 83 (1980); Nucl. Phys. B187, 184 (1981); P. Fayet and M. Mezard, Phys. Lett. 104B, 226 (1981), and in some of the later articles of Ref. 3. Fayet makes the ingenious suggestion that $\widetilde{g}$ and the $\widetilde{Z}^0$ mass are both very small, so that effects of the new neutral current are strongly suppressed in high-energy neutrino experiments. The role of the $\widetilde{U}(1)$ subgroup in allowing a symmetry-breaking solution with light quarks and leptons and heavy scalars can still be preserved, by the *ad hoc* step of letting the Fayet-Iliopoulos term for $\widetilde{U}(1)$ have a very large coefficient. As Fayet points out, even in the limit $\widetilde{g}\to 0$ the helicity-zero part of the $\widetilde{Z}^0$ boson would survive as a nearly massless Goldstone boson with semiweak couplings.

PHYSICAL REVIEW D VOLUME 27, NUMBER 11 1 JUNE 1983

Supersymmetry at ordinary energies. II. R invariance, Goldstone bosons, and gauge-fermion masses

Glennys R. Farrar
Department of Physics and Astronomy, Rutgers University, New Brunswick, New Jersey 08903

Steven Weinberg
Department of Physics, University of Texas, Austin, Texas 78712

(Received 18 October 1982)

We explore the observable consequences of supersymmetry, under the assumption that it is broken spontaneously at energies of order 300 GeV. Theories of this sort tend automatically to obey a global R symmetry, which presents us with a choice among phenomenologically unacceptable alternatives. If the R symmetry is broken by scalar vacuum expectation values of order 300 GeV, there will be a semiweakly coupled light Goldstone boson, similar to an axion. If it is not broken by such vacuum expectation values but is broken by quantum-chromodynamic (QCD) anomalies, then there will be a light ninth pseudoscalar meson. If it is not broken by QCD anomalies, then the asymptotic freedom of QCD is lost at high energies, killing the hope of an eventual meeting of the electroweak and strong couplings within the regime of validity of perturbation theory. We also confront the problem of an uncomfortably light gluino. A general analysis of gaugino masses shows that the gluino mass is at most of order 1 GeV, and in many cases much less.

I. INTRODUCTION

This paper will continue the study of supersymmetry at ordinary energies that was begun in Ref. 1.

Our theoretical framework is as follows. We assume that supersymmetry survives down to energies of order 300 GeV, where, along with the electroweak gauge symmetry, it is spontaneously broken by vacuum expectation values (VEV's) of weakly coupled scalar fields. Where relevant, we assume that these VEV's are also responsible for quark and lepton masses. In order to avoid light scalars[2] and fast proton decay,[1] we pursue the suggestion of Fayet that the gauge group at low energies should contain in addition to the usual SU(3)×SU(2)×U(1) at least an additional U(1) factor, called here $\tilde{\mathrm{U}}(1)$, with generator $\tilde{Y}$. However, most of our discussion would also apply in the alternative case[3] where the gauge group is just SU(3)×SU(2)×U(1) with light scalars avoided by having all quark, lepton, and associated scalar masses arise from radiative corrections, and we shall occasionally refer to SU(3)×SU(2)×U(1) theories. We try here to avoid basing our considerations on any specific menu of superfields, but we have in mind a model including the following left chiral superfields:

	SU(3)	SU(2)	Y	$\tilde{Y}$
$Q_L=\begin{bmatrix}U_L\\D_L\end{bmatrix}$	3	2	$-\frac{1}{6}$	1
U_R^*	$\bar{3}$	1	$\frac{2}{3}$	1
D_R^*	$\bar{3}$	1	$-\frac{1}{3}$	1
$L_L=\begin{bmatrix}N_L\\E_L\end{bmatrix}$	1	2	$\frac{1}{2}$	1
E_R^*	1	1	-1	1
$H=\begin{bmatrix}H^0\\H^-\end{bmatrix}$	1	2	$\frac{1}{2}$	-2
$H'=\begin{bmatrix}H^{+'}\\H^{0'}\end{bmatrix}$	1	2	$-\frac{1}{2}$	-2

plus additional chiral superfields like the SU(3)×SU(2)×U(1)-neutral X with $\tilde{Y}=+4$ of Ref. 2, which was introduced to allow a spontaneous violation of supersymmetry.

One severe problem with this class of theories is that they are beset with triangle anomalies in gauge currents.[1] (For instance, with just the above superfields, there is a QCD anomaly in the $\tilde{Y}$ current.) Furthermore, if enough new chiral scalar superfields are introduced to cancel these anomalies, one tends to find that the scalar VEV's either do not break supersymmetry or do break charge or color gauge invariances.[4] We will not deal with this problem here, but will simply assume that some set of chiral scalar superfields has been found which allow a realistic pattern of VEV's while at the same time canceling all anomalies in gauge currents. The considerations presented here will not be sensitive to the details of how this is done.

Our chief concern in this paper is with another problem: the phenomenological implications of a global symmetry known as R invariance. This symmetry is not put into these theories by hand, but is automatic in a wide class of supersymmetric models, including all those containing a gauge quantum number like $\tilde{Y}$ whose values for left chiral superfields are restricted to 1 (mod 3). We do not know if this symmetry is broken by scalar VEV's of order 300 GeV, or by QCD anomalies, or by QCD condensates, or by some of these mechanisms, or by none of them, so we explore all of these alternatives. Our conclusion, summarized in the last section, is that each one of these alternatives leads to a severe conflict with experiment or with current theoretical ideas.

R invariance is introduced in its several forms in Sec. II. The possible mechanisms for breaking R invariance are outlined in Sec. III. Then in Sec. IV the properties of the Goldstone bosons associated with each of these mechanisms are considered. Section V deals with the masses of the gauge fermions, in the light of the previous analysis of R invariance.

II. R INVARIANCE

A large class of renormalizable supersymmetric theories automatically have a global symmetry of the type called R invariance. By an R invariance is meant any global U(1) symmetry which acts nontrivially on the superfield coordinate θ, and therefore acts differently on the spinor and the scalar or vector components of a superfield. R symmetries were introduced by Salam and Strathdee[5] and by Fayet[6] as a means of imposing lepton conservation in supersymmetric models, and also in order to constrain the Lagrangian to rule out the possibility that the scalar fields could have vacuum expectation values which would leave supersymmetry unbroken. The discussion here will differ in that R symmetry is not imposed on the theory, but is found to arise in the theories that interest us whether we like it or not.

As one example of a large class of theories which automatically have an R symmetry, consider those renormalizable supersymmetric theories which are prohibited by gauge symmetries from including any sort of super-renormalizable linear or bilinear F terms. [For instance, this is the case if there is a U(1) gauge symmetry like that discussed by Fayet[2] and in I, for which the left chiral scalar superfields carry only the quantum numbers $1,-2,4,-5,7,\ldots$.] The Lagrangian of any such theory will contain only the kinetic terms and gauge couplings of the chiral superfields $S(x,\theta)$ [contained in the D terms $(S^*e^VS)_D$], plus the Yang-Mills F terms $(WW)_F$ (W is defined below in terms of V), and possible Fayet-Iliopoulos[7] D terms $(V)_D$ involving the gauge vector superfields $V(x,\theta)$ alone, plus trilinear F terms $(S^3)_F$ and $(S^3)^*_F$. Any Lagrangian constructed from such ingredients will automatically be invariant under a global U(1) transformation whose generator R_0 has the values $+1$ (-1) for θ_L (θ_R), $+\frac{2}{3}$ $(-\frac{2}{3})$ for all left (right) chiral superfields S (S^*), and 0 for all gauge vector superfields V.

To see this, note that the D term and F term of any function of superfields are the coefficients of $\theta_L{}^2\theta_R{}^2$ and $\theta_L{}^2$, respectively, so if we arbitrarily assign the value $R_0=+1$ to θ_L (and hence $R_0=-1$ to $\theta_R \propto \theta_L^*$), then the D terms and F terms of any function have R_0 values equal to those of the function and the function minus 2, respectively. The functions S^*e^VS and V obviously have $R_0=0$, so their D terms conserve R_0. The left chiral spinor superfield W which contains the Yang-Mills curl is given schematically by

$$W \sim \left[\frac{\partial}{\partial\theta_R}+\theta_L\partial\!\!\!/\right]^2 e^{-V}\left[\frac{\partial}{\partial\theta_L}+\theta_R\partial\!\!\!/\right]e^V ,$$

so it has $R_0=+1$; W^2 has $R_0=+2$; and so its F term conserves R_0. Finally the function S^3 has $R_0=3\times\frac{2}{3}=2$, so again its F term conserves R_0.

An R symmetry sometimes arises also in theories that do contain super-renormalizable F terms. For instance, if there are just two kinds of left chiral scalar superfields $S_\pm$ which carry values ±1 for some U(1) gauge quantum number, then the only allowed renormalizable term $[f(S)]_F$ is $(S_+S_-)_F$, and the Lagrangian is then invariant under an R_1 symmetry for which $S_\pm$ both carry the R values $R_1=1$. Where not otherwise indicated the discussion here will be restricted to theories without super-renormalizable couplings, in which all left chiral scalar superfields have $R_0=\frac{2}{3}$, but much of this discussion would also apply in more general circumstances.

The scalar and spinor component fields $\mathscr{S}$ and s_L of a left chiral scalar superfield S are the coefficients of 1 and θ_L in the expansion of $S(x,\theta)$, while the spinor and vector component fields λ_L and V^μ of a real gauge superfield V are the coefficients of $\theta_R{}^2\theta_L$ and $\theta_R\theta_L$ in the expansion of $V(x,\theta)$. Hence these component fields have the R_0 values:

$$\begin{aligned}
&\text{left chiral scalars } \mathscr{S}\text{: } R_0=\tfrac{2}{3}-0=\tfrac{2}{3}\ ,\\
&\text{left chiral spinors } s_L\text{: } R_0=\tfrac{2}{3}-1=-\tfrac{1}{3}\ ,\\
&\text{left gauge spinors } \lambda_L\text{: } R_0=0+1=1\ ,\\
&\text{vector gauge fields } V_\mu\text{: } R_0=0+0=0\ .
\end{aligned} \tag{1}$$

Any such R symmetry is surely broken by the vacuum expectation values of the $R_0=\frac{2}{3}$ Higgs scalars which break SU(2)×U(1) and give masses to the quarks and leptons. However, it is sometimes possible to combine this broken global R symmetry with a suitable broken gauge symmetry to obtain an unbroken global R symmetry. (We do not consider the possibility of combining R_0 with a broken *global* symmetry to obtain an unbroken R symmetry, because this would lead to consequences similar to those of breaking R—specifically, a semiweakly coupled Goldstone boson.) The neutral Higgs scalars whose vaccum expectation values give masses to the quarks of charge $\frac{2}{3}$ and $-\frac{1}{3}$ (and charged leptons) belong to left chiral superfields with opposite values for electroweak hypercharge and zero values for charge and color, so there is no way that the R_0 symmetry defined above could be combined with SU(3)×SU(2)×U(1) generators to yield an unbroken symmetry. On the other hand, suppose there is an additional U(1) gauge symmetry whose generator $\tilde{Y}$ has equal values for the Higgs superfields (as in the models of Fayet[2] and Secs. IV–VI of I). To keep the same notation as in I, let us take this value as $\tilde{Y}=-2$. Then we can define a new global symmetry

$$\tilde{R}=R_0+\tfrac{1}{3}\tilde{Y} \tag{2}$$

which has the value zero for the Higgs scalars, and is therefore not broken by their vacuum expectation values. Even so it is still an open question whether $\tilde{R}$ conservation is broken by other vacuum expectation values, or by Adler-Bell-Jackiw (ABJ) anomalies, or by dynamical effects of the strong interactions, or by suppressed nonrenormalizable terms in an effective Lagrangian resulting from an $\tilde{R}$-noninvariant theory at a higher energy scale. All these possibilities will be considered in following sections.

Using the values for R_0 given above and taking $\tilde{Y}=0$ and -2 for gauge and Higgs superfields, the $\tilde{R}$ values of the component fields are as follows:

$$\begin{aligned}
&\text{Higgs scalars: } \tilde{R}=0\ ,\\
&\text{left-handed Higgs spinors: } \tilde{R}=-1\ ,\\
&\text{left-handed gauge spinors: } \tilde{R}=+1\ ,\\
&\text{gauge vector bosons: } \tilde{R}=0\ .
\end{aligned} \tag{3}$$

If we suppose for simplicity that all left chiral quark and lepton superfields have equal $\tilde{Y}$ values, then in order for them to couple in pairs to the Higgs superfields they would have to have $\tilde{Y}=+1$, and their component fields would thus have the $\tilde{R}$ values

$$\begin{aligned}
&\text{quarks and leptons: } \tilde{R}=0\ ,\\
&\text{left chiral quark and lepton scalars: } R=+1\ .
\end{aligned} \tag{4}$$

Finally, in order to find a suitable supersymmetry-breaking solution it has generally been found necessary to introduce left chiral X superfields which can couple to pairs of Higgs superfields and hence have $\tilde{Y}=+4$ (see Ref. 2 and Appendix B of I). Their component fields would have $\tilde{R}$ values

$$\begin{aligned}
&\text{left chiral } X \text{ scalars: } \tilde{R}=+2\ ,\\
&\text{left chiral } X \text{ spinors: } \tilde{R}=+1\ .
\end{aligned} \tag{5}$$

It is convenient that the known particles of low mass, including quarks, leptons, gluons, and photons, all have $\tilde{R}=0$. Hence $\tilde{R}$ invariance if unbroken would rule out interactions in which exotic particles with $\tilde{R}\neq 0$ such as quark or lepton scalars or Higgs or gauge spinors or X scalars or spinors (including the Goldstone fermion) are produced singly in collisions of known low-mass particles.[8] An unbroken $\tilde{R}$ invariance would also severely restrict the mass matrices of these exotic particles, and prohibit their mixing with known low-mass particles. We will return to these masses in Sec. V but first it is necessary to study the mechanisms that might break $\tilde{R}$ invariance.

III. MECHANISMS FOR $\tilde{R}$ BREAKING

We can distinguish five different mechanisms which can either individually or jointly break R invariance in supersymmetric theories.

A. Intrinsic R_0 breaking

As explained in the previous section, R_0 invariance can only be broken in a renormalizable supersymmetric Lagrangian by super-renormalizable F terms of the form $(S_1S_2)_F$ or $(S_3)_F$, where S_i are generic left chiral scalar superfields. These are not allowed if there is a $\tilde{\text{U}}(1)$ symmetry whose generator

$\tilde{Y}$ has values equal to 1 (mod 3) (e.g., $+1, -2, +4$) for all left chiral superfields, as in the models of Fayet[2] and Secs. IV–VI of I. Bilinear F terms would be allowed if there were also left chiral superfields which belong to the complex conjugates of some of the representations of $SU(3) \times SU(2) \times U(1) \times \tilde{U}(1)$ furnished by quark, lepton, Higgs, etc., superfields; in particular these would have $\tilde{Y} = -1$ (mod 3). This would have the advantage of making it easy to cancel ABJ anomalies, and the disadvantage of making it easy to find sets of scalar vacuum expectation values which leave supersymmetry unbroken.

Alternatively it is possible to add left chiral superfields which belong to real representations of $SU(3) \times SU(2) \times U(1) \times \tilde{U}(1)$, and therefore cannot have renormalizable F-term interactions with "known" superfields, but which can have both bilinear and trilinear F-term interactions with each other. One possible addition of this sort is a superfield S_0 that is neutral under all gauge groups. This could have $(S_0{}^3)_F$, $(S_0{}^2)_F$, and $(S_0)_F$ interactions, which would break R_0, but since other superfields would have no interaction whatever with S_0, their own R_0 would still be conserved. A more interesting possibility is to add superfields which have no F-term interactions with known superfields, but belong to nontrivial real representations of gauge groups. R violation in the new F-term sector could then induce transitions of left gauge fermions ($R_0 = +1$) into their antiparticles, breaking R_0 for all particles that feel these gauge forces. For instance, a color octet $SU(2) \times U(1) \times \tilde{U}(1)$-neutral chiral superfield E could have $(E^2)_F$ and $(E^3)_F$ interactions, leading through radiative corrections to a Majorana mass term for the gluino,[9] which violates R_0.

Of course, if the gauge group at ordinary energies were just $SU(3) \times SU(2) \times U(1)$ then it would be easy to break R_0 by including an interaction of the form $(HH')_F$, which generates a Majorana Higgs-fermion mass. However if the only left chiral superfields in additon to H and H' were quark superfields Q and lepton superfields L, with F-term interactions schematically of the form $(HH')_F$, $(QQH$ (or $H'))_F$ and $(LLH)_F$, then the Lagrangian would automatically be invariant under an R symmetry for which H and H' carry the values $R_1 = 1$ while Q and L carry the values $R_1 = \frac{1}{2}$ (and θ_L still carries the value $R_1 = 1$). To break this R symmetry in the Lagrangian would require the introduction of new superfields, as, for instance, a $SU(3) \times SU(2) \times U(1)$-neutral left chiral superfield N with interactions $(NHH')_F$ and $(N^3)_F$, $(N^2)_F$, and/or $(N)_F$.

Although there are evidently many ways to break R_0 and $\tilde{R}$ in the Lagrangian, they all have a distasteful feature. One of the reasons for pursuing the possibility that supersymmetry survives down to ordinary energies is that it would help to solve the hierarchy problem. But (inverted hierarchies aside)[10] supersymmetry is not doing this job if it allows super-renormalizable F terms—without fine tuning these would have coefficients in the Lagrangian of the order of the superhigh energy scale M ($\approx 10^{17}$ GeV?) or M^2, giving some fields masses of order M. These fields would have to be integrated out in constructing the effective Lagrangian which describes physics at ordinary energies, and this effective Lagrangian would then contain no super-renormalizable F terms. Note that this argument against super-renormalizable F terms does not apply to the super-renormalizable Fayet-Iliopoulos D terms. There are general theorems[11] which prevent such terms from appearing to any finite order of perturbation theory in an effective low-energy theory which arises from an underlying grand unified theory based on a simple or semisimple gauge group. To get these D terms at all it is necessary to rely on uncertain nonperturbative effects,[12] but this is still better than having to fine-tune the Lagrangian to keep super-renormalizable F terms sufficiently small.

This is not a conclusive argument. In particular, it is possible that Fayet-Iliopoulos terms cannot be generated nonperturbatively, and that like it or not we will have to rely on an O'Raifeartaigh mechanism[13] to allow a spontaneous breakdown of supersymmetry at ordinary energies, which would necessarily require the appearance of super-renormalizable and hence R_0-noninvariant terms in the Lagrangian. However in order for the superpotential naturally to have the form required for the O'Raifeartaigh mechanism, it is generally necessary to impose some sort of R symmetry. The consequences of any such R symmetry will be much the same as those of the R_0 symmetry which characterizes theories without super-renormalizable terms.

B. Vacuum expectation values

As we have seen, in $SU(3) \times SU(2) \times U(1)$ theories any R symmetry that is formed by combining R_0 with gauge symmetries is necessarily broken by the Higgs VEV's. This is not the case in $SU(3) \times SU(2) \times U(1) \times \tilde{U}(1)$ theories, where the Higgs VEV's leave the $\tilde{R}$ of Eq. (2) unbroken. Nevertheless, it is possible that in the spontaneous breakdown of supersymmetry and $SU(2) \times U(1) \times \tilde{U}(1)$, there arise vacuum expectation values, perhaps of order 300 GeV, not only for Higgs scalars but also for other scalars with $\tilde{Y} \neq 2$. These scalars have $\tilde{R} \neq 0$, so their expectation values would produce a large spontaneous breakdown of $\tilde{R}$ in the tree approximation. One would not expect the $\tilde{R}=1$ scalar counterparts of the quarks and leptons to develop vacuum expectation values, since that would lead to an unsuppressed violation of color, charge, or lepton conservation, but the X scalars with $\tilde{Y}=+4$ and $\tilde{R}=2$ might perhaps develop large vacuum expectation values. Such vacuum expectation values would still leave unbroken an "$\tilde{R}$ parity,"[8] $\exp(i\pi\tilde{R})$, which would rule out the production of single Higgs, gauge, or X spinors or quark or lepton scalars in collisions of known low-mass particles, and would prohibit the mass mixing of these exotic particles with known low-mass particles, though not constraining their mixing with each other. Our discussion below of the consequences of breaking $\tilde{R}$ with large VEV's applies to theories with or without an extra $\tilde{\mathrm{U}}(1)$ gauge symmetry.

C. Strong ABJ anomalies

In quantum chromodynamics $\tilde{R}$ conservation can be violated by strong ABJ anomalies[14] in the presence of instantons.[15] These anomalies arise from triangle diagrams in which an $\tilde{R}$ current and two gluons are attached to a loop of colored fermions. To assess the anomalies, it is convenient to consider the R_0 currents rather than those of $\tilde{R}$; since $\tilde{Y}$ is a gauge symmetry its current must in any case be anomaly-free, so the anomalies of the R_0 and $\tilde{R}$ currents are the same, and in considering R_0 instead of $\tilde{R}$ we can lump together all chiral superfields, whatever their values of $\tilde{Y}$. Suppose that the left-handed colored fermions comprise one color octet of gauge fermions ("gluinos") with $R_0=+1$ and some number of fermionic components of various left chiral superfields, with SU(3) generator $t_{\mathrm{SU}(3)}$ and $R_0=-\frac{1}{3}$. The anomaly is then

$$A\equiv\mathrm{Tr}[R(T_{\mathrm{SU}(3)})^2]=3-\tfrac{1}{3}\mathrm{Tr}(t_{\mathrm{SU}(3)})^2 . \qquad (6)$$

The trace of $(t_{\mathrm{SU}(3)})^2$ is a half-integer, so A is a sixth-integer. The effect of this anomaly in a gluon instanton field of winding number ν is that the R_0 and $\tilde{R}$ quantum numbers change by the amounts

$$\Delta\tilde{R}=\Delta R_0=2A\nu . \qquad (7)$$

If the only colored chiral superfields were the three generations of quarks then $\mathrm{Tr}(t_{\mathrm{SU}(3)})^2=3\times4\times\frac{1}{2}$, so there would be an anomaly with $A=1$. However, additional colored chiral superfields must be added in any case in order to cancel the QCD anomaly in the $\tilde{\mathrm{U}}(1)$ current, so in general we can only conclude that $A<1$. In particular, the octet left chiral superfield O_L, which was introduced in I to cancel the QCD anomaly in the $\tilde{\mathrm{U}}(1)$ current when there are three quark generations, contributes $+3$ to $\mathrm{Tr}(t_{\mathrm{SU}(3)})^2$, which together with a gluino and the three quark generations yields $A=0$.

The case of vanishing QCD anomaly, $A=0$, has an interesting and somewhat unpleasant special feature. Recall that the divergence of the R_0 current is in the same supermultiplet as the trace of the energy-momentum tensor,[16] so if the one-loop QCD anomaly of the R_0 current vanishes, then so does the one-loop β function of QCD. This can easily be verified: for instance, one octet of Majorana fermions and their complex scalar superpartners plus the gluinos and three generations of quarks and their scalar superpartners just cancels the gluon contribution to the one-loop β function of QCD. Many of these spin-0 and spin-$\frac{1}{2}$ particles are expected to have masses of order m_W, so QCD would still appear to be asymptotically free at energies below m_W, but with a β function less negative than usually assumed. At higher energies the β function would arise only from two-loop and higher-order contributions. Detailed calculation shows that the two-loop terms[17] have the wrong sign for asymptotic freedom, and all these terms are very small for couplings of order e. Thus with $A=0$ there would be no hope that the strong and electroweak couplings could become equal at any energy below the Planck mass.[18]

Although for $A\neq0$ the ABJ anomaly would break the continuous $\tilde{R}$ symmetry, Eq. (7) shows that it would not break the discrete symmetry of multiplication by

$$Z\equiv\exp(i\pi\tilde{R}/A) . \qquad (8)$$

Whether or not this is a useful symmetry depends on the value of the sixth-integer A. For $|A|=\frac{1}{2}$ or $\frac{1}{6}$, Z invariance would have no consequences for particles of integer $\tilde{R}$, all of which would have $Z=1$. For $|A|=1$ or $\frac{1}{3}$, Z would be the same as the $\tilde{R}$ parity encountered above in Sec. III B, and would allow $|\Delta\tilde{R}|=1$ transitions. If $|A|$ were to have any value other than 1, $\frac{1}{2}$, $\frac{1}{3}$, or $\frac{1}{6}$ then transitions with $|\Delta\tilde{R}|=1$ and $|\Delta\tilde{R}|=2$ would all be forbidden, so Z invariance would lead to the same constraints on the masses of gauge, Higgs, and X fermions and quark and lepton scalars as if $\tilde{R}$ conservation were not violated by an ABJ anomaly.

For any value of A not equal to zero the $\tilde{R}$ invariance of the Lagrangian would solve the strong CP problem in the same way as the U(1) symmetry of Peccei and Quinn,[19] with the difference that $\tilde{R}$ was not invented specifically for this purpose, but is automatically encountered in a large class of supersymmetric theories. Unfortunately, as we shall see, the case $A\neq0$ runs into a variety of familiar conflicts with experiment.

D. Other strong-interaction effects

Even if $\tilde{R}$ conservation is unbroken by ABJ anomalies or by large scalar vacuum expectation values, it can still be spontaneously broken by dynamical effects of the strong interactions. However, here again we expect to retain an unbroken discrete symmetry. All strongly interacting fields that we have encountered here happen to satisfy the relation[8]

$$(-)^{2j}(-)^{3B}=(-)^{\tilde{R}} \tag{9}$$

with j the spin and B the baryon number. Hence any hadronic operator of zero spin and baryon number whose vacuum expectation value can break $\tilde{R}$ conservation would have to have $\tilde{R}$ *even*, and therefore must conserve the $\tilde{R}$ parity $\exp(i\pi\tilde{R})$, with the same consequences as described in Sec. III B.

E. Suppressed nonrenormalizable effective interactions

Since $\tilde{R}$ invariance was not imposed here as an *a priori* symmetry principle but was merely encountered as a more or less accidental consequence of renormalizability, supersymmetry, and gauge symmetries, we would not necessarily expect R invariance to be respected by the physics of much higher energy scales, and hence not by the very weak nonrenormalizable interactions in the effective interaction that describes physics at ordinary energies. These nonrenormalizable interactions were cataloged for $SU(3)\times SU(2)\times U(1)\times \tilde{U}(1)$ theories in Sec. IV of I. The lowest-dimensional allowed nonrenormalizable interactions had dimensionality $d=6$, but these are all of the form $(S^{*2}S^2)_D$ (where S is a generic left chiral superfield) and therefore conserve R_0 and hence $\tilde{R}$. However, there is a whole host of allowed $d=7$ operators of the form $(S^6)_F$ (including the F term of the square of whatever functions F term appears in the renormalizable part of the interaction), which all have $R_0=6\times\frac{2}{3}-2=2$, and hence break R_0 and $\tilde{R}$ invariance but not $\tilde{R}$ parity. These interactions are suppressed by $d-4=3$ powers of whatever superheavy mass (10^{17} GeV?) characterizes the scale of the nonrenormalizable interactions, and probably by several powers of gauge couplings as well, so their effects are extremely small.

IV. *R* GOLDSTONE BOSONS

The existence of an R symmetry of the Lagrangian which is broken spontaneously or by QCD anomalies would require the appearance of a Goldstone or pseudo-Goldstone boson. The properties of this boson depend critically on which of the various symmetry-breaking mechanisms discussed in the preceding section is in operation. Our task in dividing up the various possibilities is simplified by the observation that when R invariance is broken by a scalar VEV of order 300 GeV it is irrelevant whether it is also broken by QCD condensates with a scale of 300 MeV. Also, as we shall see, whatever the dominant mechanism for R-invariance breaking, a crucial consideration for the phenomenology is whether the QCD anomaly of the $\tilde{R}$ current vanishes or not. On the basis of these remarks, it is useful to distinguish four materially different combinations of symmetry-breaking mechanisms.

A. Large scalar VEV's, vanishing QCD anomaly

Suppose that the Lagrangian has an R symmetry but R_0 and $\tilde{R}$ and all other linear combinations of R_0 and gauge symmetries are spontaneously broken in the tree approximation by large vacuum expectation values of order $f_R\approx 300$ GeV. Then there must be a Goldstone boson (let us call it R^0) having semiweak derivative interactions with coupling constant $1/f_R$. If the R symmetry is exact and unbroken by QCD anomalies, then this Goldstone boson is strictly massless. If the symmetry is intrinsically broken only by $d=7$ operators in the effective Lagrangian, then the Goldstone boson has a squared mass inversely proportional to the cube of the superheavy mass scale M, and hence on dimensional grounds of order

$$m_{GB}{}^2\approx f_R{}^5/M^3\ . \tag{10}$$

For $f_R=300$ GeV and $M=10^{17}$ GeV, this still gives a negligible mass, of order 10^{-10} eV.

If all scalar vacuum expectation values are of the same order, then the current to which this Goldstone boson couples will be a linear combination of the currents of $\tilde{R}$ and of the ordinary weak hypercharge Y and the $\tilde{U}(1)$ generator $\tilde{Y}$, all with comparable coefficients. Hence this Goldstone boson will have semiweak couplings to ordinary quarks and leptons which are qualitatively similar to those predicted for the old axion,[20] though its mass is much less. Those experiments that have searched without success for axions through their interactions (but not their decays) thus provide evidence against this light R Goldstone boson as well.

For instance, in an $SU(3)\times SU(2)\times U(1)\times\tilde{U}(1)$ gauge theory the Goldstone boson field ϕ_{GB} has the effective interaction

$$\frac{1}{f_R}(J^\mu_{\tilde{R}}+cJ^\mu_Y+\tilde{c}J^\mu_{\tilde{Y}})_{NP}\partial_\mu\phi_{GB}\ , \tag{11}$$

where NP denotes the part of the current excluding

the Goldstone-boson pole, and c and $\tilde{c}$ are coefficients of order unity. If the complex scalar fields with nonvanishing vacuum expectation values include only Higgs fields $\mathscr{H}^0, \mathscr{H}^{0\prime}$ with $\tilde{Y}=-2$ and $Y=\pm 1$ and a field $\mathscr{X}^0$ with $Y=0$ and $\tilde{Y}=+4$, then[21]

$$f_R=\frac{2\langle\mathscr{H}\rangle\langle\mathscr{H}'\rangle\langle\mathscr{X}\rangle}{(\langle\mathscr{H}\rangle^2\langle\mathscr{H}'\rangle^2+\langle\mathscr{H}\rangle^2\langle\mathscr{X}\rangle^2+\langle\mathscr{H}'\rangle^2\langle\mathscr{X}\rangle^2)^{1/2}},$$

$$c=\frac{\langle\mathscr{X}\rangle^2(\langle\mathscr{H}'\rangle^2-\langle\mathscr{H}\rangle^2)}{(\langle\mathscr{H}\rangle^2\langle\mathscr{H}'\rangle^2+\langle\mathscr{H}\rangle^2\langle\mathscr{X}\rangle^2+\langle\mathscr{H}'\rangle^2\langle\mathscr{X}\rangle^2)},$$

$$\tilde{c}=\frac{-\langle\mathscr{X}\rangle^2(\langle\mathscr{H}\rangle^2+\langle\mathscr{H}'\rangle^2)}{2(\langle\mathscr{H}\rangle^2\langle\mathscr{H}'\rangle^2+\langle\mathscr{H}\rangle^2\langle\mathscr{X}\rangle^2+\langle\mathscr{H}'\rangle^2\langle\mathscr{X}\rangle^2)}.$$

Quarks and leptons contribute to J^μ_Y and $J^\mu_{\tilde{Y}}$, so that this Goldstone boson will in general have appreciable semiweak couplings to quarks and leptons, and probably would have been seen.

We can also consider the possibility that the scalar vacuum expectation values are of rather different magnitudes. Suppose, for instance, in the example above that $\langle\mathscr{X}\rangle$ is much larger than $\langle\mathscr{H}\rangle$ or $\langle\mathscr{H}'\rangle$. In this case the coefficients in (11) approach $\langle\mathscr{X}\rangle$-independent limits:

$$f_R\to\frac{2\langle\mathscr{H}\rangle\langle\mathscr{H}'\rangle}{(\langle\mathscr{H}\rangle^2+\langle\mathscr{H}'\rangle^2)^{1/2}},$$

$$c\to\frac{\langle\mathscr{H}'\rangle^2-\langle\mathscr{H}\rangle^2}{\langle\mathscr{H}'\rangle^2+\langle\mathscr{H}\rangle^2},\quad \tilde{c}\to-\tfrac{1}{2}.$$

Since $\langle\mathscr{H}\rangle$ and $\langle\mathscr{H}'\rangle$ break SU(2)×U(1), neither can be much greater than 300 GeV, so the Goldstone boson is still semiweakly coupled. Also $\tilde{c}$ and perhaps c are still of order unity. The semiweak couplings of the Goldstone bosons thus include interactions with quarks and leptons, which are experimentally ruled out by limits[22] on ψ and T decay to photon plus axion.

The fact that even in the large-$\langle\mathscr{X}\rangle$ limit the Goldstone boson does not decouple from quarks and leptons can be understood as follows. Since $\langle\mathscr{X}\rangle$ breaks not only $\tilde{R}$ but also the $\tilde{\mathrm{U}}(1)$ gauge symmetry, if $\langle\mathscr{H}\rangle$ and $\langle\mathscr{H}'\rangle$ were zero the Goldstone boson associated with $\langle\mathscr{X}\rangle$ would be eliminated by the Higgs mechanism. The symmetry left in this limit is the gauge SU(2)×U(1) and the global $\tilde{R}-\frac{1}{2}\tilde{Y}$. The Goldstone boson here thus arises only from the smaller vacuum expectation values $\langle\mathscr{H}\rangle$ and $\langle\mathscr{H}'\rangle$ which break $\tilde{R}-\frac{1}{2}\tilde{Y}$, so it naturally is semiweakly coupled, and coupled to quarks and leptons. The only way to make this Goldstone boson have couplings weaker than semiweak would be to suppose that $\tilde{R}$ is broken by large vacuum expectation values of scalar fields that do not carry any gauge quantum numbers, but the familiar particles would not have any renormalizable interactions with such scalars, so their $\tilde{R}$ quantum number would remain conserved.

On the other hand, suppose that $\langle\mathscr{X}\rangle$ is much *less* than $\langle\mathscr{H}\rangle$ and $\langle\mathscr{H}'\rangle$. This seems at first sight like an unpromising case, because f_R approaches $2\langle\mathscr{X}\rangle$, so this Goldstone boson couples more strongly than semiweakly. However, c and $\tilde{c}$ also vanish as $\langle\mathscr{X}\rangle\to 0$, and like $\langle\mathscr{X}\rangle^2$ instead of $\langle\mathscr{X}\rangle$. Hence the direct couplings c/f_R and $\tilde{c}/f_R$ of the Goldstone boson to quarks and leptons vanish like $\langle\mathscr{X}\rangle$ for $\langle\mathscr{X}\rangle\to 0$. The strongest limits come from beam-dump experiments[23] which give

$$[\sigma(pN\to R^0X)\sigma(R^0N\to X')]^{1/2}\lesssim 10^{-6}\text{ mb}.$$

The above estimates of R^0 quark couplings are consistent with this limit, assuming $\langle\mathscr{H}\rangle\sim\langle\mathscr{H}'\rangle$, if $\langle\mathscr{X}\rangle\lesssim 40$ GeV.[24] However, we must also take into account the indirect couplings of the Goldstone boson via gluons; this is done in Ref. 25 and in Sec. IV C below. Of course, in the limit of small $\langle\mathscr{X}\rangle$, $\tilde{R}$ invariance is only slightly broken, and we recover the consequences of an unbroken $\tilde{R}$ symmetry, to which we will come at the end of this section.

B. Large VEV's and QCD anomalies

At the same time that $\tilde{R}$ is broken by large scalar vacuum expectation values and (presumably) by suppressed nonrenormalizable effective interactions, it is possible for $\tilde{R}$ to be broken also spontaneously by hadronic vacuum expectation values and/or intrinsically by QCD anomalies. The appearance of hadronic vacuum expectation values is irrelevant as long as $\tilde{R}$ is also broken by much larger scalar-field vacuum expectation values. On the other hand, if there are QCD anomalies in the $\tilde{R}$ current, then this current is not the right place to look for Goldstone bosons. Instead we must consider a linear combination of the $\tilde{R}$ current with the U(1) axial-vector current of the light quarks, the coefficients being chosen so that the QCD anomalies of the two currents cancel. The conservation of this new current is broken, spontaneously by the large vacuum expectation values that break $\tilde{R}$ conservation, and intrinsically by the small masses of the light quarks. Hence there is a Goldstone boson here with a squared mass of order

$$m_{\mathrm{GB}}{}^2 \approx m_{u,d}\Lambda_{\mathrm{QCD}}{}^3/f_R{}^2 . \tag{12}$$

In fact, this is nothing but the old axion.

C. Hadronic VEV's, vanishing QCD anomaly

If the Lagrangian has an $\tilde{R}$ symmetry which is not broken by large vacuum expectation values but is spontaneously broken by strong interaction effects, then there will be a strongly interacting Goldstone boson. If the $\tilde{R}$ current is free of QCD anomalies then the spontaneous breakdown of $\tilde{R}$ invariance leads to a true Goldstone boson of zero mass, except that $d=7$ $\tilde{R}$-violating terms in the effective Lagrangian would give it a mass of order

$$m_{\mathrm{GB}}{}^2 \sim \Lambda_{\mathrm{QCD}}{}^5/M^3$$

for $\Lambda_{\mathrm{QCD}} \simeq 300$ MeV and $M \simeq 10^{17}$ GeV, this gives an utterly negligible mass, of order 10^{-17} eV.

The $\tilde{R}$ current does not contain quark terms, so it might be hoped that this Goldstone boson, R^0, although strongly interacting might have escaped detection. This hope unfortunately proves illusory.[25] The point is that QCD forces can only bring about a spontaneous breakdown of $\tilde{R}$ invariance if there exist $\tilde{R}$-nonneutral particles—e.g., gluinos—which escape getting masses of order m_W in the breakdown of supersymmetry and SU(2)×U(1), so that they are present in the effective theory that describes physics at energies where the QCD forces are strong. These strongly interacting $\tilde{R}$-nonneutral particles then mediate indirect interactions via gluons between the R^0 Goldstone boson and ordinary hadrons. Since the QCD anomaly of the $\tilde{R}$ current is being assumed here to vanish, the R^0 coupling to quarks via a gluino loop connected through 2, 3, or 4 gluons [Fig. 1(a)] is canceled by whatever cancels the anomaly. Thus the dominant coupling of R^0's to quarks is pairwise, as shown in Fig. 1(b).[26] We have just argued that the mass scale of the intermediate $\tilde{R}$-nonneutral state is of order of hadronic masses, and the coupling is $O(\alpha_{\mathrm{QCD}}{}^2)$, so that the R^0 pairwise interaction with hadrons is semistrong. It may not be immediately obvious that massless neutral bosons pair-produced with cross sections less than a few

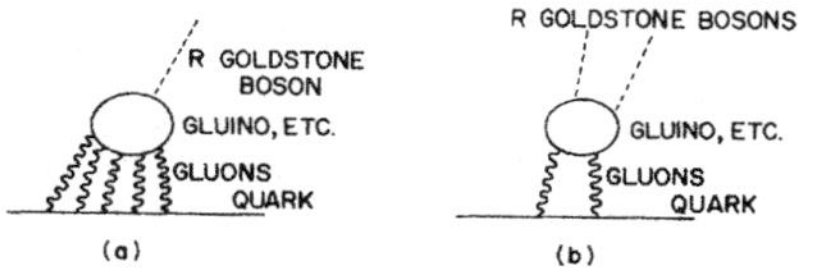

FIG. 1. Diagrams for R Goldstone boson coupling to quarks.

millibarns are experimentally excluded; certainly they would not have been observed in exclusive experiments, and conventional beam dumps with very long filters between the target and the detector are not sensitive to them because semistrongly interacting R^0's would be absorbed before reaching the detector. In fact, however, beam-dump experiments at Brookhaven[27] and at a Fermilab beam-dump test run[28] had sufficiently short filters to be able to exclude such semistrongly produced Goldstone bosons.

D. QCD anomalies, no large scalar VEV's

Now suppose that the Lagrangian has an $\tilde{R}$ symmetry that is not spontaneously broken by large scalar vacuum expectation values but is broken by QCD anomalies. This is of course only possible if there are colored fermions with $\tilde{R} \neq 0$ whose mass vanishes to all orders of perturbation theory: if all colored fermions were allowed by SU(3) and $\tilde{R}$ to get masses, then the colored left-handed fermion fields would have to form a real representation of SU(3)×$\tilde{R}$, so the anomaly A would have to vanish. Such colored massless fermions pose a problem for the consistency of theories of this sort with experiment, to be discussed in Sec. V. For the present, we have to worry about a different problem that is bad enough, the problem of pseudo-Goldstone bosons.

Although the $\tilde{R}$ current is not itself associated with a Goldstone boson, it may be combined with the U(1) axial-vector current of the light quarks to form an anomaly-free current $J^{\mu}_{\tilde{R}'}$, which is intrinsically broken only by the small masses of the light u and d quarks. This $\tilde{R}'$ current must also be spontaneously broken by dynamical effects of the strong interactions, because otherwise the theories would have to contain colorless hadrons which become massless in the limit m_u or $m_d \rightarrow 0$, and which therefore would have masses of only a few MeV. [This is intuitively plausible, and can be shown with greater rigor by adapting an argument of 't Hooft.[29] Although there is no $\tilde{R}'$-SU(3)-SU(3) triangle anomaly, since $\tilde{R}'$ receives a contribution from quarks there are a number of other nonvanishing triangle anomalies, such as $\tilde{R}'$-$\tilde{R}'$-$\tilde{R}'$, $\tilde{R}'$-Q-Q, and $\tilde{R}'$-B-B, with Q electric charge and B baryon number. If $\tilde{R}'$ were exactly conserved and not spontaneously broken by the strong interactions, then these triangle anomalies would have to be reproduced by massless untrapped color-neutral hadrons carrying nonvanishing values of $\tilde{R}'$, Q, and B.] Since $\tilde{R}'$ must be spontaneously broken by dynamical effects of strong interactions, it is irrelevant here whether also $\tilde{R}$ is or is not spontaneously broken by the strong interactions. Also, since $\tilde{R}'$ is intrinsically broken by the quark masses, it is irrelevant to us whether it is or is not also broken by suppressed nonrenormalizable terms in the effective Lagrangian.

The problem here is that the conserved anomaly-free $\tilde{R}'$ current can be used in just the same way as the axial-vector U(1) current of the quarks was used before the discovery of the effects of QCD instantons, to derive an unacceptable result[30]: The spontaneous breakdown of the $\tilde{R}'$ current leads to the same ninth pseudoscalar meson, with mass[31]

$$m_{\rm GB} < \sqrt{3}m_\pi$$

that was the crux of the old U(1) problem.

It is not obvious that the problem of the ninth pseudoscalar meson is avoided even if we add colored superfields with $\tilde{R}\neq 0$ (like the octet superfield of I) to cancel the QCD anomaly in the $\tilde{R}$ current. The point[32] is that if the scalar particles in such superfields have large masses (say $\approx m_W$) then at lower energies we have *two* approximately conserved $\tilde{R}$ currents, one for the quarks and gluinos and the other for the fermion members (e.g., color-octet fermions O) of the new superfields. Only the sum of these two currents is anomaly-free; either one individually may be combined with the axial-vector current of the quarks to deduce the existence of a ninth pseudoscalar meson. To avoid this, the scalar components of the new superfield must be supposed to be fairly light, but fortunately they do not have to be lighter than a few GeV, as it is only necessary that the scalar exchange graphs which violate the separate $\tilde{R}$ conservation for gluinos and quarks be larger than the very small up- and down-quark mass terms which give the pion its mass.

E. $\tilde{R}$ unbroken, vanishing QCD anomaly

Of course, if $\tilde{R}$ is not broken, neither spontaneously nor by QCD anomalies, then there will be no Goldstone boson to worry us. Ordinary particles have $\tilde{R}=0$, so their properties would be unconstrained by this assumption, while $\tilde{R}$-nonneutral particles would have to be massless or parity doubled. At the constituent level this would be realized by massless gluinos, while at the composite level any R hadrons would have to be parity doubled, since massless R hadrons are unacceptable. (R hadrons are color-singlet hadrons containing a gluino in addition to quarks and/or gluons, and hence coming in a variety of possible charges and flavors.) It is amusing that the 't Hooft anomaly-matching argument,[29] which can often be used to demonstrate the necessity of certain massless composite fermions, is not applicable in this case, since the $\tilde{R}$ anomaly is assumed here to vanish at the constituent level. The chief difficulty with this proposal is the absence of asymptotic freedom in QCD at high energies, which as discussed in Sec. III C above is a necessary consequence of the absence of an $\tilde{R}$ anomaly. Another potential problem is the masslessness of the gluino, to be discussed below in Sec. V.

V. GAUGINO MASSES

There are three distinct cases to be considered when discussing the masses of gauginos, the fermionic partners of gauge bosons: tree level masses, radiatively induced masses with R invariance unbroken, and radiatively induced masses with spontaneously broken R invariance.

A. Tree-level masses

Although spontaneous supersymmetry breaking permits tree-level mass splittings within chiral supermultiplets and massive gauge supermultiplets, this is not possible within multiplets containing gauge bosons of an *unbroken* gauge group. An explicit mass term for such a gaugino would violate supersymmetry since by the assumption of unbroken gauge invariance the gauge boson is massless. Furthermore, the Higgs mechanism does not produce such masses in tree approximation, since by supersymmetry gauginos couple only to fields having a nonzero charge under the corresponding gauge group, so mass generation could only occur by mixing of a gaugino with a charged fermion, which is impossible because the charged scalar superpartner has zero VEV. Thus the gluinos and photinos have zero mass at tree level even when supersymmetry is spontaneously broken.

What about the fermionic partners of massive gauge bosons, such as $w^{\pm}$, z^0, and $\tilde{z}^0$? As is well known, spontaneous supersymmetry breaking never produces pure gaugino ($\lambda\lambda$) mass terms in tree approximation, but off-diagonal mass terms (λs) connecting gauginos with chiral fermions can be produced in this way.

The isodoublet Higgs superfields H,H' contain

charged Higgs fermions h^- and $h^{+'}$ (having $\tilde{R}=-1$) that can pair with the charged gauginos w^+ and w^- (having $\tilde{R}=+1$), thereby giving them tree-approximation masses of order m_W, whether or not $\tilde{R}$ is spontaneously broken. Such particles are phenomenologically acceptable, and we need not discuss radiative corrections to their masses. However, if $\tilde{R}$ is not spontaneously broken and if there are more Higgs doublet superfields than just one H and one H', then there will be more than just one pair of charged Higgs fermions with $\tilde{R}=-1$ and some charged Higgs fermions will be left in the tree approximation with zero mass. Even if $\tilde{R}$ is broken by QCD anomalies or condensates, the mass of the left-over charged Higgs fermions will be very small, much less than 1 GeV. (See the discussion in Sec. V C below.) Charged Higgs fermions with mass below about 16 GeV are ruled out by e^+e^- colliding beam experiments,[33] so theories with extra Higgs superfields and $\tilde{R}$ unbroken by large VEV's are definitely untenable.

The case of neutral colorless gauginos and Higgs fermions is more complicated. The gauge bosons Z^0, $\tilde{Z}^0$, and γ have gaugino partners z^0, $\tilde{z}^0$, and photino, of which the photino remains massless in tree approximation, while z^0 and $\tilde{z}^0$ in general get masses of order m_W by mixing with Higgs fermions $h^0, h^{0'}$ and with the fermion member x^0 of the X chiral superfield. However, one linear combination of the z^0, $\tilde{z}^0$, h^0, h'^0, and x^0 must provide the Goldstone fermion and hence remain strictly massless.

If there are more than two neutral Higgs superfields and only one X superfield then R invariance, if unbroken, would keep all but two of the neutral Higgs fermions massless. Even if strictly massless, these extra Higgs fermions would behave at low energies like neutrinos with only neutral-current interactions, and so might have escaped detection.

B. Radiatively induced R-conserving gluino and photino masses

The issue of radiative corrections to gaugino masses is really only consequential for those particles which would have zero mass in the absence of radiative corrections, the photino and gluinos. While such masses may be compatible with gauge and supersymmetry invariances, as noted above gauginos carry $\tilde{R}=R_0=+1$, so that diagonal mass terms $\lambda\lambda$ for the photino and gluinos violate R invariance. $\tilde{R}$-conserving off-diagonal mass terms $\lambda\psi$ are possible if there are $\tilde{R}=-1$ fermions, ψ, with the same conserved quantum numbers as the photino and gluinos. However, in fact, such off-diagonal terms are often excluded by the discrete symmetry $V\to V$, $\phi\to-\phi$, where V and ϕ are the superfields containing λ and ψ, respectively. For instance, gluinos cannot mix with the color-octet fermions O introduced in I because, given the quark quantum numbers, the only O couplings allowed are O^+e^VO and OOX which respect $V\to V$, $O\to-O$. In Fayet's example[9] of gluino masses arising from off-diagonal terms he was obliged to introduce not only an additional chiral octet but also two new heavy quark fields with nonstandard quantum numbers and super-renormalizable couplings, explicitly breaking R invariance. (In fact the true vacuum of this model breaks color conservation and does not break supersymmetry.) It is doubtful that, in a model sufficiently complex to include all known particles, the additional fields necessary to generate off-diagonal gluino masses can be added while maintaining supersymmetry breaking.

In the minimal model with $SU(2)\times U(1)\times\tilde{U}(1)$ broken by two Higgs doublets, H and H', there are two neutral Higgs fermions with $\tilde{R}=-1$. However, as we have seen, there are four neutral $\tilde{R}=+1$ fermions in the three neutral gauge multiplets and the X chiral multiplet, so that in the end there are two massless fermions (the photino and the Goldstone fermion) and two massive partners for the Z^0 and $\tilde{Z}^0$. Thus for there to be enough degrees of freedom to give the photino an off-diagonal $\tilde{R}$-conserving mass an additional neutral chiral multiplet, N, must be introduced. It should be an electroweak isosinglet so that no massless charged fermion is introduced. However, if it is an isosinglet and has $\tilde{R}=-1$, its only couplings are its gauge couplings and NNX, unless nonstandard quarks or leptons are also added. Consequently the discrete symmetry $V_\gamma\to V_\gamma$ and $N\to-N$ would prevent the photino from mixing with ψ_N, even at the quantum level.

Thus, in summary, the photino and gluinos will be massless when $\tilde{R}$ invariance is unbroken unless exotic quarks or leptons are introduced.

C. Radiatively induced R-breaking gluino and photino masses

Having disposed of R-invariant gluino and photino mass terms we now examine the situation when spontaneous R-invariance breaking permits a Majorana mass $\lambda\lambda$. There are basically two cases: R breaking from a nonzero VEV such as $\langle \mathscr{X} \rangle$, or dynamical R breaking from QCD condensation giving a nonzero value to $\langle \lambda_{gl}\lambda_{gl} \rangle$ or $\langle \mathscr{Q}\mathscr{Q}^c \rangle$.

The underlying supersymmetry, R invariance, and gauge symmetries of the Lagrangian, even though spontaneously broken, result in many cancellations among the diagrams which can generate gaugino masses. Thus a method of identifying the structure associated with nonvanishing contributions is useful.

This can be accomplished[34] by considering higher order corrections as generating terms in an effective Lagrangian which necessarily has the same set of invariances as the original Lagrangian had before the fields which develop VEV's were shifted. A somewhat unconventional use is being made here of an effective Lagrangian: it is not a matter of eliminating very massive fields from a low-energy effective interaction, because all of the fields we are considering are of a relatively low ($\lesssim 300$ GeV) mass scale; rather it is being used as a device to expose the quite nontrivial constraints on possible mass terms imposed by the combined supersymmetry and internal symmetries. A term in $\mathscr{L}_{\rm eff}$ capable of generating a $\lambda\lambda$ mass term must have the following properties[34]:

(i) It must be supersymmetric and gauge invariant and respect all global symmetries.

(ii) It must be bilinear in the gauge superfield V and of such a form that the gaugino field λ enters without derivatives.

(iii) It must be at least linear in a supersymmetry breaking VEV such as $\langle \tilde{D} \rangle$ or an $\langle F \rangle$, since unbroken supersymmetry would require the gluinos and photinos to be strictly massless.

(iv) It must be linear in a $\Delta R = 2$ VEV (e.g., $\langle \lambda_g \lambda_g \rangle$, $\langle \mathscr{D}\mathscr{D}^c \rangle$, $\langle \mathscr{X} \rangle$, or $\langle F_H \rangle$).

(v) The term must not contain any additional fields with nontrivial quantum numbers under the unbroken symmetries, since to generate the mass-term structure $\lambda\lambda$ all fields in $\mathscr{L}_{\rm eff}$ except V must be given VEV's.

Finding terms in $\mathscr{L}_{\rm eff}$ with the required properties has been done in Ref. 34 and we simply take over the result here. For the minimal $\mathrm{SU(3)} \times \mathrm{SU(2)} \times \mathrm{U(1)} \times \tilde{\mathrm{U}}(1)$ model the lowest-order contributions to gluino and photino masses not resulting from R-breaking QCD condensates correspond to pieces in $\mathscr{L}_{\rm eff}$ of the form

$$(W^\alpha W_\alpha \overline{W}_{\dot\alpha} \overline{W}^{\dot\alpha} H^\dagger H)_{\theta\theta\bar\theta\bar\theta} \supset \lambda\lambda \langle \tilde{D} \rangle^2 \langle F_H \rangle \langle \mathscr{H}^* \rangle \tag{13a}$$

(and $H \to H'$) or

$$(W^\alpha W_\alpha \overline{W}_{\dot\alpha} \overline{W}^{\dot\alpha} X^\dagger X)_{\theta\theta\bar\theta\bar\theta} \supset \lambda\lambda \langle \tilde{D} \rangle^2 \langle F_X \rangle \langle \mathscr{X}^* \rangle \,. \tag{13b}$$

In fact these are equivalent when the equations of motion are used to eliminate the F's:

$$\langle F_H \rangle = -g_X^* \langle \mathscr{X} \rangle^* \langle \mathscr{H}' \rangle^* \tag{14}$$

and

$$\langle F_X \rangle = -g_X^* \langle \mathscr{H} \rangle^* \langle \mathscr{H}' \rangle^* \,. \tag{15}$$

Although we will argue below that the most important component of a gluino mass is likely to have the structure of (13), it is interesting to note that there can be a mass term independent of $\langle \tilde{D} \rangle$ coming from

$$(W^\alpha W_\alpha X^\dagger X (XHH')^\dagger)_{\theta\theta\bar\theta\bar\theta} \supset \lambda\lambda\, |\langle F_X \rangle|^2 \langle \mathscr{X}^* \rangle \langle \mathscr{H}^* \rangle \langle \mathscr{H}'^* \rangle \,. \tag{16}$$

The lowest-order diagrams yielding these structures are shown in Fig. 2 and lead to the order-of-magnitude estimates:

$$m_{\rm gl} \sim \frac{\alpha_{\rm QCD}}{2\pi} \frac{\tilde{g}^2 \langle \tilde{D} \rangle^2 g_H{}^2 \langle \mathscr{H}^* \rangle \langle F_H \rangle}{M^6} \tag{17}$$

from (13) [Fig. (2a)] and

$$m_{\rm gl} \sim \frac{\alpha_{\rm QCD}}{16\pi^3} \frac{g_U{}^2 g_D{}^2 g_X{}^3 \, |\langle F_X \rangle|^2 \langle \mathscr{X}^* \rangle \langle \mathscr{H}^* \rangle \langle \mathscr{H}'^* \rangle}{M^6} \tag{18}$$

from (16) [Fig. (2b)]. For photino masses replace α_s by $\alpha_{\rm QED}$ times the quark or lepton charges. In (17) and (18) M is the largest mass scale in the diagrams and is presumably to be identified with the spin-0 quark masses. g_U and g_D are the Yukawa couplings giving quark masses so that $g_U \langle \mathscr{H} \rangle = m_u$. Evidently the heaviest quark family makes the most important contribution to $m_{\rm gl}$. Since generally Yukawa couplings are small compared to gauge couplings we will take (17) to be the most important contribution to $m_{\rm gl}$ in the following estimates. More complicated terms in $\mathscr{L}_{\rm eff}$, containing more fields, can only be generated at the expense of higher powers of the coupling constants and thus we neglect them relative to these.

Before proceeding to bound and estimate these expressions we note several features. First, as expected, they vanish in the $\tilde{R}$-conserving limit $\langle \mathscr{X} \rangle \to 0$. Second, there is no contribution linear in $\langle \tilde{D} \rangle$. This might not have been anticipated and is an illustration of the utility of the $\mathscr{L}_{\rm eff}$ approach.

These expressions for $m_{\rm gl}$ are more amenable to approximate evaluation and bounding than one

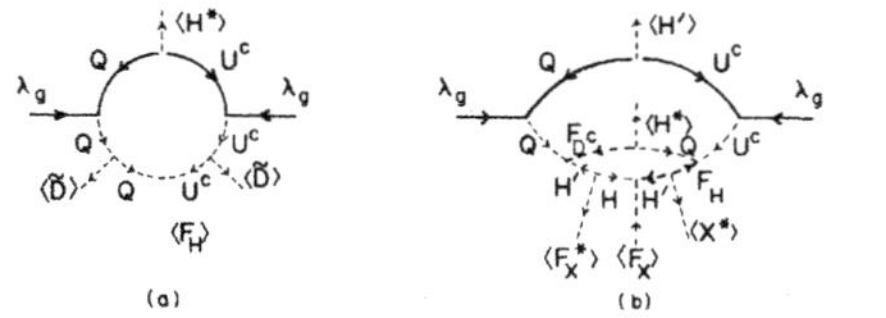

FIG. 2. Illustrative diagrams for radiatively induced gluino masses when $\tilde{R}$ is broken by large VEV's.

might at first imagine, due to the dependence of the spin-0 quark masses on $\tilde{g}\langle\tilde{D}\rangle$ and $g_X\langle\mathscr{X}\rangle$. In general the spin-0 partners of the left and right chiral quarks are not degenerate in mass: in fact, their average (mass)^2 is

$$\overline{M}^2=\tfrac{1}{2}\tilde{g}\langle\tilde{D}\rangle\ , \tag{19}$$

assuming the quark mass m is negligible in comparison, and the mass splitting is

$$\Delta M^2=4g_Q\langle F_H\rangle\ , \tag{20}$$

where g_Q is the Yukawa coupling of the quark to its Higgs boson H, i.e.,

$$m_Q=g_Q\langle\mathscr{H}\rangle\ . \tag{21}$$

Since a negative mass^2 for either spin-0 quark would cause a spontaneous breakdown of color, charge, and baryon number we know that

$$\tfrac{1}{2}\,|\,\tilde{g}\langle\tilde{D}\rangle\,|>2\,|\,g_Q\langle F_H\rangle\,|\ ,$$

or, using Eq. (14),

$$|\,\tilde{g}\langle\tilde{D}\rangle\,|>4\,|\,g_Xg_Q\langle\mathscr{H}'\rangle\langle\mathscr{X}\rangle\,|\ . \tag{22}$$

This is in accord with the prejudices that $\langle\tilde{D}\rangle\sim\langle\mathscr{H}\rangle^2$,that $0<\langle\mathscr{X}\rangle<\langle\mathscr{H}\rangle$ (the latter being also required by the Goldstone-boson analysis of Sec. IV A) and that gauge couplings be larger than Yukawa couplings.

In principle several experimental facts can further constrain the quantities we need, although in fact the constraints at present are not stronger than the guesses we would make based on the considerations mentioned above. These are the following.

(i) The bound on axion coupling. Following the discussion of Sec. IV A we have the rough upper bound $\langle\mathscr{X}\rangle\lesssim 40$ GeV.

(ii) The experimental lower limit on the masses of spin-0 quarks, $M>16$ GeV.[33] With (19)—(21) and $\langle\mathscr{H}\rangle\approx\langle\mathscr{H}'\rangle$ this gives $2g_X\langle\mathscr{X}\rangle m_Q<\frac{1}{2}\tilde{g}\langle\tilde{D}\rangle+m_Q{}^2-(16\text{GeV})^2$.

(iii) Limits on parity violation in the strong interactions. If the masses of the scalar and pseudoscalar quarks (s_q and t_q) are not degenerate, then the radiative corrections to the quark-gluon interaction coming from diagrams with gluinos and spin-0 quarks (see, e.g., Fig. 3) do not conserve parity. Suzuki[35] has recently analyzed this and finds

$$\left|\left\{\left[\left[\frac{\ln((M_{s_u}/m_{\rm gl})^2)}{M_{s_u}{}^2}\right]-(s\leftrightarrow t)\right]-(u\leftrightarrow d)\right\}\right|<\frac{1}{(380\ \text{GeV})^2}\frac{1}{\alpha_{\rm QCD}{}^2}\ . \tag{23}$$

The logarithms in Eq. (23) are of order 1 or larger, so that

$$[(M_{s_u}{}^2-M_{t_u}{}^2)-(u\leftrightarrow d)]\Big/M^4\le\frac{1}{380\ \text{GeV}^2}\frac{1}{\alpha_{\rm QCD}{}^2}\ . \tag{24}$$

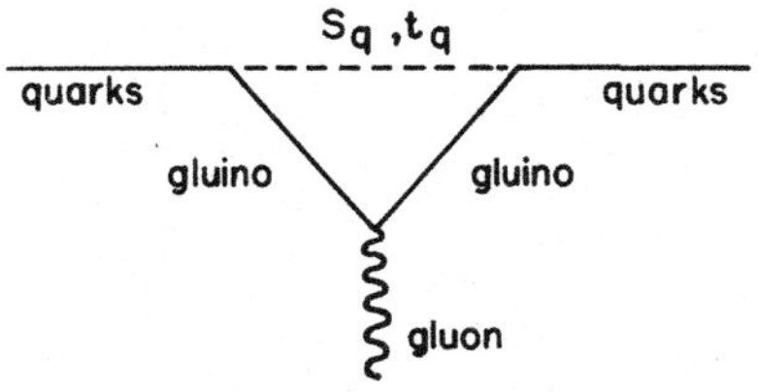

FIG. 3. Diagram responsible for parity violation in the quark-gluon coupling when the spin-0 partners of the left- and right-handed quarks are not degenerate in mass.

Using (20), (14), $\langle\mathscr{H}\rangle\approx\langle\mathscr{H}'\rangle$, (21), and $m_u-m_d=5$ MeV, this becomes (with $\alpha_{\rm QCD}\approx 1$ to be conservative, because the momentum transfers are low)

$$g_X\langle X\rangle\lesssim(\tilde{g}\langle\tilde{D}\rangle)^2\times10^{-4}\ \text{GeV}^{-3}\ . \tag{25}$$

(iv) Limits on flavor-changing "neutral currents" arising from gluino exchange: When the spin-0 quark flavor mixing is not perfectly aligned with the quark mixing, strong flavor-changing "neutral currents" are generated unless the mass splitting between spin-0 quark flavors is very small compared to the spin-0 masses. Suzuki[36] has obtained the limit $M_c{}^2-M_u{}^2<(5\times10^{-6})M^3$. Assuming $\langle F_H\rangle\ll\langle\tilde{D}\rangle$, this translates to $m_c{}^2-m_u{}^2\sim(1.5\ \text{GeV})^2<(5\times10^{-6})(\frac{1}{2}\tilde{g}\langle\tilde{D}\rangle)^{3/2}$ giving $\tilde{g}\langle\tilde{D}\rangle\gtrsim10^4\ \text{GeV}.^2$

Now we can return to the problem of estimating $m_{\rm gl}$ from (17). Using $M^2\sim\frac{1}{2}\tilde{g}\langle\tilde{D}\rangle$ and (14), assuming $\langle\mathscr{H}\rangle\sim\langle\mathscr{H}'\rangle$

$$m_{\rm gl} \sim \frac{\alpha_{\rm QCD}}{2\pi} \frac{8 g_X \langle \mathscr{X} \rangle m_t^2}{\tilde{g} \langle \tilde{D} \rangle} . \qquad (26)$$

We can use (ii) to get a (probably quite weak) bound on $g_X \langle \mathscr{X} \rangle m_t / \tilde{g} \langle \tilde{D} \rangle$ if we assume

$$m_t^2 - (16\ \text{GeV})^2 << \tfrac{1}{2} \tilde{g} \langle \tilde{D} \rangle ,$$

giving [with $m_t \sim 30$ GeV and $\alpha_s \approx 0.1$, since the loops giving Eq. (17) probe $\alpha_{\rm QCD}$ at short distances]

$$m_{\rm gl} < \frac{\alpha_{\rm QCD}}{\pi} m_t \lesssim 1\ \text{GeV} . \qquad (27)$$

Or we can use (25) and (26) to give

$$m_{\rm gl} < 8\alpha_{\rm QCD} / 2\pi (\tilde{g} \langle \tilde{D} \rangle m_t^2) \times 10^{-4}\ \text{GeV}^{-3} .$$

With $\langle \tilde{D} \rangle \sim (300\ \text{GeV})^2$ this is not as good as (27). The best bound is obtained by using (i) for $\langle \mathscr{X} \rangle$, (iv) for $\tilde{g} \langle \tilde{D} \rangle$, and assuming $g_X \leq 1$; it gives

$$m_{\rm gl} < \frac{8\alpha_{\rm QCD}}{2\pi} \frac{40 m_t^2}{10^4} \text{GeV}^{-1} \sim 400\ \text{MeV} , \qquad (28)$$

again, for $m_t \sim 30$ GeV. The limit on the photino mass corresponding to Eq. (27) is $m_{\tilde{\gamma}} \sim 4$ MeV. It should be emphasized that these are only relatively model-independent upper bounds which can be established using experimental constraints, and which we have only roughly estimated. In fact, plausible guesses for the VEV's $\langle \mathscr{X} \rangle$ and $\langle \tilde{D} \rangle$ and the couplings g_X and $\tilde{g}$ lead to much lower values of the gluino and photino masses.

These estimates have assumed that R-invariance breaking results from a nonzero value of an $R=2$ scalar VEV $\langle \mathscr{X} \rangle$. While this VEV may vanish, it is natural to expect that strong color forces can generate the $\Delta R = 2$ condensates $\langle \lambda_{\rm gl} \lambda_{\rm gl} \rangle$ and/or $\langle \mathscr{Q} \mathscr{Q}^c \rangle$ in the same way that $\langle \psi_q \psi_q^c \rangle$ forms. The lowest-order diagram of this sort is shown in Fig. 4 and comes from a term in $\mathscr{L}_{\rm eff}$ such as

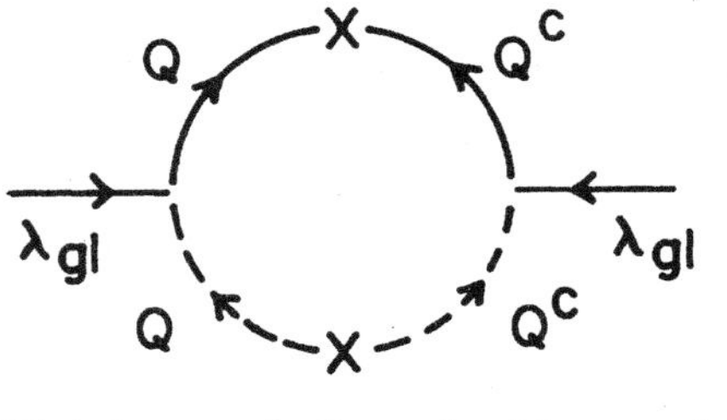

FIG. 4. Lowest-order diagram for gluino mass when R invariance is broken dynamically by a nonzero $\langle QQ^c \rangle$.

$$[D^2(Q^\dagger e^V Q) \bar{D}^2 (Q^{c\dagger} e^{-V} Q^c)]_{\theta\theta\bar\theta\bar\theta} .$$

It leads to the order-of-magnitude estimate

$$m_{\rm gl} \sim \frac{\alpha_{\rm QCD}}{2\pi} \frac{\langle \mathscr{Q}^{c*} \mathscr{Q}^* \rangle \langle \psi_q \psi_{qc} \rangle}{M^4} .$$

Taking all QCD condensates of the scale $\Lambda_{\rm QCD} \sim 300$ MeV and $M \gtrsim 16$ GeV reveals that the resulting gluino mass is much less than an MeV. Mere dimensional analysis with no recourse to any detailed discussion of $\mathscr{L}_{\rm eff}$ is sufficient to arrive at this qualitative conclusion, that dynamical R breaking by QCD forces leads to tiny gluino masses: when $\langle \lambda \lambda^c \rangle$ is responsible, gauge invariance requires quark superfields to enter in pairs and condensation requires both a quark and its charge conjugate so there will be at least a factor $\Lambda_{\rm QCD}^4 / M^3 \lesssim 10^{-5}$ GeV and when $\langle \lambda_{\rm gl} \lambda_{\rm gl} \rangle$ condensation is responsible there is minimally a factor $\Lambda_{\rm QCD}^3 / M^2 \lesssim 10^{-4}$ GeV. (Evidently a supercolor[37] dynamical breaking scheme could generate much larger gluino masses.)

We have seen that within the framework of $SU(3) \times SU(2) \times U(1) \times \tilde{U}(1)$, gluino and photino masses cannot be large, barring complicated schemes with exotic quarks or huge supersymmetry-breaking scales. Since a small (~ 15 eV) photino mass is very attractive for astrophysics[38] this result is nice. However, a small gluino mass is only barely consistent with experiments putting lower limits on R-hadron masses. R hadrons, composed of a gluino and quarks or gluons, should be pair-produced in strong interactions at a rate only inhibited by their mass. Since glueballs are thought to have masses of order 1.5 GeV this could be a reasonable guess for the mass of an R hadron assuming massless gluinos. Experiments rule out R-hadron masses of less than 1.5 or 2 GeV,[8,39] so light gluinos are only barely tolerable.[40] If R-hadron masses are in this range they should be readily observed at Tevatron energies, and if they are not it will provide still more evidence against this class of theories.

VI. CONCLUSIONS

Our conclusions are pessimistic. Using supersymmetry to solve the gauge hierarchy problem has been a very appealing if as yet unrealized possibility. However, the present analysis has revealed the seriousness of the difficulties inherent in such an approach as far as the phenomenological consequences are concerned. Supersymmetric theories tend to have an R symmetry, which if broken spontaneously *or* by QCD anomalies lends to phenomenologically unacceptable Goldstone bosons, either axionlike or mesonlike. If we instead arrange that the R current is *not* broken by QCD anomalies, then, whether or not it is spontaneously broken, we lose the asymptotic freedom of QCD above a few hundred GeV, ruling out any sort of grand unification of strong with electroweak forces below the Planck scale.[18] If we suppose that the R symmetry is not spontaneously broken by vacuum expectation values of order 300 GeV, we encounter another feature which may come into conflict with experiment: the gluino would be very light. This problem may well be with us even if R symmetry is spontaneously broken by large vacuum expectation values, since our estimates give a gluino mass in any case of order 1 GeV or less.

We can try to avoid the difficulties raised by R invariance by introducing R-noninvariant terms in the Lagrangian. However, this is difficult in existing models: for instance, R invariance is automatic in $SU(3)\times SU(2)\times U(1)\times \tilde{U}(1)$ models unless we add new superfields. Furthermore the violation of R invariance requires the introduction of superrenormalizable F terms in the Lagrangian, which seems to us to vitiate the use of low-energy supersymmetry to solve the hierarchy problem of grand unified theories.

It must be stressed that our analysis has been mainly directed at a rather conventional theoretical framework, in which supersymmetry is broken in the tree approximation by scalar vacuum expectation values of order 300 GeV. It is possible that more innovative models, with supercolor, an inverted hierarchy, or supergravity, may avoid the difficulties we have found.

ACKNOWLEDGMENTS

We have benefited from discussions with N. Cabibbo, M. Claudson, S. Dimopoulos, P. Fayet, H. Georgi, M. Wise, and E. Witten. The research of G. R. F. was supported by the National Science Foundation. The work of S. W. was supported in part by the Robert A. Welch Foundation.

---

[1]S. Weinberg, Phys. Rev. D 26, 287 (1982). This paper will be referred to here as I.

[2]P. Fayet, Phys. Lett. 69B, 489 (1977); 70B, 461 (1977); in *New Frontiers in High Energy Physics,* edited by A. Perlmutter and L. F. Scott (Plenum, New York, 1978), p. 413; in *Unification of the Fundamental Particle Interactions,* proceedings of the Europhysics Study Conference, Erice, 1980, edited by S. Ferrara, J. Ellis, and P. Van Nieuwenhuizen (Plenum, New York, 1980), p. 587; in *New Flavors and Hadron Spectroscopy,* proceedings of the XVI Rencontre de Moriond, Les Arcs, France, 1981, edited by J. Trân Thanh Vân (Editions Frontieres, Dreux, France, 1981), Vol. 1, p. 347. Also see M. Sohnius, Nucl. Phys. B122, 291 (1977).

[3]L. Ibanez and G. Ross, Phys. Lett. 110B, 215 (1982); M. Dine and W. Fischler, *ibid.* 110B, 227 (1982); C. Nappi and B. Ovrut, *ibid.* 113B, 175 (1982).

[4]L. Alvarez-Guamé, M. Claudson, and M. Wise, Nucl. Phys. B207, 96 (1982). There are models with a satisfactory pattern of supersymmetry and gauge symmetry breaking, developed by L. Hall and I. Hinchcliffe, Phys. Lett. 112B, 351 (1982). However, these models all involve superpotentials that contain linear or bilinear as well as trilinear terms. As discussed here in Sec. II, such nontrilinear terms are distasteful from the standpoint of grand unification, but as indicated by the arguments of Hall and Hinchcliffe, they may be indispensable in any realistic theory of spontaneously broken supersymmetry. In any case, although the problem of R invariance that concerns us in this paper is initially introduced in Sec. II for the case of purely trilinear superpotentials, as we shall see it arises in a similar way for a wide variety of other cases.

[5]A. Salam and J. Strathdee, Nucl. Phys. B87, 85 (1975).

[6]P. Fayet, Nucl. Phy. B90, 104 (1975).

[7]P. Fayet and J. Illiopoulos, Phys. Lett. 31B, 461 (1974).

[8]G. Farrar and P. Fayet, Phys. Lett. 76B, 575 (1978).

[9]P. Fayet, Phys. Lett. 78B, 417 (1978).

[10]E. Witten, Phys. Lett. 105B, 267 (1981); M. Dine and W. Fischler, Nucl. Phys. B204, 346, (1982); S. Dimopoulos and S. Raby, Los Alamos Report No. LA UB-82-1282 (unpublished). For a related discussion, see J. Polchinski and L. Susskind, Phys. Rev. D 26, 3661 (1982). In these models supersymmetry can resolve the original hierarchy problem by constraining the effective low-energy Lagrangian so that it is supersymmetric except for soft terms which break supersymmetry intrinsically. In this paper we are considering only theories in which the low-energy effective Lagrangian is strictly supersymmetric.

[11]M. T. Grisaru, W. Siegel, and M. Rocek, Nucl. Phys. B159, 429 (1979).

[12]E. Witten, Nucl. Phys. B188, 513 (1961); also see L. Abbott, M. T. Grisaru, and H. J. Schnitzer, Phys. Rev. D 16, 3002 (1977).

[13]L. O'Raifeartaigh, Nucl. Phys. B96, 331 (1975); also see P. Fayet, Phys. Lett. 58B, 67 (1975).

[14]S. L. Adler, Phys. Rev. 177, 2426 (1969); J. S. Bell and R. Jackiw, Nuovo Cimento 60A, 47 (1969).

[15]G. 't Hooft, Phys. Rev. Lett. 37, 8 (1976); Phys. Rev. 14, 3432 (1976); R. Jackiw and C. Rebbi, Phys. Rev. Lett. 37, 172 (1976); C. G. Callen, R. F. Dashen, and

D. J. Gross, Phys. Lett. 63B, 334 (1976).

[16]S. Ferrara and B. Zumino, Nucl. Phys. B87, 208 (1974).

[17]Formulas for the two-loop β function in supersymmetric models are given by D. R. T. Jones, Phys. Rev. D 25, 581 (1981).

[18]This possibility may still be physically relevant; see N. Cabibbo and G. Farrar, Phys. Lett. 110B, 107 (1981); L. Maiani, G. Parisi, and R. Petronzio, Nucl. Phys. B136, 115 (1978).

[19]R. D. Peccei and H. R. Quinn, Phys. Rev. Lett. 38, 1440 (1977); Phys. Rev. D 16, 1791 (1977).

[20]S. Weinberg, Phys. Rev. Lett. 40, 223 (1978); F. Wilczek, *ibid.* 40, 279 (1978).

[21]In general, c and $\tilde{c}$ can be obtained from the condition that the generator corresponding to the current in (11) must produce a variation $(\tilde{R}+cY+\tilde{c}\tilde{Y})\langle\mathscr{S}\rangle$ in the column of scalar-field vacuum expectation values $\langle\mathscr{S}\rangle$ which is orthogonal to all the variations $Y\langle\mathscr{S}\rangle, \tilde{Y}\langle\mathscr{S}\rangle$ produced by gauge generators. Also, f_R is the magnitude of the column vector $(\tilde{R}+cY+\tilde{c}\tilde{Y})\langle\mathscr{S}\rangle$.

[22]C. Edwards *et al.*, Phys. Rev. Lett. 48, 903 (1982); P. Franzini, in *Proceedings of the XVIIth Rencontre de Moriond, Les Arcs, France, 1982,* edited by J. Trân Thanh Vân (Editions Frontieres, Gif-sur-Yvette, 1982).

[23]P. C. Bosetti *et al.*, Phys. Lett. 74B, 143 (1978).

[24]A review of axion couplings and experimental results with additional references is given by G. Girardi, in *Proceedings of the XVIIth Rencontre de Moriond, Les Arcs, France, 1982* (Ref. 22).

[25]G. R. Farrar and E. Maina, Phys. Lett. B (to be published).

[26]G. R. Farrar, in *Proceedings of the XVIIth Rencontre de Moriond, Les Arcs, France, 1982* (Ref. 22).

[27]P. Jacques *et al.*, Phys. Rev. D 21, 1206 (1982); P. Coteus, Phys. Rev. Lett. 42, 1438 (1979); A. Soukas *et al.*, *ibid.* 44, 564 (1980).

[28]B. P. Roe *et al.*, in *Cosmic Rays and Particle Fields—1978,* proceedings of the Bartol Conference, edited by T. K. Gaisser (AIP, New York, 1979).

[29]G. 't Hooft, in *Recent Developments in Gauge Theories,* proceedings of the NATO Advanced Study Institute, Cargèse, 1979, edited by G. 't Hooft (Plenum, New York, 1980).

[30]This was independently realized by H. Georgi (private communication).

[31]S. Weinberg, Phys. Rev. D 11, 3583 (1975).

[32]H. Georgi (private communication).

[33]W. Bartel *et al.*, Phys. Lett. 114B, 211 (1982); H. J. Behrend *et al.*, *ibid.* 114B, 287 (1982).

[34]G. R. Farrar, Nucl. Phys. B209, 114 (1982).

[35]M. Suzuki, Phys. Lett. B115, 40 (1982).

[36]M. Suzuki, LBL Report No. UCB-PTH-82/8 (unpublished).

[37]S. Dimopoulos and S. Raby, Nucl. Phys. B192, 353 (1981); M. Dine, W. Fischler, and M. Srednicki, *ibid.* B189, 575 (1981).

[38]N. Cabibbo, G. Farrar, and L. Maiani, Phys. Lett. 105B, 155 (1981).

[39]G. Farrar and P. Fayet, Phys. Lett. 79B, 442 (1978).

[40]G. Kane and J. Leveille [Phys. Lett. 112B, 227 (1982)] have excluded the gluino mass range from about 1 to 3 GeV by use of perturbative QCD calculations.

PHYSICAL REVIEW D VOLUME 27, NUMBER 10 15 MAY 1983

Supergravity as the messenger of supersymmetry breaking

Lawrence Hall
Department of Physics, University of California, Berkeley, California 94720

Joe Lykken and Steven Weinberg
Theory Group, Department of Physics, University of Texas, Austin, Texas 78712

(Received 12 January 1983)

A systematic study is made of theories in which supergravity is spontaneously broken in a "hidden" sector of superfields that interact with ordinary matter only through supergravity. General rules are given for calculating the low-energy effective potential in such theories. This potential is given as the sum of ordinary supersymmetric terms involving a low-energy effective superpotential whose mass terms arise from integrating out the heavy particles associated with grand unification, plus supersymmetry-breaking terms that depend on the details of the hidden sector and the Kähler potential only through the values of four small complex mass parameters. The result is not the same as would be obtained by ignoring grand unification and inserting small mass parameters into the superpotential from the beginning. The general results are applied to a class of models with a pair of Higgs doublets.

I. INTRODUCTION

It was widely hoped that supersymmetry would turn out to be spontaneously broken at energies no higher than a few hundred GeV, both in order to help in understanding gauge hierarchies and also to allow some chance of confirming supersymmetry experimentally. Unhappily, it has proved difficult to construct satisfactory theories along these lines.[1] We are led to the conclusion that supersymmetry if valid at all is spontaneously broken at energies very much greater than those of $SU(2)\times U(1)$ breaking. But then if any vestige of supersymmetry is to survive at ordinary energies to help establish a gauge hierarchy, the source of supersymmetry breaking must somehow be partly isolated from ordinary particles and interactions.[2]

Recently attention has been drawn to a class of interesting models of this sort.[3–13] In these models, unextended ($N=1$) supersymmetry is broken by very large scalar-field vacuum expectation values (VEV's) of order 10^{19} GeV, but the scalars that have these large VEV's form a "hidden sector," that does not interact directly with the ordinary fields (quarks, leptons, gauge and Higgs bosons, and their superpartners) of the "observable sector." That is, the superpotential of the theory breaks up into a sum of two terms[14,15]

$$f_{\rm TOTAL}(S,\widetilde{S})=f(S)+\widetilde{f}(\widetilde{S})\ ,$$

where S^a and $\widetilde{S}^h$ are the left-chiral superfields of the observable and hidden sectors, respectively. With a minimal kinetic term and no other interactions, the potential of the scalar (nonauxiliary) components $z^a,\widetilde{z}^h$ of $S^a,\widetilde{S}^h$ would take the form

$$V(z,\widetilde{z})=\sum_{{\rm all}\,z}\left|\frac{\partial f_{\rm TOTAL}}{\partial z}\right|^2 =\sum_a\left|\frac{\partial f(z)}{\partial z^a}\right|^2+\sum_h\left|\frac{\partial \widetilde{f}(\widetilde{z})}{\partial \widetilde{z}^h}\right|^2$$

and the spontaneous breakdown of supersymmetry in the hidden sector could have no effect on the observable sector. In the models of Refs. 3–12 the news that supersymmetry is broken by the $\widetilde{z}^h$ VEV's is carried over to the observable superfields by gravity and its superpartners, which interact with both sectors.

In the papers of Ref. 3, a thorough study is presented of a model with a specific linear hidden-sector superpotential $\widetilde{f}$, and a specific grand-unified observable sector. Their results exhibit some remarkable features; in particular, the VEV's of the light Higgs scalars are of order of the gravitino mass m_g, and do not depend in any way on the grand-unified mass scale $M_{\rm GU}$, but do depend on coupling parameters of heavy fields whose masses are of order $M_{\rm GU}$. However, because the model studied was so specific, and the results were expressed in terms of values for scalar VEV's, it was difficult to see how the decoupling of heavy from light degrees of freedom works in these models, and it was difficult to know what aspects of the results would apply in general.[16]

The papers of Refs. 4 and 6 dealt with models that were in various respects more general. Reference 4 considered the same linear superpotential for the hidden sector, but put no restrictions on the form of the superpotential for the observable sector. Reference 6 considered a general superpotential for the hidden sector, and restricted the form of the observable sector only by requiring that its superpotential be purely trilinear. However, neither of these groups considered grand-unified models, in which the observable-sector superpotential involves mass scales $M_{\text{GU}} \gg m_g$. As shown by the work of Ref. 3 (and more generally in Sec. III below), the existence of a class of superheavy particles which have to be "integrated out" to construct the low-energy effective potential changes the way that the supersymmetry-breaking corrections appear in this effective potential, in a manner that (except for purely trilinear superpotentials like that of Ref. 6) cannot be simulated by inserting mass terms in an observable-sector superpotential involving only light fields.

It seemed to us that it would be useful to present a study of this class of models, with general superpotentials for both the hidden and observable sectors, and with full attention to the complications caused by the presence of heavy particles with masses of order $M_{\text{GU}} \gg m_g$. Our assumptions are spelled out in Sec. II, and in Sec. III we present our main result, a general formula [Eq. (3.11)] for the effective superpotential of the light scalars. In this formula the unknown properties of the hidden sector enter in the values of just two comparable mass parameters, m_g and m_g', one of them the gravitino mass, and all aspects of the full grand-unified theory enter only in the parameters of an effective superpotential. Section IV generalizes this result to a large class of Kähler metrics, and shows that this introduces just two more unknown mass parameters, m_g'' and m_g'''. In Sec. V we show how our results can be used to derive phenomenologically interesting predictions, even without having to make any specific assumptions about the grand-unified theory or the hidden sector.

II. ASSUMPTIONS

We assume a total superpotential of the form[14,15]

$$f_{\text{TOTAL}}(S,\tilde{S}) = f(S) + \tilde{f}(\tilde{S}) . \tag{2.1}$$

Here $f(S)$ and $\tilde{f}(\tilde{S})$ are the superpotentials for the chiral superfields S^a and $\tilde{S}^h$ of the observed and hidden sectors, respectively. The potential for the scalar field components z^a and $\tilde{z}^h$ is then[17]

$$V(z,\tilde{z}) = \exp\left[8\pi G\left[\sum_a |z^a|^2 + \sum_h |\tilde{z}^h|^2\right]\right]\left[\sum_a\left|\frac{\partial f(z)}{\partial z^a} + 8\pi G z^{a*}[f(z)+\tilde{f}(\tilde{z})]\right|^2 + \sum_h\left|\frac{\partial \tilde{f}(\tilde{z})}{\partial \tilde{z}^h} + 8\pi G \tilde{z}^{h*}[f(z)+\tilde{f}(\tilde{z})]\right|^2 - 24\pi G\,|f(z)+\tilde{f}(\tilde{z})|^2\right] + \sum_k D_k^2 . \tag{2.2}$$

This is for a quadratic d function (i.e., a flat Kähler metric); we will return to the general case later, in Sec. IV. The gauge auxiliary scalar D_k in (2.2) takes the usual form

$$D_k \equiv \sum_{a,b} (t_k)^a{}_b z^{a*} z^b , \tag{2.3}$$

where t_k is the Hermitian matrix representing the kth gauge generator, including coupling-constant factors. We assume that there are no Fayet-Iliopoulos terms, and that the hidden-sector fields are neutral with regard to all gauge symmetries, but it would be easy to include the effects of additional gauge fields that interact only with the hidden sector.

Our assumptions regarding the observable and hidden superpotentials are as follows.

A. Observable sector

It is assumed that there is a set of scalar field VEV's, z_0^a, for which, in the absence of the hidden sector, supersymmetry would be unbroken and spacetime would be flat:

$$\partial f(z)/\partial z^a = 0 , \text{ at } z = z_0 , \tag{2.4}$$

$$D_k=0\ ,\ \text{at}\ z=z_0\ , \tag{2.5}$$

$$f(z_0)=0\ . \tag{2.6}$$

[Of course, we can always make $f(z)$ vanish at the z_0 defined by (2.4) and (2.5) by shifting a constant term from $f(z)$ to $\tilde{f}(\tilde{z})$.] The tree-approximation scalar spectrum in the absence of the hidden sector then consists of a complex scalar of mass M for each eigenvalue M^2 of the Hermitian matrix

$$M^2{}_{ab}=\sum_c f^*_{ac}f_{bc} \tag{2.7}$$

with subscripts denoting differentiation with respect to z^a, z^b, etc., at $z=z_0$:

$$f_{ab}\equiv\left[\frac{\partial^2 f(z)}{\partial z^a\partial z^b}\right]_{z=z_0}\ ,$$

$$f_{abc}\equiv\left[\frac{\partial^3 f(z)}{\partial z^a\partial z^b\partial z^c}\right]_{z=z_0}\ ,\ \text{etc.}\ ,$$

plus a real scalar with mass μ for each nonzero eigenvalue μ^2 of the vector-boson mass^2 matrix:

$$\mu^2{}_{kl}=(z_0^\dagger\{t_k,t_l\}z_0)=2((t_kz_0)^\dagger(t_lz_0))\ . \tag{2.8}$$

[The second version of this formula follows from the first and Eq. (2.5).] The gauge symmetries of $f(z)$ [plus (2.4)] imply that

$$\sum_b f_{bc}(t_kz_0)^b=\sum_b M^2{}_{ab}(t_kz_0)^b=0\ . \tag{2.9}$$

These are the "Goldstone" eigenvectors of $M^2{}_{ab}$, for which the corresponding scalars are eliminated by the Higgs mechanism. We assume for reasons of naturalness that $f(z)$ depends on only a single grand-unified mass scale $M_{\rm GU}$ (presumably $M_{\rm GU}\sim10^{17}$ GeV), so that aside from coupling-constant factors, we have $z_0^a\sim M_{\rm GU}$, $f_{ab}\sim M_{\rm GU}$, $f_{abc}\sim1$, and the eigenvalues of $M^2{}_{ab}$ and $\mu^2{}_{kl}$ are either of order $M_{\rm GU}{}^2$ or zero.[18] We adopt a basis in which these matrices are diagonal, with the zero and nonzero eigenvalues of $\mu^2{}_{kl}$ labeled $\kappa,\lambda,\ldots$ and $K,L\ldots$, respectively (one nonzero eigenvalue for each linearly independent Goldstone vector t_kz_0), and the Goldstone, zero non-Goldstone, and nonzero eigenvalues of $M^2{}_{ab}$ labeled $K,L,\ldots$; $\alpha,\beta,\ldots$; and $A,B,\ldots$, respectively. That is

$$\mu^2{}_{KL}=\delta_{KL}\mu_K{}^2\ , \tag{2.10}$$

$$\mu^2{}_{\kappa L}=\mu^2{}_{\kappa\lambda}=0\ , \tag{2.11}$$

$$M^2{}_{AB}=\delta_{AB}M_A{}^2\ , \tag{2.12}$$

$$M^2{}_{\alpha B}=M^2{}_{\alpha\beta}=M^2{}_{\alpha K}=M^2{}_{KB}=M^2{}_{KL}=0\ , \tag{2.13}$$

with $\mu_K{}^2$ and $M_A{}^2$ nonzero and of order $M^2{}_{\rm GU}$. Correspondingly, the fields z^a are classified as follows:

z^α: light complex scalars, corresponding to non-Goldstone eigenvectors of $M^2{}_{ab}$ with eigenvalue zero (Higgs bosons, s-quarks, s-leptons).

z^A: superheavy complex scalars, corresponding to nonzero eigenvalues of $M^2{}_{ab}$.

z^K: superheavy real scalars, degenerate with superheavy gauge bosons, corresponding to independent Goldstone eigenvectors $(t_Kz_0)^a$ of $M^2{}_{ab}$, one for each nonzero eigenvalue of $\mu^2{}_{KL}$. (The z^K are real because the imaginary part of the coefficients of t_Kz_0 are Goldstone bosons, eliminated by the Higgs mechanism.)

Because z^α and z^A are orthogonal to z^K, we have

$$(t_Kz_0)^\alpha=(t_Kz_0)^A=0\ . \tag{2.14}$$

Also, because the z^L correspond to nonzero eigenvalues $\mu_L{}^2$ of $\mu^2{}_{kl}$, we have

$$(t_Kz_0)^L=\mu_L\delta_{KL}\ \text{nonsingular}\ . \tag{2.15}$$

Further, (2.9), (2.12), and (2.13) yield

$$f_{\alpha\beta}=f_{\alpha A}=f_{\alpha K}=f_{KA}=f_{KL}=0\ , \tag{2.16}$$

$$f_{AB}\ \text{nonsingular}\ \sim M_{\rm GU}\ . \tag{2.17}$$

We will not need to assume that $f(z)$ is a cubic polynomial, as it would be if we started with a renormalizable theory. Finally, our results will turn out to depend critically on the assumption that the light scalars do not get nonvanishing VEV's from the breakdown of the grand gauge group

$$z_0^\alpha=0\ . \tag{2.18}$$

This is an automatic consequence of symmetries like SU(3)×SU(2)×U(1) for Higgs bosons and scalar quarks and leptons, but may require fine tuning for light SU(3)×SU(2)×U(1)-neutral scalars. Also, it is an automatic consequence of the supersymmetry condition (2.5) that the scalar superpartners of the superheavy gauge bosons have zero VEV[18]:

$$z_0^K=0\ . \tag{2.19}$$

Some of the z_0^A may also vanish, but they are generally of order $M_{\rm GU}$.

B. Hidden sector

The superpotential $\tilde{f}(\tilde{z})$ is assumed to be proportional to a relatively small factor μ^3, but otherwise to depend only on $\tilde{z}^h$ and on a mass scale of order $M_{\rm PL}=1/\sqrt{G}$:

$$\tilde{f}(\tilde{z})=\mu^3\times\ \text{function of}\ \tilde{z}\sqrt{G}\ . \tag{2.20}$$

In the absence of the observable sector, the potential would take the form

$$\tilde{V}(\tilde{z})=\exp\left[8\pi G\sum_h|\tilde{z}^h|^2\right]\left[\sum_h\left|\frac{\partial\tilde{f}}{\partial\tilde{z}^h}+8\pi G\tilde{z}^{h*}\tilde{f}\right|^2-24\pi G|\tilde{f}|^2\right]. \tag{2.21}$$

We assume that there is at least a local minimum of $V(\tilde{z})$ at a point $\tilde{z}_0$, and that the additive constant in $\tilde{f}$ can be adjusted so that $\tilde{V}$ vanishes at this point

$$\tilde{V}(\tilde{z})=\partial\tilde{V}(\tilde{z})/\partial\tilde{z}^h=0 \text{ at } \tilde{z}=\tilde{z}_0 . \tag{2.22}$$

Since $\tilde{V}(\tilde{z})$ equals μ^6G times a function of $\tilde{z}\sqrt{G}$, this condition yields a μ-independent value of order $1/\sqrt{G}=M_{\rm PL}$ for $\tilde{z}_0^h$.

The supergravitational coupling between the hidden and observable sectors will introduce what appears as intrinsic supersymmetry-breaking terms in the effective Lagrangian of the observable sector. As we shall see below, the magnitude of these terms is characterized by a mass parameter

$$m_g=8\pi G\tilde{f}(\tilde{z}_0)\exp\left[4\pi G\sum_A|z_0^A|^2\right]\exp\left[4\pi G\sum_h|\tilde{z}_0^h|^2\right]. \tag{2.23}$$

It so happens that $|m_g|$ is the gravitino mass, but for us the important thing about m_g is that it sets the mass scale of particles like the $W^\pm$ and Z^0. We therefore assume that

$$|m_g|\ll M_{\rm GU} \text{ and } |m_g|\ll M_{\rm PL}\equiv 1/\sqrt{G} \tag{2.24}$$

and for orientation we may think of m_g as roughly of order 100 GeV.[19]

From now on we will use m_g rather than μ to characterize the smallness of the hidden-sector superpotential. That is, μ in (2.20) is taken to be of order $(m_g/G)^{1/3}$ (or 10^{13} GeV for $m_g\simeq 100$ GeV) so that $\tilde{f}$ is of order m_g/G, as required by (2.16). Of course, we do not at present know why μ should take this particular value, so for now m_g is simply a parameter put in by hand.

III. RESULTS

In order to characterize the breaking of supersymmetry in the observable sector in theories of the sort described in Sec. II, it seems to us most useful to calculate the complete effective potential of the light observable scalars z^α, from which we can obtain whatever information we want about scalar VEV's and masses. We do this by "integrating out" the heavy scalars z^A and z^K, expressing them as functions of z^α and $\tilde{z}^h$ by imposing the condition that

$$\partial V/\partial z^A=\partial V/\partial z^K=0 \text{ at } z^A=z^A(z^\alpha,\tilde{z}^h),\quad z^K=z^K(z^\alpha,\tilde{z}^h). \tag{3.1}$$

Leaving the hidden fields for the moment as free parameters, the effective potential of the light scalars is then

$$V_{\rm eff}(z^\alpha,\tilde{z}^h)=V(z^\alpha,z^A(z^\alpha,\tilde{z}^h),z^K(z^\alpha,\tilde{z}^h),\tilde{z}^h). \tag{3.2}$$

To render this calculation tractable, it is necessary at every point to use a power-series expansion in m_g. We take the light fields z^α to be of order m_g and the hidden fields $\tilde{z}^h$ to be of order $M_{\rm PL}$, because that is where experience teaches us to look for the minimum of V. The mass m_g also enters as the "smallness" parameter in $\tilde{f}$. Apart from m_g, the only masses in the problem are $M_{\rm GU}$ (perhaps 10^{17} GeV) and $M_{\rm PL}=1.2\times10^{19}$ GeV. These are not very different, so our expansion parameter will be taken as m_g/M, with $M_{\rm GU}$ and $M_{\rm PL}$ regarded as roughly of the same order of magnitude M. The expansion for the heavy fields then takes the form

$$z^A=z_0^A+z_1^A+z_2^A+\cdots,\quad z^K=z_0^K+z_1^K+z_2^K+\cdots, \tag{3.3}$$

with z_n^A and z_n^K of order $M(m_g/M)^n$. To repeat, M now stands for the grand-unification mass and/or the Planck mass.

The details of this calculation are presented in Appendix A. A crucial result is that the potential $V_{\rm eff}$ turns out to be independent of the light scalars z^α not only in orders M^4 and m_gM^3, but also in orders $m_g{}^2M^2$ and $m_g{}^3M$. It is therefore possible to choose a z^α-independent value of the hidden fields $\tilde{z}^h$ where the potential $V_{\rm eff}$ to this order is stationary in $\tilde{z}$, and adjust an additive constant in the superpotential $\tilde{f}$ to make the potential vanish to this order. The values of the hidden scalars and the additive constant in $\tilde{f}$ turn out to be just those that we would calculate according to Eq. (2.22) in the absence of the observable sector, plus small corrections of order m_g in $\tilde{z}^h$ and of order $m_g{}^2M$ in $\tilde{f}$. With $\tilde{z}^h$ and the constant term in $\tilde{f}$ fixed in this way, the leading terms in $V_{\rm eff}$ are of order $m_g{}^4$. As long as we are not interested in higher terms (of order $m_g{}^5/M$, etc.) it is an adequate approximation then to neglect the corrections to $\tilde{z}^h$ and to the constant term in $\tilde{f}$, and simply take them to have the values given by (2.22), which will be indicated with a subscript 0.

Our result for the $O(m_g{}^4)$ terms in $V_{\rm eff}$ can most conveniently be expressed in terms of a low-energy effective superpotential

$$f_{\rm eff}=f^{(3)}+f^{(2)}+f^{(1)}+f^{(0)} \,. \tag{3.4}$$

Here $f^{(3)}$ is proportional to the part of the original superpotential that is trilinear in *light* scalars

$$f^{(3)}\equiv\tfrac{1}{6}E_0{}^{1/2}\sum_{\alpha\beta\gamma} f_{\alpha\beta\gamma}z^{\alpha}z^{\beta}z^{\gamma} \tag{3.5}$$

while $f^{(2)}$, $f^{(1)}$, and $f^{(0)}$ result from the first-order shift in the heavy scalars

$$f^{(2)}\equiv\tfrac{1}{2}E_0{}^{1/2}\sum_{\alpha\beta A} f_{\alpha\beta A}z^{\alpha}z^{\beta}z_1^{A} \,, \tag{3.6}$$

$$f^{(1)}\equiv\tfrac{1}{2}E_0^{1/2}\sum_{\alpha AB} f_{\alpha AB}z^{\alpha}z_1^{A}z_1^{B} \,, \tag{3.7}$$

$$f^{(0)}\equiv\tfrac{1}{6}E_0{}^{1/2}\sum_{ABC} f_{ABC}z_1^{A}z_1^{B}z_1^{C} \,. \tag{3.8}$$

Also E_0 is the constant factor

$$E_0=\exp\left[8\pi G\sum_A |z_0^A|^2\right]\exp\left[8\pi G\sum_h |\tilde{z}_0^h|^2\right] \tag{3.9}$$

and z_1^A is the first-order shift in the heavy scalars

$$z_1^A=-\sum_B f^{-1}{}_{AB}z_0^{B*}m_g \,. \tag{3.10}$$

Note that f_{AB} and z_0^B are both of order $M_{\rm GU}$ and independent of G, so z_1^A is of order m_g, and otherwise independent of both $M_{\rm GU}$ and $M_{\rm PL}$, as well as of z^α. It turns out that $z_1^K=0$, so only z_1^A appears in (3.6)–(3.8).

Our main result is the formula for the $O(m_g{}^4)$ terms in the effective potential of the light scalars:

$$V_{\rm eff}=\sum_\alpha\left|\frac{\partial f_{\rm eff}}{\partial z^\alpha}\right|^2+2\,{\rm Re}(m_g'{}^*f^{(3)}) +4\,{\rm Re}(m_g^*f^{(2)})+2\,{\rm Re}[(4m_g-m_g')^*f^{(1)}] +|m_g|^2\sum_\alpha |z^\alpha|^2+\tfrac{1}{2}\sum_\kappa (z^\dagger t_\kappa z)^2+V_0 \,. \tag{3.11}$$

Here m_g is the gravitino mass (2.23), which we can write as

$$m_g=8\pi GE_0{}^{1/2}\tilde{f}_0 \tag{3.12}$$

and m_g' is a comparable mass parameter

$$m_g'=8\pi GE_0{}^{1/2}\left[\sum_h \tilde{z}_0^h\left[\frac{\partial\tilde{f}}{\partial\tilde{z}^h}\right]_0+8\pi G\tilde{f}_0\sum_h|\tilde{z}_0^h|^2\right] \tag{3.13}$$

while t_κ are the SU(3)×SU(2)×U(1) gauge generators, and V_0 is a constant of order $m_g{}^4$. (It is important to note that $f^{(2)}$ and $f^{(1)}$ are, respectively, proportional to m_g and $m_g{}^2$, and do not involve m_g'.) We can arrange to cancel the vacuum expectation value of $V_{\rm eff}$, including V_0 and all radiative corrections, by a shift in the hidden-sector superpotential $\tilde{f}$ by a constant term of order $m_g{}^3$. The first term in (3.11) is just what we would expect in a globally supersymmetric theory with superpotential $f_{\rm eff}$, while the other terms explicitly break supersymmetry.

Several features of our result are worth special mention:

(a) It is amazing how little we need to know in order to calculate the effective potential. All the unknown features of the hidden sector are embodied in just two complex mass parameters m_g and m_g', of comparable magnitude. Also, all aspects of the grand unified theory have been boiled down to the parameters in the effective superpotential. In particular, and somewhat surprisingly, there are no terms in $V_{\rm eff}$ of order $m_g{}^4(GM_{\rm GU}{}^2)^N$, so to order $m_g{}^4$ the effective potential does not even depend on the grand-unification mass scale $M_{\rm GU}$.

(b) Despite the fact that (3.11) does not depend on $M_{\rm GU}$, the supernormalizable terms in the effective potential that arise here from the shifts in the heavy scalars z^A are very different from those that would arise directly from linear and quadratic terms in the original superpotential in a theory without heavy scalars. In the latter case, the potential of the light scalars to order $m_g{}^4$ would be (as in Ref. 4)

$$V=\sum_\alpha\left|\frac{\partial f_{\rm eff}}{\partial z^\alpha}\right|^2+2\,{\rm Re}(m_g'{}^*f^{(3)})+2\,{\rm Re}[(m_g'{}^*-m_g^*)f^{(2)}]+2\,{\rm Re}[(m_g'{}^*-2m_g^*)f^{(1)}] +|m_g|^2\sum_\alpha|z^\alpha|^2+\tfrac{1}{2}\sum_\kappa(z_0^\dagger t_\kappa z_0)+V_0 \,. \tag{3.14}$$

[See Appendix C. Here $f_{\rm eff}$ is simply the constant $E_0{}^{1/2}$ times the original superpotential, with any terms of fourth or higher orders in the light fields deleted, and $f^{(n)}$ is defined as in Eqs. (3.5)—(3.8).] Comparison of (3.14) with (3.11) shows that these results are not the same, and cannot be brought into the same form by any redefinition of the constants m_g and m_g'.

(c) The bilinear and linear terms $f^{(2)}$ and $f^{(1)}$ in the potential (3.11), which distinguish our result from (3.14), may be of importance in developing realistic models.[20] The appearance of such terms in order $m_g{}^4$ in the effective superpotential is governed in part by the mechanism that is responsible for their nonappearance in order $m_g{}^3M_{\rm GU}$ or $m_g{}^2M_{\rm GU}{}^2$ in theories with a superheavy mass scale $M_{\rm GU}$. As indicated in Ref. 18, there are several possibilities for this mechanism. If the breakdown of the grand-unified gauge group leaves some scalars massless and with zero VEV because of an unbroken symmetry of the whole theory, then $f^{(1)}$ and $f^{(2)}$ will not appear even in order $m_g{}^4$, and there will be no difference between (3.11) and (3.14). If these scalar masses and VEV's vanish because of a fine tuning of the theory then $f^{(1)}$ and $f^{(2)}$ terms will in general appear in order $m_g{}^4$, but they will be unstable to tiny changes in the fine tuning. We wish to stress that it is also possible for scalar masses and VEV's to vanish in the limit $m_g\to 0$ automatically, but not because of an unbroken symmetry of the whole theory, and in this case we generally expect $f^{(1)}$ and $f^{(2)}$ terms to arise naturally in order $m_g{}^4$. One way that this can happen is for masses and VEV's of the light scalars to be kept zero in the limit $m_g\to 0$ by an R symmetry of the whole theory,[14] which is spontaneously broken in the hidden sector. The news of R-invariance breaking would then be carried to observable fields by supergravity. For instance, suppose that the chiral superfields of the observable sector comprise a set Y^n with $R=0$ plus one X with $R=2$. The observable-sector superpotential then must take the form

$$f(X,Y)=Xg(Y)\ .$$

The conditions for a supersymmetric vacuum solution are then

$$g(y)=0\ ,\quad x\,\partial g(y)/\partial y^n=0\ (\text{all } n)$$

with lower-case letters denoting scalar components of superfields. It is natural to expect that there should be a nonzero scalar field value y_0^n at which $g(y)$ vanishes, but with $\partial g(y)/\partial y^n\neq 0$ for at least some y^n, and in this case there is a supersymmetric vacuum solution with

$$y=y_0\ ;\quad x=0\ .$$

By a linear transformation we can choose the Y^n fields so that $\partial g(y)/\partial y^n$ is nonzero at y_0 for only one of the scalars, say y^1. That is

$$[\partial g(y)/\partial y^1]_0=M\ ,$$

$$[\partial g(y)/\partial y^\alpha]_0=0\ ,$$

with α running over values of $n>1$, and M nonzero and of order $M_{\rm GU}$. The y^α are the light scalars whose masses vanish for $m_g\to 0$. With suitable additional symmetries, it can also be natural for their VEV's y_0^α to vanish, while $y_0{}^1$ is nonzero and of order $M_{\rm GU}$. The matrix f_{AB} of second derivatives of the superpotential with respect to heavy scalars then has elements

$$(\partial^2 f/\partial x\,\partial x)_0=(\partial^2 f/\partial y^1\partial y^1)_0=0\ ,$$

$$(\partial^2 f/\partial y^1\partial x)_0=M\ .$$

Equation (3.10) then gives the shift in the heavy scalars x and y^1 as

$$x_1=-M^{-1}y_0{}^{1*}m_g=O(m_g)\ ,$$

$$y_1{}^1=0\ ,$$

and (3.4)—(3.8) give the effective low-energy superpotential as

$$f_{\rm eff}=\tfrac{1}{2}E_0{}^{1/2}x_1\sum_{\alpha\beta}\left[\frac{\partial^2 g(y)}{\partial y^\alpha\partial y^\beta}\right]_0 y^\alpha y^\beta\ .$$

This means that we *can* encounter bilinear mass terms in the low-energy effective superpotential without fine tuning. In a realistic model the y^α would be the Higgs doublets; we would also have to add quark and lepton superfields with $R=+1$ and perhaps additional light singlets with $R=2$.

(d) The only mass scale appearing in (3.11) is m_g (recall that z_1^A and m_g' are of order m_g) so apart from coupling-constant factors, all light scalar masses and VEV's will be of order m_g. It is for this reason that we have considered the potential for z^α values of order m_g, and have taken m_g to be of order m_W.

(e) It was crucial in the calculation of $V_{\rm eff}$ that the terms of order $m_g{}^2M^2$ and $m_g{}^3M$ turned out to be independent of light scalars, and could therefore be eliminated by an adjustment of the additive constant in the hidden-sector superpotential. The third-order terms are larger than those of order $m_g{}^4$ by a factor $M_{\rm GU}/m_g\approx 10^{15}$, so even very tiny z^α-dependent corrections to these terms could completely invalidate our results for $V_{\rm eff}$. Our calculation here shows that there are no z^α-dependent corrections of higher order in $GM_{\rm GU}{}^2$ to the $m_g{}^2$ and $m_g{}^3$ terms in the tree approximation, but it is necessary also to check both ordinary and gravitational radiative corrections to at least fifth order in α and $GM_{\rm GU}{}^2$. We have not done this, but in Appendix B we analyze what properties of a general potential are needed in order that the leading terms that depend on light scalars should be of fourth order in a perturbation. We anticipate that the no-renormalization theorems of supersymmetry can be used to show that the terms in the potential due to radiative corrections actually have these properties in theories with a natural hierarchy.[21]

Even accepting that there are no z^α-dependent radiative corrections to $V_{\rm eff}$ of order $m_g{}^2$ and $m_g{}^3$, there certainly are such corrections in order $m_g{}^4$. If we were to use $V_{\rm eff}$ to carry out calculations of quantities measured at energies of order m_g, these radiative corrections would be of order $\alpha\ln(M_{\rm GU}/m_g)$, and so could not be considered small. Instead we must interpret our results as giving the effective potential for energies of order $M_{\rm GU}$, and use (3.11) as the input to a renormalization-group calculation that would integrate the equations for the parameters in $V_{\rm eff}$ down to energies of order m_g, and only then use the results as our low-energy effective potential.[5,7,10]

IV. GENERAL KÄHLER POTENTIALS

Up to now, our results have been based on Eq. (2.2) for the potential, corresponding to a flat Kähler metric. In general, the potential would be given by a formula[17]

$$V=\exp(8\pi Gd)\left[\sum_{NM}(g^{-1})_M^N\left[\frac{\partial f_{\rm TOT}}{\partial Z^N}+8\pi G\frac{\partial d}{\partial Z^N}f_{\rm TOT}\right]\left[\frac{\partial f_{\rm TOT}}{\partial Z^M}+8\pi G\frac{\partial d}{\partial Z^M}f_{\rm TOT}\right]^* -24\pi G\,|f_{\rm TOT}|^2\right]+\text{ gauge terms },\qquad(4.1)$$

where Z^N here runs over all chiral scalars z^a, $\tilde{z}^h$, and

$$g_N^M\equiv\frac{\partial^2 d}{\partial Z^N\partial Z^{M*}}\ ,\qquad(4.2)$$

where d, the Kähler potential, is a function of both Z^N and Z^{N*}, while $f_{\rm TOT}=f+\tilde{f}$ is still a function of Z^N alone. Equation (4.1) reduces to (2.2) in the special case

$$d=\sum_N|Z^N|^2\ .\qquad(4.3)$$

It does not seem reasonable to expect that the Kähler potential will oblige us by taking a form as simple as (4.3). For one thing, this is not what we find if we start with a renormalizable theory of chiral superfields and then turn on supergravity; the Weyl rescaling that is necessary in this case yields

$$d=-\frac{3}{8\pi G}\ln\left[1-\frac{8\pi G}{3}\sum_N|Z^N|^2\right]\ .\qquad(4.4)$$

(This is the case $\phi=-3+8\pi G\sum|Z|^2$ in the notation of Cremmer *et al.*[17]) Another argument against (4.3) arises from the presence of gravitational radiative corrections, which could not be expected to preserve a simple formula like (4.3).

On the other hand, if we do not limit the form of the Kähler potential in any way, we can derive hardly any conclusions from (4.1). We may in the end be driven to such a pessimistic conclusion, but for the present it seems reasonable at least to explore the possibility that d belongs to a class of functions that is wide enough to be plausible and yet narrow enough to allow us to draw interesting conclusions.

We shall assume here that the Kähler potential takes the form

$$8\pi Gd(Z,Z^*)=P\left[8\pi G\sum_N|Z^N|^2\right]\ ,\qquad(4.5)$$

where $P(u)$ is a power series with coefficients of order unity. This includes (4.3) and (4.4) as special cases. Also, it is reasonable to expect that gravitational radiation corrections will at least approximately respect the form of (4.5), because in the absence of a superpotential these corrections possess a U(n) symmetry among the n chiral superfields[22] that would require the Kähler potential to take the form (4.5). We would expect any violations of this U(n) symmetry in $d(u)$ due to gravitational radiative corrections to be suppressed by whatever small factors ($m_g/M_{\rm PL}$ or Yukawa couplings) appear in the superpotential.

From (4.5), we obtain the Kähler metric

$$g_N^M=P'(u)\delta_N^M+8\pi GP''(u)Z^{N*}Z^M\ ,\qquad(4.6)$$

where

$$u\equiv 8\pi G\sum_N|Z^N|^2\ .\qquad(4.7)$$

This has inverse

$$(g^{-1})^M_N = P'^{-1}(u)\delta^M_N + 8\pi G Q(u) Z^{N*} Z^M , \quad (4.8)$$

where

$$Q(u) = -P''(u)/[P'^2(u) + uP'(u)P''(u)] . \quad (4.9)$$

Using (4.8) in (4.1) yields

$$V = e^{P(u)} \left[P'^{-1}(u) \sum_N \left| \frac{\partial f_{\rm TOT}}{\partial Z^N} + 8\pi G P'(u) Z^{N*} f_{\rm TOT} \right|^2 + Q(u) \left| \sum_N Z^N \frac{\partial f_{\rm TOT}}{\partial Z^N} + uP'(u) f_{\rm TOT} \right|^2 - 24\pi G \, | f_{\rm TOT} |^2 \right] + \text{gauge terms} . \quad (4.10)$$

For a superpotential of the form (2.1), Eq. (4.10) takes the form

$$V = e^{P(u)} \left[P'^{-1}(u) \sum_a \left| \frac{\partial f}{\partial z^a} + 8\pi G P'(u) z^{a*} (f + \tilde f) \right|^2 + P'^{-1}(u) \sum_h \left| \frac{\partial \tilde f}{\partial \tilde z^h} + 8\pi G P'(u) \tilde z^{h*} (f + \tilde f) \right|^2 + Q(u) \left| \sum_a z^a \frac{\partial f}{\partial z^a} + \sum_h \tilde z^h \frac{\partial \tilde f}{\partial \tilde z^h} + uP'(u)(f + \tilde f) \right|^2 - 24\pi G \, | f + \tilde f |^2 \right] + \text{gauge terms} , \quad (4.11)$$

where now

$$u = 8\pi G \left[\sum_a | z^a |^2 + \sum_h | \tilde z^h |^2 \right] . \quad (4.12)$$

We follow the same procedure as in Sec. III, integrating out the heavy scalars by setting them at values where V is stationary with respect to them. Again, we find that terms of order $m_g{}^2 M^2$ and $m_g{}^3 M$ are independent of light scalars, and can be made to be stationary with respect to the hidden-sector scalars and vanish by adjusting the value of the hidden-sector scalars and the additive constant in $\tilde f$. By a lengthy calculation just like that of Appendix A, the terms of order $m_g{}^4$ are found to take the form

$$V_{\rm eff} = \sum_\alpha \left| \frac{\partial f_{\rm eff}}{\partial z^\alpha} \right|^2 + 2\,{\rm Re}(m_g'^* f^{(3)}) + 4\,{\rm Re}(m_g^* f^{(2)}) + 2\,{\rm Re}[(4m_g - m_g'^*) f^{(1)}] + | m_g'' |^2 \sum_\alpha | z^\alpha |^2 + V_0 + \text{gauge terms} . \quad (4.13)$$

Here $f_{\rm eff}$ is an effective superpotential and $f^{(3)}$, $f^{(2)}$, and $f^{(1)}$ are its trilinear, bilinear, and linear parts, given by Eqs. (A20) or (3.4)—(3.8), but with E_0 replaced with e^P at $z^a=0$, $\tilde z^h = \tilde z^h_0$. (An additional factor $1/P'$ would appear here, but we absorb it into the normalization of z^a in order to avoid P' factors in the kinematic and gauge parts of the Lagrangian.) The first-order shift z_1^A in the heavy scalars is given here by

$$z_1^A = -m_g''' \sum_B f^{-1}{}_{AB} z_0^{B*} \quad (4.14)$$

so that $f^{(2)}$ and $f^{(1)}$ are proportional to m_g''', and $(m_g''')^2$, respectively. The constants m_g, m_g', m_g'', and m_g''' are given by complicated formulas in terms of the hidden-sector superpotential and P and its derivatives at $z^a=0$, $\tilde z^h = \tilde z^h_0$, but they are all of the same order of magnitude, roughly that of the gravitino mass. The only substantial difference between these results and those of Sec. III is that the properties of the hidden sector are now represented by four independent mass parameters m_g, m_g', m_g'', and m_g''', rather than just m_g and m_g'.

V. APPLICATIONS

We now take up some examples. Much of this analysis is already present in the articles of Refs. 3—11; we go into it here in order to illustrate the use of our results when the mass scale in the low-energy effective superpotential arises from a more fundamental theory involving superheavy particles about which nothing is explicitly known.

Consider an SU(2)×U(1) low-energy effective

gauge theory with a pair of doublet Higgs left-chiral superfields:

$$H=(H^0,H^-)\ ,\quad H'=(H'^+,H'^0)\ . \tag{5.1}$$

Additional chiral superfields will be added later. The most general effective superpotential is

$$f_{\rm eff}=\tilde{m}_g(H^T\epsilon H')\ , \tag{5.2}$$

where ϵ is the usual antisymmetric 2×2 matrix, and $\tilde{m}_g$ is a coefficient of the order of the gravitino mass, given by (3.6) and (4.14) as

$$\tilde{m}_g=-\tfrac{1}{2}m_g'''e^{P_0/2}\sum_{AB}f^{-1}{}_{AB}z_0^{A*}f_{BHH'}\ . \tag{5.3}$$

We know almost nothing about m_g''', which depends on the hidden-sector superpotential and the Kähler potential, or about the quantities appearing in the sums over heavy scalars, which depend on the underlying grand-unified model. Never mind—all these uncertainties appear here only in the value of a single unknown complex constant $\tilde{m}_g$, which will have to be taken from experiment.

From (5.2) and (4.13), we obtain the effective potential

$$\begin{aligned}V_{\rm eff}=&(\,|\tilde{m}_g|^2+|m_g''|^2)(\mathscr{H}^\dagger\mathscr{H}+\mathscr{H}^\dagger\mathscr{H}')\\&+4\,{\rm Re}(m_g^*\tilde{m}_g\mathscr{H}^T\epsilon\mathscr{H}')\\&+\tfrac{1}{2}g^2(\mathscr{H}^\dagger\vec{t}\mathscr{H}+\mathscr{H}^\dagger\vec{t}\mathscr{H}')^2\\&+\tfrac{1}{2}g'^2(\tfrac{1}{2}\mathscr{H}^\dagger\mathscr{H}-\tfrac{1}{2}\mathscr{H}'^\dagger\mathscr{H}')^2\ .\end{aligned} \tag{5.4}$$

We here use script letters for the (first) scalar components of left-chiral superfields; $\vec{t}$ denotes the electroweak isospin generator, and g and g' are the usual gauge-coupling constants.

This cannot yield a satisfactory picture of $SU(2)\times U(1)$ breaking. For charge-conserving scalar VEV's, both the gauge and $\mathscr{H}^0\mathscr{H}'^0$ terms in (5.4) are minimized on the surface of constant $|\langle\mathscr{H}^0\rangle|^2+|\langle\mathscr{H}'^0\rangle|^2$ along the direction

$$\langle\mathscr{H}^0\rangle=-e^{i\alpha}\langle\mathscr{H}'^0\rangle^* \tag{5.5}$$

with phase α chosen to minimize the $\mathscr{H}^0\mathscr{H}'^0$ term

$$\alpha={\rm Arg}(m_g\tilde{m}_g^*)\ . \tag{5.6}$$

Along this direction, the effective potential is a quadratic

$$V_{\rm eff}=2(\,|\tilde{m}_g|^2+|m_g''|^2-2|m_g|\ |\tilde{m}_g|\,)\,|\mathscr{H}^0|^2\ . \tag{5.7}$$

We see that $SU(2)\times U(1)$ is unbroken if

$$|\tilde{m}_g|^2+|m_g''|^2>2|m_g|\ |\tilde{m}_g| \tag{5.8}$$

and otherwise it can be broken only at a scale very much greater than m_g, where nonperturbative effects may halt the decrease of $V_{\rm eff}$. It is easy to show that this undesired conclusion obtains also when we add quark and lepton superfluids, or include arbitrary numbers of Higgs doublet superfields.

In Refs. 5, 7, and 10 it is noted that the symmetry between $\mathscr{H}^0$ and $\mathscr{H}^{0'}$ that is responsible for the unsatisfactory features of this model is actually broken by the different Yukawa couplings of $\mathscr{H}$ and $\mathscr{H}''$ to quarks and leptons, which enter in the renormalization-group equations used to integrate the parameters in $V_{\rm eff}$ down from grand-unification energies to ordinary energies. However, as they point out, this solves the problem only if there exist some extraordinarily heavy quarks or leptons, with masses above about 100 GeV.

An alternative possibility that has been explored by most of the authors of Refs. 3—11 is to include in the low-energy theory an $SU(3)\times SU(2)\times U(1)$-neutral left-chiral superfield J which allows trilinear terms $JH^T\epsilon H'$ in the superpotential. Usually J is identified as the "sliding singlet," needed to keep the Higgs doublets from getting very large masses like its SU(5) partners. This runs into severe difficulties, either through J mixing with the hidden sector or through its scalar component picking up a large VEV, either of which would wreck the hierarchy of mass scales. However, for us J is simply one more chiral superfield that happens like H and H' to remain massless (and with zero VEV) in the breakdown of some grand-unified symmetry, for reasons into which we do not here inquire.[23]

With J included, the most general effective superpotential is

$$\begin{aligned}f_{\rm eff}=&\,\tilde{m}_g^{(1)}(H^T\epsilon H')+(\tilde{m}_g^{(2)})^2J+\tilde{m}_g^{(3)}J^2\\&+\lambda(H^T\epsilon H')J+\lambda'J^3\ .\end{aligned} \tag{5.9}$$

We would here have to regard $\tilde{m}_g^{(n)}$ as three mass parameters of the order of the gravitino mass, which are given by formulas like (5.3), but which for practical purposes must be regarded as unknown. The constants λ and λ' are to be taken directly from the trilinear terms in the superpotential of the grand-unified theory, but for our present purposes are also just unknown dimensionless coupling constants.

The effective potential for (5.9) is

$$\begin{aligned}V_{\text{eff}}= & \,|\tilde{m}_g^{(1)}+\lambda\mathscr{J}|^2(\mathscr{H}^\dagger\mathscr{H}+\mathscr{H}'^\dagger\mathscr{H}')+|(\tilde{m}_g^{(2)})^2+2\tilde{m}_g^{(3)}\mathscr{J}+3\lambda'\mathscr{J}^2+\lambda(\mathscr{H}^T\epsilon\mathscr{H}')|^2 \\ & +2\,\mathrm{Re}[m_g'^{*}(\lambda\mathscr{H}^T\epsilon\mathscr{H}'\mathscr{J}+\lambda'\mathscr{J}^3)] \\ & +4\,\mathrm{Re}[m_g^*(\tilde{m}_g^{(1)}\mathscr{H}^T\epsilon\mathscr{H}'+\tilde{m}_g^{(3)}\mathscr{J}^2)]+2\,\mathrm{Re}[(4m_g-m_g')^*(\tilde{m}_g^{(2)})^2\mathscr{J}] \\ & +|m_g''|^2(|\mathscr{J}|^2+\mathscr{H}^\dagger\mathscr{H}+\mathscr{H}'^\dagger\mathscr{H}') \\ & +\tfrac{1}{2}g^2(\mathscr{H}^\dagger\vec{t}\,\mathscr{H}+\mathscr{H}'^\dagger\vec{t}\,\mathscr{H}')^2+\tfrac{1}{2}g'^2(\tfrac{1}{2}\mathscr{H}^\dagger\mathscr{H}-\tfrac{1}{2}\mathscr{H}'^\dagger\mathscr{H}')^2\,. \end{aligned} \tag{5.10}$$

For charge-conserving scalar VEV's, the minimum of (5.10) is again in the direction (5.5) (but with different phase α). It is well known that (5.10) has an $SU(2)\times U(1)$-breaking minimum along this direction with $\langle\mathscr{H}^0\rangle\neq0$ for a variety of special cases. For instance, if all terms in (5.9) are absent except $\lambda(H^T\epsilon H')J$, then (5.10) has an absolute minimum along the direction (5.5) with $\langle\mathscr{H}^0\rangle\neq0$, provided that

$$|m_g'|>3|m_g''|\,.$$

We will not specialize by choosing any specific values for the parameters in (5.9), but will just assume that they fall in the range where $SU(2)\times U(1)$ is broken, and consider those consequences of (5.10) that do not depend on the values of the parameters in this range.

Charged scalars: Inspection of (5.10) shows that the mass matrix of the charge-1 scalar boson has diagonal elements

$$\langle\mathscr{H}^-|M^2|\mathscr{H}^-\rangle=\langle\mathscr{H}'^{+*}|M^2|\mathscr{H}'^{+*}\rangle=\tfrac{1}{2}(m_W{}^2+\Delta^2) \tag{5.11}$$

with

$$\Delta^2/2=|\tilde{m}_g^{(1)}+\lambda\langle\mathscr{J}\rangle|^2+|m_g''|^2>0\,. \tag{5.12}$$

By the Goldstone theorem or direct calculation, we then also have

$$\langle\mathscr{H}^-|M^2|\mathscr{H}'^{+*}\rangle=\langle\mathscr{H}'^{+*}|M^2|\mathscr{H}^-\rangle=\tfrac{1}{2}(m_W{}^2+\Delta^2)e^{i\alpha}\,. \tag{5.13}$$

The eigenvalues are then 0, corresponding to the Goldstone boson eliminated by the Higgs mechanism, together with

$$m_+{}^2=m_W{}^2+\Delta^2\,. \tag{5.14}$$

Thus there is a physical charged Higgs boson heavier than the W.

Neutral scalars: There are six real scalar fields here, of which one real field is eliminated by the Higgs mechanism, leaving five real physical neutral scalars. The complete mass spectrum is quite complicated, but one of the masses is easily calculated by using the symmetry of (5.10) with $\mathscr{H}^-=\mathscr{H}'^+=0$ under the interchange of $\mathscr{H}^0$ and $\mathscr{H}'^0$. By a U(1) gauge transformation we can always choose the phase of $\langle\mathscr{H}^0\rangle$ to be $(\pi+\alpha)/2$, so that (5.5) gives $\mathscr{H}^0$ and $\mathscr{H}'^0$ equal VEV's, thus preserving this symmetry. The scalars of definite mass can therefore be classified as even or odd under the symmetry $\mathscr{H}^0\leftrightarrow\mathscr{H}'^0$: four real scalars are even, and two are odd. The Goldstone boson eliminated by the Higgs mechanism is odd (because $\mathscr{H}^0$ and $\mathscr{H}'^0$ have opposite t_3 and weak hypercharge) so there is just one physical odd neutral scalar, which does not mix with any of the other neutral scalars. Its mass is easily calculated to be

$$m_{0,\text{odd}}{}^2=m_Z{}^2+\Delta^2\,. \tag{5.15}$$

This scalar is heavier than the Z^0, and by the same amount (counting squared masses) as the charged Higgs boson is heavier than the W.

S-quarks and s-leptons: In order to account for the quark and lepton masses, we must add terms in the superpotential of the form

$$(m_u/\langle\mathscr{H}^{0\prime}\rangle)(Q_L^T\epsilon H_L')U_R^*+(m_d/\langle\mathscr{H}^0\rangle)(Q_L^T\epsilon H_L)D_R^*+(m_e/\langle\mathscr{H}^0\rangle)(L_L^T\epsilon H_L)E_R^*\,, \tag{5.16}$$

where $Q_L\equiv\{U_L,D_L\}$ and $L_L\equiv\{N_L,E_L\}$ are left-chiral quark and lepton doublets; U_R^*, D_R^*, and E_R^* are left-chiral antiquark and antilepton singlets; and we assume one generation for notational simplicity. With vanishing VEV's for the scalar counterparts of the quarks and leptons[20] (s-quarks and s-leptons) there is no change in our previous discussion of Higgs and singlet masses and VEV's. Setting the neutral scalars equal to their VEV's, the terms in the effective potential that are quadratic in the s-quarks and s-leptons are

$$\begin{aligned}& m_u{}^2(|\mathscr{U}_L|^2+|\mathscr{U}_R|^2)+m_d^2(|\mathfrak{D}_L|^2+|\mathfrak{D}_R|^2)+m_e{}^2(|\mathscr{E}_L|^2+|\mathscr{E}_R|^2) \\ & +2\,\mathrm{Re}\{[m_g'-e^{i\alpha}(\tilde{m}_g^{(1)}+\lambda\langle\mathscr{J}\rangle)]^*[m_u\mathscr{U}_L\mathscr{U}_R^*+m_d\mathfrak{D}_L\mathfrak{D}_R^*+m_e\mathscr{E}_L\mathscr{E}_R^*]\} \\ & +|m_g''|^2(|\mathscr{U}_L|^2+|\mathscr{U}_R|^2+|\mathfrak{D}_L|^2+|\mathfrak{D}_R|^2+|\mathscr{E}_L|^2+|\mathscr{E}_R|^2+|\mathscr{N}_L|^2+|\mathscr{N}_R|^2)\,. \end{aligned} \tag{5.17}$$

The up—s-quark masses are then

$$m^2_{\mathscr{U}\pm} = |m_g''|^2 + m_u^2 \pm m_u \,|m_g' - e^{i\alpha}(\widetilde{m}_g^{(1)} + \lambda\langle \mathscr{J} \rangle)| \quad (5.18)$$

and likewise for the down s-quarks and s-electrons, while the s-neutrinos have mass

$$m_{\mathscr{N}}^2 = |m_g''|^2 . \quad (5.19)$$

We note that for small quark and lepton masses, the s-quarks and s-leptons are nearly degenerate, and in any case the average mass^2 of each s-quark or s-lepton pair exceeds the corresponding quark or lepton mass^2 by the same amount, an amount *less* than the difference Δ^2 of $\mathscr{H}^\pm$ and $W^\pm$ masses^2.

Most of these results [except perhaps for Eq. (5.18)] have been obtained before in more specific models.[3–11] Our derivation here serves to emphasize that these results apply independently of the parameters of the low-energy superpotential or the details of the grand unified theory or even the details of the Kähler potential.

ACKNOWLEDGMENTS

We are grateful for valuable discussions with R. Arnowitt, M. K. Gaillard, P. Nath, H. P. Nilles, J. Polchinski, M. Wise, and B. Zumino. The research of L. H. was supported by a Miller Fellowship, while that of J. L. and S. W. was supported in part by the Robert A. Welch Foundation.

APPENDIX A: CALCULATION OF THE EFFECTIVE POTENTIAL

Under the assumptions and in the notation of Sec. II, the potential is

$$V(z,\widetilde{z}) = \exp\left[8\pi G\left(\sum_a |z^a|^2 + \sum_h |\widetilde{z}^h|^2\right)\right] \times \left[\sum_a |F_a|^2 + \sum_h |\widetilde{F}_h|^2 - 24\pi G\,|f+\widetilde{f}|^2\right] + \tfrac{1}{2}\sum_k D_k^2 , \quad (A1)$$

with

$$F_a \equiv \frac{\partial f}{\partial z^a} + 8\pi G(f+\widetilde{f})z^{a*} , \quad (A2)$$

$$\widetilde{F}_h \equiv \frac{\partial \widetilde{f}}{\partial \widetilde{z}^h} + 8\pi G(f+\widetilde{f})\widetilde{z}^{h*} , \quad (A3)$$

$$D_k \equiv \sum_{a,b} z^{a*} z^b (t_k)^a_b , \quad (A4)$$

where f and $\widetilde{f}$ are the superpotentials of the observable and hidden sectors; z^a and $\widetilde{z}^h$ are the complex scalar fields on which they, respectively, depend; and t_k are the gauge generator matrices. The scalar indices $a,b,\ldots$ run over values $A,B,\ldots$ labeling complex superheavy chiral scalars; $\alpha,\beta,\ldots$ labeling complex light scalars; and $K,L,\ldots$ labeling real scalars that would be degenerate with the superheavy gauge bosons in the limit $\widetilde{f}\to 0$. Also, the gauge indices $k,l,\ldots$ run over values $K,L,\ldots$ labeling superheavy gauge bosons, and values $\kappa,\lambda,\ldots$ labeling gauge bosons [of $SU(3)\times SU(2)\times U(1)$] that do not get masses from the breakdown of the grand-unified gauge group. These different index values are distinguished by the conditions

$$f_{\alpha\beta} = f_{\alpha A} = f_{\alpha K} = f_{KA} = f_{KL} = 0 , \quad (A5)$$

$$f_{AB} \text{ nonsingular} , \quad (A6)$$

$$(t_K z_0)^\alpha = (t_K z_0)^A = 0 , \quad (A7)$$

$$(t_K z_0)^L = \mu_{KL} \text{ nonsingular} , \quad (A8)$$

$$z_0^\alpha = z_0^K = 0 , \quad (A9)$$

where z_0 is the stationary point of $f(z)$. Recall also that $f_{abc\cdots}$ denotes the partial derivative of $f(z)$ with respect to $z^a, z^b, z^c, \ldots$ at $z = z_0$.

We will write the observable scalar fields as

$$z^a = z_0^a + \phi^a \quad (A10)$$

and take ϕ^a to be like $\widetilde{f}$ of order m_g. We are interested in calculating the potential to fourth order in m_g.

For heavy scalars, the leading term in F_A is of order m_g (recall that f as well as $\partial f/\partial z^a$ vanishes at $z=z_0$), so we need terms in F_A up to order $m_g{}^3$. Grouping terms by order in m_g, we have to third order

$$F_A \simeq \left[\sum_B f_{AB}\phi^B + 8\pi G z_0^{A*}\widetilde{f}\right] + \left[\tfrac{1}{2}\sum_{ab} f_{Aab}\phi^a\phi^b + 4\pi G z_0^{A*}\sum_{BC} f_{BC}\phi^B\phi^C + 8\pi G\phi^{A*}\widetilde{f}\right] + \left[\tfrac{1}{6}\sum_{abc} f_{Aabc}\phi^a\phi^b\phi^c + \tfrac{1}{6}8\pi G z_0^{A*}\sum_{abc} f_{abc}\phi^a\phi^b\phi^c + 4\pi G\phi^{A*}\sum_{BC} f_{BC}\phi^B\phi^C\right] . \quad (A11)$$

On the other hand, for light scalars and scalars degenerate with superheavy gauge bosons the leading terms in F_α and F_K are of order $m_g{}^2$, so we need only keep these terms:

$$F_\alpha \simeq \tfrac{1}{2}\sum_{ab} f_{\alpha ab}\phi^a\phi^b + 8\pi G\phi^{\alpha *}\tilde{f}\ , \tag{A12}$$

$$F_K \simeq \tfrac{1}{2}\sum_{ab} f_{Kab}\phi^a\phi^b + 8\pi G\phi^{K*}\tilde{f}\ . \tag{A13}$$

Similarly, for the hidden sector the leading terms in $\tilde{F}_h$ are of first order in m_g, so here we need to keep terms up to third order

$$\tilde{F}_h \simeq \left[\frac{\partial\tilde{f}}{\partial\tilde{z}^h} + 8\pi G\tilde{z}^{h*}\tilde{f}\right] + \left[4\pi G\tilde{z}^{h*}\sum_{AB} f_{AB}\phi^A\phi^B\right] + \left[\tfrac{1}{6}\times 8\pi G\tilde{z}^{h*}\sum_{abc} f_{abc}\phi^a\phi^b\phi^c\right]\ . \tag{A14}$$

Also, for superheavy gauge bosons there are terms in D_K of first and second order

$$D_K \simeq 2\sum_L \mu_{KL}\phi^L + (\phi^\dagger t_K\phi) \tag{A15}$$

while for light gauge bosons D_κ is entirely of second order in m_g

$$D_\kappa = (\phi^\dagger t_\kappa \phi)\ . \tag{A16}$$

Finally, the leading term in $f+\tilde{f}$ is of first order in m_g, so we need to keep terms up to third order

$$f+\tilde{f} = [\tilde{f}] + \left[\tfrac{1}{2}\sum_{AB}\phi^A\phi^B\right] + \left[\tfrac{1}{6}\sum_{abc} f_{abc}\phi^a\phi^b\phi^c\right]\ . \tag{A17}$$

Using these approximations in (A1), and discarding terms of fifth or sixth order in m_g, we find for the terms in V of order $m_g{}^2M^2$, $m_g{}^3M$, and $m_g{}^4$ the following expressions:

$$V_2 = N\tilde{V} + E\sum_A \left|\sum_B f_{AB}\phi^B + 8\pi G z_0^{A*}\tilde{f}\right|^2 + 2\sum_K\left[\sum_L \mu_{KL}\phi^L\right]^2 , \tag{A18}$$

$$\begin{aligned}
V_3 = {} & 16\pi G\,\mathrm{Re}\left[\sum_A z_0^{A*}\phi^A\right]V_2 + 2E\,\mathrm{Re}\left[\sum_A\left[\sum_B f_{AB}\phi^B + 8\pi G z_0^{A*}\tilde{f}\right]\right.\\
& \left.\times\left[\tfrac{1}{2}\sum_{ab} f_{Aab}\phi^a\phi^b + 4\pi G z_0^{A*}\sum_{BC} f_{BC}\phi^B\phi^C + 8\pi G\phi^{A*}\tilde{f}\right]\right]\\
& + 8\pi G E\,\mathrm{Re}\left[\sum_h\left[\tilde{z}^h\frac{\partial\tilde{f}}{\partial\tilde{z}^h} + 8\pi G\,|\tilde{z}^h|^2\tilde{f}\right]^*\sum_{AB} f_{AB}\phi^A\phi^B\right]\\
& - 24\pi G E\,\mathrm{Re}\left[\tilde{f}^*\sum_{AB} f_{AB}\phi^A\phi^B\right] + 2\sum_{KL}(\phi^\dagger t_K\phi)\mu_{KL}\phi^L\ ,
\end{aligned} \tag{A19}$$

$$\begin{aligned}
V_4 = {} & \left[-128\pi^2G^2\left[\mathrm{Re}\sum_A z_0^{A*}\phi^A\right]^2 + 8\pi G\sum_a|\phi^a|^2\right]V_2 + 16\pi G\,\mathrm{Re}\left[\sum_A z_0^{A*}\phi^A\right]V_3\\
& + E\sum_A\left|\tfrac{1}{2}\sum_{ab} f_{Aab}\phi^a\phi^b + 4\pi G z_0^{A*}\sum_{BC} f_{BC}\phi^B\phi^C + 8\pi G\tilde{f}\phi^{A*}\right|^2\\
& + 2E\,\mathrm{Re}\left[\sum_A\left[\sum_B f_{AB}\phi^B + 8\pi G z^{A*}\tilde{f}\right]\left[8\pi G z_0^{A*}\tfrac{1}{6}\sum_{abc} f_{abc}\phi^a\phi^b\phi^c + 4\pi G\phi^{A*}\sum_{AB} f_{AB}\phi^A\phi^B + \tfrac{1}{6}\sum_{abc} f_{Aabc}\phi^a\phi^b\phi^c\right]^*\right]
\end{aligned}$$

$$+E\sum_{\alpha}\left|\tfrac{1}{2}\sum_{ab}f_{\alpha ab}\phi^a\phi^b+8\pi G\phi^{\alpha *}\tilde{f}\right|^2+64\pi^2G^2E\sum_h|\tilde{z}^h|^2\left|\tfrac{1}{2}\sum_{AB}f_{AB}\phi^A\phi^B\right|^2$$

$$+16\pi GE\,\mathrm{Re}\sum_h\left[\tilde{z}^h\frac{\partial\tilde{f}}{\partial\tilde{z}^h}+8\pi G|\tilde{z}^h|^2\tilde{f}\right]\left[\tfrac{1}{6}\sum_{abc}f_{abc}\phi^a\phi^b\phi^c\right]^*-24\pi GE\left|\tfrac{1}{2}\sum_{AB}f_{AB}\phi^A\phi^B\right|^2$$

$$-48\pi GE\,\mathrm{Re}\left[\sum_{abc}\tfrac{1}{6}f_{abc}\phi^a\phi^b\phi^c\tilde{f}*\right]+\sum_K\left|\tfrac{1}{2}\sum_{ab}f_{Kab}\phi^a\phi^b+8\pi G\phi^{K*}\tilde{f}\right|^2+\tfrac{1}{2}\sum_K|\phi^\dagger t_K\phi|^2+\tfrac{1}{2}\sum_K|\phi^\dagger t_\kappa\phi|^2\,. \tag{A20}$$

Here $\tilde{V}$ is the potential of the hidden sector alone

$$\tilde{V}=\exp\left[8\pi G\sum_h|\tilde{z}^h|^2\right]\left[\sum_h\left|\frac{\partial\tilde{f}}{\partial\tilde{z}^h}+8\pi G\tilde{z}^{h*}\tilde{f}\right|^2-24\pi G|\tilde{f}|^2\right] \tag{A21}$$

and N and E are the exponential factors

$$N=\exp\left[8\pi G\sum_A|z_0^A|^2\right], \tag{A22}$$

$$E=N\exp\left[8\pi G\sum|\tilde{z}^h|^2\right]. \tag{A23}$$

Also, we remind the reader that sums over $a,b,\ldots$ run over the values $A,B,\ldots$ and $\alpha,\beta,\ldots$ and $K,L,\ldots$.

We now express the heavy scalars ϕ^A and ϕ^K as functions of $\phi^\alpha\equiv z^\alpha$ by imposing the conditions that V be stationary in heavy scalars. Expressing the heavy scalars in power series

$$\phi^A=z_1^A+z_2^A+z_3^A+\cdots\,, \tag{A24}$$

$$\phi^K=z_1^K+z_2^K+z_3^K+\cdots\,, \tag{A25}$$

with

$$z_n^A \text{ and } z_n^K\propto(m_g)^n \tag{A26}$$

the stationarity conditions become

$$0=\left[\frac{\partial V_2}{\partial\phi^A}\right]_1=\left[\frac{\partial V_2}{\partial\phi^K}\right]_1, \tag{A27}$$

$$0=\left[\frac{\partial V_3}{\partial\phi^A}\right]_1+\sum_B\left[\frac{\partial^2V_2}{\partial\phi^A\partial\phi^{B*}}\right]_1z_2^{B*}=\left[\frac{\partial V_3}{\partial\phi^K}\right]_1+\tfrac{1}{2}\sum_{KL}\left[\frac{\partial^2V_2}{\partial\phi^K\partial\phi^L}\right]_1z_2^L\,, \tag{A28}$$

and so on, the subscript indicating that ϕ^A and ϕ^K are set equal to z_1^A and z_1^K, while $\phi^\alpha\equiv z^\alpha$ is a free variable. [In writing (A28), we use the fact that the only nonvanishing second derivatives of V_2 are those shown here.] From (A27) and (A28), we find

$$z_1^A=-8\pi G\tilde{f}\sum_B z_0^{B*}\,, \tag{A29}$$

$$z_2^A=-\sum_B f^{-1}{}_{AB}\Bigg\{8\pi G\sum_C f^{-1*}{}_{BC}z_0^CE^{-1}V_2+\tfrac{1}{2}\sum_{ab}f_{Bab}z_1^az_1^b+4\pi Gz_0^{B*}\sum_{CD}f_{CD}z_1^Cz_1^D$$

$$+8\pi Gz_1^{B*}\left[\tilde{f}+\sum_h\left[\tilde{z}^h\frac{\partial\tilde{f}}{\partial\tilde{z}^h}+8\pi G|\tilde{z}^h|^2\tilde{f}\right]\right]-24\pi G\tilde{f}z_1^{B*}\Bigg\}\,, \tag{A30}$$

and

$$z_1^K = 0\,, \tag{A31}$$

$$z_2^K = -\tfrac{1}{2}\sum_L \mu^{-1}{}_{KL}(z_1^\dagger t_L z_1)\,, \tag{A32}$$

with $z_1^\alpha \equiv z^\alpha$. We will not need the formulas for the higher-order terms in z^A and z^K.

Now we can calculate the effective potential of the light scalars by inserting our results for z^A and z^K in V. To order $m_g{}^4$, this gives

$$V_{\text{eff}} = (V_2)_1 + \sum_A \left[\frac{\partial V_2}{\partial\phi^A}\right]_1 z_2^A + cc + \sum_K \left[\frac{\partial V_2}{\partial\phi^K}\right]_1 z_2^K + (V_3)_1 + \sum_A \left[\frac{\partial V_3}{\partial\phi^A}\right]_1 z_2^A + cc + \sum_K \left[\frac{\partial V_3}{\partial\phi^K}\right]_1 z_2^K$$
$$+ \sum_{AB}\left[\frac{\partial^2 V_2}{\partial\phi^A\partial\phi^{B*}}\right]_1 z_2^A z_2^{B*} + \tfrac{1}{2}\sum_{KL}\left[\frac{\partial^2 V_2}{\partial\phi^K\partial\phi^L}\right]_1 z_2^K z_2^L + \sum_A \left[\frac{\partial V_2}{\partial\phi^A}\right]_1 z_3^A + cc + \sum_K \left[\frac{\partial V_2}{\partial\phi^K}\right]_1 z_3^K + (V_4)_1\,. \tag{A33}$$

Using (A27) and (A28), this simplifies to

$$V_{\text{eff}} = (V_2)_1 + (V_3)_1 + (V_4)_1 - \sum_{AB}\left[\frac{\partial^2 V_2}{\partial\phi^A\partial\phi^{B*}}\right]_1 z_2^A z_2^{B*} - \tfrac{1}{2}\sum_{KL}\left[\frac{\partial^2 V_2}{\partial\phi^K\partial\phi^L}\right]_1 z_2^K z_2^L\,. \tag{A34}$$

First consider the effective potential to third order in m_g. Using (A29) and (A31) in (A18) and (A19), this is

$$(V_2)_1 + (V_3)_1 = \left[1 + 16\pi G\,\text{Re}\sum_h z_0^{A*} z_1^A\right] N\tilde{V} + 8\pi G E\,\text{Re}\left\{\left[\sum_h \tilde{z}^h \frac{\partial\tilde{f}}{\partial\tilde{z}^h} + \left[8\pi G\sum_h |\tilde{z}^h|^2 - 3\right]\tilde{f}\right]^* \sum_{AB} f_{AB} z_1^A z_1^A\right\}. \tag{A35}$$

We note that (A35) is completely independent of light fields, so we can find a z^α-independent value of the hidden fields $\tilde{z}^h$ where (A35) is stationary in $\tilde{z}^h$, and we can adjust an additive constant in $\tilde{f}$ to order $m_g{}^2$ so that (A35) vanishes at this point. However, the remaining terms in (A34) are already of order $m_g{}^4$, and to this order it is an adequate approximation to calculate these terms using the lowest-order values for $\tilde{z}^h$ and $f(\tilde{z}^h)$ at the stationary point. These are determined by the conditions that the second-order term $N\tilde{V}$ in (A35) vanish and be stationary, i.e., that

$$\frac{\partial\tilde{V}}{\partial\tilde{z}^h} = \tilde{V} = 0\,. \tag{A36}$$

These conditions fix $\tilde{z}^h$ to have the value denoted $\tilde{z}_0^h$ in Sec. II. From now on, we suppose that $\tilde{z}^h$ and the additive constant in $\tilde{f}$ have been determined in this way.

With $(V_2+V_3)_1$ absent, the effective potential is given by (A35) as

$$V_{\text{eff}} = (V_4)_1 - \sum_{AB}\left[\frac{\partial^2 V_2}{\partial\phi^A\partial\phi^{B*}}\right]_1 z_2^A z_2^{B*} - \tfrac{1}{2}\sum_{KL}\left[\frac{\partial^2 V_2}{\partial\phi^K\partial\phi^L}\right]_1 z_2^K z_2^L\,. \tag{A37}$$

Setting $\phi^A = z_1^A$ and $\phi^K = z_1^K = 0$ in (A20) yields

$$(V_4)_1 = E_0\sum_A \left|\tfrac{1}{2} f_{Aab} z_1^a z_1^b + 4\pi G z_0^{A*}\sum_{BC} z_1^B z_1^B + 8\pi G\tilde{f}_0 z_1^{A*}\right|^2 + E_0\sum_\alpha \left|\tfrac{1}{2} f_{\alpha ab} z^a z^b + 8\pi G z^{\alpha *}\tilde{f}_0\right|^2$$
$$+ 8\pi G E_0\left[8\pi G\sum_h |\tilde{z}^h|^2 - 3\right]\left|\tfrac{1}{2}\sum_{AB} f_{AB} z_1^A z_1^B\right|^2$$

$$+16\pi GE_0\,\mathrm{Re}\left\{\left[\sum_h\left[\tilde{z}^h\frac{\partial\tilde{f}}{\partial\tilde{z}^h}\right]_0+\left[8\pi G\sum_h|\tilde{z}_0^h|^2-3\right]\tilde{f}_0\right]\frac{1}{6}\sum_{abc}f_{abc}z_1^a z_1^b z_1^c\right\}$$

$$+\sum_K\left|\frac{1}{2}\sum_{ab}f_{Kab}z_1^a z_1^b\right|^2+\frac{1}{2}\sum_K|z_1^\dagger t_K z_1|^2+\frac{1}{2}\sum_\kappa|z_1^\dagger t_\kappa z_1|^2\,. \tag{A38}$$

Also, (A30) and (A32) give

$$\sum_{AB}\left[\frac{\partial^2 V_2}{\partial\phi^A\partial\phi^{B*}}\right]_1 z_2^A z_2^{B*}=E_0\sum_A\left|\sum_B f_{AB}z_2^B\right|^2$$

$$=E_0\sum_A\left|\frac{1}{2}\sum_{ab}f_{Aab}z_1^a z_1^b+4\pi G z_0^{A*}\sum_{CD}f_{CD}z_1^C z_1^D\right.$$

$$\left.+8\pi G z_1^{A*}\left[\sum_h\left[\tilde{z}^h\frac{\partial\tilde{f}}{\partial\tilde{z}^h}\right]_0+\left[8\pi G\sum_h|z_0^h|^2-2\right]\tilde{f}_0\right]\right|^2, \tag{A39}$$

$$\frac{1}{2}\sum_{KL}\left[\frac{\partial^2 V_2}{\partial\phi^K\partial\phi^L}\right]_1 z_2^K z_2^L=2\sum_{KL}\mu^2{}_{KL}z_2^K z_2^L$$

$$=\frac{1}{2}\sum_K(z_1^\dagger t_K z_1)^2\,. \tag{A40}$$

Part of (A39) cancels the first term in (A38), and (A40) cancels the next-to-last term in (A38), leaving us with

$$V_{\text{eff}}=-16\pi GE_0\,\mathrm{Re}\left[\sum_h\left[\tilde{z}^h\frac{\partial f}{\partial\tilde{z}^h}\right]_0+\left[8\pi G\sum_h|\tilde{z}_0^h|^2-3\right]\tilde{f}_0\right]^*$$

$$\times\left[\frac{1}{2}\sum_{abA}f_{Aab}z_1^A z_1^a z_1^b+4\pi G\sum_A z_1^A z_0^{A*}\sum_{CD}f_{CD}z_1^C z_1^D+8\pi G\tilde{f}_0\sum_A|z_1^A|^2\right]$$

$$-64\pi^2G^2\sum_A|z_1^A|^2\left|\sum_h\tilde{z}^h\frac{\partial\tilde{f}_0}{\partial\tilde{z}^h}+\left[8\pi G\sum_h|\tilde{z}_0^h|^2-3\right]\tilde{f}_0\right|^2+E_0\sum_\alpha\left|\frac{1}{2}\sum_{ab}f_{\alpha ab}z_1^a z_1^b+8\pi G z^{\alpha*}\tilde{f}_0\right|^2$$

$$+8\pi GE_0\left[8\pi G\sum_h|\tilde{z}^h|^2-3\right]\left|\frac{1}{2}\sum_{AB}f_{AB}z_1^A z_1^B\right|^2$$

$$+16\pi GE_0\,\mathrm{Re}\left[\sum_h\left[z^h\frac{\partial\tilde{f}}{\partial\tilde{z}^h}\right]_0+\left[8\pi G\sum_h|\tilde{z}^h|^2-3\right]\tilde{f}_0\right]^*$$

$$\times\frac{1}{6}\sum_{abc}f_{abc}z_1^a z_1^b z_1^c+\sum_K\left|\frac{1}{2}\sum_{ab}f_{Kab}z_1^a z_1^b\right|^2+\frac{1}{2}\sum_\kappa|z_1^\dagger t_\kappa z_1|^2\,. \tag{A41}$$

We note also that

$$f_{K\alpha\beta}=\sum_A f_{K\alpha A}z_1^A=0 \tag{A42}$$

so the sum over a and b in the next-to-last term in (A41) runs only over heavy scalar indices A and B, and this term is therefore independent of light scalars.

(*Proof*: From the gauge-invariance condition on the superpotential

$$\sum_C \frac{\partial f}{\partial z^c}(t_K z)^c = 0$$

we have by differentiating twice and setting $z = z_0$

$$\sum_c [f_{abc}(t_K z)^c + f_{ac}(t_K)^c_b + f_{bc}(t_K)^c_a] = 0 .$$

Using (A7) and (A8) allows us to solve for f_{Kab}:

$$f_{Kab} = -\sum_{Lc} \mu^{-1}{}_{KL}[f_{ac}(t_L)^c_b + f_{bc}(t_L)^c_a] .$$

Equation (A5) then yields

$$f_{K\alpha\beta} = 0$$

and

$$f_{K\alpha A} = -\sum_{LB} \mu^{-1}{}_{KL} f_{AB}(t_L)^B_\alpha$$

so

$$\sum_A f_{K\alpha A} z_1^A = +8\pi G \tilde{f} \sum_{LB} z_0^{B*}(t_L)^B_\alpha$$

which vanishes by (A7).)

It is very convenient to rewrite (A41) by introducing an effective superpotential

$$f_{\text{eff}} = \frac{E_0{}^{1/2}}{6} \sum_{abc} f_{abc} z_1^a z_1^b z_1^c . \tag{A43}$$

The other z^α-dependent quantities in (A41) can be written

$$\tfrac{1}{2} \sum_{abA} f_{Aab} z_1^A z_1^a z_1^b = 3f^{(0)} + 2f^{(1)} + f^{(2)} ,$$

$$\tfrac{1}{2} \sum_{ab\alpha} f_{\alpha ab} z^\alpha z_1^a z_1^b = f^{(1)} + 2f^{(2)} + 3f^{(3)} ,$$

$$\tfrac{1}{2} \sum_{\alpha ab} f_{\alpha ab} z_1^a z_1^b = \frac{\partial f_{\text{eff}}}{\partial z^\alpha} ,$$

where $f^{(n)}$ is the term in f_{eff} of nth order in light scalars. Equation (A41) thus takes the form

$$V_{\text{eff}} = \sum_\alpha \left| \frac{\partial f_{\text{eff}}}{\partial z^\alpha} \right|^2 + 2\,\text{Re}\{m_g'^* f^{(3)}\} + 4\,\text{Re}\{m_g^* f^{(2)}\} + 2\,\text{Re}\{(4m_g - m_g')^* f^{(1)}\} + |m_g|^2 \sum_\alpha |z^\alpha|^2$$
$$+ \tfrac{1}{2} \sum_\kappa \left| \sum_{\alpha\beta} z^{\alpha *} z^\beta (t_\kappa)^\alpha_\beta \right|^2 + V_0 , \tag{A44}$$

where m_g, m_g', and V_0 are the constants

$$m_g = 8\pi G E_0{}^{1/2} \tilde{f}_0 , \tag{A45}$$

$$m_g' = 8\pi G E_0{}^{1/2} \left| \sum_h \tilde{z}_0^h \left[\frac{\partial \tilde{f}}{\partial \tilde{z}^h} \right]_0 + 8\pi G \tilde{f}_0 \sum_h |\tilde{z}_0^h|^2 \right| , \tag{A46}$$

$$V_0=-2\,\mathrm{Re}(m_g'^*-3m_g^*)m_g\sum_A |z_1^A|^2-\sum_A |z_1^A|^2|m_g'-3m_g|^2+E_0\sum_\alpha \left|\tfrac{1}{2}\sum_{AB} f_{\alpha AB}z_1^A z_1^B\right|^2$$

$$-4E_0^{1/2}\mathrm{Re}(m_g'-3m_g)^* f^{(0)}+E_0\sum_K \left|\tfrac{1}{2}\sum_{AB} f_{KAB}z_1^A z_1^B\right|^2+v_0\,, \tag{A47}$$

where v_0 is the part of (A35) that arises from any additive constant of order $m_g{}^3$ in $\tilde{f}$. Equation (A44) is our desired result, quoted in Sec. III as Eq. (3.11).

APPENDIX B: CONDITIONS FOR LIGHT SCALAR INDEPENDENCE IN LOW ORDERS

Consider a general potential $V(z)$, expressed as a power series in a small parameter ϵ:

$$V(z)=V_0(z)+\epsilon V_1(z)+\epsilon^2 V_2(z)+\cdots\,. \tag{B1}$$

Suppose that the zeroth order potential has a minimum at $z^a=z_0^a$:

$$\frac{\partial V_0(z)}{\partial z^a}=0 \text{ at } z^a=z_0^a\,. \tag{B2}$$

Choose a basis in which a runs over values A and α, with

$$V_{0\alpha\beta}=V_{0A\alpha}=0\,, \tag{B3}$$

$$V_{0AB}\text{ nonsingular}\,, \tag{B4}$$

where subscripts denote differentiation with respect to z^a:

$$V_{nab\cdots}=\left[\frac{\partial V_n(z)}{\partial z^a\partial z^b\cdots}\right]_{z=z_0}\,. \tag{B5}$$

We write the scalars as

$$z^a=z_0^a+\epsilon\phi^a \tag{B6}$$

and expand

$$V(z)=C+\epsilon^2\left[\tfrac{1}{2}\sum_{AB}V_{0AB}\phi^A\phi^B+\sum_a V_{1a}\phi^a+\epsilon^3\left[\tfrac{1}{6}\sum_{abc}v_{0abc}\phi^a\phi^b\phi^c+\tfrac{1}{2}\sum_{ab}V_{1ab}\phi^a\phi^b\right]+\sum_a V_{2a}\phi^a\right]+O(\epsilon^4)\,, \tag{B7}$$

where C is a z^a-independent constant. We "integrate out" the heavy scalars ϕ^A, by imposing the condition that

$$\frac{\partial V}{\partial\phi^A}=0 \text{ at } \phi^A=\phi^A(\phi^\alpha)\,. \tag{B8}$$

This has a power-series solution

$$\phi^A=z_1^A+\epsilon z_2^A+\cdots\,, \tag{B9}$$

with

$$z_1^A=-\sum_B V^{-1}{}_{0AB}V_{1B}\,. \tag{B10}$$

Inserting this back into (B7) yields

$$V_{\mathrm{eff}}(\phi^\alpha)=C'+\epsilon^2\sum_\alpha V_{1\alpha}\phi^\alpha+\epsilon^3\left[\tfrac{1}{6}\sum_{\alpha\beta\gamma}V_{0\alpha\beta\gamma}\phi^\alpha\phi^\beta\phi^\gamma+\tfrac{1}{2}\sum_{\alpha\beta A}V_{0\alpha\beta A}\phi^\alpha\phi^\beta z_1^A\right.$$

$$\left.+\tfrac{1}{2}\sum_{\alpha AB}V_{0\alpha AB}\phi^\alpha z_1^A z_1^B+\tfrac{1}{2}\sum_{\alpha\beta}V_{1\alpha\beta}\phi^\alpha\phi^\beta+\sum_{\alpha A}V_{1\alpha A}\phi^\alpha z_1^A+\sum_\alpha V_{2\alpha}\phi^\alpha\right]+O(\epsilon^4) \tag{B11}$$

with C' a ϕ^α-independent constant. The conditions for the z^α-dependent terms in $V_{\rm eff}$ to vanish in orders ϵ^2 and ϵ^3 are therefore

$$V_{1\alpha}=0\ , \tag{B12}$$

$$V_{0\alpha\beta\gamma}=0\ , \tag{B13}$$

$$\sum_A V_{0\alpha\beta A}z_1^A+V_{1\alpha\beta}=0\ , \tag{B14}$$

$$\tfrac{1}{2}\sum_{AB} V_{0\alpha AB}z_1^A z_1^B+\sum V_{1\alpha A}z_1^A+V_{1\alpha}=0\ . \tag{B15}$$

In the case of the supergravity potential (A1), we can regard $\widetilde{f}$ and $\sum_n \widetilde{z}^h\partial\widetilde{f}/\partial\widetilde{f}^h$ as parameters of order ϵ (with $\widetilde{z}^h$ regarded as a fixed parameter). With $z_0^\alpha=0$, we can easily verify that $V_{1\alpha}$, $V_{0\alpha\beta\gamma}$, and $V_{2\alpha}$ all vanish, while the first two terms in (B14) and (B15) cancel. This verifies again that $V_{\rm eff}$ is independent of z^α in orders ϵ^2 and ϵ^3.

APPENDIX C: THEORIES WITHOUT SUPERHEAVY SCALARS

We suppose in this appendix that there are no superheavy scalars, but that the superpotential $f(z)$ involves a light mass scale, of order m_g. That is, the scalar field labels $a,b,\ldots$ run only over values $\alpha,\beta,\ldots$ denoting light scalars, and the superpotential now has nonvanishing but small second derivatives with respect to these scalars:

$$f_{\alpha\beta}\equiv\left[\frac{\partial^2 f}{\partial z^\alpha\partial z^\beta}\right]_0\sim m_g\ . \tag{C1}$$

To order $m_g^{\,4}$, Eq. (2.2) then gives

$$\begin{aligned}V=&\sum_\alpha\left|\frac{\partial f_{\rm eff}}{\partial z^\alpha}+8\pi GE^{1/2}z^{\alpha *}\widetilde{f}\right|^2\\&+\sum_h\left|E^{1/2}\left[\frac{\partial\widetilde{f}}{\partial\widetilde{z}^h}+8\pi G\widetilde{z}^{h*}\widetilde{f}\right]+8\pi G\widetilde{z}^{h*}f_{\rm eff}\right|^2\\&-24\pi G\,|E^{1/2}\widetilde{f}+\widetilde{f}_{\rm eff}|^2+\sum_\kappa D_\kappa^{\,2}\end{aligned} \tag{C2}$$

with z^α and $\widetilde{f}$ taken as of order m_g and

$$f_{\rm eff}=E^{1/2}\left[\tfrac{1}{2}\sum_{\alpha\beta}f_{\alpha\beta}z^\alpha z^\beta+\tfrac{1}{6}\sum_{\alpha\beta\gamma}f_{\alpha\beta\gamma}z^\alpha z^\beta z^\gamma\right]\sim m_g^{\,3}\ . \tag{C3}$$

The quantities within the absolute value signs in (C2) are purely of order $m_g^{\,2}$ in the first term, but of order m_g and $m_g^{\,3}$ in the second and third terms. Discarding the terms in V of order $m_g^{\,6}$, we have then to order $m_g^{\,4}$:

$$\begin{aligned}V=&E\widetilde{V}+\sum_\alpha\left|\frac{\partial f_{\rm eff}}{\partial z^\alpha}\right|^2+16\pi GE^{1/2}{\rm Re}\widetilde{f}^*\sum_\alpha z^\alpha\frac{\partial f}{\partial z^\alpha}\\&+64\pi^2G^2E\,|\widetilde{f}|^2\sum_\alpha|z^\alpha|^2\\&+16\pi GE^{1/2}{\rm Re}\sum_h\left[\widetilde{z}^h\frac{\partial\widetilde{f}}{\partial\widetilde{z}^h}+8\pi G\widetilde{f}\,|\widetilde{z}^h|^2\right]^*f_{\rm eff}\\&-48\pi GE^{1/2}{\rm Re}\widetilde{f}^*f_{\rm eff}+\sum_\kappa D_\kappa^{\,2}\ .\end{aligned} \tag{C4}$$

To this order the minimization of V with respect to $\widetilde{z}^h$ gives $\widetilde{z}^h=\widetilde{z}_0^h$, where $\widetilde{V}$ vanishes. Equation (C4) is then the same as the result quoted in Eq. (3.14).

---

[1]The possibilities of constructing realistic models with low-energy spontaneously broken supersymmetry were extensively explored by P. Fayet, Phys. Lett. 69B, 489 (1977); 70B, 461 (1977); in *New Frontiers in High Energy Physics,* edited by A. Perlmutter and L. F. Scott (Plenum, New York, 1978), p. 413; in *Unification of the Fundamental Particle Interactions,* proceedings of the Europhysics Study Conference, Erice, 1980, edited by S. Ferrara, J. Ellis, and P. van Nieuwenhuizen (Plenum, New York, 1980), p. 587; in *Proceedings of the XVIth Rencontre de Moriond: Vol. I, Perturbative QCD and Electroweak Interactions,* edited by J. Trân Thanh Vân (Editions Frontières, Dreux, France, 1981), p. 347. Also see M. Sohnius, Nucl. Phys. B122, 291 (1977). For more recent discussions of the difficulties encountered in these models, see E. Witten, Nucl. Phys. B188, 513 (1981); S. Dimopoulos and H. Georgi, *ibid.* B193, 150 (1981); N. Sakai, Z. Phys. C 11, 753 (1982); S. Weinberg, Phys. Rev. D 26, 287 (1982); L. Alvarez-Gaumé, M. Claudson, and M. Wise, Nucl. Phys. B207, 96 (1982); L. Hall and I. Hinchcliffe, Phys. Lett. 112B, 351 (1982); G. Farrar and S. Weinberg, Phys. Rev. D (to be published).

[2]Globally supersymmetric models with a partially isolated

supersymmetry-breaking sector have been discussed by many authors, including E. Witten, Phys. Lett. 105B, 267 (1981); L. Alvarez-Gaumé, M. Claudson, and M. Wise, Ref. 1; J. Ellis, L. Ibanez, and C. G. Ross, Phys. Lett. 113B, 283 (1982); M. Dine and W. Fischler, *ibid.* 110B, 227 (1982); C. Nappi and B. Ovrut, *ibid.* 113B, 175 (1982); T. Banks and V. Kaplunovsky, Nucl. Phys. B206, 45 (1982); S. Dimopoulos and S. Raby, *ibid.* B192, 353 (1981); R. Barbieri, S. Ferrara, and D. V. Nanopoulos, Z. Phys. C 13, 267 (1982); Phys. Lett. 113B, 219 (1982); 116B, 16 (1982); J. Polchinski and L. Susskind, Phys. Rev. D 26, 3661 (1982).

[3]A. H. Chamseddine, R. Arnowitt, and P. Nath, Phys. Rev. Lett. 49, 970 (1982); P. Nath, R. Arnowitt, and A. P. Chamseddine, Phys. Lett. 121B, 33 (1983); A. H. Chamseddine, P. Nath, and R. Arnowitt, Harvard-Northeastern Report No. HUTP-82/AO56-NUB2578 (unpublished).

[4]R. Barbieri, S. Ferrara, and C. A. Savoy, Phys. Lett. 119B, 343 (1982).

[5]L. Ibãnez, Phys. Lett. 118B, 73 (1982); Madrid Report No. FTUAM82-8 (unpublished).

[6]H. P. Nilles, M. Srednicki, and D. Wyler, Phys. Lett. 120B, 346 (1983). Also see H. P. Nilles, Phys. Lett. 115B, 193 (1982); CERN Reports Nos. TH3330 and TH3398 (unpublished).

[7]J. Ellis, D. V. Nanopoulos, and K. Tamvakis, Phys. Lett. 121B, 123 (1983).

[8]S. Ferrara, D. V. N. Nanopoulos, and C. A. Savoy, CERN Report No. TH3442 (unpublished).

[9]N. Ohta, Tokyo Report No. UT-388 (unpublished).

[10]L. Alvarez-Gaumé, J. Polchinski, and M. Wise, Harvard/CalTech Report No. HUTP-82/A063 —CALT-68-990 (unpublished).

[11]J. León, M. Quirós, and M. Ramón Medrano, Madrid report (unpublished).

[12]Forerunners of the work of Refs. 3—11 include B. A. Ovrut and J. Wess, Phys. Lett. 112B, 347 (1982) and R. Barbieri, S. Ferrara, D. V. Nanopoulos, and K. S. Stelle, *ibid.* 113B, 219 (1982). Ovrut and Wess considered a supergravity model in which supersymmetry was broken in the hidden sector not spontaneously, but through the discard of a cosmological constant. Barbieri *et al.* discussed a supergravity model in which supersymmetry is spontaneously broken by a hidden sector, but there is no observable sector superpotential.

[13]The appearance of relatively light gauge fermions in theories like those of Refs. 3—12 is discussed by S. Weinberg, Phys. Rev. Lett. 50, 387 (1983). R. Arnowitt, A. H. Chamseddine, and P. Nath, *ibid.* 50, 232 (1983).

[14]The assumption of a superpotential consisting of a sum of two terms involving different sets of superfields would be more attractive if it could be shown to follow naturally from some symmetry of the theory. This is easy to arrange if the superpotential is limited to be a cubic polynomial; for instance, if the theory has a group $G\times\widetilde{G}$ of gauge and/or global symmetries, then if all superfields S^a are non-neutral under G and neutral under $\widetilde{G}$ and all superfields $\widetilde{S}^h$ are neutral under G and non-neutral under $\widetilde{G}$, the superpotential can contain no terms involving both S^a's and $\widetilde{S}^h$'s. However, it is doubtful whether the assumption of a cubic superpotential can be justified on the grounds of renormalizability for fields like $\widetilde{S}^h$, whose scalar VEV's are of order of the Planck mass. Alternatively, it would be possible to understand the decomposition of the superpotential into $f(S)$ and $\widetilde{f}(\widetilde{S})$ if the theory had an R symmetry (a symmetry for which the superfield coordinate θ_L carries some nonvanishing quantum number, say, $+1$); for instance, if all S^a superfields have $R=\frac{2}{3}$ and all $\widetilde{S}^h$ superfields have $R=2$, then the superpotential consists of terms trilinear in the S^a plus terms linear in the $\widetilde{S}^h$. In such a theory there would be no additive constant in the superpotential that could be used to cancel the cosmological constant, but it would still be possible to arrange for a flat space by adjustment of the Kähler potential.

[15]An alternative with similar consequences is suggested by E. Cremmer, P. Fayet, and L. Girardello, Phys. Lett. 122B, 41 (1983).

[16]The independence of the physics of the light Higgs scalars on the grand-unified mass scale was subsequently discussed for a wide class of superpotentials by P. Nath, A. H. Chamseddine, and R. Arnowitt, Harvard-Northeastern Report No. HUPT82/AO57—NUB2579 (unpublished).

[17]This result was given for nongauge theories with a single chiral superfield by E. Cremmer, B. Julia, J. Scherk, S. Ferrara, L. Girardello, and P. van Nieuwenhuizen, Phys. Lett. 79B, 231 (1978); Nucl. Phys. B147, 105 (1979). The generalization to theories with arbitrary numbers of chiral scalars was given by E. Witten and J. Bagger, Phys. Lett. 115B, 202 (1982); R. Barbieri, S. Ferrara, D. V. Nanopoulos, and K. S. Stelle, *ibid.* 113B, 219 (1982). The results for general gauge theories were given by E. Cremmer, S. Ferrara, L. Girardello, and A. Van Proeyen, *ibid.* 116B, 231 (1982); J. Bagger, Princeton report (unpublished).

[18]The breakdown of the grand-gauge group can allow scalars to remain with zero mass and VEV for any one of three possible reasons: either because of unbroken symmetries, or of a fine tuning of the parameters of the theory, or as a more or less accidental property of the potential minimum that follows automatically from the symmetries and cubic polynomiality of the observable sector superpotential. In the case of fine tuning a tiny change in the input parameters could introduce additional mass parameters in the low-energy theory in addition to those produced by supergravity, so in this case it may be argued that it is not gravitation alone that sets the mass scale of known particles. In the other two cases supersymmetry can truly be said to resolve the hierarchy problem, except that the small scale of the hidden-sector superpotential still has to be put in by hand. In this paper we will consider all these cases to-

gether.

[19]This is in apparent conflict with the cosmological lower bound of about 10 TeV on the mass of heavy gravitinos; see S. Weinberg, Phys. Lett. 48, 1303 (1982). However, entropy-producing mechanisms like cosmic inflation that dilute the gravitino (and monopole) densities could reduce the entropy produced in gravitino decay sufficiently so as not to disturb calculations of cosmic nucleosynthesis; see J. Ellis, A. D. Linde, and D. V. Nanopoulos, Phys. Lett. 118B, 59 (1982); S. Dimopoulos and S. Raby, Los Alamos Report No. LA-UR-82-1282, 1982 (unpublished).

[20]Theories with a purely trilinear effective low-energy superpotential have a severe phenomenological problem: the electroweak $SU(2)\times U(1)$ symmetry is spontaneously broken only for $|m_g'/m_g|>3$, but in this case the introduction of quark and lepton superfields leads to a vacuum solution for which quark and/or lepton scalars have nonzero VEV's, as pointed out by J.-M. Frère, D. R. T. Jones, and S. Raby, Michigan Report No. UMHE82-58 (unpublished). However, $SU(2)\times U(1)$ can be spontaneously broken for $|m_g'/m_g|<3$ if bilinear terms are allowed in the effective superpotential, and in this case it is not clear that VEV's appear for quark and lepton scalars. Also it is not clear that the tunneling from the baryon- and lepton-conserving local minima of the potential to the one found by Frère *et al.* is fast enough to pose a serious problem; M. Claudson, L. Hall, and I. Hinchcliffe (unpublished).

[21]The problem of radiative corrections has been discussed by R. Arnowitt, A. H. Chamseddine, and P. Nath, Phys. Lett. 120B, 145 (1983); S. Ferrara, D. V. Nanopoulos, and C. A. Savoy, Ref. 8; H. P. Nilles, M. Srednicki, and D. Wyler, CERN Report No. TH-3461 (unpublished); A. B. Lahanas, CERN Report No. TH-3467 (unpublished); R. Barbieri and S. Cecotti, Pisa Report No. SNS9/1982 (unpublished). Also see J. Polchinski and L. Susskind, Ref. 2.

[22]This symmetry was used in a similar way by S. Weinberg, Ref. 13. Also see M. K. Gaillard, talk at the Vanderbilt University Conference on Novel Results in Particle Physics, Berkeley Report No. LBL-14647 (unpublished).

[23]The assumption that the scalar fields $\mathscr{H}$, $\mathscr{H}'$, and $\mathscr{J}$ do not get large masses or VEV's in the breakdown of the grand-gauge group is, for example, automatically satisfied if we suppose that a discrete symmetry remains unbroken, for which the superfields H, H', and J are transformed by a factor $\exp(2i\pi/3)$. In this case the only terms in (5.9) are the last two, with coefficient λ and λ'.

3.4. Gravity and Asymptotic Safety

The problem with we theoretical physicists is not that we take ourselves too seriously, but that we do not take ourselves seriously enough.
Steven Weinberg

Weinberg was among the first particle physicists to tackle the force of gravity as well as the strong, weak and electromagnetic. Exploring the similarities and differences of spin 2 gravitons and spin 1 gauge bosons provided new insights into quantum gravity not immediately apparent from the traditional geometric perspective of General Relativists. As a result, some of them regarded him as being anti-geometrical, but I think that was a bit unfair. He was simply demonstrating the advantages and disadvantages of treating Einstein's theory as a perturbation about flat space, together with all the associated apparatus of a Lorentz-invariant quantum field theory: momentum-space Feynman rules and Feynman diagrams, scattering amplitudes etc.

In fact in [3.4.1], using just S-matrix arguments, the conservation of electric charge is derived without appealing to a gauge-invariant Lagrangian and the equality of inertial and gravitational mass (aka the equivalence principle) is derived without appealing to a generally covariant Lagrangian. In fact, in a subsequent paper [3.4.2], the Maxwell equations for spin 1 and the Einstein equations for spin 2 are actually derived. It was also shown that the analogous coupling constants for spin ≥ 3 must vanish.

The purpose of [3.4.3] was to show that the infrared divergences in the quantum theory of gravitation can be treated in the same manner as in quantum electrodynamics. However, this treatment apparently does not work in other non-Abelian gauge theories, like that of Yang and Mills.

The thorny problem of "Ultraviolet divergences in the quantum theories of gravitation" was the subject of a magnum opus that Weinberg submitted to the 1979 Einstein Centenary Survey [3.4.4] and, as mentioned in Chapter 1, one on which we corresponded. See also Fig. 1. Recall from Section 3.1 that the non-renormalizability of The Einstein-Hilbert Lagrangian was not fatal when regarded as an effective low-energy approximation to some fundamental theory. There would be an infinite number of counterterms but only a finite number were required to calculate processes to any given order in energy. Moreover, since such an approximation ceased to be valid at energies where the kind of massive ghost poles catalogued by Stelle appear, unitarity was preserved. There are still two obvious drawbacks, however: (1) We still need to know the fundamental theory: string theory, M-theory or whatever it may be, if we are to understand the extreme conditions of the Big Bang or the centre of a black hole, (2) In the meantime, infinitely many counterterms superficially means infinitely many unknown parameters that can only be fixed by experiment.

In [3.4.4], however, Weinberg proposes a different alternative: Suppose the Lagrangian with all its higher order terms *is* the fundamental theory and suppose the problems of unitarity and infinitely many coefficients are illusory. What magic wand could possibly be waved to achieve this? Answer: *Asymptotic Safety.*

A theory is said to be asymptotically safe if the essential coupling parameters approach a fixed point as the momentum scale of their renormalisation point goes to infinity. Weinberg argues that from the observed properties of second order phase transitions, this condition acts much like renormalizability in fixing all but a finite number of essential coupling parameters. Furthermore, *from this point of view it is quite natural that the strong, weak and electromagnetic interactions should be described so well by renormalizable quantum field theories, and that the only force that seems to require a non-renormalizable description should be the one force that is observed purely through its macroscopic effects, the force of gravitation.*

So far, so good, but do asymptotically safe theories actually exist? While this paper presents the case for two dimensions, Weinberg concedes that the problem is still unresolved in four. The problem, as I see it, is that practical calculations require truncating the infinite number of terms to some finite subset, but then one can never be sure that the results are not just an artefact of the chosen truncation. As far as I am aware, the issue of unitarity is also still unresolved. Nevertheless, so great is the potential pay-off that Asymptotic Safety has surely earned the right to further and closer scrutiny. Indeed, it is currently enjoying a revival of interest.

In [3.4.5], Weinberg and Witten show that in all theories with a Lorentz-covariant energy-momentum tensor, such as all known renormalizable quantum field theories, composite as well as elementary massless particles with spin $s > 1$ are forbidden. A similar theorem applies to theories with a Lorentz covariant conserved four-vector current where spin $s > 1/2$ massless particles are forbidden. These theorems do not rule out gluons or gravitons because the required current or energy-momentum tensor are absent.

The final paper is this section was written by Weinberg and Candelas [3.4.6] while I was visiting UT in 1983 and on the same topic as my own research, namely Kaluza-Klein theory. Its novelty is that the energy-momentum tensor of the effective four-dimensional theory arises from the one-loop vacuum energy of the matter fields. They are then able to calculate the coupling constants associated with the Kaluza-Klein vector bosons. If one could make a realistic model along these lines, one could calculate the fine structure constant!

Such attempts were superseded by the superstring revolution of 1984 and the M-theory revolution of 1995 but calculating the dimensionless parameters of the Standard Model remains as elusive as ever. In common with some, but by no means all, members of the theoretical physics community, Weinberg came to the view that this was asking the wrong question. Just as the distance from the Sun to the Earth is merely an accident of our particular solar system, so the values of the fundamental constants may be an accident of our particular universe. This brings us to the next section on cosmology and the multiverse.

PHYSICAL REVIEW VOLUME 135, NUMBER 4B 24 AUGUST 1964

Photons and Gravitons in *S*-Matrix Theory: Derivation of Charge Conservation and Equality of Gravitational and Inertial Mass*

STEVEN WEINBERG†
Physics Department, University of California, Berkeley, California
(Received 13 April 1964)

We give a purely *S*-matrix-theoretic proof of the conservation of charge (defined by the strength of soft photon interactions) and the equality of gravitational and inertial mass. Our only assumptions are the Lorentz invariance and pole structure of the *S* matrix, and the zero mass and spins 1 and 2 of the photon and graviton. We also prove that Lorentz invariance alone requires the *S* matrix for emission of a massless particle of arbitrary integer spin to satisfy a "mass-shell gauge invariance" condition, and we explain why there are no macroscopic fields corresponding to particles of spin 3 or higher.

I. INTRODUCTION

It is not yet clear whether field theory will continue to play a role in particle physics, or whether it will ultimately be supplanted by a pure *S*-matrix theory. However, most physicists would probably agree that the place of local fields is nowhere so secure as in the theory of photons and gravitons, whose properties seem indissolubly linked with the space-time concepts of gauge invariance (of the second kind) and/or Einstein's equivalence principle.

The purpose of this article is to bring into question the need for field theory in understanding electromagnetism and gravitation. We shall show that there are no general properties of photons and gravitons, which *have* been explained by field theory, which cannot also be understood as consequences of the Lorentz invariance and pole structure of the *S* matrix for massless particles of spin 1 or 2.[1] We will also show why there can be no macroscopic fields whose quanta carry spin 3 or higher.

What are the special properties of the photon or graviton *S* matrix, which might be supposed to reflect specifically field-theoretic assumptions? Of course, the usual version of gauge invariance and the equivalence principle cannot even be stated, much less proved, in terms of the *S* matrix alone. (We decline to turn on external fields.) But there are two striking properties of the *S* matrix which *seem* to require the assumption of gauge invariance and the equivalence principle:

(1) The *S* matrix for emission of a photon or graviton can be written as the product of a polarization "vector" ϵ^μ or "tensor" $\epsilon^\mu\epsilon^\nu$ with a covariant vector or tensor amplitude, and it vanishes if any ϵ^μ is replaced by the photon or graviton momentum q^μ.

(2) Charge, defined dynamically by the strength of soft-photon interactions, is additively conserved in all reactions. Gravitational mass, defined by the strength of soft graviton interactions, is equal to inertial mass for all nonrelativistic particles (and is twice the total energy for relativistic or massless particles).

Property (1) is actually a straightforward consequence of the well-known[2,3] Lorentz transformation properties of massless particle states, and is proven in Sec. II for massless particles of arbitrary integer spin. (It has already been proven for photons by D. Zwanziger.[4])

Property (2) does not at first sight appear to be derivable from property (1). Even in field theory (1) does not prove that the photon and graviton "currents" $J_\mu(x)$ and $\theta_{\mu\nu}(x)$ are conserved, but only that their matrix elements are conserved for light-like momentum transfer, so we cannot use the usual argument that $\int d^3x J^0(x)$ and $\int d^3x\theta^{0\mu}(x)$ are time-independent. And in pure *S*-matrix theory it is not even possible to define what we mean by the operators $J^\mu(x)$ and $\theta^{\mu\nu}(x)$.

We overcome these obstacles by a trick, which replaces the operator calculus of field theory with a little simple polology. After defining charge and gravitational mass as soft photon and graviton coupling constants in Sec. III, we prove in Sec. IV that if a reaction violates charge conservation, then the same process with inner bremsstrahlung of a soft extra photon would have an *S* matrix which does not satisfy property (1), and hence would not be Lorentz invariant; similarly, the inner bremstrahlung of a soft graviton would violate Lorentz invariance if any particle taking part in the reaction has an anomalous ratio of gravitational to inertial mass.

Appendices A, B, and C are devoted to some technical problems: (A) the transformation properties of polarization vectors, (B) the construction of tensor amplitudes for massless particles of general integer spin, and (C) the presence of kinematic singularities in the conventional $(2j+1)$-component "*M* functions."

A word may be needed about our use of *S*-matrix theory for particles of zero mass. We do not know whether it will ever be possible to formulate *S*-matrix

* Research supported by the U. S. Air Force Office of Scientific Research, Grant No. AF-AFOSR-232-63.

† Alfred P. Sloan Foundation Fellow.

[1] Some of the material of this article was discussed briefly in a recent letter [S. Weinberg Phys. Letters **9**, 357 (1964)]. We will repeat a few points here, in order that the present article be completely self-contained.

[2] E. P. Wigner, in *Theoretical Physics* (International Atomic Energy Agency, Vienna, 1963), p. 59. We have repeated Wigner's work in Ref. 3.

[3] S. Weinberg, Phys. Rev. **134**, B882 (1964).

[4] D. Zwanziger, Phys. Rev. **113**, B1036 (1964). Zwanziger omits some straightforward details, which are presented here in Appendix B.

theory as a complete dynamical theory even for strong interactions alone, and the presence of massless particles will certainly add a formidable technical difficulty, since every pole sits at the beginning of an infinite number of branch cuts. All such "infrared" problems are outside the scope of the present work. We shall simply make believe that there does exist an S-matrix theory, and that one of its consequences is that the S matrix has the same poles that it has in perturbation theory, with residues that factor in the same way as in perturbation theory. (We will lapse into the language of Feynman diagrams when we do our 2π bookkeeping in Sec. IV, but the reader will recognize in this the effects of our childhood training, rather than any essential dependence on field theory.)

When we refer to the "photon" or the "graviton" in this article, we assume no properties beyond their zero mass and spin 1 or 2. We will not attempt to explain why there should exist such massless particles, but may guess from perturbation theory that zero mass has a special kind of dynamical self-consistency for spins 1 and 2, which it would not have for spin 0.

Most of our work in the present article has a counterpart in Feynman-Dyson perturbation theory. In a future paper we will show how the Lorentz invariance of the S matrix forces the coupling of the photon and graviton "potentials" to take the same form as required by gauge invariance and the equivalence principle.

II. TENSOR AMPLITUDES FOR MASSLESS PARTICLES OF INTEGER SPIN

Let us consider a process in which a massless particle is emitted with momentum $\mathbf{q}$ and helicity $\pm j$. We shall call the S-matrix element simply $S_{\pm j}(\mathbf{q},p)$, letting p stand for the momenta and helicities of all other particles participating in the reaction. The Lorentz transformation property of S can be inferred from the well-known transformation law for one-particle states[2]; we find that

$$S_{\pm j}(\mathbf{q},p)=(|\Lambda\mathbf{q}|/|\mathbf{q}|)^{1/2}\times\exp[\pm ij\Theta(\mathbf{q},\Lambda)]S_{\pm j}(\Lambda\mathbf{q},\Lambda p). \quad (2.1)$$

The angle Θ is given in Appendix A as a function of the momentum $\mathbf{q}$ and the Lorentz transformation $\Lambda^\mu{}_\nu$.

We prove in Appendix B that, in consequence of (2.1), it is always possible for integer j to write $S_{\pm j}$ as the scalar product of a "polarization tensor" and what Stapp[5] would call an "M function":

$$S_{\pm j}(\mathbf{q},p)=(2|\mathbf{q}|)^{-1/2}\epsilon_\pm{}^{\mu_1*}(\mathbf{q})\cdots\times\epsilon_\pm{}^{\mu_j*}(\mathbf{q})M_{\pm\mu_1\cdots\mu_j}(\mathbf{q},p) \quad (2.2)$$

with M a symmetric tensor,[6] in the sense that

$$M_\pm{}^{\mu_1\cdots\mu_j}(\mathbf{q},p)=\Lambda_{\nu_1}{}^{\mu_1}\cdots\Lambda_{\nu_j}{}^{\mu_j}M_\pm{}^{\nu_1\cdots\nu_j}(\Lambda\mathbf{q},\Lambda p). \quad (2.3)$$

[5] M functions for massive particles were introduced by H. Stapp, Phys. Rev. **125**, 2139 (1962). See also A. O. Barut, I. Muzinich, and D. N. Williams, Phys. Rev. **130**, 442 (1963).

[6] We use a real metric, with signature $\{+++-\}$. Indices are raised and lowered in the usual way. The inverse of the Lorentz transformation $\Lambda^\mu{}_\nu$ is $[\Lambda^{-1}]^\mu{}_\nu=\Lambda_\nu{}^\mu$.

The polarization $\epsilon_\pm{}^\mu(\hat{q})$ is defined by

$$\epsilon_\pm{}^\mu(\hat{q})\equiv R(\hat{q})^\mu{}_\nu\epsilon_\pm{}^\nu, \quad (2.4)$$

where $R(\hat{q})$ is a standard rotation that carries the z axis into the direction of $\mathbf{q}$, and $\epsilon_\pm{}^\mu$ is the polarization for momentum in the z direction:

$$\epsilon_\pm{}^\mu\equiv\{1,\pm i,0,0\}/\sqrt{2}. \quad (2.5)$$

Some properties of $\epsilon_\pm{}^\mu(\hat{q})$ are obvious:

$$\epsilon_{\pm\mu}{}^*(\hat{q})\epsilon_\pm{}^\mu(\hat{q})=1, \quad (2.6)$$

$$\epsilon_{\pm\mu}(\hat{q})\epsilon_\pm{}^\mu(\hat{q})=0, \quad (2.7)$$

$$\epsilon_\pm{}^{\mu*}(\hat{q})=\epsilon_\mp{}^\mu(\hat{q}), \quad (2.8)$$

$$\epsilon_\pm{}^0(\hat{q})=0, \quad (2.9)$$

$$q_\mu\epsilon_\pm{}^\mu(\hat{q})=0, \quad (2.10)$$

$$\sum_\pm\epsilon_\pm{}^\mu(\hat{q})\epsilon_\pm{}^{\nu*}(\hat{q})=\Pi^{\mu\nu}(\hat{q})\equiv g^{\mu\nu}+(\bar{q}^\mu q^\nu+\bar{q}^\nu q^\mu)/|\mathbf{q}|^2,\quad [\bar{q}^\mu\equiv\{-\mathbf{q},|\mathbf{q}|\}], \quad (2.11)$$

$$\sum_\pm\epsilon_\pm{}^{\mu_1}(\hat{q})\epsilon_\pm{}^{\mu_2}(\hat{q})\epsilon_\pm{}^{\nu_1*}(\hat{q})\epsilon_\pm{}^{\nu_2*}(\hat{q})=\tfrac{1}{2}\{\Pi^{\mu_1\nu_1}(\hat{q})\Pi^{\mu_2\nu_2}(\hat{q})+\Pi^{\mu_1\nu_2}(\hat{q})\Pi^{\mu_2\nu_1}(\hat{q})-\Pi^{\mu_1\mu_2}(\hat{q})\Pi^{\nu_1\nu_2}(\hat{q})\}. \quad (2.12)$$

We also note the very important transformation rule, proved in Appendix A,

$$(\Lambda_\nu{}^\mu-q^\mu\Lambda_\nu{}^0/|\mathbf{q}|)\epsilon_\pm{}^\nu(\Lambda\hat{q})=\exp\{\pm i\Theta[\mathbf{q},\Lambda]\}\epsilon_\pm{}^\mu(\hat{q}), \quad (2.13)$$

with Θ the same angle as in (2.1).

If it were not for the q^μ term in (2.13), the polarization "tensor" $\epsilon_\pm{}^{\mu_1}\cdots\epsilon_\pm{}^{\mu_j}$ would be a true tensor, and the tensor transformation law (2.3) for $M_\pm{}^{\mu_1\cdots\mu_j}$ would be sufficient to ensure the correct behavior (2.1) of the S matrix. But $\epsilon_\pm{}^\mu$ is not a vector,[7] and (2.3) and (2.13) give the S-matrix transformation rule

$$S_{\pm j}(\mathbf{q},p)=(2|\mathbf{q}|)^{-1/2}\exp\{\pm ij\Theta(\mathbf{q},\Lambda)\}\times[\epsilon_\pm{}^{\mu_1}(\Lambda\hat{q})-(\Lambda q)^{\mu_1}\Lambda_\nu{}^0\epsilon_\pm{}^\nu(\Lambda\hat{q})/|\mathbf{q}|]^*\cdots\times[\epsilon_\pm{}^{\mu_j}(\Lambda\hat{q})-(\Lambda q)^{\mu_j}\Lambda_\nu{}^0\epsilon_\pm{}^\nu(\Lambda\hat{q})/|\mathbf{q}|]^*\times M_{\pm,\mu_1\cdots\mu_j}(\Lambda\mathbf{q},\Lambda p). \quad (2.14)$$

For an infinitesimal Lorentz transformation $\Lambda^\mu{}_\nu=\delta^\mu{}_\nu+\omega^\mu{}_\nu$, we can use (2.2) and the symmetry of M to put (2.14) in the form

$$S_{\pm j}(\mathbf{q},p)=(|\Lambda\mathbf{q}|/|\mathbf{q}|)^{1/2}\exp\{\pm ij\Theta(\mathbf{q},\Lambda)\}S_{\pm j}(\Lambda\mathbf{q},\Lambda p)-j(2|q|^3)^{-1/2}(\omega_\nu{}^0\epsilon_\pm{}^{\nu*}(\hat{q}))q^{\mu_1}\epsilon_\pm{}^{\mu_2*}(\hat{q})\cdots\times\epsilon_\pm{}^{\mu_j*}(\hat{q})M_{\pm,\mu_1\cdots\mu_j}(\mathbf{q},p). \quad (2.15)$$

Hence the necessary and sufficient condition that (2.14) agree with the correct Lorentz transformation property (2.1), is that $S_\pm$ vanish when one of the $\epsilon_\pm{}^\mu$ is replaced with q^μ:

$$q^{\mu_1}\epsilon_\pm{}^{\mu_2*}(\hat{q})\cdots\epsilon_\pm{}^{\mu_j*}(\hat{q})M_{\pm,\mu_1\cdots\mu_j}(\mathbf{q},p)=0. \quad (2.16)$$

For $j=1$ this may be expressed as the conservation

[7] The transformation rule (2.13) shows that $\epsilon_\pm{}^\mu(\hat{q})$ transforms according to one of the infinite-dimensional representations of the Lorentz group discussed by V. Bargmann and E. P. Wigner, Proc. Natl. Acad. Sci. **34**, 211 (1948).

condition

$$q_\mu M_\pm{}^\mu(\mathbf{q},p)=0. \tag{2.17}$$

For $j=2$ we conclude that

$$q_\mu M_\pm{}^{\mu\nu}(\mathbf{q},p)\propto q^\nu. \tag{2.18}$$

However, (2.7) shows that the subtraction of a term proportional to $g^{\mu\nu}$ from $M_\pm{}^{\mu\nu}$ will not alter the S matrix (2.2), so $M_\pm{}^{\mu\nu}$ can always be defined in such a way that (2.18) becomes

$$q_\mu M_\pm{}^{\mu\nu}(\mathbf{q},p)=0. \tag{2.19}$$

The condition (2.16) may look empty, since it can always be satisfied by a suitable adjustment of $M_\pm{}^{0\mu_2\cdots\mu_j}$, which in light of (2.9) will have no effect on the S matrix. But we cannot play with the time-like components of $M_\pm{}^{\mu_1\cdots\mu_j}$ and still keep it a tensor in the sense of (2.3). Neither (2.3) nor (2.16) is alone sufficient for Lorentz invariance, and together they constitute a nontrivial condition on $M_\pm{}^{\mu_1\cdots\mu_j}$.

Condition (2.16) may, if we wish, be described as "mass-shell gauge invariance," because it implies that the S matrix is invariant under a regauging of the polarization vector

$$\epsilon_\pm{}^\mu(\hat{q})\rightarrow\epsilon_\pm{}^\mu(\hat{q})+\lambda_\pm(\mathbf{q})q^\mu, \tag{2.20}$$

with $\lambda_\pm(\mathbf{q})$ arbitrary. It was purely for convenience that we started with the "Coulomb gauge" in (2.4), (2.5). [However, the theorem in Sec. III of Ref. 3 shows that it is *not* possible to construct an $\epsilon_\pm{}^\mu(\hat{q})$ which would satisfy (2.13) without any q^μ term.]

The S matrix for emission and absorption of several massless particles can be treated in the same way, except that $\epsilon^{\mu *}$ is replaced by ϵ^μ when a massless particle is absorbed.

III. DYNAMIC DEFINITION OF CHARGE AND GRAVITATIONAL MASS

We are going to define the charge and gravitational mass of a particle as its coupling constants to very-low-energy photons and gravitons, with "coupling constant" understood in the same sense as the Watson-Lepore pion-nucleon coupling constant. In general, such definitions are based on the fact that the S matrix has poles, corresponding to Feynman diagrams in which a virtual particle is exchanged between two sets of A and B of incoming and outgoing particles, with four-momentum nearly on its mass shell. The residue at the pole factors into Γ_A and Γ_B, the two "vertex amplitudes" Γ_A and Γ_B depending respectively only upon the quantum numbers of the particles in sets A and B, and of the exchanged particle. Hence it is possible to give a purely S-matrix-theoretic definition of the vertex amplitude Γ for any set of physical particles, as a function of their momenta and helicities; the coupling constant or constants define the magnitude of Γ. (As discussed in the introduction, we will not be concerned in this article with whether the above remarks can be proven rigorously in S-matrix theories involving massless particles, or with the related question of whether $m=0$ poles can really be separated from the branch cuts on which they lie. Our purpose is to explore the implications of the generally accepted ideas about the pole structure.)

Let us first consider the vertex amplitude for a very-low-energy massless particle of integer helicity $\pm j$, emitted by a particle of spin $J=0$, mass m (perhaps zero), and momentum $p^\mu=\{\mathbf{p},E\}$, with $E=(\mathbf{p}^2+m^2)^{1/2}$. (We are restricting ourselves here to very soft photons and gravitons, because we only want to define the charge and gravitational mass, and not the other electromagnetic and gravitational multipole moments.) The only tensor which can be used to form $M_\pm{}^{\mu_1\cdots\mu_j}$ is $p^{\mu_1}\cdots p^{\mu_j}$ [note that terms involving $g^{\mu\mu'}$ do not contribute to the S matrix, because of (2.7)] so the tensor character of $M_\pm{}^{\mu_1\cdots\mu_j}$ dictates the form of the vertex amplitude as

$$p_{\mu_1}\cdots p_{\mu_j}\epsilon_\pm{}^{\mu_1 *}(\hat{q})\cdots\epsilon_\pm{}^{\mu_j *}(\hat{q})/2E(\mathbf{p})(2|\mathbf{q}|)^{1/2}. \tag{3.1}$$

If the emitting particle has spin $J>0$, with initial and final helicities σ and σ' then (3.1) still gives a tensor M function if we multiply it by $\delta_{\sigma\sigma'}$; this is because the unit matrix has the Lorentz transformation property

$$\delta_{\sigma\sigma'}\rightarrow D_{\sigma\sigma''}{}^{(J)}(\mathbf{p},\Lambda)D_{\sigma'\sigma'''}{}^{(J)*}(\mathbf{p},\Lambda)\delta_{\sigma''\sigma'''}=\delta_{\sigma\sigma'}, \tag{3.2}$$

where $D^{(J)}(\mathbf{p},\Lambda)$ is the unitary spin-J representation of the Wigner rotation[8] (or its analog,[2] if $m=0$) associated with momentum $\mathbf{p}$ and Lorentz transformation Λ. However, the vertex amplitude so obtained is not unique. For instance if $J=\frac{1}{2}$ and $m>0$ then we get (3.1) times $\delta_{\sigma\sigma'}$ if we use a "current"[9]

$$\bar{\psi}\{\gamma_{\mu_1}p_{\mu_2}\cdots p_{\mu_j}+\text{permutations}\}\psi, \tag{3.3}$$

while using $\gamma_5\gamma_\mu$ in place of γ_μ would give a helicity-flip vertex amplitude.

At the end of the next section we will see that these other possibilities are prohibited by the Lorentz invariance of the total S matrix. Indeed, the only allowed vertex functions for soft massless particles of spin j are of the form (3.1) times $\delta_{\sigma\sigma'}$ for $j=1$ and $j=2$ (and none at all for $j\geq 3$). We may therefore define the soft photon coupling constant e, by the statement that the $j=1$ vertex amplitude is[10]

$$\frac{2ie(2\pi)^4\delta_{\sigma\sigma'}p_\mu\epsilon_\pm{}^{\mu *}(\hat{q})}{(2\pi)^{9/2}[2E(\mathbf{p})](2|\mathbf{q}|)^{1/2}}, \tag{3.4}$$

[8] E. P. Wigner, Ann. Math. **40**, 149 (1939). For a review, see S. Weinberg, Phys. Rev. **133**, B1318 (1964).

[9] For $j=2$, see I. Y. Kobsarev and L. B. Okun, Dubna (unpublished).

[10] Proper Lorentz invariance alone would allow different charges $e_\pm$ for photon helicities ± 1. Parity conservation would normally require that $e_+=e_-$ (with an appropriate convention for the photon parity). However if space inversion takes some particle into its antiparticle then its "right charge" e_+ will be equal to the "left charge" $\bar{e}_-$ of its antiparticle, and we will see in the next section that this gives $e_+=\bar{e}_-=-e_-$. In this case we speak of a magnetic monopole rather than a charge. The same conclusions can be drawn from CP conservation. We will not consider magnetic monopoles in this paper, though in fact none of our work in Sec. IV will depend on any relation between e_+ and e_-. Time-reversal invariance allows us to take e as real.

the factors 2, i, and π being separated from e in obedience to convention. And in the same way we may define a "gravitational charge" f, by the statement that the $j=2$ vertex amplitude is[11]

$$\frac{2if(8\pi G)^{1/2}(2\pi)^4\delta_{\sigma\sigma'}(p_\mu\epsilon_\pm{}^{\mu*}(\hat{q}))^2}{(2\pi)^{9/2}[2E(\mathbf{p})](2|\mathbf{q}|)^{1/2}}, \tag{3.5}$$

the extra factor $(8\pi G)^{1/2}$ (where G is Newton's constant) being inserted to make f dimensionless.

In order to see how e and f are related to the usual charge and gravitational mss, let us consider the near forward scattering of two particles with masses m_a and m_b, spins J_a and J_b, photon coupling constants e_a and e_b, and graviton coupling constants f_a and f_b. As the invariant momentum transfer $t=-(p_a-p_a')^2$ goes to zero, the S matrix becomes dominated by its one-photon-exchange and one-graviton-exchange poles. An elementary calculation[12] using (2.11) and (2.12) shows that for $t\to 0$, the S matrix becomes

$$\frac{\delta_{\sigma_a\sigma_a'}\delta_{\sigma_b\sigma_b'}}{4\pi^2E_aE_bt}[e_ae_b(p_a\cdot p_b) +8\pi Gf_af_b\{(p_a\cdot p_b)^2-m_a{}^2m_b{}^2/2\}]. \tag{3.6}$$

If particle b is at rest, this gives

$$\frac{\delta_{\sigma_a\sigma_a'}\delta_{\sigma_b\sigma_b'}}{\pi t}\left[-\frac{e_ae_b}{4\pi}+Gf_a\left\{2E_a-\frac{m_a{}^2}{E_a}\right\}f_bm_b\right]. \tag{3.7}$$

Hence we may identify e_a as the *charge* of particle a, while its effective *gravitational mass* is

$$\tilde{m}_a=f_a\{2E_a-(m_a{}^2/E_a)\}. \tag{3.8}$$

If particle a is nonrelativistic, then $E_a\cong m_a$, and (3.8) gives its gravitational rest mass as

$$\tilde{m}_a=f_am_a. \tag{3.9}$$

[11] Proper Lorentz invariance alone would not rule out different values for the gravitational charges $f_\pm$ for gravitons of helicity ± 2. Parity conservation (with an appropriate convention for the graviton parity) requires that $f_+=f_-$. This conclusion holds even for the magnetic monopole case discussed in footnote 10, since then $f_+=\bar{f}_-$, and we will see in Sec. IV that the antiparticle has "left gravitational charge" $\bar{f}_-$ equal to f_-. The same conclusions can be drawn from CP conservation. Time-reversal invariance allows us to take f as real.

[12] The residue of the pole at $t=0$ can be most easily calculated by adopting a coordinate system in which $q\equiv p_{a'}-p_a=p_b-p_{b'}$ is a finite real light-like four-vector, while p_a, p_b, $p_{a'}$, $p_{b'}$ are on their mass shells, and hence necessarily complex. Then the gradient terms in (2.11) and (2.12) do not contribute, because $q\cdot p_a=q\cdot p_b=0$, so that $\Pi_{\mu\nu}$ may be replaced by $g_{\mu\nu}$, yielding (3.6). We are justified in using (3.6) in the physical region (where p_a, p_b, $p_{a'}$, $p_{b'}$ are real and q is small, though *not* in the direction of the light cone) because Lorentz invariance tells us that the matrix element depends only upon s and t. Lorentz invariance is actually far from trivial in a perturbation theory based on physical photons and gravitons, since then the Coulomb force and Newtonian attraction must be explicitly introduced into the interaction in order to get the invariant S matrix (3.6). (Such a perturbation theory will be discussed in an article now in preparation.) The Lorentz-invariant extrapolation of (3.6) into the physical region of small t is the analog, in S-matrix theory, of the introduction of the Coulomb and Newton forces in perturbation theory.

On the other hand, if a is massless or extremely relativistic, then $E_a\gg m_a$ and (3.8) gives

$$\tilde{m}_a=2f_aE_a. \tag{3.10}$$

[Formulas (3.8) or (3.10) should not of course be understood to mean anything more than already stated in (3.7). However, they serve to remind us that the response of a massless particle to a static gravitational field is finite, and proportional to f.]

The presence of massless particles in the initial or final state will also generate poles in the S matrix, which, like that in (3.7), lie on the edge of the physical region. It is therefore possible to measure the coupling constants e and f in a variety of process, such as Thomson scattering or soft bremsstrahlung, or their analogs for gravitons. All these different experiments will give the same value for any given particle's e or f, for purely S-matrix-theoretic reasons. The task before us is to show how the e's and f's are related for different particles.

IV. CONSERVATION OF e AND UNIVERSALITY OF f

Let $S_{\beta\alpha}$ be the S matrix for some reaction $\alpha\to\beta$, the states α and β consisting of various charged and uncharged particles, perhaps including gravitons and photons. The same reaction can also occur with emission of a very soft extra photon or graviton of momentum $\mathbf{q}$ and helicity ± 1, or ± 2, and we will denote the corresponding S-matrix element as $S_{\beta\alpha}{}^{\pm 1}(\mathbf{q})$ or $S_{\beta\alpha}{}^{\pm 2}(\mathbf{q})$.

These emission matrix elements will have poles at $\mathbf{q}=0$, corresponding to the Feynman diagrams in which the extra photon or graviton is emitted by one of the incoming or outgoing particles in states α or β. The poles arise because the virtual particle line connecting the photon or graviton vertex with the rest of the diagram gives a vanishing denominator

$$\begin{aligned}1/[(p_n+q)^2+m_n{}^2]&=1/2p_n\cdot q \quad \text{(particle } n \text{ outgoing)},\\ 1/[(p_n-q)^2+m_n{}^2]&=-1/2p_n\cdot q \quad \text{(particle } n \text{ incoming)}.\end{aligned} \tag{4.1}$$

For $|\mathbf{q}|$ sufficiently small, these poles will completely dominate the emission-matrix element. The singular factor (4.1) will be multiplied by a factor $-i(2\pi)^{-4}$ associated with the extra internal line, a factor

$$\frac{2ie[p_n\cdot\epsilon_\pm{}^*(\hat{q})](2\pi)^4}{(2\pi)^{3/2}(2|\mathbf{q}|)^{1/2}} \tag{4.2}$$

or

$$\frac{2if(8\pi G)^{1/2}[p_n\cdot\epsilon_\pm{}^*(\hat{q})]^2(2\pi)^4}{(2\pi)^{3/2}(2|\mathbf{q}|)^{1/2}} \tag{4.3}$$

arising from the vertices (3.4) or (3.5), and a factor $S_{\beta\alpha}$ for the rest of the diagram. Hence the S matrix for soft photon or graviton emission is given in the limit

$\mathbf{q}\to 0$ by[13–15]

$$S_{\beta\alpha}^{\pm1}(\mathbf{q})\to(2\pi)^{-3/2}(2|\mathbf{q}|)^{-1/2}\times\Big[\sum_n \eta_n e_n\frac{[p_n\cdot\epsilon_\pm^*(\hat{q})]}{(p_n\cdot q)}\Big]S_{\beta\alpha}\quad(4.4)$$

or

$$S_{\beta\alpha}^{\pm2}(\mathbf{q})\to(2\pi)^{-3/2}(2|\mathbf{q}|)^{-1/2}(8\pi G)^{1/2}\times\Big[\sum_n \eta_n f_n\frac{[p_n\cdot\epsilon_\pm^*(q)]^2}{(p_n\cdot q)}\Big]S_{\beta\alpha},\quad(4.5)$$

the sign η_n being $+1$ or -1 according to whether particle n is outgoing or incoming.

These emission matrices are of the general form (2.2), i.e.,

$$S_{\beta\alpha}^{\pm1}(\mathbf{q})\to(2|\mathbf{q}|)^{-1/2}\epsilon_\pm^{\mu*}(\hat{q})M_\mu(\mathbf{q},\alpha\to\beta),\quad(4.6)$$

$$S_{\beta\alpha}^{\pm2}(\mathbf{q})\to(2|\mathbf{q}|)^{-1/2}\epsilon_\pm^{\mu*}(\hat{q})\epsilon_\pm^{\nu*}(\hat{q})M_{\mu\nu}(\mathbf{q},\alpha\to\beta),\quad(4.7)$$

where M_μ and $M_{\mu\nu}$ are tensor M functions

$$M^\mu(\mathbf{q},\alpha\to\beta)=(2\pi)^{-3/2}[\sum_n \eta_n e_n p_n^\mu/(p_n\cdot q)]S_{\beta\alpha},\quad(4.8)$$

$$M^{\mu\nu}(\mathbf{q},\alpha\to\beta)=(2\pi)^{-3/2}(8\pi G)^{1/2}\times[\sum_n \eta_n f_n p_n^\mu p_n^\nu/(p_n\cdot q)]S_{\beta\alpha}.\quad(4.9)$$

However, we have learned in Sec. II that the covariance of M_μ and $M_{\mu\nu}$ is not sufficient by itself to guarantee the Lorentz invariance of the S matrix; Lorentz invariance also requires the vanishing of (2.2) when any one $\epsilon_\pm^\mu(\hat{q})$ is replaced with q^μ. For photons this implies (2.17), i.e.,

$$0=q^\mu M_\mu(\mathbf{q},\alpha\to\beta)=(2\pi)^{-3/2}[\sum_n \eta_n e_n]S_{\beta\alpha},\quad(4.10)$$

so if $S_{\beta\alpha}$ is not to vanish, *the transition* $\alpha\to\beta$ *must conserve charge*, with

$$\sum_n \eta_n e_n=0.\quad(4.11)$$

For gravitons Lorentz invariance requires (2.18), which here takes the simpler form (2.19)

$$0=q_\mu M^{\mu\nu}(\mathbf{q},\alpha\to\beta)=(2\pi)^{-3/2}(8\pi G)^{1/2}[\sum_n \eta_n f_n p_n^\nu]S_{\beta\alpha}.\quad(4.12)$$

But the p_n^μ are arbitrary four-momenta, subject only to the condition of energy momentum conservation:

$$\sum_n \eta_n p_n^\mu=0.\quad(4.13)$$

The requirement that (4.12) vanish for all such p_n^μ, can be met if and only if *all particles have the same gravitational charge.* The conventional definition of Newton's constant G is such as to make the common value of the f_n unity, so

$$f_n=1\quad(\text{all } n)\quad(4.14)$$

and (3.8) then tells us that any particle with inertial mass m and energy E has effective gravitational mass

$$\tilde{m}=2E-m^2/E.\quad(4.15)$$

In particular, a particle at rest has gravitational mass $\tilde{m}$ equal to its inertial mass m.

It seems worth emphasizing that our proof also applies when some particle n in the initial or final state is itself a graviton. Hence the graviton must emit and absorb single soft gravitons (and therefore respond to a uniform gravitational field) with gravitational mass $2E$. It would be conceivable to have a universe in which all f_n vanish, but since we know that soft gravitons interact with matter, they must also interact with gravitons.

Having reached our goal, we may look back, and see that no other vertex amplitudes could have been used for $\mathbf{q}\to 0$ except (3.4) and (3.5). A helicity-flip or helicity-dependent vertex amplitude could never give rise to the cancellations between different poles [as in (4.10) and (4.12)] needed to satisfy the Lorentz invariance conditions (2.17) and (2.19). It is also interesting that such cancellations cannot occur for massless particles of integer spin higher than 2. For suppose we take the vertex amplitude for emission of a soft massless particle of helicity $\pm j$ ($j=3, 4, \cdots$) as

$$\frac{2ig^{(j)}(2\pi)^4(\epsilon_\pm^*(\hat{q})\cdot p)^j\delta_{\sigma\sigma'}}{(2\pi)^{9/2}[2E(\mathbf{p})](2|\mathbf{q}|)^{1/2}}\quad(4.16)$$

in analogy with (3.4) and (3.5), the S matrix $S_{\beta\alpha}^{\pm j}(\mathbf{q})$ for emission of this particle in a reaction $\alpha\to\beta$ will be given in the limit $\mathbf{q}\to 0$ by

$$S_{\beta\alpha}^{\pm j}(\mathbf{q})\to(2\pi)^{-3/2}(2|\mathbf{q}|)^{-1/2}\times[\sum_n \eta_n g_n^{(j)}[p_n\cdot\epsilon_\pm^*(\hat{q})]^j/(p_n\cdot q)]S_{\beta\alpha}.\quad(4.17)$$

This is only Lorentz invariant if it vanishes when any one $\epsilon_\pm^\mu$ is replaced with q^μ, so we must have

$$\sum_n \eta_n g_n^{(j)}[p_n\cdot\epsilon_\pm^*(\hat{q})]^{j-1}=0.\quad(4.18)$$

But there is no way that this can be satisfied for all momenta p_n obeying (4.13), unless $j=1$ or $j=2$. This

---

[13] Formula (4.4) is well known to hold to all orders in quantum electrodynamic perturbation theory. See, for example, J. M. Jauch and F. Rohrlich, *Theory of Photons and Electrons* (Addison-Wesley Publishing Company, Inc., Reading, Massachusetts, 1955), p. 392, and F. E. Low, Ref. 14.

[14] It has been shown by F. E. Low, Phys. Rev. **110**, 974 (1958), that the next term in an expansion of the S matrix in powers of $|\mathbf{q}|$ is uniquely determined by the electromagnetic multipole moments of the participating particles and by $S_{\beta\alpha}$. However, this next (zeroth-order) term is Lorentz-invariant for any values of the multipole moments.

[15] Relations like (4.4) and (4.5) are also valid if $S_{\beta\alpha}^{\pm1}(\mathbf{q})$, $S_{\beta\alpha}^{\pm2}(\mathbf{q})$, and $S_{\beta\alpha}$ are interpreted as the effective matrix elements for the transition $\alpha\to\beta$, respectively, with or without one extra soft photon or graviton of momentum $\mathbf{q}$, *plus* any number of unobserved soft photons or gravitons with total energy less than some small resolution ΔE. [For a proof in quantum-electrodynamic perturbation theory, see, for example, D. R. Yennie and H. Suura, Phys. Rev. **105**, 1378 (1957). The same is undoubtedly true also for gravitons, and in pure S-matrix theory.]

is not to say that massless particles of spin 3 or higher cannot exist, but only that they cannot interact at zero frequency, and hence cannot generate macroscopic fields. And similarly, the uniqueness of the vertex amplitudes (3.4) and (3.5) does not show that electromagnetism and gravitation conserve parity, but only that parity must be conserved by zero-frequency photons and gravitons.

The crucial point in our proof is that the emission of soft photons or gravitons generates poles which individually make non-Lorentz-invariant contributions to the S matrix. Only the sum of the poles is Lorentz-invariant, and then only if e is conserved and f is universal. Just as the universality of f can be expressed as the equality of gravitational and inertial mass, the conservation of e can be stated as the equality of charge defined dynamically, with a quantum number defined by an additive conservation law. But, however, we state them, these two facts are the outstanding dynamical peculiarities of photons and gravitons, which until now have been proven only under the *a priori* assumption of a gauge-invariant or generally covariant Lagrangian density.

ACKNOWLEDGMENTS

I am very grateful for helpful conversations with N. Cabibbo, E. Leader, S. Mandelstam, H. Stapp, E. Wichmann, and C. Zemach.

APPENDIX A: POLARIZATION VECTORS AND THE LITTLE GROUP

In this Appendix we shall discuss the "little group"[2] for massless particles, with the aim of defining the angle $\Theta(\mathbf{q},\Lambda)$, and of determining the transformation properties of the polarization vectors $\epsilon_\pm(\hat{q})$.

The little group is defined as consisting of all Lorentz transformations $\mathcal{R}^\mu{}_\nu$, which leave invariant a standard light-like four-vector K^μ:

$$\mathcal{R}^\mu{}_\nu K^\nu = K^\mu\,, \tag{A1}$$

$$K^1 = K^2 = 0\,,\quad K^3 = K^0 = \kappa > 0\,. \tag{A2}$$

It is a matter of simple algebra to show that the most general such $\mathcal{R}^\mu{}_\nu$ can be written as a function of three parameters Θ, X^1, X^2:

$$\mathcal{R}^\mu{}_\nu = \begin{bmatrix} \cos\Theta & \sin\Theta & -X_1\cos\Theta - X_2\sin\Theta & X_1\cos\Theta + X_2\sin\Theta \\ -\sin\Theta & \cos\Theta & X_1\sin\Theta - X_2\cos\Theta & -X_1\sin\Theta + X_2\cos\Theta \\ X_1 & X_2 & 1 - X^2/2 & X^2/2 \\ X_1 & X_2 & -X^2/2 & 1 + X^2/2 \end{bmatrix}. \tag{A3}$$

$$X^2 \equiv X_1{}^2 + X_2{}^2\,.$$

(The rows and columns are in order 1, 2, 3, 0.) Wigner[2] has noted that this group is isomorphic to the group of rotations (by angle Θ) and translations (by vector $\{X_1,X_2\}$) in the Euclidean plane. In particular the "translations" form an invariant Abelian subgroup, defined by the condition $\Theta=0$, and are represented on the physical Hilbert space by unity. It is possible to factor any $\mathcal{R}^\mu{}_\nu$ into

$$\mathcal{R}^\mu{}_\nu = \begin{bmatrix} \cos\Theta & \sin\Theta & 0 & 0 \\ -\sin\Theta & \cos\Theta & 0 & 0 \\ 0 & 0 & 1 & 0 \\ 0 & 0 & 0 & 1 \end{bmatrix} \begin{bmatrix} 1 & 0 & -X_1 & X_1 \\ 0 & 1 & -X_2 & X_2 \\ X_1 & X_2 & 1-X^2/2 & X^2/2 \\ X_1 & X_2 & -X^2/2 & 1+X^2/2 \end{bmatrix}. \tag{A4}$$

The representation of $\mathcal{R}^\mu{}_\nu$ on physical Hilbert space is determined solely by the first factor, so

$$U[\mathcal{R}] = \exp(i\Theta[\mathcal{R}]J_3)\,. \tag{A5}$$

In discussing the transformation rules for massless particles it is necessary to consider members of the little group defined by

$$\mathcal{R}(\mathbf{q},\Lambda) = \mathcal{L}^{-1}(\mathbf{q})\Lambda^{-1}\mathcal{L}(\Lambda\mathbf{q})\,. \tag{A6}$$

Here Λ is an arbitrary Lorentz transformation, and $\mathcal{L}(\mathbf{q})$ is the Lorentz transformation:

$$\mathcal{L}^\mu{}_\nu(\mathbf{q}) = R^\mu{}_\rho(\hat{q})B^\rho{}_\nu(|\mathbf{q}|)\,, \tag{A7}$$

where $B(|\mathbf{q}|)$ is a "boost" along the z axis, with nonzero components

$$\begin{aligned} B^1{}_1 &= B^2{}_2 = 1\,, \\ B^3{}_3 &= B^0{}_0 = \cosh\varphi\,, \\ B^3{}_0 &= B^0{}_3 = \sinh\varphi\,, \\ \varphi &\equiv \log(|q|/\kappa)\,, \end{aligned} \tag{A8}$$

and $R(\hat{q})$ is the rotation introduced in (2.4), which takes the z axis into the direction of $\mathbf{q}$. The transformation $\mathcal{L}(\mathbf{q})$ takes the standard four-momentum K^μ [see (A2)] into $q^\mu \equiv \{\mathbf{q},|\mathbf{q}|\}$:

$$\mathcal{L}^\mu{}_\nu(\mathbf{q})K^\nu = q^\mu \tag{A9}$$

so therefore,

$$\begin{aligned} \mathcal{R}^\mu{}_\nu(\mathbf{q},\Lambda)K^\nu &= [\mathcal{L}^{-1}(\mathbf{q})\Lambda^{-1}]^\mu{}_\nu(\Lambda q)^\nu \\ &= [\mathcal{L}^{-1}(\mathbf{q})]^\mu{}_\nu q^\nu = K^\mu\,. \end{aligned} \tag{A10}$$

Hence $\mathcal{R}(\mathbf{q},\Lambda)$ does belong to the little group.

It was shown in Ref. 3 that, as a consequence of (A5), the S matrix obeys the transformation rule (2.1), with $\Theta(\mathbf{q},\Lambda)$ given as the Θ angle of $\mathcal{R}(\mathbf{q},\Lambda)$:

$$\Theta(\mathbf{q},\Lambda)=\Theta[\mathcal{L}^{-1}(\mathbf{q})\Lambda^{-1}\mathcal{L}(\Lambda\mathbf{q})]. \qquad (A11)$$

We now turn to the polarization "vectors" $\epsilon_\pm{}^\mu(\hat{q})$, defined in Sec. II by

$$\epsilon_\pm{}^\mu(\mathbf{q})=R^\mu{}_\nu(\hat{q})\epsilon_\pm{}^\nu, \qquad (A12)$$

$$\epsilon_\pm{}^\mu\equiv\{1,\,\pm i,\,0,\,0\}/\sqrt{2}. \qquad (A13)$$

Observe that we could just as well write (A12) as

$$\epsilon_\pm{}^\mu(q)=\mathcal{L}^\mu{}_\nu(\mathbf{q})\epsilon_\pm{}^\nu \qquad (A14)$$

since $B(|\mathbf{q}|)$ has no effect on $\epsilon_\pm$.

An arbitrary $\mathcal{R}^\mu{}_\nu$ of the form (A3) will transform $\epsilon_\pm{}^\nu$ into

$$\mathcal{R}^\mu{}_\nu\epsilon_\pm{}^\nu=\exp(\pm i\Theta[\mathcal{R}])\epsilon_\pm{}^\mu+X_\pm[\mathcal{R}]K^\mu, \qquad (A15)$$

where

$$X_\pm[\mathcal{R}]=\frac{X_1[\mathcal{R}]\pm iX_2[\mathcal{R}]}{\kappa\sqrt{2}}. \qquad (A16)$$

If we let $\mathcal{R}$ be the transformation (A6), and use (A14), then (A15) gives

$$[\mathcal{L}^{-1}(\mathbf{q})\Lambda^{-1}]^\mu{}_\nu\epsilon_\pm{}^\nu(\Lambda\mathbf{q})=\exp[\pm i\Theta(\mathbf{q},\Lambda)]\epsilon_\pm{}^\mu+X_\pm(\mathbf{q},\Lambda)K^\mu, \qquad (A17)$$

where

$$X_\pm(\mathbf{q},\Lambda)\equiv\frac{X_1[\mathcal{L}^{-1}(\mathbf{q})\Lambda^{-1}\mathcal{L}(\Lambda\mathbf{q})]\pm iX_2[\mathcal{L}^{-1}(\mathbf{q})\Lambda^{-1}\mathcal{L}(\Lambda\mathbf{q})]}{\kappa\sqrt{2}}. \qquad (A18)$$

Multiplying (A17) by $\mathcal{L}(\mathbf{q})$, we have the desired result

$$\Lambda_\nu{}^\mu\epsilon_\pm{}^\nu(\Lambda\mathbf{q})=\exp[\pm i\Theta(\mathbf{q},\Lambda)]\epsilon_\pm{}^\mu(\mathbf{q})+X_\pm(\mathbf{q},\Lambda)q^\mu. \qquad (A19)$$

Note that it is the "translations" which at the same time make the little group non-semi-simple, and which yield the gradient term in (A19).

The quantity $X_\pm(\mathbf{p},\Lambda)$ may be found in terms of $\epsilon_\pm(\mathbf{q})$ by setting $\mu=0$ in (A19):

$$X_\pm(\mathbf{q},\Lambda)|\mathbf{q}|=\Lambda_\nu{}^0e_\pm{}^\nu(\Lambda\mathbf{q}). \qquad (A20)$$

Hence we may rewrite (A19) as a homogeneous transformation rule:

$$(\Lambda_\nu{}^\mu-\Lambda_\nu{}^0q^\mu/|\mathbf{q}|)\epsilon_\pm{}^\nu(\Lambda\mathbf{q})=\exp[\pm i\Theta(\mathbf{q},\Lambda)]\epsilon_\pm{}^\mu(\hat{q}) \qquad (A21)$$

or, recalling that $\epsilon_\pm{}^0\equiv0$,

$$(\Lambda_i{}^j-\Lambda_i{}^0\hat{q}^j)e_\pm{}^i(\Lambda\mathbf{q})=\exp[\pm i\Theta(\mathbf{q},\Lambda)]\epsilon_\pm{}^j(\hat{q}).$$

This also incidentally shows that $\Theta(\mathbf{q},\Lambda)$ does not depend on $|\mathbf{q}|$.

We have not had to define the rotation $R(\hat{q})$ any further than by just specifying that it carries the z axis into the direction of q. However, the reader may wish to see explicit expressions for the polarization vectors, so we will consider one particular standardization of $R(\hat{q})$. Write $\hat{q}$ in the form

$$\hat{q}=\{-\sin\beta\cos\gamma,\,\sin\beta\sin\gamma,\,\cos\beta\} \qquad (A22)$$

and let $R(\hat{q})$ be the rotation with Euler angles 0, β, γ:

$$R^\mu{}_\nu(\hat{q})=\begin{bmatrix}\cos\beta\cos\gamma & \sin\gamma & -\sin\beta\cos\gamma & 0\\ -\cos\beta\sin\gamma & \cos\gamma & \sin\beta\sin\gamma & 0\\ \sin\beta & 0 & \cos\beta & 0\\ 0 & 0 & 0 & 1\end{bmatrix}. \qquad (A23)$$

Then (2.4) and (2.5) give

$$\epsilon_\pm{}^\mu(\hat{q})=\{\cos\beta\cos\gamma\pm i\sin\gamma,\,-\cos\beta\sin\gamma\pm i\cos\gamma,\,\sin\beta,\,0\}/\sqrt{2}\quad(\mu=1,2,3,0). \qquad (A24)$$

We can easily check (2.6)–(2.12) explicitly for (A24).

APPENDIX B: CONSTRUCTION OF TENSOR AMPLITUDES

We consider a reaction in which is emitted a massless particle of momentum $\mathbf{q}$ and integer helicity $\pm j$, all other particle variables being collected in the single symbol p. Let us first divide the set of all possible $\{\mathbf{q},p\}$ into disjoint equivalence classes, $\{\mathbf{q},p\}$ being equivalent to $\{\mathbf{q}',p'\}$ if one can be transformed into the other by a Lorentz transformation. (This is an equivalence relation, because the Lorentz group is a group.) The axiom of choice allows us to make an arbitrary selection of one set of standard values $\{\mathbf{q}_c,p_c\}$ from each equivalence class, so any $\{\mathbf{q},p\}$ determines a unique standard $\{\mathbf{q}_c,p_c\}$, such that for some Lorentz transformation $L^\mu{}_\nu$ we have

$$\mathbf{q}=L\mathbf{q}_c,\quad p=Lp_c. \qquad (B1)$$

It will invariably be the case in physical processes that the only $\Lambda^\mu{}_\nu$ leaving both $\mathbf{q}$ and p invariant is the identity $\delta^\mu{}_\nu$, so the $L^\mu{}_\nu$ in (B1) is uniquely determined by $\mathbf{q}$ and p. (This is true, for instance, if p stands for two or more general four-momenta.) Hence the arguments $\{\mathbf{q},p\}$ stand in one-to-one relation to the variables $\{\mathbf{q}_c,p_c,L\}$.

Now let us construct an $M_\pm{}^{\mu_1\cdots\mu_j}(\mathbf{q}_c,p_c)$ satisfying (2.2) for each standard $\{\mathbf{q}_c,p_c\}$. A suitable choice is

$$M_\pm{}^{\mu_1\cdots\mu_j}(\mathbf{q}_c,p_c)\equiv(2|\mathbf{q}_c|)^{1/2}\epsilon_\pm{}^{\mu_1}(\hat{q}_c)\cdots\epsilon_\pm{}^{\mu_j}(\hat{q}_c)S_{\pm j}(\mathbf{q}_c,p_c), \qquad (B2)$$

which satisfies (2.2) because of (2.6). The tensor amplitude for a general $\mathbf{q}$, p is then defined by

$$M_\pm{}^{\mu_1\cdots\mu_j}(\mathbf{q},p)\equiv L^{\mu_1}{}_{\nu_1}(\mathbf{q},p)\cdots L^{\mu_j}{}_{\nu_j}(\mathbf{q},p)M_\pm{}^{\nu_1\cdots\nu_j}(\mathbf{q}_c,p_c), \qquad (B3)$$

where $\mathbf{q}_c$, p_c, and $L(\mathbf{q},p)$ are the standard variables and Lorentz transformation defined by (B1). With this definition we can easily show that $M_\pm{}^{\mu_1\cdots\mu_j}$ is a tensor, because

$$M_\pm{}^{\mu_1\cdots\mu_j}(\mathbf{q},p)=L^{\mu_1}{}_{\nu_1}(\mathbf{q},p)\cdots L^{\mu_j}{}_{\nu_j}(\mathbf{q},p)L_{\rho_1}{}^{\nu_1}(\Lambda\mathbf{q},\Lambda p)\cdots L_{\rho_j}{}^{\nu_j}(\Lambda\mathbf{q},\Lambda p)M_\pm{}^{\rho_1\cdots\rho_j}(\Lambda\mathbf{q},\Lambda p)$$
$$=\Lambda_{\rho_1}{}^{\mu_1}\cdots\Lambda_{\rho_j}{}^{\mu_j}M_\pm{}^{\rho_1\cdots\rho_j}(\Lambda\mathbf{q},\Lambda p), \quad \text{(B4)}$$

the latter equality holding because $L(\mathbf{q},p)L^{-1}(\Lambda\mathbf{q},\Lambda p)$ induces the transformation $\{\Lambda\mathbf{q},\Lambda p\}\rightarrow\{\mathbf{q}_c,p_c\}\rightarrow\{\mathbf{q},p\}$ and hence must be just Λ^{-1}.

We must now show that (B.3) satisfies (2.2) for all $\{\mathbf{q},p\}$. The Lorentz transformation property (2.13) of $\epsilon_\pm{}^\mu$ can be written as

$$\epsilon_\pm{}^\mu(\hat{q})=\exp\{\mp i\Theta(\hat{q},L^{-1}(\mathbf{q},p))\}[L^\mu{}_\nu(\mathbf{q},p)-q^\mu L^0{}_\nu(\mathbf{q},p)/|\mathbf{q}|]\epsilon_\pm{}^\nu(\hat{q}_c).$$

Hence, (B.3) gives

$$\epsilon_\pm{}^{\mu_1*}(\hat{q})\cdots\epsilon_\pm{}^{\mu_j*}(\hat{q})M_{\mu_1\cdots\mu_j}(\hat{q},p)=\exp\{\pm ij\Theta(\hat{q},L^{-1}(\hat{q},p))\}[\epsilon_\pm{}^{\mu_1}(\hat{q}_c)-q_c{}^{\mu_1}\epsilon_\pm{}^{\nu_1}(\hat{q}_c)L^0{}_{\nu_1}(\mathbf{q},p)/|\mathbf{q}|]^*\cdots$$
$$\times[\epsilon_\pm{}^{\mu_j}(\hat{q}_c)-q_c{}^{\mu_j}\epsilon_\pm{}^{\nu_j}(\hat{q}_c)L^0{}_{\nu_j}(\mathbf{q},p)/|\mathbf{q}|]^*M_{\mu_1\cdots\mu_j}(\mathbf{q}_c,p_c). \quad \text{(B5)}$$

But (B2) and (2.10) show that all $q_c{}^\mu$ terms may be dropped, because

$$q_c{}^{\mu_r}M_{\mu_1\cdots\mu_j}(\mathbf{q}_c,p_c)=0, \quad \text{(B6)}$$

so (B5) simplifies to

$$\epsilon_\pm{}^{\mu_1*}(\hat{q})\cdots\epsilon_\pm{}^{\mu_j*}(\hat{q})M_{\mu_1\cdots\mu_j}(\mathbf{q},p)=\exp\{\pm ij\Theta(\hat{q},L^{-1}(\mathbf{q},p))\}\epsilon_\pm{}^{\mu_1*}(\hat{q}_c)\cdots\epsilon_\pm{}^{\mu_j*}(\hat{q}_c)M_{\mu_1\cdots\mu_j}(\mathbf{q}_c,p_c) \quad \text{(B7)}$$

or, using (B2) and (2.6),

$$(2|\mathbf{q}|)^{-1/2}\epsilon_\pm{}^{\mu_1*}(\hat{q})\cdots\epsilon_\pm{}^{\mu_j*}(\hat{q})M_{\mu_1\cdots\mu_j}(\mathbf{q},p)=(|\mathbf{q}_c|/|\mathbf{q}|)^{1/2}\exp\{\pm ij\Theta(\hat{q},L^{-1}(\mathbf{q},p))\}S_{\pm j}(q_c,p_c). \quad \text{(B8)}$$

The right-hand side is just the formula for $S_{\pm j}(q,p)$ obtained by setting $\Lambda=L^{-1}(\mathbf{q},p)$ in (2.1), so (B8) gives finally

$$S_{\pm j}(\mathbf{q},p)=(2|\mathbf{q}|)^{-1/2}\epsilon_\pm{}^{\mu_1*}(\hat{q})\cdots\times\epsilon_\pm{}^{\mu_j*}(\hat{q})M_{\mu_1\cdots\mu_j}(\mathbf{q},p). \quad \text{(B9)}$$

It should be noted that (B2) is *not* valid for all $\mathbf{q}$, p, since then $M_\pm{}^{0\mu_2\cdots\mu_j}(\mathbf{q},p)$ would vanish in all Lorentz frames, and $M_\pm$ could hardly then be a tensor.

APPENDIX C: $(2j+1)$-COMPONENT M FUNCTIONS

It has become customary[5] to write the S matrix for massive particles of spin j in terms of $2j+1$-component M functions, which transform under the $(j,0)$ or $(0,j)$ representation of the homogeneous Lorentz group. In contrast, the symmetric-tensor M functions used here transform according to the $(j/2, j/2)$ representation. The massless-particle S matrix could also have been written in terms of a conventional $(2j+1)$-component M function, but only at the price of giving the M function a very peculiar pole structure.

To see what sort of peculiarities can occur for zero mass, let us consider the emission of a very soft photon in a reaction like Compton scattering, in which there is only one charged particle in the initial state α and in the final state β. The S-matrix element is then given by (4.4) as

$$S_{\beta\alpha}{}^{\pm 1}(\mathbf{q})\rightarrow(2\pi)^{-3/2}(2|\mathbf{q}|)^{-1/2}\times e\left[\frac{p_\mu}{(p\cdot q)}-\frac{p'_\mu}{(p'\cdot q)}\right]\epsilon_\pm{}^{\mu*}(\hat{q})S_{\beta\alpha}, \quad \text{(C1)}$$

where p and p' are the initial and final charged-particle momenta. This may be rewritten as

$$S_{\beta\alpha}{}^\pm(\mathbf{q})\rightarrow(2|\mathbf{q}|)^{-1/2}M_{[\mu,\nu]}(\mathbf{q},\alpha\rightarrow\beta)\times\{q^{\nu*}\epsilon_\pm{}^\mu(\hat{q})-q^\mu\epsilon_\pm{}^{\nu*}(\hat{q})\}, \quad \text{(C2)}$$

where $M_{[\mu,\nu]}$ is a $(1,0)\oplus(0,1)$ M function

$$M_{[\mu,\nu]}(\mathbf{q},\alpha\rightarrow\beta)=\frac{e[p_\mu p'_\nu-p_\nu p'_\mu]S_{\beta\alpha}}{(2\pi)^{3/2}(p\cdot q)(p'\cdot q)}. \quad \text{(C3)}$$

It can be shown that $S_{\beta\alpha}{}^+$ and $S_{\beta\alpha}{}^-$ receive contributions, respectively, only from the self-dual and anti-self-dual parts of $M_{[\mu,\nu]}$, which transform according to the three-component $(0,1)$ and $(1,0)$ representations. But (C3) shows that *these conventional M functions have a double pole*, arising simultaneously from the incoming and outgoing charged particle propagators. This singularity is partly kinematic, since the S matrix (C1) involves a sum of single poles, but certainly no double pole. The presence of kinematic singularities in $M_{[\mu,\nu]}$ makes it an inappropriate covariant photon amplitude. Similar remarks apply to gravitons, but not to any other massless particles like the neutrino, for which there is no analog to charge.

PHYSICAL REVIEW VOLUME 138, NUMBER 4B 24 MAY 1965

Photons and Gravitons in Perturbation Theory: Derivation of Maxwell's and Einstein's Equations*

STEVEN WEINBERG†
Department of Physics, University of California, Berkeley, California
(Received 7 January 1965)

The S matrix for photon and graviton processes is studied in perturbation theory, under the restriction that the only creation and annihilation operators for massless particles of spin j allowed in the interaction are those for the physical states with helicity $\pm j$. The most general covariant fields that can be constructed from such operators cannot represent real photon and graviton interactions, because they give amplitudes for emission or absorption of massless particles which vanish as p^j for momentum $p \to 0$. In order to obtain long-range forces it is necessary to introduce noncovariant "potentials" in the interaction, and the Lorentz invariance of the S matrix requires that these potentials be coupled to conserved tensor currents, and also that there appear in the interaction direct current-current couplings, like the Coulomb interaction. We then find that the potentials for $j=1$ and $j=2$ must inevitably satisfy Maxwell's and Einstein's equations in the Heisenberg representation. We also show that although the existence of magnetic monopoles is consistent with parity and time-reversal invariance [provided that P and T are defined to take a monopole into its antiparticle], it is nevertheless impossible to construct a Lorentz-invariant S matrix for magnetic monopoles and charges in perturbation theory.

I. INTRODUCTION

THE classical theories of electromagnetism and gravitation were developed long before physicists discovered quantum mechanics or the S matrix. For this reason, the modern field theorist is generally content to take Maxwell's and Einstein's equations for granted as the starting point of the quantum theory of photons and gravitons.

However, the logical structure of physics is often antiparallel to its historical development. For the purposes of this article, the reader is requested to forget all he knows of electrodynamics and general relativity, and instead to take as his starting point the Lorentz invariance of the S matrix calculated by Feynman-Dyson perturbation theory. That is, we assume the S matrix to be given by

$$S=\sum_{n=0}^{\infty}\frac{(-i)^n}{n!}\int_{-\infty}^{\infty}dt_1\cdots dt_n T\{H'(t_1)\cdots H'(t_n)\} \quad (1.1)$$

with $H'(t)$ the interaction Hamiltonian in the interaction representation

$$H'(t)=\exp(iH^f t)H'\exp(-iH^f t)\,, \quad (1.2)$$

where H^f is the free-particle Hamiltonian and H' the interaction. The operator $H'(t)$ is some function of the creation and annihilation operators of free particles, and we know that these operators transform according to the various familiar representations of the inhomogeneous Lorentz group.[1] Among these representations are those characterized by mass $m=0$ and spin $j=1$ or 2, and we accord these the names of photon and graviton, with no implication intended that these particles necessarily have anything to do with gauge invariance or geometry. Our fundamental requirement on the form of $H'(t)$ is that (1.1) must yield a Lorentz-invariant S matrix.

The power of this requirement is only now beginning to be appreciated. There are strong indications,[2,3] (though as yet no careful proof) that it yields all the results usually associated with local field theories, including the existence of antiparticles, crossing symmetry, spin and statistics, CPT, the Feynman rules, etc. The purpose of this article is to explore the consequences of Lorentz invariance in perturbation theory, for the special case of zero mass and integer spin.

We shall find within this perturbative dynamical framework that *Maxwell's theory and Einstein's theory are essentially the unique Lorentz-invariant theories of massless particles with spin $j=1$ and $j=2$.* By "essentially" we mean only that the conserved current $\mathcal{J}^\mu$ and $\mathcal{J}^{\mu\nu}$ to which the photon and graviton are coupled need not be precisely equal to the electric charge current J^μ and the stress-energy tensor $\theta^{\mu\nu}$, since we can always add Pauli-type currents which vanish in the limits of zero momentum transfer, or of long range. In the same sense, we shall also find that there are *no* Lorentz-invariant theories of massless particles with $j=3$, 4 etc., that is, no theories which yield an inverse-square-law macroscopic force.

These conclusions have already been anticipated in an earlier article on pure S-matrix theory.[4] We showed

* Research supported in part by the U. S. Air Force Office of Scientific Research, Grant No. AF-AFOSR-232-63 and in part by the U. S. Atomic Energy Commission.

† Alfred P. Sloan Foundation Fellow.

[1] E. P. Wigner, Ann. Math. **40**, 149 (1939), and *Theoretical Physics* (International Atomic Energy Agency, Vienna, 1963), p. 59.

[2] For the case $m\neq 0$, see S. Weinberg, Phys. Rev. **133**, B1318 (1964).

[3] For the case $m=0$, see S. Weinberg, Phys. Rev. **134**, B882 (1964).

[4] S. Weinberg, Phys. Rev. **135**, B1049 (1964). See also D. Zwanziger, Phys. Rev. **133**, B1036 (1964). For a preliminary account of this work and that of the present article, see S. Weinberg, Phys. Letters **9**, 357 (1964). A unified treatment of Refs. 2–4 will be published in the lecture notes of the 1964 Brandeis University Summer School on theoretical physics.

there that the Lorentz invariance of the S matrix and a few elementary ideas about pole structure imply that charge is conserved and gravitational mass is equal to inertial mass, the charge and gravitational mass of a particle being defined as its coupling constants for emission of soft photons or gravitons. We also showed that the analogous coupling constants for $j \geq 3$ must vanish. But by using perturbation theory we are able to go much further here, and will in fact derive both the form of the interaction-representation Hamiltonian, and the Maxwell and Einstein field equations in the Heisenberg representation.

We start in Sec. II with a discussion of the $(2j+1)$-component free fields for massless particles of integer spin j. These fields transform according to the $(j,0)$ or $(0,j)$ representations of the homogeneous Lorentz group, and correspond for $j=1$ and $j=2$ to the left- or right-handed parts of the Maxwell field strength tensor $F^{\mu\nu}$ and the Riemann-Christoffel curvature tensor $R^{\mu\nu\lambda\eta}$. They have already been treated in detail in Ref. 3 for general spin, but we concentrate here on the tensor notation appropriate for integer spin, and we also show that *any* covariant free field may be constructed as linear combinations of these simple fields and their derivatives.

However, these simple tensor fields cannot by themselves be used to construct the interaction $H'(t)$, because the coefficients of the operators for creation or annihilation of particles of momentum p and spin j would vanish as p^j for $p \to 0$, in contradiction with the known existence of inverse-square-law forces. We are therefore forced to turn from these tensor fields to the potentials $A^{\mu_1 \cdots \mu_j}(x)$, from which they can be derived by taking a "curl" on each index. But we show in Sec. III that the potentials are not tensor fields; indeed, they cannot be, for we know from a very general theorem[3] that no symmetric tensor field or rank j can be constructed from the creation and annihilation operators of massless particles of spin j. It is for this reason that some field theorists[5] have been led to introduce fictitious photons and gravitons of helicity other than $\pm j$, as well as the indefinite metric that must accompany them.

Preferring to avoid such unphysical monstrosities, we must ask now what sort of coupling we can give our nontensor potentials without losing the Lorentz invariance of the S matrix? And it is here that the failure of manifest covariance turns out to be a blessing in disguise. In Sec. IV we remark that a Lorentz transformation will induce on $A^{\mu_1 \cdots \mu_j}(x)$ a combined tensor and gauge transformation, *so the only interactions allowed by Lorentz invariance are those satisfying gauge invariance*, i.e., those in which the potential is coupled to a conserved current.

[5] For photons, see S. N. Gupta, Proc. Phys. Soc. **63**, 681 (1950); **64**, 850 (1951); K. Bleuler, Helv. Phys. Acta **23**, 567 (1950); K. Bleuler and W. Heitler, Progr. Theoret. Phys. (Kyoto) **5**, 600 (1950). For gravitons, see S. N. Gupta, in *Recent Developments in General Relativity* (Pergamon Press, New York, 1962), p. 251, and other references quoted therein.

For example, the direct photon coupling $(A_\mu A^\mu)^2$ is forbidden, not only by gauge invariance (which we do not assume) but also by Lorentz invariance, because $A_\mu(x)$ is not a four-vector. Actually, we show in Sec. V that even gauge invariance is not sufficient for the Lorentz invariance of the S matrix; as is always the case for spins $j \geqq 1$, we must cancel a noncovariant but temporally local part of the propagator by adding an extra noncovariant interaction to $H'(t)$, which for $j=1$ is the familiar Coulomb interaction, and for $j=2$ we christen the Newton interaction. In Appendix B we present a complete proof[6] that coupling $A^\mu(x)$ to a conserved current and adding a Coulomb term to $H'(t)$ does in fact make the S matrix Lorentz-invariant for $j=1$. (The propagators for photons and gravitons are calculated in Appendix A.)

Having deduced the form of $H'(t)$, we then pass over to the Heisenberg representation, taking care to introduce extra potential components (A^0 for $j=1$; A^{00}, A^{0i}, and $A^i{}_i$ for $j=2$) to represent the effects of the direct Coulomb and Newton interactions. In Sec. VI we show that the Heisenberg representation $A^\mu(x)$ satisfies the Maxwell equations in Coulomb gauge, and in Sec. VII we show that the Heisenberg representation $A^{\mu\nu}(x)$ satisfies the Einstein field equations, in a gauge too ugly to deserve a name.

In Sec. VIII we touch briefly on an old problem: Is it possible to construct a consistent theory of magnetic monopoles? Within the dynamical framework adopted here, the answer is definitely no, because the propagator for a photon linking a charge and a monopole contains noncovariant parts which cannot be cancelled by adding direct terms like the Coulomb interaction to $H'(t)$. The behavior of monopoles under P and T is also discussed.

Although our treatment of electrodynamics is essentially complete, we are not attempting in this article to solve the really difficult problems of quantizing gravitation. In particular, we do not exhibit the conserved energy momentum tensor $\theta^{\mu\nu}$, to which $A^{\mu\nu}$ is coupled, as an explicit function of the gravitation creation and annihilation operators, and we therefore cannot complete the proof that Einstein's equations are sufficient and necessary for Lorentz invariance of the quantized theory. Needless to say, we do not touch upon the ultraviolet-divergence problem either. It is intended that the example of our treatment of photons, together with the beginning made here with gravitons, will serve as the basis for future work on the hard problems of quantum-gravitational theory.

Before setting to work, it may be instructive to compare our development with that of other authors who have also tried to derive electrodynamics or general relativity from first principles. Three different previous approaches may be distinguished.

[6] A similar result was obtained using a different approach, by J. Schwinger, Nuovo Cimento **30**, 278 (1963).

1. Extended Gauge Invariance

We may require the Lagrangian to be invariant under the extended gauge transformation

$$\psi(x) \to e^{iq\Phi(x)}\psi(x) \tag{1.3}$$

[where q is the charge destroyed by $\psi(x)$ and $\Phi(x)$ is an arbitrary c-number function of x^μ]. Then derivatives of $\psi(x)$ must always occur in the form

$$\partial_\mu\psi(x)-iqA_\mu(x)\psi(x) \tag{1.4}$$

the field $A_\mu(x)$ undergoing the gauge transformation

$$A_\mu(x)\to A_\mu(x)+\partial_\mu\Phi(x)\,. \tag{1.5}$$

Therefore,

$$\frac{\partial\mathcal{L}(x)}{\partial A_\mu(x)}=\sum_{\text{fields}}-iq\frac{\partial\mathcal{L}(x)}{\partial(\partial_\mu\psi(x))}\psi(x) \tag{1.6}$$

and this is the conserved electric current $J^\mu(x)$. Requiring the free field Lagrangian of $A_\mu(x)$ to be gauge-invariant then yields Maxwell's equations, with (1.6) as source. A similar approach has been used to derive Einstein's equations.[7]

The only criticism I can offer to this textbook approach is that no one would ever have dreamed of extended gauge invariance if he did not already know Maxwell's theory. In particular, extended gauge invariance has found no application to the strong or weak interactions, though attempts have not been lacking. In our approach (1.6) is a consequence of Lorentz invariance, and this implies that $A_\mu(x)$ enters in $\mathcal{L}(x)$ only in the combination (1.4), so that invariance under (1.3) and (1.5) appears as an incidental result rather than a mysterious postulate.

2. Geometrization

Einstein's theory rests on the identification of the gravitational field with the metric tensor of Riemannian geometry. Attempts have also been made to include electrodynamics in this geometric approach. The criticism here is, again, that the weak and strong interactions seem to have no more to do with Riemannian geometry than with extended gauge invariance. In our approach the geometric interpretation of the gravitational field arises as an incidental consequence of its coupling to the energy-momentum tensor, though we shall not go into this here.

3. Classical Fields with Definite Spin

A number of attempts[8] have been made to derive Maxwell's and/or Einstein's equations by imposing on classical vector or tensor fields the requirement that they correspond to definite spins, $j=1$ or $j=2$. In criticism of these articles we may say first that they generally seem to be based on specific Lagrangians, and secondly, that there does not seem to be much point in defining the spin of a field without being able to tie the definition to the physically relevant representations of the inhomogeneous Lorentz group, i.e., the one-particle states. In our work everything rests on the known transformation properties of the operators which destroy and create physical particles, and of course we make no use of the Lagrangian formalism.

II. THE COVARIANT FIELDS

In this section we shall show that the most general free field for a massless particle of integer helicity $\pm j$ may be constructed from the fundamental field

$$\begin{aligned}F_\pm{}^{[\mu_1\nu_1]\cdots[\mu_j\nu_j]}(x)\equiv(2\pi)^{-3/2}i^j\int d^3p(2|\mathbf{p}|)^{-1/2}\\ \times[p^{\mu_1}e_\pm{}^{\nu_1}(\mathbf{p})-p^{\nu_1}e_\pm{}^{\mu_1}(\mathbf{p})]\times\cdots\\ \times[p^{\mu_j}e_\pm{}^{\nu_j}(\mathbf{p})-p^{\nu_j}e_\pm{}^{\mu_j}(\mathbf{p})]\\ \times[a(\mathbf{p},\pm j)e^{ip\cdot x}+b^*(\mathbf{p},\mp j)e^{-ip\cdot x}],\end{aligned} \tag{2.1}$$

by taking direct sums of $F_\pm$ and/or its derivatives. In Eq. (2.1), $a(\mathbf{p},\lambda)$ and $b(\mathbf{p},\lambda)$ are the annihilation operators for a massless particle and antiparticle of momentum $\mathbf{p}$ and helicity λ; if the particle is its own antiparticle we of course set $a(\mathbf{p},\lambda)=b(\mathbf{p},\lambda)$. The "polarization vectors" $e_\pm{}^\mu(\mathbf{p})$ are defined in Ref. 4, as

$$e_\pm{}^\mu(\mathbf{p})=R^\mu{}_\nu(\hat{p})e_\pm{}^\nu \tag{2.2}$$

where

$$e_\pm{}^1=1/\sqrt{2}\,,\quad e_\pm{}^2=\pm i/\sqrt{2}\,,\quad e_\pm{}^3=e_\pm{}^0=0, \tag{2.3}$$

and $R^\mu{}_\nu(\hat{p})$ is the pure rotation that takes the z axis into the direction of $\mathbf{p}$. By a "general free field" we mean here any linear combination $\psi_n(x)$ of the $a(\mathbf{p},\lambda)$ and $b^*(\mathbf{p},\lambda)$, such that:

(1) Under an arbitrary Lorentz transformation $x^\mu\to\Lambda^\mu{}_\nu x^\nu+a^\mu$ the field transforms according to some representation $D[\Lambda]$ of the homogeneous Lorentz group

$$U[\Lambda,a]\psi_n(x)U^{-1}[\Lambda,a]=\sum_m D_{nm}[\Lambda^{-1}]\psi_m(\Lambda x+a)\,. \tag{2.4}$$

(2) The field commutes with its adjoint at space-like separations

$$[\psi_n(x),\psi_m{}^\dagger(y)]=0\quad\text{for}\quad(x-y)^2>0\,. \tag{2.5}$$

Our metric has signature $+++-$. (Fields satisfying

[7] R. Utiyama, Phys. Rev. **101**, 1597 (1956), and T. W. B. Kibble, J. Math. Phys. **2**, 212 (1961).

[8] W. E. Thirring, Ann. Phys. (N.Y.) **16**, 96 (1961); V. I. Ogievetsky and I. V. Polubarinov, Ann. Phys. (N. Y.) **25**, 358 (1963). For earlier work on similar lines see M. Fierz and W. Pauli, Proc. Roy. Soc. (London) **A173**, 211 (1939), and M. Fierz, Helv. Phys. Acta **12**, 3 (1939). One sometimes encounters a very simple version of such arguments, to the effect that the current J^μ must obviously be conserved if we *define* it as $J^\mu\equiv-\partial_\mu F^{\mu\nu}$. But this does not say that J^μ is the same as the current $\mathcal{J}^\mu\equiv-\delta H'/\delta A_\mu$ to which A_μ is coupled in the interaction Hamiltonian. In fact, we will see explicitly at the end of Sec. VI that if $\mathcal{J}^\mu$ is not conserved than J^μ is *not* equal to $\mathcal{J}^\mu$.

these two conditions can be coupled together to form a causal scalar Hamiltonian density, yielding a Lorentz-invariant S matrix.)

In order to show that all such free fields may be derived from (2.1), we shall pursue the following line of argument:

(A) We first show that $F_{\pm}(x)$ are themselves fields, by proving that they are tensors.

(B) We then study the algebraic properties of $F_{\pm}(x)$, and show that they have at most $(2j+1)$ linearly independent components.

(C) We then use the results of (A) and (B) to show that $F_{+}(x)$ and $F_{-}(x)$ are just the simple fields $\chi_\sigma(x)$ and $\varphi_\sigma(x)$ introduced in Ref. 3.

(D) We finally remark that any irreducible free field may be obtained by differentiating these simple fields a suitable number of times.

a. Tensor Behavior of $F_{\pm}(x)$

The behavior of the polarization vectors $e_{\pm}{}^{\mu}(\mathbf{p})$ under an arbitrary Lorentz transformation $\Lambda^\mu{}_\nu$ was shown in Appendix A of Ref. 4 to be

$$\{\Lambda_\nu{}^\mu - p^\mu\Lambda_\nu{}^0/|\mathbf{p}|\}e_{\pm}{}^\nu(\Lambda\mathbf{p}) = \exp[\pm i\Theta(\mathbf{p},\Lambda)]e_{\pm}{}^\mu(\mathbf{p}) \quad (2.6)$$

with Θ an angle whose precise definition need not concern us here. Furthermore, the annihilation operator $a(\mathbf{p},\lambda)$ for a massless particle of helicity λ and momentum $\mathbf{p}$ was shown in Ref. 3 to obey the transformation law

$$\begin{aligned} U[\Lambda]a(\mathbf{p},\lambda)U^{-1}[\Lambda] \\ = (|\Lambda\mathbf{p}|/|\mathbf{p}|)^{1/2}\exp[i\lambda\Theta(\mathbf{p},\Lambda)]a(\Lambda\mathbf{p},\lambda) \end{aligned} \quad (2.7)$$

with Θ the same angle as in (2.6). The creation operator $b^*(\mathbf{p},-\lambda)$ transforms like $a(\mathbf{p},\lambda)$. Using (2.6) and (2.7) in (2.1) shows instantly that $F_{\pm}(x)$ are tensors,

$$\begin{aligned} U(\Lambda,a)F_{\pm}{}^{[\mu_1\nu_1]\cdots[\mu_j\nu_j]}(x)U^{-1}[\Lambda,a] \\ = \Lambda_{\rho_1}{}^{\mu_1}\Lambda_{\eta_1}{}^{\nu_1}\cdots\Lambda_{\rho_j}{}^{\mu_j}\Lambda_{\eta_j}{}^{\nu_j}F_{\pm}{}^{[\rho_1\eta_1]\cdots[\rho_j\eta_j]}(\Lambda x+a)\,. \end{aligned} \quad (2.8)$$

The causal character of $F_{\pm}(x)$ can be deduced directly from the fact that $a(\mathbf{p},\pm j)\exp(ip\cdot x)$ and $b^*(\mathbf{p},\mp j)\times\exp(-ip\cdot x)$ enter with equal coefficients in (2.1).

b. Algebraic Properties of $F_{\pm}(x)$

It follows from (2.2) and (2.3) that the polarization vectors $e_{\pm}{}^\mu(\mathbf{p})$ have the algebraic properties

$$p_\mu e_{\pm}{}^\mu(\mathbf{p}) = 0\,, \quad (2.9)$$

$$e_{\pm\mu}(\mathbf{p})e_{\pm}{}^\mu(\mathbf{p}) = 0\,, \quad (2.10)$$

$$\epsilon^{\mu\nu\rho\eta}p_\rho e_{\pm\eta}(\mathbf{p}) = \mp i[p^\mu e_{\pm}{}^\nu(\mathbf{p}) - p^\nu e_{\pm}{}^\mu(\mathbf{p})]\,, \quad (2.11)$$

where $\epsilon^{\mu\nu\rho\eta}$ is the totally antisymmetric tensor with $\epsilon^{0123}\equiv 1$. [We prefer to use a four-vector notation, even though $e_{\pm}{}^0(\mathbf{p})\equiv 0$.]

Inspection of (2.1) and (2.9)–(2.11) shows that $F_{\pm}(x)$ obeys the algebraic conditions:

(*i*) *Symmetry*. $F_{\pm}(x)$ are symmetric under interchange of any two index pairs $[\mu_r\nu_r]\leftrightarrow[\mu_s\nu_s]$.

(*ii*) *Antisymmetry*. $F_{\pm}(x)$ are antisymmetric under interchanges $\mu_r\leftrightarrow\nu_r$ within any one index pair:

$$F_{\pm}{}^{[\mu_1\nu_1]\cdots} = -F_{\pm}{}^{[\nu_1\mu_1]\cdots}\,. \quad (2.12)$$

(*iii*) *Duality*. $F_{+}(x)$ and $F_{-}(x)$ are, respectively, self-dual or anti-self-dual with respect to each index pair $[\mu_r\nu_r]$:

$$\epsilon^{\mu\nu\mu_1\nu_1}F_{\pm[\mu_1\nu_1]}{}^{[\mu_2\nu_2]\cdots} = \mp 2iF_{\pm}{}^{[\mu\nu][\mu_2\nu_2]\cdots}\,. \quad (2.13)$$

(*iv*) *Tracelessness*. The complete contraction of any pair of indices $[\mu_r\nu_r]$, $[\mu_s\nu_s]$ gives zero:

$$g_{\mu_1\mu_2}g_{\nu_1\nu_2}F_{\pm}{}^{[\mu_1\nu_1][\mu_2\nu_2]\cdots} = 0\,. \quad (2.14)$$

It is also true that any single trace vanishes:

$$g_{\mu_1\mu_2}F_{\pm}{}^{[\mu_1\nu_1][\mu_2\nu_2]\cdots} = 0$$

but this follows from (*i*)–(*iv*), and will therefore not be listed as an independent condition. Conditions (*i*) and (*iv*) are of course empty for $j=1$.

These four conditions imply that the $F_{\pm}(x)$ each have at most $2j+1$ independent components. Condition (*ii*) lowers the number of independent values taken by each index pair $[\mu_r\nu_r]$ from 16 to 6, and (*iii*) lowers it further to 3, so under (*i*), (*ii*), and (*iii*) alone the number of independent components would be the same as for a symmetric tensor of rank j in three dimensions, i.e.,

$$N_j = \binom{j+2}{2} = \tfrac{1}{2}(j+1)(j+2)\,.$$

But condition (*iv*) imposes N_{j-2} further constraints, so the net number of independent components is at most

$$N_j - N_{j-2} = 2j+1\,.$$

c. Identification of $F_{\pm}(x)$

In Sec. III of Ref. 3, we showed that the only free fields which can be formed out of the operators

$$a(\mathbf{p},\pm j)e^{ip\cdot x} + b^*(\mathbf{p},\mp j)e^{-ip\cdot x}$$

must transform under the homogeneous Lorentz group as a direct sum of those $(2A+1)(2B+1)$-dimensional irreducible representations (A,B) with

$$B-A=\pm j\,. \quad (2.15)$$

Indeed, the irreducible fields are determined uniquely by the representation (A,B) under which they transform, as

$$\begin{aligned} \psi_{ab}{}^{AB}(x) = (2\pi)^{-3/2}\int d^3p(2|\mathbf{p}|)^{A+B-1/2} \\ \times D_{a,-A}{}^{(A)}[R(\hat{p})]D_{b,B}{}^{(B)}[R(\hat{p})] \\ \times[a(\mathbf{p},\pm j)e^{ip\cdot x} + b^*(\mathbf{p},\mp j)e^{-ip\cdot x}]\,, \end{aligned} \quad (2.16)$$

where $D^{(J)}[R(\hat{p})]$ is the usual $(2J+1)$-dimensional unitary representation of the rotation $R(\hat{p})$ which takes the z axis into the direction of **p**, and the indices a and b run by unit steps from $-A$ to $+A$ and $-B$ to $+B$, respectively. [These remarks apply for half-integer j as well as integer j.]

We have learned from Secs. IIa and b above that $F_\pm(x)$ transform according to some reducible or irreducible representation of the homogeneous Lorentz group, with dimensionality at most $2j+1$. But of the irreducible representations (A,B) satisfying (2.15), the ones with the smallest dimensionality are the $2j+1$-dimensional representations $(j,0)$ for helicity $-j$, and $(0,j)$ for helicity $+j$. Hence $F_-(x)$ and $F_+(x)$ must transform purely according to the $(j,0)$ and $(0,j)$ representations, and since the representation uniquely determines the field they must be, respectively, just the $(2j+1)$-component fields $\varphi_\sigma(x)$ and $\chi_\sigma(x)$ introduced in Ref. 3. That is, the components of $F_-^{[\mu_1\nu_1]\cdots}(x)$ and $F_+^{[\mu_1\nu_1]\cdots}(x)$ are linear combinations of those of $\varphi_\sigma(x)$ and $\chi_\sigma(x)$, and vice versa. [The fields $\varphi_a(x)$ and $\chi_b(x)$ can be obtained from the general expression (2.16) by setting $B=0$ or $A=0$.]

d. Derivation of General Fields from $F_\pm(x)$

Let us examine the Lorentz transformation properties of the $2J$th derivatives (J integer or half-integer)

$$\partial_{\lambda_1}\cdots\partial_{\lambda_{2J}}F_\pm^{[\mu_1\nu_1]\cdots[\mu_j\nu_j]}(x)\,. \tag{2.17}$$

This object is a symmetric traceless tensor with respect to the λ indices, so it transforms according to the representation

$$F_+:\ (J,J)\otimes(0,j)=(J,j+J)\oplus\cdots\oplus(J,|j-J|)\,, \tag{2.18}$$

$$F_-:\ (J,J)\otimes(j,0)=(j+J,J)\oplus\cdots\oplus(|j-J|,J)\,. \tag{2.19}$$

The only terms in these Clebsch-Gordan series that satisfy (2.15) are the first, so (2.17) transforms according to the representations $(J,j+J)$ for F_+ and $(j+J,J)$ for F_-. By letting J run over all integers and half-integers we can construct any representation (A,B) satisfying (2.15), so any free field can be built up as direct sums of (2.17). [The only possible flaw in this argument would arise if one of the $(J,j+J)$ or $(j+J,J)$ terms vanished, but then (2.17) would vanish, and this is clearly impossible.]

Incidentally, the same method of proof applies for $m\neq0$, to show that the most general (A,B) field can be obtained by projecting out the appropriate part of

$$\partial_{\lambda_1}\cdots\partial_{\lambda_{2B}}\varphi_\sigma(x)$$

or

$$\partial_{\lambda_1}\cdots\partial_{\lambda_{2A}}\chi_\sigma(x)\,,$$

where $\varphi_\sigma(x)$ and $\chi_\sigma(x)$ are the $(2j+1)$-component free fields constructed for massive particles in Ref. 2. [In this case the (A,B) part cannot vanish because $\varphi_\sigma(x)$ and $\chi_\sigma(x)$ cannot obey any homogeneous field equations.]

In contrast with the massive particle case, the massless free fields $F_\pm(x)$ obey homogeneous field equations which just express the absence of those terms in (2.18) and (2.19) which do not satisfy (2.15). The simplest such equation may be deduced directly from (2.1) and (2.9):

$$\partial_{\mu_1}F_\pm^{[\mu_1\nu_1]\cdots}=0\,. \tag{2.20}$$

For instance, for $j=1$ the algebraic properties noted under Sec. IIb let us write

$$F_\pm^{[ij]}=\epsilon_{ijk}(E^k\pm iB^k)\,,$$

$$F_\pm^{[0k]}=\pm i(E^k\pm iB^k)\,,$$

so (2.20) gives

$$\nabla\times[\mathbf{E}\pm i\mathbf{B}]=\pm i\frac{\partial}{\partial t}[\mathbf{E}\pm i\mathbf{B}]\,,$$

$$\nabla\cdot[\mathbf{E}\pm i\mathbf{B}]=0\,,$$

or

$$\nabla\times\mathbf{E}=-\partial\mathbf{B}/\partial t\,,\quad \nabla\times\mathbf{B}=\partial\mathbf{E}/\partial t\,,$$

$$\nabla\cdot\mathbf{E}=0\,,\quad \nabla\cdot\mathbf{B}=0\,,$$

justifying the identification of **E** and **B** with the free electric and magnetic fields. A similar argument allows us to identify the five independent components of $F_\pm^{[\mu\nu][\lambda\eta]}$ with the left- or right-handed parts of the source-free Riemann-Christoffel tensor.

It should perhaps be stressed that *up to this point* we have done little but put the work of Ref. 3 into tensor notation.

III. POTENTIALS

After having shown that any free field can be constructed from $F_\pm(x)$ and its derivatives, we might feel justified in trying to construct the interaction Hamiltonians for photons and gravitons out of $F_\pm^{[\mu\nu]}$ and $F_\pm^{[\mu\nu][\lambda\eta]}$. *But this does not work.* Inspection of (2.1) shows that the amplitude for emitting or absorbing a massless particle of spin j by a field $F_\pm(x)$ will vanish like $p^{j-1/2}$ for momentum $p\to0$. Hence an interaction built out of $F_\pm(x)$ could never give rise to the phenomena most closely associated with electromagnetism and gravitation, i.e., long-range forces and infrared divergences.[9] [Using other free fields would be even worse; we can see from (2.16) that the amplitudes yielded by a field of type (A,B) would vanish as $p^{A+B-1/2}$ for $p\to0$, and (2.15) gives $A+B\geqq j$. This is of course because such fields can be written as the $2A$th derivative of $F_+(x)$ or the $2B$th derivative of $F_-(x)$.]

Instead of yielding to despair at this point, let us ignore the results of Sec. II for a moment, and try to strip away the objectionable factor of p^j in $F_\pm(x)$, by writing these fields as jth derivatives of other objects. We note by inspection of Eq. (2.1) that the $F_\pm(x)$ can

[9] By "infrared divergence" here we mean that the amplitude for internal bremsstrahlung of a soft photon or graviton is dominated by a term that behaves like ω^{-1} for $\omega\to0$. See e.g., Ref. 4.

be written as generalized curls of the potentials

$$A_\pm^{\mu_1\cdots\mu_j}(x)\equiv(2\pi)^{-3/2}\int d^3p(2|\mathbf{p}|)^{-1/2}e_\pm^{\mu_1}(\mathbf{p})\cdots e_\pm^{\mu_j}(\mathbf{p})$$
$$\times[a(\mathbf{p},\pm j)e^{ip\cdot x}+(-)^j b^*(\mathbf{p},\mp j)e^{-ip\cdot x}]. \quad (3.1)$$

That is,

$$F_\pm{}^{[\mu\nu]}=\partial^\mu A_\pm{}^\nu-\partial^\nu A_\pm{}^\mu, \quad (3.2)$$

$$F_\pm{}^{[\mu\nu][\lambda\eta]}=\partial^\mu\partial^\lambda A_\pm{}^{\nu\eta}-\partial^\mu\partial^\eta A_\pm{}^{\nu\lambda}-\partial^\nu\partial^\lambda A_\pm{}^{\mu\eta}+\partial^\nu\partial^\eta A_\pm{}^{\mu\lambda}, \quad (3.3)$$

and so on. The potentials $A_\pm(x)$ which are symmetric in their μ indices, vanish if any one μ index is zero

$$A_\pm{}^{0\mu_2\cdots\mu_j}=0 \quad (3.4)$$

and have zero trace

$$g_{\mu_1\mu_2}A_\pm{}^{\mu_1\mu_2\cdots\mu_j}=0. \quad (3.5)$$

They satisfy the free-field equations

$$\partial_{\mu_1}A_\pm{}^{\mu_1\cdots\mu_j}=0 \quad (3.6)$$

$$\pm i\epsilon^{\nu_1\mu_1\nu\mu}\partial_\nu A_{\pm\mu}{}^{\mu_2\cdots\mu_j}=\partial^{\nu_1}A_\pm{}^{\mu_1\mu_2\cdots\mu_j}-\partial^{\mu_1}A_\pm{}^{\nu_1\mu_2\cdots\mu_j} \quad (3.7)$$

and of course

$$\square^2 A_\pm{}^{\mu_1\cdots\mu_j}=0. \quad (3.8)$$

The discussion of Sec. III makes it clear that the $A_\pm(x)$ cannot be fields, in the sense of (2.4) and (2.5). Indeed, we can see that they are not even tensors, because their time-like components vanish; if they were tensors then they would transform according to the $(\frac{1}{2}j,\frac{1}{2}j)$ representation of the homogeneous Lorentz group, and this representation does not satisfy the fundamental condition (2.15) for fields constructed from the operators $a(\mathbf{p},\pm j)$ and $b^*(\mathbf{p},\mp j)$.

But $F_\pm(x)$ are tensors, so the noncovariance of $A_\pm$ must be manifested in the appearance of gradient terms in the Lorentz transformation law for the potentials,[10] which do not show up when we take curls to obtain $F_\pm(x)$. In fact, this is the case. A simple calculation using (3.1), (2.6), and (2.7) shows that

$$U[\Lambda]A_\pm{}^{\mu_1\cdots\mu_j}(x)U^{-1}[\Lambda]$$
$$=\Lambda_{\nu_1}{}^{\mu_1}\cdots\Lambda_{\nu_j}{}^{\mu_j}(2\pi)^{-3/2}\int d^3p(2|\mathbf{p}|)^{-1/2}$$
$$\times[e_\pm{}^{\nu_1}(\mathbf{p})-p^{\nu_1}f_\pm(\mathbf{p},\Lambda)]\cdots[e_\pm{}^{\nu_j}(\mathbf{p})-p^{\nu_j}f_\pm(\mathbf{p},\Lambda)]$$
$$\times[a(\mathbf{p},\pm j)e^{ip\cdot\Lambda x}+(-)^j b^*(\mathbf{p},\mp j)e^{-ip\cdot\Lambda x}] \quad (3.9)$$

with

$$f_\pm(\mathbf{p},\Lambda)=\Lambda_\nu{}^0 e_\pm{}^\nu(\mathbf{p})/|\Lambda^{-1}\mathbf{p}|. \quad (3.10)$$

[10] See e.g., J. Schwinger, Phys. Rev. **74**, 1439 (1948); **127**, 324 (1964).

This can be written

$$U[\Lambda]A_\pm{}^{\mu_1\cdots\mu_j}(x)U^{-1}[\Lambda]=\Lambda_{\nu_1}{}^{\mu_1}\cdots\Lambda_{\nu_j}{}^{\mu_j}A_\pm{}^{\nu_1\cdots\nu_j}(\Lambda x)$$
$$+\sum_{r=1}^{j}\partial^{\mu_r}\Phi_\pm{}^{\mu_1\cdots\mu_{r-1}\mu_{r+1}\cdots\mu_j}(x;\Lambda). \quad (3.11)$$

We will fortunately not need the rather complicated explicit formulas for $\Phi_\pm$. However, for an infinitesimal Lorentz transformation $\Lambda^\mu{}_\nu=\delta^\mu{}_\nu+\omega^\mu{}_\nu$ the functions $f_\pm(\mathbf{p},\Lambda)$ are infinitesimal

$$f_\pm(\mathbf{p},1+\omega)=\omega_{0i}e_\pm{}^i(\mathbf{p})/|\mathbf{p}| \quad (3.12)$$

and $\Phi_\pm$ is then given by the simple expression

$$\Phi_\pm{}^{\nu_1\cdots}(x;1+\omega)=i(2\pi)^{-3/2}\int d^3p(2|\mathbf{p}|)^{-1/2}$$
$$\times f_\pm(\mathbf{p},1+\omega)e_\pm{}^{\nu_1}(\mathbf{p})\cdots$$
$$\times[a(\mathbf{p},\pm j)e^{ip\cdot x}-(-)^j b^*(\mathbf{p},\mp j)e^{-ip\cdot x}]. \quad (3.13)$$

We will find it convenient from now on to shift our attention from $A_\pm(x)$ to the potentials $A(x)$ and $B(x)$, defined by

$$A^{\mu_1\cdots\mu_j}(x)\equiv A_+{}^{\mu_1\cdots\mu_j}(x)+A_-{}^{\mu_1\cdots\mu_j}(x) \quad (3.14)$$

$$iB^{\mu_1\cdots\mu_j}(x)\equiv A_+{}^{\mu_1\cdots\mu_j}(x)-A_-{}^{\mu_1\cdots\mu_j}(x). \quad (3.15)$$

A particle that interacts with left- and right-handed particles with the same coupling constant will be coupled only to $A(x)$, while one that has coupling constants of opposite sign to left- and right-handed quanta will interact only with $B(x)$. Hence an ordinary charge will couple only to $A^\mu(x)$, while a magnetic monopole will couple to $B^\mu(x)$. The two fields can be distinguished by their different behavior [11]under parity (P) and time-reversal (T):

$$PA^{\mu_1\cdots\mu_j}(x)P^{-1}=(-)^j A^{\mu_1\cdots\mu_j}(-\mathbf{x},t), \quad (3.16)$$

$$PB^{\mu_1\cdots\mu_j}(x)P^{-1}=-(-)^j B^{\mu_1\cdots\mu_j}(-\mathbf{x},t), \quad (3.17)$$

$$TA^{\mu_1\cdots\mu_j}(x)T^{-1}=(-)^j A^{\mu_1\cdots\mu_j}(\mathbf{x},-t), \quad (3.18)$$

$$TB^{\mu_1\cdots\mu_j}(x)T^{-1}=-(-)^j B^{\mu_1\cdots\mu_j}(\mathbf{x},-t). \quad (3.19)$$

In Sec. VIII we will discuss reasons why nature has not made use of $B(x)$ in forming the interactions of massless particles.

Since the photon and graviton are both purely neutral, we will surrender a little of our extreme generality, by

[11] These are derived using the P and T behavior of the operators $a(\mathbf{p},\lambda)$ and $b^*(\mathbf{p},\lambda)$, as worked out in Sec. IX of Ref. 3. In order to obtain the particular sign changes given here for P it is necessary to adjust the relative phases of $a(\mathbf{p},+j)$ and $a(\mathbf{p},-j)$, while the sign change under T can be obtained by adjusting the over-all phase of both operators. The important thing is that the P and T sign changes are both opposite for $A(x)$ and $B(x)$, because P interchanges helicities $\pm j$, and because T is antiunitary.

restricting ourselves to the case of particles identical with their antiparticles. It will be convenient to define phases so that

$$b(\mathbf{p},\lambda)=(-)^j a(\mathbf{p},\lambda)\,. \tag{3.20}$$

We note that $e_\pm{}^{\mu *}=e_\mp{}^\mu$, so now we have

$$A_\pm{}^{\mu_1\cdots\mu_j\dagger}=A_\mp{}^{\mu_1\cdots\mu_j}\,. \tag{3.21}$$

Therefore (3.14) and (3.15) give $A(x)$ and $B(x)$ as Hermitian operators:

$$A^{\mu_1\cdots\mu_j}(x)=A^{\mu_1\cdots\mu_j\dagger}(x)=(2\pi)^{-3/2}\int d^3p(2|\mathbf{p}|)^{-1/2}\sum_\pm e_\pm{}^{\mu_1}(\mathbf{p})\cdots e_\pm{}^{\mu_j}(\mathbf{p})[a(\mathbf{p},\pm j)e^{ip\cdot x}+a^*(\mathbf{p},\mp j)e^{-ip\cdot x}] \tag{3.22}$$

$$B^{\mu_1\cdots\mu_j}(x)=B^{\mu_1\cdots\mu_j\dagger}(x)=-i(2\pi)^{-3/2}\int d^3p(2|\mathbf{p}|)^{-1/2}\sum_\pm \pm e_\pm{}^{\mu_1}(\mathbf{p})\cdots e_\pm{}^{\mu_j}(\mathbf{p})[a(\mathbf{p},\pm j)e^{ip\cdot x}+a^*(\mathbf{p},\mp j)e^{-ip\cdot x}]\,. \tag{3.23}$$

IV. LORENTZ INVARIANCE AND CURRENT CONSERVATION

The potentials $A(x)$ and $B(x)$ are not tensors, and cannot be made into tensors by redefiniton of the polarization vectors, because condition (2.15) does not allow us to construct $(\frac{1}{2}j,\frac{1}{2}j)$ tensor fields of rank j for massless particles of helicity $\pm j$. On the other hand, the true tensor fields $F_\pm(x)$ of rank $2j$ (and any other truly covariant fields) give amplitudes for emission and absorption of soft quanta which vanish at least as fast as $p^{j-1/2}$ for momentum $p\to 0$, in contradiction to our everyday experience with photons and gravitons. There seem to be just two available methods for the circumvention of this difficulty:

1. The traditional approach[5] is to introduce operators for fictitious particles of helicity other than $\pm j$ (some with negative probabilities) in such a way that $A(x)$ and $B(x)$ become true tensor fields.

2. Alternatively, we can resign ourselves to the non-tensor character of $A(x)$ and $B(x)$, but construct the interaction Hamiltonian so that the S matrix is nevertheless Lorentz-invariant.

I will follow the second path. One reason is that no one likes unphysical particles, or the indefinite metric and subsidiary state-vector conditions that they entail. But, more significant, the lack of manifest Lorentz covariance in the second approach means that we must impose powerful restrictions on the interaction Hamiltonian in order to obtain a Lorentz-invariant S matrix. This feature is a minor nuisance if we are sure we already know the correct theory of photons and gravitons, but it becomes all-important if what we want is an *a priori* derivation of electrodynamics and general relativity.

So we must ask what sort of couplings we can give $A(x)$ and $B(x)$ without violating the Lorentz invariance of the S matrix. For the moment, we will assume that only $A(x)$ enters in the interaction, (e.g., no magnetic monopoles) and will return to the more general case in Sec. VIII.

A Lorentz transformation $\Lambda^\mu{}_\nu$ induces on the potential $A(x)$ a combined tensor and "gauge" transformation

$$U[\Lambda]A^{\mu_1\cdots\mu_j}(x)U^{-1}[\Lambda]=\Lambda_{\nu_1}{}^{\mu_1}\cdots\Lambda_{\nu_j}{}^{\mu_j}A^{\nu_1\cdots\nu_j}(\Lambda x)+\sum_{r=1}^{j}\partial^{\mu_r}\Phi^{\mu_1\cdots\mu_{r-1}\mu_{r+1}\cdots\mu_j}(x;\Lambda) \tag{4.1}$$

with $\Phi=\Phi_++\Phi_-$. [See Eq. (3.11)]. The potential appears in the interaction Hamiltonian $H'(t)$ coupled to a current

$$\mathcal{J}_{i_1\cdots i_j}(x)\equiv-\delta H'(t)/\delta A^{i_1\cdots i_j}(x)\,, \tag{4.2}$$

but when we sum to all orders of perturbation theory the matrix elements for creation or annihilation of real or virtual massless particles are determined by the current in the Heisenberg representation

$$\mathcal{J}^H{}_{i_1\cdots i_j}(x)\equiv\exp(iHx^0)\mathcal{J}_{i_1\cdots i_j}(0)\exp(-iHx^0)\,. \tag{4.3}$$

The form of the two terms in (4.1) then leads us to guess that the Lorentz invariance of the S matrix requires $\mathcal{J}$ to have the properties

(a): $\mathcal{J}_{i_1\cdots i_j}(x)$ is the spatial part of a symmetric tensor $\mathcal{J}_{\mu_1\cdots\mu_j}(x)$,

$$U[\Lambda]\mathcal{J}_{\mu_1\cdots\mu_j}(x)U^{-1}[\Lambda]=\Lambda_{\mu_1}{}^{\nu_1}\cdots\Lambda_{\mu_j}{}^{\nu_j}\mathcal{J}_{\nu_1\cdots\nu_j}(\Lambda x)\,. \tag{4.4}$$

(b): $\mathcal{J}^H{}_{\mu_1\cdots\mu_j}$ is conserved

$$\partial^{\mu_1}\mathcal{J}^H{}_{\mu_1\cdots\mu_j}(x)=0\,. \tag{4.5}$$

We will remove most of the guesswork in the next section, but let us accept (*A*) and (*B*) for the moment as necessary requirements for Lorentz invariance.

There are two familiar types of conserved symmetric tensor: for $j=1$ there are the currents J^μ of additively conserved quantities such as charge and baryon number, and for $j=2$ there is the symmetric stress-energy tensor $\theta^{\mu\nu}$. In addition, it is easy to construct conserved currents of the "Pauli"-type for any j:

$$\mathcal{J}_{\text{Pauli}}{}^{\mu_1\cdots\mu_j}(x)=\partial_{\nu_1}\cdots\partial_{\nu_j}\Sigma^{[\mu_1\nu_1]\cdots[\mu_j\nu_j]}(x)\,, \tag{4.6}$$

where Σ is any tensor antisymmetric within each index pair $[\mu,\nu]$ and symmetric between different index pairs. A familiar example for $j=1$ is the Pauli-moment current

$$\mathcal{J}^\mu{}_{\text{Pauli}}\propto\partial_\nu(\bar\psi\sigma^{\mu\nu}\psi)\,.$$

However, coupling the potential $A(x)$ to the current (4.6) is equivalent to coupling the tensor field $F(x)$ to $\Sigma(x)$, and cannot by itself give finite amplitudes for producing or absorbing very soft massless particles. In particular, the "charge" carried by (4.6) vanishes, i.e.,

$$\int d^3x\,\mathcal{J}_{\text{Pauli}}{}^{0\,\mu_2\cdots\mu_j}(x)=0\,.$$

The only currents which avoid this criticism are the charge (or baryon number, etc.) current J^μ and the stress-energy tensor $\theta^{\mu\nu}$. Hence we conclude that *Lorentz invariance forces the photon potential $A^\mu(x)$ to be coupled to $J^\mu(x)$, and the graviton potential $A^{\mu\nu}(x)$ to be coupled to $\theta^{\mu\nu}(x)$*, except that in both cases there is the possibility of adding extra terms like (4.6) to J^μ and $\theta^{\mu\nu}$, or equivalently, of adding interactions involving the covariant fields $F^{[\mu\nu]}(x)$ or $F^{[\mu\nu][\lambda\eta]}(x)$.

In fact, nature does not seem to take its option of using terms like (4.6) in the interaction currents of massless particles. For the photon we have clear evidence of this in the success of Dirac's calculation of the magnetic moment of the electron. And also, the very absence of massless particles with $j\geqq 3$ is symptomatic of nature's abhorrence of Pauli-type currents, since these are the only currents with which such particles could interact. For photons the absence of Pauli couplings is sometimes referred to as the "principle of minimal electromagnetic coupling," but it remains a mystery nonetheless. Perhaps the solution will be found in considerations of high-energy behavior, since the Pauli currents are worse in this respect than J^μ and $\theta^{\mu\nu}$, and, in particular, can never give renormalizable interactions.

It seems fairly obvious that the statements that A^μ couples only to J^μ and $A^{\mu\nu}$ couples only to $\theta^{\mu\nu}$ (except in both cases for possible Pauli terms) are equivalent, respectively, to gauge invariance of the second kind and to Einstein's equivalence principle. We will not pursue this point further here, as it would lead us into the Lagrangian formalism, which we have been so far successful in avoiding. Instead, we will give a direct derivation of Maxwell's and Einstein's equations in the Heisenberg representation, in Secs. VI and VII.

V. LORENTZ INVARIANCE OF THE FEYNMAN RULES

In order to understand better what conditions are actually necessary and sufficient for the Lorentz invariance of the S matrix, we will now examine the Feynman rules generated by formula (3.22) for the potentials $A(x)$. We have already remarked in Ref. 4 that the requirements (4.4), (4.5) for a conserved tensor current are sufficient for the Lorentz invariance of S-matrix elements with external massless particle lines (provided that the covariance of matrix elements of J^μ is not spoiled by the internal massless particle lines) and that these conditions are also necessary at least on the light cone in momentum space. Our remaining task is to examine the Lorentz transformation properties of the internal massless particle lines.

The coordinate-space propagator of the field $A(x)$ is easily calculated as

$$\langle T\{A^{\mu_1\cdots\mu_j}(x)A^{\nu_1\cdots\nu_j}(y)\}\rangle_0 = (2\pi)^{-3}\int\frac{d^3p}{2|\mathbf{p}|}\Pi^{\mu_1\cdots\mu_j\nu_1\cdots\nu_j}(\mathbf{p}) \times[\theta(x-y)e^{ip\cdot(x-y)}+\theta(y-x)e^{ip\cdot(y-x)}] \quad (5.1)$$

with

$$\Pi^{\mu_1\cdots\mu_j\nu_1\cdots\nu_j}(\mathbf{p}) = \sum_\pm e_\pm^{\mu_1}(\mathbf{p})\cdots e_\pm^{\mu_j}(\mathbf{p})e_\pm^{\nu_1}(\mathbf{p})^*\cdots e_\pm^{\nu_j}(\mathbf{p})^*. \quad (5.2)$$

In momentum space the propagator is

$$\Delta_c^{\mu_1\cdots\mu_j\nu_1\cdots\nu_j}(q)\equiv i\int d^4x e^{-iq\cdot x}\langle T\{A^{\mu_1\cdots\mu_j}(x),A^{\nu_1\cdots\nu_j}(y)\}\rangle_0 = \Pi^{\mu_1\cdots\mu_j\nu_1\cdots\nu_j}(\mathbf{q})/(q^2-i\epsilon). \quad (5.3)$$

For $j=1$ we easily calculate (in Appendix A)

$$\Pi^{\mu\nu}(\mathbf{q})=g^{\mu\nu}+n^\mu\hat{q}^\nu+n^\nu\hat{q}^\mu-\hat{q}^\mu\hat{q}^\nu$$
$$n^\mu=\{0,0,0,1\} \quad (5.4)$$
$$\hat{q}^\mu=\{\hat{q},1\}.$$

In order to express (5.4) in terms of a non-light-like q^μ, we set

$$\hat{q}^\mu=[q^\mu+n^\mu(|\mathbf{q}|-q^0)]/|\mathbf{q}| \quad (5.5)$$

and we obtain

$$\Pi^{\mu\nu}(q)=g^{\mu\nu}+((n^\mu q^\nu+n^\nu q^\mu)q^0/|\mathbf{q}|^2) -(q^\mu q^\nu/|\mathbf{q}|^2)+(q^2n^\mu n^\nu/|\mathbf{q}|^2). \quad (5.6)$$

Hence the propagator may be written as the sum of three terms

$$\Delta_c^{\mu\nu}(q)=\Delta^{\mu\nu}{}_{\text{cov}}(q)+\Delta^{\mu\nu}{}_{\text{grad}}(q)+\Delta^{\mu\nu}{}_{\text{loc}}(q). \quad (5.7)$$

The first term $\Delta^{\mu\nu}{}_{\text{cov}}(q)$ is the usual covariant tensor propagator

$$\Delta^{\mu\nu}{}_{\text{cov}}(q)=g^{\mu\nu}/(q^2-i\epsilon). \quad (5.8)$$

The second term $\Delta^{\mu\nu}{}_{\text{grad}}$ is not covariant, but it is proportional to factors q^μ or q^ν which give zero[12] when multiplied into the conserved currents connected by $\Delta_c^{\mu\nu}$. The final term $\Delta^{\mu\nu}{}_{\text{loc}}$ is also not covariant, but it is characterized by the absence of the pole at $|q^0|=|\mathbf{q}|$:

$$\Delta^{\mu\nu}{}_{\text{loc}}(q)=n^\mu n^\nu/|\mathbf{q}|^2. \quad (5.9)$$

Hence it gives a coordinate-space propagator that is

---

[12] This is easy to prove in electrodynamics, where the current does not involve the potential; see R. P. Feynman, Phys. Rev. **101**, 769 (1949), Sec. 8. [This result is also implicit in the theorem proved here in Appendix B.] The situation is enormously more complicated in the case of gravitation, where the "current" must involve the potential $A^{\mu\nu}$; we will not attempt a treatment of this highly nontrivial problem here.

local in time:

$$\frac{1}{(2\pi)^4}\int d^4q\, e^{iq\cdot(x-y)}\Delta^{\mu\nu}{}_{\rm loc}(q) = \delta(x^0-y^0)\mathfrak{D}(\mathbf{x}-\mathbf{y})n^\mu n^\nu, \quad (5.10)$$

$$\mathfrak{D}(\mathbf{x}) = (2\pi)^{-3}\int d^3q\, \exp(i\mathbf{q}\cdot\mathbf{x})|\mathbf{q}|^{-2} = 1/4\pi|\mathbf{x}-\mathbf{y}| \quad (5.11)$$

and it may therefore be cancelled by the addition to $H'(t)$ of the familiar Coulomb interaction

$$H_{\rm Coul}'(t) = \frac{1}{2}\int d^3x \int d^3y\, \mathcal{J}^0(\mathbf{x},t)\mathfrak{D}(\mathbf{x}-\mathbf{y})\mathcal{J}^0(\mathbf{y},t). \quad (5.12)$$

Note that this cancellation is only possible because (5.12) is temporally local; the interaction *must* be local in time because of its definition as

$$H'(t) = e^{iH^f t}H'(0)e^{-iH^f t}. \quad (5.13)$$

In particular, $\Delta^{\mu\nu}{}_{\rm grad}(q)$ does not have a temporally local Fourier transform so it cannot be cancelled by adding a term to $H'(t)$, and it must be eliminated by requiring the current $\mathcal{J}_\mu{}^H$ to be conserved.

All this is familiar for $j=1$, and it works out much the same for $j\geqq 2$, because the general polarization sum $\Pi(q)$ is built up out of the $\Pi^{\mu\nu}$. For example

$j=2$ [see Appendix A]

$$\Pi^{\mu_1\mu_2\nu_1\nu_2}(q) = \tfrac{1}{2}[\Pi^{\mu_1\nu_1}(q)\Pi^{\mu_2\nu_2}(q) + \Pi^{\mu_1\nu_2}(q)\Pi^{\mu_2\nu_1}(q) - \Pi^{\mu_1\mu_2}(q)\Pi^{\nu_1\nu_2}(q)], \quad (5.14)$$

$j=3$

$$\begin{aligned}\Pi^{\mu_1\mu_2\mu_3\nu_1\nu_2\nu_3} = \tfrac{1}{3}[&\Pi^{\mu_1\nu_1}\Pi^{\mu_2\nu_2}\Pi^{\mu_3\nu_3} + \Pi^{\mu_1\nu_2}\Pi^{\mu_2\nu_3}\Pi^{\mu_3\nu_1} + \Pi^{\mu_1\nu_3}\Pi^{\mu_2\nu_1}\Pi^{\mu_3\nu_1} + \Pi^{\mu_1\nu_2}\Pi^{\mu_2\nu_1}\Pi^{\mu_3\nu_3} + \Pi^{\mu_1\nu_3}\Pi^{\mu_1\nu_2}\Pi^{\mu_3\nu_1}\\ &+ \Pi^{\mu_1\nu_1}\Pi^{\mu_2\nu_3}\Pi^{\mu_3\nu_2}] - \tfrac{1}{6}[\Pi^{\mu_1\nu_1}\Pi^{\mu_2\mu_3}\Pi^{\nu_2\nu_3} + \Pi^{\mu_2\nu_1}\Pi^{\mu_1\mu_3}\Pi^{\nu_2\nu_3} + \Pi^{\mu_3\nu_1}\Pi^{\mu_1\mu_2}\Pi^{\nu_2\nu_3} + \Pi^{\mu_1\nu_2}\Pi^{\mu_2\mu_3}\Pi^{\nu_1\nu_3}\\ &+ \Pi^{\mu_1\nu_3}\Pi^{\mu_1\mu_3}\Pi^{\nu_1\nu_2} + \Pi^{\mu_3\nu_2}\Pi^{\mu_1\mu_2}\Pi^{\nu_1\nu_3} + \Pi^{\mu_1\nu_2}\Pi^{\mu_2\mu_3}\Pi^{\nu_1\nu_2} + \Pi^{\mu_2\nu_2}\Pi^{\mu_1\mu_2}\Pi^{\nu_1\nu_2} + \Pi^{\mu_3\nu_3}\Pi^{\mu_1\mu_2}\Pi^{\nu_1\nu_2}], \quad (5.15)\end{aligned}$$

and so on. Evidently the propagator for any integral j can be decomposed as in (5.7), into a covariant part $\Delta_{\rm cov}$ built up out of the $g_{\mu\nu}$, plus a noncovariant part $\Delta_{\rm grad}$ proportional to one or more factors of q_μ, plus a noncovariant part $\Delta_{\rm loc}$ which lacks the pole at $q^2=0$. The last term is to be cancelled by adding a temporally local term to $H'(t)$. The second term $\Delta_{\rm grad}$ is not temporally local, so it must be eliminated by requiring that $A(x)$ be coupled to a conserved current.

For instance, Eqs. (5.3), (5.6), and (5.14) give the three parts of the $j=2$ propagator as

$$\Delta_{\rm cov}{}^{\mu_1\mu_2\nu_1\nu_2}(q) = [g^{\mu_1\nu_1}g^{\mu_2\nu_2} + g^{\mu_1\nu_2}g^{\mu_2\nu_1} - g^{\mu_1\mu_2}g^{\nu_1\nu_2}]/2(q^2-i\epsilon), \quad (5.16)$$

$$\Delta_{\rm grad}{}^{\mu_1\mu_2\nu_1\nu_2}(q) = \left[g^{\mu_1\nu_1} + \frac{n^{\mu_1}q^{\nu_1}+n^{\nu_1}q^{\mu_1}}{2|\mathbf{q}|^2}q^0 + \frac{q^2n^{\mu_1}n^{\nu_1}}{|\mathbf{q}|^2}\right] \times\frac{(n^{\mu_2}q^{\nu_2}+n^{\nu_2}q^{\mu_2})}{2|\mathbf{q}|^2(q^2-i\epsilon)}q^0 \quad (5.17)$$

$\pm$ five similar terms.

$$\begin{aligned}\Delta_{\rm loc}{}^{\mu_1\mu_2\nu_1\nu_2}(q) = [&g^{\mu_1\nu_1}n^{\mu_2}n^{\nu_2} + g^{\mu_2\nu_2}n^{\mu_1}n^{\nu_1} + g^{\mu_1\nu_2}n^{\mu_2}n^{\nu_1} + g^{\mu_2\nu_1}n^{\mu_1}n^{\nu_2} - g^{\mu_1\mu_2}n^{\nu_1}n^{\nu_2}\\ &- g^{\nu_1\nu_2}n^{\mu_1}n^{\mu_2}]/2|\mathbf{q}|^2 + n^{\mu_1}n^{\nu_1}n^{\mu_2}n^{\nu_2}q^2/2|\mathbf{q}|^4. \quad (5.18)\end{aligned}$$

The gradient term (5.17) does not contribute if we require the "current" $\mathcal{J}_{\mu\nu}{}^H(x)$ to be conserved.[12] The term (5.18) gives a temporally local contribution to the propagator

$$\begin{aligned}(2\pi)^{-4}&\int d^4q\, e^{iq\cdot(x-y)}\Delta_{\rm loc}{}^{\mu_1\mu_2\nu_1\nu_2}(q)\\ &= \tfrac{1}{2}[g^{\mu_1\nu_1}n^{\mu_2}n^{\nu_2} + g^{\mu_2\nu_2}n^{\mu_1}n^{\nu_1} + g^{\mu_1\nu_2}n^{\mu_2}n^{\nu_1} + g^{\mu_2\nu_1}n^{\mu_1}n^{\nu_2} - g^{\mu_1\mu_2}n^{\nu_1}n^{\nu_2} - g^{\nu_1\nu_2}n^{\mu_1}n^{\mu_2}\\ &\quad + n^{\mu_1}n^{\nu_1}n^{\mu_2}n^{\nu_2}]\delta(x^0-y^0)\mathfrak{D}(\mathbf{x}-\mathbf{y}) + \tfrac{1}{2}n^{\mu_1}n^{\nu_1}n^{\mu_2}n^{\nu_2}\ddot{\delta}(x^0-y^0)\mathcal{E}(\mathbf{x}-\mathbf{y}), \quad (5.19)\end{aligned}$$

where $\mathfrak{D}(\mathbf{x})$ is given by (5.11), and

$$\mathcal{E}(\mathbf{x}) \equiv (2\pi)^{-3}\int d^3q\, \exp(i\mathbf{q}\cdot\mathbf{x})|\mathbf{q}|^{-4} = \mathcal{E}(0) - \frac{|\mathbf{x}|}{8\pi}. \quad (5.20)$$

[We will see that the divergent constant $\mathcal{E}(0)$ gives no trouble.] In order to cancel (5.19) we must add to the Hamiltonian a "Newtonian" term:

$$\begin{aligned}H'_{\rm Newt}(t) = \frac{1}{2}\int d^3xd^3y[&2\mathcal{J}^\mu{}_0(\mathbf{x},t)\mathcal{J}_{\mu 0}(\mathbf{y},t) - \tfrac{1}{2}\mathcal{J}^\mu{}_\mu(\mathbf{x},t)\mathcal{J}_{00}(\mathbf{y},t) - \tfrac{1}{2}\mathcal{J}_{00}(\mathbf{x},t)\mathcal{J}^\mu{}_\mu(\mathbf{y},t)\\ &+ \tfrac{1}{2}\mathcal{J}_{00}(\mathbf{x},t)\mathcal{J}_{00}(\mathbf{y},t)]\mathfrak{D}(\mathbf{x}-\mathbf{y})\\ &+ \frac{1}{2}\int d^3xd^3y\,\mathcal{J}_{00}(\mathbf{x},t)\ddot{\mathcal{J}}_{00}(\mathbf{y},t)\mathcal{E}(\mathbf{x}-\mathbf{y}). \quad (5.21)\end{aligned}$$

In Sec. VII we shall see that this term, ugly as it seems, is precisely what is needed to generate Einstein's field equations when we pass to the Heisenberg representation.

The conclusion suggested by the above is that the conservation and covariance of the current plus the presence of direct-interaction terms like $H'_{\rm Coul}$ and $H'_{\rm Newt}$, are together the necessary and sufficient conditions for the Lorentz invariance of the S matrix. In Appendix B we show that these conditions do in fact imply the Lorentz invariance of the S matrix in quantum electrodynamics.[6] Our proof of their sufficiency makes their necessity rather evident, and can also obviously be extended to any massless particle theory in which the potential does not itself appear in the current. The rigorous treatment of Lorentz invariance in cases like the gravitational or the Yang-Mills field where the potential must appear in the current requires a much more elaborate discussion, and I reserve this for a future paper. [The problem has lost some of its urgency, because we have already seen in Ref. 4 that very simple and general arguments insure that any Lorentz-invariant theory of massless particles with $j=1$ or $j=2$ must possess the most striking dynamical features of photons and gravitons, to wit, the conservation of charge and the equality of gravitational and inertial mass.]

VI. DERIVATION OF MAXWELL'S EQUATIONS

The space-components of the vector potential $A_H{}^\mu(x)$ in the Heisenberg representation are defined, as usual, by

$$A_H{}^i(x)\equiv U(x^0)A^i(x)U^{-1}(x^0)\,, \tag{6.1}$$

$$U(t)\equiv \exp(iHt)\exp(-iH^f t) \tag{6.2}$$

with H^f the free-particle Hamiltonian and $H=H^f+H'$ the total Hamiltonian. The interaction-representation potential $A^i(x)$ is explicitly given by

$$A^i(x)=(2\pi)^{-3/2}\int d^3p(2|\mathbf{p}|)^{-1/2}\sum_\pm e_\pm{}^i(\mathbf{p}) \times[a(\mathbf{p},\pm1)e^{ip\cdot x}+a^*(\mathbf{p},\mp1)e^{-ip\cdot x}] \tag{6.3}$$

so it satisfies the field equations

$$\Box^2 A^i(x)=0\,, \tag{6.4}$$

$$\partial_i A^i(x)=0\,, \tag{6.5}$$

and the commutation reltions

$$[A^i(x),A^j(y)]=0\,, \tag{6.6}$$

$$[\dot{A}^i(x),\dot{A}^j(y)]=0\,, \tag{6.7}$$

$$[A^i(x),\dot{A}^j(y)]=i\mathfrak{D}^{ij}(\mathbf{x}-\mathbf{y})\,, \tag{6.8}$$

with

$$\mathfrak{D}^{ij}(\mathbf{x}-\mathbf{y})=(2\pi)^{-3}\int d^3p\,\Pi^{ij}(p)\exp[i\mathbf{p}\cdot(\mathbf{x}-\mathbf{y})] =\delta_{ij}\delta^3(\mathbf{x}-\mathbf{y})+\partial_i\partial_j\mathfrak{D}(\mathbf{x}-\mathbf{y})\,. \tag{6.9}$$

The lemma proved in Appendix C thus allows us immediately to write down the field equation for A^i in the Heisenberg representation:

$$\Box^2 A_H{}^i(\mathbf{x},t) =-\int d^3\mathbf{y}\;\mathfrak{D}^{ij}(\mathbf{x}-\mathbf{y})\mathfrak{J}_H{}^j(\mathbf{y},t) =-\mathfrak{J}_H{}^i(\mathbf{x},t)-\partial_i\partial_j\int d^3\mathbf{y}\;\mathfrak{D}(\mathbf{x}-\mathbf{y})\mathfrak{J}_H{}^j(\mathbf{y},t)\,. \tag{6.10}$$

However, the response of one system of charges to another system cannot be described solely in terms of the three-vector field $A_H{}^i(x)$, because there is also a direct Coulomb interaction (5.12) between the two systems. For instance, it is easy to show that the S matrix for a transition $\alpha\to\beta$ caused by an infinitesimal c-number current $\delta\mathfrak{J}^\mu(x)$ is, to first order in $\delta\mathfrak{J}$

$$S_{\beta\alpha}=i\int d^4x\langle\beta\text{ out}|A_H{}^i(x)|\alpha\text{ in}\rangle\delta\mathfrak{J}_i(x) -i\int_{-\infty}^{\infty}dt\int d^3x\int d^3y\langle\beta\text{ out}|\mathfrak{J}_H{}^0(\mathbf{x},t)|\alpha\text{ in}\rangle \times\mathfrak{D}(\mathbf{x}-\mathbf{y})\delta\mathfrak{J}^0(\mathbf{y},t)\,. \tag{6.11}$$

Therefore we invent a fourth component of $A_H{}^\mu(x)$

$$A_H{}^0(\mathbf{x},t)\equiv\int d^3\mathbf{y}\;\mathfrak{D}(\mathbf{x}-\mathbf{y})\mathfrak{J}_H{}^0(\mathbf{y},t) \tag{6.12}$$

which enables us to write an expression like (6.11) compactly as

$$S_{\beta\alpha}=i\int d^4x\langle\beta\text{ out}|A_H{}^\mu(x)|\alpha\text{ in}\rangle\delta\mathfrak{J}_\mu(x)\,. \tag{6.13}$$

The field $A_H{}^0$ obeys the Poisson equation

$$\nabla^2 A_H{}^0(x)=-\mathfrak{J}_H{}^0(x)\,. \tag{6.14}$$

Also, (6.12) and the current conservation condition (4.5) let us write (6.10) as

$$\Box^2 A_H{}^i(x)=-\mathfrak{J}_H{}^i(x)+\partial_0\partial_i A_H{}^0(x)\,. \tag{6.15}$$

Together (6.14) and (6.15) yield Maxwell's equations

$$\partial_\mu F_H{}^{\mu\nu}(x)=-\mathfrak{J}_H{}^\nu(x) \tag{6.16}$$

$$F_H{}^{\mu\nu}(x)\equiv\partial^\mu A_H{}^\nu(x)-\partial^\nu A_H{}^\mu(x)\,. \tag{6.17}$$

The particular form of (6.14) and (6.15) arises because (6.1) and (6.5) impose on $A_H{}^\mu(x)$ the Coulomb gauge condition

$$\partial_i A_H{}^i(x)=0\,. \tag{6.18}$$

It may be of interest to note that in the absence of current conservation (6.15) would become

$$\Box^2 A_H{}^i(x)=-\mathfrak{J}_H{}^i(x)+\partial_0\partial_i A_H{}^0(x) -\partial_i\int d^3\mathbf{y}\;\mathfrak{D}(\mathbf{x}-\mathbf{y})\partial_\mu\mathfrak{J}_H{}^\mu(\mathbf{y},t)$$

and Maxwell's equations would read

$$\partial_\mu F_H{}^{\mu i}(x)=-\mathfrak{J}_H{}^i(x)+\partial_i\int d^3\mathbf{y}\;\mathfrak{D}(\mathbf{x}-\mathbf{y})\partial_\mu\mathfrak{J}_H{}^\mu(\mathbf{y},t)$$

$$\partial_\mu F_H{}^{\mu 0}(x)=-\mathfrak{J}_H{}^0(x)\,.$$

The crucial importance of current conservation for Lorentz invariance is apparent again in these field equations.

VII. DERIVATION OF EINSTEIN'S EQUATIONS

The traceless part of the spatial components of the gravitational field $A_H{}^{\mu\nu}(x)$ in the Heisenberg representation are defined in the same way as we have defined $A_H{}^i$ in (6.1), i.e.,

$$A_H{}^{ij}(x)-\tfrac{1}{3}\delta^{ij}\delta_{kl}A_H{}^{kl}(x)\equiv U(x^0)A^{ij}(x)U^{-1}(x^0) \quad (7.1)$$

with the interaction representation potential $A^{ij}(x)$ given explicitly by

$$A^{ij}(x)=(2\pi)^{-3/2}\int d^3p(2|\mathbf{p}|)^{-1/2}\sum_{\pm} e_{\pm}{}^i(\mathbf{p})e_{\pm}{}^j(\mathbf{p})$$
$$\times[a(\mathbf{p},\pm 2)e^{ip\cdot x}+a^*(\mathbf{p},\mp 2)e^{-ip\cdot x}]. \quad (7.2)$$

Note that A^{ij} is traceless because $(\mathbf{e}_{\pm})^2=0$. Also, $A^{ij}(x)$ satisfies the field equations

$$\square^2 A^{ij}(x)=0, \quad (7.3)$$

$$\partial_i A^{ij}(x)=0, \quad (7.4)$$

and the commutation relations

$$[A^{ij}(x),A^{kl}(y)]=0, \quad (7.5)$$

$$[\dot{A}^{ij}(x),\dot{A}^{kl}(y)]=0, \quad (7.6)$$

$$[A^{ij}(x),\dot{A}^{kl}(y)]=i\mathfrak{D}^{ij,kl}(\mathbf{x}-\mathbf{y}), \quad (7.7)$$

with

$$\mathfrak{D}^{ij,kl}(\mathbf{x}-\mathbf{y})=(2\pi)^{-3}\int d^3p\, \Pi^{ijkl}(\mathbf{p})\exp[i\mathbf{p}\cdot(\mathbf{x}-\mathbf{y})]. \quad (7.8)$$

Equation (5.14) gives the polarization sum for $j=2$ as

$$\Pi^{ijkl}(\mathbf{p})=\tfrac{1}{2}[(\delta^{ik}-\hat{p}^i\hat{p}^k)(\delta^{jl}-\hat{p}^j\hat{p}^l)$$
$$+(\delta^{il}-\hat{p}^i\hat{p}^l)(\delta^{jk}-\hat{p}^j\hat{p}^k)-(\delta^{ij}-\hat{p}^i\hat{p}^j)(\delta^{kl}-\hat{p}^k\hat{p}^l)]$$

so

$$\mathfrak{D}^{ij,kl}(\mathbf{x}-\mathbf{y})=\tfrac{1}{2}[\delta^{ik}\delta^{jl}+\delta^{il}\delta^{jk}-\delta^{ij}\delta^{kl}]\delta^3(\mathbf{x}-\mathbf{y})$$
$$+\tfrac{1}{2}[\partial^i\partial^k\delta^{jl}+\partial^i\partial^l\delta^{jk}+\partial^i\partial^l\delta^{jk}+\partial^j\partial^k\delta^{il}$$
$$-\partial^i\partial^j\delta^{kl}-\partial^k\partial^l\delta^{ij}]\mathfrak{D}(\mathbf{x}-\mathbf{y})$$
$$+\tfrac{1}{2}\partial^i\partial^j\partial^k\partial^l\mathcal{E}(\mathbf{x}-\mathbf{y}), \quad (7.9)$$

$$\mathfrak{D}(\mathbf{x})=1/4\pi|\mathbf{x}|,$$
$$\mathcal{E}(\mathbf{x})=\mathcal{E}(0)-|\mathbf{x}|/8\pi.$$

The lemma of Appendix C thus gives us the field equations satisfied by (7.1)

$$\square^2[A_H{}^{ij}(\mathbf{x},t)-\tfrac{1}{3}\delta^{ij}\delta_{kl}A_H{}^{kl}(\mathbf{x},t)]$$
$$=-\int d^3y\, \mathfrak{D}^{ij,kl}(\mathbf{x}-\mathbf{y})\mathcal{J}_{H,kl}(\mathbf{y},t) \quad (7.10)$$

or more explicitly

$$\square^2[A_H{}^{ij}(\mathbf{x},t)-\tfrac{1}{3}\delta^{ij}\delta_{kl}A_H{}^{kl}(\mathbf{x},t)]=-\mathcal{J}_H{}^{ij}(\mathbf{x},t)+\tfrac{1}{2}\delta^{ij}\delta_{kl}\mathcal{J}_H{}^{kl}(\mathbf{x},t)-\int d^3y[\mathcal{J}_H{}^{ik}(\mathbf{y},t)\partial_k\partial^j+\mathcal{J}_H{}^{jk}(\mathbf{y},t)\partial_k\partial^i$$
$$-\tfrac{1}{2}\mathcal{J}^k{}_{Hk}(\mathbf{y},t)\partial_i\partial_j-\tfrac{1}{2}\mathcal{J}_H{}^{kl}(\mathbf{y},t)\delta^{ij}\partial_k\partial_l]\mathfrak{D}(\mathbf{x}-\mathbf{y})-\tfrac{1}{2}\partial^i\partial^j\partial_k\partial_l\int d^3y\, \mathcal{J}_H{}^{kl}(\mathbf{y},t)\mathcal{E}(\mathbf{x}-\mathbf{y}). \quad (7.11)$$

[The current $\mathcal{J}_{H,kl}$ in the Heisenberg representation is related to the interaction representation current $\mathcal{J}_{kl}=-\delta H'/\delta A^{kl}$ by the same unitary operator $U(t)$ as appears in (6.1) and (7.1).]

Just as the space components of the vector potential $A_H{}^\mu(x)$ had to be supplemented by a time component to represent the direct-Coulomb interaction, it is necessary now to invent auxiliary components of the Heisenberg representation gravitational field $A_H{}^{\mu\nu}(x)$ in order to take account of the direct Newtonian interaction (5.21). This interaction can be written

$$H'_{\text{Newt}}(t)=\frac{1}{2}\int d^3x d^3y[2\mathcal{J}^i{}_0(\mathbf{x},t)\mathcal{J}_{i0}(\mathbf{y},t)-\tfrac{1}{2}\mathcal{J}^i{}_i(\mathbf{x},t)\mathcal{J}_{00}(\mathbf{y},t)-\tfrac{1}{2}\mathcal{J}_{00}(\mathbf{x},t)\mathcal{J}^i{}_i(\mathbf{y},t)-\tfrac{1}{2}\mathcal{J}_{00}(\mathbf{x},t)\mathcal{J}_{00}(\mathbf{y},t)]\mathfrak{D}(\mathbf{x}-\mathbf{y})$$
$$+\frac{1}{2}\int d^3x d^3y\, \mathcal{J}_{00}(\mathbf{x},t)\ddot{\mathcal{J}}_{00}(\mathbf{y},t)\mathcal{E}(\mathbf{x}-\mathbf{y}).$$

Therefore, we define

$$A_H{}^{i0}(\mathbf{x},t)\equiv\int d^3y\, \mathcal{J}_H{}^{i0}(\mathbf{y},t)\mathfrak{D}(\mathbf{x}-\mathbf{y}), \quad (7.12)$$

$$A_{Hi}{}^i(\mathbf{x},t)\equiv\frac{3}{2}\int d^3y\, \mathcal{J}_H{}^{00}(\mathbf{y},t)\mathfrak{D}(\mathbf{x}-\mathbf{y}), \quad (7.13)$$

$$A_H{}^{00}(\mathbf{x},t)\equiv\frac{1}{2}\int d^3y[\mathcal{J}_{Hi}{}^i(\mathbf{y},t)+\mathcal{J}_H{}^{00}(\mathbf{y},t)\mathfrak{D}(\mathbf{x}-\mathbf{y})]$$
$$-\frac{1}{2}\int d^3y\, \ddot{\mathcal{J}}_H{}^{00}(\mathbf{y},t)\mathcal{E}(\mathbf{x}-\mathbf{y}). \quad (7.14)$$

With these definitions, the S matrix for a transition $\alpha\to\beta$ due to an infinitesimal c number $\delta\mathcal{J}_{\mu\nu}(x)$ is

$$S_{\beta\alpha}=-i\int d^4x\langle\beta\text{ out}|A_H{}^{\mu\nu}(x)|\alpha\text{ in}\rangle\delta\mathcal{J}_{\mu\nu}(x).$$

These synthetic field components obey the field equations

$$\nabla^2 A_H{}^{i0}(x)=-\mathcal{J}_H{}^{i0}(x), \quad (7.15)$$

$$\nabla^2 A_H{}^i{}_i(x)=-\tfrac{3}{2}\mathcal{J}_H{}^{00}(x), \quad (7.16)$$

$$\nabla^2 A_H{}^{00}(x)=-\tfrac{1}{2}\mathcal{J}_{Hi}{}^i(x)-\tfrac{1}{2}\mathcal{J}_H{}^{00}(x)+\tfrac{1}{3}\ddot{A}_{Hi}{}^i(x). \quad (7.17)$$

Using the current conservation condition (4.5) let us write (7.11) as

$$\Box^2 A_H{}^{ij}(x) = -\mathcal{J}_H{}^{ij}(x) + \tfrac{1}{2}\delta^{ij}\mathcal{J}_H{}^\mu{}_\mu(x) + \partial_0\partial^j A_H{}^{i0}(x) + \partial_0\partial^i A_H{}^{j0}(x) + \partial^i\partial^j[A_H{}^{00}(x) - \tfrac{1}{3}A_H{}^k{}_k(x)]. \quad (7.18)$$

To the field equations (7.15)–(7.18) we must append two first-order equations, which remind us that we have defined the traceless part (7.1) of $A_H{}^{ij}$ to be divergenceless

$$\partial_i A_H{}^{ij}(x) = \tfrac{1}{3}\partial_j A_H{}^i{}_i(x) \quad (7.19)$$

and have defined $A_H{}^{i0}$ and $A_H{}^i{}_i$ in (7.12) and (7.13) so that the conservation of $\mathcal{J}_H{}^{\mu\nu}$ relates them by

$$\partial_i A_H{}^{i0}(x) = -\tfrac{2}{3}\partial_0 A_H{}^i{}_i(x). \quad (7.20)$$

Equations (7.15)–(7.20) can be put together compactly as

$$R_H{}^{\mu\nu}(x) = -\mathcal{J}_H{}^{\mu\nu}(x) + \tfrac{1}{2}g^{\mu\nu}\mathcal{J}_H{}^\lambda{}_\lambda(x), \quad (7.21)$$

where

$$R_H{}^{\mu\nu}(x) \equiv \Box^2 A_H{}^{\mu\nu}(x) - \partial^\mu\partial_\lambda A_H{}^{\lambda\nu}(x) - \partial^\nu\partial_\lambda A_H{}^{\mu\lambda}(x) + \partial^\mu\partial^\nu A_H{}^\lambda{}_\lambda(x). \quad (7.22)$$

The complicated form of (7.15)–(7.18) just arises from the fact that we happen to have defined $A_H{}^{\mu\nu}$ in the peculiar gauge characterized by (7.19) and (7.20). We might have avoided some algebra along the way had we chosen a different polarization "tensor" in forming the potential (7.2), but the choice we made was the most obvious generalization of the Coulomb gauge used for $j=1$, and at any rate has brought us safely to our goal.

Equation (7.21) can also be put in the familiar form

$$R_H{}^{\mu\nu}(x) - \tfrac{1}{2}g^{\mu\nu}R_H{}^\lambda{}_\lambda(x) = -\mathcal{J}_H{}^{\mu\nu}(x). \quad (7.23)$$

If $\mathcal{J}_H{}^{\mu\nu}(x)$ were proportional to the energy-momentum tensor of matter alone then (7.23) would be identical with Einstein's equations in the weak field limit, where we set the Einstein metric tensor equal to the Minkowski $g^{\mu\nu}$ plus our $A_H{}^{\mu\nu}$, and keep only terms of first order in $A_H{}^{\mu\nu}$. However, such a theory would not be Lorentz invariant, because Lorentz invariance requires that $\partial_\mu\mathcal{J}_H{}^{\mu\nu}=0$, and this condition is fulfilled only if the current $\mathcal{J}_H{}^{\mu\nu}$ contains terms involving $A_H{}^{\mu\nu}$, representing the energy and momentum density of gravitation. If we therefore identify $\mathcal{J}_H{}^{\mu\nu}$ with the full-energy momentum tensor $\theta^{\mu\nu}$ of matter plus gravitation, Eq. (7.23) becomes highly nonlinear. As remarked by Gupta,[13] there is obviously one choice of a conserved $\theta^{\mu\nu}$ which makes (7.23) equivalent to Einstein's nonlinear equations, namely that obtained by identifying the nonlinear terms on the left-hand-side of Einstein's equations with the negative of the gravitational part of $\theta^{\mu\nu}$. In fact, Feynman[14] has shown that this is the only choice which works. In Feynman's Lagrangian approach, Lorentz invariance is built in, but other desiderata of perturbation theory such as unitarity can be lost by making the wrong choice of $\theta^{\mu\nu}$, while in our approach unitarity and the particle interpretation are built in, and only Lorentz invariance can go wrong; therefore we may presume that the sort of covariance proof given in Appendix B for photons will only work for gravitons if we choose $\theta^{\mu\nu}$ in agreement with Einstein's theory. However, this still leaves an ambiguity in the matter part of $\theta^{\mu\nu}$, because we can always add "Pauli" terms such as (4.6).

[13] S. N. Gupta, Proc. Phys. Soc. **A65**, 608 (1952).

[14] R. P. Feynman (private communication). I am indebted to Professor Feynman for a discussion of this point.

VIII. MAGNETIC AND OTHER MONOPOLES

We saw in Sec. III that the particle operators for mass zero and integer spin j can be used to construct two different Hermitian potentials, a normal one $A^{\mu_1\cdots\mu_j}(x)$ with parity and time-reversal-phase $(-)^j$, and an abnormal one $B^{\mu_1\cdots\mu_j}(x)$ with P and T phases equal to $-(-)^j$. [See Eqs. (3.14)–(3.19).] Both $A(x)$ and $B(x)$ must be coupled to conserved tensor currents. However, the Hermitian current J^μ of charge (or baryon number, etc.) and the Hermitian energy-momentum tensor $\theta^{\mu\nu}(x)$ both have normal P and T phases, by which we mean that their spatial components obey the same P and T transformation rules (3.16) and (3.18) as for $A^i(x)$ and $A^{ij}(x)$. Therefore, both P and T invariance do not allow $B^\mu(x)$ and $B^{\mu\nu}(x)$ to be coupled to $J^\mu(x)$ or $\theta^{\mu\nu}(x)$. We could, of course, couple $B(x)$ to a Pauli current (4.6), but such interactions can be rewritten in terms of $A(x)$; for instance the coupling $B^\mu\partial^\nu[\bar\psi\gamma_5\sigma_{\mu\nu}\psi]$ is equivalent [using (3.7)] to $A^\mu\partial^\nu[\bar\psi\sigma_{\mu\nu}\psi]$. Hence, we would normally conclude from P or T invariance that all interactions may be expressed in terms of the normal potential $A(x)$, and in particular that there can be no magnetic monopoles.[15]

But there *is* one way that magnetic monopoles can occur without violating P or T. Suppose there is a particle which turns into its antiparticle under the operation of either parity[16] or time-reversal, and that the number of such particles is conserved. Then the Hermitian current $M^\mu(x)$ of the particle would undergo an extra sign change under P and T, and hence could be coupled to $B^\mu(x)$. Note that in this case P or T would forbid $A^\mu(x)$ from being coupled to $M^\mu(x)$; that is, a magnetic monopole cannot also carry a normal charge. Note also that we are *defining* P and T so that they act as usual on familiar particles like electrons and photons,

[15] The apparent violation of time-reversal invariance by magnetic monopoles has been noted by L. I. Schiff, Am. J. Phys., **32**, 812 (1964).

[16] This is sometimes expressed in the statement that the true symmetry is not P but PM, where M changes the sign of all magnetic monopole moments. See N. F. Ramsey, Phys. Rev. **109**, 225 (1959). We would prefer to say that M takes magnetic monopoles into their antiparticles, and include this in the definition of C, P, and T. (The product CPT takes all particles into their antiparticles, including magnetic monopoles.) This redefinition of T resolves the contradiction noted in Ref. 15.

and it is these ordinary inversions that take magnetic monopoles into their antiparticles; if all particles had this abnormal behavior under P and T we would just interchange the definitions of P and CP, T and CT, $A^\mu(x)$ and $B^\mu(x)$, charge and magnetic pole strength, etc.

In contrast, P or T do not allow the abnormal gravitational potential $B^{\mu\nu}(x)$ to interact with anything. Even if there were magnetic monopoles which went into their antiparticles under P and T, they would still make a contribution to the energy-momentum tensor $\theta^{\mu\nu}(x)$ which behaved normally under P and T, and which therefore could only be coupled to the normal potential $A^{\mu\nu}(x)$.

Since magnetic monopoles are allowed by C, P, and T, but are not observed in nature, we must ask if there is any other reason why they should not exist. Zwanziger[17] has noted that their existence would give the charge-monopole scattering amplitude $A(s,t)$ two very peculiar branch points in s near $t=0$. This suggests that field theories of photons, charges, and monopoles might be unavoidably acausal, and therefore, not Lorentz invariant. We now show that this is the case, at least within the interaction-representation dynamical framework used here.

The trouble arises in diagrams in which a photon is exchanged between a charge and monopole. Since the charge current $J_\mu(x)$ is coupled to $A^\mu(x)$ and the monopole current $M_\nu(y)$ is coupled to $B^\nu(y)$, the photon propagator will be

$$-i\Delta_{AB}{}^{\mu\nu}(q)=\int d^4x\, e^{-iq\cdot(x-y)}\langle T\{A^\mu(x),B^\nu(y)\}\rangle_0. \quad (8.1)$$

This can be easily calculated using (3.22) and (3.23) and the results of Appendix A we find

$$\Delta_{AB}{}^{\mu\nu}(q)=\frac{\Xi^{\mu\nu}(q)(q^0/|\mathbf{q}|)}{q^2-i\epsilon} \quad (8.2)$$

$$\Xi^{\mu\nu}(q)=i\textstyle\sum_\pm(\pm)e_\pm{}^\mu(\mathbf{q})e_\pm{}^\nu(\mathbf{q})^*$$
$$=\epsilon^{\mu\nu\lambda\rho}q_\lambda n_\rho/|\mathbf{q}|\,. \quad (8.3)$$

This is not covariant, but more important, it cannot be split up as in (5.7) into a covariant part, a noncovariant gradient part which vanishes between conserved currents, and a noncovariant "local" part which can be cancelled by adding a temporally local term to $H'(t)$.

To see that this crucial decomposition is impossible for (8.3), note that the one-photon-exchange matrix element for scattering of a charge, with conserved current J_μ, and a monopole, with conserved current M_μ, is

$$J_\mu\Delta_{AB}{}^{\mu\nu}M_\nu=(q^0/|\mathbf{q}|)[(J_\mu J^\mu+\alpha J_0{}^2)(M_\mu M^\mu+\alpha M_0{}^2)$$
$$-(J_\mu M^\mu+\alpha J_0M_0)^2]^{1/2}/(q^2-i\epsilon) \quad (8.4)$$

with

$$\alpha\equiv q^2/|\mathbf{q}|^2. \quad (8.5)$$

[17] D. Zwanziger (to be published).

This may be compared with one-photon-exchange between two charges (or two monopoles)

$$J_\mu\Delta_{AA}{}^{\mu\nu}J_\nu'=[J_\mu J^{\mu\prime}+\alpha J_0J_0']/(q^2-i\epsilon). \quad (8.6)$$

In both cases the matrix element is invariant for q^μ precisely on the light cone ($\alpha=0$) but not otherwise. The great difference between (8.4) and (8.6) is that the α term in (8.6) can be cancelled by a temporally local interaction $J_0J_0'/|\mathbf{q}|^2$, while no similar cancellation is possible in (8.4).

Incidentally, the square root in (8.4) would yield Zwanziger's branch points[17] if we set $\alpha=0$. But the failure of analyticity is academic if the theory of monopoles isn't even Lorentz invariant.

There is one possible hope for saving Lorentz invariance. According to Dirac,[18] the coupling constant ge for charge-monopole interactions must be an integer or a half-integer. Perhaps the exact S matrix is Lorentz-invariant for these particular large values of ge, though not in any finite order of perturbation theory. However, preliminary examination of the ladder series by A. Goldhaber indicates that this is unlikely.

There is a possibility that time-reversal as well as parity is violated by the weak interactions. In this case, some of the conclusions reached earlier in this section might need revision. In particular, CPT alone would not prevent a particle from carrying a magnetic monopole moment as well as an ordinary charge, or in other words, of coupling with different strength to the left- and and right-handed parts of the electromagnetic field. And in the same way, all particles might respond with different coupling constants $f_\pm$ (the ratio of gravitational to inertial mass) to the left and right-handed parts of the gravitational field. However, this still would not produce observable anomalies in gravitational interactions, for Lorentz invariance tells us[4] that all particles must have the same f_+ and the same f_- (perhaps $\neq f_+$). The contribution of virtual graviton lines in Feynman diagrams would therefore be proportional to

$$\textstyle\sum_\pm f_\pm f_\mp\langle T\{A_\pm{}^\mu(x),A_\mp{}^\nu(y)\}\rangle_0=f_+f_-\langle T\{A^\mu(x),A^\nu(y)\}\rangle_0$$

and this has the same form as if the coupling constants $f_\pm$ for right- and left-handed gravitons were the same.

ACKNOWLEDGMENTS

I am very grateful for valuable discussions with S. Deser, A. Goldhaber, S. Mandelstam, L. Schiff, and J. Schwinger.

APPENDIX A: THE POLARIZATION SUMS

We wish to evaluate the sums

$$\Pi^{\mu\nu}(\hat{q})=\textstyle\sum_\pm e_\pm{}^\mu(\hat{q})e_\pm{}^{\nu*}(\hat{q})\,, \quad (A.1)$$

$$\Xi^{\mu\nu}(\hat{q})=i\textstyle\sum_\pm(\pm)e_\pm{}^\mu(\hat{q})e_\pm{}^{\nu*}(\hat{q})\,, \quad (A.2)$$

$$\Pi^{\mu_1\mu_2\nu_1\nu_2}(\hat{q})=\textstyle\sum_\pm e_\pm{}^{\mu_1}(\hat{q})e_\pm{}^{\mu_2}(\hat{q})e_\pm{}^{\nu_1*}(\hat{q})e_\pm{}^{\nu_2*}(\hat{q})\,. \quad (A.3)$$

[18] P. A. M. Dirac, Proc. Roy. Soc. (London) **133**, 60 (1931).

These are the numerators, respectively, of the photon propagator linking two charges or two monopoles, the photon propagator linking a charge and a monopole, and the graviton propagator.

First take $\hat{q}=\hat{k}$, defined as the unit vector in the z direction. Then the polarization is

$$e_\pm^{1}(\hat{k})=1/\sqrt{2}\,,\quad e_\pm^{2}(\hat{k})=\pm i/\sqrt{2}\,,\quad e_\pm^{3}(\hat{k})=e_\pm^{0}(\hat{k})=0\,,$$

so the only nonvanishing components of (A.1)–(A.3) are

$$\Pi^{11}=\Pi^{22}=1\,, \tag{A.4}$$

$$\Xi^{12}=-\Xi^{21}=1\,, \tag{A.5}$$

$$\Pi^{1111}=\Pi^{2222}=\Pi^{1212}=\Pi^{2112}=\Pi^{2121}=\Pi^{1221}=-\Pi^{1122}=-\Pi^{2211}=1/2\,. \tag{A.6}$$

For $\hat{q}=\hat{k}$, (A.4) agrees with (5.4) or (5.6), so it agrees with them for all $\hat{q}$, because $\Pi^{\mu\nu}(\hat{q})$ is related to $\Pi^{\mu\nu}(\hat{k})$ by the rotation $R^\mu{}_\lambda(\hat{q})$ which takes $\hat{k}$ into $\hat{q}$:

$$\Pi^{\mu\nu}(\hat{q})=R^\mu{}_\lambda(\hat{q})R^\nu{}_\rho(\hat{q})\Pi^{\lambda\rho}(\hat{k})\,. \tag{A.7}$$

A similar argument verifies Eq. (5.14) for $\Pi^{\mu_1\mu_2\nu_1\nu_2}(\hat{q})$ and verifies Eq. (8.3) for $\Xi^{\mu\nu}(\hat{q})$.

APPENDIX B: LORENTZ INVARIANCE OF THE QUANTUM-ELECTRODYNAMICAL S MATRIX

We shall show in a separate paper that if the interaction is translation and rotation invariant then the S matrix will be Lorentz invariant *if and only if* the behavior of the interaction under infinitesimal "boosts" takes the form

$$[\mathbf{K}^f,H'(t)]=-[\mathbf{K}'(t),H^f+H'(t)]\,. \tag{B.1}$$

Here $\mathbf{K}^f$ is the generator of pure Lorentz transformations on the free-particle states; H^f is the free-particle Hamiltonian, and $H'(t)$ is the interaction in the interaction representation

$$H'(t)=\exp(iH^ft)H'\exp(-iH^ft)\,.$$

The operator $\mathbf{K}'(t)$ is unrestricted, except that it must have the same t dependence as $H'(t)$:

$$\mathbf{K}'(t)=\exp(iH^ft)\mathbf{K}'\exp(-iH^ft) \tag{B.2}$$

with the free-particle matrix elements of K' sufficiently smooth functions of energy so that, effectively,

$$\mathbf{K}'(t)\to 0\text{ for }t\to\pm\infty \tag{B.3}$$

this limit being understood in the same sense as the usual "adiabatic switching on and off" of $H'(t)$.

We will prove here that (B.1) is satisfied in the simplest case, i.e., quantum electrodynamics with an A-independent current:

$$H'(t)=-\int d^3x\,J_i(\mathbf{x},t)A^i(\mathbf{x},t)+\frac{1}{2}\int d^3xd^3y\,J^0(\mathbf{x},t)\mathfrak{D}(\mathbf{x}-\mathbf{y})J^0(\mathbf{y},t) \tag{B.4}$$

provided that the current is a vector and conserved in both the interaction and Heisenberg representations, i.e.,

$$\partial_\mu J^\mu(x)=0 \tag{B.5}$$

$$[H'(t),J^0(\mathbf{x},t)]=0\,. \tag{B.6}$$

(This is the case in spinor electrodynamics, and it can always be arranged by introducing enough auxiliary fields to make the free-field Lagrangian linear in space-time derivatives.)

The interaction (B.4) is manifestly translation- and rotation-invariant, so we need only check that it satisfies (B.1). The product $\mathbf{J}\cdot\mathbf{A}=J_\mu A^\mu$ is scalar except for the extra Φ term in Eq. (4.1), which for infinitesimal Lorentz transformations is given by (3.12) and (3.13) as

$$\Phi(x)=\Phi_+(x)+\Phi_-(x)=-i\omega_{i0}C^i(x)$$

$$C^i(x)\equiv(2\pi)^{-3/2}\int d^3p(2|\mathbf{p}|^3)^{-1/2}\sum_\pm e_\pm^{i}(\mathbf{p})\times[a(\mathbf{p},\pm1)e^{ip\cdot x}-a^*(\mathbf{p},\mp1)e^{-ip\cdot x}]\,. \tag{B.7}$$

Hence $\mathbf{J}\cdot\mathbf{A}$ transforms under infinitesimal boosts according to

$$i[\mathbf{K}^f,J_i(x)A^i(x)]=(x_0\boldsymbol{\nabla}-\mathbf{x}\partial_0)J_i(x)A^i(x)+iJ_\mu(x)\partial^\mu\mathbf{C}(x)\,. \tag{B.8}$$

Also, since $J^\mu(x)$ is a vector we have

$$i[\mathbf{K}^f,J^0(x)]=\mathbf{J}(x)+(x_0\boldsymbol{\nabla}-\mathbf{x}\partial_0)J^0(x)\,. \tag{B.9}$$

The $\boldsymbol{\nabla}$ terms drop out when we integrate over 3 space, leaving us with

$$[\mathbf{K}^f,H'(t)]=-i\partial_0\int d^3x\,\mathbf{x}J_i(\mathbf{x},t)A^i(\mathbf{x},t)-\int d^3x\,J_\mu(\mathbf{x},t)\partial^\mu\mathbf{C}(\mathbf{x},t)-i\int d^3xd^3y\,\mathbf{J}(\mathbf{x},t)\mathfrak{D}(\mathbf{x}-\mathbf{y})J^0(\mathbf{y},t)+i\int d^3xd^3y[\partial_0J^0(\mathbf{x},t)]\mathbf{x}\mathfrak{D}(\mathbf{x}-\mathbf{y})J^0(\mathbf{y},t)\,.$$

Using (B.5), and writing $\mathbf{x}$ in the last term as $\frac{1}{2}(\mathbf{x}+\mathbf{y})+\frac{1}{2}(\mathbf{x}-\mathbf{y})$, we can put this in the form

$$[\mathbf{K}^f,H'(t)]=-i(d\mathbf{K}'(t)/dt)-L(t) \tag{B.10}$$

with

$$\mathbf{K}'(t)\equiv\int d^3x\,\mathbf{x}J_\mu(\mathbf{x},t)A^\mu(\mathbf{x},t)-i\int d^3x\,J^0(\mathbf{x},t)\mathbf{C}(\mathbf{x},t)-\frac{1}{2}\int d^3xd^3y\,J^0(\mathbf{x},t)\mathbf{x}\mathfrak{D}(\mathbf{x}-\mathbf{y})J^0(\mathbf{y},t)\,, \tag{B.11}$$

$$L_i(t)\equiv-i\int d^3xd^3y\,J_j(\mathbf{x},t)\mathfrak{F}_{ij}(\mathbf{x}-\mathbf{y})J^0(\mathbf{y},t)\,, \tag{B.12}$$

$$\mathfrak{F}_{ij}(\mathbf{x})\equiv\delta_{ij}\mathfrak{D}(\mathbf{x})-\tfrac{1}{2}\partial_j(x_i\mathfrak{D}(\mathbf{x}))\,. \qquad (B.13)$$

Because of (B.6), the only term in $\mathbf{K}'(t)$ that does not commute with $H'(t)$ is that containing $\mathbf{C}(x)$, and therefore

$$[\mathbf{K}'(t),H'(t)]=i\int d^3xd^3y\, J_j(\mathbf{x},t)J^0(\mathbf{y},t) \times[A_j(\mathbf{x},t),\mathbf{C}(\mathbf{y},t)]\,. \qquad (B.14)$$

But we can calculate directly from (B.7) and (6.3) that

$$[A_j(\mathbf{x},t),C_i(\mathbf{y},t)] =(2\pi)^{-3}\int d^3p|\mathbf{p}|^{-2}(\delta_{ij}-\hat{p}_i\hat{p}_j)\exp[i\mathbf{p}\cdot(\mathbf{x}-\mathbf{y})]$$

$$=\delta_{ij}\mathfrak{D}(\mathbf{x}-\mathbf{y})+\partial_i\partial_j\mathcal{E}(\mathbf{x}-\mathbf{y})=\mathfrak{F}_{ij}(\mathbf{x}-\mathbf{y}) \qquad (B.15)$$

so (B.14) gives

$$[\mathbf{K}'(t),H'(t)]=-L(t)\,. \qquad (B.16)$$

Also, $\mathbf{K}'(t)$ evidently has the t dependence (B.2), so (B.10) and (B.16) give (B.1). Condition (B.3) is satisfied because (B.11) is as "smooth" an operator as the interaction (B.4).

APPENDIX C: DERIVATION OF FIELD EQUATIONS IN THE HEISENBERG REPRESENTATION

Suppose the interaction representation fields $\phi_n(x)$ have the properties

$$\Box^2\phi_n(x)=0\,, \qquad (C.1)$$

$$[\phi_n(\mathbf{x},t),\phi_m(\mathbf{y},t)]=0\,, \qquad (C.2)$$

$$[\dot{\phi}_m(\mathbf{x},t),\dot{\phi}_m(\mathbf{y},t)]=0\,, \qquad (C.3)$$

$$[\phi_n(\mathbf{x},t),\dot{\phi}_m(\mathbf{y},t)]=i\mathfrak{D}_{nm}(\mathbf{x}-\mathbf{y})\,. \qquad (C.4)$$

[This is the case for the potentials $A(x)$ defined for general j in Sec. III.] Suppose that the interaction $H'(t)$ does not involve any derivatives of ϕ_n higher than the first, and define the "partial currents" $\mathcal{S}_n$, $\mathcal{S}_n{}^\mu$ by the statement

$$\delta H'(t)=-\int d^3x\sum_n[\mathcal{S}_n(\mathbf{x},t)\delta\phi_n(\mathbf{x},t) +\mathcal{S}_n{}^\mu(\mathbf{x},t)\partial_\mu\delta\partial_\mu\phi_n(\mathbf{x},t)] \qquad (C.5)$$

where $\delta\phi_n$ and $\delta\partial_\mu\phi_n$ are arbitrary infinitesimal c-number variations of ϕ_n and $\partial_\mu\phi_n$. [If $H'(t)$ is the space integral of a local $\mathfrak{H}(x)$, then $\mathcal{S}_n\equiv\partial\mathfrak{H}/\partial\phi_n$ and $\mathcal{S}_n{}^\mu\equiv\partial\mathfrak{H}/\partial(\partial_\mu\phi_n)$.] Define the Heisenberg representation field $\phi_{nH}(x)$ by

$$\phi_{nH}(\mathbf{x},t)\equiv U(t)\phi_n(\mathbf{x},t)U^{-1}(t) \qquad (C.6)$$

$$U(t)\equiv\exp(iHt)\exp(-iH't)\,. \qquad (C.7)$$

Then ϕ_H will obey the field equation

$$\Box^2\phi_{nH}(\mathbf{x},t)=-\int d^3\mathbf{y}\,\mathfrak{D}_{nm}(\mathbf{x}-\mathbf{y})J_{mH}(\mathbf{y},t) \qquad (C.8)$$

with J_H defined as the total current

$$J_{nH}(x)\equiv\mathcal{S}_{nH}(x)-\partial_\mu\mathcal{S}_{nH}{}^\mu(x)\,, \qquad (C.9)$$

$$\mathcal{S}_{nH}(\mathbf{x},t)\equiv U(t)\mathcal{S}_n(x)U^{-1}(t)\,, \qquad (C.10)$$

$$\mathcal{S}_{nH}{}^\mu(\mathbf{x},t)\equiv U(t)\mathcal{S}_n{}^\mu(x)U^{-1}(t)\,. \qquad (C.11)$$

Proof: We note first that

$$dU(t)/dt=iU(t)H'(t)$$

$$dU^{-1}(t)/dt=-iH'(t)U^{-1}(t)\,.$$

Therefore the time derivative of (C.6) gives

$$\dot{\phi}_{nH}(\mathbf{x},t)=U(t)\{\dot{\phi}_n(\mathbf{x},t)+i[H'(t),\phi_n(\mathbf{x},t)]\}U^{-1}(t)\,.$$

But (C.5), (C.2), and (C.4) give the commutator

$$[H'(t),\phi_n(\mathbf{x},t)]=+i\int d^3y\sum_m\mathcal{S}_m{}^0(\mathbf{y},t)\mathfrak{D}_{nm}(\mathbf{x}-\mathbf{y}) \qquad (C.12)$$

so

$$\dot{\phi}_{nH}(\mathbf{x},t)=U(t)\dot{\phi}_n(\mathbf{x},t)U^{-1}(t) -\sum_m\int d^3y\,\mathcal{S}_{mH}{}^0(\mathbf{y},t)\mathfrak{D}_{nm}(\mathbf{x}-\mathbf{y})\,.$$

A second time derivative gives

$$\ddot{\phi}_{nH}(\mathbf{x},t)=U(t)\{\ddot{\phi}(\mathbf{x},t)+i[H'(t),\dot{\phi}_n(\mathbf{x},t)]\}U^{-1}(t) -\sum_m\int d^3y\,\partial_0\mathcal{S}_{mH}{}^0(\mathbf{y},t)\mathfrak{D}_{nm}(\mathbf{x}-\mathbf{y})\,. \qquad (C.13)$$

But (C.5), (C.3), and (C.4) give the commutator

$$[H'(t),\dot{\phi}_n(\mathbf{x},t)]=-i\int d^3y\sum_m[\mathcal{S}_m(\mathbf{y},t)\mathfrak{D}_{nm}(\mathbf{y}-\mathbf{x}) +\mathcal{S}_m{}^i(\mathbf{y},t)\partial_i\mathfrak{D}_{mn}(\mathbf{y}-\mathbf{x})]\,. \qquad (C.14)$$

Using (C.14) and (C.1) and integrating by parts let us write (C.13) as

$$\ddot{\phi}_{nH}(\mathbf{x},t)=\nabla^2\phi_{nH}(\mathbf{x},t)+\int d^3y\sum_m[\mathfrak{D}_{mn}(\mathbf{y}-\mathbf{x})\mathcal{S}_{mH}(\mathbf{y},t) -\mathfrak{D}_{mn}(\mathbf{y}-\mathbf{x})\partial_i\mathcal{S}_{mH}(\mathbf{y},t) \mathfrak{D}_{nm}(\mathbf{x}-\mathbf{y})\partial_0\mathcal{S}_{mH}{}^0(\mathbf{y},t)]\,. \qquad (C.15)$$

But differentiating (C.2) with respect to t gives

$$\mathfrak{D}_{nm}(\mathbf{y}-\mathbf{x})=\mathfrak{D}_{mn}(\mathbf{x}-\mathbf{y}) \qquad (C.16)$$

so (C.15) and (C.9) yield the desired Eq. (C.8).

PHYSICAL REVIEW VOLUME 140, NUMBER 2B 25 OCTOBER 1965

Infrared Photons and Gravitons*

STEVEN WEINBERG†
Department of Physics, University of California, Berkeley, California
(Received 1 June 1965)

It is shown that the infrared divergences arising in the quantum theory of gravitation can be removed by the familiar methods used in quantum electrodynamics. An additional divergence appears when infrared photons or gravitons are emitted from noninfrared external lines of zero mass, but it is proved that for infrared gravitons this divergence cancels in the sum of all such diagrams. (The cancellation does not occur in massless electrodynamics.) The formula derived for graviton bremsstrahlung is then used to estimate the gravitational radiation emitted during thermal collisions in the sun, and we find this to be a stronger source of gravitational radiation (though still very weak) than classical sources such as planetary motion. We also verify the conjecture of Dalitz that divergences in the Coulomb-scattering Born series may be summed to an innocuous phase factor, and we show how this result may be extended to processes involving arbitrary numbers of relativistic or nonrelativistic particles with arbitrary spin.

I. INTRODUCTION

THE chief purpose of this article is to show that the infrared divergences in the quantum theory of gravitation can be treated in the same manner as in quantum electrodynamics. However, this treatment apparently does not work in other non-Abelian gauge theories, like that of Yang and Mills. The divergent phases encountered in Coulomb scattering will incidentally be explained and generalized.

It would be difficult to pretend that the gravitational infrared divergence problem is very urgent. My reasons for now attacking this question are:

(1) Because I can. There still does not exist any satisfactory quantum theory of gravitation, and in lieu of such a theory it would seem well to gain what experience we can by solving any problems that can be solved with the limited formal apparatus already at our disposal. The infrared divergences are an ideal case of this sort, because we already know all about the coupling of a very soft graviton to any other particle,[1] and about the external graviton line wave functions[1] and internal graviton line propagators.[2]

(2) Because something might go wrong, and that would be interesting. Unfortunately, nothing does go wrong. In Sec. II we see that the dependence on the infrared cutoffs of real and virtual gravitons cancels just as in electrodynamics.

However, there is a more subtle difficulty that might have been expected. Ordinary quantum electrodynamics would contain unremovable logarithmic divergences if the electron mass were zero, due to diagrams in which a soft photon is emitted from an external electron line with momentum parallel to the electron's.[3] There are no charged massless particles in the real world, but hard neutrinos, photons, and gravitons do carry a gravitational "charge," in that they can emit soft gravitons. In Sec. III we show that diagrams in which a soft graviton is emitted from some other hard massless particle line do contain divergences like the $\ln m_e$ terms in massless electrodynamics, but that these divergences cancel when we sum all such diagrams.[4] However, this cancellation is definitely due to the details of gravitational coupling, and does not save theories (like Yang and Mills's) in which massless particles can emit soft massless particles of spin one.

(3) Because in solving the infrared divergence problem we obtain a formula for the emission rate and spectrum of soft gravitons in arbitrary collision processes, which may (if our experience in electrodynamics is a guide) be numerically the most important gravitational radiative correction. In Sec. IV this formula is used to calculate the soft gravitational inner bremsstrahlung in an arbitrary nonrelativistic collision, and the result is then used to estimate the thermal gravitational radiation from the sun. The answer is several

* Research supported in part by the Air Force Office of Scientific Research, Grant No. AF-AFOSR-232-65.
† Alfred P. Sloan Foundation Fellow.

[1] S. Weinberg, Phys. Rev. **135**, B1049 (1965).

[2] See, e.g., S. Weinberg, Phys. Rev. **138**, B988 (1965). The graviton propagator given in Eq. (2.20) of the present article is not just the vacuum expectation value of a time-ordered product, but includes the effects of instantaneous "Newton" interactions that must be added to the interaction to maintain Lorentz invariance, and further, it does *not* include certain non-Lorentz-invariant gradient terms which disappear because the gravitational field is coupled to a conserved source. This disappearance has so far only been proved for graviton lines linking particles on their mass shells, and in fact this is the one impediment which keeps us from claiming that we possess a completely satisfactory quantum theory of gravitation. In using (2.20) we are to some extent relying on an act of faith, but this faith seems particularly well-founded in our present context because we use (2.20) here to link particle lines with momenta only infinitesimally far from their mass shells. See also S. Weinberg, in *Brandeis 1964 Summer Lectures on Theoretical Physics* (Prentice-Hall, Inc., New York, 1965).

[3] The extra divergences in massless quantum electrodynamics have long been known to many theorists. Recently, it has been noted by T. D. Lee and M. Nauenberg, Phys. Rev. **133**, B1549 (1964), that these divergences cancel if transition rates are computed only between suitable ensembles of final *and initial* states. [See also T. Kinoshita, J. Math. Phys. **3**, 650 (1962)]. However, these ensembles include not only indefinite numbers of very soft quanta but also hard massless particles with indefinite energies, and I remain unconvinced that transition rates between such ensembles are the only ones that can be measured and need be finite.

[4] I understand that this cancellation has also been found by R. P. Feynman.

orders of magnitude greater than from more usually considered sources, like planetary motion.

The formalism derived in Sec. II is used in Sec. V to calculate the divergent final- and initial-state interaction phases in arbitrary scattering processes.

II. REAL AND VIRTUAL INFRARED DIVERGENCES

This section shows how to treat the infrared divergences arising from very soft real and virtual gravitons. In order to keep the discussion as perspicuous as possible, I repeat the conventional treatment[5,6] of infrared divergences in electrodynamics (correcting a mistake in Ref. 5), with explanations at each step of how the same arguments apply to gravitation. It is not really quite correct to treat gravitation and electromagnetism as mutually exclusive phenomena, but it will be made obvious that in a combined theory the infrared photons and gravitons simply supply independent correction factors to transition rates.

1. One Soft Photon or Graviton

If we attach a soft-photon line with momentum q to an outgoing charged-particle line in a Feynman diagram, we must supply one extra charged-particle propagator with momentum $p+q$ and one extra vertex for the transition $p+q \to p$. If the soft-photon line is attached to an incoming charged-particle line, the extra propagator is for momentum $p-q$ and the transition is $p \to p-q$. For instance, if the charged particle has zero spin, these factors are[7]

$$i(2\pi)^4 e(2p^\mu+\eta q^\mu)[-i(2\pi)^{-4}] \times[(p+\eta q)^2+m^2-i\epsilon]^{-1}, \quad (2.1)$$

where $\eta=+1$ or -1 for an outgoing or incoming charged particle. In the limit $q\to 0$ Eq. (2.1) becomes (because $p^2+m^2=0$):

$$e\eta p^\mu/[p\cdot q-i\eta\epsilon]. \quad (2.2)$$

Although (2.1) applies only for zero spin, the limiting form (2.2) is well known[8] to hold for any spin.

Diagrams with the soft-photon line attached to an internal charged-particle line lack the denominator $p\cdot q$, and therefore are negligible for $q\to 0$. Hence the effect of attaching one soft-photon line to an arbitrary diagram is simply to supply an extra factor,

$$\sum_n e_n\eta_n p_n{}^\mu/[p_n\cdot q-i\eta_n\epsilon], \quad (2.3)$$

the sum running over all external lines in the original diagram.

If we attach a soft-graviton line to an external spin-zero line, the extra factors are[2]

$$\tfrac{1}{2}i(2\pi)^4(8\pi G)^{1/2}(2p^\mu+\eta q^\mu)(2p^\nu+\eta q^\nu) \times[-i(2\pi)^{-4}][(p+\eta q)^2+m^2-i\epsilon]^{-1}, \quad (2.4)$$

where μ, ν are the graviton polarization indices. For $q\to 0$ this gives

$$(8\pi G)^{1/2}\eta p^\mu p^\nu/[p\cdot q-i\eta\epsilon]. \quad (2.5)$$

The limiting form (2.5) is actually valid whatever the spin of the external line to which we attach the graviton.[1] For example, if this line is outgoing and has spin $\frac{1}{2}$, then we have instead of Eq. (2 4) the factor

$$-\tfrac{1}{4}(2\pi)^4(8\pi G)^{1/2}\{(2p^\mu+q^\mu)\gamma^\nu+(2p^\nu+q^\nu)\gamma^\mu\} \times[-i(2\pi)^{-4}]\left[\frac{-i(p^\lambda+q^\lambda)\gamma_\lambda+m}{(p+q)^2+m^2-i\epsilon}\right]. \quad (2.6)$$

But (2.6) appears multiplied on the left with a Dirac spinor $\bar{u}$ such that

$$\bar{u}[ip^\lambda\gamma_\lambda+m]=0.$$

Thus, moving the propagator numerator to the left of the vertex function, we are left with an anticommutator equal to (2.5) in the limit $q\to 0$. For general spin the same conclusion can be reached on grounds of Lorentz invariance,[1] without embroiling oneself in higher spin formalisms. The normalization factor $(8\pi G)^{1/2}$ is chosen so that an arbitrary nonrelativistic two-particle scattering amplitude will have a one-graviton-exchange pole with the correct residue to correspond to a potential Gm_1m_2/r.

The dominance of the $1/(p\cdot q)$ pole in (2.5) implies that the effect of attaching one soft-graviton line to an arbitrary diagram is to supply a factor equal to the sum of (2.5) over all external lines in the diagram

$$(8\pi G)^{1/2}\sum_n \eta_n p_n{}^\mu p_n{}^\nu/[p_n\cdot q-i\eta_n\epsilon]. \quad (2.7)$$

2. Many Soft Photons or Gravitons

It is well known that the effect of attaching several soft-photon lines to an arbitrary diagram is to supply a product of factors of the form (2.3), one for each soft photon. For we note that if N soft photons are emitted from an outgoing (incoming) charged-particle line with photon r last (first), photon s next-to-last (second), etc., then the charged particle propagators will contribute a multiple pole factor

$$[p\cdot q_r-i\eta\epsilon]^{-1}[p\cdot(q_r+q_s)-i\eta\epsilon]^{-1}\cdots,$$

but this must be summed over the $N!$ permutations $12\cdots N\to rs\cdots$, and the sum is just

$$[p\cdot q_1-i\eta\epsilon]^{-1}[p\cdot q_2-i\eta\epsilon]^{-1}\cdots.$$

---

[5] J. M. Jauch and F. Rohrlich, *Theory of Photons and Electrons* (Addison-Wesley Publishing Company, Inc., Reading, Massachusetts, 1955), Chap. 16.

[6] D. R. Yennie, S. C. Frautschi, and H. Suura, Ann. Phys. (N. Y.) **13**, 379 (1961).

[7] The notation used is that of Ref. 5. In particular, $\hbar=c=1$, and $p\cdot q\equiv\mathbf{p}\cdot\mathbf{q}-p^0q^0$.

[8] For $j=\frac{1}{2}$ see Ref. 5 or 6. For general spin see Ref. 1.

For example, for $N=2$

$$[p\cdot q_1-i\eta\epsilon]^{-1}[p\cdot(q_1+q_2)-i\eta\epsilon]^{-1}+[p\cdot q_2-i\eta\epsilon]^{-1}[p\cdot(q_2+q_1)-i\eta\epsilon]^{-1}=[p\cdot q_1-i\eta\epsilon]^{-1}[p\cdot q_2-i\eta\epsilon]^{-1}.$$

The result may be proved for general N by an easy mathematical induction. The same factorization occurs trivially when soft photons are emitted from different legs.

The pole structure created by inserting the soft-graviton factors (2.7) is the same as for the soft-photon factors (2.3), so by precisely the same reasoning the effect of attaching N soft-graviton lines to an arbitrary Feynman diagram is just to multiply the matrix element by N factors (2.7). It is this factorizability that will allow us to obtain the sum of an unlimited number of very complicated Feynman diagrams.

3. Virtual Infrared Divergences

We will define an infrared virtual photon or graviton as one which connects two external lines and carries energy less than Λ, where Λ is some convenient dividing point chosen low enough to justify the approximations made above in subsections (1) and (2). By "connecting external lines" we mean that the infrared line may join onto a line that has already emitted soft real quanta or virtual infrared quanta, but not onto one which, by real or virtual emission, has acquired a momentum far off its mass shell. In addition to the cutoff $|\mathbf{q}|\leqq\Lambda$ which just defines the infrared lines, we will also impose a cutoff $|\mathbf{q}|\geqq\lambda$ in order to display the logarithmic divergences as powers of $\ln\lambda$. We take λ very small (in particular, $\lambda\ll\Lambda$) so this cutoff only affects the infrared lines because it is only these that give infrared divergences for $\lambda=0$.

The effect of adding N virtual infrared-photon lines to a diagram that does not already involve any infrared lines is to multiply the matrix element by N pairs of the factors (2.3), each pair connected by a photon propagator

$$\frac{-i}{(2\pi)^4}\frac{g_{\mu\nu}}{q^2-i\epsilon}, \tag{2.8}$$

and then sum over polarization indices and integrate over q's. In addition we must divide by $2^N N!$, because we saw in Subsec. 2 that the external-line poles factor only if we sum over all places to which we may attach the two ends of each infrared virtual-photon line, and this includes spurious sums over the $N!$ permutations of the lines and over the two directions that each line might be thought to flow. The result is then

$$\frac{1}{N!}\left[\frac{1}{2}\int_\lambda^\Lambda d^4q\,A(q)\right]^N, \tag{2.9}$$

where

$$A(q)=\frac{-i}{(2\pi)^4[q^2-i\epsilon]}\times\sum_{nm}\frac{e_ne_m\eta_n\eta_m(p_n\cdot p_m)}{[p_n\cdot q-i\eta_n\epsilon][-p_m\cdot q-i\eta_m\epsilon]}. \tag{2.10}$$

The limits on the integral in (2.9) refer to $|\mathbf{q}|$. Note that we have changed the sign of $p_m\cdot q$ in the second denominator in (2.10), because if we define q as the momentum emitted by line n then q must be absorbed by line m. Summing over N, we conclude that the S matrix for an arbitrary process may be expressed as

$$S_{\beta\alpha}=S_{\beta\alpha}{}^0\exp\left(\frac{1}{2}\int_\lambda^\Lambda d^4q\,A(q)\right), \tag{2.11}$$

where $S_{\beta\alpha}{}^0$ is the S matrix without infrared virtual photons.

The rate for $\alpha\to\beta$ is given by the absolute square of (2.11),

$$\Gamma_{\beta\alpha}=\Gamma_{\beta\alpha}{}^0\exp\left\{\mathrm{Re}\int_\lambda^\Lambda d^4q\,A(q)\right\}. \tag{2.12}$$

The real part of the integral arises wholly from the $i\pi\delta(q^2)$ term in the photon propagator [for details, see Sec. V], so

$$\mathrm{Re}\int_\lambda^\Lambda d^4q\,A(q)=-\frac{1}{2(2\pi)^3}\int_\lambda^\Lambda d^4q\,\delta(q^2)\times\sum_{nm}\frac{e_ne_m\eta_n\eta_m(p_n\cdot p_m)}{(p_n\cdot q)(p_m\cdot q)}=-A\ln(\Lambda/\lambda), \tag{2.13}$$

where A is the positive dimensionless constant

$$A\equiv\int d^2\Omega\,A(\hat{q}), \tag{2.14}$$

$$A(\hat{q})\equiv\frac{1}{2(2\pi)^3}\sum_{nm}\frac{e_ne_m\eta_n\eta_m(p_n\cdot p_m)}{[E_n-\mathbf{p}_n\cdot\hat{q}][E_m-\mathbf{p}_m\cdot\hat{q}]}. \tag{2.15}$$

The integral in (2.14) is elementary, and yields

$$A=-\frac{1}{8\pi^2}\sum_{mn}\eta_n\eta_m e_ne_m\beta_{nm}{}^{-1}\ln\left(\frac{1+\beta_{nm}}{1-\beta_{nm}}\right), \tag{2.16}$$

where β_{nm} is the relative velocity of particles n and m in the rest frame of either:

$$\beta_{nm}\equiv\left[1-\frac{m_n{}^2m_m{}^2}{(p_n\cdot p_m)^2}\right]^{1/2}. \tag{2.17}$$

Using (2.13) in (2.12), we find

$$\Gamma_{\beta\alpha}=\Gamma_{\beta\alpha}{}^0(\lambda/\Lambda)^A. \tag{2.18}$$

The same combinatorics apply to gravitons, and yield an expression for the infrared virtual-graviton corrections to any process $\alpha \to \beta$

$$S_{\beta\alpha}=S_{\beta\alpha}{}^0 \exp\left\{\frac{1}{2}\int_\lambda^\Lambda d^4q\, B(q)\right\}, \tag{2.19}$$

where S^0 is the S matrix without virtual infrared gravitons, and $B(q)$ is the result of joining a pair of factors (2.7) with a graviton propagator. The effective graviton propagator joining a $(\mu\nu)$ vertex with a $(\rho\sigma)$ vertex is known[2] to be

$$\frac{-i}{2(2\pi)^4}\frac{\{g_{\mu\rho}g_{\nu\sigma}+g_{\mu\sigma}g_{\nu\rho}-g_{\mu\nu}g_{\rho\sigma}\}}{q^2-i\epsilon}. \tag{2.20}$$

Therefore we find

$$B(q)=\frac{-8\pi G i}{(2\pi)^4[q^2-i\epsilon]} \times\sum_{nm}\frac{\eta_n\eta_m\{(p_n\cdot p_m)^2-\frac{1}{2}m_n{}^2m_m{}^2\}}{[p_n\cdot q-i\eta_n\epsilon][-p_m\cdot q-i\eta_m\epsilon]}. \tag{2.21}$$

The rate for $\alpha\to\beta$ is the absolute square of (2.19)

$$\Gamma_{\beta\alpha}=\Gamma_{\beta\alpha}{}^0 \exp\left\{\mathrm{Re}\int_\lambda^\Lambda d^4q\, B(q)\right\}. \tag{2.22}$$

The real part of the integral comes only from the $+i\pi\delta(q^2)$ term in the graviton propagator, so

$$\mathrm{Re}\int_\lambda^\Lambda d^4q\, B(q)=-\frac{-8\pi G}{2(2\pi)^3}\int_\lambda^\Lambda d^4q\,\delta(q^2) \times\sum_{nm}\frac{\eta_n\eta_m\{(p_n\cdot p_m)^2-\frac{1}{2}m_n{}^2m_m{}^2\}}{[p_n\cdot q][p_m\cdot q]} =-B\ln(\Lambda/\lambda), \tag{2.23}$$

where B is the positive dimensionless constant

$$B\equiv\int d^2\Omega\, B(\hat{q}) \tag{2.24}$$

$$B(\hat{q})\equiv\frac{8\pi G}{2(2\pi)^3}\sum_{nm}\frac{\eta_n\eta_m\{(p_n\cdot p_m)^2-\frac{1}{2}m_n{}^2m_m{}^2\}}{[E_n-\mathbf{p}_n\cdot\hat{q}][E_m-\mathbf{p}_m\cdot\hat{q}]}. \tag{2.25}$$

The solid-angle integration (2.24) yields

$$B=\frac{G}{2\pi}\sum_{nm}\eta_n\eta_m m_n m_m\frac{1+\beta_{nm}{}^2}{\beta_{nm}(1-\beta_{nm}{}^2)^{1/2}}\ln\left(\frac{1+\beta_{nm}}{1-\beta_{nm}}\right) \tag{2.26}$$

with β_{nm} the relative velocity (2.17). Using (2.23) in (2.22), we find the cutoff dependence of the rate is

$$\Gamma_{\beta\alpha}=\Gamma_{\beta\alpha}{}^0(\lambda/\Lambda)^B. \tag{2.27}$$

Since $B>0$ this shows that *all* processes have zero rate in the limit $\lambda\to 0$, just as all charged-particle processes have zero rate for $\lambda\to 0$ in electrodynamics. The paradox is resolved in both cases by taking into account the infrared divergences attributable to emission of real soft photons and gravitons.

4. Real Infrared Divergences

The S-matrix element for emitting N real soft photons in a process $\alpha\to\beta$ is given by multiplying the nonradiative S matrix by N factors of form (2.3), and then contracting each of these factors with the appropriate "wave function"

$$(2\pi)^{-3/2}(2|\mathbf{q}|)^{-1/2}\epsilon_\mu{}^*(\mathbf{q},h), \tag{2.28}$$

where $\mathbf{q}$ is the photon momentum, $h=\pm 1$ is its helicity, and ϵ_μ is the corresponding polarization vector.[1] We therefore find for the radiative transition amplitude

$$S_{\beta\alpha}{}^{ph}(12\cdots N)=S_{\beta\alpha}\prod_{r=1}^N (2\pi)^{-3/2}(2|\mathbf{q}_r|)^{-1/2} \times\sum_n\frac{\eta_n e_n[p_n\cdot\epsilon^*(\mathbf{q}_r,h_r)]}{[p_n\cdot q_r]}. \tag{2.29}$$

The S-matrix element for emitting N real soft gravitons in a process $\alpha\to\beta$ is similarly obtained by multiplying the S matrix for $\alpha\to\beta$ by N factors of form (2.7) and then contracting each of these factors with the appropriate graviton "wave function"[1]

$$(2\pi)^{-3/2}(2|\mathbf{q}|)^{-1/2}\epsilon_\mu{}^*(\mathbf{q},\pm 1)\epsilon_\nu{}^*(\mathbf{q},\pm 1), \tag{2.30}$$

where $\mathbf{q}$ is the graviton momentum, $h=\pm 2$ is its helicity, and ϵ_μ is the same as in (2.28). We therefore find the graviton emission-matrix element

$$S_{\beta\alpha}{}^{gr}(12\cdots N)=S_{\beta\alpha}\prod_{r=1}^N (2\pi)^{-3/2}(2|\mathbf{q}_r|)^{-1/2}(8\pi G)^{1/2} \times\sum_n\frac{\eta_n[p_n\cdot\epsilon^*(\mathbf{q}_r,\frac{1}{2}h_r)]^2}{[p_n\cdot q_r]}. \tag{2.31}$$

The rate for emitting N soft photons or gravitons with momenta near $\mathbf{q}_1\cdots\mathbf{q}_N$ is given by squaring (2.29) or (2.31), summing over helicities, and dividing by $N!$ because photons and gravitons are bosons. This gives

$$\Gamma_{\beta\alpha}{}^{ph}(\mathbf{q}_1\cdots\mathbf{q}_N)d^3q_1\cdots d^3q_N =(1/N!)\Gamma_{\beta\alpha}\prod_{r=1}^N \mathcal{A}(\mathbf{q}_r)d^3q_r, \tag{2.32}$$

$$\Gamma_{\beta\alpha}{}^{gr}(\mathbf{q}_1\cdots\mathbf{q}_N)d^3q_1\cdots d^3q_N =(1/N!)\Gamma_{\beta\alpha}\prod_{r=1}^N \mathcal{B}(\mathbf{q}_r)d^3q_r, \tag{2.33}$$

where $\Gamma_{\beta\alpha}=|S_{\beta\alpha}|^2$, and

$$\mathcal{A}(\mathbf{q})=(2\pi)^{-3}(2|\mathbf{q}|)^{-1}\sum_{nm}\frac{\eta_n\eta_m e_n e_m p_n{}^\mu p_m{}^\nu \Pi_{\mu\nu}(\mathbf{q})}{(p_n\cdot q)(p_m\cdot q)}, \tag{2.34}$$

$$\mathcal{B}(\mathbf{q}) = (2\pi)^{-3}(2|\mathbf{q}|)^{-1}(8\pi G)$$
$$\times \sum_{nm} \frac{\eta_n \eta_m p_n^\mu p_n^\nu p_m^\rho p_m^\sigma \Pi_{\mu\nu\rho\sigma}(\mathbf{q})}{(p_n \cdot q)(p_m \cdot q)}. \quad (2.35)$$

Here $\Pi_{\mu\nu}$ and $\Pi_{\mu\nu\rho\sigma}$ are the polarization sums

$$\Pi_{\mu\nu}(\mathbf{q}) = \sum_\pm \epsilon_\mu(\mathbf{q}, \pm)\epsilon_\nu^*(\mathbf{q}, \pm), \quad (2.36)$$

$$\Pi_{\mu\nu\rho\sigma}(\mathbf{q}) = \sum_\pm \epsilon_\mu(\mathbf{q}, \pm)\epsilon_\nu(\mathbf{q}, \pm)\epsilon_\rho^*(\mathbf{q}, \pm) \times \epsilon_\sigma^*(\mathbf{q}, \pm). \quad (2.37)$$

We recall that[2]

$$\Pi_{\mu\nu}(\mathbf{q}) = g_{\mu\nu} + q_\mu \lambda_\nu + q_\nu \lambda_\mu, \quad \lambda^\mu \equiv \{-\mathbf{q}, |\mathbf{q}|\}/2|\mathbf{q}|^2. \quad (2.38)$$

The $q\lambda$ terms do not contribute to $\mathcal{A}(\mathbf{q})$ because charge is conserved:

$$q_\mu \sum_n \eta_n e_n p_n^\mu / (p_n \cdot q) = \sum_n \eta_n e_n = 0,$$

so (2.34) becomes

$$\mathcal{A}(\mathbf{q}) = (2\pi)^{-3}(2|\mathbf{q}|)^{-1} \sum_{nm} \frac{\eta_n \eta_m e_n e_m (p_n \cdot p_m)}{(p_n \cdot q)(p_m \cdot q)}$$

or

$$\mathcal{A}(\mathbf{q}) = A(\hat{q})/|\mathbf{q}|^3, \quad (2.39)$$

where $A(\hat{q})$ is given by (2.15). We also recall that[2]

$$\Pi_{\mu\nu\rho\sigma}(\mathbf{q}) = \tfrac{1}{2}\{\Pi_{\mu\rho}(\mathbf{q})\Pi_{\nu\sigma}(\mathbf{q}) + \Pi_{\mu\sigma}(\mathbf{q})\Pi_{\nu\rho}(\mathbf{q}) - \Pi_{\mu\nu}(\mathbf{q})\Pi_{\rho\sigma}(\mathbf{q})\}. \quad (2.40)$$

But again the λq terms in Π do not contribute, this time because energy and momentum are conserved:

$$q_\mu \sum_n \eta_n p_n^\mu p_n^\nu / (p_n \cdot q) = \sum_n \eta_n p_n^\nu = 0,$$

so $\Pi_{\mu\nu}$ in (2.40) is effectively just $g_{\mu\nu}$, and (2.35) becomes

$$\mathcal{B}(\mathbf{q}) = (2\pi)^{-3}(2|\mathbf{q}|)^{-1}(8\pi G) \times \sum_{nm} \frac{\eta_n \eta_m \{(p_n \cdot p_m)^2 - \frac{1}{2} m_n^2 m_m^2\}}{(p_n \cdot q)(p_m \cdot q)}$$

or

$$\mathcal{B}(\mathbf{q}) = B(\hat{q})/|\mathbf{q}|^3, \quad (2.41)$$

where $B(\hat{q})$ is given by (2.25).

The rates for emission of N photons or gravitons with energies near $\omega_1 \cdots \omega_N$ are given by integrating (2.32) and (2.33) over solid angles. Using (2.39), (2.41), (2.14), and (2.24), we find

$$\Gamma_{\beta\alpha}^{ph}(\omega_1 \cdots \omega_N) d\omega_1 \cdots d\omega_N = \frac{A^N}{N!} \Gamma_{\beta\alpha} \frac{d\omega_1}{\omega_1} \cdots \frac{d\omega_N}{\omega_N}, \quad (2.42)$$

$$\Gamma_{\beta\alpha}^{gr}(\omega_1 \cdots \omega_N) d\omega_1 \cdots d\omega_N = \frac{B^N}{N!} \Gamma_{\beta\alpha} \frac{d\omega_1}{\omega_1} \cdots \frac{d\omega_N}{\omega_N}. \quad (2.43)$$

These formulas show that the integrated photon or graviton emission rate will contain logarithmic infrared divergences. In order to display these divergences quantitatively, it is convenient to calculate the rate for the transition $\alpha \to \beta$ accompanied by any number of soft photons with total energy less than E, and with each individual ω_r greater than the infrared cutoff λ. We use the well-known representation of the step function to write this rate as

$$\Gamma_{\beta\alpha}(\leq E) = \frac{1}{\pi} \sum_{N=0}^{\infty} \int_\lambda^E d\omega_1 \cdots \int_\lambda^E d\omega_N \int_{-\infty}^{\infty} d\sigma \frac{\sin E\sigma}{\sigma} \times \exp\{i\sigma \textstyle\sum_r \omega_r\} \Gamma_{\beta\alpha}(\omega_1 \cdots \omega_N). \quad (2.44)$$

Applying this to (2.42) and (2.43), the photon and graviton emission rates are

$$\Gamma_{\beta\alpha}^{ph}(\leq E) = \frac{1}{\pi} \int_{-\infty}^{\infty} \frac{\sin E\sigma}{\sigma} \exp\left\{A \int_\lambda^E \frac{d\omega}{\omega} e^{i\omega\sigma}\right\} d\sigma, \quad (2.45)$$

$$\Gamma_{\beta\alpha}^{gr}(\leq E) = \frac{1}{\pi} \int_{-\infty}^{\infty} \frac{\sin E\sigma}{\sigma} \exp\left\{B \int_\lambda^E \frac{d\omega}{\omega} e^{i\omega\sigma}\right\} d\sigma. \quad (2.46)$$

For $\lambda \to 0$ the ω integrals become

$$\int_\lambda^E \frac{d\omega}{\omega} e^{i\omega\sigma} \to \ln(E/\lambda) + \int_0^E \frac{d\omega}{\omega}(e^{i\omega\sigma} - 1) + \mathcal{O}(\lambda). \quad (2.47)$$

Hence (2.45) and (2.46) give for $\lambda \to 0$

$$\Gamma_{\beta\alpha}^{ph}(\leq E) = (E/\lambda)^A b(A) \Gamma_{\beta\alpha}, \quad (2.48)$$

$$\Gamma_{\beta\alpha}^{gr}(\leq E) = (E/\lambda)^B b(B) \Gamma_{\beta\alpha}, \quad (2.49)$$

where $b(x)$ is the real function[5]

$$b(x) = \frac{1}{\pi} \int_{-\infty}^{\infty} d\sigma \frac{\sin \sigma}{\sigma} \exp\left\{x \int_0^1 \frac{d\omega}{\omega}(e^{i\omega\sigma} - 1)\right\}$$
$$b(x) \simeq 1 - \tfrac{1}{12}\pi^2 x^2 + \cdots. \quad (2.50)$$

Since A and B are positive, the factors $(E/\lambda)^A$ and $(E/\lambda)^B$ become infinite for $\lambda \to 0$.

5. Cancellation of Divergences

It now only remains to insert the formulas (2.18) and (2.27) which display the virtual infrared divergences into (2.48) and (2.49). As promised, all dependence on the infrared cutoff λ disappears, leaving us with

$$\Gamma_{\beta\alpha}^{ph}(\leq E) = (E/\Lambda)^A b(A) \Gamma_{\beta\alpha}^0, \quad (2.51)$$

$$\Gamma_{\beta\alpha}^{gr}(\leq E) = (E/\Lambda)^B b(B) \Gamma_{\beta\alpha}^0. \quad (2.52)$$

We repeat that A is given by (2.16), B by (2.26), $b(x)$ by (2.50), and $\Gamma_{\beta\alpha}^0$ is the rate for the process $\alpha \to \beta$ without soft photon and graviton emission and without including virtual infrared photons or gravitons. The quantity Λ is an ultraviolet cutoff that has been used to define what we mean by "infrared," but (2.51) and (2.52) show that a change $\Lambda \to \Lambda'$ just renormalizes

$\Gamma_{\beta\alpha}{}^0$ by a factor $(\Lambda'/\Lambda)^A$ in electrodynamics or $(\Lambda'/\Lambda)^B$ for gravitation theory. Thus it makes no difference how we fix Λ, except that if we wish to estimate $\Gamma_{\beta\alpha}{}^0$ by ignoring *all* radiative corrections it will usually be a good strategy to fix Λ equal to some mass typical of the particles in the reaction $\alpha\to\beta$.

For reasons beyond my ken, Jauch and Rohrlich[6] did not fix Λ, but instead took $\Lambda=E$. They therefore missed the energy-dependent factor $(E/\Lambda)^A$ in (2.51). The factors E^A in (2.51) and E^B in (2.52) correctly represent the shape of the energy spectrum for E ranging from zero (where Γ vanishes) up to some maximum smaller (though not necessarily much smaller) than any energy characterizing the process $\alpha\to\beta$.

6. Remark

It was crucial in the above that the infrared divergences arise only from diagrams in which the soft real or virtual photon or graviton is attached to an external line, with "external line" *not* including the soft real photons or gravitons themselves. In electrodynamics this is true because photons are electrically neutral. In gravitation theory it is justified because the effective coupling constant for emission of a very soft graviton from a graviton (or photon) line with energy E is proportional to E, and the vanishing of this factor prevents simultaneous infrared divergences from a graviton and the line to which it is attached.

But these remarks do not apply to theories involving charged massless particles. In such theories (including the Yang-Mills theory) a soft photon emitted from an external line can itself emit a pair of soft charged massless particles, which themselves emit soft photons, and so on, building up a cascade of soft massless particles each of which contributes an infrared divergence. The elimination of such complicated interlocking infrared divergences would certainly be a Herculean task, and might even not be possible.

We may be thankful that the zero charge of soft photons and the zero gravitational mass of soft gravitons saves the real world from this mess. Perhaps it would not be too much to suggest that it is the infrared divergences that prohibit the existence of Yang-Mills quanta or other charged massless particles. See Sec. III for further remarks in this direction.

7. Another Remark

To lowest order in G, Eq. (2.52) gives the power spectrum of soft gravitons accompanying a reaction $\alpha\to\beta$ as

$$Ed\Gamma_{\beta\alpha}(\leqq E)=B\Gamma_{\beta\alpha}{}^0dE. \tag{2.53}$$

This formula could also have been derived in classical weak-field gravitational radiation theory. [Note that in cgs units (2.26) will give a dimensionless B only if we divide the right-hand side by $\hbar c$. But $dE=\hbar d\omega$, so $\hbar$ does not appear in (2.53) if written as a formula for power per unit frequency interval.]

There is no infrared divergence in (2.53), but this is because it gives the energy rather than the number of gravitons emitted per second with energy between E and $E+dE$. The power spectrum formula (2.53) is both classical and quantum-mechanical, but the infrared divergences are purely quantum mechanical, because it is only in quantum mechanics that we count individual gravitons as well as their total energy.

III. PHOTON AND GRAVITON EMISSION FROM MASSLESS-PARTICLE LINES

Equations (2.16) and (2.26) seem to indicate that the soft photon and graviton emission rates become logarithmically divergent when the mass of one of the *noninfrared* particles in the reaction is allowed to vanish. This divergence occurs because the denominator factors $(p\cdot q)$ in (2.15) and (2.25) will vanish for $\mathbf{q}$ parallel to $\mathbf{p}$ if p^2 is zero. However, we will show that there is a cancellation of these divergences for gravitons, though *not* for photons.[3]

Suppose we let the mass m_1 of particle one vanish, but hold its momentum $\mathbf{p}_1$ fixed. Then Eq. (2.17) becomes

$$\beta_{1n}=1-m_1{}^2m_n{}^2/2(p_1\cdot p_n)^2+O(m_1{}^4)\,,\quad (n\neq 1) \tag{3.1}$$

where $p_1=\{\mathbf{p}_1,|\mathbf{p}_1|\}$. For $m_1\to 0$ Eq. (2.16) gives the infrared-photon spectral index as

$$A=-\frac{e_1{}^2}{4\pi^2}-\frac{\eta_1e_1}{4\pi^2}\sum_{n\neq 1}\eta_ne_n\ln\left(\frac{4(p_1\cdot p_n)^2}{m_1{}^2m_n{}^2}\right)$$
$$-\frac{1}{8\pi^2}\sum_{n,m\neq 1}\eta_n\eta_me_ne_m\beta_{nm}{}^{-1}\ln\left(\frac{1+\beta_{nm}}{1-\beta_{nm}}\right). \tag{3.2}$$

Using charge conservation, we may write the divergent term as

$$+\frac{\eta_1e_1}{2\pi^2}\ln m_1\sum_{n\neq 1}\eta_ne_n=-\frac{e_1{}^2}{2\pi^2}\ln m_1. \tag{3.3}$$

Hence quantum electrodynamics would be in serious trouble if any charged particle had zero mass.[3] The Yang-Mills theory is for these purposes just the electrodynamics of charged massless vector mesons, so it also shares this trouble. (We have already remarked that a complete treatment of infrared divergences in massless electrodynamics would be enormously more difficult than for ordinary electrodynamics or gravitation, but we are now only considering processes with infrared photons and no infrared charged particles, and for these our present formalism is adequate.)

There are no charged massless particles, so this divergence in A is of only academic interest. But any massless particles can contribute to the infrared-graviton spectral index B. When $m_1\to 0$ Eq. (2.26) gives

$$B=-\frac{\eta_1 G}{\pi}\sum_{n\neq 1}\eta_n(p_n\cdot p_1)\ln\left(\frac{4(p_1\cdot p_n)^2}{m_1^2 m_n^2}\right)$$
$$-\frac{G}{2\pi}\sum_{n,m\neq 1}\eta_n\eta_m(p_n\cdot p_m)(1+\beta_{nm}^2)\ln\left(\frac{1+\beta_{nm}}{1-\beta_{nm}}\right). \quad (3.4)$$

Using energy and momentum conservation, we may write the divergent term as

$$+\frac{2\eta_1 G}{\pi}(\ln m_1)\sum_{n\neq 1}\eta_n(p_n\cdot p_1)=-\frac{2G}{\pi}(\ln m_1)p_1^2=0. \quad (3.5)$$

Since this vanishes, m_1 in Eq. (3.4) may be replaced with any convenient mass; for instance the first logarithm in (3.4) could be written

$$\ln[4(p_1\cdot p_n)^2/m_n^4].$$

I leave it to the reader to show that the same cancellations occur when several particles have vanishing mass.

IV. GRAVITATIONAL RADIATION IN NONRELATIVISTIC COLLISIONS

The rate of emission of energy in soft gravitational radiation during collisions is

$$P(\leqq\Lambda)=\int_0^\Lambda E\,d\Gamma(\leqq E). \quad (4.1)$$

Here "soft" means that the emitted energy E is less than some cutoff Λ chosen smaller than the energies characteristic of the collision process. The rate $\Gamma(\leqq E)$ for a collision with radiated energy $\leqq E$ was calculated in Sec. II as

$$\Gamma(\leqq E)=(E/\Lambda)^B b(B)\Gamma_0, \quad (4.2)$$

with B given by (2.26), $b(B)$ by (2.50), and Γ_0 equal to the collision rate without real or virtual infrared gravitons. Hence the power (4.1) is

$$P(\leqq\Lambda)=(B/(1+B))b(B)\Lambda\Gamma_0. \quad (4.3)$$

Since B is always very tiny ($\leqq 10^{-38}$) both $1+B$ and $b(B)$ are extremely close to one, and we may write

$$P(\leqq\Lambda)=B\Lambda\Gamma_0. \quad (4.4)$$

Also, to a very good approximation Γ_0 may be calculated as the collision rate ignoring gravitation altogether.

For our present purposes it will be convenient to write Eq. (2.26) for B as

$$B=(G/\pi)\sum_{nm}\eta_n m_n\eta_m m_m f(\beta_{nm}), \quad (4.5)$$

where

$$f(\beta)\equiv\frac{1+\beta^2}{2\beta(1-\beta^2)^{1/2}}\ln\left[\frac{1+\beta}{1-\beta}\right]^{1/2} \quad (4.6)$$

and $\eta_n=+1$ or -1 if n is a final or initial particle; m_n is the mass of particle n, and β_{nm} is the relative velocity of particles n and m:

$$\beta_{nm}\equiv[1-m_n^2m_m^2/(p_n\cdot p_m)^2]^{1/2}. \quad (4.7)$$

If all particles involved in the collision are nonrelativistic then β_{nm} may be expanded in powers of $\mathbf{v}_n$ and $\mathbf{v}_m$, with $\mathbf{v}\equiv\mathbf{p}/E$. We find

$$\beta_{nm}^2=\mathbf{v}_n^2+\mathbf{v}_m^2-2\mathbf{v}_n\cdot\mathbf{v}_m-\mathbf{v}_n^2\mathbf{v}_m^2-3(\mathbf{v}_n\cdot\mathbf{v}_m)^2$$
$$+2(\mathbf{v}_n^2+\mathbf{v}_m^2)(\mathbf{v}_n\cdot\mathbf{v}_m)+\cdots. \quad (4.8)$$

Also, $f(\beta)$ may be expanded in powers of β^2:

$$f(\beta)=1+(11/6)\beta^2+(63/40)\beta^4+\cdots. \quad (4.9)$$

Using (4.8) in (4.9) gives

$$f(\beta_{nm})=1+(11/6)(\mathbf{v}_n^2+\mathbf{v}_m^2-2\mathbf{v}_n\cdot\mathbf{v}_m)$$
$$+(63/40)(\mathbf{v}_n^2+\mathbf{v}_m^2)^2-(79/30)(\mathbf{v}_n^2+\mathbf{v}_m^2)$$
$$\times(\mathbf{v}_n\cdot\mathbf{v}_m)+(4/5)(\mathbf{v}_n\cdot\mathbf{v}_m)^2+\cdots. \quad (4.10)$$

We are keeping terms up to order v^4 in (4.8)–(4.10), since the lower order terms contribute only to order v^4 in B because of the energy and momentum conservation equations:

$$\sum_n\eta_n m_n(1+\tfrac{1}{2}\mathbf{v}_n^2+\tfrac{3}{8}\mathbf{v}_n^4+\cdots)=0, \quad (4.11)$$

$$\sum_n\eta_n m_n\mathbf{v}_n(1+\tfrac{1}{2}\mathbf{v}_n^2+\cdots)=0. \quad (4.12)$$

Using (4.10), (4.11), and (4.12) in (4.5), we find to lowest nonvanishing order in v

$$B=(G/\pi)[(16/5)Q_{ij}Q_{ij}+(94/15)(Q_{ii})^2], \quad (4.13)$$

where

$$Q_{ij}=\tfrac{1}{2}\sum_n\eta_n m_n v_{ni}v_{nj} \quad (4.14)$$

and repeated Latin indices are summed over 1, 2, 3. Since (4.13) is only correct to order v^4, the velocities in Eq. (4.14) must be subjected to the nonrelativistic conservation laws

$$\sum_n\eta_n m_n(1+\tfrac{1}{2}\mathbf{v}_n^2)=0, \quad (4.15)$$

$$\sum_n\eta_n m_n\mathbf{v}_n=0. \quad (4.16)$$

Hence Q_{ii} is just the usual Q value

$$Q_{ii}=-\sum_n\eta_n m_n. \quad (4.17)$$

Also, (4.16) makes Q_{ij} invariant under the Galilean transformation $\mathbf{v}_n\to\mathbf{v}_n+\mathbf{u}$, so B may be computed in any convenient reference frame.

As an example, consider nonrelativistic elastic two-body scattering. We find here

$$Q_{ij}Q_{ij}=\tfrac{1}{2}\mu^2v^4\sin^2\theta_c,$$
$$Q_{ii}=0,$$

where μ is the reduced mass, $v=|\mathbf{v}_1-\mathbf{v}_2|$ is the relative velocity, and θ_c is the scattering angle in the center-of-mass system. Thus Eq. (4.13) gives

$$B=(8G/5\pi)\mu^2v^4\sin^2\theta_c. \quad (4.18)$$

The rate for such collisions per cm^3 per sec is $vn_1n_2(d\sigma/d\Omega)$, where n_1 and n_2 are the number densities of particles 1 and 2. Hence (4.4) and (4.8) give the total power emitted in soft gravitational radiation attributable to 1–2 collisions as

$$P(\leqq\Lambda)=\frac{8G}{5\pi}\mu^2v^5n_1n_2V\Lambda\int\left(\frac{d\sigma}{d\Omega}\right)\sin^2\theta_c d\Omega\,, \quad (4.19)$$

with V the volume of the source. For practical purposes we may generally define "soft" radiation by taking the cutoff Λ at half the relative kinetic energy

$$\Lambda\approx\tfrac{1}{4}\mu v^2. \quad (4.20)$$

Everything in the universe is transparent to gravitons, so (4.19) may be used directly to compute the thermal gravitational radiation from any hot body.

We will use these results to estimate the thermal gravitational radiation from the sun. By far the most frequent collisions are the Coulomb collisions between electrons and protons or electrons. In this case we may take

$$\begin{aligned}\mu&=m_e, & v&=(3KT/m_e)^{1/2}\\ n_1&=n_e, & n_2&=n_e+n_p=2n_e.\end{aligned} \quad (4.21)$$

Also, the integral in Eq (4 19) is nothing but the familiar diffusion coefficient, and as is well known[9] can be estimated as

$$\int\left(\frac{d\sigma}{d\Omega}\right)\sin^2\theta_c d\Omega=\frac{8\pi e^2}{(3KT)^2}\ln\Lambda_D\,, \quad (4.22)$$

where e is now unrationalized, and Λ_D is the ratio of the Debye shielding radius (used to cutoff the integral) to the average impact parameter. Putting together (4.19)–(4.22), we find

$$P_\odot=(32/5)G(3KT)^{3/2}m_e^{-1/2}n_e^2V_\odot e^4(\hbar c^5)^{-1}\ln\Lambda_D. \quad (4.23)$$

We did our calculation in natural units with $\hbar=c=1$, but we have supplied a factor $(\hbar c^5)^{-1}$ in Eq. (4.23) to convert it to cgs units. In the sun's core the parameters in (4.23) take the values[10]

$$\begin{aligned}T&\simeq10^7\,^\circ\text{K},\\ n_e&\simeq3\times10^{25}\text{ cm}^{-3},\\ V_\odot&\simeq2\times10^{31}\text{ cm}^3,\\ \ln\Lambda_D&\simeq4.\end{aligned}$$

The solar gravitational radiation power is then

$$P_\odot\simeq6\times10^{14}\text{ erg/sec}. \quad (4.24)$$

Although this is not much more power than used by the city of Berkeley, it nevertheless compares favorably with the gravitational radiation from previously considered classical sources like planetary motion. A planet of mass m moving in a circular orbit of radius R around a star of mass M emits gravitational radiation with power

$$P=(32/5)Gc^{-5}m^2R^4(GM/R^3)^3. \quad (4.25)$$

For the Jupiter-Sun system this is 7.6×10^{11} erg/sec. Venus and the Earth radiate comparable amounts, and the other planets considerably less, so the thermal gravitational radiation (4.14) from the Sun appears to be the dominant source of gravitational radiation from the solar system. A binary star like Sirius A and B radiates more classically—in this case Eq. (4.25) gives 8×10^{14} erg/sec—but it also radiates more thermally. Thus thermal collisions possibly may provide the most important source of gravitational radiation in the universe. I have no idea what it is good for.

V. PHASE DIVERGENCES

It is well known that the Born series for the Coulomb scattering amplitude gives divergent integrals beyond the first order. Dalitz[11] has studied the scattering by a screened potential

$$V(r)=(e_1e_2/4\pi r)e^{-\lambda r} \quad (5.1)$$

and conjectured that the $\ln\lambda$ term found in second Born approximation might represent the beginning of a series whose sum is a phase factor

$$\exp\left\{\frac{ie_1e_2}{2\pi\beta_{12}}\ln\lambda\right\}, \quad (5.2)$$

which would correctly represent the cutoff dependence for $\lambda\to0$, and would not affect the cross section. This is very reasonable, but to my knowledge it has not been proved. I will show here that this conjecture is correct, and has an almost trivial extension to any process involving any number of relativistic or nonrelativistic particles of arbitrary spin.

The divergences in the nonrelativistic Born series for Coulomb scattering are obviously not the same as the ordinary infrared divergences, which depend on retardation effects, i.e., on the $i\pi\delta(q^2)$ term in the photon propagator. However, Eq. (2.11) shows that the full effect of virtual infrared photons is to contribute to the S matrix for any process $\alpha\to\beta$ a factor

$$\frac{S_{\beta\alpha}}{S_{\beta\alpha}{}^0}=\exp\left\{\frac{1}{2}\int_\lambda^\Lambda d^4q\,A(q)\right\}. \quad (5.3)$$

The real part of $A(q)$ gives the familiar infrared-divergence factor $(\lambda/\Lambda)^{A/2}$ which is eventually cancelled by real soft-photon emission processes. But $A(q)$ also has an imaginary part, and we shall find that it is this

[9] See, e.g., L. Spitzer, Jr., *Physics of Fully Ionized Gases* (Interscience Publishers, Inc., New York, 1956), Chap. 5.

[10] These parameters apply to the inner $\frac{1}{64}$ of the sun's volume. The first two are from M. Schwarzschild, *Structure and Evolution of the Stars* (Princeton University Press, Princeton, New Jersey, 1958), p. 259. The last is from Table 5.1 of Ref. 9.

[11] R. H. Dalitz, Proc. Roy. Soc. (London) **206**, 509 (1951).

that accounts for the $\ln\lambda$ terms in the nonrelativistic Born series.

Referring back to (2.10), we see that (5.3) may be written as

$$\frac{S_{\beta\alpha}}{S_{\beta\alpha}{}^0}=\exp\left\{\frac{1}{2(2\pi)^3}\sum_{nm} e_n e_m \eta_n \eta_m (p_n\cdot p_m) J_{nm}\right\}, \quad (5.4)$$

$$J_{nm}\equiv i\int^{\Lambda}\frac{d^4q}{[q^2+\lambda^2-i\epsilon][p_n\cdot q-i\eta_n\epsilon][p_m\cdot q+i\eta_m\epsilon]}. \quad (5.5)$$

We are still using an ultraviolet cutoff $|\mathbf{q}|\leqq\Lambda$ to separate the infrared from the noninfrared virtual photons, but in order to facilitate the comparison with (5.2) we are now using a photon mass λ in place of the infrared cutoff $\lambda\leqq|\mathbf{q}|$.

The integrand of (5.5) is analytic in q^0 except at the four poles

$$q^0=\omega-i\epsilon,\qquad q^0=-\omega+i\epsilon,$$
$$q^0=\mathbf{v}_n\cdot\mathbf{q}-i\eta_n\epsilon,\quad q^0=\mathbf{v}_m\cdot\mathbf{q}+i\eta_m\epsilon,$$

where $\omega\equiv(\mathbf{q}^2+\lambda^2)^{1/2}$ and $\mathbf{v}\equiv\mathbf{p}/E$. Also, we may close the q^0 contour with a large semicircle in either the upper or the lower half-planes. If particle n is outgoing and m is incoming then $\eta_n=+1$, $\eta_m=-1$, so by closing the contour in the upper half-plane we avoid the contributions from the poles $q^0=\mathbf{v}_n\cdot\mathbf{q}-i\eta_n\epsilon$ or $q^0=\mathbf{v}_m\cdot\mathbf{q}+i\eta_m\epsilon$. Similarly, if n is incoming and m is outgoing we can avoid these two poles by closing the contour in the lower half-plane. In these two cases it is only the poles at $q^0=\pm(\omega-i\epsilon)$ that contribute, and we find J_{nm} purely real:

$$J_{nm}=-\pi\int^{\Lambda}\frac{d^3q}{\omega(\omega E_n-\mathbf{q}\cdot\mathbf{p}_n)(\omega E_m-\mathbf{q}\cdot\mathbf{p}_m)}\qquad (\text{for }\ \eta_n=-\eta_m=\pm1). \quad (5.6)$$

On the other hand, if particles n and m are both outgoing or both incoming, then the poles at $\mathbf{v}_n\cdot\mathbf{q}-i\eta_n\epsilon$ and $\mathbf{v}_m\cdot\mathbf{q}+i\eta_m\epsilon$ lie on opposite sides of the real q^0 axis, and we cannot avoid a contribution from one of them whichever way we close the q^0 contour. We now have, after some elementary integrations,

$$J_{nm}=-\pi\int^{\Lambda}\frac{d^3q}{\omega(\omega E_n-\mathbf{q}\cdot\mathbf{p}_n)(\omega E_m-\mathbf{q}\cdot\mathbf{p}_m)}+\frac{2i\pi^3}{[(p_n\cdot p_m)^2-m_n^2m_m^2]^{1/2}}\ln\left(\frac{\Lambda^2}{\lambda^2}+1\right)\qquad (\text{for }\ \eta_n=\eta_m=\pm1). \quad (5.7)$$

In both (5.6) and (5.7) we find a real term which, if we now reconverted from a photon-mass cutoff to an energy cutoff, would contribute to (5.4) the real factor $(\lambda/\Lambda)^{1/2}$ [see (2.14) and (2.15)]. This λ dependence is then cancelled by real-photon emission, and does not interest us in this section. But (5.7) shows that (5.4) will also contain a divergent phase factor

$$\frac{S_{\beta\alpha}}{S_{\beta\alpha}{}^0}=\prod_{nm}{}'\exp\left\{\frac{-i\,e_ne_m}{16\pi\,\beta_{nm}}\ln\left(\frac{\Lambda^2}{\lambda^2}+1\right)\right\}, \quad (5.8)$$

where β_{nm} is the relative velocity

$$\beta_{nm}=[(p_n\cdot p_m)^2-m_n^2m_m^2]^{1/2}/(-p_n\cdot p_m), \quad (5.9)$$

and the prime reminds us that the product runs over particle pairs with $\eta_n=\eta_m$, i.e., both in the initial or both in the final state. In (5.8) the pairs nm and mn are counted separately, so *each different pair of particles in the initial or final state contributes to the S matrix a phase factor which for $\lambda\ll\Lambda$ may be written*

$$\exp\left\{\frac{i}{4\pi}\frac{e_ne_m}{\beta_{nm}}\ln(\lambda/\Lambda)\right\}. \quad (5.10)$$

In two-particle elastic Coulomb scattering this factor occurs in both the initial and final states, and therefore accounts for Dalitz's conjectured phase factor (5.2). *The phase factor (5.10) is correct even if particles n and m are relativistic and/or have spin.*

It hardly needs to be said that a similar result holds for gravitation. Each different particle pair in the initial or final state contributes to the S matrix a divergent phase factor

$$\exp\left\{-i\frac{Gm_nm_m(1+\beta_{nm}^2)}{\beta_{nm}[1-\beta_{nm}^2]^{1/2}}\ln\frac{\lambda}{\Lambda}\right\}. \quad (5.11)$$

These results might have some practical application to the calculation of scattering by potentials with Coulomb tails. Such potentials may be cut off as in (5.2), and we will then find $\ln\lambda$ terms in each order beyond the first. But however complicated the potential is at small distances, these $\ln\lambda$ terms will always sum to phase factors (5.10), and can therefore be removed in a systematic way.

ACKNOWLEDGMENT

I am grateful for valuable conversations with S. Mandelstam.

Reproduced from Chapter 16 of General Relativity: an Einstein Centenary Survey, S. Hawking and W. Israel (Eds.), Cambridge University Press, 1980. ISBN 9780521299282.

16. Ultraviolet divergences in quantum theories of gravitation

STEVEN WEINBERG†

16.1 Introduction

Ever since physicists began to think about a quantum theory of gravitation it has been clear that ultraviolet divergences would present problems. Simple dimensional analysis[1] tells us that if the coupling constant of a field theory has dimensions $[\text{mass}]^d$ (taking $\hbar = c = 1$), then the integral for a Feynman diagram of order N will behave at large momenta like $\int p^{A-Nd}\, dp$, where A depends on the process in question but not on N. Hence, the dangerous interactions are those with $d < 0$; in this case, the integrals for any process will diverge at sufficiently high order. But Newton's constant has dimensionality $d = -2$ (for $\hbar = c = 1$, $G = 6.7 \times 10^{-39}\ \text{GeV}^{-2}$) so general relativity is a prime example of a theory with dangerous interactions.

This intuitive argument became precise with the development of covariant rules for calculating Feynman diagrams in the quantum theory of gravitation.[2] Inspection of these rules showed immediately that general relativity fails the usual tests for renormalizability. This was confirmed by detailed calculations:[3] the one-loop integral for vacuum fluctuations in a classical background field $g_{\mu\nu}$ was found to have divergent terms[4] proportional to R^2 and $R^{\mu\nu}R_{\mu\nu}$. (These of course vanish if $g_{\mu\nu}$ satisfies the vacuum Einstein equation, $R_{\mu\nu} = 0$, but not in the presence of matter fields with $T_{\mu\nu} \neq 0$.) More generally, on dimensional grounds one would expect that diagrams with L loops should have divergent terms proportional to $L+1$ powers of the curvature tensor. As in any non-renormalizable theory, the cancellation of these ultraviolet divergences would require introduction of an infinite number of terms in the Lagrangian, proportional to arbitrarily high powers of the curvature tensor and its covariant derivatives, and if these terms were introduced the divergences would be encountered even earlier.

† This research was supported in part by the National Science Foundation under Grant No. PHY77-22864.

Of course, most of the business of physics can go on perfectly well without an understanding of quantum gravity and its ultraviolet divergences. However, it is deeply disturbing that the fusion of two such fundamental theories as quantum mechanics and general relativity should lead to an apparent contradiction. Further, the effort to construct a unified gauge theory of weak, electromagnetic, *and* strong interactions may force us to consider energies as high as 10^{19} GeV,[5] energies at which gravitation is a strong interaction. If so, then progress in developing such a 'superunified' theory will have to wait until we understand how to deal with the ultraviolet divergences in quantum gravity.

It is possible that this problem has arisen because the usual flat-space formalism of quantum field theory simply cannot be applied to gravitation. After all, gravitation is a very special phenomenon, involving as it does the very topology of space and time. It is also possible that a way may yet be found to describe gravitation, together with suitable matter fields, by an ordinary renormalizable quantum field theory. Several approaches along this line are reviewed in section 16.2. However, this paper will be chiefly concerned with another possibility, that a quantum field theory which incorporates gravitation may satisfy a generalized version of the condition of renormalizability known as *asymptotic safety*.

A theory is said to be asymptotically safe if the 'essential' coupling parameters approach a fixed point as the momentum scale of their renormalization point goes to infinity. This condition is introduced in section 16.3 as a means of avoiding unphysical singularities at very high energy. From the observed properties of second-order phase transitions, one can infer that for some, and probably all, fixed points the condition of asymptotic safety acts much like the condition of renormalizability, in fixing all but a finite number of the essential coupling parameters of a theory. For some fixed points, asymptotic safety even requires that the theory be renormalizable in the usual sense.

In section 16.4 we consider the behaviour of an asymptotically safe theory at ordinary energies, far below the Planck mass[6] $M_P = G^{-1/2} = 1.2\times 10^{19}$ GeV. It appears that such a theory would be effectively renormalizable in the usual sense, with all non-renormalizable interactions suppressed by powers if E/M_P. The one exception is gravitation which, although incredibly weak, can nevertheless be detected because its coherence and long range allow us to observe its macroscopic effects; gravitation would be described at distances much longer than the Planck length, $M_P^{-1} = 1.6\times 10^{-33}$ cm, by the Einstein Lagrangian,[7] $-\sqrt{g}R/16\pi G$. Thus from this point of view it is quite natural that the weak, electromagnetic, and strong interactions of microscopic physics should be described so well by renormalizable quantum field theories, and that the only force that seems to require a non-renormalizable description should be the one force that is observed purely through its macroscopic effects, the force of gravitation.

The question remains, whether there actually are any theories of gravitation which are asymptotically safe. In section 16.5 we show how to use dimensional continuation to give a tentative answer to this question. This method is applied to gravitation in section 16.6, and it is concluded that there is an asymptotically safe theory of pure gravity in $2+\varepsilon$ dimensions (here $0<\varepsilon \ll 1$). Matter fields may or may not change this conclusion, depending on their type and number. Unfortunately, the problem of continuation from $2+\varepsilon$ to 4 dimensions remains unsolved.

As the reader will discover, the analysis presented here is extremely 'soft' – much is conjectured, but little is actually proved about quantum gravity. I am taking the opportunity afforded by this volume to present my own point of view on the most likely direction for progress in dealing with the ultraviolet divergences in the quantum theory of gravitation, rather than to exhibit a finished formalism.

One other purpose that I had here was to review some mathematical techniques that have been developed in the theories of critical phenomena and of elementary particles, which go under the general rubric of the 'renormalization group'.[8] Very little of this material is new, and almost none of it is due to me, but I hoped that my bringing it together in this article would provide some readers with a background to the renormalization group that they might find useful in future work on the quantum theory of gravitation.

16.2 Renormalizable theories of gravitation

As far as I know, there are just three approaches by which one might hope to construct a renormalizable theory of gravitation:

(a) Extended theories of gravitation

It could be that by combining gravitation with suitable matter fields, and imposing suitable symmetries on the theory, all ultraviolet divergences would be found to cancel, aside from a finite number which could be absorbed into a renormalization of the parameters of the theory. The extended theory of matter and gravitation would then be renormalizable in the usual sense.

As remarked in section 16.1 dimensional analysis suggests that a quantum field theory of gravitation will have ultraviolet divergencies everywhere. Experience has generally shown that if an ultraviolet divergence is expected on dimensional grounds, and is not ruled out by some symmetry of the theory, then the ultraviolet divergence will actually occur. The question of whether an ultraviolet divergence is ruled out by a symmetry is equivalent to the question of whether the term in the Lagrangian which would be needed to provide a counterterm for this divergence is forbidden by the symmetry. Thus it seems that the only hope for an extended theory of gravitation and matter that is renormalizable in the usual sense is that the symmetries of the theory should exclude all but a finite number of types of interaction from the Lagrangian.

This is certainly not possible for ordinary symmetries. If the symmetries allow the construction of a set of invariant interactions $\sqrt{g}\mathscr{L}_i$, then we can construct an infinite number of other allowed interactions by forming the products $\sqrt{g}\mathscr{L}_i\mathscr{L}_j$, $\sqrt{g}\mathscr{L}_i\mathscr{L}_j\mathscr{L}_k$, etc. and every one of the infinite number of corresponding ultraviolet divergences may be expected to occur.

However, it may be that the symmetries of the theory do not allow the construction of *any* invariant interactions, but only interactions $\sqrt{g}\mathscr{L}_i$ whose variations $\delta\sqrt{g}\mathscr{L}_i$ under the transformations of the symmetry group are total derivatives. Then the contributions $\int d^4x\sqrt{g}\mathscr{L}_i$ of these interactions to the action are invariant, so these are allowed interactions. However, the variations in the products $\sqrt{g}\mathscr{L}_i\mathscr{L}_j$, $\sqrt{g}\mathscr{L}_i\mathscr{L}_j\mathscr{L}_k$, etc. are in general not total derivatives, so these interactions could be forbidden by the symmetries of the theory, and the corresponding ultraviolet divergences would not occur.

The most interesting examples of a theory of this type are provided by supersymmetry theories,[9] in which particles of different spin are put together in irreducible representations. The Lagrangian is not actually invariant under supersymmetry transformations, but is a particular member of a supermultiplet which transforms into a total derivative. Supersymmetry imposes such stringent conditions on the Lagrangian that in some 'supergravity' theories[10] there are no allowed interactions of the sort that would be needed to cancel divergences in diagrams with one or two loops;[11] in consequence, no ultraviolet divergences are encountered in these theories in low orders of perturbation theory. However, there is as yet no reason to expect that there would not be an infinite number of allowed

Chapter 16. Ultraviolet divergences in quantum gravity

interactions in these theories, which will be needed as counter-terms in higher orders of perturbation theory.

(*b*) Resummation

It is an old hope that a non-renormalizable theory could be made effectively renormalizable by a rearrangement of the perturbation series.[12] Certain infinite subclasses of graphs would have to be summed first, and if these sums vanished sufficiently rapidly at large momenta, then they could be used as building blocks in a new set of more convergent Feynman rules. However, it is difficult to show that such a resummation leads to a more reliable perturbation series, or indeed, to justify it on any grounds other than the elimination of ultraviolet divergences. In fact, in many cases we can see directly that terms of finite order in the resummed perturbation series contain what appear to be unphysical singularities.

The simplest example of this sort of resummation in the quantum theory of gravitation is provided by a theory based on the Lagrangian[13]

$$\mathscr{L} = -\frac{1}{16\pi G}\sqrt{g}R - f\sqrt{g}R^2 - f'\sqrt{g}R_{\mu\nu}R^{\mu\nu}, \tag{16.1}$$

Normally one would treat the last two terms as a perturbation, and expand in powers of f and f' as well as G. In this way, of course, one would encounter horrible ultraviolet divergences, but there would be no unphysical singularities in any finite order of perturbation theory. However, we can sum up the contributions of the terms in R^2 and $R^{\mu\nu}R_{\mu\nu}$ which are quadratic in the gravitational field $h_{\mu\nu}$ (where as usual $g_{\mu\nu} = \eta_{\mu\nu} + \sqrt{32\pi G}h_{\mu\nu}$). This gives a modified graviton propagator, in which the usual $1/k^2$ is replaced with

$$\frac{1}{k^2}+\frac{1}{k^2}\alpha fGk^4\frac{1}{k^2}+\frac{1}{k^2}\alpha fGk^4\frac{1}{k^2}\alpha fGk^4\frac{1}{k^2}+\cdots=(k^2-\alpha fGk^4)^{-1}, \tag{16.2}$$

where α is a dimensionless function of f'/f. (Alternatively, we could simply regard the quadratic part of the R^2 and $R_{\mu\nu}R^{\mu\nu}$ terms as part of the zeroth-order Lagrangian, and obtain (16.2) directly as the 'bare' propagator for this theory.) This vanishes fast enough at infinity so that the Lagrangian (16.1) now passes the usual power-counting tests for renormalizability. The trouble is, of course, that (16.2) now has a pole at $k^2 = -1/\alpha fG$, with a residue, $-1/\alpha^2 f^2 G^2$, of the wrong sign to be consistent with unitarity.[14]

The problem encountered here is a special case of a more general impasse. Under the usual analyticity assumptions, any resummation would in general replace the $1/k^2$ in the graviton propagator with an integral[15]

$$\int \frac{\rho(\mu)}{\mu^2+k^2}\, d\mu^2. \tag{16.3}$$

Unitarity requires that $\rho(\mu) \geqslant 0$ (except possibly for the contribution of loop graphs containing gravitons or Faddeev–Popov ghosts in Lorentz covariant gauges). As long as $\rho(\mu) \geqslant 0$, the integral (16.3) cannot vanish faster than $1/k^2$ as $k^2 \to \infty$.

We can see this again in a resummation based in the '$1/N$' approximation.[16] Suppose that gravity is coupled to N types of matter field, with gravitational constant, G, of the order of $1/N$. For large N, the dominant diagrams for the complete propagator consists of a chain of self-energy bubbles, each constructed from a single loop of matter lines, and connected with bare graviton propagators. (The two factors of $1/\sqrt{N}$ introduced by the graviton coupling to each matter loop are cancelled by a factor N from the sum over types of fields in the matter loop, so these diagrams are of zeroth order in $1/N$.) The matter-loop graphs are logarithmically divergent, so to cancel these divergences we must add terms proportional to $\sqrt{g}R^2$ and $\sqrt{g}R^{\mu\nu}R_{\mu\nu}$ in the Lagrangian, and the quadratic parts of these terms must be summed along with the matter loops to all orders. When this is done, the $1/k^2$ in the graviton propagator is found to be replaced with[17]

$$k^{-2} \to [k^2 + GNk^4\alpha \ln(k^2/\mu^2)]^{-1}, \tag{16.4}$$

where α is a dimensionless constant, and μ is a renormalization mass which determines the values of f and f'. This vanishes fast enough as $k \to \infty$ so that no new ultraviolet divergences are encountered in graphs of higher order in $1/N$. However, the propagator (16.4) now has singularities at complex values of k^2. This certainly violates our usual ideas of analyticity, though it is not clear that complex singularities of this sort necessarily lead to a conflict with fundamental physical principles.[18]

(*c*) Composite gravitons

There is no renormalizable theory of the 3–3 resonance field, but no one worries about this – the 3–3 resonance is believed to be a three-quark bound state arising in an underlying renormalizable theory of strong interactions known as quantum chromodynamics. In the same way, the graviton may be a composite particle of mass zero and spin two which arises in a renormalizable quantum field theory of some sort.

If the graviton is a mere bound state, then why should it be described by a theory which is so elegantly geometrical as general relativity? A possible answer can be found within the framework of flat-space special relativity and quantum mechanics. It is very difficult to incorporate massless particles of spin two into any Lorentz-invariant quantum theory of long-range forces; it is necessary to couple such a particle to a conserved energy–momentum tensor which includes the gravitons themselves. It has in fact been shown[19] that massless particles of spin two must be described by an effective field theory satisfying the Principle of Equivalence. To the extent that we are only interested in effects of gravitation at long distances, the unique such theory is general relativity.[20]

An illuminating example of such effective field theories is provided by the theory of soft pions known as chiral dynamics.[21] In the limit of vanishing u and d quark masses, quantum chromodynamics tells us that the strong interactions have a global symmetry, chiral $SU(2)\otimes SU(2)$. This symmetry is spontaneously broken, giving rise to a massless Goldstone boson, the pion. Even though the pion is a quark–antiquark composite, its interactions are described by an effective field theory,[22] in which the pion is represented by a field $\vec{\pi}$ which transforms according to the nonlinear three-dimensional realization of $SU(2)\otimes SU(2)$. With one convenient definition of the pion field, the chiral dynamics Lagrangian takes the form

$$\mathscr{L}=-\frac{1}{2F_\pi^2}D_\mu\vec{\pi}\cdot D^\mu\vec{\pi}-f(D_\mu\vec{\pi}\cdot D^\mu\vec{\pi})^2\cdots, \tag{16.5}$$

where $D_\mu\vec{\pi}$ is a 'covariant derivative',

$$D_\mu\vec{\pi}=\partial_\mu\vec{\pi}/(1+\vec{\pi}^2), \tag{16.6}$$

and $F_\pi\simeq 190$ MeV and $f=O(1)$ are constants whose value must be taken from experiment. Suppose we use this theory to calculate any one of the invariant amplitudes $M(E)$ for pion–pion scattering at fixed angle and energy E. For E sufficiently small, the matrix element is given by a term of first order in the quartic part of the first term in (16.5), so $M(E)\propto E^2/F_\pi{}^2$. To next order in E we have to take into account the one-loop graphs constructed from the first term in (16.5), and the first-order effects of the second term in (16.5). The one-loop graphs are of course divergent, but the divergence can be absorbed in a renormalization of f, yielding a matrix element

$$M(E)=\frac{aE^2}{F_\pi^2}+\frac{bE^4}{F_\pi^4}\ln\left(\frac{E}{\mu}\right)+cf(\mu)\frac{E^4}{F_\pi^4}+O(E^6\ln^2 E). \qquad (16.7)$$

Here a, b, and c are known dimensionless quantities, which depend on the angle and isospin variables; $f(\mu)$ is a renormalized value of f; and μ is a unit of mass used in the definition of $f(\mu)$. Even though the effective theory is non-renormalizable, and in principle involves an infinite number of unknown parameters, we can calculate the E^2 and $E^4 \ln E$ terms in pion–pion scattering in terms of the single constant, F_π.

In much the same way, whether or not the graviton is a bound particle, the requirements of Lorentz invariance and quantum mechanics constrain the effective gravitational Lagrangian to take the form[19]

$$\mathscr{L}_G=-\frac{1}{16\pi G}\sqrt{g}R-f\sqrt{g}R^2-f'\sqrt{g}R_{\mu\nu}R^{\mu\nu}-16\pi Gf''\sqrt{g}R^3-\cdots \qquad (16.8)$$

with a matter Lagrangian which depends on $g_{\mu\nu}$ in such a way as to be generally covariant. This is to be used to generate a perturbation series in powers of GE^2 or G/r^2 (where E and r are an energy and a length that are characteristic of the process under study), so the unperturbed Lagrangian must be taken as the quadratic part of only the first term, $-\sqrt{g}R/16\pi G$. Since the quadratic parts of the second and third terms must here be regarded as higher-order perturbations, we do not encounter the unphysical singularities discussed in section 16.2(*b*). We shall see in section 16.4 that the full Einstein term, $-\sqrt{g}R/16\pi G$, can be used in the tree approximation to calculate the leading terms in all multi-graviton Green's functions, and can even be used in the one-loop approximation to calculate the quantum corrections of relative order $GE^2 \ln E$ or $(G/r^2)\ln r$.

The main objection to viewing the graviton as a composite particle is simply that it is not clear why a bound state of spin two should have precisely zero mass. It is possible that this could be understood either dynamically,[23] or in terms of a supersymmetry[10] which puts the graviton in a multiplet with other composite particles which have to be massless because of chiral or other symmetries. However, this discussion has shown that if all we want is to study the low-energy or long-range properties of pions or gravitons, it is not necessary to know anything about the mechanism by which these particles are bound, or even whether pions and gravitons are composite particles at all. Section 16.4 will deal in

Chapter 16. Ultraviolet divergences in quantum gravity

a somewhat more systematic way with the low-energy limit of quantum theories of gravitation, without needing to raise the question whether gravitons are composite or elementary.

16.3 Asymptotic safety

Let us suppose for the sake of argument that none of the approaches outlined in the previous section can be successfully applied to gravitation. Suppose that no combination of gravity with matter fields yields a theory which is renormalizable in the usual sense, and that no resummation procedure yields a renormalizable perturbation theory consistent with unitarity, and that the graviton cannot be reinterpreted as a composite particle arising in an underlying renormalizable field theory. If gravitation is to be described by a flat-space quantum field theory at all, we would then have to face the prospect of dealing with a theory that is not renormalizable in the usual sense. The present section will describe a generalized version of the condition of renormalizability, which might still be applicable to gravitation in this case.

In any non-renormalizable theory, ultraviolet divergences occur in all Green's functions (and to all orders in their external momenta) so that in order to provide counterterms for these infinities, we must include in the Lagrangian all possible interactions allowed by the symmetries of the theory. Thus the gravitational Lagrangian would have to include not only the Einstein term $-\sqrt{g}R/16\pi G$, but also terms proportional to $\sqrt{g}R^2$, $\sqrt{g}R^{\mu}{}_{\nu}R^{\nu}{}_{\mu}$, $\sqrt{g}R^3$, etc., plus terms involving arbitrary powers of matter as well as gravitational fields.

It might be thought that the inclusion of terms proportional to $\sqrt{g}R^2$ and $\sqrt{g}R^{\mu}{}_{\nu}R^{\nu}{}_{\mu}$ would lead to unphysical singularities, of the sort discussed in section 16.2(*b*). However, these unphysical singularities are not encountered in perturbation theory unless the parts of these interactions quadratic in the fields are summed to all orders, as in (16.2). There is in general no justification for this partial summation – the apparent poles occur at momenta of the order of the Planck mass,[6] $G^{-1/2} = 1.2\times 10^{19}$ GeV, and at such momenta gravitation is so strong that perturbation theory breaks down altogether. A Lagrangian with $\sqrt{g}R^2$ and $\sqrt{g}R^{\mu}{}_{\nu}R^{\nu}{}_{\mu}$ terms will *not* lead to unphysical singularities at energies $E \ll 10^{19}$ GeV, where perturbation theory can be trusted, and we cannot use perturbation theory to tell whether there will be unphysical singularities at energies of order 10^{19} GeV in such a theory, or indeed in general relativity itself. The question of possible unphysical singularities at very high energy is nevertheless an important one, and it is in fact the key issue which will lead us to the requirement of asymptotic safety.

In order to search for such singularities, we use the method of the renormalization group.[8] Let $g_i(\mu)$ denote the full set of all renormalized coupling parameters of a theory, defined at a renormalization point with momenta characterized by an energy scale μ. If $g_i(\mu)$ has the dimensions of $[\text{mass}]^{d_i}$ we replace it with a dimensionless coupling,

$$\bar{g}_i(\mu) = \mu^{-d_i} g_i(\mu). \tag{16.9}$$

Any sort of partial or total reaction rate R may be written in the form

$$R = \mu^D f\left(\frac{E}{\mu}, X, \bar{g}(\mu)\right) \tag{16.10}$$

where D is the ordinary dimensionality of R (e.g., for total cross sections, $D = -2$); E is some energy characterizing the process; and X stands for all other dimensionless physical variables, including all ratios of energies. The central idea of the renormalization group method is simply to recognize that the physical quantity R cannot depend on the arbitrary choice of renormalization point μ at which the couplings are defined, so that we can take μ in (16.10) to be anything like, and in particular to be $\mu = E$, in which case (16.10) becomes

$$R = E^D f(1, X, \bar{g}(E)). \tag{16.11}$$

Thus, apart from the trivial scaling factor E^D, the high-energy behavior of reaction rates depends on the behavior of the couplings $\bar{g}(\mu)$ as $\mu \to \infty$.

There is a technicality here which deserves a word of explanation at this point. Among the coupling parameters $g_i(\mu)$ are the particle masses $m(\mu)$, with dimensionality $d = +1$. The corresponding dimensionless parameters (16.9) are of the form $m(\mu)/\mu$, and may reasonably be expected to vanish[24] for $\mu \to \infty$. However, reaction rates will in general contain singularities for zero mass, corresponding to the infrared divergences present in a massless quantum field theory. Thus, it is not possible to evaluate the high-energy behavior of arbitrary reaction rates by simply setting the masses equal to zero. It is for this reason that the renormalization group method has historically been used to calculate the high-energy behavior of Green's functions far off the mass shell, where no infrared divergences occur even for zero mass. However, mass singularities can be eliminated from physical reaction rates themselves by performing suitable sums over certain sets of initial and final states.[25] In what follows, we will tacitly assume that this has been done.

Chapter 16. Ultraviolet divergences in quantum gravity

Our emphasis here on reaction rates rather than off-shell Green's functions has a very important advantage. Mass-shell matrix elements and reaction rates do not depend on how the fields are defined, so they are functions only of 'essential' coupling parameters, i.e. of those combinations of the coupling parameters in the Lagrangian that do not change when we subject the fields to a point transformation (as for instance $\phi \to \phi + \phi^2$ for a scalar field ϕ). In contrast, the off-shell Green's functions will of course reflect the definition of the fields involved, and will therefore be functions of all the coupling parameters in the Lagrangian, including those 'inessential' coupling parameters (like the field renormalization constants, Z) which are not invariant under redefinitions of the fields. It will always be understood here that the $g_i(\mu)$ comprise only the *essential* coupling parameters of the theory.

(There is a well-known test which can be used to identify the inessential coupling parameters in any theory. When we change any unrenormalized coupling parameter γ_0 by an infinitesimal amount ε the whole Lagrangian changes by

$$\mathscr{L} \to \mathscr{L} + \varepsilon \frac{\partial \mathscr{L}}{\partial \gamma_0}.$$

Suppose we try to produce this change by a mere redefinition of fields

$$\psi_n(x) \to \psi_n(x) + \varepsilon F_n(\psi(x), \partial_\mu \psi(x), \ldots).$$

The change in $\mathscr{L}$ induced thereby is

$$\delta\mathscr{L} = \varepsilon \sum_n \left[\frac{\partial \mathscr{L}}{\partial \psi_n} F_n + \frac{\partial \mathscr{L}}{\partial(\partial_n \psi_n)} \partial_n F_n + \cdots \right]$$

$$= \varepsilon \sum_n \left[\frac{\partial \mathscr{L}}{\partial \psi_n} - \partial_\mu \left(\frac{\partial \mathscr{L}}{\partial(\partial_\mu \psi_n)} \right) + \cdots \right] F_n + \text{total derivative terms.}$$

Thus a change $\delta\mathscr{L} = \varepsilon\, \partial\mathscr{L}/\partial\gamma_0$ in the Lagrangian can be brought about by a redefinition of the fields if and only if we can find functions, F_n, of the fields and their derivatives such that

$$\frac{\partial \mathscr{L}}{\partial \gamma_0} = \sum_n \left[\frac{\partial \mathscr{L}}{\partial \psi_n} - \partial_\mu \frac{\partial \mathscr{L}}{\partial(\partial_\mu \psi_n)} + \cdots \right] F_n(\psi, \partial_\mu \psi, \ldots) + \text{total derivative terms.}$$

In other words, *the coupling parameter γ_0 is inessential if and only if $\partial\mathscr{L}/\partial\gamma_0$ vanishes or is a total derivative when we use the Euler–Lagrange equations*

$$0 = \frac{\partial \mathscr{L}}{\partial \psi_n} - \partial_\mu \frac{\partial \mathscr{L}}{\partial(\partial_\mu \psi_n)} + \cdots.$$

For instance, in the renormalizable scalar field theory with Lagrangian

$$\mathscr{L} = -\tfrac{1}{2}Z(\partial_\mu\phi\ \partial^\mu\phi + m^2\phi^2) - \tfrac{1}{24}\lambda Z^2\phi^4$$

the field renormalization constant is an inessential coupling, because we can write

$$\frac{\partial\mathscr{L}}{\partial Z} = \tfrac{1}{2}\phi(\Box^2\phi - m^2\phi - \tfrac{1}{6}\lambda Z\phi^3) - \tfrac{1}{2}\partial_\mu(\phi\ \partial^\mu\phi)$$

and the first term vanishes when we use the field equation

$$\Box^2\phi = m^2\phi + \tfrac{1}{6}\lambda Z\phi^3.$$

On the other hand, neither the mass, m, nor the coupling, λ, nor any combination of m and λ are inessential. In this example, the field redefinition, $\phi \to \phi + \varepsilon F$, associated with the single inessential coupling Z is a simple rescaling, with $F \propto \phi$, but this is just because the theory is constrained to be renormalizable; more complicated transformations with F a nonlinear function of ϕ and its derivatives would have produced non-renormalizable terms in $\mathscr{L}$. In non-renormalizable theories, we have to consider all possible redefinitions of the fields consistent with their symmetry properties, and in consequence there is an infinite number of inessential as well as essential couplings.)

As we shall see, by working only with essential couplings we will be able to formulate the condition of asymptotic safety in a concise way. In addition, it fits in well with the 'background field method',[26] in which the renormalization of the coupling parameters is determined by calculating the matrix element between the 'in' and 'out' vacua in a classical background field which satisfies the Euler–Lagrange field equations.

Now let us return to the problem of determining the behavior of the essential couplings $\bar{g}_i(\mu)$. The change in $\bar{g}_i(\mu)$ under a given fractional change in μ is a dimensionless quantity, and can therefore depend on all the $\bar{g}_i(\mu)$, but not separately on μ itself, because there are no other dimensional parameters here with which μ can be compared. Thus the rate of change of $\bar{g}_i(\mu)$ with respect to rescaling of the renormalization point μ may be rewritten as a generalized Gell-Mann–Low equation[8]

$$\mu\frac{d}{d\mu}\bar{g}_i(\mu) = \beta_i(\bar{g}(\mu)). \tag{16.12}$$

Each specific theory is characterized by a *trajectory* in coupling constant space, generated by the solution of (16.12) with given initial conditions.

Chapter 16. Ultraviolet divergences in quantum gravity

The function $\beta(\bar{g})$ can be calculated as a power series in $\bar{g}_i$, but in general this will not help us determining the behavior of $\bar{g}(\mu)$ for $\mu \to \infty$. However, we can identify one general class of theories in which unphysical singularities are almost certainly absent. If the couplings $\bar{g}_i(\mu)$ approach a 'fixed point', g^*, as $\mu \to \infty$, then (16.11) gives a simple scaling behavior, $R \propto E^D$, for $E \to \infty$. In order for $\bar{g}(\mu)$ to approach g_i^* as $\mu \to \infty$, it is necessary that $\beta_i(\bar{g})$ vanish at this point

$$\beta_i(g^*) = 0 \tag{16.13}$$

and also that the couplings lie on a trajectory $\bar{g}_i(\mu)$ which actually hits the fixed point. The surface formed of such trajectories will be called the *ultraviolet critical surface*. The generalized version of renormalizability that we wish to propose for the quantum theory of gravitation is that the coupling constants must lie on the ultraviolet critical surface of some fixed point. Such theories will be called *asymptotically safe*.[27]

(Incidentally, the demand that the coupling parameters approach a fixed point for $\mu \to \infty$ cannot, in general, be met if we include inessential as well as essential coupling parameters. For instance, the renormalization group equations cannot change their form when we multiply each field by an independent constant; hence, if the field-renormalization constants $Z_r(\mu)$ satisfy these equations, then so do the $Z_r(\mu)$ times arbitrary constants, and therefore the equations for the $Z_r(\mu)$ must take the form

$$\mu \frac{\mathrm{d}}{\mathrm{d}\mu} Z_r(\mu) = Z_r(\mu)\gamma_r(\bar{g}(\mu)).$$

In general, there is no reason why $\gamma_r(g^*)$ should vanish or diverge, so the solution for $\mu \to \infty$ will be of the form

$$Z_r(\mu) \propto \mu^{\gamma_r(g^*)}.$$

This introduces corrections to scaling in off-shell Green's functions. However, reaction rates do not depend on the Zs, so they can exhibit 'naive' scaling $R \sim E^D$ even if $Z_r(\mu) \to \infty$ for $\mu \to \infty$.)

We do not really know that a theory which is not asymptotically safe will have unphysical singularities – the assumption of asymptotic safety is just one way of being reasonably sure that unphysical singularities do not occur. As an example of what can happen when a theory is *not* asymptotically safe, let us consider the sample differential equation[28]

$$\mu \frac{\mathrm{d}}{\mathrm{d}\mu} \bar{g}_i = a_i \sum_j (\bar{g}_j - g_j^*)^2 \tag{16.14}$$

where a_i and g_i^* are a set of arbitrary constants. The trajectories which hit the fixed point g^* are evidently those with initial values along the line

$$\bar{g}_i = g_i^* - a_i\xi, \quad \xi > 0, \tag{16.15}$$

so here the critical surface is one-dimensional. If $\bar{g}(\mu)$ is on this surface, with $\xi = \xi_0 > 0$ at $\mu = \mu_0$, then at greater values of μ, $\bar{g}$ is given by (16.15), with

$$\xi = \xi_0\left[1 + \xi_0 \sum_i a_i^2 \ln(\mu/\mu_0)\right]^{-1}.$$

We see that in this case $\bar{g}_i$ smoothly approaches g_i^* as $\mu \to \infty$, and the theory is asymptotically safe. On the other hand, if $\bar{g}_i$ does not lie on the line (16.15), then it will diverge at a *finite* value of μ. For $\bar{g}_i \to \infty$, the solution to (16.14) becomes

$$\bar{g}_i \to a_i[-\sum a_j^2 \ln(\mu/\mu_\infty)]^{-1},$$

and we see that $\bar{g}_i \to \infty$ as μ approaches μ_∞. If we assume that an infinity in coupling parameters would produce an unphysical singularity in reaction rates, then we can conclude here that a theory which does not lie on the ultraviolet critical surface (16.15) will develop unphysical singularities at the energy $E = \mu_\infty$.

Of course, the question of whether or not an infinity in coupling constants betokens a singularity in reaction rates depends on how the coupling constants are defined. We could always adopt a perverse definition (e.g. $\bar{g}' \equiv (\bar{g} - g^*)^{-1}$) such that reaction rates are finite even at an infinity of the coupling parameters. This problem is avoided if we define the coupling constants as coefficients in a power series expansion of the reaction rates themselves around some physical renormalization point. In lowest-order perturbation theory, this procedure is indistinguishable from the usual renormalization procedure, in which the $\bar{g}_i(\mu)$ are defined in terms of a power series expansion of the Green's functions around some off-shell renormalization point. It seems worth exploring whether it would be possible to carry out a consistent definition of renormalized coupling parameters in terms of reaction rates rather than Green's functions, but this will not be attempted here.

The number of free parameters in an asymptotically safe theory is simply equal to the dimensionality of the ultraviolet critical surface. If the critical surface is infinite-dimensional, then the demand that the physical theory lie on this surface leaves us with an infinite number of undetermined parameters, and little has been learned. At the other extreme, if

Chapter 16. Ultraviolet divergences in quantum gravity

the critical surface is zero-dimensional, then the demand for asymptotic safety cannot be satisfied at all. The hope is that the dimensionality of the critical surface is some finite number, C; in this case, the theory will have C free parameters, of which $C-1$ are dimensionless parameters which identify a particular trajectory on the C-dimensional critical surface, and one is a dimensional parameter, which tells us the value of μ at which some given point on this trajectory is reached. Best of all of course would be the case $C=1$; in this case, physics would have no free parameters aside from one dimensional constant which merely defines our units of mass or length. As long as C is finite, the condition of asymptotic safety will play a role for us similar to that of the condition of renormalizability in quantum electrodynamics: it serves to fix all but a finite number of the parameters of the theory. In fact, as we shall see, the condition of asymptotic safety will in some cases require that a theory be renormalizable, in the usual sense.

The dimensionality of the critical surface can be determined from the behavior of the $\beta_i(\bar{g})$ functions near the fixed point. In the neighborhood of g^*, (16.12) may be written

$$\mu\frac{\mathrm{d}}{\mathrm{d}\mu}\bar{g}_i(\mu)\to\sum_j B_{ij}[\bar{g}_j(\mu)-g_j^*], \tag{16.16}$$

where

$$B_{ij}\equiv(\partial\beta_i(\bar{g})/\partial\bar{g}_j)_{\bar{g}=g^*}. \tag{16.17}$$

The general solution is

$$\bar{g}_i(\mu)\to\sum_K C_K V_i^K\mu^{\lambda_K}+g_i^* \tag{16.18}$$

where V^K is the eigenvector of B_{ij} with eigenvalue λ_K,

$$\sum_j B_{ij}V_j^K=\lambda_K V_i^K, \tag{16.19}$$

and the C_K are arbitrary coefficients. Clearly, the condition for $\bar{g}_i(\mu)$ to approach g_i^* as $\mu\to\infty$ is that C_K should vanish for all positive eigenvalues $\lambda_K>0$. (The possibility of zero eigenvalues is a nuisance here, to which we shall return later.) The dimensionality of the ultraviolet critical surface then is the number of remaining C_K parameters, i.e. the number of negative eigenvalues of B_{ij}.

The crucial problem then is to determine how many eigenvalues of the B-matrix are negative. In all cases of which I know, this number is finite. This may be understood by a suggestive, though highly non-rigorous, argument. Recall that

$$\bar{g}_i(\mu) \equiv \mu^{-d_i} g_i(\mu),$$

where d_i is the dimensionality of the un-rescaled renormalized coupling $g_i(\mu)$ in powers of mass. The μ-dependence of $g_i(\mu)$ arises from the dependence of loop graphs on the momenta at the renormalization point, so that

$$\beta_i = -d_i \bar{g}_i + \text{loop contributions} \tag{16.20}$$

and

$$B_{ij} = -d_i \delta_{ij} + \text{loop contributions.} \tag{16.21}$$

Now, adding derivatives or powers of fields to an interaction will always *lower* the dimensionality d_i, so at most a finite number of d_i can be positive, and all but a finite number are below any given negative value. In the absence of the 'loop contributions', the eigenvalues of B_{ij} would be just the quantities $-d_i$, of which all but a finite number are positive. The 'loop contributions' can of course change the signs of some of the eigenvalues of B_{ij}, but if these contributions are bounded then they cannot change the sign of the infinite number of large positive eigenvalues, and only a finite number can be negative. We thus may guess that the ultraviolet critical surface will in general be of finite dimensionality.

This conclusion is in some cases empirically verified by the observed existence of second-order phase transitions. A second-order phase transition will in general occur at values of the parameters in a theory at which masses vanish or correlation lengths diverge, so that physical quantities can exhibit scaling at large distances or small momenta. By repeating the arguments of this section, we see that such scaling must be associated with a fixed point g^* where $\beta_i(g^*)$ vanishes; the phase transition occurs when the parameters of the theory take values on an *infrared* critical surface,[8] consisting of trajectories (16.12) which hit the fixed point for $\mu \to 0$. From (16.18), we see that the number of parameters which have to be adjusted in order to put the theory on the infrared critical surface is equal to the number of negative eigenvalues of B_{ij}, and hence is just equal to the dimensionality of the *ultraviolet* critical surface. But in all cases we know, this number is finite: only a finite number of parameters (temperature, pressure, magnetic field) have to be adjusted to induce a second-order phase transition. ***Thus, at least for those fixed points associated with second-order phase transitions, we can be sure that the ultraviolet critical surface is of finite dimensionality.***

Chapter 16. Ultraviolet divergences in quantum gravity

We have already remarked that for a finite-dimensional critical surface, the condition of asymptotic safety works much like the condition of renormalizability in limiting the free parameters in physical theories. In fact, we can now see that the connection between asymptotic safety and renormalizability is even closer. Any theory will always have a fixed point at the origin, $g^* = 0$. (If the essential couplings vanish at one renormalization scale μ, they will vanish at all μ, so $\beta_i(\bar{g})$ always vanishes at $\bar{g} = 0$.) Suppose that for some theory this is the only suitable fixed point with an ultraviolet critical surface of nonzero dimensionality, so that asymptotic safety requires that the couplings lie on this surface. The 'loop contribution' in (16.21) vanishes at $\bar{g}_i = 0$, so the B-matrix for this fixed point is just

$$B_{ij} = -d_i \delta_{ij}. \tag{16.22}$$

In order for the trajectory $\bar{g}i(\mu)$ to hit $g^* = 0$ as $\mu \to \infty$, it is thus necessary that all g_j with $d_j < 0$ should vanish. But these are precisely the non-renormalizable interactions,[1] so this sort of theory must be renormalizable in the usual sense.

Strictly speaking, renormalizability may not be enough. The B-matrix for $g^* = 0$ will in general have some zero eigenvalues, corresponding to those 'strictly' renormalizable interactions with $d_i = 0$, and it is also necessary that these interactions have $\beta_i / g_i < 0$ near $\bar{g}_i = 0$, in order that $\bar{g}_i(\mu)$ should vanish for $\mu \to \infty$. Hence the ultraviolet critical surface of the fixed point $g^* = 0$ actually consists of all theories that are renormalizable *and* asymptotically free.[29] However, from a practical point of view a renormalizable theory like quantum electrodynamics can be regarded as asymptotically safe even though it is not asymptotically free, because the growth of a strictly renormalizable coupling like $e(\mu)$ is only logarithmic, and any unphysical singularities would only be encountered at exponentially high energies. This would not be the case if the theory involved non-renormalizable interactions, like Fermi interactions.

In some cases, asymptotic safety can lead to renormalizability even though the fixed point is not at $g^* = 0$. The seminal paper[8] of Gell–Mann and Low suggested the existence of a possible fixed point $e^* \neq 0$ in quantum electrodynamics, at which $\beta_e(e^*) = 0$. If there is such a fixed point in quantum electrodynamics, then it is a fixed point for the most general field theory of photons and electrons, but there is no reason to expect that trajectories with non-vanishing values for non-renormalizable couplings would hit this fixed point. Thus, barring other fixed points, the effect of the condition of asymptotic safety here would be just to require renormalizability of the usual sort.

The problem that we face in dealing with quantum gravity is that there may not be any theories that are renormalizable, let alone asymptotically free. Therefore we must search for other fixed points, away from $g^* = 0$. In general, there is no particular reason why a fixed point with $g^* \neq 0$ should have g^* small, so perturbation theory will not be of much use to us in searching for such fixed points, or in exploring their properties.

This is much the same as the problem encountered in the theory of critical phenomena: the failure of mean field theory showed that the phase transitions are not associated with a fixed point at zero coupling, and it became necessary to search for other fixed points. In this case, the problem could be solved[30] by a continuation in the spatial dimensionality of the system, the 'ε-expansion'. It therefore seems reasonable to try a continuation in spacetime dimensionality in our present problem.

Suppose we could find some spacetime dimensionality $D_r < 4$ at which there exists a renormalizable and asymptotically free theory of gravitation. As we have seen, this would imply that the fixed point at $g^* = 0$ has an ultraviolet critical surface of finite dimensionality. If we then increase the dimensionality, D, the theory will become non-renormalizable, but continuity would lead us to expect that the fixed point which is at $g^* = 0$ for $D = D_r$ will move smoothly away from the origin, and that for at least a finite range of D above D_r its ultraviolet critical surface will retain the same dimensionality. Hence for $D = D_r + \varepsilon$, we can expect to find a fixed point with g^* of order ε, and to learn the properties of this fixed point by an expansion in powers of ε.

This approach has already been applied[31] to the nonlinear σ-model (which is renormalizable and asymptotically free at $D_r = 2$), though from a somewhat different point of view. Its application to gravitation will be described in section 16.6, using techniques to be discussed in section 16.5.

In carrying out actual calculations, it is very useful to recognize that although the β_i-functions and the B_{ij}-matrix depend on the details of our renormalization procedure and on how the coupling parameters g_i are defined, the *eigenvalues* of the B_{ij}-matrix do not.[32] There is a wide variety of ways in which we might change the definition of the $g_i(\mu)$, as for example:

(*a*) We might simply choose a different set of renormalization points, in which case the rescaled coupling parameters (16.9) would become functions, $\tilde{g}_i(\tilde{\mu})$, of the scale, $\tilde{\mu}$, of the momenta at the new renormalization points.

Chapter 16. Ultraviolet divergences in quantum gravity

(b) We can use dimensional regularization,[33] in which case the 'renormalized' couplings can be taken as the constant terms in a Laurent expansion of the unrenormalized couplings, g_{i0}, around spacetime dimensionality $D=4$:

$$g_{i0} \xrightarrow[D\to 4]{} \tilde{\mu}^{d_i(D)}\left[\tilde{g}_i(\tilde{\mu}) + \sum_{\nu=1}^{\infty} \frac{b_{\nu i}(\tilde{g}(\tilde{\mu}))}{(D-4)^{\nu}}\right], \tag{16.23}$$

with $\tilde{\mu}$ a 'unit of mass' introduced to make $\tilde{g}_i(\tilde{\mu})$ dimensionless.

(c) We can introduce an ordinary ultraviolet cut-off at momentum $\tilde{\mu}$, and take the rescaled unrenormalized couplings as functions $\tilde{g}_i(\tilde{\mu})$, with a cut-off dependence chosen so that the reaction rates are cut-off independent.[34]

In these or other cases, the new couplings $\tilde{g}_i$ must be expressible as functions of the old couplings $\bar{g}_j(\mu)$ and of the only other dimensionless quantity $\tilde{\mu}/\mu$:

$$\tilde{g}_i(\tilde{\mu}) = \tilde{g}_i\left(\frac{\tilde{\mu}}{\mu}, \bar{g}_i(\mu)\right).$$

But the new couplings do not depend on how the old couplings are defined, so they are independent of μ:

$$0 = \mu\frac{\mathrm{d}\tilde{g}_i}{\mathrm{d}\mu} = -\tilde{\mu}\frac{\partial \tilde{g}_i}{\partial \tilde{\mu}} + \sum_j \frac{\partial \tilde{g}_i}{\partial \bar{g}_j}\mu\frac{\mathrm{d}\bar{g}_j}{\mathrm{d}\mu}.$$

In other words, we may define a new β_i-function

$$\tilde{\beta}_i(\tilde{g}(\tilde{\mu})) \equiv \tilde{\mu}\frac{\partial \tilde{g}_i}{\partial \tilde{\mu}} \tag{16.24}$$

which is related to the old β-function by the transformation rule of a contravariant vector

$$\tilde{\beta}_i(\tilde{g}) = \sum_J \frac{\partial \tilde{g}_i}{\partial \bar{g}_j}\beta_j(\bar{g}). \tag{16.25}$$

We see that the existence of a fixed point is an invariant concept: if $\beta_i(g^*)=0$ then $\tilde{\beta}_i(\tilde{g}(g^*))=0$. The β-functions themselves are not invariant, and neither are their derivatives

$$\frac{\partial \tilde{\beta}_i}{\partial \tilde{g}_j} = \sum_{kl} \frac{\partial^2 \tilde{g}_i}{\partial \bar{g}_k \partial \bar{g}_l}\frac{\partial \bar{g}_l}{\partial \tilde{g}_j}\beta_k(\bar{g}) + \sum_{kl}\frac{\partial \tilde{g}_i}{\partial \bar{g}_k}\frac{\partial \beta_k(\bar{g})}{\partial \bar{g}_l}\frac{\partial \bar{g}_l}{\partial \tilde{g}_j}.$$

But at a fixed point the first term vanishes, so the new B-matrix is related to the old one by

$$\tilde{B}_{ij} = \sum_{kl} A_{ik} B_{kl} A^{-1}{}_{lj} \tag{16.26}$$

$$A_{ik} \equiv \left(\frac{\partial \tilde{g}_i(\bar{g})}{\partial \bar{g}_k} \right)_{\bar{g}=g^*}. \tag{16.27}$$

This is a similarity transformation, so the eigenvalues of $\tilde{B}$ are the same as the eigenvalues of B. (These eigenvalues are known as *critical exponents*; they depend only on the nature of the degrees of freedom of a system, and on no other physical inputs.) In particular, the question of asymptotic safety is one that can be addressed in any of the formalisms (a), (b), (c) outlined above, with confidence that the answer will be the same.

16.4 Physics at ordinary energies

The condition of asymptotic safety entails the appearance of a fundamental energy scale M: it is the value of μ at which the trajectory $\bar{g}_i(\mu)$ approaches to within some definite distance of the fixed point $g_i{}^*$. We will see below that this characteristic energy should be of the order of the Planck mass,[6] $M_P = 1.2 \times 10^{19}$ GeV. The problem to which we now turn is the description of physical phenomena at ordinary energies, which are vastly less than M. In particular, we want to inquire why gravitational interactions are so well described at macroscopic distances by the Einstein Lagrangian, $-\sqrt{g}R/16\pi G$, and why the weak, strong, and electromagnetic interactions should be so well described at ordinary energies by renormalizable quantum field theories.[35]

Let us first consider the case of pure gravity, with a Lagrangian of form (16.8). We assume that the theory is asymptotically safe, so the infinite set of couplings are constrained to lie on a trajectory $\bar{g}_i(\mu)$ which hits some fixed point $g_i{}^*$ for $\mu \to \infty$. The renormalization-group arguments of the previous section show that physics at low energies is governed by the behavior of the couplings $\bar{g}_i(\mu)$ as $\mu \to 0$. The simplest possibility is for $\bar{g}_i(\mu)$ to approach another fixed point in this limit. Now, we saw in the previous section that in addition to $g_i{}^*$, there is always a fixed point at the origin. Also, every term in the Lagrangian (16.8) is non-renormalizable, so every eigenvalue of the matrix $\partial\beta_i/\partial\bar{g}_j$ at $\bar{g}_i = 0$ is positive. This means that the fixed point at the origin is entirely repulsive for $\mu \to \infty$, but by the same token, it is entirely attractive for $\mu \to 0$: whatever direction we go from the origin, we remain in the infrared critical surface. It follows that there is at least a finite region around the origin, within which *every*

Chapter 16. Ultraviolet divergences in quantum gravity

trajectory is attracted to the origin for $\mu \to 0$. We do not know whether the ultraviolet fixed point, $g_i{}^*$, lies within this region, but there is nothing unreasonable in supposing that it does. Under this assumption, we can conclude that all $\bar{g}_i(\mu)$ vanish for $\mu \to 0$.

(We have tacitly assumed here that the Lagrangian does not contain a cosmological constant term $\Lambda\sqrt{g}$. Such a term would be super-renormalizable, and would therefore correspond to an infrared-repulsive eigenvector of $\partial\beta_i/\partial\bar{g}_j$ at $\bar{g}_i = 0$. A cosmological constant is not needed as a counterterm in pure gravity, at least in the dimensional regularization scheme. However, theories with massive particles will in general require a cosmological constant counterterm, and it is somewhat of a mystery why Λ is not some forty orders of magnitude larger than the observed upper limit.)

For $\bar{g}_i(\mu)$ near zero, the 'loop contribution' in (16.20) is negligible, and the solution of (16.12) becomes simply

$$\bar{g}_i(\mu) \approx (M_i/\mu)^{d_i} \quad \text{for } \mu \ll M_i,$$

with M_i a set of unknown integration constants. The M_i are related to each other by the asymptotic safety condition, that $\bar{g}(\mu)$ must lie on the ultraviolet critical surface of a fixed point g^*. For a one-dimensional critical surface, the condition of asymptotic safety leaves us with only one free parameter, the fundamental energy scale M, so all M_i must be of order M. The same is true even if the dimensionality C of the critical surface is greater than unity, provided that the $C-1$ parameters which determine the orbit of $\bar{g}_i(\mu)$ do not take exceptionally large or exceptionally small values. We conclude then that the original couplings, before the rescaling (16.9), have the order of magnitude

$$g_i(\mu) \approx M^{d_i} \quad \text{for } \mu \ll M. \tag{16.28}$$

In particular, (16.28) gives a gravitational constant G of order M^{-2}, so M must be of the order of the Planck mass

$$M \approx M_{\rm P} \equiv G^{-1/2} = 1.2 \times 10^{19}\,\text{GeV}. \tag{16.29}$$

We can now see why gravitational phenomena are so well described by Einstein's theory at macroscopic distances. Consider a connected Green's function for a set of gravitational fields at points with typical spacetime separations r. Ultraviolet divergencies are removed by the renormalization of the infinite set of coupling parameters, but it is essential here that the renormalized coupling parameters $g_i(\mu)$ be defined at renormalization points with momenta of order $1/r$, not $M_{\rm P}$. In this way, the integrals in the Feynman diagrams will begin to converge at momenta of order $1/r$, so r will be the only dimensional parameter in the theory other than the coupling parameters themselves. Equations (16.28) and (16.29) show that the coupling constants in a graph with $\mathcal{N}_i$ vertices of type i yield a factor proportional to N powers of $G^{1/2}$ or $M_P{}^{-1}$, where

$$N = -\sum_i \mathcal{N}_i d_i, \tag{16.30}$$

so ordinary dimensional analysis tells us that the contribution of such a graph will be suppressed by a factor

$$(G^{1/2}/r)^N = (rM_P)^{-N}. \tag{16.31}$$

The leading graphs for r much larger than the Planck length $M_P{}^{-1} = 1.6 \times 10^{-33}$ cm will thus be those with the smallest values of N. The dimensionality, d_i, is given by

$$d_i = 4 - p_i - g_i, \tag{16.32}$$

where p_i is the number of derivatives and g_i the number of graviton fields in the interaction of type i. To calculate N, we use the well known topological relations

$$\mathcal{L} = \ell - \sum_i \mathcal{N}_i + 1 \tag{16.33}$$

$$2\ell + \mathcal{E} = \sum_i \mathcal{N}_i g_i \tag{16.34}$$

where $\mathcal{L}$ is the number of loops; ℓ is the number of internal lines; and $\mathcal{E}$ is the number of external lines in the graph. Together with (16.30) and (16.32), these give

$$N = \sum_i \mathcal{N}_i(p_i - 2) + 2\mathcal{L} + \mathcal{E} - 2. \tag{16.35}$$

Leaving aside any super-renormalizable cosmological constant term, $\Lambda\sqrt{g}$, the gravitational interactions with the smallest number of derivatives are those derived from the Einstein term, $-\sqrt{g}R/16\pi G$, all of which have $p_i = 2$. Hence for any given Green's function with a given number $\mathcal{E}$ of external lines, the leading graphs for $r \gg M_P{}^{-1}$ will be the tree graphs ($\mathcal{L} = 0$) constructed purely from the Einstein Lagrangian. Summing these tree graphs is tantamount to solving the classical field equations;[36] in particular, the one-graviton Green's function in the presence of a classical background distribution of energy and momentum satisfies the classical Einstein field equations for the gravitational field produced by this energy–momentum tensor. The next corrections will arise both from the graphs in pure general relativity with one loop, and equally from the tree graphs containing *one* vertex arising from the higher interactions, $\sqrt{g}R^2$ or $\sqrt{g}R^{\mu\nu}R_{\mu\nu}$, for which $p_i = 4$. (The ultraviolet divergence in the loop graph is cancelled by the counterterm provided by the $\sqrt{g}R^2$ and $\sqrt{g}R^{\mu\nu}R_{\mu\nu}$ interactions.) Equation (16.35) shows that these quantum corrections to classical general relativity are suppressed by a factor of order $(rM_P)^{-2}$, and hence are utterly negligible at macroscopic distances.

Incidentally, (16.35) shows that even the classical tree-graph contributions to a given Green's function are suppressed by a factor proportional to $G^{1/2}$ for each extra external graviton line. When we calculate the metric produced by a mass, m, we also pick up another factor of $G^{1/2}m$ for each coupling of these external lines to the mass. The reason why the exchange of *trees* of graviton lines with $\mathscr{E} > 2$ external lines has a detectable effect on planetary motion is just that the solar mass, $m_\odot$, is so large that $Gm_\odot/r$ is not an utterly negligible quantity.

The above discussion can be immediately extended to theories of gravitation and matter, at least in the case that there are no masses and no super-renormalizable or asymptotically free renormalizable interactions among the matter fields. (For instance, this would be the case for the theory of gravitons, photons, and massless electrons.) In such a theory, the fixed point at zero coupling is still entirely attractive for $\mu \to 0$, so according to our previous arguments, the couplings at ordinary energy will have the order of magnitude

$$g_i(\mu) = O(M_P{}^{d_i}) \quad \text{for } \mu \ll M_P.$$

Physical phenomena at ordinary energies $E \ll M_P$ will be entirely governed by the renormalizable interactions, for which $d_i = 0$, because all non-renormalizable interactions with $d_i < 0$ are suppressed by powers of E/M_P and hence entirely undetectable. Only gravitation is an exception: as we have seen, the fact that gravitation couples coherently to every particle in a large body like the sun allows us to observe its macroscopic effects despite its intrinsic weakness.

Theories of gravitation and matter that involves masses or asymptotically free renormalizable interactions require a bit more attention. First, there may be particles with mass of order M_P – for instance, there are intermediate vector bosons almost this heavy in some superunified theories of weak, electromagnetic, and strong interactions.[5] Such particles cause no problems here; in Green's functions at ordinary energy, $E \ll M_P$, any internal line with mass M of order M_P may be replaced with a series of non-renormalizable interactions, obtained by expanding the heavy-particle propagators

$$\frac{1}{q^2+M^2}=\frac{1}{M^2}-\frac{q^2}{M^4}+\frac{(q^2)^2}{M^6}-\ldots.$$

(Again, it is essential here that the momentum scales, μ, of all renormalization points be taken of order E, not M, so that integrals converge at momenta $q \approx E \ll M$.) In this way, we obtain an effective field theory,[37] involving only the light particles with masses much less than M_P. These of course include all the particles with which we are familiar – gravitons, photons, leptons, quarks, intermediate vector bosons, Higgs bosons, and gluons.[38]

Now suppose that this effective field theory involves either some small masses $m \ll M_P$, or some asymptotically free, renormalizable interactions acting among the light particles, or both. In this case, the fixed point at the origin in coupling constant space is no longer entirely attractive for $\mu \to 0$. However, we can now identify an infrared-attractive *area*, if not an infrared-attractive *point*. Let us denote rescaled renormalizable couplings (including masses) as $\bar{f}_a(\mu)$, and non-renormalizable couplings as $\bar{F}_A(\mu)$. Since renormalizable couplings never generate infinities which require non-renormalizable counterterms, the renormalization group equations for $\bar{F}_A(\mu) \ll 1$ must take the form

$$\mu\frac{d}{d\mu}\bar{f}_a(\mu)\approx\beta_a(\bar{f}(\mu))$$

$$\mu\frac{d}{d\mu}\bar{F}_A(\mu)\approx\sum_B B_{AB}(\bar{f}(\mu))\bar{F}_B(\mu).$$

Furthermore, for $\bar{f}(\mu) \ll 1$ the loop contributions to B_{AB} become negligible, so that

$$B_{AB}(\bar{f})\approx -d_A\delta_{AB} \quad (\bar{f}\ll 1).$$

By definition, F_A is a non-renormalizable interaction, so $d_A > 0$. Thus there is at least a finite area $\mathscr{A}$ on the surface $\bar{F}_A = 0$ within which all eigenvalues of the matrix B_{AB} are positive. This is an infrared-attractive area – every trajectory in at least a finite slab $\mathscr{R}_{\mathscr{A}}$ around $\mathscr{A}$ is attracted to it for $\mu \to 0$. Once again, we do not know whether the ultraviolet fixed point $g_i{}^*$ lies within the region $\mathscr{R}_{\mathscr{A}}$, but it is not unreasonable to suppose that it does. Indeed, this seems to be experimentally verified: there is an enormous range of renormalization scales, $1\,\text{GeV} \ll \mu \ll 10^{19}\,\text{GeV}$, within which the rescaled non-renormalizable gravitational couplings,

Chapter 16. Ultraviolet divergences in quantum gravity

$\bar{F}_A(\mu)$, are all quite small, and the renormalizable weak, electromagnetic, and 'strong' interactions also have fairly small couplings, so that B_{AB} has positive eigenvalues $\approx -d_A$. On the assumption that the ultraviolet fixed point $g_i{}^*$ does lie within the region $\mathscr{R}_{\mathscr{A}}$, we can conclude again that the effect of non-renormalizable couplings becomes negligible at ordinary energies. More specifically, within the range of renormalization scales in which all $\bar{g}_i(\mu)$ are small, the unrescaled couplings $g_i(\mu)$ will again be of order M^{d_i}, as in (16.28), with M a common integration constant related to the renormalization scale at which $\bar{g}_i(\mu)$ approaches the fixed point, $g_i{}^*$. The conventional values of the gravitational couplings, G, and other non-renormalizable couplings are specified at a renormalization point with μ of the order of the mass of a typical light particle, where the rescaled renormalizable couplings (including m/μ) are just beginning to be of order unity, so these conventional couplings will also be of order M^{d_i}. We can thus identify M with the Planck mass, as in (16.29), and conclude again that effects of non-renormalizable couplings at ordinary energies, $E \ll M_P$, will be suppressed by powers of E/M_P.

16.5 Dimensional continuation

It was emphasized in section 16.3 that the existence of a fixed point and the dimensionality of its ultraviolet critical surface do not depend on whether we define the coupling parameters by ordinary renormalization, or by dimensional regularization, or by a 'floating' ultraviolet cut-off. However, experience has shown that the method of dimensional regularization[33] is by far the most convenient for actual calculation.[39] Somewhat suprisingly, dimensional regularization also turns out to provide a very convenient basis for the study of fixed points at arbitrary, non-integral dimensionality;[40] we remarked in section 16.3 that such dimensional continuation provides one method by which perturbation theory can be used in the study of fixed points.

Dimensional regularization allows us to calculate all Feynman integrals in finite form for non-rational spacetime dimensionality D, but the integrals will have poles as D approaches various rational values. Let us concentrate on some particular set of rational values, D_s, of the spacetime dimensionality, and suppose that the unrenormalized coupling constants $g_{i0}(D)$ have poles which will cancel the poles at $D = D_s$ in Feynman integrals, yielding reaction rates which are finite at $D = D_s$. (In the original formulation[39] of this approach, the set of D_s consisted of the single physical spacetime dimensionality $D = 4$; the generalization to several D_s is presented here because it requires essentially no extra work, and might turn out to be useful.) In order to express $g_{i0}(D)$ in terms of renormalized coupling parameters which are dimensionless for all D, we must introduce a unit of mass μ and write a Laurent expansion for the rescaled coupling, $g_{i0}(D)\mu^{-d_i(D)}$, instead of the ordinary unrenormalized coupling, $g_{i0}(D)$. (As usual, $d_i(D)$ is the dimensionality of $g_{i0}(D)$ in powers of mass, at a space-time dimensionality D). In addition to the poles at $D=D_s$, this expansion will have a remainder term which is analytic in D, and which we simply *define* as the dimensionless renormalized coupling, $g_i(\mu, D)$. The coefficients, $b_{\nu i}^{(s)}$, of the poles of order ν in $g_{i0}(D)\mu^{-d_i(D)}$ at $D=D_s$ will depend on μ and D only through their dependence on the renormalized coupling $g_i(\mu, D)$, because there is no dimensional parameter with which μ could be compared, and because any separate dependence of $b_{\nu i}^{(s)}$ on D would merely change the analytic terms and the lower-order poles in D. Thus the Laurent expansion may be written in the form

$$g_{i0}(D)\mu^{-d_i(D)} = g_i(\mu, D) + \sum_s \sum_{\nu=1}^{\infty} (D-D_s)^{-\nu} b_{\nu i}^{(s)}(g(\mu, D)). \tag{16.36}$$

(We now drop the tilde used in section 16.3 to distinguish $g_i(\mu, D)$ from the conventional rescaled renormalized coupling $\bar{g}_i(\mu)$.)

To calculate the Gell-Mann–Low functions, $\beta_i(g, D)$, we first differentiate with respect to μ, and find

$$-d_i(D)\Big[g_i + \sum_{s,\nu} (D-D_s)^{-\nu} b_{\nu i}^{(s)}(g)\Big]$$
$$= \beta_i(g, D) + \sum_{s,\nu,j} (D-D_s)^{-\nu} b_{\nu ij}^{(s)}(g)\beta_j(g, D), \tag{16.37}$$

where

$$\beta_i(g(\mu, D), D) \equiv \mu \frac{\partial}{\partial \mu} g_i(\mu, D) \tag{16.38}$$

and

$$b_{\nu ij}^{(s)}(g) \equiv \partial b_{\nu i}^{(s)}(g)/\partial g_j. \tag{16.39}$$

Again, we write β_i as a function of all the $g_j(\mu, D)$ and also of D, but not of μ, because there is no dimensional quantity here with which μ could be compared.

Now, the dimensionality, $d_i(D)$, of g_{i0} is always a linear function of the spacetime dimensionality D, which we shall write as[39]

$$d_i(D) = \sigma_i + \rho_i D. \tag{16.40}$$

Chapter 16. Ultraviolet divergences in quantum gravity

The left-hand side of (16.37) can then be re-written as

$$-\rho_i g_i D - \left[\sigma_i g_i + \sum_s b_{1j}^{(s)}(g)\rho_i\right]$$
$$-\sum_{s,\nu} (D-D_s)^{-\nu}[\rho_i b_{\nu+1,i}^{(s)}(g) + (\sigma_i + \rho_i D_s) b_{\nu,i}^{(s)}(g)] \quad (16.41)$$

the sum over ν running as before from 1 to ∞. Since the highest power of D in the analytic part here is of first order, the same must be true on the right-hand side of (16.37). However, all poles in D are supposed to be eliminated when we express g_{i0} in terms of g_i, so $\beta_i(g, D)$ should be analytic in D. In order that the analytic part of the right-hand side of (16.37) contain no terms of higher than first order in D, it is necessary then that $\beta_i(g, D)$ be *linear* in D:

$$\beta_i(g, D) = \beta_i^{(1)}(g)D + \beta_i^{(0)}(g).$$

Equating terms of first and zeroth order in D in (16.41) and the right-hand side of (16.37) gives then

$$-\rho_i g_i = \beta_i^{(1)}(g)$$

$$-\sigma_i g_i - \sum_s b_{1i}^{(s)}(g)\rho_i = \beta_i^{(0)}(g) + \sum_{sj} b_{1ij}^{(s)}(g)\beta_j^{(1)}(g),$$

and therefore

$$\beta_i(g, D) = -\rho_i g_i D - \sigma_i g_i - \sum_s b_{1i}^{(s)}(g)\rho_i + \sum_{sj} b_{1ij}^{(s)}(g)\rho_j g_j. \quad (16.42)$$

It is both remarkable and convenient that the β_i-functions are linear in D; the D-dependence arises entirely from the term $-d_i(D)g_i$ in β_i, and the loop contributions in (16.20) are D-independent.

(It should be mentioned in passing that the comparison of pole terms in (16.37) leads to further relations,[39] which determine the $b_{\nu i}^{(s)}$ for $\nu > 1$ in terms of $b_{1i}^{(s)}$. Using (16.42), the right-hand side of (16.37) may be put in the form

$$-\rho_i g_i D - \sigma_i g_i - \sum_s b_{1i}^{(s)}(g)\rho_i$$
$$+\sum_{s,\nu,j} (D-D_s)^{-\nu}[b_{\nu ij}^{(s)}(g)\beta_j(g, D_s) - b_{\nu+1,ij}^{(s)}(g)\rho_j g_j].$$

Equating the pole terms here with those in (16.41), we obtain the recursion relation

$$\rho_i b_{\nu+1,i}^{(s)}(g) - \sum_j \rho_j g_j b_{\nu+1,ij}^{(s)}(g)$$
$$= -(\sigma_i + \rho_i D_s) b_{\nu i}^{(s)}(g) - \sum_j b_{\nu ij}^{(s)}(g)\beta_j(g, D_s) \qquad (16.43)$$

for all $\nu \geq 1$.)

Further useful information about the residue functions can be obtained from dimensional considerations. One of the pecularities of dimensional regularization is that the poles in Green's functions arise only from logarithmic ultraviolet divergences, not from quadratic or higher divergences. But a logarithmic divergence can occur in a given quantity only if the dimensionality of this quantity just equals the dimensionality of the coupling constants to which the quantity is proportional. It follows that $b_i^{(s)}(g)$ can contain a term of order $g_a g_b g_c \cdots$ only if the dimensionality of g_{i0} equals the total dimensionality of $g_{a0} g_{b0} g_{c0} \cdots$ at spacetime dimensionality D_s:

$$d_i(D_s) = d_a(D_s) + d_b(D_s) + \cdots. \qquad (16.44)$$

From (16.42) we see that the same is then true of the D_s terms in β_i. For instance, in a Euclidean scalar field theory with symmetry under the transformation $\phi \to -\phi$ the general Lagrangian is

$$\mathscr{L} = -\frac{1}{2}\partial_\mu\phi\ \partial_\mu\phi - \frac{1}{2}g_{10}\phi^2 - \frac{1}{4!}g_{20}\phi^4 - \frac{1}{6!}g_{30}\phi^6 - \frac{1}{8!}g_{40}\phi^8 \cdots. \qquad (16.45)$$

(We use our freedom to make point transformations with $\delta\phi$ a linear combination of ϕ, ϕ^3, $\Box\phi$, ϕ^5, $\phi^2\Box\phi$, $\Box^2\phi$, etc. to eliminate terms such as $\phi^3\Box\phi$, $(\Box\phi)^2$, $\phi^5\Box\phi$, $\phi^2(\Box\phi)^2$, $(\Box^2\phi)^2$, etc., and to adjust the coefficient of $-\frac{1}{2}\partial_\mu\phi\ \partial_\mu\phi$ to unity.) In four dimensions, the g_{i0} have dimensionalities

$$d_1 = 2, \quad d_2 = 0, \quad d_3 = -2, \quad d_4 = -4, \cdots. \qquad (16.46)$$

Hence if we define the renormalized couplings to eliminate only the poles at $D_s = 4$, the β-functions will have the structure

$$\beta_1 = g_1 F_1 \quad \beta_2 = F_2$$
$$\beta_3 = g_3 F_3 + g_1 g_4 F_{14} + \cdots \qquad (16.47)$$
$$\beta_4 = g_4 F_4 + g_3^2 F_{33} + \cdots$$

where the Fs are functions only of the variables g_2, $g_1 g_3$, $g_1^2 g_4, \ldots$.

This formalism allows a very compact and convenient analysis of the properties of certain fixed points. Let us suppose from now on that the g_i

are defined to eliminate only the poles at a single spacetime dimensionality D_s. The β_i for any non-renormalizable or super-renormalizable coupling must always be proportional to one or more powers of non-renormalizable or super-renormalizable couplings, respectively. Hence a set of couplings $g_i{}^*$ with vanishing values for all interactions that are non-renormalizable or super-renormalizable at D_s will represent a fixed point, provided only that the $\beta_i(g^*)$ corresponding to the renormalizable couplings should vanish. Further, the only terms in $\partial\beta_i/\partial g_j$ which do not vanish at this fixed point are those with $d_i = d_j$, so the matrix B_{ij} is diagonal in the dimensionality of the couplings, and the eigenvalues of this matrix can be obtained by diagonalizing the submatrices connecting couplings of the same dimensionality. For instance, in the theory described by the Lagrangian (16.45), all the β-functions (16.46) will vanish at the point

$$g_1 = 0, \quad g_2 = g_2{}^*, \quad g_3 = 0, \quad g_4 = 0, \ldots \tag{16.48}$$

provided only that $g_2{}^*$ satisfies the condition

$$\beta_2(0, g_2{}^*, 0, \ldots) = 0. \tag{16.49}$$

In addition, the B-matrix for this fixed point is diagonal, with non-vanishing elements

$$B_{11} = F_1(g^*) \quad B_{22} = (\partial F_2/\partial g_2)^* \quad B_{33} = F_3(g^*) \cdots. \tag{16.50}$$

These diagonal elements are then all eigenvalues of the B-matrix. In general, a fixed point of this sort may be found by working in a strictly renormalizable theory (even though the actual theory may not be renormalizable at all) and the eigenvalues of the B-matrix, the 'critical exponents', can be obtained by treating all super-renormalizable and non-renormalizable interactions as first-order perturbations.

It should perhaps be emphasized that even though the definition (16.36) of renormalized couplings gives the β-function a trival dependence on the spacetime dimensionality, D, the critical exponents have precisely the same complicated D-dependence that they would have with any other definition of the renormalized couplings. To illustrate this point, let us return to the Lagrangian (16.45), with $D_s = 4$. For reasons discussed above, we can find a fixed point by working with a truncated, strictly-renormalizable theory

$$\mathscr{L} = -\tfrac{1}{2}\partial_\mu\phi\, \partial_\mu\phi - \tfrac{1}{24}\lambda_0\phi^4. \tag{16.51}$$

Collins[41] has calculated the poles in λ_0 that are needed to cancel the poles in Green's functions as $D\to 4$ from below; to the two-loop order he finds

$$\lambda_0\mu^{D-4}=\lambda+(D-4)^{-1}\left(\frac{-3\lambda^2}{16\pi^2}+\frac{17\lambda^3}{6(16\pi^2)^2}+\cdots\right)$$
$$+(D-4)^{-2}\left(\frac{\lambda^3}{(16\pi^2)^2}+\cdots\right)+\cdots. \tag{16.52}$$

The residue function for $\nu=1$ is thus

$$b_{1\lambda}=\frac{-3\lambda^2}{16\pi^2}+\frac{17}{6}\frac{\lambda^3}{(16\pi^2)^2}+\cdots. \tag{16.53}$$

The dimensionality, $\sigma_\lambda+\rho_\lambda D$, of λ_0 is $4-D$, so $\sigma_\lambda=4$, $\rho_\lambda=-1$, and (16.42) gives the β-function here as

$$\beta_\lambda(\lambda,D)=(D-4)\lambda+b_{1\lambda}-\lambda\;\partial b_{1\lambda}/\partial\lambda \tag{16.54}$$

or, using (16.53),

$$\beta_\lambda(\lambda,D)=(D-4)\lambda+\frac{3\lambda^2}{16\pi^2}-\frac{17}{3}\frac{\lambda^3}{(16\pi^2)^2}+\cdots. \tag{16.55}$$

The fixed point where $\beta_\lambda(\lambda^*,D)$ vanishes can therefore be obtained as a power series in $D-4$:

$$\lambda^*=16\pi^2|\tfrac{1}{3}(4-D)+\tfrac{17}{9}(4-D)^2+\cdots|. \tag{16.56}$$

(This is known as the Wilson–Fisher fixed point.[30] It has a physical sign $\lambda^*>0$ only for $D<4$.) Also, we easily calculate the critical exponent

$$\left(\frac{\partial\beta_\lambda}{\partial\lambda}\right)_{\lambda=\lambda^*}=-(4-D)+\frac{6\lambda^*}{16\pi^2}-\frac{17\lambda^{*2}}{(16\pi^2)^2}+\cdots$$
$$=(4-D)+\tfrac{85}{9}(4-D)^2+\cdots. \tag{16.57}$$

We note that this is positive for at least a finite range of D below $D=4$. All other critical exponents are also positive in a neighborhood of $D=4$, except for the one corresponding to the super-renormalizable coupling $-\frac{1}{2}m_0^2\phi^2$. Collins[41] has also calculated the poles in m_0^2 needed to cancel the singularities at $D=4$ introduced by this coupling; to two-loop order, his result is

$$m_0^2\mu^{-2}=m^2\left[1+(D-4)^{-1}\left(\frac{-\lambda}{16\pi^2}+\frac{5}{12}\frac{\lambda^2}{(16\pi^2)^2}+\cdots\right)\right.$$
$$\left.+(D-4)^{-2}\frac{2\lambda^2}{(16\pi^2)^2}+\cdots\right]. \tag{16.58}$$

Chapter 16. Ultraviolet divergences in quantum gravity

Hence the residue function for $\nu = 1$ and $D_s = 4$ is

$$b_{1,m^2} = m^2\left[\frac{-\lambda}{16\pi^2} + \frac{5}{12}\frac{\lambda^2}{(16\pi^2)^2} + \cdots\right]. \tag{16.59}$$

The dimensionality, $\sigma + \rho D$, of m_0^2 is $+2$, so $\sigma_{m^2} = +2$, $\rho_{m^2} = 0$, and (16.42) gives the β-function for m^2 as

$$\beta_{m^2} = -2m^2 - \lambda\frac{\partial b_{1,m^2}}{\partial\lambda} \tag{16.60}$$

or using (16.59),

$$\beta_{m^2} = m^2\left[-2 + \frac{\lambda}{16\pi^2} - \frac{5}{6}\frac{\lambda^2}{(16\pi^2)^2} + \cdots\right]. \tag{16.61}$$

The corresponding critical exponent is usually denoted $-\nu^{-1}$:

$$-\nu^{-1} \equiv \left(\frac{\partial\beta_{m^2}}{\partial m^2}\right)_{\lambda=\lambda^*, m^2=0} = -2 + \frac{\lambda^*}{16\pi^2} - \frac{5}{6}\frac{\lambda^{*2}}{(16\pi^2)^2} + \cdots \tag{16.62}$$

or, using (16.56),

$$\nu = \tfrac{1}{2} + \tfrac{1}{12}(4-D) + \tfrac{7}{162}(4-D)^2 + \cdots \tag{16.63}$$

in agreement with known results.[42] The fact that only one eigenvalue, $-\nu^{-1}$, of B_{ij} is negative for $D = 4 - \varepsilon$ implies that the Wislon–Fisher fixed point has a one-dimensional ultraviolet critical surface, and that there is just one parameter that need be adjusted to produce a second-order phase transition.

We saw in section 16.3 that the existence of a theory which is renormalizable and asymptotically safe at a spacetime dimensionality D_r indicates the existence of a fixed point near $g^* = 0$ with a finite-dimensional critical surface for at least a finite range of spacetime dimensionalities D above D_r. This can be seen very conveniently by using the method of dimensional continuation described in this section. Suppose for simplicity that the theory which is (strictly) renormalizable at $D = D_r$ has just a single coupling parameter, λ_0, with dimensionality $-(D - D_r)\rho$, where $\rho > 0$. Define the dimensionless renormalized coupling parameter $\lambda(\mu)$ so as to eliminate all poles in reaction rates at $D = D_r$. For the reasons discussed above in this section, we can find a fixed point of the whole theory for any D by setting all couplings equal to zero which would be non-renormalizable or super-renormalizable for $D = D_r$, and looking for a fixed point of the truncated theory. Having done this, the renormalization group equation satisfied by the coupling $\lambda(\mu)$ will be of the form

$$\mu\frac{d}{d\mu}\lambda(\mu)=\beta(\lambda(\mu),D)=[D-D_r]\rho\lambda(\mu)+\beta(\lambda(\mu),D_r). \quad (16.64)$$

The second term on the right arises from loop diagrams, so its power series will generally begin with terms of second order

$$\beta(\lambda(\mu),D_r)=-b\lambda^2(\mu)+O(\lambda^3(\mu)). \quad (16.65)$$

In order for the theory to be asymptotically free when $D=D_r$ (and $\lambda(\mu)>0$) it is necessary that b be *positive*. For $D-D_r$ positive and sufficiently small, there will then be a fixed point

$$\lambda^*=(D-D_r)\rho/b+O((D-D_r)^2). \quad (16.66)$$

All critical exponents are positive, except for the one associated with λ

$$\left(\frac{\partial\beta_\lambda}{\partial\lambda}\right)_{\lambda=\lambda^*}=(D-D_r)\rho-2b\lambda^*+O(\lambda^{*2})$$

$$=-(D-D_r)\rho+O((D-D_r)^2)<0, \quad (16.67)$$

and those associated with any masses or couplings that would be super-renormalizable at $D=D_r$. Thus the ultraviolet critical surface is finite-dimensional, consisting of just those theories which would be renormalizable at $D=D_r$.

This should not be interpreted as a statement that these asymptotically safe theories are renormalizable as usual for $D>D_r$. The method of dimensional regularization is a bit misleading here – it eliminates ultraviolet divergences in *all* theories at non-rational values of the spacetime dimensionality D, at the price of introducing poles at rational values of D. With any other regularization scheme, there is a plethora of ultraviolet divergences at $D>D_r$, and these must be eliminated by including in the Lagrangian all possible interactions allowed by the symmetries of the theory. The occurrence of a fixed point (16.66) with a finite-dimensional critical surface in the dimensional regularization formalism ensures that there *is* a fixed point, with an ultraviolet critical surface of the same dimensionality, in the more conventional renormalization schemes, but the fixed point there will in general have non-vanishing values for *all* couplings, renormalizable *and* non-renormalizable, and the asymptotically safe theories will not even appear to be renormalizable in the usual sense.

Why then should we leave the elegant formalism of dimensional regularization, in which asymptotically safe theories appear so simple?

Chapter 16. Ultraviolet divergences in quantum gravity

The reason is just that we must in the end concern ourselves with a physical spacetime dimensionality $D = 4$ which is greater than D_r, and in continuing from $D = D_r$ up to $D = 4$ we must avoid the poles at intervening rational values of D and at $D = 4$ itself, which would be present with dimensional regularization. The conventional renormalization scheme offers a possible way of carrying out this continuation, at the price of giving up the appearance of renormalizability. However, the first step in assuring ourselves that there is a fixed point with a finite dimensional critical surface for D just above D_r is one that can be most easily accomplished by the method of dimensional regularization.

16.6 Gravity in $2+\varepsilon$ dimensions

At last we come back to gravitation. We want to know whether it is possible to demand that the quantum theory of gravitation is asymptotically safe, and how many free parameters there would be in such a theory. This depends on whether there is a fixed point g^*, and on the dimensionality of its critical surface.

To address this question, we use the technique of dimensional continuation discussed in the previous section. In two dimensions there is a unique, strictly renormalizable theory of pure gravity, based on the Einstein Lagrangian $-\sqrt{g}R/16\pi G$. (The integral $\int d^2x\sqrt{g}R$ is dimensionless in two dimensions, so that G must be dimensionless in order for the action $\int d^2x\mathscr{L}$ to have no dimensions.) The theory remains renormalizable if we add matter fields with a minimal coupling to gravitation, though it may then be necessary to add couplings of matter fields with each other.

Of course, general relativity is not much of a theory in two dimensions. The Lagrangian $\sqrt{g}R$ is a total derivative for $D = 2$, and in consequence the left-hand side of the Einstein field equations, $R_{\mu\nu} - \frac{1}{2}g_{\mu\nu}R$, vanishes identically.[43] This raises a problem in using the method of dimensional regularization. Suppose that when we calculate invariant amplitudes in $2+\varepsilon$ dimensions, we find that some invariant amplitude has a pole at $\varepsilon = 0$, and that in order to cancel this pole we have to add a term to the Lagrangian proportional to $\sqrt{g}R/\varepsilon$. Can we ignore such counterterms, on the grounds that $\int d^2x\sqrt{g}R$ vanishes for $\varepsilon = 0$? The answer appears to be no:[44] if we did not include a counterterm proportional to $\sqrt{g}R/\varepsilon$ where needed to cancel poles in invariant amplitudes at $\varepsilon = 0$, the Green's functions we calculate might be finite at $\varepsilon = 0$, but they would not be analytic functions of ε at $\varepsilon = 0$ in $2+\varepsilon$ dimensions, as assumed in the dimensional-continuation formulation of the renormalization group.

We conclude then that the gravitational coupling constant which appears in the Einstein Lagrangian, $-\sqrt{g}R/16\pi G$, is subject to renormalization in $2+\varepsilon$ dimensions, even for $\varepsilon \to 0$. We will return later to the question of whether G is an *essential* coupling, which cannot be altered by a suitable redefinition of fields.

The unrenormalized gravitational constant, $G_0(\varepsilon)$, in $2+\varepsilon$ dimensions has dimensionality $[\text{mass}]^{-\varepsilon}$, so in the notation of section 16.5, $\rho_G=-1$, $\sigma_G=0$. The singularity structure of $G_0(\varepsilon)$ for $\varepsilon \to 0$ is given here by (16.36) as

$$G_0(\varepsilon)\mu^{\varepsilon} \to G(\mu)+\sum_{\nu=1}^{\infty} \varepsilon^{-\nu} b_\nu(G(\mu)). \tag{16.68}$$

Also (16.38) and (16.42) yield a renormalization group equation for *finite* ε

$$\mu \frac{\mathrm{d}}{\mathrm{d}\mu} G(\mu)=\beta(G(\mu), \varepsilon) \tag{16.69}$$

with

$$\beta(G, \varepsilon)=\varepsilon G+b_1(G)-G b_1'(G). \tag{16.70}$$

For small G, we expect

$$b_1(G)=bG^2+O(G^3) \tag{16.71}$$

so (16.70) gives

$$\beta(G, \varepsilon)=\varepsilon G-bG^2+O(G^3). \tag{16.72}$$

The crucial question is whether b is positive; if so, then there is a fixed point

$$G^*=\varepsilon/b+O(\varepsilon^2) \tag{16.73}$$

and as shown in section 16.5, it has an ultraviolet critical surface of finite dimensionality.

The calculation of b was carried out for a number of special cases by several different groups.[45] Their results up to mid-1977 can be summarized in the statement that the singularities in all purely gravitational Green's functions in $2+\varepsilon$ dimensions at $\varepsilon=0$ are canceled in one-loop order if we suppose that the bare gravitational constant has the pole

$$G_0\mu^{\varepsilon} \to G+bG^2/\varepsilon, \tag{16.74}$$

with

$$b=\tfrac{38}{3}+4N_{\mathrm{V}}-\tfrac{1}{3}N_{\mathrm{F}}-\tfrac{2}{3}N_{\mathrm{S}}. \tag{16.75}$$

Chapter 16. Ultraviolet divergences in quantum gravity

The four terms here arise from one-loop graphs, whose internal lines are respectively either graviton lines or matter lines of spin $1, \frac{1}{2}$, or 0; N_V and N_S are the number of real vector and scalar fields, and N_F is the number of Majorana fermion fields. It therefore appears at first sight that $b > 0$ and hence general relativity is asymptotically safe in $2 + \varepsilon$ dimensions, provided there are enough gauge fields to balance any scalar or fermion fields.

However, before reaching any such conclusion, we have to pay careful attention to the physical interpretation of (16.75). In pure general relativity, the trace of the vacuum Einstein field equations gives $R = 0$ for any spacetime dimensionality, so as explained in section 16.3, the coefficient $1/16\pi G$ of this Lagrangian is not an essential coupling, and there is no reason why it should be required to approach a fixed point for $\mu \to \infty$. The same is true if we add to the theory any number of 'photon' fields with purely gravitational interactions; in this case the Einstein field equations give $\sqrt{g}R$ proportional to $\sqrt{g}\sum F_{\mu\nu}F^{\mu\nu}$, but Maxwell's equations allow us to re-write this as the total derivative $\partial_\mu[\sqrt{g}\sum A_\nu F^{\mu\nu}]$, so here $\partial\mathscr{L}/\partial G$ is a total derivative, and G is again not an essential coupling. (The simplest theory which is renormalizable in two dimensions and which *does* have essential couplings is the Einstein–Yang–Mills[46] theory. In this case, reaction rates depend on the single essential coupling $e^{2\varepsilon}G^{2-\varepsilon}$, where e is the gauge coupling constant. However, this quantity is dimensionless for all ε, and therefore satisfies the trivial renormalization-group equation $\mathrm{d}(e^{2\varepsilon}G^{2-\varepsilon})/\mathrm{d}\mu = 0$.)

Recently a new interpretation of these calculations has been proposed by Gastmans, Kallosh and Truffin[47] (GKT). Their starting point is a reconsideration of the structure of the Lagrangian of general relativity. Gibbons and Hawking[48] have emphasized that in applying the functional formalism to general relativity, we do not actually use the Einstein Lagrangian $-\sqrt{g}R/16\pi G_0$, but rather

$$\mathscr{L}_G = -\frac{1}{16\pi G_0}[\sqrt{g}R - \Phi], \tag{16.76}$$

with Φ a total derivative designed so that $\mathscr{L}_G$ is a function only of $g_{\mu\nu}$ and its *first* derivatives

$$\Phi = \frac{\partial}{\partial X^\mu}\left\{\frac{\sqrt{g}}{2}\left[g^{\lambda\nu}g^{\mu\kappa}\frac{\partial g_{\lambda\nu}}{\partial X^\kappa} - g^{\lambda\nu}g^{\mu\kappa}\frac{\partial g_{\kappa\nu}}{\partial X^\lambda} - g^{\lambda\nu}g^{\mu\kappa}\frac{\partial g_{\lambda\kappa}}{\partial X^\nu} + g^{\lambda\mu}g^{\nu\kappa}\frac{\partial g_{\nu\kappa}}{\partial X^\lambda}\right]\right\}. \tag{16.77}$$

Of course, adding a total derivative has no effect if we restrict our attention to metrics which vanish sufficiently fast for $|X|\to\infty$, but in the functional formulation of quantum gravity we must sum over *all* metrics in Euclidean spacetime, and Φ makes an important contribution to the action for some of these metrics, such as the Euclidean Schwarzschild metric.

Now, when we calculate one-loop graphs in $2+\varepsilon$ dimensions, we must expect to find $1/\varepsilon$ poles as $\varepsilon\to 0$ which require independent counterterms proportional both to $\sqrt{g}R$ and to Φ. GKT argue that the Lagrangian should therefore then be written as the sum of two independent terms

$$\mathscr{L}_G = -\frac{1}{16\pi G_0}[\sqrt{g}R-\Phi]-\frac{1}{16\pi F_0}\sqrt{g}R. \tag{16.78}$$

By using the trace of the Einstein field equations, we can express $\sqrt{g}R$ in terms of matter fields, so its coefficient is not an independent essential coupling. However, the gravitational coupling, G_0, appears in the coefficient of a different term $\sqrt{g}R-\Phi$, which is *not* given by the field equations in terms of matter fields, so G_0 is an independent essential coupling. In other words, the earlier calculations[45] gave the counterterms in $1/G_0+1/F_0$ correctly, but the physically interesting essential coupling is $1/G_0$, and its counterterms must be calculated anew.

GKT have calculated the poles in G_0 required to cancel the $1/\varepsilon$ poles in gravitational Green's functions in a theory with N_S and N_V real scalar and vector fields and N_F and N_Δ Majorana fields of spin $\frac{1}{2}$ and $\frac{3}{2}$. Their results can be summarized in the statement that G_0 must have a pole (16.74), with b now given by

$$b=\tfrac{2}{3}[1+\tfrac{15}{2}N_\Delta-N_F-N_S], \tag{16.79}$$

the first term arising from graviton loops. Very recently, Christensen and Duff[49] (CD) have carried out a calculation along the same lines, and find a formula for b with different fermion contributions:

$$b=\tfrac{2}{3}[1-N_\Delta+N_F-N_S]. \tag{16.80}$$

In either case, there is an asymptotically safe theory of pure gravity in $2+\varepsilon$ dimensions, with a one-dimensional critical surface. Asymptotic safety is also preserved when we add matter fields, provided we add fields of spin $\frac{3}{2}$ (GKT) or $\frac{1}{2}$ (CD) to balance the contributions of fields of spin zero and $\frac{1}{2}$ (GKT) or $\frac{3}{2}$ (CD), and providing also that the couplings of the matter fields with themselves do not raise problems.

It may be noted that (16.79) or (16.80) and (16.75) give the same result for the contribution of scalar particles to b, so in this case the earlier calculations[45] did actually give the counterterms proportional to $\sqrt{g}R-\Phi$, not $\sqrt{g}R$. According to both GKT and CD, the new feature introduced by the distinction between $\sqrt{g}R-\Phi$ and $\sqrt{g}R$ is that the contribution of particles of arbitrary spin to b is simply proportional to the number of degrees of freedom of their fields, but (according to CD) with an extra minus sign for fermions. (The factor $\frac{15}{2}$ in (16.79) is puzzling here.) For instance, a symmetric traceless tensor field $h_{\mu\nu}$ has $\frac{1}{2}D(D+1)-1$ independent components, of which D are eliminated by the gauge condition which specifies $\partial_\mu h^\mu{}_\nu$, and $D-1$ are eliminated by our freedom to make further gauge transformations $\delta h_{\mu\nu}=\partial_\mu\phi_\nu+\partial_\nu\phi_\mu$ with $\partial_\mu\phi^\mu=0$ and $\Box\phi^\mu=0$; hence the number of degrees of freedom of the gravitational field in D dimensions is

$$\tfrac{1}{2}D(D+1)-1-D-(D-1)=\tfrac{1}{2}D(D-3).$$

This is -1 for $D=2$, so the contribution of the graviton to b should be equal and opposite to the contribution of a single spinless particle. On the other hand, a vector field A_μ has D components, of which one is eliminated by the gauge condition which specifies $\partial_\mu A^\mu$, and one is eliminated by our freedom to make further gauge transformations $\delta A_\mu=\partial_\mu\phi$ with $\Box\phi=0$; hence the number of degrees of freedom of the photon field in D dimensions is $D-2$. This vanishes for $D=2$, so photons make no contributions to b.

Since the evaluation of the contribution of scalar fields to b does not apparently raise problems of distinguishing between $\sqrt{g}R$ and $\sqrt{g}R-\phi$, and since this contribution sets the scale for the contributions of particles of nonzero spin, it may be of interest to see one more calculation of this quantity. A calculation based on the methods of reference 15 is presented in the Appendix.

It is amusing to apply these results to extended supergravity theories,[10] in which the graviton appears in a multiplet with fields of lower spin. In a four-dimensional theory with $n\leqslant 7$ supersymmetry generators, the helicity $+2$ graviton will appear in a multiplet with $\begin{bmatrix} n \\ r \end{bmatrix}$ massless particles of helicity $2-r/2$, with $r=1,2,\ldots n$, and there is a separate multiplet containing the helicity -2 graviton and $\begin{bmatrix} n \\ r \end{bmatrix}$ massless particles of helicity $-2+r/2$. For $n=8$, there is a single multiplet containing gravitons of helicity ± 2, and $\begin{bmatrix} 8 \\ r \end{bmatrix}$ massless particles of helicity $2-r/2$. We suppose that for arbitrary spacetime dimensionality, the number of fields of a given spin S is equal to the number of fields of that spin in four dimensions, and hence equal to the number of states in four dimensions with helicity $+S$ (or $-S$, but not both), even though the supersymmetry is actually present only for four dimensions. The numbers N_Δ, N_V, N_F, N_S of fields of spin $\frac{3}{2}$, $1, \frac{1}{2}, 0$ in $O(n)$-extended supergravity are given in Table 16.1, along with values of b obtained from (16.79) or (16.80): According to GKT's results, $b>0$ for pure supergravity ($n=1$), and also for $0(n)$-extended supergravity with $n \leqslant 5$. However, if CD are correct, then a two-loop calculation is needed to settle the question of asymptotic safety in pure supergravity, although we will always have $b>0$ for $n=1$ if we add enough 'vector' supermultiplets with spins 1 and $\frac{1}{2}$. According to CD's results, it appears to be impossible to have $b>0$ in $0(n)$-extended supergravity with $n \geqslant 2$, whether or not we add additional matter supermultiplets.

Table 16.1. *Numbers of field types and values of b in extended supersymmetry theories with n supersymmetry generators*

n	N_Δ	N_V	N_F	N_S	$3b$ (GKT)	$3b$ (CD)
0	0	0	0	0	2	2
1	1	0	0	0	17	0
2	2	1	0	0	32	−2
3	3	3	1	0	45	−2
4	4	6	4	2	50	−2
5	5	10	11	10	35	−6
6	6	16	26	30	−20	−18
7	8	28	56	70	−130	−42
8	8	28	56	70	−130	−42

The really important question here is that of continuation to four dimensions. In this respect, the dimensional regularization formalism in $2+\varepsilon$ dimensions may be somewhat misleading. It is true that the asymptotically safe theory of gravitation in this formalism is based on Lagrangians that must be renormalizable in two dimensions, and these Lagrangians do not contain any counterterms that would cancel the poles in Feynman diagrams at $D=4$ spacetime dimensions. However, the presence of these poles indicates that the perturbation expansion in powers of ε will have broken down long before we reach $\varepsilon=2$.

Chapter 16. Ultraviolet divergences in quantum gravity

A better view of the possibilities of continuation from $2+\varepsilon$ to 4 dimensions may be provided by the conventional renormalization scheme with which we started in section 16.3. In this formalism, integrals are regulated with some sort of large ultraviolet cut-off Λ, and the cut-off dependence is removed for $\Lambda \to \infty$ by cancellation with the counterterms provided by the unrenormalized coupling constants. There are an infinite number of counterterms required in $D > 2$ dimensions, and such a theory of gravitation cannot be said to be renormalizable in the usual sense, but there are also no new singularities encountered when D approaches 4. The one result of the dimensional regularization method that can be taken over directly into the conventional renormalization scheme is that in $2+\varepsilon$ dimensions for sufficiently small ε there exists a fixed point with an ultraviolet critical surface of finite dimensionality.

Acknowledgements

It is a pleasure to thank S. Coleman, S. Deser, and M. J. Duff for their frequent and valuable help in the preparation of this report. I am also grateful to L. Brown, S. Christensen, J. C. Collins, B. DeWitt, E. S. Fradkin, G. 't Hooft, B. Lee, P. Martin, D. Nelson, A. Salam, H. Schnitzer, L. Smollin, H.-S. Tsao, P. van Nieuwenhuizen, K. Wilson, E. Witten and B. Zumino for informative conversations on various special topics.

16.7 Appendix. Calculation of *b*

This appendix will present a calculation of the residue of the pole in the bare gravitational constant at $D=2$, using the method of reference 15.

We introduce a gravitational field $h_{\mu\nu}$ with

$$g_{\mu\nu} = \eta_{\mu\nu} + (32\pi G)^{1/2} h_{\mu\nu} \tag{A.1}$$

and work in a gauge with

$$\partial_\mu h^{\mu\nu} = 0. \tag{A.2}$$

(Indices here are raised and lowered with $\eta_{\mu\nu}$ not $g_{\mu\nu}$.) The bare graviton propagator in D spacetime dimensions is then

$$\langle T\{h_{\mu\nu}(x), h_{\lambda\rho}(0)\}\rangle_0 = \int \frac{d^D q}{(2\pi)^D} e^{iq\cdot x} \Delta_{\mu\nu,\lambda\rho}(q) \tag{A.3}$$

Appendix

$$\Delta_{\mu\nu,\lambda\rho}(q)=\frac{1}{2q^2}\left[L_{\mu\rho}(q)L_{\nu\lambda}(q)+L_{\mu\lambda}(q)L_{\nu\rho}(q)-\frac{2}{D-2}L_{\mu\nu}(q)L_{\lambda\rho}(q)\right] \tag{A.4}$$

$$L_{\mu\nu}(q)\equiv\eta_{\mu\nu}-q_\mu q_\nu/q^2. \tag{A.5}$$

In one-loop order, the propagator becomes

$$\Delta'_{\mu\nu,\lambda\rho}(q)=\Delta_{\mu\nu,\lambda\rho}(q)+\Delta_{\mu\nu,\mu'\nu'}(q)\Pi^{\mu'\nu',\lambda'\rho'}(q)\Delta_{\lambda'\rho',\lambda\rho}(q), \tag{A.6}$$

where Π is the graviton vacuum polarization tensor, defined by

$$\langle T\{T^{\mu\nu}(x),\,T^{\lambda\rho}(0)\}\rangle_0=\frac{-i}{8\pi G}\int\frac{d^Dq}{(2\pi)^D}\,e^{iq\cdot x}\Pi^{\mu\nu,\lambda\rho}(q). \tag{A.7}$$

Since the energy–momentum tensor is conserved, Π can be written

$$\Pi^{\mu\nu,\lambda\rho}(q)=(q^2)^2A(q^2)L^{\mu\nu}(q)L^{\lambda\rho}(q)$$
$$-q^2B(q^2)[L^{\mu\lambda}(q)L^{\nu\rho}(q)+L^{\mu\rho}(q)L^{\nu\lambda}(q)-2L^{\mu\nu}(q)L^{\lambda\rho}(q)] \tag{A.8}$$

with $A(q^2)$ and $B(q^2)$ free of poles at $q^2=0$. It is straightforward then to calculate the corrected propagator as

$$\Delta'^{\mu\nu,\lambda\rho}(q)=\frac{1}{2q^2}(1-2B(q^2))\Big[L^{\mu\rho}(q)L^{\mu\lambda}(q)$$
$$+L^{\mu\lambda}(q)L^{\nu\rho}(q)-\frac{2}{D-2}L^{\mu\nu}(q)L^{\lambda\rho}(q)\Big]$$
$$+\frac{A(q^2)}{(D-2)^2}L^{\mu\nu}(q)L^{\lambda\rho}(q). \tag{A.9}$$

We see that the renormalized gravitational constant which measures the strength of long-range graviton exchange is

$$G=G_0(1-2B(0)), \tag{A.10}$$

where G_0 is the bare gravitational constant. Hence if $B(0)$ has a pole at $D=2$ of the form

$$B(0,D)\to\frac{bG}{2}\left(\frac{1}{D-2}\right) \tag{A.11}$$

we must introduce a pole in G_0 of the form

$$\mu^{D-2}G_0\xrightarrow[D\to 2]{}G+bG^2\left(\frac{1}{D-2}\right). \tag{A.12}$$

Chapter 16. Ultraviolet divergences in quantum gravity

Thus the quantity b in (A.11) is the same as the b in (16.71)–(16.73). Our task is to calculate the residue of the pole in $B(q^2, D)$ at $D=2$ and determine b by comparison with (A.11).

For this purpose, we introduce the spectral functions ρ_A, ρ_B by the formula

$$\sum_n \delta^D(p_n - p)\langle 0|T^{\mu\nu}(0)|n\rangle\langle 0|T^{\rho\sigma}(0)|n\rangle^*$$
$$= (2\pi)^{-D+1}\theta(p^0)[\rho_A(-p^2)(p^2)^2 L^{\mu\nu}(p)L^{\rho\sigma}(p)$$
$$-p^2\rho_B(-p^2)\{L^{\mu\rho}(p)L^{\nu\sigma}(p)+L^{\mu\sigma}(p)L^{\nu\rho}(p)-2L^{\mu\nu}(p)L^{\mu\rho}(p)\}]. \tag{A.13}$$

Apart from possible subtractions, (A.7), (A.8), and (A.13) give

$$A(q^2) = 8\pi G\int_0^\infty [q^2+\mu^2-i\varepsilon]^{-1}\rho_A(\mu^2)\,d\mu^2 \tag{A.14}$$

$$B(q^2) = 8\pi G\int_0^\infty [q^2+\mu^2-i\varepsilon]^{-1}\rho_B(\mu^2)\,d\mu^2. \tag{A.15}$$

Now let us consider the contribution of a state consisting of a pair of identical neutral spinless particles of mass m and momenta $\vec{k}$, $\vec{k}'$ to ρ_A and ρ_B. In lowest-order perturbation theory, for any D,[50]

$$\langle 0|T^{\mu\nu}(0)|\vec{k}, \vec{k}'\rangle$$
$$= -(2\pi)^{-(D-1)}(2\omega)^{-1/2}(2\omega')^{-1/2}[k^\mu k'^\nu + k'^\mu k^\nu + \eta^{\mu\nu}(-k^\lambda k'_\lambda + m^2)$$
$$-\frac{(D-2)}{2(D-1)}\{(k+k')^\mu(k+k')^\nu - (k+k')^2\eta^{\mu\nu}\}], \tag{A.16}$$

where

$$k^0 = \omega = (\vec{k}^2 + m^2)^{1/2} \quad k^{0\prime} = \omega' = (\vec{k}'^2 + m^2)^{1/2}.$$

Equation (A.13) then gives

$$\rho_A(\mu^2) = \tfrac{1}{8}(2\pi)^{-D+1}\Omega_D k^{D-3}\mu^{-5}\left[\frac{12k^4}{D^2-1} - \frac{2k^2\mu^2}{(D-1)^2} + \frac{\mu^4}{4(D-1)^2}\right] \tag{A.17}$$

$$\rho_B(\mu^2) = \tfrac{1}{2}(2\pi)^{-D+1}\Omega_D k^{D+1}\mu^{-3}/(D^2-1), \tag{A.18}$$

where Ω_D is the surface area of a unit sphere in $D-1$ spatial dimensions,

$$\Omega_D = 2\pi^{(D-1)/2}/\Gamma\left(\frac{D-1}{2}\right),$$

Appendix

and

$$k \equiv \left(\frac{\mu^2}{4} - m^2\right)^{1/2}.$$

Now we can calculate b. The function $B(q^2, D)$ is given by (A.15) and (A.18) as

$$B(q^2, D) = \frac{4\pi G \Omega_D}{(D^2-1)(2\pi)^{D-1}} \int_{4m^2}^{\infty} \frac{k^{D+1}\, d\mu^2}{\mu^3(\mu^2+q^2)}. \quad \text{(A.19)}$$

The integral is well defined for $D < 2$, and can be analytically continued to $D > 2$, with a pole at $D = 2$

$$B(q^2, D) \underset{D\to 2}{\longrightarrow} \frac{G}{3}\left(\frac{1}{2-D}\right). \quad \text{(A.20)}$$

Comparing with (A.11), we see that

$$b = -2/3 \quad \text{(A.21)}$$

in agreement with (6.79) and (6.80) for $N_s = 1$.

This method of calculation has the advantage of allowing us to draw general conclusions about the sign and other properties of spectral functions and pole residues at various dimensionalities for intermediate states of arbitrary spin. We take p in (A.13) to lie in the 'time' direction $p = (0, \ldots, 0, \mu)$, and contract with $a_\mu b_\nu a_\rho b_\sigma$, where a and b are purely spatial vectors separated by an angle ϕ; this gives

$$(1+\tan\phi)^2 \rho_A(\mu^2)\mu^4 - 4\tan\phi\, \rho_B(\mu^2)\mu^2 \geqslant 0 \quad \text{(A.22)}$$

for all ϕ and μ. For $D = 2$ we must of course take $\phi = 0$, so this gives only the condition

$$\rho_A(\mu^2) \geqslant 0. \quad \text{(A.23)}$$

For integer dimensionalities $D \geqslant 3$, we can pick ϕ freely; by choosing it to minimize the left-hand side of (A.22), we find

$$0 \leqslant \rho_B(\mu^2) \leq \rho_A(\mu^2)\mu^2. \quad \text{(A.24)}$$

Finally, for a traceless energy–momentum tensor, (A.13) gives

$$0 = \rho_A(\mu^2)\mu^2(D-1) + \rho_B(\mu^2)(4-2D). \quad \text{(A.25)}$$

In particular, $\rho_A(\mu^2) = 0$ for $D = 2$. Even if finite masses give the energy–momentum tensor a non-vanishing trace, (A.25) will be asymptotically valid for $\mu \to \infty$, so the integral (A.14) for $A(q^2)$ will not have a pole at $D = 2$.

Chapter 16. References

Chapter 16 Ultraviolet divergences in quantum gravity

(1) Long before the modern development of renormalization theory, Heisenberg proposed a classification of elementary-particle interactions into those with dimensionless couplings and those whose couplings have the dimensions of a negative power of mass, and he suggested that the mass scale which enters in the latter couplings would set a bound to the applicability of existing theories (see W. Heisenberg (1938) *Z. Physik*, **110**, 251). He also noted that in Fermi's theory of beta decay, the coupling constant had the dimensions of $[\text{mass}]^{-2}$, and he suggested that dynamical effects might be associated with energies of order $G_F^{-1/2}$, as for instance in cosmic ray showers (see W. Heisenberg (1938) *Z. Physik*, **101**, 251; (1939) *Z. Physik*, **113**, 61). After the development of renormalization theory, it was noted that non-renormalizable theories are in general just those whose couplings have the dimensionality of negative powers of mass, and that reaction rates would grow rapidly with energy in such theories (see S. Sakata, H. Umezawa and S. Kamefuchi (1952) *Prog. Theor. Phys.*, **7**, 327).

(2) R. P. Feynman (1963) *Acta Phys. Polon.*, **24**, 697; B. S. DeWitt (1967) *Phys. Rev.*, **162**, 1195, 1239 (erratum (1968) *Phys. Rev.*, **171**, 1834); L. D. Faddeev and V. N. Popov (1967) *Phys. Lett.*, **B25** (1967) 29; S. Mandelstam (1968) *Phys. Rev.*, **175**, 1604; E. S. Fradkin and I. V. Tyutin (1970), *Phys. Rev.*, **2**, 2841. For a derivation of covariant rules from the canonical formalism see E. S. Fradkin and G. A. Vilkovsky (1975) *Phys. Lett.*, **55B**, 224; *Nuovo Cimento*, **13**, 187. For a review see M. J. Duff (1975) in

Chapter 16. References

Quantum Gravity, eds. C. J. Isham, R. Penrose and D. W. Sciama (Oxford University Press).

(3) G. 't Hooft (1973) *Nucl. Phys.*, **B62**, 444; G. 't Hooft and M. Veltman (1974) *Ann. Inst. Poincaré*, **20**, 69; S. Deser and P. van Nieuwenhuizen (1974) *Phys. Rev. Lett.*, **32**, 245; S. Deser and P. van Nieuwenhuizen (1974) *Phys. Rev.*, **D10**, 401, 411; S. Deser, H-S Tsao and P. van Nieuwenhuizen (1974) *Phys. Lett.*, **50B**, 491. Infrared divergences pose no problem here; see S. Weinberg (1965) *Phys. Rev.*, **140**, B516.

(4) The need for counterterms proportional to $R_{\mu\nu}R^{\mu\nu}$ and R^2 was pointed out very early by R. Utiyama and B. S. DeWitt (1962) *J. Math. Phys.*, **3**, 608.

(5) H. Georgi, H. Quinn and S. Weinberg (1974). *Phys. Rev. Lett.*, **33**, 451.

(6) The Planck mass is given (for $\hbar = c = 1$) in terms of the Newton Constant G, as $G^{-1/2} = 1.2 \times 10^{19}$ GeV. See M. Planck (1899) *Sitz. Deut. Akad. Wiss.* (Berlin), 440.

(7) In the notation used in this article, $\hbar = c = 1$; the flat-space metric $\eta_{\mu\nu}$ has diagonal elements +, +, +,−; and sign conventions for $R, R_{\mu\nu}$, etc. are the same as in S. Weinberg (1972) *Gravitation and Cosmology – Principles and Applications of the General Theory of Relativity* (Wiley: New York).

(8) The renormalization group idea was introduced in its modern form into elementary particle physics by M. Gell-Mann and F. E. Low (1954) *Phys. Rev.*, **95**, 1300. Similar concepts were also discussed by E. C. G. Stueckelberg and A. Petermann (1953) *Helv. Phys. Acta*, **26**, 499. For surveys of the applications to the theory of critical phenomena see the following reviews: K. G. Wilson and J. Kogut (1974) *Phys. Rep.*, **12C**, No. 2; M. E. Fisher (1974) *Rev. Mod. Phys.*, **46**, 597; E. Brézin, J. C. LeGuillou and J. Zinn-Justin (1975) in *Phase Transitions and Critical Phenomena*, eds. C. Domb and M. S. Green (Academic Press: London, New York); F. J. Wegner (1975) in *Trends in Elementary Particle Theory*, p. 171 (Springer-Verlag, Berlin); K. Wilson (1975) *Rev. Mod. Phys.*, **47**, 773; Shang-Keng Ma (1976) *Modern Theory of Critical Phenomena* (W. A. Benjamin, New York).

(9) Supersymmetry was introduced by J. Wess and B. Zumino (1974) *Nucl. Phys.*, **B70**, 34; *Nucl. Phys.*, **B78**, 1; *Phys. Lett.*, **49B**, 52. Similar ideas had also been explored by Yu. A. Gol'fand and E. P. Likhtman (1971) *Sov. Phys.: JETP Lett.*, **13**, 323; D. V. Volkov and V. P. Akulov (1973) *Phys. Lett.*, **46B**, 109. A 'superspace' formulation was given by A. Salam and J. Strathdee (1974) *Nucl. Phys.*, **B76**, 477. For a review, see P. Fayet and S. Ferrara (1977) *Phys. Rep.*, **32C**, 249.

(10) 'Supergravity' is the supersymmetric theory of the multiplet consisting of the graviton plus a massless Majorana particle of spin 3/2. It was introduced by D. Z. Freedman, P. van Nieuwenhuizen and S. Ferrara (1976) *Phys. Rev.*, **D13**, 3214; S. Deser and B. Zumino (1976) *Phys. Lett.*, **62B**, 335. For reviews with discussions of 'extended supergravity' theories see B. Zumino (1977) CERN preprint; S. Deser (1978) Brandeis preprint; D. Z. Freedman and P. van Nieuwenhuizen (1978) *Rev. Mod. Phys.*, to be published. For supergravity theories based on a generalization of the Lagrangian $\sqrt{g}R^2$, see M. Kaku, P. K. Townsend and P. van Nieuwenhuizen (1977) *Phys. Rev. Lett.*, **39**, 1109; S. Ferrara and P. van Nieuwenhuizen (1978) Ecole Normale Superieur preprint 78/14.

Chapter 16. References

1) The cancellation of some divergences in globally supersymmetric field theories of matter was noted by B. W. Lee (1974) (unpublished; cited in Wess and Zumino, *infra*); J. Wess and B. Zumino (1974) *Phys. Lett.*, **49B**, 52; J. Iliopoulos and B. Zumino (1974) *Nucl. Phys.*, **B76**, 310 (1974); S. Ferrara, J. Iliopoulos and B. Zumino (1974) *Nucl. Phys.*, **B77**, 413. The cancellation of divergences in gravitational Green's functions in a theory of supersymmetric matter fields was pointed out by B. Zumino (1975) *Nucl. Phys.*, **B89**, 535. This led to the conjecture that a supersymmetric theory of gravitation could be renormalizable; see B. Zumino (1974) in *Proceedings of the XVIII International Conference on High Energy Physics*, ed. J. R. Smith (Rutherford Laboratory, Chilton, Didcot, Oxfordshire). Cancellation of one-loop divergences in pure supergravity theory were noted by D. Z. Freedman, P. van Nieuwenhuizen and S. Ferrara (1976) *Phys. Rev.* **D13**, 3214; *Phys. Rev.*, **D14**, 912; S. Deser and B. Zumino (1976) *Phys. Lett.*, **62B**, 335; M. T. Grisaru, P. van Nieuwenhuizen and J. A. M. Vermaseren (1976) *Phys. Rev. Lett.*, **37**, 1662; S. Deser, J. H. Kay and K. S. Stelle (1977) *Phys. Rev. Lett.*, **38**, 527. One-loop divergences in theories with supergravity coupled to matter were found in explicit calculations by P. van Nieuwenhuizen and J. A. M. Vermaseren (1976) *Phys. Lett.*, **65B**, 263, and were explained in general terms by Deser, Kay and Stelle (1977) *Phys. Rev. Lett.*, **38**, 527. However, the finiteness of one-loop graphs for $O(n)$ extended supergravity has been shown by Grisaru, van Nieuwenhuizen and Vermaseren (1976) *Phys. Rev. Lett.*, **37**, 1662, for $n=2$; by P. van Nieuwenhuizen and J. A. M. Vermaseren (1977) *Phys. Rev.*, **D16**, 298, for $n=3$ and $n=4$; and by M. Fischler and P. van Nieuwenhuizen (1978) to be published, for $n=8$. In two-loop order, pure supergravity was shown to be free of ultraviolet divergences by M. T. Grisaru (1977) *Phys. Lett.*, **66B**, 75; E. Tomboulis (1977) *Phys. Lett.*, **67B**, 417; Deser, Kay, and Stelle (1977) *Phys. Rev. Lett.* (The same was found to be true in one- and two-loop order for $O(n)$-extended supergravity with n supersymmetry generators.) However, it was pointed out by Deser, Kay, and Stelle (1977), *op. cit.*, and Ferrara and Zumino (1978), CERN preprint, that there are possible ultraviolet divergences in pure supergravity in three-loop order which cannot yet be ruled out by symmetry arguments. Similar problems have been found in three-loop order in $O(2)$-extended supergravity; S. Deser and J. H. Kay (1978), private communication. The possible ultraviolet divergences in supergravity have been analyzed for all orders of perturbation theory, by S. Ferrara and P. van Nieuwenhuizen (1978), Ecole Normale Superieur preprint 78/14. Possible divergences are found in every order beyond two loops, but it is not yet known whether any of their coefficients are non-zero.

2) Various methods of resummation in general relativity or analogous theories have been studied by B. S. DeWitt (1964) *Phys. Rev. Lett.*, **13**, 114; I. B. Khriplovich (1966) *Yadernaya Fizika*, **3**, 575 (English translation in *Sov. J. Nucl. Phys.*, **3**, 415); A. Salam and J. Strathdee (1970) *Nuovo Cimento Lett.*, **4**, 101; C. J. Isham, A. Salam and J. Strathdee (1971) *Phys. Rev.*, **D3**, 867; (1972) *Phys. Rev.*, **D5**, 2548; A. Salam (1963), *Phys. Rev.*, **130B**, 1287; J. Strathdee (1964) *Phys. Rev.*, **135B**, 1428; A. Salam and R. Delbourgo (1964) *Phys. Rev.*, **135B**, 1398; R. Delbourgo (1977), University of Tasmania preprint; R. Delbourgo and P. West (1977) *J. Phys.*, **A10**, 1049.

Chapter 16. References

(13) Renormalizable theories of gravitation based on Lagrangians which include $\sqrt{g}R^2$ and $\sqrt{g}R^{\mu}{}_{\nu}R^{\nu}{}_{\mu}$ terms were proposed by S. Deser (1975) in *Proceedings of the Conference on Gauge Theories and Modern Field Theory*, eds. R. Arnowitt and P. Nath (MIT Press, Cambridge, Massachusetts); S. Weinberg (1974) in *Proceedings of the XVII International Conference on High Energy Physics*, ed. J. R. Smith (Rutherford Laboratory, Chilton, Didcot, Oxfordshire), III-59. A general study of such theories has been carried out by K. S. Stelle (1977) *Phys. Rev.* **D16**, 953.
(14) A resummation of the theory which might eliminate this pole has been suggested by A. Salam and J. Strathdee (1978) Trieste preprint. Similar idea: have been pursued by S. Deser (1978), private communication.
(15) G. Källén (1952) *Helv. Phys. Acta*, **25**, 417; H. Lehmann (1954) *Nuovo Cimento*, **11**, 342. (Subtractions might be needed in the dispersion relation for the propagator, but this would make its asymptotic behavior even worse than (16.3).)
(16) The 'large *N*' approximation was developed in statistical physics by H. E. Stanley (1968) *Phys. Rev.*, **176**, 718; E. Brézin and D. J. Wallace (1973), *Phys. Rev.*, **B7**, 1967; K. G. Wilson (1973) *Phys. Rev.*, **D7**, 2911; L. Dolan and R. Jackiw (1974) *Phys. Rev.*, **D9**, 3320; and in relativistic quantum field theory by R. Abe (1972) *Prog. Theor. Phys.*, **48**, 1414; G. Parisi and L. Peliti (1972) *Phys. Lett.*, **41A**, 331; M. Suzuki (1972) *Phys. Lett.*, **54A**, 5; R. A. Ferrel and D. J. Scalapino (1972) *Phys. Rev. Lett.*, **29**, 413; S. Ma (1972) *Phys. Rev. Lett.*, **29**, 1361. Unphysical 'tachyon' poles were found in some theories by S. Coleman, R. Jackiw, and H. D. Politzer (1974) *Phys. Rev.*, **D10**, 2491; D. J. Gross and A. Neveu (1974) *Phys. Rev.*, **D10**, 3235. However, it was pointed out that these poles are absent in the stable solutions of the theory, by L. F. Abbott, J. S. Kang and H. J. Schnitze (1976) *Phys. Rev.*, **D13**, 2212. Fixed points in non-renormalizable field theories were studied in the large *N* approximation by G. Parisi (1975) *Nucl Phys.*, **B100**, 368.
(17) E. Tomboulis (1977) Princeton University preprint.
(18) T. D. Lee and G. C. Wick (1969) *Nucl. Phys.*, **B9**, 209; *Nucl. Phys.* **B10**, 1; (1970) *Phys. Rev.*, **D2**, 1033.
(19) S. Weinberg (1964) *Phys. Lett.*, **9**, 357; (1964) *Phys. Rev.*, **B135**, 1049; (1965) *Phys. Rev.*, **B138**, 988; and (1965) in *Lectures on Particles and Field Theory*, eds. S. Deser and K. Ford, p. 988 (Prentice-Hall: New Jersey). The program of deriving classical general relativity from quantum mechanics and special relativity was completed by D. Boulware and S. Deser (1975) *Ann. Phys.*, **89**, 173. I understand that similar ideas were developed by R. Feynman in unpublished lectures at Cal Tech.
(20) Einstein derived his field equations as the unique generally covariant equations in which each term (on the left) would involve just two derivatives of the metric. General covariance is just a convenient way of implementing the Principle of Equivalence, and the limitation to two derivatives picks out those terms in the most general generally covariant field equations which are relevant at long distances. For a discussion of the Einstein field equations along these lines, see S. Weinberg (1972) *Gravitation and Cosmology – Principles and Applications of the General Theory of Relativity*, Section 7.1 (Wiley: New York).

Chapter 16. Reference

(21) For reviews, see S. Weinberg (1968) *Proceedings of the XIV International Conference on High Energy Physics* (CERN, Geneva), 253; S. Weinberg (1970) in *Lectures on Elementary Particles and Quantum Field Theory – 1970 Brandeis Summer Institute in Theoretical Physics*, eds. S. Deser, M. Grisaru and H. Pendleton (MIT Press: Cambridge, Mass.); B. W. Lee (1972) *Chiral Dynamics* (Gordon and Breach: New York).

(22) S. Weinberg (1967) *Phys. Rev. Lett.*, **18**, 507; J. Schwinger (1967) *Phys. Lett.*, **24B**, 473; S. Weinberg (1968) *Phys. Rev.*, **166**, 1568; S. Coleman, J. Wess, and B. Zumino (1968), *Phys. Rev.* **177**, 2239; C. Callan, S. Coleman, J. Wess, and B. Zumino (1968), *Phys. Rev.*, **177**, 2247.

(23) Photon pairing instabilities have been proposed as a possible origin for gravitation by S. L. Adler, J. Lieberman, Y. J. Ng and H.-S. Tsao (1976) *Phys. Rev.*, **D14**, 359; S. L. Adler (1976) *Phys. Rev.*, **D14**, 379. For other theories with composite gravitons, see P. R. Phillips (1966) *Phys. Rev.*, **146**, 966; A. D. Sakharov (1967) *Dokl. Akad. Nauk. SSSR*, **177**, 70; H. C. Ohanian (1969) *Phys. Rev.*, **184**, 1305; H. P. Dürr (1973) *Gen. Relativ. Grav.*, **4**, 29; D. Atkatz (1977) Stony Brook preprint ITP–SB–77–59; H. Teragawa, Y. Chikashige, K. Akama and T. Matsuki (1977) *Phys. Rev.*, **D15**, 1181; T. Matsuki (1978) *Prog. Theor. Phys.*, **59**, 235; K. Akama, Y. Chikashige and T. Matsuki (1978) *Prog. Theor. Phys.*, **59**, 653; K. Akama, Y. Chikashige, T. Matsuki and H. Terazawa (1977), INS–Report–304; K. Akama (1978), Saitama preprint.

(24) The condition that $m(\mu)/\mu$ should vanish for $\mu \to \infty$ is discussed by S. Weinberg (1973) *Phys. Rev.*, **D8**, 3497.

(25) T. Kinoshita (1962) *J. Math. Phys.*, **3**, 650; T. D. Lee and M. Nauenberg (1964) *Phys. Rev.*, **133**, B1549. For a recent application see G. Sterman and S. Weinberg (1977) *Phys. Rev. Lett.*, **39**, 1436.

(26) B. S. DeWitt (1967) *Phys. Rev.*, **162**, 1195, 1239; (1975) *Phys. Rep.*, **19**, 295; G. 't Hooft and M. Veltman (1974) ref. 3; J. Honerkamp (1972) *Nucl. Phys.*, **B48**, 269; R. Kallosh (1974) *Nucl. Phys.*, **B78**, 293; M. T. Grisaru, P. van Nieuwenhuizen and C. C. Wu (1975) *Phys. Rev.*, **D12**, 3203.

(27) For earlier discussions, see S. Weinberg (1977), invited talk at the Eighth International Conference on General Relativity, Waterloo, Ontario, Canada (unpublished). Related ideas are discussed by E. S. Fradkin and G. A. Vilkovsky (1976) *Proceedings of the XVIII International Conference on High Energy Physics*, vol. 2, Sec. T28 (JINR, Dubna, 1977); (1976), Berne preprint; (1978), I.A.S. preprint 778–IPP.

(28) This is a generalization of an example suggested to me by E. Witten. Of course, this example is specifically chosen to have trajectories with the desired behavior, that they either hit a fixed point for $\mu \to \infty$ or go to infinity for finite values of μ. However, this example does show that this behavior can arise for β-functions that are not at all pathological.

(29) The asymptotic freedom of non-Abelian gauge theories was discovered by H. D. Politzer (1973) *Phys. Rev. Lett.*, **30**, 1346; D. J. Gross and F. Wilczek (1973) *Phys. Rev. Lett.*, **30**, 1343.

(30) K. G. Wilson and M. E. Fisher (1972) *Phys. Rev. Lett.*, **28**, 240; K. G. Wilson (1972) *Phys. Rev. Lett.*, **28**, 548.

(31) W. A. Bardeen, B. W. Lee and R. E. Shrock (1976) *Phys. Rev.*, **D14**, 985; E. Brézin, J. Zinn-Justin and J. C. Le Guillou (1976) *Phys. Rev.*, **D14**,

Chapter 16. References

2615; also see A. M. Polyakov (1975) *Phys. Rev. Lett.*, **59B**, 79; A. Midgal (1975) *Zh. Eksp. Teor. Fiz.*, **69**, 1457.

(32) This is shown for the theory of critical phenomena by F. J. Wegner (1974) *J. Phys., C. Solid State Physics*, **7**, 2098. This work emphasizes that the invariance of the critical exponents applies only to those eigenvectors associated with essential couplings. In statistical physics, couplings that are not essential are called 'redundant'. I use the term 'inessential' here instead, because in referring to its opposite, it is easier to say 'essential' than 'irredundant'.

(33) G. 't Hooft and M. Veltman (1972) *Nucl. Phys.*, **B48**, 189; C. G. Bollini and J. J. Giambiagi (1972) *Phys. Lett.*, **40B**, 566; J. F. Ashmore (1972) *Nuovo Cimento Lett.*, **4**, 289.

(34) This is the method used in most applications of the renormalization group to critical phenomena. See, for example, Kogut and Wilson (1974) ref. 8.

(35) For a review of modern renormalizable gauge theory of weak and electromagnetic interactions, see E. S. Abers and B. W. Lee (1973) *Phys. Rep.*, **9**, 1; J. C. Taylor (1976) *Gauge Theories of Weak Interactions* (Cambridge University Press).

(36) This is shown for general field theories by Y. Nambu (1968) *Phys. Lett.*, **26B**, 626; D. G. Boulware and L. S. Brown (1968) *Phys. Rev.*, **172**, 1628; L. V. Prokhorov (1969) *Phys. Rev.*, **183**, 1515, and specifically for general relativity in low orders by M. J. Duff (1973) *Phys. Rev.*, **D7**, 2317. Quantum loop connections are considered by M. J. Duff (1974) *Phys. Rev.*, **D9**, 1837. The applications of these results to the motion of large masses is considered by D. Boulware and S. Deser (1975) *Ann. Phys.*, **89**, 173.

(37) This is a generalization of a result of T. Applequist and J. Carazzone (1975) *Phys. Rev.*, **D11**, 2856. They showed that, in a renormalizable theory of 'light' and (much heavier) 'heavy' particles, the interactions among the 'light' particles is given by an effective renormalizable field theory in which 'heavy' particles do not appear.

(38) For a discussion, see pp. 49–50 of S. Weinberg (1977) *Physics Today*, **30**, No. 4; also E. Gildener and S. Weinberg (1976) *Phys. Rev.*, **D13**, 3333.

(39) The use of dimensional regularization to construct a new set of renormalization group equations is due to G. t'Hooft (1973), *Nucl. Phys.*, **B61**, 455; *Nucl. Phys.*, **B82**, 444. The derivation given here is a somewhat simplified version of the one given by 't Hooft.

(40) In ref. 39, the renormalized coupling constant was defined strictly at $D=4$, as the constant term in a Laurent expansion around $D=4$. The extension of this definition and the corresponding renormalization group equation to arbitrary irrational spacetime dimensions has been implicity used for some time by the Saclay group (see Brézin *et al.* (1976), ref. 42) and was given explicitly bt D. J. Gross (1976), in *Methods in Field Theory*, eds. R. Balian and J. Zinn-Justin, section 4 (North-Holland: Amsterdam).

(41) J. C. Collins (1974) *Phys. Rev.*, **D10**, 1213; and University of Cambridge thesis (unpublished.)

(42) E. Brézin, J. C. LeGuillou, J. Zinn-Justin and B. G. Nickel (1973) *Phys. Lett.*, **44A**, 227; E. Brézin, J. C. LeGuillou and J. Zinn-Justin (1973) *Phys. Rev.*, **D8**, 2418; Kogut and Wilson (1974), ref. 8, Table 8.1.

Chapter 16. References

(43) See, for example, S. Weinberg (1972), ref. 6, Section 6.7.

(44) A. Duncan (1977) *Phys. Lett.*, **66B**, 170.

(45) For spin 0 contributions see L. Brown (1977) *Phys. Rev.*, **D15**, 1469. For spin $\frac{1}{2}$ contributions see D. M. Capper and M. J. Duff (1974) *Nucl. Phys.*, **B82**, 147. For spin 1 contributions see D. M. Capper, M. J. Duff and L. Halpern (1974) *Phys. Rev.*, **D10**, 461. For graviton contributions, see H.-S. Tsao (1977) *Phys. Lett.*, **66B**, 79.

(46) C. N. Yang and R. L. Mills (1954) *Phys. Rev.*, **96**, 141.

(47) R. Gastmans, R. Kallosh and C. Truffin (1977), Lebedev Institute preprint, to be published. Also see E. S. Fradkin and G. A. Vilkovsky, ref. 27.

(48) G. W. Gibbons and S. W. Hawking (1977) *Phys. Rev.*, **D15**, 2752. Also see S. W. Hawking (1977) *Phys. Lett.*, **60A**, 81 (1977). In ref. 49, Christensen and Duff relate the extra term in (16.76) back to early work of R. Arnowitt, S. Deser, and C. W. Misner (1962), in *Gravitation: An Introduction to Current Research*, ed. L. Witten (Wiley: New York); J. W. York (1972) *Phys. Rev. Lett.*, **28**, 1082; C. W. Misner, K. S. Thorne and J. A. Wheeler (1973) *Gravitation*, Chapter 21 (Freeman: San Francisco).

(49) S. M. Christensen and M. J. Duff (1978), Brandeis preprint.

(50) In this formula, it is assumed that the form of the energy–momentum tensor is ‘improved’ for arbitrary D in the manner suggested by C. Callan, S. Coleman and R. Jackiw (1970), *Ann. Phys.*, **59**, 42, for $D = 4$. However, this only affects $A(q^2)$, not $B(q^2)$.

Volume 96B, number 1,2 PHYSICS LETTERS 20 October 1980

LIMITS ON MASSLESS PARTICLES

Steven WEINBERG and Edward WITTEN
Lyman Laboratory of Physics, Harvard University, Cambridge, MA 02138, USA

Received 6 August 1980

We show that in all theories with a Lorentz-covariant energy–momentum tensor, such as all known renormalizable quantum field theories, composite as well as elementary massless particles with $j > 1$ are forbidden. Also, in all theories with a Lorentz-covariant conserved current, such as renormalizable theories with a symmetry that commutes with all local symmetries, there cannot exist composite or elementary particles with nonvanishing values of the corresponding charge and $j > 1/2$.

It has been known for many years that there are problems in the construction of lagrangian field theories for massless particles of higher spin. The difficulty is definitely not one of constructing free field theories. There are physically acceptable free-field lagrangians [1] for massless particles of arbitrary spin j, in which these particles are represented by tensor or tensor–spinor fields with j or $j - 1/2$ Lorentz indices, and the lagrangian satisfies a gauge invariance principle which eliminates unphysical degrees of freedom. Nor is there any difficulty in giving these particles interactions of some sort. For instance, we can construct interactions that trivially satisfy the gauge invariance conditions for higher spin, by simply requiring that a "curl" be taken on each Lorentz index. (Thus a spin-5/2 field $\Psi_{\mu\nu}$ would have to appear in the interaction in the form $\partial_\mu\partial_\nu\Psi_{\rho\sigma} - \partial_\mu\partial_\sigma\Psi_{\rho\nu} - \partial_\rho\partial_\nu\Psi_{\mu\sigma} + \partial_\rho\partial_\sigma\Psi_{\mu\nu}$.) Rather, the problem appears to do specifically with giving massless higher spin particles electromagnetic or gravitational interactions. The replacement of the derivatives in the lagrangians of ref. [1] with gauge-covariant or generally covariant derivatives yields a lagrangian that for high spin does not satisfy the appropriate higher-spin gauge invariance conditions, and leads to field equations that are not even algebraically consistent [2].

We find it difficult to believe that this problem is solely a limitation on the properties of *elementary* massless particles whose fields appear in the lagrangian, and that this limitation need to apply to composite particles. If there did exist massless higher spin composite particles with electromagnetic or gravitational interactions, then could we not describe their interactions by an effective lagrangian? There may well be a distinction between elementary and composite particles, but this distinction must have to do with whether or not they appear in the fundamental lagrangian, and not with whether they can appear in any lagrangian at all.

It seems likely to us instead that whenever it proves impossible to construct a lagrangian field theory for certain kinds of massless particles, then such particles simply cannot exist, whether elementary or composite ‡1. In support of this view, we offer a pair of very simple theorems ‡2, which rule out higher-spin massless particles in certain contexts.

Theorem 1. A theory that allows the construction of a Lorentz-covariant conserved four-vector current J^μ cannot contain massless particles of spin $j > 1/2$ with nonvanishing values of the conserved charge $\int J^0 \, d^3x$.

‡1 There is an S-matrix-theoretical argument against the possibility of giving gravitational interactions to massless particles with $j = 5/2$, by Grisaru et al. [3].

‡2 A result similar to our theorem 1 has been derived by Coleman [4].

Volume 96B, number 1,2 PHYSICS LETTERS 20 October 1980

Theorem 2. A theory that allows the construction of a conserved Lorentz covariant energy–momentum tensor $\theta^{\mu\nu}$ for which $\int \theta^{0\nu}\, d^3x$ is the energy–momentum four-vector cannot contain massless particles of spin $j > 1$.

These theorems are proved by studying the matrix elements $\langle p', \pm j \,|\, J^\mu \,| p, \pm j\rangle$ and $\langle p', \pm j \,|\, \theta^{\mu\nu} | p, \pm j\rangle$ of J^μ and $\theta^{\mu\nu}$ between one-massless-particle states of helicity $\pm j$ and four-momenta p' and p. We first note that under the assumptions of these theorems, the matrix elements cannot vanish in the limit $p' \to p$, and we then show that for all p' and p with $(p - p')^2 \neq 0$, the matrix element of J^μ must vanish for $j > 1/2$, and the matrix element of $\theta^{\mu\nu}$ must vanish for $j > 1$.

To see that the matrix elements do not vanish for $p' \to p$, note that Lorentz invariance dictates their form in this limit to be [‡3]

$$\langle p' | J^\mu | p\rangle \to g p^\mu / E(2\pi)^3 \,, \tag{1}$$

$$\langle p' | \theta^{\mu\nu} | p\rangle \to f p^\mu p^\nu / E(2\pi)^3 \,. \tag{2}$$

The coefficient g is the one-particle value of the charge $\int J^0\, d^3x$, and does not vanish by hypothesis. The quantity fp^μ is the one-particle value of the energy–momentum four-vector $\int \theta^{0\nu}\, d^3x$, and so $f = 1$.

To show that the matrix elements *do* vanish for high spin, we adopt a Lorentz frame in which

$$p = (\boldsymbol{p}, |\boldsymbol{p}|)\,, \quad p' = (-\boldsymbol{p}, |\boldsymbol{p}|)\,. \tag{3}$$

(This is always possible for $(p' - p)^2 \neq 0$, because then $p' + p$ is timelike, and we need simply choose a frame in which $p' + p$ has no space component.) Consider the effect of a rotation $R(\theta)$ by an angle θ around the $\boldsymbol{p}$ direction. The one-particle states undergo the transformations

$$| p, \pm j\rangle \to \exp(\pm i\theta j)\, | p, \pm j\rangle \,, \tag{4}$$

$$| p', \pm j\rangle \to \exp(\mp i\theta j)\, | p', \pm j\rangle \,. \tag{5}$$

(The difference of the signs in the exponents arises because $R(\theta)$ is a rotation of $+\theta$ around $\boldsymbol{p}$ but of $-\theta$ around $\boldsymbol{p}' = -\boldsymbol{p}$.) Rotational invariance tells us then that

$$\exp(\pm 2i\theta j)\, \langle p', \pm j \,|\, J^\mu \,|\, p, \pm j\rangle = R(\theta)^\mu{}_\rho \langle p', \pm j \,|\, J^\rho \,|\, p, \pm j\rangle \,, \tag{6}$$

$$\exp(\pm 2i\theta j)\, \langle p', \pm j \,|\, \theta^{\mu\nu} \,|\, p, \pm j\rangle = R(\theta)^\mu{}_\rho R(\theta)^\nu{}_\sigma \langle p', \pm j \,|\, \theta^{\rho\sigma} \,|\, p, \pm j\rangle \,. \tag{7}$$

But the rotation matrix $R(\theta)$ has Fourier components $e^{i\theta}$, 1, and $e^{-i\theta}$ only, so these equations require the matrix element of J^μ to vanish unless $2j = 0$ or 1, and the matrix element of $\theta^{\mu\nu}$ to vanish unless $2j = 0, 1$ or 2. Since these matrix elements thus vanish for $j > 1/2$ or $j > 1$ in the special Lorentz frame defined by eq. (3), and the helicities of massless particles are Lorentz invariant while J^μ and $\theta^{\mu\nu}$ are assumed to be Lorentz covariant, the matrix elements would have to vanish in all frames, and hence for all p' and p with $(p' - p)^2 \neq 0$. This concludes the proof.

Of course, there are acceptable theories that have massless charged particles with spin $j > 1/2$ (such as the massless version of the original Yang–Mills theory), and also theories that have massless particles with spin $j > 1$ (such as supersymmetry theories or general relativity). Our theorem does not apply to these theories because they do not have Lorentz-covariant conserved currents or energy–momentum tensors, respectively. For instance, interpreting the Yang–Mills theory as a theory of charged massless spin-one particles and photons, the electric current is

$$J^\mu = e\, \mathrm{Im}[B^{\mu\dagger}(\partial_\mu B_\nu - \partial_\nu B_\mu) - \partial^\mu (B^\dagger_\mu B_\nu)] \,,$$

where B^μ is the complex field of the charged bosons. This is not a four-vector, because under a Lorentz transformation $x^\mu \to \Lambda^\mu{}_\nu x^\nu$ the boson field B^μ transforms into $\Lambda^\mu{}_\nu B^\nu$ plus a term proportional to a derivative $\partial^\mu \Phi$. Similarly, the energy–momentum pseudotensor in general relativity is not a Lorentz tensor,

[‡3] It is important that we define g and f in terms of limits of matrix elements as the momentum transfer $p' - p$ approaches zero, and not in terms of the values of matrix elements at $p' - p = 0$. Our definition corresponds to the method by which charges, energies, and momenta are actually determined: by measuring the nearly forward scattering caused by exchange of spacelike but nearly lightlike massless vector bosons or gravitons, or else by evaluating $\int_V J^0 \times d^3x$ or $\int_V \theta^{0\mu}\, d^3x$ for a large but finite volume V. Of we had defined g and f in terms of matrix elements with $p' - p = 0$, we could not have used the vanishing of the higher-spin matrix elements for $(p - p')^2 \neq 0$ to conclude that $g = 0$ for $j > 1/2$ and $f = 0$ for $j > 1$, without invoking an assumption of continuity between spacelike and lightlike momentum transfers. Such a continuity assumption seems entirely plausible, but with our definition of f and g, it is not necessary.

Volume 96B, number 1,2 PHYSICS LETTERS 20 October 1980

because it involves the gravitational field $h_{\mu\nu}$, which has a non-tensor behavior under Lorentz transformations. (Of course, we can make J^μ or $\theta^{\mu\nu}$ into tensors by introducing unphysical helicity states, such as longitudinal and timelike charged bosons in the Yang–Mills theory, but then the proof of our theorem would be invalid because the helicities of physical states would not be Lorentz invariant.)

Our theorem does apply to all known renormalizable quantum field theories such as quantum chromodynamics, with the qualification that theorem 1 only applies to conserved currents associated with symmetries that commute with any local symmetries. It can be shown by direct construction that these theories have a Lorentz-covariant energy–momentum tensor $\theta^{\mu\nu}$, and a Lorentz-covariant Noether current J^μ for all symmetries that commute with local symmetries. Thus we conclude that all these theories have no massless bound states with $j > 1$, and quantum chromodynamics has no flavor nonsinglet massless bound states with $j > 1/2$.

It is perhaps not surprising that ordinary field theories like quantum chromodynamics do not have massless bound states of high spin. What is somewhat surprising is that this result can be proved so easily, and with such generality.

We close by noting some applications of this result to current research.

Because of the difficulties with general relativity as a quantum theory, it has occasionally been suggested that the graviton is not an elementary particle, but a massless spin-two bound state that arises in an ordinary renormalizable field theory [5]. (The couplings of a massless spin-two particle, composite or not, must at low energies mimic general relativity.) However, our theorems rule out this possibility [‡4].

Our theorems also remove one objection to the idea that the familiar quarks and leptons are bound states of more nearly fundamental particles. It might be thought that in this case, quarks and leptons would have to form closely lying states of differing angular momenta (like atoms and nuclei) in disagreement with the observation that quarks and leptons all seem to have spin 1/2. However, the absence of any detectable quark or lepton structure indicates that quark and lepton masses are very much less than the characteristic energy scale of the binding forces, so it seems reasonable to suppose that in the absence of electroweak or other perturbations, the quark and lepton masses would vanish. Assuming that the constituent particles and binding forces are described by an ordinary renormalizable theory, our theorem then rules out any other particles with $j > 1$ that would also be massless in the absence of electroweak or other perturbations to the binding forces. There may well exist excited quarks and leptons with high spin, but they are likely to have masses of the order of the characteristic scale of the binding forces, and hence to be much heavier than the ordinary quarks and leptons.

In particular, our theorems also close a loophole in recent discussions [6] of the implications of Adler–Bell–Jackiw triangle anomalies. It has been argued that theories with triangle anomalies in the amplitudes of conserved currents must involve massless untrapped physical particles. Our theorems confirm the conjecture by 't Hooft [6] that these massless particles would have to have spin zero or one-half. Our theorems do leave open the possibility that there are massless spin-one particles that are neutral under all symmetries that commute with gauge symmetries, but there is an argument by Frishman et al. [6] that such particles could not produce the anomaly, so that there would still have to be massless particles of spin zero or one-half.

‡4 However, the theorem clearly does not apply to theories [7] in which the gravitational field is a basic degree of freedom but the Einstein action is induced by quantum effects.

We are grateful for valuable conversations with S. Coleman, S. Deser, M. Grisaru, Y. Ne'eman and H. Schnitzer. This research is supported in part by the National Science Foundation under Grant No. PHY77-22864.

References

[1] M. Fierz and W. Pauli, Proc. Roy. Soc. (London) 173 (1939) 211;
W. Rarita and J. Schwinger, Phys. Rev. 60 (1941) 61;
J. Schwinger, Particles, sources, and fields (Addison-Wesley, Reading, MA, 1970);
L.P.S. Singh and C.R. Hagen, Phys. Rev. D9 (1974) 898, 910;
A. Kawakami and S. Kamefuchi, Nuovo Cimento 48A (1976) 239;
C. Fronsdal, Phys. Rev. D18 (1978) 3624;
J. Fang and C. Fronsdal, Phys. Rev. D18 (1978) 3630;

Volume 96B, number 1,2 PHYSICS LETTERS 20 October 1980

F.A. Berends, J.W. van Holten, P. van Nieuwenhuizen and B. de Wit, Phys. Lett. 83B (1979) 188; Nucl. Phys. B54 (1979) 261;
T. Curtright, Phys. Lett. 85B (1979) 219;
C. Aragone and S. Deser, Phys. Rev. D21 (1980) 352; and to be published in Nucl. Phys. B (1980);
B. de Wit and D. Freedman, Phys. Rev. D21 (1980) 358.

[2] C. Aragone and S. Deser, Phys. Lett. 85B (1979) 161;
F.A. Berends, J.W. van Holten, B. de Wit and P. van Nieuwenhuizen, J. Phys. A Math. Gen. 13 (1980) 1643;
B. de Wit and D. Freedman, ref. [1];
for earlier discussions of related difficulties, see K. Johnson and E.C.G. Sudarshan, Ann. Phys. (NY) 13 (1961) 126;
G. Velo and D. Zwanziger, Phys. Rev. 186 (1967) 1337.

[3] M.T. Grisaru, H.N. Pendleton and P. van Nieuwenhuizen, Phys. Rev. D15 (1977) 996.

[4] S. Coleman, unpublished.

[5] G. Papini, Nuovo Cimento 39 (1965) 716;
H. Terazawa, Y. Chikashigi, K. Ikama, and T. Matsuki, Phys. Rev. D15 (1977) 1181;
F. Cooper, G. Guralnik and N. Snyderman, Phys. Rev. Lett. 40 (1978) 1620.

[6] G. 't Hooft, Cargese lectures (1979);
Y. Frishman, A. Schwimmer, T. Banks and S. Yankielowicz Weizmann Institute preprint WIS-80/27 (1980);
S. Coleman and E. Witten, Phys. Rev. Lett. 45 (1980) 100.

[7] A.D. Sakharov, Dokl. Akad. Nauk SSSR 177 (1967) 70 [Sov. Phys. Dokl. 12 (1968) 1040];
O. Klein, Phys. Scr. 9 (1974) 69;
P. Minkowski, Phys. Lett. 71B (1977) 419;
A. Zee, Phys. Rev. Lett. 42 (1979) 417;
L. Smolin, Nucl. Phys. B160 (1979) 253;
K. Ikama, Y. Chikashigi, T. Matsuki and H. Terazawa, Prog. Theor. Phys. 60 (1978) 868;
S. Adler, Phys. Rev. Lett. 44 (1980) 1567; Phys. Lett. 95B (1980) 241;
B. Hasslacher and E. Mottola, Phys. Lett. 95B (1980) 237.

Nuclear Physics B237 (1984) 397–441

CALCULATION OF GAUGE COUPLINGS AND COMPACT CIRCUMFERENCES FROM SELF-CONSISTENT DIMENSIONAL REDUCTION*

Philip CANDELAS and Steven WEINBERG

Center for theoretical Physics,
and
Theory Group, Department of Physics, The University of Texas at Austin, Austin, Texas 78712, USA

Received 19 May 1983
(Corrected version received 24 January 1984)

We consider a system of gravity plus free massless matter fields in $4+N$ dimensions, and look for solutions in which N dimensions form a compact curved manifold, with the energy-momentum tensor responsible for the curvature produced by quantum fluctuations in the matter fields. For manifolds of sufficient symmetry (including spheres, CP^N, and manifolds of simple Lie groups) the metric depends on only a single multiplicative parameter ρ^2, and the field equations reduce to an algebraic equation for ρ, involving the potential of the matter fields in the metric of the manifold. With a large number of species of matter fields, the manifold will be larger than the Planck length, and the potential can be calculated using just one-loop graphs. In odd dimensions these are finite, and give a potential of form C_N/ρ^4. Also there are induced Yang-Mills and Einstein–Hilbert terms in the effective 4-dimensional action, proportional to additional numerical coefficients, D_N and E_N. General formulas are given for the gauge coupling g^2 in terms of C_N and D_N, and the ratio $\rho^2/8\pi G$ in terms of C_N and E_N. Numerical values for C_N, D_N, and E_N are obtained for scalar and spinor fields on spheres of odd dimensionality N. It is found that the potential, g^2 and $\rho^2/8\pi G$ can all be positive but only when the compact manifold has $N=3+4k$ dimensions. (The positivity of the potential is needed for stability of the sphere against uniform dilations or contractions.) In this case, solutions exist either for spinor fields alone or for suitable mixes of spinor and scalar fields provided the ratio of the number of scalar fields to the number of fermion fields is not too large. Numerical values of the $O(N+1)$ gauge couplings and $8\pi G/\rho^2$ are calculated for illustrative values of the numbers of spinor fields. It turns out that large numbers of matter fields are needed to make these parameters reasonably small.

1. Introduction

The calculation of the fine-structure constant has presented a challenge to theoretical physicists ever since the measurement of e and $\hbar$ in the early years of this century. We do not presume in this paper to offer a calculation of *the* fine-structure constant, $\frac{1}{137}$. Rather, we wish to show how fine-structure constants in general can be calculated in certain theories, in which the gauge fields arise from the metric of a higher-dimensional space.

* Research supported in part by the Robert A. Welch Foundation, the National Science Foundation under Grant PHY8205717 and the University of Texas Center for Theoretical Physics.

In the original Kaluza–Klein model [1], all electric charges are integer multiples of the ratio of $2\pi(16\pi G)^{1/2}$ and the circumference of the compact fifth dimension. However, this circumference is a free parameter of the model, one that is not determined even when we specify all parameters in the lagrangian. Thus, although this model accounts for the quantization of electric charge, it does not lead to a prediction of the basic unit of charge.

There are more general $(4+N)$-dimensional models [2, 3], in which N dimensions form a compact manifold, and a massless gauge field appears in four dimensions for each Killing vector of this manifold. A general prescription [4] has recently been given for calculating the various gauge couplings in such models in terms of the ratio of $2\pi(16\pi G)^{1/2}$ and various r.m.s. circumferences. Some of these models [3] exhibit a so-called "spontaneous compactification": the compact manifold is a solution of Einstein's field equations with non-zero curvature, with the energy-momentum tensor arising from monopole-like configurations of classical fields. Unfortunately, the "monopole" strength of these configurations is again a free parameter, unrelated to the parameters in the lagrangian. Hence, at least on a classical level, these models do not lead to a determination of the scale of the compact manifold, or of the gauge coupling constants.

In this paper we consider dynamical compactification of a different sort, proposed in ref. [4]. The $(4+N)$-dimensional space is again supposed to break up into a 4-dimensional Minkowski space and a curved compact N-dimensional manifold, with the curvature governed by Einstein's field equations. However, the energy-momentum tensor on the right-hand side of these equations is now supposed to arise not from some topologically non-trivial classical field configuration, but rather from the one-loop quantum fluctuations in various matter fields.

One naturally approaches any juxtaposition of quantum mechanics and general relativity with caution. This may explain why previous models of dynamical compactification have sought a solution in purely classical terms*. However, there is one class of models of quantum spontaneous compactification in which multi-loop gravitational corrections can be systematically neglected. These are the models outlined in ref. [4], with a large number f of light matter fields.

For an N-dimensional compact manifold whose linear dimensions are of order ρ, the one-loop energy density of f light matter fields is of order $f\rho^{-4-N}$. The $(4+N)$-dimensional gravitational constant $\bar{G}$ is of order $G\rho^N$, and the Einstein tensor

* Quantum matter fluctuations have been considered as a source of an energy-momentum tensor in higher-dimensional theories by the authors of ref. [5]. However, they consider the extra dimensions to form a circle or torus, so that there is no curvature to balance the energy-momentum tensor, and the field equations require that the matter potential itself be stationary. This is only possible for a mix of bosons and fermions, some of which must be massive, and the circumference of the compact manifold will depend on the values assumed for these masses. In contrast, in the present work the energy-momentum tensor is balanced by the curvature, and solutions are possible without mass parameters in the lagrangian, and with the scale of the compact manifold set by the gravitational constant.

$G_{\mu\nu}$ is of order ρ^{-2}. Hence the Einstein field equations give

$$\rho^{-2} \approx \bar{G} f \rho^{-4-N} \approx G f \rho^{-4} ,$$

and so

$$\rho^2 \approx Gf . \tag{1}$$

The L-loop gravitational corrections to the one-loop matter energy density are proportional to $f\bar{G}^{L-1}$, and hence on dimensional grounds are less for $L \geq 2$ than the one-loop matter terms by a factor

$$(\bar{G}\rho^{-N-2})^{L-1} \approx (G\rho^{-2})^{L-1} \approx (1/f)^{L-1} .$$

Also, the L-loop purely gravitational contributions to the energy density do not carry an over-all factor f, and are proportional to $\bar{G}^{L-1}$, so on dimensional grounds are less for $L \geq 1$ than the one-loop matter terms by a factor

$$\frac{1}{f}(\bar{G}\rho^{-N-2})^{L-1} \approx (1/f)^L .$$

Adding matter loops within virtual graviton lines adds equal numbers of additional $\bar{G}$ and f factors, and hence does not change the order in $1/f$.

We conclude that for f sufficiently large the scale of the compact manifold is of order $\sqrt{Gf}$, and can be calculated by including only one-loop matter contributions to the energy-momentum tensor*.

For an odd number N of compact dimensions, these models offer an added bonus. If we define our integrals by dimensional regularization then in the limit of 4 spacetime dimensions the one-loop energy density is actually finite [7]. (This is because dimensional regularization makes all divergences logarithmic, and the one-loop graphs are independent of dimensional couplings like $\bar{G}$, so any divergences would have to correspond to local counterterms with the dimensionality of the energy density in $4+N$ dimensions, i.e. $[\text{length}]^{-4-N}$. But the only local functions of the $(4+N)$-dimensional metric and its derivatives that are invariant under general coordinate transformations (and inversions) in the full space are those containing even numbers of derivatives, such as powers of the curvature scalar, and none of these can have the odd dimensionality $4+N$). The general formal considerations of this paper will apply to compact manifolds of arbitrary dimensionality N, but our numerical calculations will be limited to the case of odd N.

* The approximation used in our paper, based on the classical Einstein field equations with an energy-momentum tensor generated by one-loop quantum fluctuations in a large number of matter fields had previously been employed by Hartle and Horowitz [6]. Their work dealt only with 4-dimensional spacetimes.

2. Solution of the field equations

We consider a $(4+N)$-dimensional theory of gravity plus a number of massless matter fields. The Einstein field equations are*

$$\bar{R}^{NM}-\tfrac{1}{2}\bar{g}^{NM}(\bar{R}+\bar{\Lambda})=-8\pi\bar{G}\bar{T}^{NM}\,, \tag{2}$$

where $\bar{g}_{NM}$ is the $(4+N)$-dimensional metric; $\bar{R}^{NM}$ and $\bar{R}$ are the Ricci tensor and curvature scalar calculated from $\bar{g}_{NM}$; $\bar{G}$ and $\bar{\Lambda}$ are the gravitational and cosmological constants in $4+N$ dimensions; and $\bar{T}^{NM}$ is the $(4+N)$-dimensional energy-momentum tensor, given as usual by

$$\tfrac{1}{2}\sqrt{\bar{g}}\,\bar{T}^{NM}=\delta\bar{I}/\delta\bar{g}_{NM}\,, \tag{3}$$

with $\bar{I}$ the effective matter action, produced here by one-loop quantum fluctuations.

We seek a vacuum solution with Poincaré invariance in 4 dimensions. That is, the metric is taken to have the components

$$\bar{g}_{\mu\nu}=\eta_{\mu\nu}\,,\qquad \bar{g}_{\mu n}=0\,,\qquad \bar{g}_{nm}=\tilde{g}_{nm}(y)\,, \tag{4}$$

where $\eta_{\mu\nu}$ is the usual Minkowski metric for spacetime coordinates x^{μ}, and $\tilde{g}_{nm}(y)$ is the metric of some compact manifold with coordinates y^{m}. The Ricci tensor and curvature scalar appearing in (2) are then

$$\bar{R}^{\mu\nu}=\bar{R}^{\mu n}=\bar{R}^{n\mu}=0\,, \tag{5}$$

$$\bar{R}^{nm}=\tilde{R}^{nm}(y)\,, \tag{6}$$

$$\bar{R}=\tilde{R}(y)\,, \tag{7}$$

where $\tilde{R}^{nm}$ and $\tilde{R}$ are the N-dimensional Ricci tensor and curvature scalar calculated from $\tilde{g}_{nm}(y)$. Also, the vacuum Poincaré invariance gives the energy-momentum tensor the structure

$$\bar{T}^{\mu\nu}=B(y)\eta^{\mu\nu}\,, \tag{8}$$

$$\bar{T}^{\mu n}=\bar{T}^{n\mu}=0\,, \tag{9}$$

$$\bar{T}^{nm}=\tilde{T}^{nm}(y)\,. \tag{10}$$

The field equations (2) thus reduce to

$$\tfrac{1}{2}(\tilde{R}(y)+\bar{\Lambda})=-8\pi\bar{G}B(y)\,, \tag{11}$$

$$\tilde{R}^{nm}(y)-\tfrac{1}{2}\tilde{g}^{nm}(y)[\tilde{R}(y)+\bar{\Lambda}]=-8\pi\bar{G}\tilde{T}^{nm}(y)\,. \tag{12}$$

In calculating $\tilde{T}^{nm}(y)$, we must evaluate a variational derivative with respect to $\bar{g}_{nm}$, so we can fix $\bar{g}_{\mu n}$ and $\bar{g}_{\mu\nu}$ at the values given by (4). The matter action is then

$$\bar{I}=-\int \mathrm{d}^{4}x\; V[\tilde{g}_{nm}]\,, \tag{13}$$

* We must include a cosmological constant in $4+N$ dimensions and tune its value to ensure that the field equations have a solution with a flat 4-dimensional Minkowski space. This of course is an unpleasant feature of our theory, and all other existing theories.

with V the "potential" for the classical field $\tilde{g}_{nm}(x, y)$. The nm energy-momentum components are given by eq. (3) for $\tilde{g}_{nm}(x, y) = \tilde{g}_{nm}(y)$ as

$$\tfrac{1}{2}\sqrt{\tilde{g}(y)}\,\tilde{T}^{nm}(y) = -\frac{\delta V[\tilde{g}]}{\delta \tilde{g}_{nm}(y)}. \tag{14}$$

We will return later to the calculation of the other component, $B(y)$.

The nm field equations (12) are in general not easy to solve. However, as discussed in appendix A, there is a certain class of manifolds whose symmetries dictate (i) that $\tilde{g}_{nm}(y)$ is fixed up to an over-all factor ρ^2, and (ii) all other tensors like $\tilde{T}_{nm}(y)$ and $\tilde{R}_{nm}(y)$ formed by variational derivatives of scalars with respect to $\tilde{g}_{nm}(y)$ are proportional to $\tilde{g}_{nm}(y)$, with constant coefficients. These manifolds consist of the homogeneous coset spaces G/H for which generators of G that are not in H form a representation of H that is either real and irreducible or the direct sum of a complex irreducible representation and its complex conjugate. (This includes spheres, CP^N, and manifolds of simple groups.) For manifolds of this class, we can normalize the one free parameter ρ^2 in the metric by specifying that

$$\tilde{R}_{nm}(y) = -(N-1)\rho^{-2}\tilde{g}_{nm}(y). \tag{15}$$

(The factor $-(N-1)$ is inserted here so that for spheres ρ will simply be the radius.) Also, the energy-momentum tensor here takes the form

$$\tilde{T}_{nm}(y) = A\tilde{g}_{nm}(y), \tag{16}$$

and homogeneity implies that B is y independent.

We can calculate A by contracting (14) with $\tilde{g}_{nm}(y)$ and integrating over the manifold:

$$-\tfrac{1}{2}AN\Omega_N = \int \mathrm{d}^N y\, \tilde{g}_{nm}(y)\frac{\delta V[\tilde{g}]}{\delta \tilde{g}_{nm}(y)}, \tag{17}$$

where Ω_N is the volume of the manifold

$$\Omega_N \equiv \int \mathrm{d}^N y \sqrt{\tilde{g}(y)}. \tag{18}$$

The right-hand side of (17) is just the logarithmic rate of change of $V[\tilde{g}]$ when we vary the over-all scale of $\tilde{g}_{nm}(y)$, so we can write this as

$$-\tfrac{1}{2}N\Omega_N A = \rho^2\frac{\mathrm{d}V}{\mathrm{d}\rho^2}. \tag{19}$$

To calculate B, it is only necessary to notice that if we had used a spacetime metric with $g_{\mu\nu} = \lambda^2\eta_{\mu\nu}$, then a factor $\sqrt{g} = \lambda^4$ would have appeared in (13), so

$$\tfrac{1}{2}\tilde{T}^{\mu}{}_{\mu} = \left(g_{\mu\nu}\frac{\delta \bar{I}}{\delta g_{\mu\nu}}\right)_{\lambda=1} = -\frac{V}{\Omega_N}\left(\lambda^2\frac{d}{d\lambda^2}\lambda^4\right)_{\lambda=1} = -\frac{2V}{\Omega_N},$$

and therefore

$$\Omega_N B = -V. \tag{20}$$

Note that this gives the one-loop energy density $\bar{T}^{00}$ per unit volume in $(4+N)$-dimensional space as $-B = V/\Omega_N$, in accord with the usual interpretation of V as the energy density in 4-dimensional space. This result would apply to any homogeneous manifold, not only those with the irreducibility property discussed in appendix A.

The factors Ω_N in the formulas (19) and (20) for A and B will be cancelled by the Ω_N factor in the well-known relation between the gravitational constants G_0 and $\bar{G}$ in 4 and $4+N$ dimensions

$$\bar{G} = G_0 \Omega_N. \tag{21}$$

Using (15), (16), (19), (20), and (21), the field equations (11) and (12) now become

$$-\frac{1}{2}\left[\frac{-N(N-1)}{\rho^2} + \bar{\Lambda}\right] = 8\pi G_0 V, \tag{22}$$

$$\frac{-(N-1)}{\rho^2} - \frac{1}{2}\left[\frac{-N(N-1)}{\rho^2} + \bar{\Lambda}\right] = \frac{8\pi G_0}{N}\rho\frac{\mathrm{d}V}{\mathrm{d}\rho}. \tag{23}$$

Using the first equation to determine $\bar{\Lambda}$, the second becomes

$$\frac{N-1}{8\pi G_0 \rho^2} = V(\rho) - \frac{\rho}{N}\frac{\mathrm{d}V(\rho)}{\mathrm{d}\rho}. \tag{24}$$

Eq. (24) can be solved for the radius parameter ρ, and given the symmetries of the manifold, this completely determines its metric.

We now make the further assumptions that the matter fields are massless in $4+N$ dimensions and that N is odd. Otherwise the radius we calculate from (24) would depend on the mass parameters in the lagrangian, and there would be no hope of obtaining a definite value for the scale of gauge couplings without further information. Since there is no one-loop conformal anomaly in odd dimensions [7], naive dimensional analysis tells us that the potential here has the form

$$V = \frac{C_N}{\rho^4}, \tag{25}$$

with C_N a pure number, depending only on the symmetries of the manifold, on N, and on the numbers of massless fields of various types. Eq. (24) gives immediately*

$$\rho^2 = \frac{8\pi G_0 (N+4) C_N}{N(N-1)}, \tag{26}$$

* Note that the value for the radius ρ given by eq. (26) blows up for $N = 1$. This just reflects the fact that the compact one-dimensional manifold S^1 is flat, so there is no solution of the field equations in the original Kaluza–Klein model with a non-vanishing energy-momentum tensor.

while (22) gives the cosmological constant

$$\bar{\Lambda} = \frac{N^2(N-1)^2(N+2)}{8\pi G_0(N+4)^2 C_N} . \tag{27}$$

As we will see in sect. 7, the stability of the compact manifold requires that G_0 be positive. This being the case, a solution can be found for ρ if and only if C_N is *positive*, and in this case $\bar{\Lambda}$ must be tuned to a definite positive value to obtain a flat spacetime.

3. Calculation of the potential

We consider here the potential V due to b real massless scalar fields and f Dirac massless spin-$\frac{1}{2}$ fields in $4+N$ dimensions. Each of these fields will appear in 4 dimensions as an infinite tower of spin-0 and spin-$\frac{1}{2}$ particles with masses given by the eigenvalues M_l^2 and m_l of the N-dimensional laplacian and Dirac differential operators. The potential produced by a single real spin-0 particle of mass M is known to be [8]

$$V_0(M) = \frac{-i}{2(2\pi)^4} \int \mathrm{d}^4k \ln(k^2 + M^2 - i\varepsilon) ,$$

while for spin-$\frac{1}{2}$, we just replace M with m and multiply by -4. The total potential is then

$$V = b \sum_l D_l V_0(M_l) - 4f \sum_l d_l V_0(m_l) , \tag{28}$$

where D_l and d_l are the degeneracies of the eigenvalues M_l and m_l of the laplacian and Dirac operators.

We will use dimensional regularization to evaluate $V_0(M)$. An elementary calculation gives

$$V_0(M) = -\tfrac{1}{2}(4\pi)^{-n/2}\Gamma(-\tfrac{1}{2}n)M^n , \tag{29}$$

where n is the spacetime dimensionality, later to be set equal to 4. For manifolds whose metric is characterized by a single over-all scale factor ρ^2 such as those discussed in appendix A, the eigenvalues are

$$M_l = \Lambda_l/\rho , \tag{30}$$

$$m_l = \lambda_l/\rho \tag{31}$$

where Λ_l and λ_l as well as the degeneracies D_l and d_l are ρ independent, depending only on the symmetries and dimensionality of the compact manifold. For instance, for massless scalars on a sphere S^N we have [9]*

$$\Lambda_l^2 = l(l+N-1) ,$$

$$D_l = \frac{(2l+N-1)(l+N-2)!}{(N-1)!\, l!} .$$

* Our conventions regarding standard functions follow those of this reference.

Using (29)–(31) in (28), the potential is

$$V=\lim_{n\to 4}[-\tfrac{1}{2}(4\pi)^{-n/2}\Gamma(-\tfrac{1}{2}n)\rho^{-n}\{b\zeta_N^{(0)}(-n)-4f\zeta_N^{(1/2)}(-n)\}]\,, \tag{32}$$

where the ζ_N are generalized zeta functions

$$\zeta_N^{(0)}(z)=\sum_l D_l\Lambda_l^{-z}\,, \tag{33}$$

$$\zeta_N^{(1/2)}(z)=\sum_l d_l\lambda_l^{-z}\,. \tag{34}$$

All terms in the power series for $\Gamma(-\frac{1}{2}n)$ and $\zeta_N(-n)$ are positive, so at first sight it looks like the condition $C_N>0$ found necessary in sect. 2 can be satisfied if and only if there is a sufficiently large ratio of fermion to boson field types. However, both of these series diverge for $n>0$, and must therefore be defined by analytic continuation for $n\to 4$. Only then can we tell the signs.

The analytic continuation is carried out by two different methods in appendices B, D in the special case where the compact manifold is a sphere S^N of odd dimensionality N. The results confirm the general expectation that there should be no one-loop ultraviolet divergences in odd dimensions: the generalized zeta functions $\zeta_N^{(0)}(z)$ and $\zeta_N^{(1/2)}(z)$ are found like the ordinary Riemann zeta function $\zeta(z)$ to have a simple zero at negative even integer arguments, which cancels the pole in $\Gamma(-\frac{1}{2}n)$ as $n\to 4$. From (32) we see then that $V=C_N/\rho^4$, with

$$C_N=bC_N^{(0)}+fC_N^{(1/2)}\,, \tag{35}$$

$$C_N^{(0)}=-\frac{1}{32\pi^2}\zeta_N^{(0)\prime}(-4)\,, \tag{36}$$

$$C_N^{(1/2)}=\frac{1}{8\pi^2}\zeta_N^{(1/2)\prime}(-4)\,. \tag{37}$$

The most convenient way of calculating C_N numerically is provided by the integral representations developed in appendix B. For spheres and for the case of minimally coupled scalar fields these yield

$$\begin{aligned}C_N^{(0)}=\frac{1}{4\pi\nu}\int_0^{\pi/2}\frac{\mathrm{d}\theta}{[2\cosh(\frac{1}{2}\pi\tan\theta)]^{2\nu}}&\left\{\left(\frac{\nu^3}{\pi^3}\cos\theta\sin 3\theta-\frac{15\nu}{\pi^5}\cos^3\theta\sin 5\theta\right)\right.\\ &\times\sinh(\nu\pi\tan\theta)+\left(\frac{6\nu^2}{\pi^4}\cos^2\theta\cos 4\theta-\frac{15}{\pi^6}\cos^4\theta\cos 6\theta\right)\\ &\left.\times\cosh(\nu\pi\tan\theta)\right\}\,,\end{aligned} \tag{38}$$

while for fermions we find

$$C_N^{1/2}=\frac{(-1)^{\nu+1}3\cdot 2^{\nu+1}}{\pi^6}\int_0^{\pi/2}\frac{\mathrm{d}\theta\cos^3\theta\cos 5\theta}{[2\cosh(\frac{1}{2}\pi\tan\theta)]^{2\nu+1}}\,, \tag{39}$$

TABLE 1

Values of $C_N^{(0)}$, the one-loop potential for a single massless real spin-0 field on an N-dimensional sphere of unit radius

N	$C_N^{(0)}$
1 (a)	-5.05576×10^{-5}
(b)	4.73982×10^{-5}
3	7.56870×10^{-5}
5	4.28304×10^{-4}
7	8.15883×10^{-4}
9	1.13389×10^{-3}
11	1.32932×10^{-3}
13	1.37403×10^{-3}
15	1.25249×10^{-3}
17	9.55916×10^{-4}
19	4.79352×10^{-4}
21	-1.79909×10^{-4}

Since the circle S^1 is multiply connected, for $N=1$ we have the choice of whether or not to allow the field to change sign around the circle. These two cases of untwisted and twisted fields are labelled here as (a) and (b), respectively.

TABLE 2

Values of $C_N^{(1/2)}$, the one-loop potential for a single massless Dirac spin-$\frac{1}{2}$-field on an N-dimensional sphere of unit radius

N	$C_N^{(1/2)}$
1 (a)	$+2.022304\times10^{-4}$
(b)	-1.895928×10^{-4}
3	$+1.945058\times10^{-4}$
5	-1.140405×10^{-4}
7	$+5.958744\times10^{-5}$
9	-2.99172×10^{-5}
11	$+1.477709\times10^{-5}$
13	-7.242740×10^{-6}
15	$+3.537614\times10^{-6}$
17	-1.725405×10^{-6}
19	$+8.412070\times10^{-7}$
21	-4.101970×10^{-7}

As in the scalar case (a) and (b) denote the values appropriate to the field being, respectively, untwisted and twisted.

where

$$\nu \equiv \tfrac{1}{2}(N-1)\,. \tag{40}$$

Numerical results are presented in tables 1 and 2 (the corresponding results for conformally coupled scalar fields are presented in table 3). The asymptotic values of the $C_N^{(j)}$ for large N are derived in appendix C:

$$C_N^{(0)} = -\frac{\nu^2}{16\pi^2\mu^2}\left(1+\mathrm{O}\left(\frac{1}{\mu}\right)\right), \tag{41}$$

where, in this context, μ is the root of

$$\mathrm{e}^{\mu}/\mu = \tfrac{2}{3}\nu\,, \tag{42}$$

$$C_N^{(1/2)} = \frac{(-)^{\nu+1}\cdot 3}{2^{\nu}\pi^{13/2}(\nu+\frac{1}{2})^{1/2}}\left\{1-\left(\frac{30}{\pi^2}-\frac{1}{8}\right)\frac{1}{(\nu+\frac{1}{2})}+\mathrm{O}\left(\frac{1}{(\nu+\frac{1}{2})^2}\right)\right\}. \tag{43}$$

We believe that the sign change in $C_N^{(0)}$ that occurs between $N=19$ and $N=21$ is the one "predicted" by the asymptotic form (41) and we therefore expect that the sign of $C_N^{(0)}$ will remain negative for all $N>21$. On the other hand for fermions the alternation of sign, expected asymptotically for large-N, sets in immediately at $N=3$, and this alternation will continue for all N.

We see that a solution can be found (i.e. $C_N>0$) with any numbers of massless scalars and fermions only for spaces of total dimensionality $N+4$ equal to 7, 11, 15, 19, and 23. There is no solution at all if the total dimensionality is 25, 29, 33, $\cdots$. A solution can be found provided the ratio of fermions to bosons is sufficiently large for $N+4=27$, 31, 35, $\cdots$ and provided the ratio of bosons to fermions is sufficiently large for $N+4=9$, 13, 17, and 21. However, as we shall see, there are additional positivity conditions that exclude most of these possibilities.

TABLE 3

Values of $C_N^{\text{conformal}}$, the one-loop potential for a single massless conformally coupled scalar field on an N-dimensional sphere of unit radius

N	$C_N^{\text{conformal}}$
3	$+7.145890\times10^{-6}$
5	-7.857102×10^{-7}
7	$+7.049427\times10^{-8}$
9	-1.821234×10^{-9}
11	-1.566188×10^{-9}
13	$+6.501802\times10^{-10}$
15	-1.943404×10^{-10}
17	$+5.201765\times10^{-11}$
19	-1.322957×10^{-11}
21	$+3.274453\times10^{-12}$

4. Gauge and gravitational couplings

When a $(4+N)$-dimensional space breaks up into a 4-dimensional Minkowski space and a compact manifold, the perturbations of this metric appear in 4 dimensions in part as a set of massless fields: the Yang-Mills fields $A^{\mu}_{\alpha}(x)$ and the gravitational field $g_{\mu\nu}(x)$. In ref. [4] it was shown that the classical Einstein–Hilbert action of pure gravity in $4+N$ dimensions yields in 4 dimensions an action for these fields of the form:

$$\bar{I} \equiv -\frac{1}{16\pi\bar{G}}\int \bar{R}\sqrt{\bar{g}}\,\mathrm{d}^{4+N}\bar{x}$$
$$= -\frac{1}{16\pi G_0}\left[\int R\sqrt{g}\,\mathrm{d}^4x + \frac{1}{4}\sum_e \frac{g_e^2 N_e^2 s_e^2}{4\pi^2}\int F_e^{\mu\nu}F_{e\mu\nu}\sqrt{g}\,\mathrm{d}^4x + \cdots\right], \tag{44}$$

where in our present notation $F_e^{\mu\nu}$ are the Yang–Mills curls of those gauge fields A_e^{μ} that correspond to closed Killing curves of the compact manifold; s_e is the r.m.s. circumference of the manifold along these curves; g_e is the corresponding gauge coupling, defined as the lowest eigenvalue of the generator t_e in the faithful N_e-valued representation whose structure constants $C_{ef}{}^{g}$ appear in $F_e^{\mu\nu}$; and (as in sect. 3)

$$G_0 \equiv \bar{G}\Big/\int \mathrm{d}^N y\sqrt{\bar{g}(y)}\,.$$

If this were the whole story, then G_0 could be identified as the Newton gravitational coupling constant G, and the normalization condition for the Yang–Mills fields would yield the formula for g_e derived in ref. [4]:

$$g_e = \frac{2\pi(16\pi G)^{1/2}}{N_e s_e}\,.$$

However, we are now assuming that there are many species of matter fields, so that one-loop radiative corrections can be important* despite the smallness of G/ρ^2.

In general, radiative corrections generate induced Einstein–Hilbert and Yang–Mills terms in the effective four-dimensional action, of the form

$$I_{\text{induced}} = -\frac{1}{4}\sum_e D_{Ne}g_e^2\int F_e^{\mu\nu}F_{e\mu\nu}\sqrt{g}\,\mathrm{d}^4x - \frac{E_N}{\rho^2}\int R\sqrt{g}\,\mathrm{d}^4x + \cdots. \tag{45}$$

The D_{Ne} and E_N are new dimensionless coefficients, which for massless matter fields on one-parameter manifolds (see appendix A) of odd dimensionality are (like C_N) ρ-independent numerical quantities roughly of the order of the number of species of matter fields. Combining (44) and (45), we see that the true Newton constant G is given by

$$\frac{1}{16\pi G} = \frac{1}{16\pi G_0} + \frac{E_N}{\rho^2}, \tag{46}$$

* The induced Einstein–Hilbert and Yang–Mills terms are calculated in the original Kaluza–Klein model with $N=1$ by Toms, ref. [10]. Toms' preprint was instrumental in reminding us of the importance of these terms.

and the normalization condition for the gauge fields is

$$1=\frac{g_e^2 N_e^2 s_e^2}{4\pi^2\cdot 16\pi G_0}+D_{Ne}g_e^2\,. \tag{47}$$

In particular, for spheres S^N, all gauge couplings and D-coefficients have common values g and D_N; the r.m.s. circumference is

$$s_e=2\pi\rho\sqrt{\frac{2}{N+1}}\,,$$

and $N_e=1$; so (47) here reads:

$$1=g^2\left[\frac{\rho^2}{16\pi G_0}\,\frac{2}{N+1}+D_N\right]. \tag{48}$$

The previous results of ref. [4] gave ρ^2/G and $1/g^2$ of the order of the number of species of matter fields, and hence of the order of E_N and D_N, so the radiative corrections represented by E_N and D_N in (46) and (47)–(48), are of the same order as the other, classical, terms. However higher radiative corrections are suppressed by factors $1/f$ and $1/b$.

Now let us apply what we have previously learned about the size of the compact manifold. Recall that eq. (26) gives ρ^2 in terms of G_0, not G:

$$\rho^2=\frac{8\pi G_0(N+4)C_N}{N(N-1)}\,, \tag{49}$$

so eq. (48) gives the $O(N+1)$ coupling as

$$g^2=\left[\frac{(N+4)C_N}{N(N^2-1)}+D_N\right]^{-1}. \tag{50}$$

This is what we wanted: a formula for the gauge coupling that depends only on the dimensionality and shape of the compact manifold and the numbers of various types of massless matter fields, and that gives a small gauge coupling if these numbers are large. However, we now encounter a new positivity condition

$$\frac{(N+4)C_N}{N(N^2-1)}+D_N>0\,. \tag{51}$$

If this is not satisfied, then the Yang–Mills term in the effective $(4+N)$-dimensional action has the "wrong" sign, and the vacuum is unstable with regard to exponentially growing Yang–Mills waves.

Similarly, we can use (49) to eliminate G_0 in eq. (46), and solve for the squared radius ρ^2 in terms of the observed gravitational constant G:

$$\frac{\rho^2}{8\pi G}=\frac{(N+4)C_N}{N(N-1)}+2E_N\,. \tag{52}$$

Thus as before the size of the compact manifold is larger than the Planck length if there are many matter fields. The immediate consequence of (52) is another positivity condition

$$\frac{(N+4)C_N}{N(N-1)}+2E_N>0\,. \tag{53}$$

If this is not satisfied, then the Einstein–Hilbert term in the effective action has the "wrong" sign, and the vacuum is unstable with regard to exponentially growing gravitational waves.

5. Calculation of D_N and E_N

We now turn to a calculation of the new coefficients D_N and E_N. We note first that in four dimensions the one-loop graphs involving a single minimally coupled particle of mass M and gauge coupling g_eQ_e will produce the induced Einstein–Hilbert and Yang–Mills terms*

$$-\tfrac{1}{4}\sum_e \mathscr{D}_e(M)g_e^2\int F_e^{\mu\nu}F_{e\mu\nu}\sqrt{g}\,\mathrm{d}^4x$$

$$-\mathscr{E}(M)\int R\sqrt{g}\,\mathrm{d}^4x\,,$$

where for scalars

$$\mathscr{D}_e(M)=\frac{1}{96\pi^2}\lim_{n\to 4}\Gamma(2-\tfrac{1}{2}n)M^{n-4}Q_e^2\,, \tag{54}$$

$$\mathscr{E}(M)=\frac{1}{192\pi^2}\lim_{n\to 4}\Gamma(1-\tfrac{1}{2}n)M^{n-2}\,. \tag{55}$$

If in $4+N$ dimensions we have b massless scalars, then in 4 dimensions we will observe b infinite towers of massive scalars: say bD_l scalars of mass M_l. Thus the D_N and E_N coefficients for each species will be given by

$$D_N^{(0)}=\frac{1}{96\pi^2}\lim_{n\to 4}\Gamma\left(2-\frac{n}{2}\right)\sum_l D_lM_l^{n-4}Q_{le}^2\,, \tag{56}$$

$$E_N^{(0)}=\frac{1}{192\pi^2}\lim_{n\to 4}\Gamma\left(1-\frac{n}{2}\right)\sum_l D_lM_l^{n-2}\rho^2\,, \tag{57}$$

where the Q_{le}^2 are the mean square charges (in units of g_e^2) of the scalars of mass M_l.

For spinors induced gravity and Yang–Mills terms arise similarly although the situation with regard to the induced Yang–Mills term is complicated by the fact

* The contributions of individual massive particles to the induced Einstein–Hilbert and Yang–Mills actions in 4 dimensions can be inferred from the work of deWitt, Toms, and Awada and Toms [10], or taken for spin-0 (for E_N) from S. Weinberg, ref. [11] and for spin-$\frac{1}{2}$ from Zee, ref. [12].

that, through the process of dimensional reduction, the Dirac equation acquires a non-minimal coupling to the background gauge field. Fortunately these effects have been considered by Awada and Toms [10] who find that for each species

$$D_N^{(1/2)} = -\frac{1}{192\pi^2}\frac{(20\nu^2+28\nu+5)}{(\nu+1)^2}\lim_{n\to 4}\Gamma\left(2-\frac{n}{2}\right)\sum_l d_l m_l^{n-4} q_{le}^2\,, \tag{58}$$

$$E_N^{(1/2)} = \frac{1}{96\pi^2}\lim_{n\to 4}\Gamma\left(1-\frac{n}{2}\right)\sum_l d_l m_l^{n-2}\rho^2\,, \tag{59}$$

where d_l and q_{le}^2 are the number and mean square charges of spinors of mass m_l and

$$\nu \equiv \tfrac{1}{2}(N-1)\,.$$

Now let us specialize to the case where the compact manifold is a sphere S^N of radius ρ. As we saw in the calculation of C_N, the masses and degeneracies for spin-0 are

$$M_l^2 = l(l+N-1)/\rho^2\,, \tag{60}$$

$$D_l = \frac{(2l+N-1)(l+N-2)!}{(N-1)!\,l!}\,, \tag{61}$$

and for spin-$\frac{1}{2}$

$$m_l = (l+\tfrac{1}{2}N)/\rho\,, \tag{62}$$

$$d_l = \frac{2^{(N+1)/2}(l+N-1)!}{(N-1)!\,l!}\,. \tag{63}$$

The mean square charges Q_{le}^2 and q_{le}^2 are the same for all $O(N+1)$ generators; they are calculated in appendix E as

$$Q_l^2 = [l(l+N-1)]/\tfrac{1}{2}N(N+1)\,, \tag{64}$$

$$q_l^2 = [l^2 + lN + N(N+1)/8]/\tfrac{1}{2}N(N+1)\,. \tag{65}$$

Using methods like those previously applied to the calculation of C_N, we show in appendix F that eqs. (60)–(65) allow us to write (56)–(59) as the integrals:

$$D_N^{(0)} = -\frac{4}{N(N+1)}E_N^{(0)}\,, \tag{66}$$

$$E_N^{(0)} = \frac{1}{48\pi^4\nu}\int_0^{\pi/2} d\theta\,[2\cosh(\tfrac{1}{2}\pi\tan\theta)]^{-2\nu}$$
$$\times\{3\nu\cos\theta\sin 3\theta\sinh(\nu\pi\tan\theta)$$
$$+\left(\frac{3}{\pi}\cos^2\theta\cos 4\theta - \nu^2\pi\cos 2\theta\right)\cosh(\nu\pi\tan\theta)\}\,, \tag{67}$$

$$D_N^{(1/2)} = \frac{(20\nu^2+28\nu+5)}{16(\nu+1)^2}\left\{\frac{16}{N(N+1)}E_N^{(1/2)}\right.$$
$$\left.+\left(\frac{N-1}{N+1}\right)\frac{(-1)^{\nu+1}2^{\nu}}{12\pi^2}\int_0^{\pi/2} d\theta[2\cosh(\tfrac{1}{2}\pi\tan\theta)]^{-2\nu-1}\right\}\,, \tag{68}$$

$$E_N^{(1/2)} = \frac{(-1)^{\nu+1}2^\nu}{12\pi^2} \int_0^{\pi/2} d\theta [2 \cosh (\tfrac{1}{2}\pi \tan \theta)]^{-2\nu-1} \cos \theta \cos 3\theta . \tag{69}$$

where as before, $\nu \equiv \frac{1}{2}(N-1)$.

Instead of giving the values of D_N and E_N, we present in tables 4 and 5 the values of the combinations

$$\beta_N^{(j)} \equiv \frac{(N+4)}{N(N-1)} C_N^{(j)} + 2E_N^{(j)} , \tag{70}$$

$$\gamma_N^{(j)} \equiv \frac{(N+4)}{N(N^2-1)} C_N^{(j)} + D_N^{(j)} , \tag{71}$$

whch appear in the formulas for ρ^2/G and g^2.

6. Results

For spherical compact manifolds, the radius ρ and $O(N+1)$ coupling g are given by eqs. (52) and (50) as

$$\rho^2 = 8\pi G \beta_N , \tag{72}$$

$$g^2 = 1/\gamma_N , \tag{73}$$

where according to (54) and (55)

$$\beta_N = b\beta_N^{(0)} + f\beta_N^{(1/2)} , \tag{74}$$

$$\gamma_N = b\gamma_N^{(0)} + f\gamma_N^{(1/2)} , \tag{75}$$

with $\beta_N^{(j)}$ and $\gamma_N^{(j)}$ the coefficients (70) and (71). As mentioned earlier it is crucial that β_N and γ_N as well as C_N be positive. A glance at tables 4 and 5 shows that for $N = 5$ (mod 4), both $\gamma_N^{(0)}$ and $\gamma_N^{(1/2)}$ are negative, so in this case there is no way

TABLE 4

Values of $\beta_N^{(0)}$ and $\gamma_N^{(0)}$

N	$\beta_N^{(0)}$	$\gamma_N^{(0)}$
3	3.9281071×10^{-4}	$-2.8676136 \times 10^{-5}$
5	7.8068082×10^{-4}	$-7.0734863 \times 10^{-6}$
7	1.1159701×10^{-3}	$-5.5139621 \times 10^{-6}$
9	1.4314887×10^{-3}	$-6.7881832 \times 10^{-6}$
11	1.7353643×10^{-3}	$-8.4410152 \times 10^{-6}$
13	2.0310803×10^{-3}	$-9.9788801 \times 10^{-6}$
15	2.3205490×10^{-3}	$-1.1311056 \times 10^{-5}$
17	2.6049752×10^{-3}	$-1.2443484 \times 10^{-5}$
19	2.8851865×10^{-3}	$-1.3403666 \times 10^{-5}$
21	3.1617857×10^{-3}	$-1.4220508 \times 10^{-5}$

TABLE 5
Values of $\beta_N^{(1/2)}$ and $\gamma_N^{(1/2)}$

N	$\beta_N^{(1/2)}$	$\gamma_N^{(1/2)}$
3	$+3.3832530\times10^{-4}$	$+5.0098703\times10^{-4}$
5	-1.0389980×10^{-4}	-2.6068446×10^{-4}
7	$+4.0049456\times10^{-5}$	$+1.2830668\times10^{-4}$
9	-1.6805816×10^{-5}	-6.1937531×10^{-5}
11	$+7.3681781\times10^{-6}$	$+2.9734556\times10^{-5}$
13	-3.3174183×10^{-6}	-1.4265962×10^{-5}
15	$+1.5203817\times10^{-6}$	$+6.8526826\times10^{-6}$
17	-7.0568431×10^{-7}	-3.2977732×10^{-6}
19	$+3.3066589\times10^{-7}$	$+1.5902169\times10^{-6}$
21	-1.5608669×10^{-7}	-7.6834453×10^{-7}

to avoid a negative value for g^2, which would imply a vacuum that is not even a local minimum of the energy. We therefore must restrict our attention to the cases $N=3\ (\mathrm{mod}\ 4)$. For these cases, all three of $C_N^{(1/2)}$, $\beta_N^{(1/2)}$ and $\gamma_N^{(1/2)}$ are positive so fermions alone furnish a solution. If $N=3\ (\mathrm{mod}\ 4)$ and $N<21$, then $C_N^{(0)}$ and $\beta_N^{(0)}$ are positive while $\gamma_N^{(0)}$ is negative, so solutions exist for which there is a mix of bosons and fermions provided the boson to fermion ratio is not too large, the bound being

$$\frac{b}{f}<\frac{\gamma_N^{(1/2)}}{|\gamma_N^{(0)}|}, \qquad N=3\ (\mathrm{mod}\ 4), \qquad N<21. \tag{76}$$

For $N>21$, the $C_N^{(0)}$ are also negative so the bound is

$$\frac{b}{f}<\min\left\{\frac{C_N^{(1/2)}}{|C_N^{(0)}|}, \frac{\gamma_N^{(1/2)}}{|\gamma_N^{(0)}|}\right\}, \qquad N=3\ (\mathrm{mod}\ 4), \qquad N>21. \tag{77}$$

The bound (76) is given in table 6.

TABLE 6
Upper bounds on the ratio of bosons to fermions

N	Upper bound on b/f: $\dfrac{\gamma_N^{(1/2)}}{\lvert\gamma_N^{(0)}\rvert}$
3	17.47
7	23.27
11	3.523
15	0.6058
19	0.1186

What is not so satisfactory is the *value* of $\beta_N^{(j)}$ and $\gamma_N^{(j)}$. These are so small that we need an enormous number of matter fields to get reasonably small values for the gauge couplings and for $8\pi G/\rho^2$. For instance, suppose (to take a round number) that there are $f = 1000$ species of fermions. The resulting values of the O($N+1$) and gravitational coupling parameters $g^2/8\pi^2$ and $8\pi G/8\pi^2\rho^2$ (which indicate the strength of radiative corrections) are shown in table 7. We see that even for such large numbers of matter fields, these coupling parameters are not much less than unity for any N.

We are not sure how seriously we ought to take this problem. It may be that the smallness of the coefficients $\beta_N^{(j)}$ and $\gamma_N^{(j)}$ is a special property of the sphere, which will not reappear when we turn to more realistic models with less symmetry*. Then again, maybe there really are thousands of species of matter fields.

Even if $g^2/8\pi^2$ turns out small, we are not justified in neglecting higher-order corrections if they would involve very large logarithms of energy ratios. This would be the case for instance if we used our methods to calculate the gauge coupling constants observed at ordinary energies; loops would then be associated with factors like $(g^2/8\pi^2)\ln(m_W\rho)$. The renormalization group approach teaches us that to avoid these large logarithms we should use the full $(4+N)$-dimensional theory to calculate gauge coupling constants that are renormalized at energies of order $1/\rho$, then form an effective gauge field theory by integrating out all particles of mass $\approx 1/\rho$, and only then calculate the gauge coupling constants at ordinary energies by integrating the Gell-Mann–Low equations of the effective theory down from energies

TABLE 7

Values of the O($N+1$) and gravitational coupling parameters for the case $f = 1000$, $b = 0$

N	Coupling parameters $\frac{g^2}{8\pi^2}$	Coupling parameters $\frac{G}{\pi\rho^2}$
3	0.253	0.0374
7	0.0987	0.316
11	0.426	1.72
15	1.85	8.33
19	7.96	38.3

* It is possible that the potential may get large contributions from one-loop graphs involving the towers of massive particles of spin 2, 1, and 0 associated with the metric itself. The numbers of these towers with spin 1 or 0 grow with N like N and N^2, respectively, so we may be able to exploit a large N instead of large b or f approximation to justify the neglect of higher graphs. Also, Duff informs us that the one-loop graphs involving massive spin-2 particles typically carry very large numerical factors, and suggests that these may dominate the one-loop potential.

of order $1/\rho$. It is unlikely that any specially simple results should be expected for fine structure constants like $\frac{1}{137}$ that appear as the end-product of this long chain of calculation.

7. Stability

We have so far only found a set of equilibrium configurations of our models, and considered their instability against growing Yang–Mills and gravitational waves. We must also ask whether the solution is stable against perturbations of the compact manifold itself.

This question can most conveniently be studied in terms of an effective potential for the whole system of matter and gravitation. In order to study stability against spacetime-independent perturbations, it is enough to consider the total action under the constraint of Poincaré invariance, which allows it to be written as

$$I_{\text{total}} = -\int d^4x\, V_{\text{eff}}\,, \tag{78}$$

$$V_{\text{eff}} = \frac{1}{16\pi\bar{G}}\int d^N y\sqrt{\tilde{g}(y)}[\tilde{R}(y)+\bar{\Lambda}] + V[\tilde{g}]\,. \tag{79}$$

As is well-known [13], a potential V_{eff} constructed in this way is equal to the minimum energy of any state in which the expectation value of the fields has the value indicated by the argument of the potential—in our case, $\tilde{g}_{NM}(x, y)$. Therefore, provided that there are no negative-energy perturbations, a stable solution is one associated with a *minimum* of V_{eff}.

In studying perturbations which also preserve those symmetries of the manifold that keep it of the type described in appendix A (e.g. S^N, CP^N, etc.), we can regard the metric g_{nm} here as depending on a single radius parameter ρ. Eqs. (15) and (18) then let us write (79) as

$$V_{\text{eff}} = \frac{1}{16\pi G_0}\left[\frac{-N(N-1)}{\rho^2}+\bar{\Lambda}\right] + V(\rho)\,. \tag{80}$$

However it should be kept in mind that it is $\bar{G}$ that is ρ-independent, with $G_0(\rho)$ given by (18) as proportional to ρ^{-N}. Hence the derivative of V_{eff} is given by

$$\rho\frac{dV_{\text{eff}}}{d\rho} = \frac{N}{16\pi G_0}\left[\frac{-N(N-1)}{\rho^2}+\bar{\Lambda}\right] + \frac{2N(N-1)}{16\pi G_0\rho^2} + \rho\frac{dV}{d\rho}\,.$$

We see that the field equation (23) is simply the statement that V_{eff} is stationary in ρ. (The other field equation (22) just says that $\bar{\Lambda}$ takes a value that makes V_{eff} vanish at its stationary point.) We expect that V_{eff} can also be used to judge the stability of the solution against spacetime-independent uniform dilations or contractions of the compact manifold: it is stable if and only if the stationary point of V_{eff} is a *minimum*.

Strictly speaking, this is only a guess. We really should consider the contribution to the action of terms quadratic in the time derivative of infinitesimal perturbations. We are anticipating here that these terms will as usual be negative, so that the field equations for a perturbation $\delta\rho$ will give $\delta\ddot{\rho}$ of opposite sign to $\delta\rho\, \mathrm{d}^2V/\mathrm{d}\rho^2$, and hence $\delta\rho$ will oscillate stably if $\mathrm{d}^2V/\mathrm{d}\rho^2$ is positive. This should be checked.

Returning now to $V_{\text{eff}}(\rho)$, for very small ρ we know that V_{eff} for N odd grows like C_N/ρ^4. This is the behavior when all fields have vanishing $(4+N)$-dimensional mass, and even if they do not the potential becomes independent of their masses for sufficiently small ρ. On the other hand, the curvature and cosmological constant terms in (80) go as ρ^{N-2} and ρ^N, so are negligible compared with $V(\rho)$ as $\rho \to 0$, and therefore

$$V_{\text{eff}}(\rho) \to \frac{C_N}{\rho^4}, \qquad \text{as } \rho \to 0\,.$$

For very large ρ we expect $V(\rho)$ to vanish as ρ^{-4} if all fields in $4+N$ dimensions are massless, or at worst approach a constant if there are $(4+N)$-dimensional masses in the theory. Here the dominant term is the cosmological term, which grows as ρ^N:

$$V_{\text{eff}}(\rho) \to \frac{\bar{\Lambda}}{16\pi G_0} \propto \rho^N, \qquad \text{as } \rho \to \infty\,.$$

It is clear that for C_N and $\bar{\Lambda}$ positive, the potential $V_{\text{eff}}(\rho)$ must have a minimum at some finite ρ. We found that for C_N and $\bar{\Lambda}$ positive, there is only one stationary point, and it must therefore be a minimum. Our tentative conclusion is that our solution is stable against uniform dilations or contractions. We could have found an equilibrium solution with C_N, G_0, and $\bar{\Lambda}$ all negative, but it would have been unstable.

We still have to consider stability against deformation. Page [14] has calculated the change in the C_N coefficients, for massless scalar fields, when the spherical compact manifold is subject to an aspherical but homogeneous deformation. We understand that work on the case of fermions is in progress. In fact, we would not be unhappy to learn that spheres are unstable since the high energy gauge group may very well not be an $\mathrm{SO}(N+1)$.

Even a manifold that is stable against all deformations will become unstable if the temperature is raised above a critical value T_c, which is independent of the shape of the manifold. To see this, it is sufficient to examine the form of the free energy $F(\rho)$ for large values of the radius ρ. Suppose that the temperature has some finite value T, and consider the behavior of the total free energy when the linear scale of the manifold is much larger than $1/kT$. For definiteness, suppose that there are b species of massless scalars in $4+\mathrm{N}$ dimensions. In 4 dimensions, these appear as b towers of massive particles, yielding a free energy

$$F = V_{\text{eff}} + b \sum_l D_l f(T, M_l) \,, \tag{81}$$

where V_{eff} is the zero temperature free energy discussed previously, and $f(T, M)$ is the thermal free energy of a scalar of mass M

$$f(T, M) = -\frac{4\pi}{3(2\pi)^3} \int_0^\infty \frac{p^4 \, \mathrm{d}p}{E_M(p)[\exp(E_M(p)/kT) - 1]} \,. \tag{82}$$

For a manifold whose linear scale is much larger than $1/kT$, the sum in (81) is dominated by very high eigenmodes, and can be replaced with an integral

$$F = V_{\text{eff}} + b \int_0^\infty D(M) f(T, M) \, \mathrm{d}M \,, \tag{83}$$

where $D(M)\,\mathrm{d}M$ is the number of eigenmodes with M_l between M and $M + \mathrm{d}M$. For instance, for spheres of radius ρ we have

$$M_l \to l/\rho \,,$$

$$D_l \to 2l^{N-1}/(N-1)!$$

and so

$$D(M) = \frac{2\rho^N M^{N-1}}{(N-1)!} \,.$$

We can also write this in terms of the volume of the N-sphere

$$\Omega_N = \frac{2\pi^{(N+1)/2}}{\Gamma[\frac{1}{2}(N+1)]} \rho^N \,,$$

so that

$$D(M) = \frac{\Omega_N M^{N-1}}{2^{N-1} \pi^{N/2} \Gamma(\frac{1}{2}N)} \,. \tag{84}$$

The point of writing the formula this way is that in this form it is valid for manifolds of arbitrary shape, not just spheres [15]. It is now elementary to substitute (82) and (84) in (83), and do the integrals over M and p/M (in that order). The result is

$$F = V_{\text{eff}} - \frac{b(N+3)!\,\zeta(N+4)\,\Omega_N (kT)^{N+4}}{2^{N+3} \pi^{(N+3)/2} \Gamma[\frac{1}{2}(N+5)]} \,. \tag{85}$$

This formula applies (for large manifolds) even to the free energy generated by the massive excitations of the metric itself, with

$$b_{\text{grav}} = \tfrac{1}{2}(N+4)(N+5) - 2(N+4) = \tfrac{1}{2}(N+1)(N+4) \,.$$

The negative $(kT)^5\rho$ term contributed by the excitations of the metric has already been found in the original Kaluza–Klein model (where $b_{\text{grav}} = 5$) by Rubin and Roth [16]. It is easy to see that (85) holds as long as the total dimensionality is $N+4$, whatever the dimensionalities of the compact manifold and Minkowski subspace.

Comparing (85) with (80) and recalling that $G_0 = \bar{G}/\Omega_N$, we see that the cosmological constant term dominates over the thermal term for large manifolds only for T less than a critical temperature T_c, given by

$$\frac{\bar{\Lambda}}{16\pi\bar{G}} = \frac{b(N+3)!\,\zeta(N+4)(kT_c)^{N+4}}{2^{N+3}\pi^{(N+3)/2}\Gamma[\frac{1}{2}(N+5)]}\,. \tag{86}$$

This gives kT_c roughly of order $1/\sqrt{bG}$. For T greater than T_c, the free energy (85) becomes increasingly negative for $\rho \to \infty$, and we no longer have even an equilibrium solution.

This suggests that there is a dramatic phase transition at $T = T_c$, in which the compactified dimensions explode outward. To see what happens then, it will clearly be necessary to give up our assumption of Poincaré invariance, and include at least a time dependence in our ansatz for the $(4+N)$-dimensional metric. This phase transition presumably will come into play in the interior of black holes (though perhaps not in any finite time as seen by outside observers) and perhaps also in the last stages of black hole evaporation and cosmic contraction. One wonders also whether our universe started with equal circumferences in all $3+N$ spatial directions, and became tightly contracted in N of those dimensions only when the temperature fell below the critical temperature T_c. Work on these problems is in progress.

We have been greatly helped throughout the course of this work by frequent enlightening conversations with M. Duff, D. Page, and M. Rubin. Thanks for valuable discussions are also due to B. DeWitt, H. Georgi, P. Ginsparg, C. Pope, D. Toms and E. Witten. We wish to thank M. Awada and D. Toms for bringing to our attention errors with regard to the calculation of the $D_N^{(1/2)}$ and $E_N^{(1/2)}$ coefficients that were present in an earlier version of this article and for communicating their work prior to publication.

Appendix A

ONE-PARAMETER MANIFOLDS

In this appendix we identify a class of Riemmanian manifolds whose symmetries dictate that:

(i) the metric has only a single free parameter ρ, which appears (in a suitable coordinate system) as a multiplicative factor ρ^2;

(ii) all symmetric second-rank tensors that share the symmetries of the metric are proportional to the metric, with constant coefficients.

By a tensor S_{nm} sharing the symmetries of the metric, we mean that for any Killing vector $\xi^n(y)$ that satisfies the usual condition

$$\xi_{n;m} + \xi_{m;n} = 0 , \qquad (A.1)$$

the Lie derivative of S with respect to ξ vanishes [17]:

$$\xi^p S_{nm;p} + S_{pm}\xi^p{}_{;n} + S_{np}\xi^p{}_{;m} = 0 . \qquad (A.2)$$

(Tildes are dropped in this appendix.) We will see that (as is pretty obvious) the matter energy-momentum tensor T^{nm} and the Ricci tensor R^{nm} share the symmetries of the metric in this sense, so property (ii) allows us to replace the $\frac{1}{2}N(N+1)$ differential equations (12) with a single algebraic equation.

We understand that the conditions for a manifold to satisfy properties (i) and (ii) are well known to mathematicians [18]. However, we have not been able to find a reference for this, and we felt that a discussion here might be of some use to physicists working on this sort of theory. (See added note below.)

The manifolds to be considered here are homogeneous, with a group of isometries G. By "homogeneous" we mean that any point is carried into any other point by some element of G. For any specific point y_0, there may be some subgroup H of G that leaves y_0 fixed. (Homogeneity ensures that any other point is left fixed by a subgroup isomorphic to H.) Any other point y can be obtained from y_0 by some transformation g in G, so points in the manifold can be labelled by the elements of G, two elements g_1, g_2 of G corresponding to the same point of the manifold if $g_1^{-1}g_2$ is in H. That is, elements of the manifold are in one-to-one correspondence with the cosets that make up the coset space G/H. It is often convenient to split the generators of G into generators T_i of H and a linearly independent set X_n outside the algebra of H, and define the coordinates y^n of the manifold (at least locally) by specifying that y_0 is carried into y by the G-element $\exp(iy^n X_n)$.

Let us first apply the invariance of the metric and other tensors S_{nm} under the isotropy subgroup H. An infinitesimal element $h = 1 + i\varepsilon^i T_i$ of H induces a coordinate transformation

$$y^n \to y'^n = y^n + \varepsilon^i \xi^n{}_i(y) , \qquad (A.3)$$

with ξ^n_i a set of Killing vectors corresponding to the generators T_i of H. Since H leaves y_0 fixed, these Killing vectors vanish as y_0, so (A.2) becomes at y_0:

$$S_{pm}(y_0)(t_i)^p{}_n + S_{np}(y_0)(t_i)^p{}_m = 0 , \qquad (A.4)$$

where

$$(t_i)^p{}_n \equiv (\partial_n \xi^p_i)_{y=y_0} . \qquad (A.5)$$

The matrices t_i are just the generators of H in the representation D_X furnished by the coordinates, i.e. by the generators X_n of G that are not in the subgroup H. If the symmetric direct product $(D_X \otimes D_X)_s$ contains the identity representation just once, then (A.4) gives

$$S_{nm}(y^0) = \mathscr{S}\Delta_{nm}\,, \tag{A.6}$$

where Δ_{nm} is the Clebsch–Gordan coefficient for forming a singlet from $(D_X \otimes D_X)_s$, and $\mathscr{S}$ is an unknown constant.

Now we apply the invariance of these tensors under finite transformations in G that do not leave y_0 fixed. In general, the integrated form of (A.2) is

$$S_{nm}(y') = S'_{nm}(y') \equiv \frac{\partial y^p}{\partial y'^n}\frac{\partial v^q}{\partial y'^m} S_{pq}(y)\,, \tag{A.7}$$

where $y \to Y'(y, g)$ is the transformation induced by some element g of G. Setting $y = y_0$ and taking g as $\exp(iy^n X_n)$ (with a new y^n), we find

$$S_{nm}(y) = \mathscr{S}(B^{-1}(y))^p{}_n (B^{-1}(y))^q{}_m \Delta_{pq} \tag{A.8}$$

where $B(y)$ is the "boost" matrix

$$B^n{}_m(y) = \frac{\partial}{\partial y_0^m} y'^n(y_0, \exp(iy^n X_n))\,. \tag{A.9}$$

The important thing about (A.8) is that it shows that all symmetric tensors that satisfy the invariance conditions (A.2) are equal, up to multiplicative constant factors. This is in particular true of the metric tensor (for which (A.1) and (A.2) are the same) so the metric tensor is uniquely specified by its symmetries and our choice of coordinates up to a multiplicative constant, and all other tensors satisfying (A.2) (including the Ricci tensor) are proportional to the metric.

The special assumption that led to this result was that the manifold is a coset space G/H for which the generators of G not in H form a representation D_X of H such that the symmetric direct product $(D_X \times D_X)_s$ contains the identity just once. The real representation D_X will have this property if and only if it is either irreducible, or the sum of an irreducible complex representation and its complex conjugate. Let us see how this works in some familiar cases.

Spheres: The sphere S^N can be described as the coset space SO(N+1)/SO(N). The generators X_n of SO(N+1) which are not in SO(N) for a single N-vector representation of SO(N). Since this is irreducible, properties (i) and (ii) hold for spheres, as is of course well known.

CP^N: The space CP^N can be described as the coset space SU(N+1)/SU(N)× U(1). The generators of SU(N+1) which are not in SU(N)×U(1) form two representations $\boldsymbol{N}$ and $\bar{\boldsymbol{N}}$ of SU(N), with opposite values for the U(1) quantum number. Since $\boldsymbol{N}$ is an irreducible complex representation of SU(N)×U(1), properties (i) and (ii) hold also for CP^N.

Group manifolds: We can endow the manifold of a compact Lie group G with a symmetry under the group G×G, an element g of G being carried by an element

(a, b) of G×G into agb^{-1}. The isotropy group H that leaves the particular element $g = g_0$ of G invariant is $\mathrm{G}_{\mathrm{diag}}$, consisting of elements (a, b) of G×G with $b = g_0^{-1} a g_0$, so the manifold of G can be described as the coset space $\mathrm{G}\times\mathrm{G}/\mathrm{G}_{\mathrm{diag}}$. If A_n and B_n are the generators of the left and right G factors of G×G, then the generators of $\mathrm{H}=\mathrm{G}_{\mathrm{diag}}$ are $A_n + g_0^{-1} B_n g_0$, and the generators X_n of G×G that are not in $\mathrm{G}_{\mathrm{diag}}$ can be taken as just A_n. The A_n transform according to the adjoint representation of $\mathrm{G}_{\mathrm{diag}}$, which for a simple group G is real and irreducible, so properties (i) and (ii) hold for simple group manifolds. On the other hand, if the group G is the direct product of $K > 1$ simple subgroups, then $(\mathrm{D}_X \otimes \mathrm{D}_X)_{\mathrm{s}}$ contains the identity K times, and properties (i) and (ii) do not hold.

The squashed 7-sphere: The squashed 7-sphere can be described as the coset space SO(5)×SO(3)/SO(3)×SO(3), with the generators of SO(5)×SO(3) that are not in SO(3)×SO(3) transforming as the SO(3)×SO(3) representation (1,3)+(2,2) [19]. There are here two ways of forming a singlet, from (1,3)×(1,3) or (2,2)×(2,2), so (as is well known) properties (i) and (ii) do not hold for the squashed 7-sphere.

There is a generalization of properties (i) and (ii), which applies to all homogeneous spaces G/H. If the elements of G that are not in H form a representation D_X for which the symmetric direct product $(\mathrm{D}_X \otimes \mathrm{D}_X)_{\mathrm{s}}$ contains the identity K times, then (A.8) is replaced with

$$S_{nm}(y) = \sum_{j=1}^{K} \mathcal{S}_j (B^{-1}(y))^p{}_n (B^{-1}(y))^q{}_m \Delta^{(j)}_{pq} ,$$

with $\Delta^{(j)}$ a set of Clebsch–Gordan coefficients that depend only on the manifold and our coordinate system, and $\mathcal{S}_j$ a set of constant coefficients. The metric and all other tensors satisfying (A.2) are thus linear combinations of just K standard tensors. This is in particular true for the manifolds of semi-simple groups, for which K is the number of simple factors, and for the squashed 7-sphere, for which $K = 2$.

Finally we note that one class of tensors that automatically share the symmetries of the metric are those formed by variational derivatives of scalar functionals $F[g]$ with respect to the metric:

$$\tfrac{1}{2}\sqrt{g(y)}\, S^{nm}(y) \equiv \frac{\delta F[g]}{\delta g_{nm}(y)} . \tag{A.10}$$

This can be proved formally by noting that the general covariance of $F[g]$ implies that S^{nm} is conserved for all metrics, and therefore

$$\int \xi_n S^{nm}{}_{;m} \sqrt{g}\, \mathrm{d}^N y = 0 ,$$

for *all* g_{nm}, whether or not ξ_n is a Killing vector of g_{nm}. A small variation in g_{nm}

(with ξ_n fixed) thus satisfies

$$\int \xi_n(\delta S^{nm})_{;m}\sqrt{g}\,\mathrm{d}^N y + \int \xi_n S^{pm}\sqrt{g}\,\delta\Gamma^n_{pm}\,\mathrm{d}^N y = 0\,.$$

Now fix g_{nm} to be a metric with Killing vector ξ_n. Integrating the first term here by parts and using (A.1) and the symmetry of S^{nm}, we see that this term vanishes. Also, the variation in the affine connection can be expressed as

$$\delta\Gamma^n_{pm} = \tfrac{1}{2}g^{nq}[(\delta g_{qp})_{;m} + (\delta g_{qm})_{;p} - (\delta g_{pm})_{;q}]\,.$$

Using this in the second term above and integrating by parts then yields

$$0 = \int [\xi^m S^{pq}{}_{;m} - \xi^q{}_{;m}S^{mp} - \xi^p{}_{;m}S^{mq}]\sqrt{g}\,\delta g_{qp}\,\mathrm{d}^N y\,.$$

Since this holds for all δg_{qp}, the coefficient of δg_{qp} must vanish which (using (A.1) again) is equivalent to (A.2).

The Einstein tensor $R^{nm} - \frac{1}{2}g^{nm}R$ and the matter-fluctuation energy-momentum tensor T^{nm} are of the type (A.10), with $F[g]$ respectively the action $\int \mathrm{d}^N y\sqrt{g}R(y)$ and the potential V. Thus, without any calculation, we know that S^N, CP^N, etc. are Einstein spaces, and that the field equations reduce for each such manifold to a single algebraic equation for the radius ρ.

Added Note. The results regarding one-parameter manifolds discussed in appendix A were earlier derived by Zee [20], and Boulware and Brown [21]. It may be useful to note further that a homogeneous space G/H can satisfy the irreducibility condition for a compact manifold (i.e. that the generators of G not in the algebra of H form a representation of H that is irreducible into real representations) only if H is a maximal subgroup of G (i.e., there is no subgroup of G that contains H, other than G or H). Also, the irreducibility condition *will* be satisfied if H is maximal and G/H is a symmetric space (in the sense that G has an automorphism that leaves the generators of H unchanged and reverses the sign of the other generators of G); see Helgason [22], Peskin [23].

Appendix B

THE EFFECTIVE POTENTIAL FOR $\mathrm{M}^4 \times \mathrm{S}^N$

In this appendix we shall compute the effective potential for the scalar and Dirac fields for a background geometry which is the direct product of a four-dimensional Minkowski spacetime with the N-sphere.

Minimally coupled scalar fields

We shall take the action for a minimally coupled scalar field to be

$$S[\phi] = -\tfrac{1}{2}\mu^{n-4}\int \mathrm{d}^n x \int \mathrm{d}^N y\, g^{1/2}(\phi_{;A}\phi^{;A}_{;} + m^2\phi^2)\,, \qquad \text{(B.1)}$$

where n, in the sense of dimensional regularization, denotes the dimension of spacetime, upper case indices run over the values $0, 1, \ldots, n+N-1$, and μ is an arbitrary constant with the dimension of mass.

The effective potential V is defined through the relation

$$\exp\left(-i\mu^{n-4}\int \mathrm{d}^n x\, V\right) = \int \mathscr{D}\phi\, \mathrm{e}^{iS[\phi]}. \tag{B.2}$$

It follows that

$$\exp\left(-i\mu^{n-4}\int \mathrm{d}^n x\, V\right) = \det{}^{-1/2}(-\Box + m^2), \tag{B.3}$$

and hence that

$$V = -\tfrac{1}{2}i\mu^{-(n-4)}\int \frac{\mathrm{d}^n k}{(2\pi)^n}\sum_{l=0}^{\infty} D_l \log(k^2 + \Lambda_l^2\rho^{-2} + m^2), \tag{B.4}$$

where $-\Lambda_l^2$ and D_l denote the eigenvalues and the degeneracies of the laplacian on the unit N-sphere. These are known to be [9]

$$\Lambda_l^2 = l(l+N-1), \qquad D_l = (2l+N-1)\frac{\Gamma(l+N-1)}{\Gamma(N)l!}, \tag{B.5}$$

where we express the D_l in a form that anticipates that it will be convenient in the following to regard N as a continuous variable. A short calculation reveals that

$$\int \frac{\mathrm{d}^n k}{(2\pi)^n}\log(k^2 + m_l^2) = -\frac{i\Gamma(-\frac{1}{2}n)m_l^n}{(4\pi)^{n/2}}. \tag{B.6}$$

On using this result in (4) we find

$$V = -\frac{\mu^{-(n-4)}\Gamma(-\frac{1}{2}n)}{(4\pi)^{n/2}\rho^n}\sum_{l=0}^{\infty}\frac{\Gamma(l+N-1)}{l!\,\Gamma(N)}(l+\tfrac{1}{2}(N-1))(l(l+N-1)+m^2\rho^2)^{n/2}, \tag{B.7}$$

subject to the restriction $\mathrm{Re}\,(N+n) < 0$ which is required to secure the convergence of the sum.

We seek now to simplify the sum and to analytically continue V to the required values of n and N. To this end we note that the summand in (B.7) is a binomial coefficient multiplied by the troublesome term

$$\begin{aligned}&(l+\tfrac{1}{2}(N-1))[l(l+N-1)+m^2\rho^2]^{n/2}\\&= (l+\tfrac{1}{2}(N-1))[(l+\tfrac{1}{2}(N-1))^2 - ((\tfrac{1}{2}(N-1))^2 - m^2\rho^2)]^{n/2}.\end{aligned} \tag{B.8}$$

Our strategy will be to replace this quantity by its representation as a Laplace integral; this will enable us to perform the sum. The following identity proves useful in this regard

$$z(z^2-b^2)^{n/2}=\frac{\pi^{1/2}2^{n/2+1/2}b^{n/2+3/2}}{\Gamma(-\frac{1}{2}n)}\int_0^\infty \mathrm{d}t\, \mathrm{e}^{-zt}t^{-n/2-1/2}I_{-n/2-3/2}(bt)\,, \qquad \mathrm{Re}\, n<-1 \tag{B.9}$$

with I a modified Bessel function [9]. On using (B.9) in (B.7) and taking note of the fact that

$$\begin{aligned}\sum_{l=0}^{\infty}\frac{\Gamma(l+N-1)}{l!}\mathrm{e}^{-(l+\frac{1}{2}(N-1))t}&=\Gamma(N-1)\,\mathrm{e}^{-(N-1)t/2}(1-\mathrm{e}^{-t})^{-(N-1)}\\&=\Gamma(N-1)(2\,\mathrm{sh}\,\tfrac{1}{2}t)^{-(N-1)}\,,\end{aligned} \tag{B.10}$$

we find

$$V=-\frac{\mu^{-(n-4)}}{(2\pi)^{n/2-1/2}(N-1)\rho^n}\int_0^\infty \mathrm{d}t\, t^{-N-n-1}f(t^2)\,, \qquad \mathrm{Re}\,(N+n)<0\,, \tag{B.11}$$

where we have defined

$$\begin{aligned}f(t^2)&=\left(\frac{t}{2\mathrm{sh}\frac{1}{2}t}\right)^{N-1}(bt)^{n/2+3/2}I_{-n/2-3/2}(bt)\,,\\ b^2&=[\tfrac{1}{2}(N-1)]^2-m^2\rho^2\,,\end{aligned} \tag{B.12}$$

which we observe to be a function analytic in t^2 about $t^2=0$.

It remains now to continue the integral in (B.11) to the physical values of N and n. One way to proceed would be to integrate by parts repeatedly. However a simpler method presents itself. Consider the integral

$$\int_C \mathrm{d}t\, t^{-N-n-1}f(t^2)\,,$$

with C the contour of fig. 1. By displacing the contour to C′, which runs just above the branch cut associated with t^{-N-n-1}, we see that

$$\int_C \mathrm{d}t\, t^{-N-n-1}f(t^2)=(1-\mathrm{e}^{-i\pi(N+n)})\int_0^\infty \mathrm{d}t\, t^{-N-n-1}f(t^2)\,, \tag{B.13}$$

i.e.

$$\int_0^\infty \mathrm{d}t\, t^{-N-n-1}f(t^2)=\frac{1}{(1-\mathrm{e}^{-i\pi(N+n)})}\int_C \mathrm{d}t\, t^{-N-n-1}f(t^2)\,. \tag{B.14}$$

We observe that since the contour C does not run through the point $t=0$ the integral over C exists and is analytic for all values of $N+n$, the poles of the left-hand side

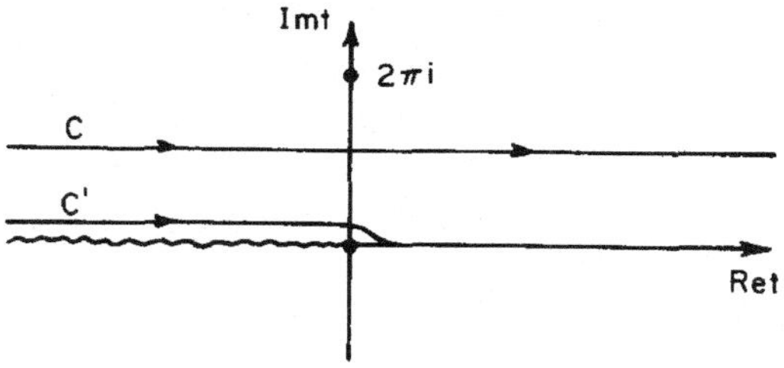

FIG. 1.

being contained in the factor

$$\frac{1}{(1-e^{-i\pi(N+n)})}\,.$$

The right-hand side of (B.14) represents, therefore, the analytic continuation of the left-hand side to the entire $N+n$ plane with the exception of the points $N+n=0, 2, 4, \ldots$. In order to avoid the poles for $N+n$ even we shall set $n=4$ and take N to be an odd integer. It follows from (B.12) and (B.14) that

$$V=-\frac{1}{2(2\pi)^{3/2}(N-1)\rho^4}\int_C \frac{dt}{t^6}\frac{(bt)^{7/2}I_{-7/2}(bt)}{(2\mathrm{sh}\frac{1}{2}t)^{N-1}}\,, \qquad n=4\,, \qquad N \text{ odd}\,. \tag{B.15}$$

We specialize now to the case $m=0$ for which

$$b=\tfrac{1}{2}(N-1)\equiv\nu\,, \qquad m=0\,. \tag{B.16}$$

Note also that since the Bessel function is of half odd-integer order it is related to hyperbolic functions. Specifically

$$z^{7/2}I_{-7/2}(z)=\sqrt{2/\pi}\{(15z+z^3)\,\mathrm{sh}\,z-(15+6z^2)\,\mathrm{ch}\,z\}\,, \tag{B.17}$$

and hence

$$V=-\frac{1}{(2\pi)^2 2\nu\rho^4}\int_C \frac{dt}{t^6}\frac{1}{(2\mathrm{sh}\frac{1}{2}t)^{2\nu}} \times\{(15\nu t+(\nu t)^3)\,\mathrm{sh}\,(\nu t)-(15+6(\nu t)^2)\,\mathrm{ch}\,(\nu t)\}\,. \tag{B.18}$$

The integrand has poles corresponding to the zeros of $\mathrm{sh}\,\frac{1}{2}t$ i.e. when $t=\pm 2p\pi i$, $p=1,2,\ldots$ and may, in principle, be evaluated by residues which leads to sums of Riemann ζ functions. Thus, for example, we find for $\nu=1$ (i.e. $N=3$)

$$\rho^4 V=\frac{1}{2}\left(\frac{\zeta(3)}{(2\pi)^4}-\frac{39\zeta(5)}{(2\pi)^6}+\frac{90\zeta(7)}{(2\pi)^8}\right)=7.56870\times 10^{-5}\,, \qquad n=4\,, \qquad N=3\,, \qquad m=0\,. \tag{B.19}$$

However the computations become arduous if ν is large. We prefer, therefore, to cast (B.18) in a form that is amenable to numerical integration. We take C to be the contour $(-\infty+\pi i, \infty+\pi i)$. On setting

$$t=\pi(i+\tan\theta)\,, \tag{B.20}$$

and availing ourselves of the elementary identity

$$\int_{-\pi/2}^{\pi/2} d\theta\, h(\theta)=\int_0^{\pi/2} d\theta\,(h(\theta)+h(-\theta))\,, \tag{B.21}$$

we obtain our final expression for the effective potential

$$V=\frac{1}{4\pi\nu\rho^4}\int_0^{\pi/2}\frac{d\theta}{[2\,\mathrm{ch}\,(\frac{1}{2}\pi\tan\theta)]^{2\nu}}\left\{\left(\frac{\nu^3}{\pi^3}\cos\theta\sin 3\theta-\frac{15\nu}{\pi^5}\cos^3\theta\sin 5\theta\right)\right.$$
$$\left.\times\,\mathrm{sh}\,(\nu\pi\tan\theta)+\left(\frac{6\nu^2}{\pi^4}\cos^2\theta\cos 4\theta-\frac{15}{\pi^6}\cos^4\theta\cos 6\theta\right)\mathrm{ch}\,(\nu\pi\tan\theta)\right\}\,,$$
$$n=4\,,\qquad N\ \text{odd}\,,\qquad m=0\,. \tag{B.22}$$

The case $\nu=0$ (i.e. $N=1$) is best dealt with by returning to (B18) and taking the limit $\nu\to 0$. Since

$$\int_C \frac{dt}{t^6}=0\,, \tag{B.23}$$

we find in this case that

$$V=-\frac{15}{4\pi^2\rho^4}\int_C\frac{dt}{t^6}\log(\mathrm{sh}\,(\tfrac{1}{2}t))\quad . \tag{B.24}$$

Integration by parts and evaluation by residues immediately yields

$$V=-\frac{3\zeta(5)}{(2\pi)^6\rho^4}=-\frac{5.05576\times 10^{-5}}{\rho^4}\,,\qquad n=4\,,\qquad N=1\,,\qquad m=0 \tag{B.25}$$

in agreement with the result of ref. [5].

Conformally coupled scalar fields

The case of conformal coupling is dealt with by making the replacement

$$m^2\to\frac{N(N+2)(N-1)}{4(N+3)\rho^2}\,, \tag{B.26}$$

or equivalently, the replacement

$$b\to i\gamma\,,\qquad \gamma^2=\frac{3}{4}\left(\frac{N-1}{N+3}\right)\,, \tag{B.27}$$

in (15). This leads rapidly to the expression

$$C_N^{\text{conformal}} = \frac{(-1)^{\nu+1}}{8\pi\nu}\int_0^{\pi/2} \frac{d\theta}{[2\,\text{ch}\,(\frac{1}{2}\pi\tan\theta)]^{2\nu}}$$

$$\times\left\{\frac{15\gamma}{\pi^5}\cos^3\theta(e^{-\pi\gamma}\cos(5\theta+\pi\gamma\tan\theta)-e^{\pi\gamma}\cos(5\theta-\pi\gamma\tan\theta))\right.$$

$$+\frac{\gamma^3}{\pi^3}\cos\theta(e^{-\pi\gamma}\cos(3\theta+\pi\gamma\tan\theta)-e^{\pi\gamma}\cos(3\theta-\pi\gamma\tan\theta))$$

$$+\frac{15}{\pi^6}\cos^4\theta(e^{-\pi\gamma}\cos(6\theta+\pi\gamma\tan\theta)+e^{\pi\gamma}\cos(6\theta-\pi\gamma\tan\theta))$$

$$\left.+\frac{6\gamma^2}{\pi^4}\cos^2\theta(e^{-\pi\gamma}\cos(4\theta+\pi\gamma\tan\theta)+e^{\pi\gamma}\cos(4\theta-\pi\gamma\tan\theta))\right\},$$

$$n=4,\ N \text{ odd},\ m=0,\ \text{conformal coupling}. \qquad \text{(B.28)}$$

Numerical integration yields the results of table 3. A check on the numerical work is provided by evaluating the integral by residues for the case $\nu = 1$ (i.e. $N = 3$) this yields the expression

$$C_3 = -\frac{1}{107520\pi} + \frac{1}{512\pi}\sum_{p=1}^{\infty}\frac{e^{-p\pi}}{(p\pi)^3}\left(1+\frac{9}{p\pi}+\frac{39}{(p\pi)^2}+\frac{90}{(p\pi)^3}+\frac{90}{(p\pi)^4}\right)$$

$$= 7.145890079\times 10^{-6}.$$

$$n=4,\ N=3,\ m=0, \qquad \text{conformal coupling}. \qquad \text{(B.29)}$$

These values of C_n can be used to calculate the potential V for the case of conformal coupling, but in this case there are other changes in the gravitational field equations, which we shall not consider in this paper.

Fermions

The calculation of the effective potential for fermions is a problem of a slightly more recondite nature than the corresponding scalar problem since, in order to find the expression analogous to (B.3), we must first compute the eigenvalues and the degeneracies of the Dirac laplacian on S^N. We shall therefore calculate V via the Green function method and discover the eigenvalues and their degeneracies as we proceed.

Our conventions regarding the Dirac matrices, $\hat{\gamma}^A$, for $M^4 \times S^N$ are as follows

$$\{\hat{\gamma}^A, \hat{\gamma}^B\} = 2g^{AB},$$

$$\hat{\gamma}^\mu = \gamma^\mu \otimes \mathbb{1}, \qquad \hat{\gamma}^p = \gamma^5 \otimes \Gamma^p, \qquad (\gamma^5)^2 = +1, \qquad \text{(B.30)}$$

where the $\gamma^\mu(\mu = 0, 1, 2, 3)$ are Dirac matrices appropriate to M^4 while the $\Gamma^p(p = 4, 5, \ldots, 4+N-1)$ are those appropriate to S^N.

We take the Dirac action to be

$$S[\psi]=i\mu^{n-4}\int d^n x\int d^N y\, g^{1/2}\bar{\psi}(\hat{\gamma}^A\psi_{;A}+m)\psi\,, \tag{B.31}$$

and define the Green function and the effective potential by

$$G(x,y;x',y')=i\mu^{n-4}\frac{\int \mathscr{D}\psi\mathscr{D}\bar{\psi}\psi(x,y)\bar{\psi}(x',y')\,e^{iS[\psi]}}{\int \mathscr{D}\psi\mathscr{D}\bar{\psi}\,e^{iS[\psi]}}\,, \tag{B.32}$$

$$\exp\left(-i\mu^{n-4}\int d^n x\, V\right)=\int \mathscr{D}\psi\mathscr{D}\bar{\psi}\,e^{iS[\psi]}\,. \tag{B.33}$$

By differentiating V with respect to m we see that the effective potential is related to the coincident limit of the Green function by the equation

$$\frac{\partial V}{\partial m}=\mu^{-(n-4)}\Omega_N \operatorname{Tr} G(x,y;x,y)\,, \tag{B.34}$$

where Tr indicates a trace over spinor indices.

It is convenient to resolve the x, x' dependence of G by means of Fourier integral

$$G(x,y;x',y')=\int\frac{d^n k}{(2\pi)^n}\,e^{ik(x-x')}G(y,y'|k)\,, \tag{B.35}$$

and hence we have for the coincident limit

$$G(x,y;x,y)=\int\frac{d^n k}{(2\pi)^n}\,G(y,y|k)\,. \tag{B.36}$$

We wish now to relate the Green function on $M^n\times S^N$ to that on S^N. $G(x,y;x',y')$ satisfies the equation

$$-i(\hat{\gamma}^A\mathscr{D}_A+m)G(x,y;x',y')=-g^{-1/2}\delta(x,x')\delta(y,y')\,, \tag{B.37}$$

and hence $G(y,y'|k)$ satisfies

$$(-i(m+i\not{k})\otimes\mathbb{1}+\gamma^5\otimes(-i\Gamma^p\mathscr{D}_p))G(y,y'|k)=-g^{-1/2}\delta(y,y')\,. \tag{B.38}$$

To solve this equation we shall decompose $G(y,y'|k)$ according to the action of γ^5. In order to do this we introduce the matrices

$$U_\pm=\frac{1}{2\sqrt{k^2+m^2}}(\gamma^5\sqrt{k^2+m^2}\pm(m-i\not{k}))\,. \tag{B.39}$$

These have the properties that

$$(m+i\not{k})U_\pm=\pm\sqrt{k^2+m^2}\,\gamma^5U_\pm\,, \tag{B.40}$$

$$\gamma^5(U_++U_-)=\mathbb{1}\,. \tag{B.41}$$

In view of the above we set

$$G(y, y'|k) = U_+ \otimes G_+(y, y'|k) + U_- \otimes G_-(y, y'|k) . \qquad \text{(B.42)}$$

On substituting this ansatz into (B.38) we find that

$$(\gamma^5 U_+)\otimes[-i(\Gamma^p \mathscr{D}_p + \sqrt{k^2+m^2})G_+]$$
$$+(\gamma^5 U_-)\otimes[-i(\Gamma^p \mathscr{D}_p - \sqrt{k^2+m^2})G_-] = -g^{-1/2}\delta(y, y') . \qquad \text{(B.43)}$$

Therefore the ansatz (B.42) solves (B.38) provided the $G_\pm$ satisfy the equations

$$-i(\Gamma^p \mathscr{D}_p \pm \sqrt{k^2+m^2})G_\pm(y, y'|k) = -g^{-1/2}\delta(y, y') , \qquad \text{(B.44)}$$

i.e. the $G_\pm$ are the spinor Green functions on S^N.

Referring back to (B.34) and noting that

$$\mathrm{Tr}\, U_\pm = \frac{m \,\mathrm{tr}\, 1}{2\sqrt{k^2+m^2}} , \qquad \text{(B.45)}$$

where tr 1 indicates the dimensionality of the γ^μ, we see that

$$\frac{\partial V}{\partial m} = \tfrac{1}{2}\mu^{-(n-4)}\Omega_N(\mathrm{tr}\, 1)\int \frac{\mathrm{d}^n k}{(2\pi)^n}\frac{m}{\sqrt{k^2+m^2}}(\mathrm{tr}\, G_+(y, y|k) - \mathrm{tr}\, G_-(y, y|k)) . \qquad \text{(B.46)}$$

The coincident limits of the $G_\pm$ are known from the results of ref. [24] to be given by

$$\mathrm{tr}\, G_\pm = \pm i(\mathrm{tr}\, \mathbb{1})\frac{1}{(4\pi)^{N/2}\rho^{N-1}}\beta\frac{\Gamma(\frac{1}{2}N+i\beta)\Gamma(\frac{1}{2}N-i\beta)}{\Gamma(1+i\beta)\Gamma(1-i\beta)}\Gamma(1-\tfrac{1}{2}N) , \qquad \mathrm{Re}\, N<2 \qquad \text{(B.47)}$$

where tr $\mathbb{1}$ indicates the dimensionality of the Γ^p and

$$\beta = \rho\sqrt{k^2+m^2} . \qquad \text{(B.48)}$$

If we substitute (B.47) into (B.46) and recall the duplication formula

$$\Gamma(z)\Gamma(z+\tfrac{1}{2}) = \pi^{1/2}2^{-2z+1}\Gamma(2z) , \qquad \text{(B.49)}$$

we find

$$V = \frac{\mu^{-(n-4)}}{(4\pi)^{n/2}}\mathrm{Tr}\,(1\otimes\mathbb{1})\frac{\sin\frac{1}{2}n\pi}{\sin\frac{1}{2}N\pi}\frac{\Gamma(1-\frac{1}{2}n)}{\Gamma(N)}$$
$$\times\int_0^\infty \mathrm{d}k\, k^{n-1}\int_{(k^2+m^2)\rho^2}^\infty \mathrm{d}\beta^2\frac{\Gamma(\frac{1}{2}N+i\beta)\Gamma(\frac{1}{2}N-i\beta)}{\Gamma(1+i\beta)\Gamma(1-i\beta)} \qquad \text{(B.50)}$$

$$= -\frac{\mu^{-(n-4)}}{(4\pi)^{n/2}}\mathrm{Tr}\,(1\otimes\mathbb{1})\frac{\sin\frac{1}{2}n\pi}{\sin\frac{1}{2}N\pi}\frac{\Gamma(-\frac{1}{2}n)}{\Gamma(N)}\rho^2$$
$$\times\int_0^\infty \mathrm{d}k\, k^{n+1}\frac{\Gamma(\frac{1}{2}N+i\beta)\Gamma(\frac{1}{2}N-i\beta)}{\Gamma(1+i\beta)\Gamma(1(-i\beta)} , \qquad \mathrm{Re}\, N+n<0 , \qquad \mathrm{Re}\, n>-2 \qquad \text{(B.51)}$$

$$= \tfrac{1}{2} i \frac{\mu^{-(n-4)}}{(4\pi)^{n/2}} \operatorname{Tr}(1 \otimes \mathbb{1}) \frac{e^{-in\pi/2}}{\sin \frac{1}{2} N\pi} \frac{\Gamma(-\frac{1}{2}n)}{\Gamma(N)} \rho^2$$

$$\times \int_C \mathrm{d}k\, k^{n+1} \frac{\Gamma(\frac{1}{2}N + i\beta)\Gamma(\frac{1}{2}N - i\beta)}{\Gamma(1+i\beta)\Gamma(1-i\beta)}, \qquad \operatorname{Re} N + n < 0\,. \tag{B.52}$$

The second equality following by an integration by parts and the third by replacing the contour of integration by one similar to that of fig. 1. The simplest way to proceed seems to be to note that the integrand in (B.52) has poles when

$$i\beta = \pm(\tfrac{1}{2}N + l)\,, \qquad l = 0, 1, 2, \ldots, \tag{B.53}$$

and to evaluate the integral by residues. This yields an expression analogous to (B.7)

$$V = \frac{\mu^{-(n-4)}}{(4\pi)^{n/2}\rho^n} \operatorname{Tr}(1 \otimes \mathbb{1}) \Gamma(-\tfrac{1}{2}n) \sum_{l=0}^{\infty} \frac{\Gamma(l+N)}{l!\,\Gamma(N)} (m^2\rho^2 + (l + \tfrac{1}{2}N)^2)^{n/2},$$

$$\operatorname{Re}(N+n) < 0\,. \tag{B.54}$$

Comparing this sum with (B.7) we see that the eigenvalues of the spinor laplacian on S^N are, for $l = 0, 1, 2, \ldots,$

$$-\lambda_l^2 = -(l + \tfrac{1}{2}N)^2\,, \tag{B.55}$$

with degeneracies (recalling that $\operatorname{Tr} \mathbb{1} = 2^{(N-1)/2}$)

$$d_l = 2^{(N+1)/2} \frac{\Gamma(l+N)}{l!\,\Gamma(N)}\,. \tag{B.56}$$

The calculation proceeds now in close analogy to the scalar case. The identity that allows us to sum the series is in this case

$$[(l + \tfrac{1}{2}N)^2 + m^2\rho^2]^{n/2} = \frac{\pi^{1/2}}{\Gamma(-\frac{1}{2}n)} (2m\rho)^{n/2+1/2}$$

$$\times \int_0^\infty \mathrm{d}t\, e^{-(l+N/2)t} t^{-n/2-1/2} J_{-n/2-1/2}(m\rho t)\,,$$

$$\operatorname{Re} n > 0\,. \tag{B.57}$$

Proceeding now in precise analogy to the scalar case yields finally

$$V = (-1)^{\nu+1} \frac{3 \cdot 2^{\nu+1}}{\pi^6 \rho^4} \int_0^{\pi/2} \frac{\mathrm{d}\theta \cos^3\theta \cos 5\theta}{[2 \operatorname{ch}(\frac{1}{2}\pi \tan\theta)]^{2\nu+1}}, \qquad N \text{ odd}, n = 4, m = 0\,. \tag{B.58}$$

As in the scalar cases this integral may be evaluated by residues for low values of ν. This expresses V in terms of η functions where

$$\eta(s) = \sum_{p=1}^{\infty} \frac{(-1)^{p-1}}{p^s} = \left(1 - \frac{1}{2^{s-1}}\right) \zeta(s)\,; \tag{B.59}$$

for $\nu = 0, 1$ (i.e. $N = 1, 3$) we find

$$\rho^4 V_1 = -\frac{12\eta(5)}{(2\pi)^6} = -1.895927872 \times 10^{-4}, \tag{B.60}$$

$$\rho^4 V_3 = 3\frac{\eta(5)}{(2\pi)^6} + 360\frac{\eta(7)}{(2\pi)^8} = 1.945058002 \times 10^{-4}. \tag{B.61}$$

Appendix C

THE ASYMPTOTIC FORM OF C FOR LARGE N

Minimally coupled scalars

As $\nu \to \infty$, we make the replacements

$$\left.\begin{array}{l}\mathrm{ch}\,(\nu\pi\lambda)\\ \mathrm{sh}\,(\nu\pi\lambda)\end{array}\right\} \to \tfrac{1}{2}\,\mathrm{e}^{\nu\pi\lambda}, \tag{C.1}$$

in (B.21) at the cost of an exponentially small error. It will become clear as we proceed that the dominant term in (B.22) is indeed the one with coefficient ν^2. In fact we shall see that

$$C_\nu = -\frac{\nu^2}{8\pi}\int_0^\infty \frac{\mathrm{d}\lambda}{(1+\mathrm{e}^{-\pi\lambda})^{2\nu}}\frac{(\lambda^3 - 3\lambda)}{(\lambda^2+1)^3\pi^3} + \mathrm{O}\left(\frac{\nu}{\log\nu}\right). \tag{C.2}$$

The integrand in (C.2) has its maximum when $\pi\lambda$ is very close to μ where $\mu(\nu)$ is the large root of the equation

$$\frac{\mathrm{e}^\mu}{\mu} = \tfrac{2}{3}\nu. \tag{C.3}$$

Specifically

$$\mu = \log[\tfrac{2}{3}\nu\log[\tfrac{2}{3}\nu\log[\tfrac{2}{3}\nu\log[\cdots]]]]. \tag{C.4}$$

It is evident that

$$\mu = \log\tfrac{2}{3}\nu + \log\mu, \tag{C.5}$$

and hence also that

$$\frac{\mu}{\log\frac{2}{3}\nu} \to 1, \tag{C.6}$$

as $\nu \to \infty$.

In (C.2) set

$$\lambda = \frac{\mu x}{\pi}. \tag{C.7}$$

This keeps the maximum near $x=1$ and yields the expression

$$C_\nu = -\frac{\nu^2}{8\pi^2\mu^2}\int_0^\infty dx \exp[-2\nu\log(1+e^{-\mu x})]\frac{(x^3-(3\pi^2/\mu^2)x)}{(x^2+\pi^2/\mu^2)^3}+O(\nu^{-1}\mu^{-1})\,. \tag{C.8}$$

Now

$$2\nu\log(1+e^{-\mu x}) = 2\nu\, e^{-\mu x}+O(\nu\, e^{-2\mu})$$

$$= \frac{3}{\mu}e^{-\mu(x-1)}+O(\nu^{-1}\mu^{-2})\,, \tag{C.9}$$

in virtue of (B.3).

$$\text{As } \nu\to\infty\,, \qquad \exp\left(-\frac{3}{\mu}e^{-\mu(x-1)}\right)\to\begin{cases}1, & x>1\\ 0, & x<1,\end{cases} \tag{C.10}$$

however the scale over which the function changes from zero to unity is of order $\mu^{-1}\log\mu$. If we, therefore, make the replacement

$$\exp(-2\nu\log(1+e^{-\mu x}))\to\theta(x-1)\,, \tag{C.11}$$

in (C.8) and retain only the leading term in $1/\mu$ we obtain

$$C_N = -\left(\frac{\nu}{4\pi\mu}\right)^2(1+O(\mu^{-1}\log\mu))\,, \qquad m=0\,, \qquad N\text{ odd}\,, \qquad N\to\infty\,. \tag{C.12}$$

Unfortunately this result, apart from showing that C_N eventually becomes negative, is of little practical use owing to the fact that μ increases only slowly with ν.

Fermions

It is much simpler to examine the asymptotic behavior of the Dirac field owing to the fact that (B.53) may be analyzed by means of the method of steepest descent. A short calculation reveals that

$$C_\nu = \frac{(-1)^{\nu+1}3}{2^\nu\pi^{13/2}(\nu+\frac{1}{2})^{1/2}}\left\{1-\left(\frac{30}{\pi^2}-\frac{1}{8}\right)\frac{1}{(\nu+\frac{1}{2})}+O\left(\frac{1}{(\nu+\frac{1}{2})^2}\right)\right\},$$

$$m=0\,, \qquad N\text{ odd}\,, \qquad N\to\infty\,. \tag{C.13}$$

Appendix D

We present here a different approach to the evaluation of the quantity

$$C_N^{(0)} = -\frac{1}{32\pi^2}\lim_{n\to4}\Gamma(-\tfrac{1}{2}n)\sum_l D_l\Lambda_l^n\,, \tag{D.1}$$

for spheres of odd dimensionality N. Following ref. [25] we write the eigenvalue Λ_l^2 as

$$\Lambda_l^2 = l(l+N-1) = L^2-\nu^2\,, \tag{D.2}$$

$$L \equiv l + \nu\,, \qquad \nu \equiv \tfrac{1}{2}(N-1)\,, \tag{D.3}$$

and expand

$$\Lambda_l^n = \sum_{r=0}^{\infty} \frac{\Gamma(r-\frac{1}{2}n)}{\Gamma(-\frac{1}{2}n)r!}\,\nu^{2r}L^{n-2r}\,. \tag{D.4}$$

Also, we write the degeneracy D_l as

$$D_l = \frac{(2l+N-1)(l+N-2)!}{(N-1)!\,l!}$$
$$= \frac{2L^2}{(2\nu)!}[L^2-(\nu-1)^2][L^2-(\nu-2)^2]\cdots[L^2-1]\,. \tag{D.5}$$

Eq. (D.5) can be written as a polynomial in L^2:

$$D_l = \sum_{m=0}^{\nu-i} a_{\nu m} L^{2m+2}\,, \tag{D.6}$$

with $a_{\nu m}$ independent of L (i.e. of l):

$$a_{10} = 1\,, \qquad a_{20} = -\tfrac{1}{12}\,, \qquad a_{21} = \tfrac{1}{12}\,,$$
$$a_{30} = \tfrac{1}{90}\,, \qquad a_{31} = -\tfrac{1}{72}\,, \qquad a_{32} = \tfrac{1}{360} \qquad \text{etc.} \tag{D.7}$$

Using (D.6) and (D.5) in (D.1), we have now

$$C_N^{(0)} = -\frac{1}{32\pi^2}\sum_{m=0}^{\nu-1} a_{\nu m} \lim_{n\to 4} \sum_{r=0}^{\infty} \frac{\Gamma(r-\frac{1}{2}n)}{r!}\,\nu^{2r}\zeta(2r-2m-n-2,\,\nu)\,, \tag{D.8}$$

where ζ is the incomplete Riemann zeta function

$$\zeta(z,\,\nu) = \sum_{L=\nu}^{\infty} L^{-z}\,. \tag{D.9}$$

The factor $\Gamma(-\frac{1}{2}n)$ in (D.1), which blows up for $n\to 4$, has been cancelled by the $\Gamma(-\frac{1}{2}n)$ in the denominator of eq. (D.4). In its place we now have a factor $\Gamma(r-\frac{1}{2}n)$, which blows up for $n\to 4$ only in the cases $r=0$, 1, and 2. To see how this pole is cancelled, note first that because (D.5) vanishes for $L=1, 2, \ldots \nu-1$, the sum over L can be taken to run from 1 to ∞ instead of ν to ∞ in any of the incomplete zeta functions in (D.8), and in particular for the singular terms with $r=0$, 1, and 3, so that in these terms in place of an incomplete zeta function we have the ordinary zeta function $\zeta(2r-2m-n-2)$. But for $r\leqslant m+2$, the argument $2r-2m-n-2$ approaches a negative even integer as $n\to 4$, where the zeta function has a simple zero that cancels the pole in $\Gamma(r-\frac{1}{2}n)$. Thus there is no singularity here for $n\to 4$.

In evaluating the terms in (D.8) with $r=0, 1, 2$, it is convenient to use the identity [26]

$$\Gamma(\tfrac{1}{2}z)\zeta(z) = \pi^{z-1/2}\Gamma\!\left(\frac{1-z}{2}\right)\zeta(1-z)\,, \tag{D.10}$$

which yields

$$\Gamma(r-\tfrac{1}{2}n)\zeta(2r-2m-n-2) \xrightarrow[n\to 4]{} (-)^{m+1}\pi^{2r-2m-13/2}$$

$$\times(3+m-r)!\,\Gamma(\tfrac{7}{2}+m-r)\zeta(7+2m-2r)/(2-r)! \tag{D.11}$$

Inserting these terms in (D.8), we have now

$$C_N^{(0)} = -\frac{1}{32\pi^2}\sum_{m=0}^{\nu-1} a_{\nu m}$$

$$\times\left[(-)^{m+1}\pi^{-2m-13/2}\sum_{r=0,1,2}\frac{(\pi\nu)^{2r}(3+m-r)!\,\Gamma(\tfrac{7}{2}+m-r)\zeta(7+2m-2r)}{r!\,(2-r)!}\right.$$

$$\left.+\sum_{r=m+3}^{\infty}\frac{\nu^{2r}}{r(r-1)(r-2)}\zeta(2r-2m-6,\nu)\right]. \tag{D.12}$$

The convergence of this sum is apparent: the incomplete zeta function $\zeta(2r-2m-6,\ \nu)$ behaves for $r\to\infty$ like $\nu^{-2r+2m+6}$, so the series converges like $\sum 1/r^3$. Eq. (D.12) can therefore be used in numerical calculations of $C_N^{(0)}$.

As a check on our calculation of C_N in appendix B, we have used the more cumbersome series (D.12) to calculate $C_N^{(0)}$ in the case $N=3$ (i.e. $\nu=1$). Here (D.12) becomes

$$C_3^{(0)} = \frac{-1}{32\pi^2}\left[-\pi^{-13/2}\sum_{r=0,1,2}\frac{\pi^{2r}(3-r)!\,\Gamma(\tfrac{7}{2}-r)\zeta(7-2r)}{r!\,(2-r)!}\right.$$

$$\left.+\sum_{r=3}^{\infty}\frac{1}{r(r-1)(r-2)}\zeta(2r-6)\right]. \tag{D.13}$$

Since $\zeta(2r-6)$ approaches unity as $r\to\infty$, it is very convenient to improve the convergence of the series in (D.13) by subtracting the elementary series

$$\sum_{r=3}^{\infty}\frac{1}{r(r-1)(r-2)} = \frac{1}{4}.$$

We then have

$$C_3^{(0)} = -\frac{1}{32\pi^2}\left[\frac{-45\zeta(7)}{8\pi^6} - \frac{3\zeta(5)}{2\pi^4} - \frac{\zeta(3)}{4\pi^2} + \tfrac{1}{4} + \sum_{r=3}^{\infty}\frac{1}{r(r-1)(r-2)}(\zeta(2r-6)-1)\right].$$

To show the speed of convergence of the series, here are the terms up to $r=8$:

$$C_3^{(0)} = -\frac{1}{32\pi^2}[-0.00589976-0.01596762-0.03044846+\tfrac{1}{4}$$

$$-\tfrac{1}{4}+0.02687225+0.00137205+0.00014453$$

$$+0.00001942+0.00000296+\cdots] = 7.568(9)\times 10^{-5}.$$

This agrees with the more precise result derived by the methods of appendix B, shown in table 1.

For *spinors* we have

$$C_N^{(1/2)} = +\frac{1}{8\pi^2}\lim_{n\to 4}\Gamma(-\tfrac{1}{2}n)\sum_{l=0}^{\infty} d_l\lambda_l^n\,. \tag{D.14}$$

The degeneracies and masses derived in appendix B can be written

$$\lambda_l = l+\nu+\tfrac{1}{2}\,, \tag{D.15}$$

$$d_l = \frac{2^{(N+1)/2}}{(N-1)!}\frac{(l+N-1)!}{l!} = \frac{2^{\nu+1}}{(2\nu)!}\sum_{m=0}^{\nu}\mathcal{B}_{\nu m}(l+\nu+\tfrac{1}{2})^{2m}\,, \tag{D.16}$$

where again $\nu \equiv \frac{1}{2}(N-1)$, and now

$$\mathcal{B}_{00}=1\,,\qquad \mathcal{B}_{11}=1\,,\qquad \mathcal{B}_{10}=-\tfrac{1}{4}\,,$$
$$\mathcal{B}_{22}=1\,,\qquad \mathcal{B}_{21}=-\tfrac{5}{2}\,,\qquad \mathcal{B}_{20}=\tfrac{9}{16}\cdots\,. \tag{D.17}$$

Since the sum over m evidently vanishes when $l=-1,-2,\ldots,-(N-1)$, we can let the sum over l run from $l=-\nu$ to ∞ instead of from $l=0$ to ∞; setting $L=l+\nu$, this gives

$$C_N^{(1/2)} = \frac{2^\nu}{4\pi^2(2\nu)!}\sum_{m=0}^{\nu}\mathcal{B}_{\nu m}\lim_{n\to 4}\Gamma(-\tfrac{1}{2}n)\sum_{L=0}^{\infty}(L+\tfrac{1}{2})^{n+2m}\,. \tag{D.18}$$

The sum over L can easily be expressed in terms of ordinary Riemann zeta functions

$$\sum_{L=0}^{\infty}(L+\tfrac{1}{2})^{n+2m} = -(1-2^{-n-2m})\zeta(-n-2m)\,.$$

Using (D.10) again, the limit $n\to 4$ gives

$$\Gamma(-\tfrac{1}{2}n)\zeta(-n-2m)\xrightarrow[n\to 4]{}\tfrac{1}{2}(-)^m\pi^{-2m-9/2}(2+m)!\,\Gamma(\tfrac{5}{2}+m)\zeta(5+2m)\,.$$

Putting this all together, we find

$$C_N^{(1/2)} = \frac{-2^\nu}{8\pi^2(2\nu)!}\sum_{m=0}^{\nu}(1-2^{-4-2m})\mathcal{B}_{\nu m}(-)^m\pi^{-2m-9/2}(2+m)!\,\Gamma(\tfrac{5}{2}+m)\zeta(5+2m)\,. \tag{D.19}$$

For instance, in the case $N=3$ $(\nu=1)$, this gives simply

$$C_3^{(1/2)} = \frac{45\zeta(5)}{1024\pi^6}+\frac{2835\zeta(7)}{2048\pi^8} = 1.9451\times 10^{-4}\,.$$

This reproduces the value derived in appendix B, shown in table 3. At least for fermions, the present method seems an especially convenient way of obtaining expressions for C_N in closed form.

Appendix E

MEAN SQUARE O($N+1$) GENERATORS

In this appendix we calculate Q_l^2 and q_l^2, the eigenvalues of the squared O($N+1$) generators, averaged over the eigenmodes of the laplacian and Dirac operators on a sphere S^N of radius ρ with eigenvalues $M_l^2 = l(l+N-1)/\rho^2$ and $m_l = (l+\frac{1}{2}N)/\rho$, respectively. These mean eigenvalues are the same for all O($N+1$) generators, so they can be expressed as the eigenvalues $\boldsymbol{J}_l^2$ of the O($N+1$) Casimir operators, which are the same for all eigenmodes of definite mass, divided by the number $\frac{1}{2}N(N+1)$ of O($N+1$) generators.

For spin-0, the O($N+1$) Casimir operator is trivially related to the laplacian. Embedding S^N in an $(N+1)$-dimensional euclidean space with coordinates $Z^i (i = 1, 2, \ldots, N+1)$, we have SO($N+1$) generators

$$J_{ij} = L_{ij} \equiv -i\left(Z^i \frac{\partial}{\partial Z^j} - Z^j \frac{\partial}{\partial Z^i}\right), \tag{E.1}$$

so the Casimir operator is

$$\boldsymbol{J}^2 \equiv \tfrac{1}{2}\sum_{ij} J_{ij}^2 = -\rho^2\nabla^2 + \rho^2 \frac{\partial^2}{\partial \rho^2} + N\rho \frac{\partial}{\partial \rho}, \tag{E.2}$$

where

$$\rho^2 \equiv \sum_i (Z^i)^2 \tag{E.3}$$

$$\nabla^2 \equiv \sum_i \frac{\partial^2}{\partial Z^{i2}}. \tag{E.4}$$

Acting on a function of the unit vector Z^i/ρ, for which the laplacian ∇^2 has eigenvalue $-M_l^2$, the Casimir operator has eigenvalues

$$\boldsymbol{J}_l^2 = \rho^2 M_l^2 = l(l+N-1), \tag{E.5}$$

and therefore

$$Q_l^2 = \frac{\rho^2 M_l^2}{\frac{1}{2}N(N+1)} = \frac{l(l+N-1)}{\frac{1}{2}N(N+1)}. \tag{E.6}$$

An immediate consequence of the relation between Q_l^2 and M_l^2 is that, comparing (56) and (58),

$$D_N^{(0)} = -\frac{4}{N(N+1)} E_N^{(0)}. \tag{E.7}$$

For spin-$\frac{1}{2}$ the O$(N+1)$ generators involve an intrinsic as well as an orbital term

$$J_{ij} = L_{ij} + S_{ij}\,, \tag{E.8}$$

where L_{ij} is as before

$$L_{ij} = -i\left(Z^i \frac{\partial}{\partial Z^j} - Z^j \frac{\partial}{\partial Z^i}\right),$$

and now

$$S_{ij} = -\tfrac{1}{4} i[\Gamma_i, \Gamma_j]\,, \tag{E.9}$$

where Γ_i are the gamma matrices in $(N+1)$-dimensional euclidean space

$$\{\Gamma_i, \Gamma_j\} = 2\delta_{ij}\,. \tag{E.10}$$

The Casimir operator is then

$$\boldsymbol{J}^2 \equiv \tfrac{1}{2} \sum_{ij} (L_{ij} + S_{ij})^2\,. \tag{E.11}$$

Following the approach of ref. [27], an elementary manipulation of the commutation relations for Z^i and $\partial/\partial Z^j$ and the anticommutation relations for Γ^i and Γ^j shows that

$$(\boldsymbol{J}^2 - \boldsymbol{L}^2 - \boldsymbol{S}^2)^2 - (N-1)(\boldsymbol{J}^2 - \boldsymbol{L}^2 - \boldsymbol{S}^2) - \boldsymbol{L}^2 = 0\,, \tag{E.12}$$

with $\boldsymbol{J}^2$, $\boldsymbol{L}^2$, and $\boldsymbol{S}^2$ interpreted as

$$\boldsymbol{A}^2 \equiv \tfrac{1}{2} \sum_{ij} (A_{ij})^2\,.$$

Also,

$$\boldsymbol{S}^2 = \tfrac{1}{8} N(N+1)\,, \tag{E.13}$$

and, acting on the kth spherical harmonic on S^N,

$$\boldsymbol{L}^2 = k(k+N-1)\,. \tag{E.14}$$

It follows that the direct product of a Dirac spinor and a spherical harmonic of order k is a mixture of eigenmodes of $\boldsymbol{J}^2$ with eigenvalues given by the two roots of (E.12):

$$\boldsymbol{J}^2 = \begin{cases} k^2 + kN + \tfrac{1}{8} N(N+1) \\ (k-1)^2 + (k-1)N + \tfrac{1}{8} N(N+1)\,. \end{cases} \tag{E.15}$$

In other words, the eigenvalues of $\boldsymbol{J}^2$ take the form

$$\boldsymbol{J}_l^2 = l^2 + lN + \tfrac{1}{8} N(N+1)\,,$$
$$l = 0, 1, \dots, \tag{E.16}$$

with eigenmodes consisting of mixtures of spherical harmonics with $k = l$ and

$k=l+1$. The eigenmodes of the Dirac operator are of course also eigenmodes of $\boldsymbol{J}^2$, and the integer l can be identified as the integer used to label the eigenvalues $m_l=(l+\frac{1}{2}N)/\rho$ of the Dirac operator. In accordance with our earlier remarks, the mean square $O(N+1)$ generator is given by

$$q_l^2=\boldsymbol{J}_l^2/\tfrac{1}{2}N(N+1)=\frac{l^2+lN+\frac{1}{8}N(N+1)}{\frac{1}{2}N(N+1)}. \tag{E.17}$$

We can easily check this for the familiar case of O(3), i.e. $N=2$. In this case (E.5) gives the familiar result for spinless fields, $\boldsymbol{J}_l^2=l(l+1)$. Also, for spin-$\frac{1}{2}$ a spherical harmonic of order k will have total spin $j=k\pm\frac{1}{2}$, and hence the Casimir operator

$$\boldsymbol{J}^2=j(j+1)=(k\pm\tfrac{1}{2})(k\pm\tfrac{1}{2}+1)$$

in agreement with (E.15) for $N=2$.

Appendix F

EVALUATION OF THE COEFFICIENTS D_N AND E_N

We shall calculate, in this appendix, the coefficients that renormalize the gravitational constant and the Yang–Mills coupling.

Scalar fields

For scalar fields it is sufficient to compute $E_N^{(0)}$ since $D_N^{(0)}$ and $E_N^{(0)}$ are related by eq. (E.7). From eq. (58) we have

$$\frac{E_N^{(0)}}{\rho^2}=\frac{1}{192\pi^2}\lim_{n\to4}\Gamma(1-\tfrac{1}{2}n)\sum_{l=0}^{\infty}D_l(M_l^2)^{n/2-1}, \tag{F.1}$$

while $V(m^2)$, the effective potential for a scalar field of mass m, is given by (B.7) in the form

$$V(m^2)=-\frac{1}{32\pi^2}\lim_{n\to4}\Gamma(-\tfrac{1}{2}n)\sum_{l=0}^{\infty}D_l(M_l^2+m^2)^{n/2}. \tag{F.2}$$

Comparing (F.1) and (F.2) and setting $n=4$ we obtain

$$\frac{E_N}{\rho^2}=\tfrac{1}{6}V'(0), \qquad \text{scalars, } n=4, \tag{F.3}$$

where a prime denotes $\partial/\partial m^2$.

We shall calculate $V'(m^2)$ from (B.15). Recalling that

$$b^2=\nu^2-m^2\rho^2, \tag{F.4}$$

we see that the m dependence of V is contained in the quantity

$$(bt)^{7/2}I_{-7/2}(bt).$$

Setting $z = bt$ we have

$$\frac{\partial}{\partial m^2}(z^{7/2}I_{-7/2}(z)) = -\tfrac{1}{2}\rho^2 t^2 \frac{1}{z}\frac{\mathrm{d}}{\mathrm{d}z}(z^{7/2}I_{-7/2}(z))$$

$$= -\tfrac{1}{2}\rho^2 t^2 z^{5/2} I_{-5/2}(z)\,, \tag{F.5}$$

where, in the last equality, we have used the identity

$$\frac{1}{z}\frac{\mathrm{d}}{\mathrm{d}z}(z^{-\nu}I_\nu(z)) = z^{-(\nu+1)}I_{\nu+1}(z)\,. \tag{F.6}$$

Noting also that

$$z^{5/2}I_{-5/2}(z) = \sqrt{\frac{2}{\pi}}\{-3z\,\mathrm{sh}\,z + (3+z^2)\,\mathrm{ch}\,z\}\,, \tag{F.7}$$

we obtain E_N in the form

$$E_N^{(0)} = \frac{1}{96\pi^2\nu}\int_c \frac{\mathrm{d}t}{t^4(2\,\mathrm{sh}\,\frac{1}{2}t)^{2\nu}}[-3\nu t\,\mathrm{sh}\,(\nu t) + (3+\nu^2 t^2)\,\mathrm{ch}\,(\nu t)]\,. \tag{F.8}$$

As in previous cases evaluation by residues yields exact results. Thus, by way of example, we find

$$E_3^{(0)} = \frac{1}{24}\left(\frac{5\zeta(3)}{(2\pi)^4} - \frac{12\zeta(5)}{(2\pi)^6}\right)$$

$$= 1.522542\times 10^{-4}\,. \tag{F.9}$$

To facilitate numerical computation for general values of N, we again set $t = \pi i + \pi\tan\theta$ and obtain as our final expression

$$E_N^{(0)} = \frac{1}{48\pi^2\nu}\int_0^{\pi/2}\frac{\mathrm{d}\theta}{[2\,\mathrm{ch}\,(\frac{1}{2}\pi\tan\theta)]^{2\nu}}\left\{3\nu\cos\theta\sin 3\theta\,\mathrm{sh}\,(\nu\pi\tan\theta)\right.$$

$$\left. + \left(\frac{3}{\pi}\cos^2\theta\cos 4\theta - \nu^2\pi\cos 2\theta\right)\mathrm{ch}\,(\nu\pi\tan\theta)\right\}\,. \tag{F.10}$$

Numerical values of this integral are used (along with the values of $C_N^{(0)}$ obtained in appendix B) to calculate the values of the coefficients $\beta_N^{(0)}$ and $\gamma_N^{(0)}$ presented in table 4.

Fermions

For massless fermions we have

$$\frac{E_N^{(1/2)}}{\rho^2} = \frac{1}{96\pi^2}\lim_{n\to 4}\Gamma(1-\tfrac{1}{2}n)\sum_{l=0}^{\infty} d_l(m_l^2)^{n/2-1}\,, \tag{F.11}$$

$$D_N^{(1/2)} = -\frac{1}{192\pi^2}\frac{20\nu^2+28\nu+5}{(\nu+1)^2}\lim_{n\to 4}\Gamma\left(2-\frac{n}{2}\right)\sum_l d_l (m_l^2)^{n/2-2}$$
$$\times\frac{[l^2+lN+\frac{1}{8}N(N+1)]}{\frac{1}{2}N(N+1)}. \tag{F.12}$$

As in the scalar case we proceed by relating $E_N^{(1/2)}$ and $D_N^{(1/2)}$ to the derivative of $V(m^2)$,. which is given for spin-$\frac{1}{2}$ by (B.54) as

$$V(m^2) = \frac{1}{8\pi^2}\lim_{n\to 4}\Gamma(-\tfrac{1}{2}n)\sum_{l=0}^{\infty} d_l (m_l^2+m^2)^{n/2}. \tag{F.13}$$

We see that

$$\frac{E_N}{\rho^2} = -\tfrac{1}{12}V'(0), \qquad \text{fermions, } n=4, \tag{F.14}$$

and by noting that we may write

$$l^2+lN+\tfrac{1}{8}N(N+1) = m_l^2\rho^2 - \tfrac{1}{8}N(N-1), \tag{F.15}$$

we find

$$D_N^{(1/2)} = \frac{20\nu^2+28\nu+5}{(\nu+1)^2}\left\{\frac{E_N^{(1/2)}}{N(N+1)} + \frac{1}{96}\left(\frac{N-1}{N+1}\right)V''(0)\right\}, \qquad \text{fermions,} \qquad n=4. \tag{F.16}$$

Proceeding as in the scalar case we make use of a representation that follows from (B.57)

$$V(m^2) = \frac{2^{\nu-1/2}}{\pi^{3/2}\rho^4}\int_C \frac{\mathrm{d}t}{t^5}\frac{(mt\rho)^{5/2}J_{-5/2}(mt\rho)}{(2\,\mathrm{sh}\,\frac{1}{2}t)^{2\nu+1}}. \tag{F.17}$$

The derivatives of V are readily calculated in virtue of the relations

$$\left(\frac{1}{z}\frac{\mathrm{d}}{\mathrm{d}z}\right)^k (z^{-\nu}J_\nu(z)) = (-1)^k z^{-(\nu+k)}J_{\nu+k}(z), \tag{F.18}$$

and

$$\lim_{z\to 0} z^{-\nu}J_\nu(z) = \frac{2^{-\nu}}{\Gamma(1+\nu)}. \tag{F.19}$$

Thus we find

$$E_N^{(1/2)} = \frac{(-1)^{\nu+1}2^\nu}{12\pi^4}\int_0^{\pi/2}\frac{\mathrm{d}\theta\cos\theta\cos 3\theta}{[2\,\mathrm{ch}\,(\frac{1}{2}\pi\tan\theta)]^{2\nu+1}}, \qquad \text{fermions, } n=4, \tag{F.20}$$

$$V''(0) = \frac{(-2)^{\nu-1}}{\pi^2}\int_0^{\pi/2}\frac{\mathrm{d}\theta}{[2\,\mathrm{ch}\,(\frac{1}{2}\pi\tan\theta)]^{2\nu+1}}, \qquad \text{fermions, } n=4, \tag{F.21}$$

from which $D_N^{(1/2)}$ is obtained using (F.16). Numerical values of these integrals for $D_N^{(1/2)}$ and $E_N^{(1/2)}$ are used (along with the values of $C_N^{(1/2)}$ obtained in appendix B) to calculate the values of the coefficients $\beta_N^{(1/2)}$ and $\gamma_N^{(1/2)}$ presented in table 5.

As a check, we have also calculated $D_N^{(j)}$ and $E_N^{(j)}$ for $j=0$ and $\frac{1}{2}$ and low N by the summation method of appendix B. The results confirm those obtained in this appendix.

References

[1] Th. Kaluza, Sitz. Preuss. Akad. Wiss. K1 (1921) 966;
O. Klein, Nature 118 (1926) 516; Z. Phys. 37 (1926) 895;
H. Mandel, Z. Phys. 39 (1926) 136;
A. Einstein and P. Bergmann, Ann. of Math. 39 (1938) 683

[2] B. DeWitt, in Relativity, groups, and topology (Gordon and Breach, New York and London, 1964) p. 725;
J. Scherk and J.H. Schwarz, Phys. Lett. 57B (1975) 463;
Y.M. Cho and P.G.O. Freund, Phys. Rev. D12 (1975) 1711;
E. Witten, Nucl. Phys. B186 (1981) 412;
E. Cremmer, in Supergravity 81, eds. S. Ferrara and J. Taylor (Cambridge University Press, 1982);
A. Salam and J. Strathdee, Ann. of Phys. 141 (1982) 316;
M.J. Duff and D. Toms, in Unification of fundamental interactions 2, eds. S. Ferrara and J. Ellis (Plenum, 1982);
C. Wetterich, Phys. Lett. 110B (1982) 379,
F. Englert, Phys. Lett. 119B (1982) 339

[3] E. Cremmer and J. Scherk, Nucl. Phys. B108 (1976) 408;
Z. Horvath, L. Palla, E. Cremmer and J. Scherk, Nucl. Phys. B127 (1977) 57;
J.F. Luciani, Nucl. Phys. B135 (1978), 111; P.G.O. Freund and M. Rubin, Phys. Lett. 97B (1980) 233;
M. Duff, in Supergravity 81, eds. S. Ferrara and J. Taylor (Cambridge University Press, 1982);
S. Randjbar-Daemi, A. Salam and J. Strathdee, ICTP preprints IC/82/208 (1982), IC/83/6 (1983);
M.J. Duff and C.N. Pope, Imperial College preprint ICTP/82–83/7 (1983);
M.J. Duff, B. Nilsson, and C. Pope, Texas preprint UTTG-5 (1983)

[4] S. Weinberg, Phys. Lett. 125B (1983) 265

[5] T. Appelquist and A. Chodos, Phys. Rev. Lett. 50 (1983) 141;
D. Pollard, ICTP preprint 82–83/11;
M.A. Rubin and B.D. Roth, Texas preprint UTTG-3 (1983), Phys. Lett. B, to appear;
T. Appelquist, A. Chodos and E. Myers, Yale preprint YTP 83–04 (1983);
T. Appelquist and A. Chodos, Yale preprint YTP 83-05 (1983)

[6] J.B. Hartle and G.T. Horowitz, Phys. Rev. D24 (1981) 257;
Also see E. Tomboulis, Phys. Lett. 70B (1977) 361; 97B (1980) 77

[7] M.J. Duff and D. Toms, in Unification of fundamental interactions, eds. S. Ferrara and J. Ellis (Plenum, 1982)

[8] S. Coleman and E. Weinberg, Phys. Rev. D7 (1973) 1888;
S. Weinberg, Phys. Rev. D7 (1973) 1068

[9] A. Erdelyi, W. Magnus, F. Oberhettinger and F.R. Tricomi, Higher transcendental functions (New York, McGraw-Hill, 1953)

[10] B.S. DeWitt, *Dynamical theory of groups and fields* (Gordon and Breach, 1964);
D.J. Toms, University of Wisconsin at Milwaukee preprint;
M. Awada and D.J. Toms, University of Wisconsin preprint.

[11] S. Weinberg, in Gravitational theories since Einstein, eds. S. Hawking and W. Israel (Cambridge University Press, 1978), sect. 16.7

[12] A. Zee, Phys. Rev. D23 (1981) 858
[13] K. Symanzik, Comm. Math. Phys. 16 (1970) 48;
S. Coleman, in Laws of hadronic matter, ed. A. Zichichi (Academic Press, NY, 1975) 139
[14] D. Page, paper in preparation
[15] H. Weyl, Rend. Circ. Mat. Palermo 31 (1915) 1
[16] M.A. Rubin and B.D. Roth, Nucl. Phys. B226 (1983) 444
[17] S. Weinberg, Gravitation and cosmology (Wiley, New York, 1972) Chap. 13
[18] E. Witten, private communication
[19] M.J. Duff, C.N. Pope and B.E.W. Nilsson, private communication
[20] A. Zee, in Proc. 4th Summer Institute on grand unified theories and related topics, eds. M. Konuma and T. Maskawa (World Scientific, Singapore, 1981) appendix VI-5
[21] D.G. Boulware and L.S. Brown, Ann. of Phys. 138 (1982) 392
[22] S. Helgason, Differential geometry and symmetric spaces (Academic Press, New York, 1962)
[23] M.E. Peskin, Nucl. Phys. B175 (1981) 197
[24] P. Candelas and D.J. Raine, Phys. Rev. D12 (1975) 965
[25] S.M. Christensen and M.J. Duff, Nucl. Phys. B170 (1980) 480
[26] I.S. Gradshteyn and I.M. Ryzik, Table of integrals, series, and products (Academic Press, New York, 1980)
[27] I.T. Drummond and G.M. Shore, Ann. of Phys. 117 (1979) 89

3.5. Cosmology and the Multiverse

Just as Darwin and Wallace explained how the wonderful adaptations of living forms could arise without supernatural intervention, so the string landscape may explain how the constants of nature that we observe can take values suitable for life without being fine-tuned by a benevolent creator.

Steven Weinberg

Although Weinberg worked for a while on string theory and continued to defend research in this area in the face of criticism, in later life he focused more on cosmology, which he thought to be as exciting now as particle physics was in the 1960s and 1970s. He had already written a standard text book in 1971, *Gravitation and Cosmology*, and a very influential (and very controversial) paper [3.5.1] on the cosmological constant Λ. He invoked "anthropic" arguments to place an upper bound on Λ: it must not be so large as to prevent the formation of gravitationally bound states, and a lower bound: it must not be so negative as to lead to a closed universe. He devoted his 1988 Loeb Lectures at Harvard to reviewing the history of the cosmological constant problem and surveying the various attempts to solve it, including anthropic ones [3.5.2].

Many were surprised that Steven Weinberg, the epitome of the level-headed physicist, should embrace a fringe idea like the anthropic principle. But subsequent developments in astrophysics, in particular eternal inflation, and in high-energy physics, in particular the string landscape, were to change people's perspectives so that Weinberg's view seemed more compelling. In *Living in the multiverse* [3.5.3] he says:

> *Now we may be at a new turning point, a radical change in what we accept as a legitimate foundation for a physical theory. The current excitement is of course a consequence of the discovery of a vast number of solutions of string theory, beginning in 2000 with the work of Bousso and Polchinski... Susskind gave the name "string landscape" to this multiplicity of vacua.*

So he had not given up on string theory.

When it was realized in 1998 that the universe was not only expanding but accelerating, the simplest explanation was a small positive cosmological constant of about the size predicted by Weinberg. The theoretical physics community continues to be divided on whether this is evidence for the multiverse.

Three of his favorite topics — cosmology, effective field theory and asymptotic safety — were brought together in [3.5.4] and [3.5.5].

VOLUME 59, NUMBER 22 PHYSICAL REVIEW LETTERS 30 NOVEMBER 1987

Anthropic Bound on the Cosmological Constant

Steven Weinberg

Theory Group, Department of Physics, University of Texas, Austin, Texas 78712

(Received 5 August 1987)

In recent cosmological models, there is an "anthropic" upper bound on the cosmological constant Λ. It is argued here that in universes that do not recollapse, the only such bound on Λ is that it should not be so large as to prevent the formation of gravitationally bound states. It turns out that the bound is quite large. A cosmological constant that is within 1 or 2 orders of magnitude of its upper bound would help with the missing-mass and age problems, but may be ruled out by galaxy number counts. If so, we may conclude that anthropic considerations do not explain the smallness of the cosmological constant.

PACS numbers: 98.80.Dr, 04.20.Cv

Our knowledge of the present expansion rate of the Universe indicates that the effective value Λ of the cosmological constant is vastly less than what would be produced by quantum fluctuations[1] in any known realistic theory of elementary particles. In view of the continued failure to find a microscopic explanation of the smallness of the cosmological constant, it seems worthwhile to look for a solution in other, "anthropic," directions.[2] Perhaps Λ must be small enough to allow the Universe to evolve to its present nearly empty and flat state, because otherwise there would be no scientists to worry about it. Without having a definite framework for such reasoning, one can at least point to four lines of current cosmological speculation, in which anthropic considerations could set bounds on the value we observe for the effective cosmological constant:

(a) The effective cosmological constant may evolve very slowly, perhaps because of slow changes in the value of some scalar field, as in the model of Banks.[3] In this case, it would be natural to expect that for some very long epoch the cosmological constant would remain near zero. The question then is, why do we find ourselves in such an epoch? As remarked by Banks, the answer may be anthropic: Perhaps only in such epochs is life possible.

(b) The Universe may evolve through a very large number of first-order phase transitions, in which bubbles form within bubbles within bubbles . . . , each bubble having within it a smaller value of the vacuum energy, and hence of the effective cosmological constant. If the steps in vacuum energy are very small, then it would be natural to expect that there would be some phase in which the effective cosmological constant is correspondingly small. Abbott[4] has suggested a scalar-field theory with a potential that has an infinite number of closely spaced local minima; bubbles form within bubbles as the scalar-field value jumps from one minimum to the next. Recently Brown and Teitelboim[5] have proposed a model in which a similar sequence of phase transitions occurs, but in which the bubble walls are elementary membranes coupled to a three-form gauge field, with the difference in cosmological constants between the inside and outside of each membrane caused by the differences in the values of the four-form field strength.[6] In models of the type discussed in Refs. 4 and 5 it may not be strictly necessary to invoke the anthropic principle because gravitational effects can stop the process of bubble formation when the vacuum energy is about to become negative.[7] However, it takes an enormously long time to reach this final stage, and anthropic arguments may be needed to explain why we are not still in an earlier stage, with large effective cosmological constant.

(c) Fluctuations in scalar fields can trigger cosmic inflation in regions of the Universe where the fields happen to be large. Except near the edges, the inflationary region would appear to its inhabitants as a separate subuniverse. In this region further fluctuations can produce new inflations, and so on. This has been studied by Linde,[8] who remarks that the physical constants of the subuniverse in which we live may be in part constrained by the requirement that life could arise in such a subuniverse.

(d) Quantum fluctuations in the very early Universe may cause incoherence between different terms in the state vector of the Universe; each term would then in effect represent a separate universe. Such a picture has been considered by Hawking.[9] Our own Universe could correspond to any one of the terms in the state vector, subject only to the anthropic condition, that it be a universe in which life could develop.

Without committing ourselves to any one of these cosmological models, it seems appropriate at least to ask, just what limit does the anthropic principle place on the effective cosmological constant Λ?

Fortunately, at least for $\Lambda > 0$, the anthropic principle provides a rather sharp upper bound on Λ. This is because in a continually expanding universe, the cosmological constant (unlike charges, masses, etc.) can affect the evolution of life in only one way. Without undue anthropocentrism, it seems safe to assume that in order for any sort of life to arise in an initially homogeneous and isotropic universe, it is necessary for sufficiently large gravi-

VOLUME 59, NUMBER 22 PHYSICAL REVIEW LETTERS 30 NOVEMBER 1987

tationally bound systems to form first. (By "sufficiently large" is meant large enough to form stars, and large enough also to contain the heavy elements produced by early generations of stars so that planets can form around subsequent generations of stars. Galaxies and probably also the larger globular clusters are sufficiently large in this sense.[10]) However, once a sufficiently large gravitationally bound system has formed, a cosmological constant would have no further effect on its dynamics, or on the eventual evolution of life. In particular, it makes no difference if the e-folding time of the cosmic expansion is much shorter than the time required for the evolution of intelligent life.[11] (Note that I am here *not* requiring that the cosmological constant have a value consistent with the astronomical knowledge, but only that it have a value consistent with the appearance of beings that could measure it.) Thus, irrespective of what we think are the possible forms of intelligent life, the necessary and sufficient anthropic condition on the cosmological constant is that it should not be so large as to prevent the appearance of gravitationally bound states.

I evaluate this bound here in the context of conventional big-bang cosmologies, described by a Robertson-Walker metric with initially small perturbations. For definiteness I will concentrate here on the case of positive cosmological constant and zero spatial curvature, $k=0$, and with the energy density of the Universe dominated since recombination by nonrelativistic matter. However, it would not be hard to adapt the arguments to any other case in which the Universe does not recollapse.

Let us consider then the fate of a density perturbation in a Universe with $\Lambda > 0$. Such a perturbation can be modeled[12] as a sphere within which there is a uniform excess density $\Delta\rho(t)$, and a gravitational field described by a Robertson-Walker (RW) metric with positive curvature constant $\Delta k > 0$, and with RW scale factor $a(t)$. The evolution of the perturbation is governed by a Friedmann equation

$$(da/dt)^2+\Delta k = \tfrac{8}{3}\pi G a^2(\rho+\Delta\rho+\rho_V), \qquad (1)$$

where $\rho(t)$ is the unperturbed cosmic mass density, ρ_V is the constant vacuum energy density

$$\rho_V \equiv \Lambda/8\pi G, \qquad (2)$$

and $\Delta\rho(t)$ is the perturbation, satisfying the equation of mass conservation

$$a^3(\rho+\Delta\rho)=\text{const}. \qquad (3)$$

I am *not* assuming here that $\Delta\rho$ or Δk is small, but there is a branch of the solutions of Eq. (1) for which, as $t\to 0$, $\Delta\rho \propto t^{-4/3}$ while $\rho \propto t^{-2}$, so that in this limit $\rho_V \ll \Delta\rho \ll \rho$. I assume that all perturbations are on this branch, so that the universe looks smooth for $t\to 0$. The strength of such a perturbation can then be characterized by the parameter

$$\tilde\rho \equiv \lim_{t\to 0}\{[\Delta\rho(t)]^3/\rho^2(t)\}. \qquad (4)$$

Treating Δk and $\Delta\rho$ to first order for $t\to 0$ in Eq. (1), we find for the perturbed curvature constant

$$\Delta k = \tfrac{40}{9}\pi G a^2(\rho+\Delta\rho)^{2/3}\tilde\rho^{1/3}. \qquad (5)$$

Equation (1) shows that the perturbed scale factor $a(t)$ will increase to a maximum and then collapse back to $a=0$, provided that there is a value of $a(t)$ where the right-hand side of Eq. (1) is equal to Δk. The right-hand side of Eq. (1) reaches a minimum at a value of $a(t)$ such that $\rho+\Delta\rho=2\rho_V$, and so the condition for a given perturbation to undergo gravitational condensation is that at this minimum, the right-hand side of Eq. (1) should be less than Δk. This condition can be written

$$8\pi G\rho_V^{1/3}[\tfrac{1}{2}(\rho+\Delta\rho)]^{2/3}a^2 < \Delta k. \qquad (6)$$

With use of Eq. (5) to express the perturbed curvature constant in terms of the parameter $\tilde\rho$, this becomes

$$\rho_V < \tfrac{500}{729}\tilde\rho. \qquad (7)$$

If there is an upper bound on the parameter $\tilde\rho$, and if this upper bound is independent of Λ (because it refers to very early times, when ρ_V is negligible), then Eq. (7) provides our anthropic bound on the vacuum energy density.

[It is instructive to compare this with the result of a linear analysis, in which Δk and $\Delta\rho$ are treated throughout as first-order perturbations. The solution of the second-order differential equation[13] for $\Delta\rho/\rho$ takes the form

$$\Delta\rho/\rho \propto (\sinh\tau)^{-2/3}Q_{1/3}^{2/3}(\coth\tau),$$

where

$$\tau \equiv (6\pi G\rho_V)^{1/2}t,$$

and $Q_\nu^\mu(z)$ is the associated Legendre function of the second kind.[14] This behaves for $\tau \ll 1$ like $\tau^{2/3}$, and then rises monotonically[15] to a constant limit as $\tau\to\infty$. By comparing the asymptotic behavior of $\Delta\rho/\rho$ for $\tau\to 0$ and $\tau\to\infty$, we can see that if we normalize so that $\Delta\rho/\rho \to \epsilon\tau^{2/3}$ for $\tau\to 0$, then for $\tau\to\infty$

$$\Delta\rho/\rho \to (2/\sqrt{\pi})\Gamma(\tfrac{11}{6})\Gamma(\tfrac{2}{3})\epsilon = 1.437\epsilon.$$

With this normalization, the parameter $\tilde\rho$ is just $\epsilon^3\rho_V$, so that $\Delta\rho/\rho \to 1.437(\tilde\rho/\rho_V)^{1/3}$. One might guess that the necessary and sufficient condition for gravitational condensation is that the linear analysis should give an asymptotic value of $\Delta\rho/\rho$ at least of order unity. If this were correct, then the upper bound on ρ_V for gravitational condensation to occur would be

$$\rho_V < (1.437)^3\tilde\rho = 2.97\tilde\rho.$$

This is quite different from the result (7), showing the

VOLUME 59, NUMBER 22 PHYSICAL REVIEW LETTERS 30 NOVEMBER 1987

inadequacy here of linear methods.]

Now, what is the distribution of $\tilde{\rho}$ values for actual perturbations? Eventually, the theory of the very early Universe may develop to the point that the distribution of $\tilde{\rho}$ will be known in terms of fundamental constants. Alternatively, it may some day become possible to read off the distribution of values of $\tilde{\rho}$ from observations of small-angle fluctuations in the cosmic microwave background temperature.

For the present, it seems that the best we can do is to use the existence of quasars at high red shifts to set an empirical lower bound on any maximum value for $\tilde{\rho}$, and hence a lower bound on the anthropic upper bound (6) on ρ_V. This is not as pointless as it may seem. If it turns out that the empirical lower bound on the maximum value of the right-hand side of Eq. (6) is much larger than empirically allowed values for ρ_V, then we would have to conclude that the anthropic principle does not explain why the cosmological constant is as small as it is.

In the model we have been considering, the time required for the scale factor $a(t)$ of a perturbation to reach its maximum and then collapse back to $a=0$ is an increasing function of ρ_V, so that it is bounded below by its value when $\rho_V=0$:

$$t_c > \tfrac{9}{2}\pi(250\pi G\tilde{\rho})^{-1/2}. \tag{8}$$

Actually, we do not measure the age of early gravitational condensations, but we do observe their red shifts. The time corresponding to a red shift z_c is bounded *above* by its value for $\rho_V=0$:

$$t_c < \tfrac{2}{3}(3/8\pi G\rho_0)^{1/2}(1+z_c)^{-3/2}, \tag{9}$$

where ρ_0 is the present cosmic mass density. Thus, putting together (8) and (9), we see that the observation of gravitational condensations (e.g., quasars) of red shift z_c or greater sets an empirical lower bound on the maximum value of the parameter $\tilde{\rho}$:

$$\tfrac{500}{729}\tilde{\rho}_{\max} > \tfrac{1}{3}\pi^2\rho_0(1+z_c)^3. \tag{10}$$

For instance, we know that quasars exist with red shifts up to about $z=4.4$, so taking $z_c=4.5$ in (10) gives a lower bound of $550\rho_0$ on the anthropic upper bound on ρ_V.

Now, if the intrinsic distribution of ρ_V values is smooth and featureless below the anthropic bound (as would be expected in the models of Refs. 3–5, 8, and 9 if the natural scale for ρ_V is set by the Planck mass), then it seems likely that ρ_V would be within 1 or 2 orders of magnitude of its upper bound. (This can be made more precise by calculations like those of Carter in Ref. 2.) With a lower bound of $550\rho_0$ on the anthropic upper bound on ρ_V, we would then conclude that ρ_V must be much greater than the present mass density ρ_0. Is this plausible? The answer unfortunately depends on data whose interpretation is far from settled.

On one hand, a vacuum energy density much larger than ρ_0 would resolve a problem that has been posed by measurements of ρ_0 for those who believe, on grounds of either inflation or aesthetics, that (as assumed here) the Universe has a vanishing unperturbed curvature constant k. This implies that the total energy density $\rho_V+\rho_0$ is equal to its critical value $3H_0^2/8\pi G$; that is $\Omega_V+\Omega_0=1$, where $\Omega_V=8\pi G\rho_V/3H_0^2$ and $\Omega_0=8\pi G\rho_0/3H_0^2$. The deuterium abundance indicates[16] that the contribution of baryons to Ω_0 is no greater than about 0.03 (for $H_0=100$ km/sec Mpc), while observations of galaxies suggest[17] that they contribute about 0.02 to Ω_0. Even allowing for nonbaryonic extra-galactic matter, the dynamics of clusters of galaxies lead to estimated values[17] of Ω_0 only about 0.1 to 0.2. It would be difficult to satisfy the condition $\Omega_0+\Omega_V=1$ if Ω_V were much smaller than Ω_0. However, if for instance we suppose that ρ_V is at least 1% of its anthropic upper bound (10), then Ω_V/Ω_0 must be at least 5.5, so that we would not need Ω_0 to be greater than 0.15.

The assumption of a vacuum energy density roughly comparable with the anthropic upper bound would also help[18] with the problem of cosmic ages that arises if the Hubble constant H_0 is as large as 100 km/sec Mpc. In this case, if $\Lambda=0$ and $k=0$, then the age of the Universe is about $\tfrac{2}{3}H_0^{-1}=6.5\times10^9$ yr, less than the (10–20) $\times10^9$ yr usually given[19] for the globular clusters. However, if at present ρ_V is much greater than ρ_0, then over most of the history of the Universe $R(t)$ behaves as $\exp(Ht)$ rather than $t^{2/3}$, and the age of the Universe is greater than $\tfrac{2}{3}H_0^{-1}$. Specifically, the present age of a gravitational condensation which forms at a red shift z_c is

$$t_0-t_c=\tfrac{2}{3}(1+\rho_0/\rho_V)^{1/2}H_0^{-1}\{\operatorname{invsinh}(\rho_V/\rho_0)^{1/2}-\operatorname{invsinh}[(\rho_V/\rho_0)^{1/2}(1+z_c)^{-3/2}]\}. \tag{11}$$

If for instance $z_c=4$ and $\rho_V/\rho_0=9$ (i.e., $\Omega_0=0.1$), then the age (11) is $1.1H_0^{-1}$, which even for H_0 close to 100 km/sec Mpc leaves adequate time for the evolution of globular clusters.

On the other hand, counts of galaxies as a function of red shift indicate[20] (for $k=0$) that $\Lambda/3H_0^2=0.1^{+0.2}_{-0.4}$ or in other words $\rho_V/\rho_0=0.1\pm0.3$. According to (10), this is at least 3 orders of magnitude less than the anthropic upper bound (7).

A similar conclusion would be reached if it turns out that gravitational condensations occur at red shifts larger than 4.5. For instance, if gravitational condensations occur at a red shift $z_c=30$, then according to Eq. (10), the anthropic upper bound on ρ_V is at least $10^5\rho_0$. But Ω_0 cannot be less than about 0.01, indicating (for $k=0$)

VOLUME 59, NUMBER 22 PHYSICAL REVIEW LETTERS 30 NOVEMBER 1987

a vacuum energy density ρ_V no greater than $100\rho_0$, which is again at least 3 orders of magnitude less than the anthropic upper bound.

Thus if the interpretation of galaxy number counts in Ref. 20 holds up, or if gravitational condensations are found at red shifts $z \gg 4$, we will be able to conclude that the cosmological constant is so small that even the anthropic principle could not explain its smallness.

I am grateful for conversations with T. Banks, G. Field, F. Wilczek, E. Witten, A. Zee, and with many colleagues at the University of Texas, and especially to P. Shapiro for helpful discussions of the data and references in observational cosmology. This work is supported in part by the Robert A. Welch Foundation and National Science Foundation Grant No. 8605978.

[1]See, e.g., Ya. B. Zeldovich, Pis'ma Zh. Eksp. Teor. Fiz. **6**, 883 (1967) [JETP Lett. **6**, 316 (1967)].

[2]For a comprehensive survey of the anthropic principle, see J. D. Barrow and F. J. Tipler, *The Anthropic Cosmological Principle* (Clarendon, Oxford, 1986). Also see P. C. W. Davies, *The Accidental Universe* (Cambridge Univ. Press, Cambridge, 1982), Chap. 5; B. Carter, Philos. Trans. Roy. Soc. London A **310** 347 (1983).

[3]T. Banks, Nucl. Phys. **B249**, 332 (1985).

[4]L. F. Abbott, Phys. Lett. **150B**, 427 (1985).

[5]J. D. Brown and C. Teitelboim, to be published.

[6]The introduction of a three-form gauge field has the effect of making the effective cosmological constant a constant of integration, as shown by E. Witten, in *Shelter Island II: Proceedings of the 1983 Shelter Island Conference on Quantum Field Theory and the Fundamental Problems of Physics, Shelter Island, New York, 1983*, edited by R. Jackiw *et al.* (M.I.T. Press, Cambridge, MA, 1985); M. Henneaux and C. Teitelboim, Phys. Lett. **143B**, 415 (1984).

[7]S. Coleman and F. De Luccia, Phys. Rev. D **21**, 3305 (1980).

[8]A. D. Linde, to be published.

[9]S. Hawking, as quoted by M. Gell-Mann, Phys. Scr. **T15**, 202 (1987).

[10]This remark about globular clusters is due to H. Smith. He points out that the observed low gas abundance in globular clusters presumably results from their inevitable frequent passages through the plane of our galaxy with consequent sweeping out of the gas from the clusters during these passages. If globular clusters formed first, and then did not associate with galaxies, there could still be heavy elements formed by the evolving stars in a cluster, condensing to the plane of rotation of the cluster, perhaps reaching densities high enough to allow the formation of second-generation stars with planets. This could work for large enough globular clusters which thus effectively become small galaxies. A small globular cluster would probably not have sufficient gravitational attraction to hold the gas against the blast effects of supernovae stripping it from the cluster.

[11]In this respect, I differ from Barrow and Tipler (Ref. 2) and consequently obtain a different anthropic bound on Λ. They require that $|\Lambda|^{-1/2}$ (roughly the vacuum Hubble time) should be at least as large as the main-sequence stellar lifetime t_*. This seems to me correct for $\Lambda < 0$ [where the Universe recollapses in a time $\pi(3|\Lambda|)^{-1/2}$], but not for $\Lambda > 0$, the case under consideration here. It is true that for $\Lambda > 0$, gravitational condensation must occur within a time of order $\Lambda^{-1/2}$, because after that time the matter density drops below the vacuum density, and gravitational condensation becomes impossible. However, once gravitationally bound systems form, their subsequent evolution is unaffected by the cosmological constant, and it makes no difference how long it takes for stars to evolve on the main sequence.

[12]P. J. E. Peebles, Astrophys. J. **147**, 859 (1967).

[13]See, e.g., S. Weinberg, *Gravitation and Cosmology* (Wiley, New York, 1972), Eq. (15.10.57).

[14]I use the notation of the *Handbook of Mathematical Functions*, edited by M. Abramowitz and I. A. Stegun (Dover, New York, 1965), Sec. 8.

[15]I am very grateful to P. Candelas for a numerical integration of the differential equation for $\Delta\rho/\rho$.

[16]A. M. Boesgaard and G. Steigman, Annu. Rev. Astron. Astrophys. **23**, 310 (1985).

[17]*Dark Matter in the Universe*, edited by G. R. Knapp and J. Kormendy, Proceedings of the International Astronomical Union Symposium No. 117 (Reidel, Dordrecht, 1987).

[18]G. de Vaucouleurs has long advocated the introduction of a cosmological constant to reconcile the apparent inconsistency of his measurement of the Hubble constant with globular cluster ages; see Astrophys. J. **268**, 468 (1983), Appendix B, and Nature (London) **299**, 303 (1982), etc. Also see P. J. E. Peebles, Astrophys. J. **284**, 439 (1984).

[19]A. Renzini, in *Galaxy Distances and Deviations from Universal Expansion*, edited by B. F. Madore and R. B. Tully (Reidel, Dordrecht, 1986), pp. 177–184.

[20]E. D. Loh, Phys. Rev. Lett. **57**, 2865 (1986).

The cosmological constant problem*

Steven Weinberg

Theory Group, Department of Physics, University of Texas, Austin, Texas 78712

Astronomical observations indicate that the cosmological constant is many orders of magnitude smaller than estimated in modern theories of elementary particles. After a brief review of the history of this problem, five different approaches to its solution are described.

CONTENTS

As I was going up the stair,
I met a man who wasn't there.
He wasn't there again today,
I wish, I wish he'd stay away.

Hughes Mearns

I. INTRODUCTION

Physics thrives on crisis. We all recall the great progress made while finding a way out of various crises of the past: the failure to detect a motion of the Earth through the ether, the discovery of the continuous spectrum of beta decay, the τ-θ problem, the ultraviolet divergences in electromagnetic and then weak interactions, and so on. Unfortunately, we have run short of crises lately. The "standard model" of electroweak and strong interactions currently faces neither internal inconsistencies nor conflicts with experiment. It has plenty of loose ends; we know no reason why the quarks and leptons should have the masses they have, but then we know no reason why they should not.

Perhaps it is for want of other crises to worry about that interest is increasingly centered on one veritable crisis: theoretical expectations for the cosmological constant exceed observational limits by some 120 orders of magnitude.[1] In these lectures I will first review the history of this problem and then survey the various attempts that have been made at a solution.

*Morris Loeb Lectures in Physics, Harvard University, May 2, 3, 5, and 10, 1988.

[1]For a good nonmathematical description of the cosmological constant problem, see Abbott (1988).

II. EARLY HISTORY

After completing his formulation of general relativity in 1915–1916, Einstein (1917) attempted to apply his new theory to the whole universe. His guiding principle was that the universe is static: "The most important fact that we draw from experience is that the relative velocities of the stars are very small as compared with the velocity of light." No such static solution of his original equations could be found (any more than for Newtonian gravitation), so he modified them by adding a new term involving a free parameter λ, the cosmological constant:[2]

$$R_{\mu\nu}-\tfrac{1}{2}g_{\mu\nu}R-\lambda g_{\mu\nu}=-8\pi G T_{\mu\nu}\ . \tag{2.1}$$

Now, for $\lambda>0$, there was a static solution for a universe filled with dust of zero pressure and mass density

$$\rho=\frac{\lambda}{8\pi G}\ . \tag{2.2}$$

Its geometry was that of a sphere S_3, with proper circumference $2\pi r$, where

$$r=1/\sqrt{8\pi\rho G}\ , \tag{2.3}$$

so the mass of the universe was

$$M=2\pi^2 r^3\rho=\frac{\pi}{4}\lambda^{-1/2}G^{-1}\ . \tag{2.4}$$

In some popular history accounts, it was Hubble's discovery of the expansion of the universe that led Einstein to retract his proposal of a cosmological constant. The real story is more complicated, and more interesting.

One disappointment came almost immediately. Einstein had been pleased at the connection in his model between the mass density of the universe and its geometry, because, following Mach's lead, he expected that the mass distribution of the universe should set inertial frames. It was therefore unpleasant when his friend de Sitter, with whom Einstein remained in touch during the war, in 1917 proposed another apparently static cosmological model with no matter at all. (See de Sitter, 1917.) Its line element (using the same coordinate system as de Sitter, but in a different notation) was

$$d\tau^2=\frac{1}{\cosh^2 Hr}[dt^2-dr^2 - H^{-2}\tanh^2 Hr(d\theta^2+\sin^2\theta\, d\varphi^2)]\ , \tag{2.5}$$

[2]The notation used here for metrics, curvatures, etc., is the same as in Weinberg (1972).

with H related to the cosmological constant by

$$H=\sqrt{\lambda/3} \tag{2.6}$$

and $\rho=p=0$. Clearly matter was not needed to produce inertia.

At about this time, the redshift of distant objects was being discovered by Slipher. Over the period from 1910 to the mid-1920s, Slipher (1924) observed that galaxies (or, as then known, spiral nebulae) have redshifts $z\equiv\Delta\lambda/\lambda$ ranging up to 6%, and only a few have blueshifts. Weyl pointed out in 1923 that de Sitter's model would exhibit such a redshift, increasing with distance, because although the metric in de Sitter's coordinate system is time independent, test bodies are not at rest; there is a nonvanishing component of the affine connection

$$\Gamma^r_{tt}=-H\sinh Hr\tanh Hr \tag{2.7}$$

giving a redshift proportional to distance

$$z\simeq Hr \quad \text{for } Hr\ll 1\ . \tag{2.8}$$

In his influential textbook, Eddington (1924) interpreted Slipher's redshifts in terms of de Sitter's "static" universe.

But of course, although the cosmological constant *was* needed for a static universe, it was not needed for an expanding one. Already in 1922, Friedmann (1924) had described a class of cosmological models, with line element (in modern notation)

$$d\tau^2=dt^2-R^2(t)\left[\frac{dr^2}{1-kr^2}+r^2(d\theta^2+\sin^2\theta\, d\varphi^2)\right]. \tag{2.9}$$

These are comoving coordinates; the universe expands or contracts as $R(t)$ increases or decreases, but the galaxies keep fixed coordinates r,θ,φ. The motion of the cosmic scale factor is governed by an energy-conservation equation

$$\left[\frac{dR}{dt}\right]^2=-k+\tfrac{1}{3}R^2(8\pi G\rho+\lambda)\ . \tag{2.10}$$

The de Sitter model is just the special case with $k=0$ and $\rho=0$; in order to put the line element (2.5) in the more general form (2.9), it is necessary to introduce new coordinates,

$$t'=t-H^{-1}\ln\cosh Hr\ ,$$
$$r'=H^{-1}\exp(-Ht)\sinh Hr\ , \tag{2.11}$$
$$\theta'=\theta,\quad \varphi'=\varphi\ ,$$

and then drop the primes. However, we can also easily find expanding solutions with $\lambda=0$ and $\rho>0$. Pais (1982) quotes a 1923 letter of Einstein to Weyl, giving his reaction to the discovery of the expansion of the universe: "If there is no quasi-static world, then away with the cosmological term!"

III. THE PROBLEM

Unfortunately, it was not so easy simply to drop the cosmological constant, because anything that contributes to the energy density of the vacuum acts just like a cosmological constant. Lorentz invariance tells us that in the vacuum the energy-momentum tensor must take the form

$$\langle T_{\mu\nu}\rangle=-\langle\rho\rangle g_{\mu\nu}\ . \tag{3.1}$$

(A minus sign appears here because we use a metric which for flat space-time has $g_{00}=-1$.) Inspection of Eq. (2.1) shows that this has the same effect as adding a term $8\pi G\langle\rho\rangle$ to the effective cosmological constant

$$\lambda_{\rm eff}=\lambda+8\pi G\langle\rho\rangle\ . \tag{3.2}$$

Equivalently we can say that the Einstein cosmological constant contributes a term $\lambda/8\pi G$ to the total effective vacuum energy

$$\rho_V=\langle\rho\rangle+\lambda/8\pi G=\lambda_{\rm eff}/8\pi G\ . \tag{3.3}$$

A crude experimental upper bound on $\lambda_{\rm eff}$ or ρ_V is provided by measurements of cosmological redshifts as a function of distance, the program begun by Hubble in the late 1920s. The present expansion rate is today estimated as

$$\left[\frac{1}{R}\frac{dR}{dt}\right]_{\rm now}\equiv H_0\simeq 50-100\ \text{km/sec Mpc}$$
$$\simeq(\tfrac{1}{2}-1)\times10^{-10}/\text{yr}\ .$$

Furthermore, we do not gross effects of the curvature of the universe, so very roughly

$$|k|/R^2_{\rm now}\lesssim H_0^2\ .$$

Finally, the ordinary nonvacuum mass density of the universe is not much greater than its critical value

$$|\rho-\langle\rho\rangle|\lesssim 3H_0^2/8\pi G\ .$$

Hence (2.10) shows that

$$|\lambda_{\rm eff}|\lesssim H_0^2$$

or, in physicists' units,

$$|\rho_V|\lesssim10^{-29}\ \text{g/cm}^3\approx10^{-47}\ \text{GeV}^4\ . \tag{3.4}$$

A more precise observational bound will be discussed in Sec. V, but this one will be good enough for our present purposes.

As everyone knows, the trouble with this is that the energy density $\langle\rho\rangle$ of empty space is likely to be enormously larger than 10^{-47} GeV^4. For one thing, summing the zero-point energies of all normal modes of some field of mass m up to a wave number cutoff $\Lambda\gg m$ yields a vacuum energy density (with $\hbar=c=1$)

$$\langle\rho\rangle=\int_0^\Lambda\frac{4\pi k^2dk}{(2\pi)^3}\frac{1}{2}\sqrt{k^2+m^2}\simeq\frac{\Lambda^4}{16\pi^2}\ . \tag{3.5}$$

If we believe general relativity up to the Planck scale, then we might take $\Lambda \simeq (8\pi G)^{-1/2}$, which would give

$$\langle \rho \rangle \approx 2^{-10}\pi^{-4}G^{-2} = 2\times 10^{71}\ \mathrm{GeV}^4 \ . \quad (3.6)$$

But we saw that $|\langle \rho \rangle + \lambda/8\pi G|$ is less than about 10^{-47} GeV^4, so the two terms here must cancel to better than 118 decimal places. Even if we only worry about zero-point energies in quantum chromodynamics, we would expect $\langle \rho \rangle$ to be of order $\Lambda_{\mathrm{QCD}}^4/16\pi^2$, or 10^{-6} GeV^4, requiring $\lambda/8\pi G$ to cancel this term to about 41 decimal places.

Perhaps surprisingly, it was a long time before particle physicists began seriously to worry about this problem, despite the demonstration in the Casimir effect of the reality of zero-point energies.[3] Since the cosmological upper bound on $|\langle \rho \rangle + \lambda/8\pi G|$ was vastly less than any value expected from particle theory, most particle theorists simply assumed that for some unknown reason this quantity was zero. But cosmologists generally continued to keep an open mind, analyzing cosmological data in terms of models with a possibly nonvanishing cosmological constant.

In fact, as far as I know, the first published discussion of the contribution of quantum fluctuations to the effective cosmological constant was triggered by astronomical observations. In the late 1960s it seemed that an excessively large number of quasars were being observed with redshifts clustered about $z = 1.95$. Since $1+z$ is the ratio of the cosmic scale factor $R(t)$ at present to its value at the time the light now observed was emitted, this could be explained if the universe loitered for a while at a value of $R(t)$ equal to $1/2.95$ times the present value. A number of authors [Petrosian, Salpeter, and Szekeres (1967); Shklovsky (1967); Rowan-Robinson (1968)] proposed that such a loitering could be accounted for in a model proposed by Lemaître (1927, 1931). In this model there is a positive cosmological constant λ_{eff} and positive curvature $k = +1$, just as in the static Einstein model, while the mass of the universe is taken close to the Einstein value (2.4). The scale factor $R(t)$ starts at $R = 0$ and then increases; however, when the mass density drops to near the Einstein value (2.2), the universe behaves for a while like a static Einstein universe, until the instability of this model takes over and the universe starts expanding again. In order for this idea to explain a preponderance of redshifts at $z = 1.95$, the vacuum energy density ρ_V would have to be $(2.95)^3$ times the present nonvacuum mass density ρ_0.

These considerations led Zeldovich (1967) to attempt to account for a nonzero vacuum energy density in terms of quantum fluctuations. As we have seen, the zero-point energies themselves gave far too large a value for $\langle \rho \rangle$, so Zeldovich assumed that these were canceled by $\lambda/8\pi G$, leaving only higher-order effects: in particular, the gravitational force between the particles in the vacuum fluctuations. (In Feynman diagram terms, this corresponds to throwing away the one-loop vacuum graphs, but keeping those with two loops.) Taking Λ^3 particles of energy Λ per unit volume gives the gravitational self-energy density of order

$$\langle \rho \rangle \approx (G\Lambda^2/\Lambda^{-1})\Lambda^3 = G\Lambda^6 \ . \quad (3.7)$$

For no clear reason, Zeldovich took the cutoff Λ as 1 GeV, which yields a density $\langle \rho \rangle \approx 10^{-38}$ GeV^4, much smaller than from zero-point energies themselves, but still larger than the observational bound (3.4) on $|\langle \rho \rangle + \lambda/8\pi G|$ by some 9 orders of magnitude. Neither Zeldovich nor anyone else felt encouraged to pursue these ideas.

The real beginning of serious worry about the vacuum energy seems to date from the success of the idea of spontaneous symmetry breaking in the electroweak theory. In this theory, the scalar field potential takes the form (with $\mu^2 > 0$, $g > 0$)

$$V = V_0 - \mu^2\phi^\dagger\phi + g(\phi^\dagger\phi)^2 \ . \quad (3.8)$$

At its minimum this takes the value

$$\langle \rho \rangle = V_{\mathrm{min}} = V_0 - \frac{\mu^4}{4g} \ . \quad (3.9)$$

Apparently some theorists felt that V should vanish at $\phi = 0$, which would give $V_0 = 0$, so that $\langle \rho \rangle$ would be negative definite.[4] In the electroweak theory this would give $\langle \rho \rangle \simeq -g(300\ \mathrm{GeV})^4$, which even for g as small as α^2 would yield $|\langle \rho \rangle| \simeq 10^6$ GeV^4, larger than the bound on ρ_V by a factor 10^{53}. Of course we know of no reason why V_0 or λ must vanish, and it is entirely possible that V_0 or λ cancels the term $-\mu^4/4g$ (and higher-order corrections), but this example shows vividly how unnatural it is to get a reasonably small effective cosmological constant. Moreover, at early times the effective temperature-dependent potential has a positive coefficient for $\phi^\dagger\phi$, so the minimum then is at $\phi = 0$, where $V(\phi) = V_0$. Thus, in order to get a zero cosmological constant today, we have to put up with an enormous cosmological constant at times before the electroweak phase transition. [This is *not* in conflict with experiment; in fact, the phase transition occurs at a temperature T of order $\mu/\sqrt{g}$, so the black-body radiation present at that

[3] Casimir (1948) showed that quantum fluctuations in the space between two flat conducting plates with separation d would produce a force per unit area equal to $\hbar c\pi^2/240d^4$, or 1.30×10^{-18} $\mathrm{dyn\,cm}^2/d^4$. This was measured by Sparnaay (1957), who found a force per area of $(1-4)\times 10^{-18}$ $\mathrm{dyn\,cm}^2/d^4$, when d was varied between 2 and 10 μm.

[4] Veltman (1975) attributes this view to Linde (1974), himself (quoted as "to be published"), and Dreitlein (1974). However, Linde's paper does not seem to me to take this position. Dreitlein's paper proposed that Eq. (3.9) could give an acceptably small value of $\langle \rho \rangle$, with $\mu/\sqrt{g}$ fixed by the Fermi coupling constant of weak interactions, if μ is very small, of order 10^{-27} MeV. Veltman's paper gives experimental arguments against this possibility.

time has an energy density of order μ^4/g^2, larger than the vacuum energy by a factor $1/g$ (Bludman and Ruderman, 1977).] At even earlier times there were other transitions, implying an even larger early value for the effective cosmological constant. This is currently regarded as a good thing; the large early cosmological constant would drive cosmic inflation, solving several of the long-standing problems of cosmological theory (Guth, 1981; Albrecht and Steinhardt, 1982; Linde, 1982). We want to explain why the effective cosmological constant is small now, *not* why it was always small.

Before closing this section, I want to take up a peculiar aspect of the problem of the cosmological constant. The appearance of an effective cosmological constant makes it impossible to find any solutions of the Einstein field equations in which $g_{\mu\nu}$ is the constant Minkowski term $\eta_{\mu\nu}$. That is, the original symmetry of general covariance, which is always broken by the appearance of any given metric $g_{\mu\nu}$, cannot, without fine-tuning, be broken in such a way as to preserve the subgroup of space-time translations.

This situation is unusual. Usually if a theory is invariant under some group G, we would not expect to have to fine-tune the parameters of the theory in order to find vacuum solutions that preserve any given subgroup $H \subset G$. For instance, in the electroweak theory, there is a finite range of parameters in which any number of doublet scalars will get vacuum expectation values that preserve a U(1) subgroup of $SU(2)\times U(1)$. So why will this not work for the translational subgroup of the group of general coordinate transformations? Suppose we look for a solution of the field equations that preserves translational invariance. With all fields constant, the field equations for matter and gravity are

$$\frac{\partial \mathcal{L}}{\partial \psi_i}=0 , \tag{3.10}$$

$$\frac{\partial \mathcal{L}}{\partial g_{\mu\nu}}=0 . \tag{3.11}$$

With N ψ's, these are $N+6$ equations for $N+6$ unknowns, so one might expect a solution without fine-tuning. The problem is that when (3.10) is satisfied, the dependence of $\mathcal{L}$ on $g_{\mu\nu}$ is too simple to allow a solution of (3.11). There is a GL(4) symmetry that survives as a vestige of general covariance even when we constrain the fields to be constants: under the GL(4) transformation

$$g_{\mu\nu} \rightarrow A^{\rho}{}_{\mu} A^{\sigma}{}_{\nu} g_{\rho\sigma} , \tag{3.12}$$

$$\psi_i \rightarrow D_{ij}(A)\psi_j ; \tag{3.13}$$

the Lagrangian transforms as a density,

$$\mathcal{L} \rightarrow \mathrm{Det}\, A \mathcal{L} . \tag{3.14}$$

When Eq. (3.10) is satisfied, this implies that $\mathcal{L}$ transforms as in (3.14) under (3.12) *alone*. This has the unique solution

$$\mathcal{L}=c(\mathrm{Det} g)^{1/2} , \tag{3.15}$$

with c independent of $g_{\mu\nu}$. With this $\mathcal{L}$, there are no solutions of Eq. (3.11), unless for some reason the coefficient c vanishes when (3.10) is satisfied.

Now that the problem has been posed, we turn to its possible solution. The next five sections will describe five directions that have been taken in trying to solve the problem of the cosmological constant.

IV. SUPERSYMMETRY, SUPERGRAVITY, SUPERSTRINGS

Shortly after the development of four-dimensional globally supersymmetric field theories, Zumino (1975) pointed out that supersymmetry in these theories would, if unbroken, imply a vanishing vacuum energy. The argument is very simple: the supersymmetry generators Q_α satisfy an anticommutation relation

$$\{Q_\alpha, Q_\beta^\dagger\}=(\sigma_\mu)_{\alpha\beta}P^\mu , \tag{4.1}$$

where α and β are two-component spin indices; σ_1, σ_2, and σ_3 are the Pauli matrices; $\sigma_0=1$; and P^μ is the energy-momentum 4-vector operator. If supersymmetry is unbroken, then the vacuum state $|0\rangle$ satisfies

$$Q_\alpha|0\rangle=Q_\alpha^\dagger|0\rangle=0 , \tag{4.2}$$

and from (4.1) and (4.2) we infer that the vacuum has vanishing energy and momentum

$$\langle 0|P^\mu|0\rangle=0 .$$

This result can also be obtained by considering the potential $V(\phi,\phi^*)$ for the chiral scalar fields ϕ^i of a globally supersymmetric theory:

$$V(\phi,\phi^*)=\sum_i \left|\frac{\partial W(\phi)}{\partial \phi^i}\right|^2 , \tag{4.3}$$

where $W(\phi)$ is the so-called superpotential. (Gauge degrees of freedom are ignored here, but they would not change the argument.) The condition for unbroken supersymmetry is that W be stationary in ϕ, which would imply that V take its minimum value,

$$\langle \rho \rangle = V_{\min}=0 . \tag{4.4}$$

Quantum effects do not change this conclusion, because with boson-fermion symmetry, the fermion loops cancel the boson ones.

The trouble with this result is that supersymmetry is broken in the real world, and in this case either (4.1) or (4.3) shows that the vacuum energy is positive-definite. *If* this vacuum energy were the sole contribution to the effective cosmological constant, then the effect of supersymmetry would be to convert the problem of the cosmological constant from a crisis into a disaster.

Fortunately this is not the whole story. It is not possible to decide the value of the effective cosmological constant unless we explicitly introduce gravitation into the theory. Any globally supersymmetric theory that involves gravity is inevitably a locally supersymmetric supergravity theory. In such a theory the effective cosmological constant *is* given by the expectation value of the potential, but the potential is now given by (Cremmer *et al.*, 1978, 1979; Barbieri *et al.*, 1982; Witten and Bagger, 1982)

$$V(\phi,\phi^*)=\exp(8\pi GK)[D_i W(\mathcal{G}^{-1})^i{}_j(D_j W)^* -24\pi G|W|^2] , \qquad (4.5)$$

where $K(\phi,\phi^*)$ is a real function of both ϕ and ϕ^* known as the Kahler potential, $D_i W$ is a sort of covariant derivative

$$D_i W \equiv \frac{\partial W}{\partial \phi^i}+8\pi G\frac{\partial K}{\partial \phi^i} , \qquad (4.6)$$

and $(\mathcal{G}^{-1})^i{}_j$ is the inverse of a metric

$$\mathcal{G}^i{}_j \equiv \frac{\partial^2 K}{\partial \phi^{i*}\partial \phi^j} . \qquad (4.7)$$

The condition for unbroken supersymmetry is now $D_i W=0$. This again yields a stationary point of the potential, but now it is one at which V is generally negative. In fact, even if we fine-tuned W so that there were a supersymmetric stationary point at which $W=0$ and hence $V=0$, such a solution would not, in general, be the state of lowest energy, though it would be stable [Coleman and de Luccia (1980), Weinberg (1982)]. It should, however, be mentioned that if there is a set of field values at which $W=0$ and $D_i W=0$ for all i in lowest order of perturbation theory, then the theory has a supersymmetric equilibrium configuration with $V=0$ to all orders of perturbation theory, though not necessarily beyond perturbation theory (Grisaru, Siegel, and Rocek, 1979). The same is believed to be true in superstring perturbation theory (Dine and Seiberg, 1986; Friedan, Martinec, and Shenker, 1986; Martinec, 1986; Attick, Moore, and Sen, 1987; Morozov and Perelomov, 1987).

Without fine-tuning, we can generally find a nonsupersymmetric set of scalar field values at which $V=0$ and $D_i W\neq 0$, but this would not normally be a stationary point of V. Thus in supergravity the problem of the cosmological constant is no more a disaster, but just as much a crisis, as in nonsupersymmetric theories.

On the other hand, supergravity theories offer opportunities for changing the context of the cosmological constant problem, if not yet for solving it. Cremmer *et al.* (1983) have noted that there is a class of Kahler potentials and superpotentials that, for a broad range of most parameters, automatically yield an equilibrium scalar field configuration in which $V=0$, even though supersymmetry is broken. Here is a somewhat generalized version: the Kahler potential is

$$K=-3\ln|T+T^*-h(C^a,C^{a*})|/8\pi G +\tilde{K}(S^n,S^{n*}) \qquad (4.8)$$

while the superpotential is

$$W=W_1(C^a)+W_2(S^n) , \qquad (4.9)$$

and T,C^a,S^n are all chiral scalar fields. No constraints are placed on the functions $h(C^a,C^{a*})$, $\tilde{K}(S^n,S^{n*})$, $W_1(C^a)$, or $W_2(S^n)$, except that h and $\tilde{K}$ are real, and functions all depend only on the fields indicated; in particular, the superpotential must be independent of the single chiral scalar T.

With these conditions the potential (4.5) takes the form

$$V=\exp(8\pi G\tilde{K})\left[\frac{1}{3(T+T^*+h)^3}\left[\frac{\partial W}{\partial C^a}\right](\mathcal{N}^{-1})^a{}_b \times\left[\frac{\partial W}{\partial C^b}\right]^* +(D_n W)(\mathcal{G}^{-1})^n{}_m(D_m W)^*\right] , \qquad (4.10)$$

where $(\mathcal{N}^{-1})^a{}_b$ is the reciprocal of the matrix

$$\mathcal{N}^a{}_b=\frac{\partial^2 h}{\partial C^{a*}\partial C^b} . \qquad (4.11)$$

The matrices $\mathcal{N}^a{}_b$ and $\mathcal{G}^n{}_m$ are necessarily positive-definite, because of their role in the kinetic part of the scalar Lagrangian

$$\begin{aligned}\mathcal{L}_{\rm kin}&=-\mathcal{G}^i{}_j\frac{\partial\phi^{i*}}{\partial x^\mu}\frac{\partial\phi^j}{\partial x_\mu}\\ &=-\frac{3}{(T+T^*+h)^2}\left[\frac{\partial T}{\partial x^\mu}+\frac{\partial h}{\partial C^a}\frac{\partial C^a}{\partial x^\mu}\right]\left[\frac{\partial T}{\partial x_\mu}+\frac{\partial h}{\partial C^b}\frac{\partial C^b}{\partial x_\mu}\right]^* -\frac{3}{|T+T^*-h|}\mathcal{N}^a{}_b\frac{\partial C^{a*}}{\partial x^\mu}\frac{\partial C^b}{\partial x_\mu}-\mathcal{G}^n{}_m\frac{\partial S^{n*}}{\partial x^\mu}\frac{\partial S^m}{\partial x_\mu} .\end{aligned} \qquad (4.12)$$

Hence Eq. (4.10) is positive and therefore, without further fine-tuning, may be expected to have a stationary point with $V=0$, specified by the conditions

$$\frac{\partial W}{\partial C^a}=D_n W=0 . \qquad (4.13)$$

But this is not necessarily a supersymmetric configuration, because here

$$D_a W \equiv \frac{\partial W}{\partial C^a}+8\pi G\frac{\partial K}{\partial C^a}W = -\frac{3}{|T+T^*+h|}\frac{\partial h}{\partial C^a}W , \qquad (4.14)$$

and this does not necessarily vanish. (However, to have supersymmetry broken, it is essential that the superpotential actually depend on all of the chiral scalars S^n, because otherwise the conditions $D_n W=0$ would require $W=0$ and hence $D_a W=0$.)

The superpotential W depends on C^a and S^n, but not on T, so the conditions (4.13) will generally fix the values of C^a and S^n at the minimum of V, while leaving T undetermined. The field T enters the potential only in the overall scale of the part that depends on the C^a, so such theories are called "no-scale" models. An intensive phenomenological study of these models was carried out at CERN for several years following 1983 (Ellis, Lahanas, *et al.*, 1984; Ellis, Kounnas, *et al.*, 1984; Barbieri *et al.*, 1985).

Of course, these models do not solve the cosmological constant problem, because neither Eq. (4.8) nor Eq. (4.9) is dictated by any known physical principle. In particular, in order to cancel the second term in Eq. (4.5), it is essential that the coefficient of the logarithm in the first term in (4.8) be given the apparently arbitrary value $-3/8\pi G$.

It was therefore exciting when, in some of the first work on the physical implications of superstring theory, it was found that compactification of six of the ten original dimensions yielded a four-dimensional supergravity theory with Kahler potential and superpotential of the form (4.8) and (4.9). Specifically, Witten (1985) found a Kahler potential of the form (4.8), with h quadratic in the C's and $\tilde{K}=-\ln(S+S^*)/8\pi G$, but with a superpotential that depended solely on the C's. By including nonperturbative gaugino condensation effects, Dine *et al.* (1985) were able to give the superpotential a dependence on S (though they did not treat the dependence of the Kahler potential or superpotential on the C^a fields). In this work, the S field is a complex function (now often called Y) of four-dimensional dilaton and axion fields, while the T field represents the scale of the compactified six-dimensional manifold. The factor 3 in Eq. (4.8) arises in these models because one compactifies on a complex manifold with $(10-4)/2=3$ complex dimensions (Chang *et al.*, 1988).

Intriguing as these results are, they have not been taken seriously (even by the original authors) as a solution of the cosmological constant problem. The trouble is that no one expects the simple structures (4.8) and (4.9) to survive beyond the lowest order of perturbation theory, because they are not protected by any symmetry that survives down to accessible energies.

Recently Moore (1987a, 1987b) has attempted a more specifically "stringy" attack on the problem. Early work by Rohm (1984) and Polchinski (1986) had shown that in the calculation of the vacuum energy density, the sum over zero-point energies can be converted into an integral over a complex "modular parameter" τ. (In string theories, two-dimensional conformal symmetry makes the tree-level vacuum energy vanish.) Last year Moore pointed out that for some special compactifications there is a discrete symmetry of modular space, known as Atkin-Lehner symmetry, that makes the integral over τ vanish despite the absence of space-time supersymmetry. So far, the only examples where this occurs entail a compactification to two rather than four space-time dimensions, but it does not seem unlikely that four-dimensional examples could be found. A more serious obstacle is that the Atkin-Lehner symmetry seems irretrievably tied to one-loop order.

Indeed, it is very hard to see how any property of supergravity or superstring theory could make the effective cosmological constant sufficiently small. It is not enough that the vacuum energy density cancel in lowest order, or to all finite orders of perturbative theory; even nonperturbative effects like ordinary QCD instantons would give far too large a contribution to the effective cosmological constant if not canceled by something else. According to our modern theories, properties of elementary particles, like approximate baryon and lepton conservation, are dictated by gauge symmetries of the standard model, which survive down to accessible energies. We know of no such symmetry (aside from the unrealistic example of unbroken supersymmetry) that could keep the effective cosmological constant sufficiently small. It is conceivable that in supergravity the property of having zero effective cosmological constant does survive to low energies without any symmetry to guard it, but this would run counter to all our experience in physics.

V. ANTHROPIC CONSIDERATIONS

I now turn to a very different approach to the cosmological constant, based on what Carter (1974) has called the anthropic principle.[5] Briefly stated, the anthropic principle has it that the world is the way it is, at least in part, because otherwise there would be no one to ask why it is the way it is. There are a number of different versions of this principle, ranging from those that are so weak as to be trivial to those that are so strong as to be absurd. Three of these versions seem worth distinguishing here.

(i) In one very weak version, the anthropic principle amounts simply to the use of the fact that we are here as one more experimental datum. For instance, recall M. Goldhaber's joke that "we know in our bones" that the lifetime of the proton must be greater than about 10^{16} yr, because otherwise we would not survive the ionizing particles produced by proton decay in our own bodies. No one can argue with this version, but it does not help us to explain anything, such as *why* the proton lives so long. Nor does it give very useful experimental information; certainly experimental physicists (including Goldhaber) have provided us with better limits on the proton lifetime.

[5]Recent discussions of the anthropic principle are given in the books by Davies (1982) and Barrow and Tipler (1986), and in articles by Carter (1983), Page (1987), and Rees (1987).

(ii) In one rather strong version, the anthropic principle states that the laws of nature, which are otherwise incomplete, are completed by the requirement that conditions must allow intelligent life to arise, the reason being that science (and quantum mechanics in particular) is meaningless without observers. I do not know how to reach a decision about such matters and will simply state my own view, that although science is clearly impossible without scientists, it is not clear that the universe is impossible without science.

(iii) A moderate version of the anthropic principle, sometimes known as the "weak anthropic principle," amounts to an explanation of which of the various possible eras or parts of the universe we inhabit, by calculating which eras or parts of the universe we *could* inhabit. An example is provided by what I think is the first use of anthropic arguments in modern physics, by Dicke (1961), in response to a problem posed by Dirac (1937). In effect, Dirac had noted that a combination of fundamental constants with the dimensions of a time turns out to be roughly of the order of the present age of the universe:

$$\hbar/Gcm_\pi^3=4.5\times10^{10}\ \text{yr}\ . \tag{5.1}$$

[There are various other ways of writing this relation, such as replacing m_π with various combinations of particle masses and introducing powers of $e^2/\hbar c$. Dirac's original "large-number" coincidence is equivalent to using Eq. (5.1) as a formula for the age of the universe, with m_π replaced by $(137m_pm_e^2)^{1/3}=183$ MeV. In fact, there are so many different possibilities that one may doubt whether there is any coincidence that needs explaining.] Dirac reasoned that if this connection were a real one, then, since the age of the universe increases (linearly) with time, some of the constants on the left side of (5.1) must change with time. He guessed that it is G that changes, like $1/t$. [Zee (1985) has applied similar arguments to the cosmological constant itself.] In response to Dirac, Dicke pointed out that the question of the age of the universe could only arise when the conditions are right for the existence of life. Specifically, the universe must be old enough so that some stars will have completed their time on the main sequence to produce the heavy elements necessary for life, and it must be young enough so that some stars would still be providing energy through nuclear reactions. Both the upper and lower bounds on the ages of the universe at which life can exist turn out to be roughly (very roughly) given by just the quantity (5.1). Hence there is no need to suppose that any of the fundamental constants vary with time to account for the rough agreement of the quantity (5.1) with the present age of the universe.

It is this "weak anthropic principle" that will be applied here. Its relevance arises from the fact that, in some modern cosmological models, the universe does have parts or eras in which the effective cosmological constant takes a wide variety of values. Here are some examples.

(1) The vacuum energy may depend on a scalar field vacuum expectation value that changes slowly as the universe expands, as in a model of Banks (1985).

(2) In a model of Linde (1986, 1987, 1988b), fluctuations in scalar fields produce exponentially expanding regions of the universe, within which further fluctuations produce further subuniverses, and so on. Since these subuniverses arise from fluctuations in the fields, they have differing values of various "constants" of nature.

(3) The universe may go through a very large number of first-order phase transitions in which bubbles of smaller vacuum energy form; within these bubbles there form further bubbles of even smaller vacuum energy, and so on. This can happen if the potential for some scalar field has a large number of small bumps, as in a model of Abbott (1985). Alternatively, the bubble walls may be elementary membranes coupled to a 3-form gauge potential $A_{\mu\nu\lambda}$, as in the work of Brown and Teitelboim (1987a, 1987b).

(4) The universe may start in a quantum state in which the cosmological constant does not have a precise value. Any "measurement" of the properties of the universe yields a variety of possible values for the cosmological constant, with *a priori* probabilities determined by the initial state (Hawking, 1987a). We will see examples of this in Secs. VII and VIII.

In models of these types, it is perfectly sensible to apply anthropic considerations to decide which era or part of the universe we could inhabit, and hence which values of the cosmological constant we could observe.

A large cosmological constant would interfere with the appearance of life in different ways, depending on the sign of λ_{eff}. For a large *positive* λ_{eff}, the universe very early enters an exponentially expanding de Sitter phase, which then lasts forever. The exponential expansion interferes with the formation of gravitational condensations, but once a clump of matter becomes gravitationally bound, its subsequent evolution is unaffected by the cosmological constant. Now, we do not know what weird forms life may take, but it is hard to imagine that it could develop at all without gravitational condensations out of an initially smooth universe. Therefore the anthropic principle makes a rather crisp prediction: λ_{eff} must be small enough to allow the formation of sufficiently large gravitational condensations (Weinberg, 1987).

This has been worked out quantitatively, but we can easily understand the main result without detailed calculations. We know that in our universe gravitational condensation had already begun at a redshift $z_c \geq 4$. At this time, the energy density was greater than the present mass density ρ_{M_0} by a factor $(1+z_c)^3 \geq 125$. A cosmological constant has little effect as long as the nonvacuum energy density is larger than ρ_V, so one can conclude that a vacuum energy density ρ_V no larger than, say $100\rho_{M_0}$ would not be large enough to prevent gravitational condensations. [The quantitative analysis of Weinberg

(1987) shows that for $k=0$, a vacuum energy density no greater than $\pi^2(1+z_c)^3\rho_{M_0}/3$ would not prevent gravitational condensation at a redshift z_c; this is $410\rho_{M_0}$ for $z_c=4$.]

This result suggests strongly that if it is the anthropic principle that accounts for the smallness of the cosmological constant, then we would expect a vacuum energy density $\rho_V \sim (10-100)\rho_{M_0}$, because there is no anthropic reason for it to be any smaller.

Is such a large vacuum energy density observationally allowed? There are a number of different types of astronomical data that indicate differing answers to this question.

A. Mass density

If, as often assumed, the universe now has negligible spatial curvature, then

$$\Omega_V+\Omega_{M_0}=1\ , \tag{5.2}$$

where Ω_V and Ω_{M_0} are the ratios of the vacuum energy density and the present mass density to the critical density

$$\Omega_V \equiv \frac{8\pi G\rho_V}{3H_0^2}=\frac{\lambda_{\text{eff}}}{3H_0^2}\ ,\qquad \Omega_{M_0}=\frac{8\pi G\rho_{M_0}}{3H_0^2}\ . \tag{5.3}$$

The dynamics of clusters of galaxies seems to indicate that Ω_{M_0} is in the range $0.1-0.2$ (Knapp and Kormendy, 1987), which with these assumptions would indicate a value for ρ_V/ρ_{M_0} in the range 4–9. If we discount the evidence from the dynamics of clusters of galaxies, then Ω_{M_0} could be as small as 0.02 (Knapp and Kormendy, 1987), corresponding to a value of $\rho_V/\rho_{M_0}\simeq 50$. [See also Bahcall *et al.* (1987).]

B. Ages

In a dust-dominated universe with $k=0$ and $\rho_V=0$, the age of the universe is $t_0=2/3H_0$. For $H_0=100$ km/sec Mpc, this is 7×10^9 yr, considerably less than the ages usually estimated for globular clusters (Renzini, 1986). On the other hand, for a dust-dominated universe with $k=0$ and $\rho_V\neq0$, the present age of an object that formed at a redshift z_c is

$$t_0(z_c)=\frac{2}{3}\left[1+\frac{\rho_{M_0}}{\rho_V}\right]^{1/2}H_0^{-1}\left\{\sinh^{-1}\left[\frac{\rho_V}{\rho_{M_0}}\right]^{1/2}-\sinh^{-1}\left[\left[\frac{\rho_V}{\rho_{M_0}}\right]^{1/2}(1+z_c)^{-3/2}\right]\right\}\ . \tag{5.4}$$

For instance, for $z_c=4$ and $\rho_V/\rho_{M_0}=9$ (i.e., $\Omega_{M_0}=0.1$), this gives an age $1.1H_0^{-1}$ in place of $\frac{2}{3}H_0^{-1}$. This is not in conflict with globular cluster ages even for Hubble constants near 100 km/sec Mpc.

These considerations of cosmic age and density have led a number of astronomers to suggest a fairly large positive cosmological constant, with $\rho_V>\rho_{M_0}$ [de Vaucouleurs (1982, 1983); Peebles (1984, 1987a, 1987b); Turner, Steigman, and Krauss (1984)]. However, there recently has appeared a strong argument against this view, which we shall now consider.

C. Number counts

Loh and Spillar (1986) have carried out a survey of numbers of galaxies as a function of redshift, subsequently analyzed by Loh (1986). For a uniformly distributed class of objects that are all bright enough to be detectable at redshifts $\le z_{\max}$, the number of objects observed at redshift less than $z\le z_{\max}$ in a dust-dominated universe with $k=0$ is

$$N(<z)\propto\int_{(1+z)^{-3/2}}^{1}ds\, s^{4/3}(1+\rho_V s^2/\rho_{M_0})^{-1/2}\left[\int_0^1 ds' s'^{-2/3}(1+\rho_V s'^2/\rho_{M_0})^{-1/2}\right]^2\ . \tag{5.5}$$

Of course, in the real world there are always some objects too dim to be seen. Loh's analysis allowed for an unknown luminosity distribution, assuming only that its shape does not evolve with time. Under these assumptions, he found that the vacuum energy must be quite small: specifically,

$$\rho_V/\rho_{M_0}=0.1^{-0.4}_{+0.2}\ .$$

This is more than 3 orders of magnitude below the anthropic upper bound discussed earlier. If the effective cosmological constant is really this small, then we would have to conclude that the anthropic principle does *not* explain why it is so small. [However, there are reasons to be cautious in reaching this conclusion. Bahcall and Tremaine (1988) have recently reanalyzed the data of Loh and Spillar, using a plausible model of galaxy evolution in which the shape of the luminosity distribution does change with time. They considered only the case $\rho_V=0$, leaving Ω_{M_0} undetermined, and found that evolution in this model could increase or decrease the inferred value of Ω_{M_0} by as much as unity. Presumably it would also have a similarly large effect on the inferred value of ρ_V/ρ_{M_0} when $\Omega_{M_0}+\Omega_V$ is constrained to be unity. In addition, the redshifts of Loh and Spillar are photometric and therefore less certain than those obtained from shifts of individual spectral lines.]

Now let us consider a cosmological constant of the other sign, $\lambda_{\text{eff}}<0$. Here the cosmological constant does not interfere with the formation of gravitational condensations. Instead (for $k=0$ or $k=+1$), the whole universe collapses to a singularity in a finite time T. The anthropic constraint here is simply that the universe last long enough for the appearance of life (Barrow and Tipler, 1986), say, $T \gtrsim 0.5 H_0^{-1}$, where H_0^{-1} is the Hubble time in *our* universe. For a dust-dominated universe with $k=0$, we have

$$T=\pi(8\pi G|\rho_V|)^{1/2} ,$$
$$H_0^{-1}=(8\pi G\rho_{M_0}/3)^{1/2} , \tag{5.6}$$

so the anthropic constraint here is just

$$|\rho_V| \lesssim \rho_{M_0} . \tag{5.7}$$

In this case the anthropic principle can explain why the cosmological constant is as small as found by Loh (1986), but not much smaller. On the other hand, a negative cosmological constant would not help with the cosmic mass and age problems.

Before closing this section, let me take up one possibility that may confront us in a few years. Suppose it really is confirmed that, as suggested by cosmic ages and densities, there is a cosmological constant with ρ_V of order ρ_{M_0}. Would we then have any alternative to an anthropic explanation for this value of ρ_V? The mass density ρ_M changes with time, so without anthropic considerations it is very hard to explain why a constant ρ_V should equal the value that ρ_M happens to have at present. But perhaps ρ_V really is not constant. For instance, Peebles and Ratra (1988) and Ratra and Peebles (1988) have considered a model in which the vacuum energy depends on a scalar field that changes as the universe expands. In order to qualify as a vacuum energy, it is only necessary for ρ_V to be accompanied with a pressure $p_V=-\rho_V$; the value of ρ_V can change if the vacuum exchanges energy with matter and radiation. The conservation of energy then relates the change of ρ_V to the change in the densities of matter (with $p_M=0$) and radiation (with $p_R=\rho_R/3$):

$$\frac{d}{dt}\rho_V+R^{-3}\frac{d}{dt}(R^3\rho_M)+R^{-4}\frac{d}{dt}(R^4\rho_R)=0 . \tag{5.8}$$

Freese *et al.* (1987) have considered the possibility that energy is exchanged only between the vacuum and matter, or the vacuum and radiation, in such a way that either ρ_V/ρ_M or ρ_V/ρ_R remain constant, respectively (see also Reuter and Wetterich, 1987). In order for the vacuum to transfer energy to ordinary matter in such a way that ρ_V/ρ_M remains fixed, and if baryon number is conserved, then it would be necessary to create baryon-antibaryon pairs at a sufficient rate to produce a troublesome γ-ray background. Alternatively, if the vacuum transfers energy to radiation in such a way that ρ_V/ρ_R remains constant, and if ρ_V is comparable with the present mass density ρ_{M_0}, then ρ_V/ρ_R must be rather large, completely changing the results of cosmological nucleosynthesis.

One more possibility that was not considered by Freese *et al.* is that the vacuum transfers energy to radiation, avoiding the problems of baryon-antibaryon annihilation, but in such a way as to keep a fixed ratio ρ_V/ρ_M rather than ρ_V/ρ_R. However, this also does not work. With $\rho_V=c\rho_M$ and $R^3\rho_M$ constant, Eq. (5.8) yields

$$\rho_R=\rho_{R_0}\left[\frac{R_0}{R}\right]^4+3c\rho_{M_0}\left[\left[\frac{R_0}{R}\right]^3-\left[\frac{R_0}{R}\right]^4\right]$$
$$\underset{R\ll R_0}{\longrightarrow}(\rho_{R_0}-3c\rho_{M_0})\left[\frac{R_0}{R}\right]^4 . \tag{5.9}$$

So that there is no interference with calculations of cosmological nucleosynthesis, we need

$$\rho_R \approx \rho_{R_0}\left[\frac{R_0}{R}\right]^4 ,$$

and therefore

$$|\rho_V|/\rho_M \equiv |c| \lesssim \frac{\rho_{R_0}}{3\rho_{M_0}} \ll 1 . \tag{5.10}$$

Thus, even if we are willing to suppose that the vacuum energy changes with time, a vacuum energy density comparable with the present mass density seems very difficult to explain on other than anthropic grounds.

VI. ADJUSTMENT MECHANISMS

I now turn to an idea that has been tried by virtually everyone who has worried about the cosmological constant [see, e.g., Dolgov (1982); Wilczek and Zee (1983); Wilczek (1984, 1985); Peccei, Solá, and Wetterich (1987); Barr and Hochberg (1988)]. Suppose there is some scalar ϕ whose source is proportional to the trace of the energy-momentum tensor

$$\Box^2\phi \propto T^\mu{}_\mu \propto R . \tag{6.1}$$

(Here $T^{\mu\nu}$ is the total energy-momentum tensor that includes a possible cosmological constant term $-\lambda g^{\mu\nu}/8\pi G$.) Suppose also that $T^\mu{}_\mu$ depends on ϕ and vanishes at some field value ϕ_0. Then ϕ will evolve until it reaches an equilibrium value ϕ_0, where $T^\mu{}_\mu=0$, and the

Einstein field equations have a flat-space solution.

Of course, we do not observe such a scalar field, but for these purposes it can couple as weakly as we like; a weak coupling simply implies that the equilibrium value ϕ_0 is very large. In this respect the scalar ϕ is analogous to the axion, especially in its later "invisible" version [Kim (1979); Dine, Fischler, and Srednicki (1981)].

Even very weakly coupled, it is possible that the ϕ field could have interesting effects, because it must have very small mass. If it has any nonzero mass M_ϕ, then at energies below m_ϕ we can work with an effective Lagrangian in which ϕ has been "integrated out," and so does not appear explicitly. But massless fields like the gravitational and electromagnetic field will still appear in this effective Lagrangian, and their vacuum fluctuations will contribute to the effective cosmological constant. In order to keep $\rho_V < 10^{-48}$ GeV^4, we need the scalar field adjustment to cancel the effect of gravitational and electromagnetic field fluctuations down to frequencies 10^{-12} GeV; for this purpose we must have $m_\phi < 10^{-12}$ GeV. A field this light will have a macroscopic range: $\hbar/m_\phi c \gtrsim 0.01$ cm.

Unfortunately it seems to be impossible to construct a theory with one or more scalar fields having the assumed properties. This can be seen in very general terms. What we want is to find an equilibrium solution of the field equations in which $g_{\mu\nu}$ and all matter fields ψ_n (perhaps tensors as well as scalars) are constant in space-time. For such constant fields the Euler-Lagrange equations are simply

$$\frac{\partial \mathcal{L}}{\partial g_{\mu\nu}} = 0 \, , \tag{6.2}$$

$$\frac{\partial \mathcal{L}}{\partial \psi_n} = 0 \, . \tag{6.3}$$

As we saw in Sec. III, the problem is in satisfying the trace of the gravitational field equation. To make a solution natural, we would like this trace to be a linear combination of the ψ_n field equations; that is, we want

$$g_{\lambda\nu}\frac{\partial \mathcal{L}(g,\psi)}{\partial g_{\lambda\nu}} = \sum_n \frac{\partial \mathcal{L}(g,\psi)}{\partial \psi_n} f_n(\psi) \tag{6.4}$$

for all constant $g_{\mu\nu}$ and ψ_n. This can be restated as a symmetry condition: for constant fields the Lagrangian must be invariant under the transformation

$$\delta g_{\lambda\nu} = 2\varepsilon g_{\lambda\nu}, \quad \delta\psi_n = -\varepsilon f_n(\psi) \, . \tag{6.5}$$

With this condition, if we find a solution $\psi^{(0)}$ of the Euler-Lagrange equations for ψ_n,

$$\frac{\partial \mathcal{L}}{\partial \psi_n} = 0 \quad \text{at } \psi_n = \psi_n^{(0)} \, , \tag{6.6}$$

then the trace of the field equation for $g_{\mu\nu}$ is automatically satisfied.

The problem is that under these assumptions, it is impossible (without fine-tuning $\mathcal{L}$) to find a solution to the field equations (6.3) for the ψ_n. To see this, we replace the N fields ψ_n with $N-1$ fields σ_a (not necessarily scalars) and one scalar ϕ, in such a way that the symmetry transformation (6.5) takes the form

$$\delta g_{\lambda\nu} = 2\varepsilon g_{\lambda\nu}, \quad \delta\sigma_a = 0, \quad \delta\phi = -\varepsilon \, . \tag{6.7}$$

[To do this, we first define a "transverse" surface S in field space by an equation $T(\psi)=0$, where $T(\psi)$ is any function on which $\sum_n (\partial T/\partial\psi_n) f_n(\psi)$ does not vanish. We take σ_a as any set of coordinates on this $(N-1)$-dimensional surface, and define $\psi_n(\sigma;\phi)$ as the solution of the ordinary differential equation $d\psi_n/d\phi = f_n(\psi)$ subject to the condition that at $\phi=0$, ψ_n is at the point on S with coordinates σ. The condition that S be a transverse surface ensures that, at least within a finite region of field space, any point ψ_n is on just one of these trajectories.] This symmetry simply ensures that for constant fields the Lagrangian can depend on $g_{\lambda\nu}$ and ϕ only in the combination $e^{2\phi} g_{\lambda\nu}$. The general arguments of Sec. III then show that when the field equations for σ are satisfied, the Lagrangian must take the form

$$\mathcal{L} = e^{4\phi}(\mathrm{Det}g)^{1/2}\mathcal{L}_0(\sigma) \, . \tag{6.8}$$

We see that the source of ϕ is the trace of the energy-momentum tensor

$$\frac{\partial \mathcal{L}}{\partial \phi} = T^\mu{}_\mu (\mathrm{Det}g)^{1/2} \, , \tag{6.9}$$

$$T^{\mu\nu} = g^{\mu\nu} e^{4\phi}\mathcal{L}_0(\sigma) \, . \tag{6.10}$$

It is true that if there were a value of ϕ where $\mathcal{L}$ is stationary in ϕ, then the trace of the Einstein field equations would automatically be satisfied at this point, but clearly there is no such stationary field value (unless, of course, we fine-tune $\mathcal{L}_0$ so that it vanishes at its stationary point). To put this another way, since $\mathcal{L}$ depends only on ϕ and $g_{\mu\nu}$ only in the combination $\hat{g}_{\mu\nu} \equiv e^{2\phi} g_{\mu\nu}$ (and derivatives of ϕ and $g_{\mu\nu}$), we might as well redefine the metric as $\hat{g}_{\mu\nu}$ instead of $g_{\mu\nu}$. Then ϕ is just a scalar with only derivative couplings and clearly cannot help with our problem.[6]

As one example of many failed attempts along this line, let us consider a proposal of Peccei, Solà, and Wetterich (1987). They observed that the symmetry (6.5) or (6.7) may be broken by conformal anomalies, such as those that produce the β function of quantum chromodynamics, in such a way that the effective Lagrangian becomes[7]

$$\mathcal{L}_{\mathrm{eff}} = (\mathrm{Det}g)^{1/2}[e^{4\phi}\mathcal{L}_0(\sigma) + \phi\Theta^\mu{}_\mu] \, , \tag{6.11}$$

where $\Theta^\mu{}_\mu$ represents the effect of the conformal anoma-

[6]This remark is due to Polchinski (1987).

[7]An equation essentially equivalent to (6.11) appeared in the preprint version of the paper by Peccei, Solá, and Wetterich (1987). In the published version this equation was removed, and it was acknowledged that fine-tuning is still needed to make the cosmological constant vanish. However, this equation was quoted in the meantime in a paper by Ellis, Tsamis, and Voloshin (1987), which mostly deals with the observable consequences of the light scalar particle in this model.

ly. The source of the ϕ field is now

$$\frac{\partial\mathcal{L}}{\partial\phi}=(T^{\mu}{}_{\mu}+\Theta^{\mu}{}_{\mu})(\mathrm{Det} g)^{1/2}\ , \qquad (6.12)$$

with $T^{\mu\nu}$ the previous energy-momentum tensor (6.10). Now we can find an equilibrium solution for the ϕ field, at a value ϕ_0 such that

$$4e^{4\phi_0}\mathcal{L}_0+\Theta^{\mu}{}_{\mu}=0\ . \qquad (6.13)$$

The trouble is that this is *not* the condition for a flat-space solution; the Einstein equation for a constant metric is

$$0=\frac{\partial\mathcal{L}_{\mathrm{eff}}}{\partial g_{\mu\nu}}\propto e^{4\phi}\mathcal{L}_0+\phi\Theta^{\mu}{}_{\mu}\ , \qquad (6.14)$$

which is not the same as (6.13). The point is that just calling the anomalous term in (6.11) $\Theta^{\mu}{}_{\mu}$ does not make it a term in the trace of the energy-momentum tensor to which $g_{\mu\nu}$ is coupled. This result is not surprising, since (6.11) does not obey the symmetry (6.7). One cannot have it both ways: either we preserve the symmetry, in which case there is no equilibrium solution for ϕ, or we break the symmetry, in which case such an equilibrium solution does not imply a solution of the field equations for a constant metric. (Also see Coughlan *et al.*, 1988; Wetterich, 1988.)

In a slightly different version of this general class of models, we can try coupling a scalar field so that it is the curvature scalar R rather than the trace of the energy-momentum tensor that directly serves as the source of the scalar field. [See, e.g., Dolgov (1982); Barr (1987); Ford (1987).] For instance, we might take the Lagrangian as

$$\mathcal{L}=\sqrt{g}\left[-\tfrac{1}{2}\partial_{\mu}\phi\partial^{\mu}\phi-\frac{\lambda}{8\pi G}-\frac{1}{16\pi G}R-U(\phi)R\right]\ . \qquad (6.15)$$

This has a flat-space solution with $g_{\mu\nu}=\eta_{\mu\nu}$ and $\phi=\phi_0$ (a constant), provided

$$U(\phi_0)=\infty\ . \qquad (6.16)$$

However, as the above authors observed, the effective gravitational coupling in this theory is given by

$$G_{\mathrm{eff}}=\frac{G}{1+16\pi G U(\phi_0)}=0\ . \qquad (6.17)$$

This is not much progress; we always knew that a nonzero vacuum energy does not prevent a flat-space solution if the gravitational constant is zero.

The "no-go" theorem proved in this section should not be regarded as closing off all hope in this direction. No-go theorems have a way of relying on apparently technical assumptions[8] that later turn out to have exceptions of great physical interest. (A famous example is the Coleman-Mandula theorem.) More discouraging than any theorem is the fact that many theorists have tried to invent adjustment mechanisms to cancel the cosmological constant, but without any success so far.

VII. CHANGING GRAVITY

A number of authors have suggested changing the rules of classical general relativity in such a way that the cosmological constant appears as a constant of integration, unrelated to any parameters in the action [Van der Bij *et al.* (1982); Weinberg (1983); Wilczek and Zee (1983); Buchmüller and Dragon (1988a, 1988b)]. This does not solve the cosmological constant problem, but it does change it in a suggestive way.

I will describe one version of this idea, in which one maintains general covariance, but reinterprets the formalism so that the determinant of the metric is not a dynamical field. *Any* theory can be written in a way that is formally generally covariant, so by the usual arguments we can take the action for gravity and matter as

$$I[\psi,g]=\frac{-1}{16\pi G}\int d^4x\sqrt{g}R+I_M[\psi,g]\ , \qquad (7.1)$$

where ψ are a set of matter fields appearing in the matter action I_M. (I_M includes a possible cosmological constant term $-\lambda\int\sqrt{g}\,d^4x/8\pi G$.) The variational derivative of Eq. (7.1) with respect to the metric is

$$\frac{\delta I}{\delta g_{\mu\nu}}=\frac{1}{8\pi G}(R^{\mu\nu}-\tfrac{1}{2}g^{\mu\nu}R)+T^{\mu\nu}\ , \qquad (7.2)$$

where, as usual, $T^{\mu\nu}$ is the variational derivative of I_M with respect to $g_{\mu\nu}$. In ordinary general relativity all components of the metric are dynamical fields, so Eq. (7.2) vanishes for all μ,ν, yielding the usual Einstein field equations. However, just because we use a generally covariant formalism does not mean that we are committed to treating all components of the metric as dynamical fields. For instance, we all learn in childhood how to write the equations of Newtonian mechanics in general curvilinear spatial coordinate systems, without supposing that the 3-metric has to obey any field equations at all.

In particular, if the determinant g is not dynamical, then the action only has to be stationary with respect to variations in the metric that keep the determinant fixed,

---

[8]For instance, we assumed that in the solution for flat space all fields are constant, but it might be that this solution preserves only some combination of translation and gauge invariance, in which case some gauge-noninvariant fields might vary with space-time position. (This is the case for the 3-form gauge field model discussed at the end of Sec. VII and in Sec. VIII.) Furthermore, it is possible that the foliation of field space, which allows us to replace the ψ_n with σ_a and ϕ, does not work throughout the whole of field space.

i.e., for which $g^{\mu\nu}\delta g_{\mu\nu}=0$; hence only the traceless part of (7.2) needs to vanish, yielding the field equation

$$R^{\mu\nu}-\tfrac{1}{4}g^{\mu\nu}R=-8\pi G(T^{\mu\nu}-\tfrac{1}{4}g^{\mu\nu}T^{\lambda}{}_{\lambda}) . \quad (7.3)$$

This is just the traceless part of the Einstein field equations; these equations evidently contain less information than Einstein's, but as we shall see, not much less. Because the whole formalism is generally covariant, the energy-momentum tensor satisfies the usual conservation law

$$T^{\mu\nu}{}_{;\mu}=0 , \quad (7.4)$$

and of course the Bianchi identities still hold,

$$(R^{\mu\nu}-\tfrac{1}{2}g^{\mu\nu}R)_{;\mu}=0 . \quad (7.5)$$

The full Einstein field equations are automatically consistent with (7.4) and (7.5), but for the traceless part we get a nontrivial consistency condition. Taking the covariant derivative of Eq. (7.3) with respect to x^{μ} yields

$$\tfrac{1}{4}\partial_{\mu}R=8\pi G\tfrac{1}{4}\partial_{\mu}T^{\lambda}{}_{\lambda} ,$$

or, in other words, $R-8\pi GT^{\lambda}{}_{\lambda}$ is a constant, which we will call -4Λ:

$$R-8\pi GT^{\lambda}{}_{\lambda}=-4\Lambda \text{ (constant)} . \quad (7.6)$$

From (7.3) and (7.6), we obtain

$$R^{\mu\nu}-\tfrac{1}{2}g^{\mu\nu}R-\Lambda g^{\mu\nu}=-8\pi GT^{\mu\nu} . \quad (7.7)$$

Thus we recover the Einstein field equations, but with a cosmological constant that has nothing to do with any terms in the action or vacuum fluctuations, arising, instead, as a mere integration constant. To put this another way, Eq. (7.3) does not involve a cosmological constant; the contribution of vacuum fluctuations automatically cancel on the right-hand side of Eq. (7.3), so this equation does have flat-space solutions in the absence of matter and radiation. The remaining problem in this formulation is: why should we choose the flat-space solutions?

Before proceeding with this theory, I should pause to mention that it is closely related[9] to a proposal made long ago by Einstein (1919). After his formulation of general relativity and its application to cosmology, Einstein turned to the old problem of a field theory of matter. In a paper titled "Do Gravitational Fields Play an Essential Part in the Structure of the Elementary Particles of Matter?" he proposed to replace the original gravitational field equation with the equation

$$R_{\mu\nu}-\tfrac{1}{4}g_{\mu\nu}R=-8\pi Gt_{\mu\nu} . \quad (7.8)$$

This is consistent only if $t_{\mu\nu}$ is traceless; however, Einstein took for $t_{\mu\nu}$ not the full energy-momentum tensor of matter and radiation, but just the traceless tensor of radiation alone. This is, of course, conserved only outside matter. In such regions there is no difference between Eqs. (7.8) and (7.3), so by the same calculation as shown here, Einstein was able to recover Eq. (7.7), with Λ a constant of integration. However, inside matter, Eq. (7.8) is different from (7.3), the difference being that the right-hand side of Eq. (7.3) includes the traceless part of the energy-momentum tensor of matter. A consequence of this difference is that in charged matter R is an undetermined function, except that it is constant along world lines.

I will also take the opportunity of this pause to comment on the connection between the formulation described here and that of Zee (1985) and Buchmüller and Dragon (1988a, 1988b). These authors take as their starting point the assumption that the action is invariant not under the group of all coordinate transformations, but only under the subgroup of transformations $x^{\mu}\to x'^{\mu}$ with $\mathrm{Det}(\partial x'^{\mu}/\partial x^{\nu})=1$. This is not really in conflict with the formulation presented here; the general covariance of Eq. (7.1) is achieved at the cost of introducing a metric that is partly nondynamical (just as we can make Newtonian mechanics formally Lorentz invariant by introducing a nondynamical quantity, the velocity of the reference frame). However, in giving up general covariance, one may be led to a theory with unnecessary elements. Under transformations with $\mathrm{Det}(\partial x'/\partial x)=1$, the determinant of the metric g behaves just like any scalar field, so one can introduce arbitrary functions of g here and there in the action. There is nothing wrong with this, but it is not necessary, no different from inserting a new scalar field into the theory.

Now let us return to the theory described by the field equations (7.3). In my view, the key question in deciding whether this is a plausible classical theory of gravitation is whether it can be obtained as the classical limit of any physically satisfactory quantum theory of gravitation. To help in answering this, and also to illuminate the points raised in the previous paragraph, let us look at a simple model (Teitelboim, 1982) that shares several features with the theory of gravitation studied here.

Consider a free relativistic particle, with space-time trajectory $x^{\mu}(s)$ parametrized by a variable s. In order for the action to be invariant under arbitrary reparametrizations $s\to s'(s)$, we must introduce an "einbein" $g(s)$, with transformation rule

$$g(s)\to g'(s')=g(s)\left[\frac{ds'}{ds}\right]^{-1} . \quad (7.9)$$

The action may then be taken as

$$I[x,g]=\tfrac{1}{2}\int ds\, g^{-1}(s)\frac{dx^{\mu}(s)}{ds}\frac{dx_{\mu}(s)}{ds} -\frac{m^2}{2}\int ds\, g(s) . \quad (7.10)$$

[9]This was pointed out to me by someone in the audience of the lectures at Harvard. I thank my informant for this interesting historical reference.

The conditions that I be stationary with respect to variations in $x^{\mu}(s)$ and $g(s)$ are, respectively,

$$\frac{dp^{\mu}}{ds}=0 \ , \tag{7.11}$$

$$p^{\mu}p_{\mu}=-m^2 \ , \tag{7.12}$$

where p_{μ} is the canonical conjugate to x^{μ}:

$$p_{\mu}(s)=g^{-1}(s)\frac{dx_{\mu}(s)}{ds} \ . \tag{7.13}$$

However, just because we choose to write the action in a reparametrization-invariant way does not necessarily mean that we must treat the einbein $g(s)$ as a dynamical quantity. If we treat $x^{\mu}(s)$, but not $g(s)$, as dynamical variables, then we obtain Eq. (7.11), but not (7.12). Of course, Eq. (7.11) implies that $p_{\mu}p^{\mu}$ is a constant [just as Eq. (7.3) implies that $R-8\pi GT^{\lambda}{}_{\lambda}$ is constant]. If we like, we can call this constant $-m^2$, but this is now a mere integration constant, unrelated to anything in the original action.

Now to quantization. The Hamiltonian here is

$$H(s)=p_{\mu}\frac{dx^{\mu}}{ds}-L=\tfrac{1}{2}g(p^{\mu}p_{\mu}+m^2) \ , \tag{7.14}$$

so in quantum mechanics we calculate amplitudes by the functional integral

$$A=\int[dx][dp][dg] \times\exp\left[i\int ds\left[p_{\mu}(s)\frac{dx^{\mu}(s)}{ds}-H(s)\right]\right] \ . \tag{7.15}$$

The einbein $g(s)$ has no canonical conjugate, and so appears here only as a Lagrange multiplier, whose integral yields a factor

$$\prod_s \delta(p^{\mu}p_{\mu}+m^2) \ . \tag{7.16}$$

Presumably the classical theory in which g is not dynamical would be obtained as the classical limit of a quantum theory in which we do not do a functional integral over $g(s)$, and hence do not get the factor (7.16). But then there would be nothing to keep p^{μ} timelike. This is such a trivial theory that it is hard to say that anything goes wrong physically; but we may anticipate that in less trivial theories, we need a field to serve as a Lagrange multiplier for every negative norm degree of freedom like p^0. This is the case, for instance, in string theories, where the integration over the world-sheet metric is needed to enforce the Virasoro conditions on physical states.

The quantum theory of gravitation can be put in similar terms. Using the Arnowitt-Deser-Misner (1962) formalism, we calculate amplitudes as functional integrals,

$$Z=\int[dh_{ij}][d\pi^{ij}][d\tilde{N}][dN^i] \times\exp\left[i\int\left[\pi^{ij}\frac{\partial h_{ij}}{\partial t}-(\tilde{\mathcal{H}}-2\lambda)\tilde{N}-\mathcal{H}_iN^i\right]d^4x\right] \ . \tag{7.17}$$

Here h_{ij}, $\tilde{N}$, and N^i parametrize the 4-metric, with line element given by

$$d\tau^2=(h^{-1}\tilde{N}^2-N^iN^jh_{ij})dt^2 -2h_{ij}N^idx^jdt-h_{ij}dx^idx^j \ , \tag{7.18}$$

$$h\equiv\mathrm{Det}(h_{ij}) \ . \tag{7.19}$$

Furthermore, π^{ij} is the canonical conjugate to h_{ij}, and $\tilde{\mathcal{H}}$ and $\mathcal{H}_i$ are functions of h_{ij} and π^{ij} and their space derivatives, given by

$$\tilde{\mathcal{H}}=\tfrac{1}{2}\mathcal{G}_{ij,kl}\pi^{ij}\pi^{kl}-{}^{(3)}R \ , \tag{7.20}$$

$$\mathcal{H}_i=-2h_{ij}\nabla_k\pi^{jk} \ , \tag{7.21}$$

where ${}^{(3)}R$ is the scalar curvature and ∇_k is the covariant derivative, both calculated using the 3-metric h_{ij}, and

$$\mathcal{G}_{ij,kl}\equiv h_{ik}h_{jl}+h_{il}h_{jk}-h_{ij}h_{kl} \ . \tag{7.22}$$

We see that $\tilde{N}$ and N^i just act as Lagrange multipliers for $\tilde{\mathcal{H}}$ and $\mathcal{H}_i$, respectively. Moreover, from (7.18), we see that $\tilde{N}^2$ is just the quantity whose status is under question here, the determinant of the 4-metric[10]

$$\tilde{N}=(\mathrm{Det}g_{\mu\nu})^{1/2} \ . \tag{7.23}$$

Thus, just as the integral over the einbein $g(s)$ enforced the constraint $p^{\mu}p_{\mu}=-m^2$, the integral over Detg enforces the constraint

$$\tilde{\mathcal{H}}=2\lambda \ . \tag{7.24}$$

The two conditions are quite similar. Just as $\eta_{\mu\nu}$ has signature $+++-$, the quantity (7.22), viewed as a 6×6 matrix, has signature $+,+,+,+,+,-$. Hence the integration over $\mathrm{Detg}_{\mu\nu}$ has the effect of eliminating one negative norm degree of freedom for each $\mathbf{x}$, $\pi^{ij}\propto(h^{-1})^{ij}$, just as the integral over the einbein $g(s)$ allows one to eliminate the variable p^0. However, for gravity there is a "potential" term in $\tilde{\mathcal{H}}$, proportional to the 3-curvature, and it is not entirely clear to me that it really is necessary to constrain $\tilde{\mathcal{H}}$ to take a fixed value. For the present, the question of whether it is necessary to integrate over $\mathrm{Detg}_{\mu\nu}$ must be left open. [Recent work by Henneaux and Teitelboim (1988) shows that there is a sensible generally covariant quantum version of the classical theory described by Eq. (7.3).]

Before closing this section, I should note that several authors have made a rather different suggestion, which also has the effect of converting the cosmological constant from a function of parameters in the action into a constant of the motion (Aurilia *et al.*, 1980; Witten, 1983; Henneaux and Teitelboim, 1984). They proposed adding to the action a term

[10]In order to obtain this result, I have defined $\tilde{\mathcal{H}}$ and $\tilde{N}$ differently from the usual $\mathcal{H}$ and N, by moving a factor $h^{1/2}$ from N to $\mathcal{H}$.

$$I_F = -\tfrac{1}{48}\int d^4x\sqrt{g}\,F_{\mu\nu\rho\sigma}F^{\mu\nu\rho\sigma}\ , \tag{7.25}$$

where $F_{\mu\nu\rho\sigma}$ is the exterior derivative of a 3-form gauge field $A_{\nu\rho\sigma}$,

$$F_{\mu\nu\rho\sigma} = \partial_{[\mu}A_{\nu\rho\sigma]}\ , \tag{7.26}$$

and $g \equiv -\mathrm{Det}\,g_{\mu\nu}$. Since $F^{\mu\nu\rho\sigma}$ is totally antisymmetric, it can be expressed as

$$F^{\mu\nu\rho\sigma} = c\,\varepsilon^{\mu\nu\rho\sigma}/\sqrt{g}\ , \tag{7.27}$$

where $\varepsilon^{\mu\nu\rho\sigma}$ is the Levi-Cività tensor density, with $\varepsilon^{0123}\equiv 1$, and c is a scalar field. The field equation for A is

$$F^{\mu\nu\rho\sigma}{}_{;\mu} = 0\ , \tag{7.28}$$

so, using (7.27)

$$\frac{\partial c}{\partial x^\mu} = 0\ . \tag{7.29}$$

But the action (7.25) then takes the form

$$I_F = +\tfrac{1}{2}c^2\int d^4x\sqrt{g}\ . \tag{7.30}$$

In other words, whatever else contributes to the cosmological constant, there is one term that depends on the integration constant c,

$$\lambda_F = 4\pi G c^2\ . \tag{7.31}$$

Again, this does not solve the cosmological constant problem, but it does change the way it arises.

If λ is a constant of integration, then in a quantum theory we expect the state vector of the universe to be a superposition of states with different values of λ, in which case the anthropic considerations of Sec. V would set a bound on th effective cosmological constant.

VIII. QUANTUM COSMOLOGY

The last approach to the cosmological constant problem that I shall describe here is based on the application of quantum mechanics to the whole universe. In 1984 Hawking (1984b) described how in quantum cosmology there could arise a distribution of values for the effective cosmological constant, with an enormous peak at $\lambda_{\text{eff}}=0$. Very recently, this approach has been revived in an exciting paper by Coleman (1988b), using a new mechanism for producing a distribution of values for the cosmological constant (that rests in part on other work of Hawking and Coleman) and finding an even sharper peak. Related ideas have also been recently discussed by Banks (1988). Before describing the work of Coleman and Hawking, I will have to say something about quantum cosmology in general.

Most treatments of quantum cosmology are based on the "wave function of the universe," a function $\Psi[h,\phi]$ of the 3-metric and matter fields on a spacelike surface. [The 3-metric h_{ij} can be conveniently defined by adapting the space-time coordinate system so that the spacelike surface has constant t, and then decomposing the 4-metric $g_{\mu\nu}$ as in Eq. (7.19).] This wave function satisfies a sort of Schrödinger equation, known as the Wheeler-DeWitt equation [DeWitt (1967); Wheeler (1968)]:

$$\left[\frac{1}{2h^{1/2}}\frac{\delta}{\delta h_{ij}}h^{1/2}g_{ij,kl}\frac{\delta}{\delta h_{kl}} - {}^{(3)}R - 2\lambda + 8\pi G T_{00}\right]\Psi = 0\ , \tag{8.1}$$

with notation explained in Sec. VII (except that we now include a matter energy density T_{00}, in which the canonical conjugate of a matter field Φ is replaced with $\delta/\delta\Phi$). It will be very important in what follows that we express the solution as a *Euclidean* path integral

$$\Psi \propto \int [dg][d\Phi]\exp(-S[g,\Phi])\ , \tag{8.2}$$

where we integrate over all Euclidean-signature 4-metrics $g_{\mu\nu}$ and matter fields Φ defined on a 4-manifold M_4, that have the 3-manifold $M_3[h,\phi]$ with 3-metric h_{ij} and matter fields ϕ as a boundary. (The Wheeler-DeWitt equation is the constraint obtained from integrating the Lagrange multiplier $\tilde{N}$ as discussed in Sec. VII.) Here S is the Euclidean action[11]

$$S = \frac{1}{16\pi G}\int_{M_4}\sqrt{g}\,(R+2\lambda) + \text{matter terms} + \text{surface terms}\ . \tag{8.3}$$

Since Eq. (8.1) is a differential equation in an infinite-dimensional space [the set of $h_{ij}(\mathbf{x})$ and $\phi(\mathbf{x})$ for all $\mathbf{x}$], it has an infinite variety of solutions, which can be specified by giving the 4-manifold in Eq. (8.2) other boundaries, besides the $M_3[h,\phi]$ on which the 3-metric and matter fields are specified. Hartle and Hawking (1983) proposed as a cosmological initial condition that the manifold M_4 should have no boundaries other than $M_3(h,\phi)$. We will see that Coleman's (1988b) approach does not depend critically on the choice of initial conditions.

There are technical problems associated with this formalism. One is an operator-ordering ambiguity: there are various ways of ordering[12] the h_{ij} fields and $\delta/\delta h_{ij}$ operators in (8.1), all of which have (8.2) as solution, but

[11]The Euclidean action S is opposite in sign to what we would get if we replaced the metric $g_{\mu\nu}$ in the action I in Eq. (7.1) with one of signature $+,+,+,+$. This sign of S is chosen so that ordinary matter makes a positive contribution to S.

[12]The insertion of factors $h^{-1/2}$ and $h^{1/2}$ in Eq. (8.1) represents one choice of operator ordering, which is made in order to allow the derivation of the conservation equation (8.8).

with different ways of calculating the measure $[dg][d\Phi]$ (Hawking and Page, 1986). Another problem, potentially more worrisome, is that for gravity the Euclidean action (8.3) is not bounded below. Gibbons, Hawking, and Perry (1978) have proposed rotating the contour of integration for the overall scale of the 4-metric so that it runs parallel to the imaginary axis. We will not need to go into these technicalities here, because it will turn out that we only need to deal with the effective action at its equilibrium point.

A problem that is more relevant to us here has to do with the probabilistic interpretation of the wave function Ψ and of Euclidean path integrals like (8.2). Hawking has proposed (1984a, 1984c) that $\exp(-S[g,\Phi])$ should be regarded as proportional to the probability of a particular metric and matter field history. It is not immediately clear what is meant by this—even supposing that we had the godlike ability to measure the gravitational and matter fields throughout space-time, it would be in a space-time of Lorentzian rather than Euclidean signature. However, since we can (sometimes) go from one signature to another by a complex coordinate transformation, it may be that a Euclidean history $g_{\mu\nu}(x)$, $\Phi(x)$ can be interpreted in terms of correlations of scalar quantities, just as if the space-time were Lorentzian. In much of Hawking's work (e.g., Hawking, 1979), these questions are avoided by using the formalism only to calculate the probability that, in the space-time history of the universe, there is a spacelike 3-surface with a given 3-metric $h_{ij}(\mathbf{x})$ and matter fields $\phi(\mathbf{x})$. For instance, with Hartle-Hawking (1983) initial conditions, we would integrate over all closed 4-manifolds that contain such a 3-surface. If this surface bisects the 4-manifold, then it can be regarded as the boundary of the two halves of the 4-manifold, and so the integral is (with some qualifications) just the square of the wave function (8.2). But questions still arise concerning the probabilistic interpretation of Ψ, particularly with regard to normalization. If $|\Psi[h,\phi]|^2$ is the probability density that there exists *some* 3-surface on which the 3-metric is $h_{ij}(\mathbf{x})$ and the matter fields are $\phi(\mathbf{x})$, then we would not simply want to set the functional integral of $|\Psi[h,\phi]|^2$ over $h_{ij}(\mathbf{x})$ and $\phi(\mathbf{x})$ equal to unity, because in this functional integral we are summing up possibilities that are not exclusive; if the universe has some $h_{ij}(\mathbf{x})$ and $\phi(\mathbf{x})$ on one 3-surface, then it may also have some other $h'_{ij}(\mathbf{x})$ and $\phi'(\mathbf{x})$ on some other 3-surface. After all, you would not expect that the probabilities that you ever in your life have flipped a coin and gotten heads, and that you ever in your life have flipped a coin and gotten tails, should add up to unity.

I would like to offer an interpretation of what is meant by treating $|\Psi[h,\Phi]|^2$ as a probability density, which seems to me implicit in Hawking's writings (and may already be stated explicitly somewhere in the literature). As everyone has recognized, the problem has to do with the role of time in quantum gravity. [See, e.g., Hartle (1987).] The problems raised here do not arise in asymptotically flat cosmologies, because in such theories there is a natural definition of time, and we generally ask for the probabilities that the fields have certain values at a definite time. However, here time is a coordinate with no objective significance, and this coordinate time is even imaginary. As Augustine (398) warned, "I must not allow my mind to insist that time is something objective."[13] Heeding this warning, suppose we choose some "time-keeping" field $a(\mathbf{x},t)$, for instance, the trace of the energy-momentum tensor, and use its value to define a local time α. Each value of α defines a 3-surface, on which the coordinate time t is a function $t(\mathbf{x},\alpha)$ defined implicitly by

$$a(\mathbf{x},t(\mathbf{x},\alpha))\equiv\alpha\ . \tag{8.4}$$

We are then interested in the probability that the tangential components of the metric and all matter fields other than $a(\mathbf{x},t)$ have specified values on this surface. Calling these quantities $b_n(\mathbf{x},t)$, we see that the probability density for the $b_n(\mathbf{x},t)$ to have the values $\beta_n(\mathbf{x})$ at local time α is

$$P_\alpha[\beta]=N\int_{M_4}[dg][d\Phi]\exp(-S[g,\Phi])\times\prod_{n,\mathbf{x}}\delta(b_n(\mathbf{x},t(\mathbf{x},\alpha))-\beta_n(\mathbf{x}))\ , \tag{8.5}$$

with N a normalization factor, determined by the condition that the total probability of finding *any* value for the $b_n(\mathbf{x})$ at local time α should be unity:

$$1=\int P_\alpha[\beta][d\beta] =N\int_{M_4}[dg][d\Phi]\exp(-S[g,\Phi])\ . \tag{8.6}$$

[This usually makes N a function of α, because in (8.5) and (8.6) we integrate only over matter and metric histories for which Eq. (8.4) is satisfied on some 3-surface. With some boundary conditions, this condition is automatically satisfied, and then N is α independent. For instance, if M_4 has two boundaries, on which $a(x)$ is required to take values α_1 and α_2, then there are 3-surfaces on which (8.4) is satisfied for all α in the range $\alpha_1<\alpha<\alpha_2$.] Where the surface of constant a bisects the 4-space, $P_\alpha[\beta]$ can be written as proportional to the square of the wave function $\Psi[\alpha,\beta]$, but with α constant in 3-space.

[13]This quote is not merely a display of useless erudition. Book XI of Augustine's *Confessions* contains a famous discussion of the nature of time, and it seems to have become a tradition to quote from this chapter in writing about quantum cosmology. Thus Hawking (1979) quotes "What did God do before He made Heaven and Earth? I do not answer as one did merrily: He was preparing a Hell for those that ask such questions. For at no time had God not made anything, for time itself was made by God." Coleman (1988a) quotes "The past is present memory." To this, I can add one more very relevant quote: "I confess to you, Lord, that I still do not know what time is. Yet I confess too that I do know that I am saying this in time, that I have been talking about time for a long time, ..."

Coleman (1988b) short-circuits many of the problems that arise in giving a probabilistic interpretation to Euclidean path integrals by using such integrals only to calculate expectation values: the expectation value of an arbitrary scalar field $A_{g,\Phi}(x)$, which may depend on the metric and matter fields and their derivatives, is taken as

$$\langle A \rangle = \frac{\int [dg][d\Phi] A_{g,\Phi}(x) \exp(-S[g,\Phi])}{\int [dg][d\Phi] \exp(-S[g,\Phi])} . \quad (8.7)$$

The general covariance of the theory makes $\langle A \rangle$ independent of x. In fact, it should be emphasized that this sort of expectation values includes an average over the time in the history of the universe that A is measured. On the other hand, the probability $P_\alpha[\beta]$ discussed above is the expectation value of a nonlocal operator, the delta function in (8.5), and refers to a specific local time α.

(I should mention here that there is a very different and apparently unrelated approach to the problem of giving a probabilistic interpretation to the wave function Ψ. The Wheeler-DeWitt equation (8.1) is somewhat like the Klein-Gordon equation for a particle in a scalar potential and leads immediately to a somewhat similar conservation law (now given for pure gravity):

$$0 = \frac{\delta}{\delta h_{ij}(\mathbf{x})} \left[h^{1/2}(\mathbf{x}) \mathcal{G}_{ij,kl}(\mathbf{x}) \times \mathrm{Im} \left[\Psi^*[h] \frac{\delta}{\delta h_{kl}(\mathbf{x})} \Psi[h] \right] \right] . \quad (8.8)$$

Since the beginning, it was hoped that such a conservation law could be used to construct a suitable probability density (DeWitt, 1967). Usually (8.8) is stated in a minisuperspace context, where $h_{ij}(\mathbf{x})$ is constrained to depend on only a finite number of parameters. Since $\mathcal{G}_{ij,kl} h^{kl} = -h_{ij}$, it is natural to treat the overall scale of h_{ij} as a sort of global time coordinate, and take as a probability density the corresponding component of the conserved "current" in (8.8). I wish to point out here that such a construction is not limited to any particular minisuperspace formulation, but can be carried out in the general case. Take Ψ to depend on a "global time"

$$T[h] = \left[\int d^3x \, h^{1/2}(\mathbf{x}) \right]^{1/2} \quad (8.9)$$

and an arbitrary (in fact, infinite) number of other parameters $\zeta_n[h]$, all ζ_n independent of the overall scale of $h_{ij}(\mathbf{x})$:

$$0 = \int d^3x \, h^{1/2}(\mathbf{x}) h_{kl}(\mathbf{x}) \frac{\delta \zeta_n[h]}{\delta h_{kl}(\mathbf{x})} . \quad (8.10)$$

We also introduce a Jacobian $J(\zeta, T)$ and write the functional measure as

$$[dh] = J[d\zeta] dT . \quad (8.11)$$

Multiplying Eq. (8.8) with $\delta(T[h]-T)$ and doing an integral over $\mathbf{x}$ and a functional integral over $h_{ij}(\mathbf{x})$, we easily find a constancy condition

$$0 = \frac{d}{dT} \int J[d\zeta] \mathrm{Im} \left[\Psi^* \frac{\partial \Psi}{\partial T} \right] . \quad (8.12)$$

The trouble here is, of course, the same as that encountered in giving a probabilistic interpretation to the Klein-Gordon equation: the integrand in (8.12) is not, in general, positive. Banks, Fischler, and Susskind (1985) and Vilenkin (1986, 1988a), have considered minisuperspace models in which Ψ is complex, with increasing phase, for which the integrand of Eq. (8.12) is positive-definite; however, this is not the case in general, and, in particular, not for Hartle-Hawking boundary conditions. For a recent more general discussion, see Vilenkin (1988b).)

I now want to give a simplified description of Hawking's (1984b) proposed solution of the cosmological constant problem, using for this purpose parts of Coleman's (1988b) analysis. In order to make the cosmological constant into a dynamical variable, Hawking introduces a 3-form gauge field $A_{\mu\nu\lambda}$ of the sort described at the end of Sec. VII. According to the general ideas of Euclidean quantum cosmology, the probability distribution for the scalar $c(x)$ defined by Eq. (7.27) at any one point $x = x_1$ is

$$P(c) = \langle \delta(c(x_1)-c) \rangle \propto \int [dA][dg][d\Phi] \delta(c(x_1)-c) \times \exp(-S[A,g,\Phi]) . \quad (8.13)$$

It is well known that such functional integrals can be expressed as exponentials of the effective action at its stationary point.[14] In the present case, we have

$$P(c) \propto \exp(-\Gamma[A_c, g_c, \Phi_c]) , \quad (8.14)$$

where $\Gamma[A,g,\Phi]$ is the total action (the sum of one-particle irreducible graphs with external lines replaced with fields A,g,Φ) and the subscript c indicates that this quantity is to be evaluated at a point where Γ is stationary with respect to any variations in $A_{\mu\nu\lambda}(x)$, $g_{\mu\nu}(x)$, or $\Phi(x)$ that leave $c(x_1)=c$ fixed. Now, among all the possible stationary points of Γ, there is one that can be found knowing only the effective action relevant to large

[14]The usual proof, for the case without a delta function in the integrand, proceeds by adding a term $\int J\Omega$ to the action, where Ω denotes the various fields, and J is a set of corresponding currents. The path integral is then $\exp[-W(J)] \equiv \int d\Omega \exp(-S - \int J\Omega)$. The effective action is defined by the Legendre transformation $\Gamma(\Omega) = W(J_\Omega) - \int J_\Omega \Omega$, where J_Ω is the current that produces a given expectation value $\Omega = \delta W / \delta J$. The condition for zero current is that $\Gamma(\Omega)$ be stationary with respect to Ω, and at this point $\Gamma(\Omega) = W(0)$. The delta function in (8.13) can be dealt with by writing it as an integral $\int d\omega \exp\{i\omega[c(x_1)-c]\}$. One can then use the above theorem to evaluate the functional integral before integrating over ω, now with no restriction on $c(x)$, and then doing the ω integral.

4-manifolds. In this case, it is convenient to set all fields except $A_{\mu\nu\lambda}$ and $g_{\mu\nu}$ equal to their (A- and g-dependent) stationary values, in which case the effective action can be expanded in inverse powers of the size of the manifold[15]

$$\Gamma_{\rm eff}[A,g]=\frac{\lambda}{8\pi G}\int\sqrt{g}\,d^4x+\frac{1}{16\pi G}\int\sqrt{g}R\,d^4x + \tfrac{1}{48}\int d^4x\sqrt{g}\,F_{\mu\nu\lambda\rho}F^{\mu\nu\lambda\rho}+\cdots\,, \tag{8.15}$$

the omitted terms involving more than two derivatives of g and/or A. As we saw in Sec. VII, the condition that this be stationary in $A_{\mu\nu\lambda}$ [for variations that keep $c(x_1)$ fixed] is that $F_{\mu\nu\lambda\rho}$ have vanishing covariant divergence, from which it follows that c in Eq. (7.27) is constant; hence

$$\Gamma_{\rm eff}=\frac{\lambda(c)}{8\pi G}\int\sqrt{g}\,d^4x+\frac{1}{16\pi G}\int\sqrt{g}R\,d^4x+\cdots\,, \tag{8.16}$$

where

$$\lambda(c)=\frac{c^2}{2}+\lambda\,. \tag{8.17}$$

The condition that this be stationary in $g_{\mu\nu}$ is, of course, that $g_{\mu\nu}$ satisfy the Einstein field equations with cosmological constant $\lambda(c)$. For any such solution, $R=-4\lambda(c)$, so at the stationary point

$$\Gamma_{\rm eff}=-\frac{\lambda(c)}{8\pi G}\int\sqrt{g}\,d^4x\,. \tag{8.18}$$

With Hartle-Hawking boundary conditions, the solution of the Einstein equations for $\lambda(c)>0$ is a 4-sphere of proper circumference $2\pi r$, where

$$r=\sqrt{3/\lambda(c)}\,, \tag{8.19}$$

yielding a probability density proportional to

$$\exp(-\Gamma_{\rm eff})=\exp[3\pi/G\lambda(c)]\,. \tag{8.20}$$

On the other hand, for $\lambda(c)<0$ the solutions can be made compact by imposing periodicity conditions, but they all have $\Gamma_{\rm eff}\geq 0$. Hawking's conclusion is that the probability density has an infinite peak for $\lambda(c)\to 0+$; hence, after normalizing P,

$$P(c)=\delta(c-c_0)\,, \tag{8.21}$$

where c_0 is the value of c (assuming there is one), for which $\lambda(c)=0$.

It is important that the quantity $\lambda(c)$ is the true effective cosmological constant, previously called $\lambda_{\rm eff}$, that would be measured in gravitational phenomena at long ranges.[16] The constant λ in Eq. (8.15) includes all effects of fields other than $g_{\mu\nu}$ and $A_{\mu\nu\lambda}$, including all quantum fluctuations. Hence the result (8.21), if valid, really does solve the cosmological constant problem.

We can check that this result is not invalidated by the terms neglected in Eq. (8.16). For a large radius r, the exhibited terms in (8.16) are of order $\lambda r^4/G$ and r^2/G, respectively, while a term with $D\geq 4$ derivatives would yield a contribution to $\Gamma_{\rm eff}$ of order $(mr)^{4-D}$, where m is some combination of the Planck mass and elementary-particle masses. For $\lambda(c)\lesssim m^2$, this shifts the size of the manifold by

$$\delta r/r\approx G\lambda(c)[\lambda(c)/m^2]^{(D-4)/2}\ll 1\,.$$

The change in the stationary value of the action is then

$$\delta\Gamma_{\rm eff}\approx[\lambda(c)/m^2]^{(D-4)/2}\lesssim 1\,,$$

so these higher-derivative terms have no effect on the singularity (8.20).

Coleman (1988b) does not need to introduce a 3-form gauge field $A_{\mu\nu\lambda}$; rather, in order to make the cosmological constant into a dynamical variable, he considers the effect of topological fixtures known as wormholes.[17] An explicit example of a wormhole is provided by the metric (Hawking, 1987b, 1988)

$$ds^2=(1+b^2/x^\mu x^\mu)^2dx^\mu dx^\mu\,. \tag{8.22}$$

This appears to have a singularity at $x^\mu=0$, but the line element is invariant under the transformation

$$x^\mu\to x'^\mu=x^\mu b^2/x^\nu x^\nu\,, \tag{8.23}$$

so the region $x^\mu x^\mu<b^2$ actually has the same geometry as that with $x^\mu x^\mu>b^2$. The space described by Eq. (8.22) therefore consists of two asymptotically flat 4-spaces, joined together at the 3-surface with $x^\mu x^\mu=b^2$, a 3-sphere known as a "baby universe." This 4-metric is not a solution of the classical Einstein equations (though it does have $R=0$), but this is not very relevant; the action is

$$S=3\pi b^2/G\,, \tag{8.24}$$

so the factor $\exp(-S)$ suppresses the effects of all

[15]Such an effective action may be used as the input for calculations in which we include quantum effects only from virtual massless particles with $|q^2|$ less than some cutoff Λ^2. Such effects are, of course, finite, and their Λ dependence is to be canceled by giving the coefficients in $\Gamma_{\rm eff}$ a suitable Λ dependence. (This point of view is described by Weinberg, 1979b.) In order to prevent these quantum effects from generating an unacceptable cosmological constant, the cutoff Λ must be taken very small.

[16]This property is shared by an imaginative solution to the cosmological constant problem proposed by Linde (1988a).

[17]The importance of quantum fluctuations in space-time topology at small scales has been emphasized for many years by Wheeler (e.g., 1964), and more recently by Hawking (1978) and Strominger (1984). Such "space-time foam" was considered as a mechanism for canceling a cosmological constant by Hawking (1983).

wormholes except those of Planck dimensions or less, for which quantum effects are surely important. [A model with classical wormhole solutions, based on a 2-form axion, has been presented by Giddings and Strominger (1988a).]

If Planck-sized wormholes can connect asymptotically flat 4-spaces, then they can connect any 4-spaces that are large compared to the Planck scale. We are therefore led to consider contributions to the Euclidean path integral from large 4-spaces [like the 4-sphere in Hawking's (1984b) theory] connected to themselves and each other with Planck-sized wormholes. Each wormhole can be regarded as the creation and subsequent destruction of a baby universe [like the 3-sphere of proper circumference $4\pi b$ in Hawking's (1987b, 1988) wormhole model], and such baby universes may also appear as part of the boundary of the 4-manifold.

What are the effects of these wormholes and baby universes? At scales large compared with the scale of the baby universe, the creation or destruction of a baby universe can only show up through the insertion of a local operator in the path integral. The various types of baby universes can be classified according to the form of these local operators. The effect of creating and destroying arbitrary numbers of baby universes of all types can thus be expressed by adding a suitable term in the action

$$\tilde{S}=S+\sum_i (a_i+a_i^\dagger)\int d^4x\, O_i(x)\ , \tag{8.25}$$

where a_i and $a_i^\dagger$ are the annihilation and creation operators for a baby universe of type i, and $O_i(x)$ is the corresponding local operator. [This was first stated by Hawking (1987b). Creation and annihilation operators for baby universes were earlier used by Strominger (1984). For a proof of Eq. (8.25), see Coleman (1988a) and Giddings and Strominger (1988b).] The path integral over all 4-manifolds with given boundary conditions is to be calculated as

$$\int [dg][d\Phi]e^{-S}=\int_{\mathrm{No}}[dg][d\Phi]\langle B|e^{-\tilde{S}}|B\rangle\ , \tag{8.26}$$

where No means that wormholes and baby universes are excluded, and $|B\rangle$ is a normalized baby-universe state depending on the boundary conditions. For instance, with Hartle-Hawking boundary conditions, $|B\rangle$ is the empty state

$$a_i|B\rangle=0\ . \tag{8.27}$$

These baby universes have an important effect even if none of them appear as part of the boundary of the 4-manifold, as would be the case for Hartle-Hawking boundary conditions. Hawking (1987b, 1988) has suggested that since the baby universes are unobservable, their effect is an effective loss of quantum coherence. [See also Hawking (1982); Teitelboim (1982); Strominger (1984); Lavrelashvili, Rubakov, and Tinyakov (1987, 1988); Giddings and Strominger (1988b). A contrary view was taken by Gross (1984).] Recently Coleman (1988a) has argued (convincingly, in my view) for a different interpretation [see also Giddings and Strominger (1988b)]. The state $|B\rangle$ in Eq. (8.26) may always be expanded in eigenstates of the operators $a_i+a_i^\dagger$:

$$|B\rangle=\int f_B(\alpha)\prod_i d\alpha_i|\alpha\rangle\ , \tag{8.28}$$

$$(a_i+a_i^\dagger)|\alpha\rangle=\alpha_i|\alpha\rangle\ , \tag{8.29}$$

$$\langle\alpha'|\alpha\rangle=\prod_i\delta(\alpha_i'-\alpha_i)\ , \tag{8.30}$$

the function $f_B(\alpha)$ depending on the boundary conditions. For instance, for Hartle-Hawking conditions, $|B\rangle$ satisfies Eq. (8.27), and so

$$f_B(\alpha)=\prod_i\pi^{-1/4}\exp(-\alpha_i^2/2)\ . \tag{8.31}$$

(With n baby universes on the boundary of the 4-space, this would be multiplied with a Hermite polynomial of order n.) In the state $|\alpha\rangle$, the effect of the creation and annihilation of baby universes is to change the action S to

$$S_\alpha=S+\sum_i\alpha_i\int O_i(x)d^4x\ . \tag{8.32}$$

That is, the coupling constant multiplying each possible local term $\int O_i d^4x$ is changed by an amount α_i. As soon as we start to make any sort of measurements, the state of the universe breaks up into an incoherent superposition of these $|\alpha\rangle$'s, each appearing with *a priori* probability $|f_B(\alpha)|^2$; but for each term we have an ordinary wormhole-free quantum theory, with α-dependent action (8.32).

If all we want is to explain why the cosmological constant is not enormous, then our work is essentially done. The effective cosmological constant is a function of the α_i, because among the O_i there is a simple operator $O_1=\sqrt{g}$, whose coefficient contributes a term $8\pi G\alpha_1$ to λ, and also because the vacuum energy $\langle\rho\rangle$ depends on the couplings of all interactions, each of which has a term proportional to one of the α_i. Now, generic baby-universe states $|B\rangle$ will have components $|\alpha\rangle$ for which $\lambda_{\mathrm{eff}}(\alpha)$ is very small, as well as others for which it is enormous. The anthropic considerations of Sec. VI tell us that any scientist who asks about the value of the cosmological constants can only be living in components $|\alpha\rangle$ for which λ_{eff} is quite small, for otherwise galaxies and stars could never have formed (for $\lambda_{\mathrm{eff}}>0$), or else there would not be time for life to evolve (for $\lambda_{\mathrm{eff}}<0$).

However, it is of great interest to ask whether the effective cosmological constant is really zero, or just small enough to satisfy anthropic bounds, in which case it should show up observationally. The probability of getting any particular value of the α_i, and hence of finding a value $\lambda_{\mathrm{eff}}(\alpha)$, is not just given by the function $|f_B(\alpha)|^2$ arising from the boundary conditions, but is also affected by the functional integral itself.

In calculating this effect, Coleman (1988b) observed that although we are to integrate only over connected 4-manifolds, on a scale much large than the wormhole scale those manifolds that appear disconnected will really be connected by wormholes. Hence any sort of probability density or expectation value will contain as a factor a sum over disconnected manifolds consisting of arbitrary numbers of closed connected wormhole-free components.[18] Just as for Feynman diagrams, this sum is the exponential of the path integral for a single closed connected wormhole-free manifold

$$F(\alpha)=\exp\left[\int_{CC}[dg]\exp(-S_\alpha[g])\right], \tag{8.33}$$

where CC indicates that we include only closed connected wormhole-free manifolds, and $S_\alpha[g]$ is the action (8.32) with all fields other than $g_{\mu\nu}(x)$ integrated out.

The path integral in (8.33) can be evaluated by precisely the same methods as described above in connection with Hawking's (1984b) model [and used for this purpose by Coleman (1988b)]. The result is that the probability density for λ_{eff} contains a factor (for $\lambda_{\text{eff}}>0$)

$$F=\exp\left[\exp\left[\frac{3\pi}{G\lambda_{\text{eff}}}+O(1)\right]\right]. \tag{8.34}$$

The fact that this is now an exponential of an exponential, instead of a mere exponential, is not essential in solving the cosmological constant problem (though it is important in fixing other constants, as described at the end of this section). Either way, the probability distribution has an infinite peak at $\lambda_{\text{eff}}\to 0+$, which, after normalizing so that the total probability is unity, means that $P(\alpha)$ has a factor

$$P(\alpha)\propto\delta(\lambda_{\text{eff}}(\alpha)). \tag{8.35}$$

In addition, as in Hawking's case, from the way that F has been calculated it is clear that this λ_{eff} is the constant that appears in the effective action for pure gravity with all high-energy fluctuations integrated out; hence it is the cosmological constant relevant to astronomical observation.

Has the cosmological constant problem been solved? Perhaps so, but there are still some things to worry about in Coleman's approach, as also in the earlier work of Hawking. Here is a short list of qualms.

(1) Does Euclidean quantum cosmology have anything to do with the real world? It is essential to both Coleman and Hawking that the path integral be given by a stationary point of the Euclideanized action—the conclusion would be completely wiped out if in place of $\exp(3\pi/G\lambda_{\text{eff}})$ we had found $\exp(3\pi i/G\lambda_{\text{eff}})$. Some of the technical and conceptual difficulties of Euclidean quantum cosmology were discussed at the beginning of this section.

(2) What are the boundary conditions? It is always possible that the essential singularity in $\exp\{\exp[3\pi/G\lambda(\alpha)]\}$ is canceled by an essential zero in the *a priori* probability $|f_B(\alpha)|^2$. However, this is not the case for Hartle-Hawking boundary conditions, where $|f_B(\alpha)|^2$ is a simple Gaussian. Moreover, Coleman (1988b) has shown that in his theory such an essential zero would be destroyed by almost any perturbation of the boundary conditions; instead of its being unnatural to have zero cosmological constant, it would be highly unnatural not to. Still, the problem of boundary conditions is disturbing, because it reminds us that quantum cosmology is an incomplete theory.

(3) Are wormholes real? Coleman's calculation depends on there being a clear separation between the very large 4-manifolds, for which the long-range effective action is stationary (and large and negative), and very small wormholes, whose contribution to the action is of order unity (and generally positive). Furthermore, the wormholes have been assumed to be so well separated that we can ignore their interactions (the "dilute gas" approximation). It may be possible to construct a theory in which the wormhole scale [like b in Eq. (8.22)] is somewhat larger than the Planck scale, large enough to allow the wormhole metric to be calculated classically, but we would still have to ask whether this is actually the case. Hawking (1984b) does not need to worry about wormholes, but how do we know that the 3-form gauge field is real? A related question for both authors: even granting the existence of the stationary point of the action at which $\Gamma_{\text{eff}}=-3\pi/\lambda G$, how do we know that this is the dominant stationary point?

(4) What about the other terms in the effective action? For instance, suppose we include the 6-derivative term[19]

$$\Gamma_{\text{eff}}=\int d^4x\sqrt{g}\left[\frac{\lambda(\alpha)}{8\pi G(\alpha)}+\frac{R}{16\pi G(\alpha)}\right.$$
$$\left.+\zeta(\alpha)G(\alpha)R_{\mu\nu}{}^{\lambda\rho}R_{\lambda\rho}{}^{\sigma\tau}R_{\sigma\tau}{}^{\mu\nu}\right], \tag{8.36}$$

with $\zeta(\alpha)$ a dimensionless coefficient that, like λ and G, depends on the baby-universe parameters α_i. Hawking and Coleman found a stationary point of this action for which $\Gamma_{\text{eff}}\to-\infty$ when $\lambda(\alpha)G(\alpha)\to 0$, but for this purpose it is essential that $\zeta(\alpha)$ remain bounded in this limit. (We recall that in our previous discussion of the higher-derivative terms in Γ_{eff}, we assumed that the coefficient m^{4-D} of terms with $D\geq 4$ derivatives remains less than $\lambda^{(D-4)/2}$ as $\lambda\to 0$.) But if we can let $1/\lambda G$ go to infinity, then why not let ζ go to infinity also? In particular, why not use a dimensional factor $1/\lambda(\alpha)$ in place of $G(\alpha)$ in

[18]This sum actually includes manifolds that are truly not connected by wormholes or anything else, but their contribution is a harmless multiplicative factor, which will cancel out anyway in normalizing $P(\alpha)$.

[19]Terms involving the Ricci tensor $R_{\mu\nu}$ or its trace R are not included here, because they represent merely a redefinition of the metric; see, e.g., Weinberg (1979a). The 4-derivative term $R_{\mu\nu\lambda\rho}R^{\mu\nu\lambda\rho}$ is not included, because it can be combined with terms involving $R_{\mu\nu}$ or R to make a topological invariant and is therefore physically unmeasureable for fixed large-scale topology.

the last term of Eq. (8.36)? This would completely invalidate the analysis of the singularity in the probability density $P(\alpha)$, and could well wipe it out.

The last of these four qualms suggests some interesting possibilities. Suppose we do assume that for some reason constants like $\zeta(\alpha)$ in Eq. (8.36) are bounded. Then the effect of wormholes is not only to fix $\lambda(\alpha)$ at zero, but also to fix these other constants at their lower or upper bounds. [I think this is the correct interpretation of what Coleman (1988b) calls "the big fix."] For instance, for $\zeta(\alpha)$ bounded and $|\lambda(\alpha)G(\alpha)| \ll 1$, the action (8.36) is stationary for a sphere of proper circumference $2\pi r$, where

$$r^2 = \frac{3}{\lambda} - \frac{64\zeta\pi G^2\lambda}{3} , \tag{8.37}$$

for which the effective action takes the value

$$\Gamma_{\text{eff}} = -\frac{3\pi}{G\lambda} - \frac{128\zeta G\lambda\pi^2}{3} . \tag{8.38}$$

Thus the probability distribution $\exp[\exp(-\Gamma_{\text{eff}})]$ not only has an infinite peak at $\lambda(\alpha)=0$, but also contains a factor

$$\exp\left[\frac{128\zeta G\lambda\pi^2}{3}\exp\left[\frac{3\pi}{G\lambda}\right]\right] . \tag{8.39}$$

For $G\lambda \to 0$, the quantity $G\lambda\exp(3\pi/G\lambda)$ becomes infinite, so the normalized probability will have a delta function at the upper bound of $\zeta(\alpha)$. All constants in the effective action for gravitation, including terms with any numbers of derivatives, can be calculated in this way,[20] but they all have to be bounded as $\lambda(\alpha)G(\alpha) \to 0$ for any of this to make sense.

It may be that the bounds (if any) on parameters like $\zeta(\alpha)$ arise from the details of wormhole physics, in which case these remarks are not going to be useful numerically for some time. However, there is another more exciting possibility, that there are just unitarity bounds, which could be calculated working only with low-energy effective theory itself. Of course, we are not likely to be able to measure parameters like $\zeta(\alpha)$, but it would still be nice to be able to calculate them, because up to now the only really unsatisfactory feature of the quantum theory of gravitation has been the apparent arbitrariness of this infinite set of parameters.

[20]To the extent that it will become possible to calculate functions like $\lambda(\alpha)$, $G(\alpha)$, $\zeta(\alpha)$ etc., in terms of the parameters in an underlying fundamental theory, such as a string theory, the location of the delta functions in F may allow us to infer something about the values of the α_i and of the parameters in the underlying theory. However, without such an underlying theory, it is impossible to use calculations of λG, ζ, etc., to infer anything about the observed parameters of some intermediate theory like the standard model. This is because, in addition to charges, masses, etc., the standard model implicitly also involves parameters $\lambda_0, G_0, \zeta_0, \ldots$ appearing in the effective action for gravitation. When we integrate out the quarks, leptons, and gauge and Higgs bosons, we obtain new values for λ, G, ζ, etc.; but these new values depend on an equal number of unknowns λ_0, G_0, ζ_0, etc., as well as on charges and masses.

IX. OUTLOOK

All of the five approaches to the cosmological constant problem described in Secs. IV–VIII remain interesting. At present, the fifth, based on quantum cosmology, appears the most promising. However, if wormholes (or 3-form gauge fields) do produce a distribution of values for the cosmological constant, but without an infinite peak at $\lambda_{\text{eff}}=0$, then we will have to fall back on the anthropic principle to explain why λ_{eff} is not enormously larger than allowed by observation. Alternatively, it may be some change in the theory of gravity, like that described here in Sec. VII, that produces the distribution in values for λ_{eff}. The approaches based on supersymmetry and adjustment mechanisms described in Secs. IV and VI seem least promising at present, but this may change.

All five approaches have one other thing in common: They show that any solution of the cosmological constant problem is likely to have a much wider impact on other areas of physics or astronomy. One does not need to explain the potential importance of supergravity and superstrings. A light scalar like that needed for adjustment mechanisms could show up macroscopically, as a "fifth force." Changing gravity by making $\text{Det}g_{\mu\nu}$ not dynamical would make us rethink our quantum theories of gravitation, and wormholes might force all the constants in these theories to their outer bounds. Finally, and of greatest interest to astronomy, if it is only anthropic constraints that keep the effective cosmological constant within empirical limits, then this constant should be rather large, large enough to show up before long in astronomical observations.

Note added in proof. As might have been expected, in the time since this report was submitted for publication there have appeared a large number of preprints that follow up on various aspects of the work of Coleman (1988b) and Banks (1988). Here is a partial list: Accetta *et al.* (1988); Adler (1988); Fischler and Susskind (1988); Giddings and Strominger (1988c); Gilbert (1988); Grinstein and Wise (1988); Gupta and Wise (1988); Hosoya (1988); Klebanov, Susskind, and Banks (1988); Myers and Periwal (1988); Polchinski (1988); Rubakov (1988). I am not able to review all of these papers here. However, I do want to mention two further qualms, regarding Coleman's proposed solution of the cosmological constant problem, that are raised by some of these papers. First, Fischler and Susskind (1988), partly on the basis of conversations with V. Kaplunovsky, have pointed out that the exponential damping of large wormholes may be overcome by Coleman's double exponential. If this were the case, we would be confronted with closely packed wormholes of macroscopic as well as Planck scales. This would be a disaster for Coleman's proposed solution of the cosmological constant problem, and would also indicate that we do not fully understand how to use Euclidean path integrals in quantum cosmology. Next, Polchinski (1988) has found that the Euclidean path integral over closed, connected, wormhole-free manifolds inside the exponential in (8.33) has a phase that might eliminate the peak in the probability distribution at zero cosmological constant. As pointed out here in footnote 15, when we use an effective action Γ_{eff} to evaluate such path integrals, the effective action must be taken as an input to calculations in which we include quantum fluctuations in massless particle fields with momenta up to some ultraviolet cutoff Λ. This cutoff must be taken as the same as the *infrared* cutoff that was used in calculating Γ_{eff}, so that all fluctuations are taken into account. It was remarked in footnote 15 that Λ must be taken very small, to avoid reintroducing a cosmological constant, but as Polchinski now remarks, no matter how small we take Λ, the integral over fluctuations in the gravitational field with momenta less than Λ produces a phase in the integral. Since this phase appears inside the exponential in Eq. (8.33), if its real part is not positive definite there would be no exponential peak at zero cosmological constant. On the other hand, in the absence of wormholes this phase would appear as an overall factor in front of a single exponential, so it would not affect the peaking at zero cosmological constant found by Hawking (1984b).

ACKNOWLEDGMENTS

I have been greatly helped in preparing this review by conversations with many colleagues. Here is a list of a few of those to whom my thanks are especially due. Section II: G. Holton; Sec. III: E. Witten; Sec. IV: S. de Alwis, J. Polchinski, E. Witten; Sec. V: P. J. E. Peebles, P. Shapiro, E. Vishniac; Sec. VI: J. Polchinski; Sec. VII: C. Teitelboim, F. Wilczek; Sec. VIII: L. Abbott, S. Coleman, B. DeWitt, W. Fischler, S. Giddings, L. Susskind, C. Teitelboim, F. Wilczek. Of course, they take no responsibility for anything that I may have gotten wrong. Research was supported in part by the Robert A. Welch Foundation and NSF Grant No. PHY 8605978.

REFERENCES

Abbott, L., 1985, Phys. Lett. **B150**, 427.

Abbott, L., 1988, Sci. Am. **258** (No. 5), 106.

Accetta, F. S., A. Chodos, F. Cooper, and B. Shao, 1988, "Fun with the wormhole calculus," Yale University Preprint No. YCTP-P20-88.

Adler, S. L., 1988, "On the Banks-Coleman-Hawking argument for the vanishing of the cosmological constant," Institute for Advanced Study Preprint No. IASSNS-HEP-88/35.

Albrecht, A., and P. J. Steinhardt, 1982, Phys. Rev. Lett. **48**, 120.

Arnowitt, R., S. Deser, and C. W. Misner, 1962, in *Gravitation: An Introduction to Current Research*, edited by L. Witten (Wiley, New York) p. 227.

Attick, J., G. Moore, and A. Sen, 1987, Institute for Advanced Studies preprint.

Augustine, 398, *Confessions,* translated by R. S. Pine-Coffin (Penguin Books, Harmondsworth, Middlesex, 1961), Book XI.

Bahcall, J., T. Piran, and S. Weinberg, 1987, Eds., *Dark Matter in the Universe: 4th Jerusalem Winter School for Theoretical Physics* (World Scientific, Singapore).

Bahcall, S. R., and S. Tremaine, 1988, Astrophys. J. **326**, L1.

Banks, T., 1985, Nucl. Phys. B **249**, 332.

Banks, T., 1988, University of California, Santa Cruz, Preprint No. SCIPP 88/09.

Banks, T., W. Fischler, and L. Susskind, 1985, Nucl. Phys. B **262**, 159.

Barbieri, R., E. Cremmer, and S. Ferrara, 1985, Phys. Lett. B **163**, 143.

Barbieri, R., S. Ferrara, D. V. Nanopoulos, and K. S. Stelle, 1982, Phys. Lett. B **113**, 219.

Barr, S. M., 1987, Phys. Rev. D **36**, 1691.

Barr, S. M., and D. Hochberg, 1988, Phys. Lett. B **211**, 49.

Barrow, J. D., and F. J. Tipler, 1986, *The Anthropic Cosmological Principle* (Clarendon, Oxford).

Bernstein, J., and G. Feinberg, 1986, Eds., *Cosmological Constants* (Columbia University, New York).

Bludman, S. A., and M. Ruderman, 1977, Phys. Rev. Lett. **38**, 255.

Brown, J. D., and C. Teitelboim, 1987a, Phys. Lett. B **195**, 177.

Brown, J. D., and C. Teitelboim, 1987b, Nucl. Phys. B **297**, 787.

Buchmüller, W., and N. Dragon, 1988a, University of Hannover Preprint No. ITP-UH 1/88.

Buchmüller, W., and N. Dragon, 1988b, Phys. Lett. B **207**, 292.

Carter, B., 1974, in *International Astronomical Union Symposium 63: Confrontation of Cosmological Theories with Observational Data,* edited by M. S. Longair (Reidel, Dordrecht), p. 291.

Carter, B., 1983, in *The Constants of Physics, Proceedings of a Royal Society Discussion Meeting, 1983,* edited by W. H. McCrea and M. J. Rees (printed for The Royal Society, London, at the University Press, Cambridge), p. 137.

Casimir, H. B. G., 1948, Proc. K. Ned. Akad. Wet. **51**, 635.

Chang, N.-P., D.-X. Li, and J. Pérez-Mercader, 1988, Phys. Rev. Lett. **60**, 882.

Coleman, S., 1988a, Nucl. Phys. B **307**, 867.

Coleman, S., 1988b, "Why there is nothing rather than something: A theory of the cosmological constant," Harvard University Preprint No. HUTP-88/A022.

Coleman, S., and F. de Luccia, 1980, Phys. Rev. D **21**, 3305.

Coughlan, G. D., I. Kani, G. G. Ross, and G. Segré, 1988, CERN Preprint No. TH. 5014/88.

Cremmer, E., S. Ferrara, C. Kounnas, and D. V. Nanopoulos, 1983, Phys. Lett. B **133**, 61.

Cremmer, E., B. Julia, J. Scherk, S. Ferrara, L. Girardello, and P. van Nieuwenhuizen, 1978, Phys. Lett. B **79**, 231.

Cremmer, E., B. Julia, J. Scherk, S. Ferrara, L. Girardello, and P. van Nieuwenhuizen, 1979, Nucl. Phys. B **147**, 105.

Davies, P. C. W., 1982, *The Accidental Universe* (Cambridge University, Cambridge).

de Sitter, W., 1917, Mon. Not. R. Astron. Soc. **78**, 3 (reprinted in Bernstein and Feinberg, 1986).

de Vaucouleurs, G., 1982, Nature (London) **299**, 303.

de Vaucouleurs, G., 1983, Astrophys. J. **268**, 468, Appendix B.

DeWitt, B., 1967, Phys. Rev. **160**, 1113.

Dicke, R. H., 1961, Nature (London) **192**, 440.

Dine, M., W. Fischler, and M. Srednicki, 1981, Phys. Lett. B **104**, 199.

Dine, M., R. Rohm, N. Seiberg, and E. Witten, 1985, Phys. Lett. B **156**, 55.
Dine, M., and N. Seiberg, 1986, Phys. Rev. Lett. **57**, 2625.
Dirac, P. A. M., 1937, Nature (London) **139**, 323.
Dolgov, A. D., 1982, in *The Very Early Universe: Proceedings of the 1982 Nuffield Workshop at Cambridge,* edited by G. W. Gibbons, S. W. Hawking, and S. T. C. Siklos (Cambridge University, Cambridge), p. 449.
Dreitlein, J., 1974, Phys. Rev. Lett. **34**, 777.
Eddington, A. S., 1924, *The Mathematical Theory of Relativity,* 2nd Ed. (Cambridge University, London).
Einstein, A., 1917, Sitzungsber. Preuss. Akad. Wiss. Phys.-Math. Kl. 142 [English Translation in *The Principle of Relativity* (Methuen, 1923, reprinted by Dover Publications), p. 177; and in Bernstein and Feinberg, 1986].
Einstein, A., 1919, Sitzungsber. Preuss. Akad. Wiss., Phys.-Math. Kl. [English translation in *The Principle of Relativity* (Methuen, 1923, reprinted by Dover Publications), p. 191].
Ellis, J., C. Kounnas, and D. V. Nanopoulos, 1984, Nucl. Phys. B **247**, 373.
Ellis, J., A. B. Lahanas, D. V. Nanopoulos, and K. Tamvakis, 1984, Phys. Lett. B **134**, 429.
Ellis, J., N. C. Tsamis, and M. Voloshin, 1987, Phys. Lett. B **194**, A 291.
Fischler, W., and L. Susskind, 1988, "A wormhole catastrophe," Texas Preprint No. UTTG-26-88.
Ford, L. H., 1987, Phys. Rev. D **35**, 2339.
Freese, K., F. C. Adams, J. A. Frieman, and E. Mottola, 1987, Nucl. Phys. B **287**, 797.
Friedan, D., E. Martinec, and S. Shenker, 1986, Nucl. Phys. B **271**, 93.
Friedmann, A., 1924, Z. Phys. **21**, 326 [English translation in Bernstein and Feinberg, 1986, Eds., *Cosmological Constants* (Columbia University, New York)].
Gibbons, G. W., S. W. Hawking, and M. J. Perry, 1978, Nucl. Phys. B **138**, 141.
Giddings, S. B., and A. Strominger, 1988a, Nucl. Phys. B **306**, 890.
Giddings, S. B., and A. Strominger, 1988b, Nucl. Phys. B **307**, 854.
Giddings, S. B., and A. Strominger, 1988c, "Baby universes, third quantization, and the cosmological constant," Harvard Preprint No. HUTP-88/A036.
Gilbert, G., 1988, "Wormhole induced proton decay," Caltech Preprint No. CALT-68-1524.
Grinstein, B., and M. B. Wise, 1988, "Light scalars in quantum gravity," Caltech Preprint No. CALT-68-1505.
Grisaru, M. T., W. Siegel, and M. Rocek, 1979, Nucl. Phys. B **159**, 429.
Gross, D. J., 1984, Nucl. Phys. B **236**, 349.
Gupta, A. K., and M. B. Wise, 1988, "Comment on wormhole correlations," Caltech Preprint No. CALT-68-1520.
Guth, A. H., 1981, Phys. Rev. D **23**, 347.
Hartle, J. B., 1987, in *Gravitation in Astrophysics: Cargese 1986,* edited by B. Carter and J. B. Hartle (Plenum, New York), p. 329.
Hartle, J. B., and S. W. Hawking, 1983, Phys. Rev. D **28**, 2960.
Hawking, S. W., 1978, Nucl. Phys. B **144**, 349.
Hawking, S. W., 1979, in *Three Hundred Years of Gravitation,* edited by S. W. Hawking and W. Israel (Cambridge University, Cambridge).
Hawking, S. W., 1982, Commun. Math. Phys. **87**, 395.
Hawking, S. W., 1983, Philos. Trans. R. Soc. London, Ser. A **310**, 303.
Hawking, S. W., 1984a, Nucl. Phys. B **239**, 257.
Hawking, S. W., 1984b, Phys. Lett. B **134**, 403.
Hawking, S. W., 1984c, in *Relativity, Groups and Topology II,* NATO Advanced Study Institute Session XL . . . Les Houches 1983, edited by B. S. DeWitt and Raymond Stora (Elsevier, Amsterdam), p. 336.
Hawking, S. W., 1987a, remarks quoted by M. Gell-Mann, Phys. Scr. **T15**, 202 (1987).
Hawking, S. W., 1987b, Phys. Lett. B **195**, 337.
Hawking, S. W., 1988, Phys. Rev. D **37**, 904.
Hawking, S., and D. Page, 1986, Nucl. Phys. B **264**, 185.
Henneaux, M., and C. Teitelboim, 1984, Phys. Lett. B **143**, 415.
Henneaux, M., and C. Teitelboim, 1988, "The cosmological constant and general covariance," University of Texas preprint.
Hosoya, A., 1988, "A diagrammatic derivation of Coleman's vanishing cosmological constant," Hiroshima Preprint No. RRK-88-28.
Kim, J., 1979, Phys. Rev. Lett. **43**, 103.
Klebanov, I., L. Susskind, and T. Banks, 1988, "Wormholes and cosmological constant," SLAC Preprint No. SLAC-Pub.-4705.
Knapp, G. R., and J. Kormendy, 1987, Eds., *Dark Matter in the Universe: I.A.U. Symposium No. 117* (Reidel, Dordrecht).
Lavrelashvili, G. V., V. A. Rubakov, and P. G. Tinyakov, 1987, Pis'ma Zh. Eksp. Teor. Fiz. **46**, 134 [JETP Lett. **46**, 167 (1987)].
Lavrelashvili, G. V., V. A. Rubakov, and P. G. Tinyakov, 1988, Nucl. Phys. B **299**, 757.
Lemaître, G., 1927, Ann. Soc. Sci. Bruxelles, Ser. 1 **47**, 49.
Lemaître, G., 1931, Mon. Not. R. Astron. Soc. **91**, 483.
Linde, A. D., 1974, Pis'ma Zh. Eksp. Teor. Fiz. **19**, 320 [JETP Lett. **19**, 183 (1974)].
Linde, A. D., 1982, Phys. Lett. B **129**, 389.
Linde, A. D., 1986, Phys. Lett. B **175**, 395.
Linde, A. D., 1987, Phys. Scri. **T15**, 169.
Linde, A. D., 1988a, Phys. Lett. B **200**, 272.
Linde, A. D., 1988b, Phys. Lett. B **202**, 194.
Loh, E. D., 1986, Phys. Rev. Lett. **57**, 2865.
Loh, E. D., and E. J. Spillar, 1986, Astrophys. J. **303**, 154.
Martinec, E., 1986, Phys. Lett. B **171**, 189.
Moore, G., 1987a, Nucl. Phys. B **293**, 139.
Moore, G., 1987b, Institute for Advanced Study Preprint No. IASSNS-HEP-87/59, to be published in the proceedings of the Cargèse School on Nonperturbative Quantum Field Theory.
Morozov, A., and A. Perelomov, 1987, Phys. Lett. B **199**, 209.
Myers, R. C., and V. Periwal, 1988, "Constants and correlations in the Coleman calculus," Santa Barbara Preprint No. NSF-ITP-88-151.
Page, D., 1987, in *The World and I* (in press).
Pais, A., 1982, *'Subtle is the Lord . . .': The Science and the Life of Albert Einstein* (Oxford University, New York).
Peccei, R. D., J. Solà, and C. Wetterich, 1987, Phys. Lett. B **195**, 183.
Peebles, P. J. E., 1984, Astrophys. J. **284**, 439.
Peebles, P. J. E., 1987a, in *Proceedings of the Summer Study on the Physics of the Superconducting Super Collider,* edited by R. Donaldson and J. Marx (Division of Particles and Fields of the APS, New York).
Peebles, P. J. E., 1978b, Publ. Astron. Soc. Pac., in press.
Peebles, P. J. E., and B. Ratra, 1988, Astrophys. J. Lett. **325**, L17.
Petrosian, V., E. E. Salpeter, and P. Szekeres, 1967, Astrophys. J. **147**, 1222.
Polchinski, J., 1986, Commun. Math. Phys. **104**, 37.

Polchinski, J., 1987, private communication.
Polchinski, J., 1988, in preparation.
Ratra, B., and P. J. E. Peebles, 1988, Phys. Rev. D **37**, 3406.
Rees, M. J., 1987, New Sci. August 6, 1987, p. 43.
Renzini, A., 1986, in *Galaxy Distances and Deviations from Universal Expansion*, edited by B. F. Madore and R. B. Tully (Reidel, Dordrecht), p. 177.
Reuter, M., and C. Wetterich, 1987, Phys. Lett. **188**, 38.
Rohm, R., 1984, Nucl. Phys. B **237**, 553.
Rowan-Robinson, M., 1968, Mon. Not. R. Astron. Soc. **141**, 445.
Rubakov, V. A., 1988, "On the third quantization and the cosmological constant," DESY preprint.
Shklovsky, I., 1967, Astrophys. J. **150**, L1.
Slipher, V. M., 1924, table in Eddington (1924), *The Mathematical Theory of Relativity*, 2nd Ed. (Cambridge University, London), p. 162
Sparnaay, M. J., 1957, Nature (London) **180**, 334.
Strominger, A., 1984, Phys. Rev. Lett. **52**, 1733.
Teitelboim, C., 1982, Phys. Rev. D **25**, 3159.
Turner, M. S., G. Steigman, and L. M. Krauss, 1984, Phys. Rev. Lett. **52**, 2090.
Van der Bij, J. J., H. Van Dam, and Y. J. Ng, 1982, Physica **116A**, 307.
Veltman, M., 1975, Phys. Rev. Lett. **34**, 777.
Vilenkin, A., 1986, Phys. Rev. D **33**, 3560.
Vilenkin, A., 1988a, Phys. Rev. D **37**, 888.
Vilenkin, A., 1988b, Tufts Preprint No. TUTP-88-3.
Weinberg, S., 1972, *Gravitation and Cosmology* (Wiley, New York).
Weinberg, S., 1979a, in *General Relativity: An Einstein Centenary Survey*, edited by S. W. Hawking and W. Israel (Cambridge University, Cambridge), p. 800.
Weinberg, S., 1979b, Physica **96A**, 327.
Weinberg, S., 1982, Phys. Rev. Lett. **48**, 1776.
Weinberg, S., 1983, unpublished remarks at the workshop on "Problems in Unification and Supergravity," La Jolla Institute, 1983.
Weinberg, S., 1987, Phys. Rev. Lett. **59**, 2607.
Wetterich, C., 1988, Nucl. Phys. B **302**, 668.
Wheeler, J. A., 1964, in *Relativity, Groups and Topology*, Lectures Delivered at Les Houches, 1963 . . . , edited by B. DeWitt and C. DeWitt (Gordon and Breach, New York), p. 317.
Wheeler, J. A., 1968, in *Battelle Rencontres*, edited by C. DeWitt and J. A. Wheeler (Benjamin, New York).
Wilczek, F., 1984, Phys. Rep. **104**, 143.
Wilczek, F., 1985, in *How Far Are We from the Gauge Forces: Proceedings of the 1983 Erice Conference*, edited by A. Zichichi (Plenum, New York), p. 208.
Wilczek, F., and A. Zee, 1983, unpublished work quote by Zee (1985), in *High Energy Physics: Proceedings of the Annual Orbis Scientiae*, edited by S. L. Mintz and A. Perlmutter (Plenum, New York).
Witten, E., 1983, in *Proceedings of the 1983 Shelter Island Conference on Quantum Field Theory and the Fundamental Problems of Physics*, edited by R. Jackiw, N. N. Khuri, S. Weinberg, and E. Witten (MIT, Cambridge, Massachusetts), p. 273.
Witten, E., 1985, Phys. Lett. B **155**, 151.
Witten, E., and J. Bagger, 1982, Phys. Lett. B **115**, 202.
Zee, A., 1985, in *High Energy Physics: Proceedings of the 20th Annual Orbis Scientiae, 1983*, edited by S. L. Mintz and A. Perlmutter (Plenum, New York).
Zeldovich, Ya., B., 1967, Pis'ma Zh. Eksp. Teor. Fiz. **6**, 883 [JETP Lett. **6**, 316 (1967)].
Zumino, B., 1975, Nucl. Phys. B **89**, 535.

2

Living in the multiverse

Steven Weinberg
Physics Department, University of Texas at Austin

Opening talk at the symposium *Expectations of a Final Theory* at Trinity College, Cambridge, 2 September 2005

2.1 Introduction

We usually mark advances in the history of science by what we learn about nature, but at certain critical moments the most important thing is what we discover about science itself. These discoveries lead to changes in how we score our work, in what we consider to be an acceptable theory.

For an example, look back to a discovery made just one hundred years ago. Before 1905 there had been numerous unsuccessful efforts to detect changes in the speed of light, due to the motion of the Earth through the ether. Attempts were made by Fitzgerald, Lorentz and others to construct a mathematical model of the electron (which was then conceived to be the chief constituent of all matter) that would explain how rulers contract when moving through the ether in just the right way to keep the apparent speed of light unchanged. Einstein instead offered a symmetry principle, which stated that not just the speed of light, but all the laws of nature are unaffected by a transformation to a frame of reference in uniform motion. Lorentz grumbled that Einstein was simply assuming what he and others had been trying to prove. But history was on Einstein's side. The 1905 Special Theory of Relativity was the beginning of a general acceptance of symmetry principles as a valid basis for physical theories.

This was how Special Relativity made a change in science itself. From one point of view, Special Relativity was no big thing – it just amounted to the replacement of one 10-parameter spacetime symmetry group, the Galileo group, with another 10-parameter group, the Lorentz group. But never

Universe or Multiverse?, ed. Bernard Carr. Published by Cambridge University Press.

before had a symmetry principle been taken as a legitimate hypothesis on which to base a physical theory.

As usually happens with this sort of revolution, Einstein's advance came with a retreat in another direction: the effort to construct a classical model of the electron was permanently abandoned. Instead, symmetry principles increasingly became the dominant foundation for physical theories. This tendency was accelerated after the advent of quantum mechanics in the 1920s, because the survival of symmetry principles in quantum theories imposes highly restrictive consistency conditions (existence of antiparticles, connection between spin and statistics, cancellation of infinities and anomalies) on physically acceptable theories. Our present Standard Model of elementary particle interactions can be regarded as simply the consequence of certain gauge symmetries and the associated quantum mechanical consistency conditions.

The development of the Standard Model did not involve any changes in our conception of what was acceptable as a basis for physical theories. Indeed, the Standard Model can be regarded as just quantum electrodynamics writ large. Similarly, when the effort to extend the Standard Model to include gravity led to widespread interest in string theory, we expected to score the success or failure of this theory in the same way as for the Standard Model: string theory would be a success if its symmetry principles and consistency conditions led to a successful prediction of the free parameters of the Standard Model.

Now we may be at a new turning point, a radical change in what we accept as a legitimate foundation for a physical theory. The current excitement is, of course, a consequence of the discovery of a vast number of solutions of string theory, beginning in 2000 with the work of Bousso and Polchinski [1].[1] The compactified six dimensions in Type II string theories typically have a large number (tens or hundreds) of topological fixtures (3-cycles), each of which can be threaded by a variety of fluxes. The logarithm of the number of allowed sets of values of these fluxes is proportional to the number of topological fixtures. Further, for each set of fluxes one obtains a different effective field theory for the modular parameters that describe the compactified 6-manifold, and for each effective field theory the number of local minima of the potential for these parameters is again proportional to

1 Smolin [2] had noted earlier that string theory has a large number of vacuum solutions, and explored an imaginative possible consequence of this multiplicity. Even earlier, in the 1980s, Duff, Nilsson and Pope had noted that $D = 11$ supergravity has an infinite number of possible compactifications, but of course it was not then known that this theory is a version of string theory. For a summary, see ref. [3].

https://doi.org/10.1017/CBO9781107050990.004 Published online by Cambridge University Press

the number of topological fixtures. Each local minimum corresponds to the vacuum of a possible stable or metastable universe.

Subsequent work by Giddings, Kachru, Kallosh, Linde, Maloney, Polchinski, Silverstein, Strominger and Trivedi (in various combinations) [4–6] established the existence of a large number of vacua with positive energy densities. Ashok and Douglas [7] estimated the number of these vacua to be of order 10^{100} to 10^{500}. String theorists have picked up the term 'string landscape' for this multiplicity of solutions from Susskind [8], who took the term from biochemistry, where the possible choices of orientation of each chemical bond in large molecules lead to a vast number of possible configurations. Unless one can find a reason to reject all but a few of the string theory vacua, we may have to accept that much of what we had hoped to calculate are environmental parameters, like the distance of the Earth from the Sun, whose values we will never be able to deduce from first principles.

We lose some and win some. The larger the number of possible values of physical parameters provided by the string landscape, the more string theory legitimates anthropic reasoning as a new basis for physical theories. Any scientists who study nature must live in a part of the landscape where physical parameters take values suitable for the appearance of life and its evolution into scientists.

An apparently successful example of anthropic reasoning was already at hand by the time the string landscape was discovered. For decades there seemed to be something peculiar about the value of the vacuum energy density ρ_V. Quantum fluctuations in known fields at well understood energies (say, less than 100 GeV) give a value of ρ_V larger than observationally allowed by a factor 10^{56}. This contribution to the vacuum energy might be cancelled by quantum fluctuations of higher energy, or by simply including a suitable cosmological constant term in the Einstein field equations, but the cancellation would have to be exact to fifty-six decimal places. No symmetry argument or adjustment mechanism could be found that would explain such a cancellation. Even if such an explanation could be found, there would be no reason to suppose that the remaining net vacuum energy would be comparable to the *present* value of the matter density, and since it is certainly not very much larger, it was natural to suppose that it is very much less, too small to be detected.

On the other hand, if ρ_V takes a broad range of values in the multiverse, then it is natural for scientists to find themselves in a subuniverse in which ρ_V takes a value suitable for the appearance of scientists. I pointed out in 1987 that this value for ρ_V cannot be too large and positive, because then galaxies and stars would not form [9]. Roughly, this limit is that ρ_V should be less than the mass density of the universe at the time when galaxies first condense. Since this was in the past, when the mass density was larger than at present, the anthropic upper limit on the vacuum energy density is larger than the present mass density, but not many orders of magnitude greater.

But anthropic arguments provide not just a bound on ρ_V, they give us some idea of the value to be expected: ρ_V should be not very different from the mean of the values suitable for life. This is what Vilenkin [10] calls the 'Principle of Mediocrity'. This mean is positive, because if ρ_V were negative it would have to be less in absolute value than the mass density of the universe during the whole time that life evolves (otherwise the universe would collapse before any astronomers come on the scene [11]), while if ρ_V were positive, it would only have to be less than the mass density of the universe at the time when most galaxies form, giving a much broader range of possible positive than negative values. In 1997–98 Martel, Shapiro and I [12] carried out a detailed calculation of the probability distribution of values of ρ_V seen by astronomers throughout the multiverse, under the assumption that the *a priori* probability distribution is flat in the relatively very narrow range that is anthropically allowed (for earlier calculations, see refs. [13] and [14]). At that time, the value of the primordial root-mean-square (rms) fractional density fluctuation σ was not well known, since the value inferred from observations of the cosmic microwave background depended on what one assumed for ρ_V. It was therefore not possible to calculate a mean expected value of ρ_V, but for any assumed value of ρ_V we could estimate σ and use the result to calculate the fraction of astronomers that would observe a value of ρ_V as small as the assumed value. In this way, we concluded that if Ω_Λ (the dimensionless density parameter associated with ρ_V) turned out to be much less than 0.6, anthropic reasoning could not explain why it was so small. The editor of the *Astrophysical Journal* objected to publishing papers about anthropic calculations, and we had to sell our article by pointing out that we had provided a strong argument for abandoning an anthropic explanation of a small value of ρ_V if it turned out to be too small.

Of course, it turned out that ρ_V is not too small. Soon after this work, observations of type Ia supernovae revealed that the cosmic expansion is accelerating [15,16] and gave the result that $\Omega_\Lambda \simeq 0.7$. In other words, the ratio of the vacuum energy density to the present mass density ρ_{M0} in *our* subuniverse (which I use just as a convenient measure of density) is about 2.3, a conclusion subsequently confirmed by observations of the microwave background [17].

This is still a bit low. Martel, Shapiro and I had found that the probability of a vacuum energy density this small was 12%. I have now recalculated the probability distribution, using WMAP data and a better transfer function, with the result that the probability of a random astronomer seeing a value as small as $2.3\rho_{M0}$ is increased to 15.6%.[2] Now that we know σ, we can also calculate that the median vacuum energy density is $13.3\rho_{M0}$.

I should mention a complication in these calculations. The average of the product of density fluctuations at different points becomes infinite as these points approach each other, so the rms fractional density fluctuation σ is actually infinite. Fortunately, it is not σ itself that is really needed in these calculations, but the rms fractional density fluctuation averaged over a sphere of comoving radius R taken large enough so that the density fluctuation is able to hold on efficiently to the heavy elements produced in the first generation of stars. The results mentioned above were calculated for R (projected to the present) equal to 2 Mpc. These results are rather sensitive to the value of R; for $R = 1$ Mpc, the probability of finding a vacuum energy as small as $2.3\rho_{M0}$ is only 7.2%. The estimate of the required value of R involves complicated astrophysics, and needs to be better understood.

2.2 Problems

Now I want to take up four problems we have to face in working out the anthropic implications of the string landscape.

What is the shape of the string landscape?

Douglas [18] and Dine and co-workers [19, 20] have taken the first steps in finding the statistical rules governing different string vacua. I cannot comment usefully on this, except to say that it would not hurt in this work if we knew what string theory is.

What constants scan?

Anthropic reasoning makes sense for a given constant if the range over which the constant varies in the landscape is large compared with the anthropically allowed range of values of the constant; for then it is reasonable to assume that the *a priori* probability distribution is flat in the anthropically allowed range. We need to know what constants actually 'scan' in this sense. Physicists would like to be able to calculate as much as possible, so we hope that not too many constants scan.

2 This situation has improved since the release of the second and third year WMAP results. Assuming flat space, the ratio of the vacuum energy density to the matter density is now found to be about 3.2 rather than 2.3.

https://doi.org/10.1017/CBO9781107050990.004 Published online by Cambridge University Press

The most optimistic hypothesis is that the only constants that scan are the few whose dimensionality is a positive power of mass: the vacuum energy and whatever scalar mass or masses set the scale of electroweak symmetry-breaking. With all other parameters of the Standard Model fixed, the scale of electroweak symmetry-breaking is bounded above by about 1.4 to 2.7 times its value in our subuniverse, by the condition that the pion mass should be small enough to make the nuclear force strong enough to keep the deuteron stable against fission [21]. (The condition that the deuteron be stable against beta decay, which yields a tighter bound, does not seem to me to be necessary. Even a beta-unstable deuteron would live long enough to allow cosmological helium synthesis; helium would be burned to heavy elements in the first generation of very massive stars; and then subsequent generations could have long lifetimes burning hydrogen through the carbon cycle.) But the mere fact that the electroweak symmetry-breaking scale is only a few orders of magnitude larger than the QCD scale should not in itself lead us to conclude that it must be anthropically fixed. There is always the possibility that the electroweak symmetry-breaking scale is determined by the energy at which some gauge coupling constant becomes strong, and if that coupling happens to grow with decreasing energy a little faster than the QCD coupling, then the electroweak breaking scale will naturally be a few orders of magnitude larger than the QCD scale.

If the electroweak symmetry-breaking scale is anthropically fixed, then we can give up the decades long search for a natural solution of the hierarchy problem. This is a very attractive prospect, because none of the 'natural' solutions that have been proposed, such as technicolor or low-energy supersymmetry, were ever free of difficulties. In particular, giving up low-energy supersymmetry can restore some of the most attractive features of the non-supersymmetric Standard Model: automatic conservation of baryon and lepton number in interactions up to dimension 5 and 4, respectively; natural conservation of flavors in neutral currents; and a small neutron electric dipole moment. Arkani-Hamed and Dimopoulos [22] and others [23–25] have even shown how it is possible to keep the good features of supersymmetry, such as a more accurate convergence of the $SU(3) \times SU(2) \times U(1)$ couplings to a single value, and the presence of candidates for dark matter, WIMPs. The idea of this 'split supersymmetry' is that, although supersymmetry is broken at some very high energy, the gauginos and higgsinos are kept light by a chiral symmetry. (An additional discrete symmetry is needed to prevent lepton-number violation in higgsino–lepton mixing, and to keep the lightest supersymmetric particle stable.) One of the nice things about split supersymmetry is that, unlike many of the things we talk about these days,

https://doi.org/10.1017/CBO9781107050990.004 Published online by Cambridge University Press

it makes predictions that can be checked when the LHC starts operation. One expects a single neutral Higgs with a mass in the range 120 to 165 GeV, possible winos and binos, but no squarks or sleptons, and a long-lived gluino. (Incidentally, a Stanford group [26] has recently used considerations of Big Bang nucleosynthesis to argue that a 1 TeV gluino must have a lifetime less than 100 seconds, indicating a supersymmetry breaking scale less than 10^{10} GeV. But I wonder whether, even if the gluino has a longer lifetime and decays after nucleosynthesis, the universe might not thereby be reheated above the temperature of helium dissociation, giving Big Bang nucleosynthesis a second chance to produce the observed helium abundance.)

What about the dimensionless Yukawa couplings of the Standard Model? If these couplings are very tightly constrained anthropically, then we might reasonably suspect that they take a wide range of values in the multiverse, so that anthropic considerations can have a chance to affect the values we observe. Hogan [27, 28] has analyzed the anthropic constraints on these couplings, with the electroweak symmetry-breaking scale and the sum of the u and d Yukawa couplings held fixed, to avoid complications due to the dependence of nuclear forces on the pion mass. He imposes the following conditions: (1) $m_d - m_u - m_e > 1.2$ MeV, so that the early universe does not become all neutrons; (2) $m_d - m_u + m_e < 3.4$ MeV, so that the pp reaction is exothermic; and (3) $m_e > 0$. With three conditions on the two parameters $m_u - m_d$ and m_e, he naturally finds these parameters are limited to a finite region, which turns out to be quite small. At first sight, this gives the impression that the quark and lepton Yukawa couplings are subject to stringent anthropic constraints, in which case we might infer that the Yukawa couplings probably scan.

I have two reservations about this conclusion. The first is that the pp reaction is not necessary for life. For one thing, the pep reaction $p + p + e^- \to d + \nu$ can keep stars burning hydrogen for a long time. For this, we do not need $m_d - m_u + m_e < 3.4$ MeV, but only the weaker condition $m_d - m_u - m_e < 3.4$ MeV. The three conditions then do not constrain $m_d - m_u$ and m_e separately to any finite region, but only constrain the single parameter $m_d - m_u - m_e$ to lie between 1.2 MeV and 3.4 MeV, not a very tight anthropic constraint. (In fact, He^4 will be stable as long as $m_d - m_u - m_e$ is less than about 13 MeV, so stellar nucleosynthesis can begin with helium burning in the heavy stars of Population III, followed by hydrogen burning in later generations of stars.) My second reservation is that the anthropic constraints on the Yukawa couplings are alleviated if we suppose (as discussed above) that the electroweak symmetry-breaking scale is not fixed, but free to take whatever value is anthropically necessary. For instance, according to the results of ref. [21], the deuteron binding energy could be made as large as about 3.5 MeV by taking the electroweak breaking scale much less than it is in our universe, in which case even the condition that the *pp* reaction be exothermic becomes much looser.

Incidentally, I do not set much store by the famous 'coincidence', emphasized by Hoyle, that there is an excited state of C^{12} with just the right energy to allow carbon production via α–Be^8 reactions in stars. We know that even–even nuclei have states that are well described as composites of α-particles. One such state is the ground state of Be^8, which is unstable against fission into two α-particles. The same α–α potential that produces that sort of unstable state in Be^8 could naturally be expected to produce an unstable state in C^{12} that is essentially a composite of three α-particles, and that therefore appears as a low-energy resonance in α–Be^8 reactions. So the existence of this state does not seem to me to provide any evidence of fine tuning.

What else scans? Tegmark and Rees [29] have raised the question of whether the rms density fluctuation σ may itself scan. If it does, then the anthropic constraint on the vacuum energy becomes weaker, resuscitating to some extent the problem of why ρ_V is so small. But Garriga and Vilenkin [30] have pointed out that it is really ρ_V/σ^3 that is constrained anthropically, so that, even if σ does scan, the anthropic prediction of this ratio remains robust.

Arkani-Hamed, Dimopoulos and Kachru [31], referred to below as ADK, have offered a possible reason for supposing that most constants do not scan. If there are a large number N of decoupled modular fields, each taking a few possible values, then the probability distribution of quantities that depend on all these fields will be sharply peaked, with a width proportional to $1/\sqrt{N}$. According to Distler and Varadarajan [32], it is not really necessary here to make arbitrary assumptions about the decoupling of the various scalar fields; it is enough to adopt the most general polynomial superpotential that is stable, in the sense that radiative corrections do not change the effective couplings for large N by amounts larger than the couplings themselves. Distler and Varadarajan emphasize cubic superpotentials, because polynomial superpotentials of order higher than cubic presumably make no physical sense. But it is not clear that even cubic superpotentials can be plausible approximations, or that peaks will occur at reasonable values in the distribution of dimensionless couplings rather than of some combinations of these couplings.[3] It also is not clear that the multiplicity of vacua in this kind of effective scalar field theory can properly represent the multiplicity of flux values in string theories [33], but even if it cannot, it presumably can represent the variety of minima of the potential for a given set of flux vacua.

3 M. Douglas, private communication.

https://doi.org/10.1017/CBO9781107050990.004 Published online by Cambridge University Press

If most constants do not effectively scan, then why should anthropic arguments work for the vacuum energy and the electroweak breaking scale? ADK point out that, even if some constant has a relatively narrow distribution, anthropic arguments will still apply if the anthropically allowed range is even narrower and near a point around which the distribution is symmetric. (ADK suppose that this point would be at zero, but this is not necessary.) This is the case, for instance, for the vacuum energy if the superpotential W is the sum of the superpotentials W_n for a large number of decoupled scalar fields, for each of which there is a separate broken R symmetry, so that the possible values of each W_n are equal and opposite. The probability distribution of the total superpotential $W = \sum_{n=1}^{N} W_n$ will then be a Gaussian peaked at $W = 0$ with a width proportional to $1/\sqrt{N}$, and the probability distribution of the supersymmetric vacuum energy $-8\pi G|W|^2$ will extend over a correspondingly narrow range of negative values, with a maximum at zero. When supersymmetry-breaking is taken into account, the probability distribution widens to include positive values of the vacuum energy, extending out to a positive value depending on the scale of supersymmetry-breaking. For any reasonable supersymmetry-breaking scale, this probability distribution, though narrow compared with the Planck scale, will be very wide compared with the very narrow anthropically allowed range around $\rho_V = 0$, so within this range the probability distribution can be expected to be flat, and anthropic arguments should work. Similar remarks apply to the μ-term of the supersymmetric Standard Model, which sets the scale of electroweak symmetry-breaking.

How should we calculate anthropically conditioned probabilities?

We would expect the anthropically conditioned probability distribution for a given value of any constant that scans to be proportional to the number of scientific civilizations that observe that value. In the calculations described above, Martel, Shapiro and I took this number to be proportional to the *fraction* of baryons that find themselves in galaxies, but what if the total number of baryons itself scans? What if it is infinite?

How is the landscape populated?

There are at least four ways in which we might imagine the different 'universes' described by the string landscape actually to exist.

(i) The various subuniverses may be simply different regions of space. This is most simply realized in the chaotic inflation theory [34–38]. The scalar fields in different inflating patches may take different values, giving rise to different values for various effective coupling constants. Indeed, Linde speculated about the application of the

https://doi.org/10.1017/CBO9781107050990.004 Published online by Cambridge University Press

anthropic principle to cosmology soon after the proposal of chaotic inflation [39, 40].

(ii) The subuniverses may be different eras of time in a single Big Bang. For instance, what appear to be constants of nature might actually depend on scalar fields that change very slowly as the universe expands [41].

(iii) The subuniverses may be different regions of spacetime. This can happen if, instead of changing smoothly with time, various scalar fields on which the 'constants' of nature depend change in a sequence of first-order phase transitions [42–44]. In these transitions, metastable bubbles form within a region of higher vacuum energy; then within each bubble there form further bubbles of even lower vacuum energy; and so on. In recent years this idea has been revived in the context of the string landscape [45, 46]. In particular, it has been suggested [47] that in this scenario the curvature of our universe is small for anthropic reasons, and hence possibly large enough to be detected.

(iv) The subuniverses could be different parts of quantum mechanical Hilbert space. In a reinterpretation of Hawking's earlier work on the wave-function of the universe [48, 49],[4] Coleman [51] showed that certain topological fixtures known as wormholes in the path integral for the Euclidean wave-function of the universe would lead to a superposition of wave-functions in which any coupling constant not constrained by symmetry principles would take any possible value.[5] Ooguri, Vafa and Verlinde [57] have argued for a particular wave-function of the universe, but it escapes me how anyone can tell whether this or any other proposed wave-function is *the* wave-function of the universe.

These alternatives are by no means mutually exclusive. In particular, it seems to me that, whatever one concludes about the first three alternatives, we will still have the possibility that the wave-function of the universe is a superposition of different terms representing different ways of populating the landscape in space and/or time.

4 Some of this work is based on an initial condition for the origin of the universe proposed by Hartle and Hawking [50].

5 It has been argued by Hawking and others that the wave-function of the universe is sharply peaked at values of the constants that yield a zero vacuum energy at late times [52–55]. This view has been challenged in ref. [56]. I am assuming here that there are no such peaks.

2.3 Conclusion

In closing, I would like to comment on the impact of anthropic reasoning within and beyond the physics community. Some physicists have expressed a strong distaste for anthropic arguments. (I have heard David Gross say 'I hate it'.) This is understandable. Theories based on anthropic calculations certainly represent a retreat from what we had hoped for: the calculation of all fundamental parameters from first principles. It is too soon to give up on this hope, but without loving it we may just have to resign ourselves to a retreat, just as Newton had to give up Kepler's hope of a calculation of the relative sizes of planetary orbits from first principles.

There is also a less creditable reason for hostility to the idea of a multiverse, based on the fact that we will never be able to observe any subuniverses except our own. Livio and Rees [58] and Tegmark [59] have given thorough discussions of various other ingredients of accepted theories that we will never be able to observe, without our being led to reject these theories. The test of a physical theory is not that everything in it should be observable and every prediction it makes should be testable, but rather that enough is observable and enough predictions are testable to give us confidence that the theory is right.

Finally, I have heard the objection that, in trying to explain why the laws of nature are so well suited for the appearance and evolution of life, anthropic arguments take on some of the flavour of religion. I think that just the opposite is the case. Just as Darwin and Wallace explained how the wonderful adaptations of living forms could arise without supernatural intervention, so the string landscape may explain how the constants of nature that we observe can take values suitable for life without being fine-tuned by a benevolent creator. I found this parallel well understood in a surprising place, a *New York Times* article by Christoph Schönborn, Cardinal Archbishop of Vienna [60]. His article concludes as follows.

> Now, at the beginning of the 21st century, faced with scientific claims like neo-Darwinism and the multiverse hypothesis in cosmology invented to avoid the overwhelming evidence for purpose and design found in modern science, the Catholic Church will again defend human nature by proclaiming that the immanent design evident in nature is real. Scientific theories that try to explain away the appearance of design as the result of 'chance and necessity' are not scientific at all, but, as John Paul put it, an abdication of human intelligence.

It is nice to see work in cosmology get some of the attention given these days to evolution, but of course it is not religious preconceptions like these that can decide any issues in science.

It must be acknowledged that there is a big difference in the degree of confidence we can have in neo-Darwinism and in the multiverse. It is settled, as well as anything in science is ever settled, that the adaptations of living things on Earth have come into being through natural selection acting on random undirected inheritable variations. About the multiverse, it is appropriate to keep an open mind, and opinions among scientists differ widely. In the Austin airport on the way to this meeting I noticed for sale the October issue of a magazine called *Astronomy*, having on the cover the headline 'Why You Live in Multiple Universes'. Inside I found a report of a discussion at a conference at Stanford, at which Martin Rees said that he was sufficiently confident about the multiverse to bet his dog's life on it, while Andrei Linde said he would bet his own life. As for me, I have just enough confidence about the multiverse to bet the lives of both Andrei Linde *and* Martin Rees's dog.

Acknowledgements

This material is based upon work supported by the National Science Foundation under Grants nos. PHY-0071512 and PHY-0455649 and with support from The Robert A. Welch Foundation, Grant no. F-0014, and also grant support from the US Navy, Office of Naval Research, Grant nos. N00014-03-1-0639 and N00014-04-1-0336, Quantum Optics Initiative.

References

[1] R. Bousso and J. Polchinski. *JHEP*, **0006** (2000), 006.
[2] L. Smolin. *Life of the Cosmos* (New York: Oxford University Press, 1997).
[3] M. J. Duff, B. E. W. Nilsson and C. N. Pope. *Phys. Rep.* **130** (1986), 1.
[4] S. B. Giddings, S. Kachru and J. Polchinski. *Phys. Rev.* **D 66** (2002), 106006.
[5] A. Strominger, A. Maloney and E. Silverstein. In *The Future of Theoretical Physics and Cosmology*, eds. G. W. Gibbons, E. P. S. Shellard and S. J. Ranken (Cambridge: Cambridge University Press, 2003), pp. 570–91.
[6] S. Kachru, R. Kallosh, A. D. Linde and S. P. Trivedi. *Phys. Rev.* **D 68** (2003), 046005.
[7] S. K. Ashok and M. Douglas. *JHEP*, **0401** (2004), 060.
[8] L. Susskind. This volume (2007) [hep-th/0302219].
[9] S. Weinberg. *Phys. Rev. Lett.* **59** (1987), 2607.
[10] A. Vilenkin. *Phys. Rev. Lett.* **74** (1995), 846.
[11] J. D. Barrow and F. J. Tipler. *The Anthropic Cosmological Principle* (Oxford: Clarendon, 1986).
[12] H. Martel, P. Shapiro and S. Weinberg. *Astrophys. J.* **492** (1998), 29.
[13] G. Efstathiou. *Mon. Not. Roy. Astron. Soc.* **274** (1995), L73.

[14] S. Weinberg. In *Critical Dialogues in Cosmology*, ed. N. Turok (Singapore: World Scientific, 1997).
[15] A. G. Riess, A. V. Filippenko, P. Challis *et al. Astron. J.* **116** (1998), 1009.
[16] S. Perlmutter, G. Aldering, G. Goldhaber *et al. Astrophys. J.* **517** (1999), 565.
[17] D. N. Spergel, L. Verde, H. V. Peiris *et al. Astrophys. J. Supp.* **148**, (2003), 175.
[18] M. R. Douglas. *Compt. Rend. Phys.* **5** (2004), 965 [hep-ph/0401004].
[19] M. Dine, D. O'Neil and Z. Sun. *JHEP*, **0507** (2005), 014.
[20] M. Dine and Z. Sun. *JHEP* **0601** (2006), 129 [hep-th/0506246].
[21] V. Agrawal, S. M. Barr, J. F. Donoghue and D. Seckel. *Phys. Rev.* **D 57** (1998), 5480.
[22] N. Arkani-Hamed and S. Dimopoulos. *JHEP*, **0506** (2005), 073.
[23] G. F. Giudice and A. Romanino. *Nucl. Phys.* **B 699** (2004), 65.
[24] N. Arkani-Hamed, S. Dimopoulos, G. F. Giudice and A. Romanino. *Nucl. Phys.* **B 709** (2005), 3.
[25] A. Delgado and G. F. Giudice. *Phys. Lett.* **B 627** (2005), 155 [hep-ph/0506217].
[26] A. Arvanitaki, C. Davis, P. W. Graham, A. Pierce and J. G. Wacker. *Phys. Rev.* **D 72** (2005), 075011 [hep-ph/0504210].
[27] C. Hogan. *Rev. Mod. Phys.* **72**, 1149 (2000).
[28] C. Hogan. This volume (2007) [astro-ph/0407086].
[29] M. Tegmark and M. J. Rees. *Astrophys. J.* **499** (1998), 526.
[30] J. Garriga and A. Vilenkin. *Prog. Theor. Phys. Suppl.* **163** (2006), 245 [hep-th/0508005].
[31] N. Arkani-Hamed, S. Dimopoulos and S. Kachru [hep-th/0501082].
[32] J. Distler and U. Varadarajan (2005) [hep-th/0507090].
[33] T. Banks (2000) [hep-th/0011255].
[34] A. D. Linde. *Phys. Lett.* **129 B** (1983), 177.
[35] A. Vilenkin. *Phys. Rev.* **D 27** (1983), 2848.
[36] A. D. Linde. *Phys. Lett.* **B 175** (1986), 305.
[37] A. D. Linde. *Phys. Script.* **T 15** (1987), 100.
[38] A. D. Linde. *Phys. Lett.* **B 202** (1988), 194.
[39] A. D. Linde. In *The Very Early Universe*, eds. G. W. Gibbons, S. W. Hawking and S. Siklos (Cambridge: Cambridge University Press, 1983).
[40] A. D. Linde. *Rep. Prog. Phys.* **47** (1984), 925.
[41] T. Banks. *Nucl. Phys.* **B 249** (1985), 332.
[42] L. Abbott. *Phys. Lett.* **B 150** (1985), 427.
[43] J. D. Brown and C. Teitelboim. *Phys. Lett.* **B 195** (1987), 177.
[44] J. D. Brown and C. Teitelboim. *Nucl. Phys.* **B 297** (1987), 787.
[45] J. L. Feng, J. March-Russel, S. Sethi and F. Wilczek. *Nucl. Phys.* **B 602** (2001), 307.
[46] H. Firouzjahi, S. Sarangji and S.-H. Tye. *JHEP*, **0409** (2004), 060.
[47] B. Freivogel, M. Kleban, M. R. Martinez and L. Susskind. *JHEP* **3** (2006), 39 [hep-th/0505232].
[48] S. W. Hawking. *Nucl. Phys.* **B 239** (1984), 257.
[49] S. W. Hawking. *Relativity, Groups, and Topology II*, NATO Advanced Study Institute Session XL, Les Houches, 1983, eds. B. S. DeWitt and R. Stora (Amsterdam: Elsevier, 1984), p. 336.

[50] J. Hartle and S. W. Hawking. *Phys. Rev.* **D 28** (1983), 2960.
[51] S. Coleman. *Nucl. Phys.* **B 307** (1988), 867.
[52] S. W. Hawking. In *Shelter Island II – Proceedings of the 1983 Shelter Island Conference on Quantum Field Theory and the Fundamental Problems of Physics*, eds. R. Jackiw *et al.* (Cambridge: MIT Press, 1985).
[53] S. W. Hawking. *Phys. Lett.* **B 134** (1984), 403.
[54] E. Baum. *Phys. Lett.* **B 133** (1984), 185.
[55] S. Coleman. *Nucl. Phys.* **B 310** (1985), 643.
[56] W. Fischler, I. Klebanov, J. Polchinski and L. Susskind. *Nucl. Phys.* **B 237** (1989), 157.
[57] H. Ooguri, C. Vafa and E. Verlinde. *Lett. Math. Phys.* **74** (2005), 311 [hep-th/0502211].
[58] M. Livio and M. J. Rees. *Science*, **309** (2003), 1022.
[59] M. Tegmark. *Ann. Phys.* **270** (1998), 1.
[60] C. Schönborn. *N.Y. Times* (7 July 2005), p. A23.

PHYSICAL REVIEW D **77,** 123541 (2008)

Effective field theory for inflation

Steven Weinberg*

Theory Group, Department of Physics, University of Texas, Austin, Texas, 78712, USA

(Received 15 May 2008; published 27 June 2008)

The methods of effective field theory are used to study generic theories of inflation with a single inflaton field. For scalar modes, the leading corrections to the $\mathcal{R}$ correlation function are found to be purely of the k-inflation type. For tensor modes the leading corrections to the correlation function arise from terms in the action that are quadratic in the curvature, including a parity-violating term that makes the propagation of these modes depend on their helicity. These methods are also briefly applied to nongeneric theories of inflation with an extra shift symmetry, as in so-called ghost inflation.

DOI: 10.1103/PhysRevD.77.123541 PACS numbers: 98.80.−k, 04.60.−m, 04.62.+v, 98.80.Jk

I. GENERIC THEORIES OF INFLATION

Observations of the cosmic microwave background and large scale structure are consistent with a simple theory of inflation [1] with a single canonically normalized inflaton field $\varphi_c(x)$, described by a Lagrangian

$$\mathcal{L}_0 = \sqrt{g}\left[-\frac{M_P^2}{2}R - \frac{1}{2}g^{\mu\nu}\partial_\mu\varphi_c\partial_\nu\varphi_c - V(\varphi_c)\right], \quad (1)$$

where $g \equiv -\mathrm{Det}g_{\mu\nu}$, $M_P \equiv 1/\sqrt{8\pi G}$ is the reduced Planck mass, and $V(\varphi_c)$ is a potential down which the scalar field rolls more-or-less slowly. With this theory, the strength of observed fluctuations in the microwave background matter density indicates that the cosmic expansion rate $H \equiv \dot{a}/a$ and the physical wave number k/a at horizon exit, when these are equal, have the value [2] $H = k/a \approx \sqrt{\epsilon} \times 2 \times 10^{14}$ GeV where ϵ is the value of $-\dot{H}/H^2$ at this time, and a is the Robertson–Walker scale factor. Hence H and k/a at horizon exit are likely to be much less than $M_P \simeq 2.4 \times 10^{18}$ GeV, and even considerably less than a plausible grand unification scale $\approx 10^{16}$ GeV. This provides a justification after the fact for using a Lagrangian (1) with a minimum number of spacetime derivatives. [As is well known, (1) is the most general Lagrangian density for gravitation and a single scalar field with no more than two spacetime derivatives. An arbitrary function of φ multiplying the first term could be eliminated by a redefinition of the metric, and an arbitrary function of φ multiplying the second term could be eliminated by a redefinition of φ.]

But H and k/a at horizon exit are not entirely negligible compared with whatever fundamental scale characterizes the theory underlying inflation, and at earlier times k/a is exponentially larger than at horizon exit, so it is worth considering the next corrections to (1). We assume that (1) is just the first term in a generic effective field theory, in which terms with higher derivatives are suppressed by negative powers of some large mass M, characterizing whatever fundamental theory underlies this effective field theory. Rather than committing ourselves to any particular underlying theory, we will simply assume that all constants in the higher derivative terms of the effective Lagrangian take values that are powers of M indicated by dimensional analysis, with coefficients roughly of order unity. Because H and k/a are so large during inflation, observations of fluctuations produced during inflation provide a unique opportunity for detecting effects of higher derivative terms in the gravitational action.

To get some idea of the value of M, we note that the unperturbed canonically normalized scalar field $\bar{\varphi}_c$ described by the Lagrangian (1) has a time derivative $\dot{\bar{\varphi}}_c = \sqrt{2\epsilon}M_PH$, so the change in $\bar{\varphi}_c$ during a Hubble time $1/H$ at around the time of horizon exit is of order $\dot{\bar{\varphi}}_c/H = \sqrt{2\epsilon}M_P$. If we are to use effective field theory to study fluctuations at about the time of horizon exit in generic theories in which the dependence of the action on φ_c is unconstrained by symmetry principles or by other consequences of an underlying theory, and if (1) is at least a fair first approximation to the full theory, then the mass M that is characteristic of the effective field theory of inflation cannot be much smaller than $\sqrt{2\epsilon}M_P$, for if it were then there would be no limit on the size of higher-derivative terms containing many powers of φ_c/M. It follows that the expansion parameter H/M in this class of theories is no greater than $H/\sqrt{2\epsilon}M_P \simeq 6 \times 10^{-5}$, whatever the value of ϵ.

We will tentatively assume here that M is of order $\sqrt{2\epsilon}M_P$, in which case the coefficients of the higher-derivative terms in the effective Lagrangian have to be taken as arbitrary functions of φ_c/M. This is likely to be the case if ϵ is not too small, say of order 0.02, since then there is not much difference between $\sqrt{2\epsilon}M_P$ and M_P, and M is unlikely to be much larger than M_P. (The considerations presented below would still be valid if M were instead much larger than $\sqrt{2\epsilon}M_P$, as for instance if $M \approx M_P$ and ϵ is very small, but then we would have to count powers of φ_c/M as well as numbers of derivatives in judging how much the various higher-derivative terms are

*weinberg@physics.utexas.edu

suppressed, and some of the coefficient functions in the effective Lagrangian derived below would be negligible, and all others very simple.) From now on we will work with a dimensionless scalar field $\varphi \equiv \varphi_c/M$, and write (1) as

$$\mathcal{L}_0 = \sqrt{g}\left[-\frac{M_P^2}{2}R - \frac{M^2}{2}g^{\mu\nu}\partial_\mu\varphi\partial_\nu\varphi - M_P^2 U(\varphi)\right], \tag{2}$$

where $U(\varphi) \equiv V(M\varphi)/M_P^2$. Note that the unperturbed value of U is $(3-\epsilon)H^2$, so we can think of U as well as $\partial_\mu\varphi\partial^\mu\varphi$ as both being of order H^2 at horizon exit.

The leading correction to (2) will consist of a sum of all generally covariant terms with four spacetime derivatives and coefficients of order unity [3]. By a judicious weeding out of total derivatives, the most general such correction term can be put in the form [4]

$$\begin{aligned}\Delta\mathcal{L} = \sqrt{g}[&f_1(\varphi)(g^{\mu\nu}\varphi_{,\mu}\varphi_{,\nu})^2 + f_2(\varphi)g^{\rho\sigma}\varphi_{,\rho}\varphi_{,\sigma}\Box\varphi \\ &+ f_3(\varphi)(\Box\varphi)^2 + f_4(\varphi)R^{\mu\nu}\varphi_{,\mu}\varphi_{,\nu} \\ &+ f_5(\varphi)Rg^{\mu\nu}\varphi_{,\mu}\varphi_{,\nu} + f_6(\varphi)R\Box\varphi + f_7(\varphi)R^2 \\ &+ f_8(\varphi)R^{\mu\nu}R_{\mu\nu} + f_9(\varphi)C^{\mu\nu\rho\sigma}C_{\mu\nu\rho\sigma}] \\ &+ f_{10}(\varphi)\epsilon^{\mu\nu\rho\sigma}C_{\mu\nu}{}^{\kappa\lambda}C_{\rho\sigma\kappa\lambda},\end{aligned} \tag{3}$$

where as usual commas denote ordinary derivatives and semicolons denote covariant derivatives; $\Box\varphi \equiv g^{\mu\nu}\varphi_{,\mu;\nu}$ is the invariant d'Alembertian of φ; $\epsilon^{\mu\nu\rho\sigma}$ is the totally antisymmetric tensor density with $\epsilon^{1230} \equiv +1$; and the $f_n(\varphi)$ are dimensionless functions, treated here as of order unity. In the last two terms, instead of the Riemann–Christoffel tensor $R_{\mu\nu\rho\sigma}$, we have used the Weyl tensor

$$\begin{aligned}C_{\mu\nu\rho\sigma} \equiv R_{\mu\nu\rho\sigma} &- \frac{1}{2}(g_{\mu\rho}R_{\nu\sigma} - g_{\mu\sigma}R_{\nu\rho} - g_{\nu\rho}R_{\mu\sigma} \\ &+ g_{\nu\sigma}R_{\mu\rho}) + \frac{R}{6}(g_{\mu\rho}g_{\nu\sigma} - g_{\nu\rho}g_{\mu\sigma}).\end{aligned} \tag{4}$$

Writing the last two terms in Eq. (3) as bilinears in $C_{\mu\nu\rho\sigma}$ rather than $R_{\mu\nu\rho\sigma}$ has no effect in the last term, and in the penultimate term of course just amounts to a different definition of f_7 and f_8. (Similarly, instead of writing the penultimate term as a bilinear in $C_{\mu\nu\rho\sigma}$ or $R_{\mu\nu\rho\sigma}$, we could have written it as the linear combination of curvature bilinears that appears in the Gauss–Bonnet identity; even though this linear combination is a total derivative, it would affect the field equations because its coefficient $f_9(\varphi)$ is not constant.) Our reason for choosing to use the Weyl tensor in the last two terms will become apparent soon.

The correction term (3) involves second time derivatives, as well as fields and their first time derivatives. If we took such a theory literally, we would find more than just the usual two adiabatic modes for single-field inflation, and the commutation relations (as given by the Ostrogradski formalism [5]) would be bizarre, with φ commuting with $\dot\varphi$. (For instance, Kallosh, Kang, Linde, and Mukhanov [6] encounter such additional modes when the Ostrogradski formalism is applied to a scalar field Lagrangian involving second time derivatives.) Similarly, there are metric components (such as g^{00} and g^{0i} in the ADM formalism [7]) whose time derivatives do not appear in $\mathcal{L}_0$, but that do appear in $\Delta\mathcal{L}$. If we were to take $\mathcal{L}_0 + \Delta\mathcal{L}$ as the full Lagrangian, then the correction term $\Delta\mathcal{L}$ would cause these auxiliary fields to become dynamical, with a further expansion of the modes of the system.

Instead, we should remember that from the point of view of effective field theory, Eqs. (2) and (3) represent just the lowest two terms in an expansion in inverse powers of M, so we must rule out any modes that cannot be expanded in this way [8]. This means, in particular, that we *must* eliminate all second time derivatives and time derivatives of auxiliary fields in the first correction terms in the effective action by using the field equations derived from the leading terms in the action.[1] In the present case, we must

---

[1]This is equivalent to what is generally done in deriving Feynman rules in effective flat-space quantum field theories. Consider for instance the very simple effective Lagrangian

$$\mathcal{L} = -\tfrac{1}{2}[\partial_\mu\varphi\partial^\mu\varphi + m^2\varphi^2 + M^{-2}(\Box\varphi)^2] + J\varphi$$

where $M \gg m$ is some very large mass, and J is a c-number external current. We can easily find the connected part Γ of the vacuum persistence amplitude:

$$\Gamma = i\int d^4k\frac{|J(k)|^2}{k^2 + m^2 + k^4/M^2}.$$

If we took this result seriously, then we would conclude that in addition to the usual particle with mass $m + O(m^3/M^2)$, the theory contains an unphysical one particle state with mass $M + O(m^2/M)$. But if we regard $\mathcal{L}$ as just the first two terms in a power series in $1/M^2$, then we must treat the term $M^{-2}(\Box\varphi)^2$ as a first-order perturbation, so that the vacuum persistence amplitude is

$$\Gamma = i\int d^4k|J(k)|^2\left[\frac{1}{k^2+m^2} - \frac{k^4}{M^2(k^2+m^2)^2} + \ldots\right],$$

and the only pole is at $k^2 = -m^2$. This is just the same result for Γ that we would find if we were to eliminate the second time derivatives in the $O(M^{-2})$ term in $\mathcal{L}$ by using the field equation derived from the leading term in the Lagrangian

$$\Box\varphi = m^2\varphi - J.$$

In this case the effective Lagrangian becomes

$$\mathcal{L} = -\frac{1}{2}[\partial_\mu\varphi\partial^\mu\varphi + m^2\varphi^2 + m^4M^{-2}\varphi^2] + (1 + m^2/M^2)J\varphi - J^2/2M^2.$$

Taking into account all J-dependent terms, it is straightforward to see that with this Lagrangian we get the same vacuum persistence amplitude as found above for the original Lagrangian, when $M^{-2}(\Box\varphi)^2$ is treated as a first-order perturbation.

eliminate second time derivatives and time derivatives of auxiliary fields in (3) by using the zeroth order field equations derived from (2):

$$M^2\Box\varphi = M_P^2 U'(\varphi),$$
$$R_{\mu\nu} = -(M^2/M_P^2)\varphi_{,\mu}\varphi_{,\nu} - U(\varphi)g_{\mu\nu}. \tag{5}$$

Using these field equations in Eq. (3) allows us, with some redefinitions, to eliminate all of the terms in (3) except the first one and the last two. Specifically, the second term in Eq. (3) just provides a field-dependent correction to the kinematic term in (2), which can be eliminated by a redefinition of the inflaton field; the third term just provides a correction $f_3 U'^2 M_P^4/M^4$ to the potential in (2), which can be absorbed into a redefinition of $U(\varphi)$; the fourth and fifth terms supply corrections to both $f_1(\varphi)$ and the kinematic term in (2); the sixth term provides corrections to the kinematic term and the potential in (2); and the seventh and eighth terms provide corrections to the kinematic term and potential in (2) and to $f_1(\varphi)$. That is, with suitable redefinitions of φ, $U(\varphi)$, and $f_1(\varphi)$, and with various total derivatives dropped, the Lagrangian is the sum of (2) and a correction term of the form

$$\Delta\mathcal{L} = \sqrt{g}f_1(\varphi)(g^{\mu\nu}\varphi_{,\mu}\varphi_{,\nu})^2 + \sqrt{g}f_9(\varphi)C^{\mu\nu\rho\sigma}C_{\mu\nu\rho\sigma} + f_{10}(\varphi)\epsilon^{\mu\nu\rho\sigma}C_{\mu\nu}{}^{\kappa\lambda}C_{\rho\sigma\kappa\lambda}, \tag{6}$$

The first term is of the type encountered in theories of "k-inflation" [9]. This term must be included in the Lagrangian, as a counterterm to ultraviolet divergences encountered when the leading terms in (2) are used in one-loop order. The second term (or an equivalent Gauss–Bonnet term) has been considered in connection with inflation and the evolution of dark energy [10].

For a general function $f_{10}(\varphi)$ the final term in Eq. (6) violates parity conservation [11]. That is, although the action is invariant under coordinate transformations $x^\mu \to x'^\mu$ that are "small," in the sense that $\mathrm{Det}(\partial x'/\partial x) > 0$, it is not invariant under *inversions*, that is, under coordinate transformations with $\mathrm{Det}(\partial x'/\partial x) < 0$. It is only invariance under "small" coordinate transformations that is needed to ensure the conservation of the energy-momentum tensor, and no sequence of small coordinate transformations can ever add up to an inversion, so there is no *a priori* reason to impose invariance under inversions, including space inversion. The fact that parity has always been observed to be conserved in gravitational interactions is sufficiently explained by the fact that terms in the effective action for gravity and scalars with no more than two spacetime derivatives that are invariant under small coordinate transformations cannot be complicated enough to violate invariance under inversions.

From now on we shall work in perturbation theory, writing

$$g_{\mu\nu}(\mathbf{x},t) = \bar{g}_{\mu\nu}(t) + h_{\mu\nu}(\mathbf{x},t),$$
$$\varphi(\mathbf{x},t) = \bar{\varphi}(t) + \delta\varphi(\mathbf{x},t), \tag{7}$$

where $\bar{g}_{\mu\nu}(t)$ is the flat-space Robertson–Walker metric with $\bar{g}_{00} = -1$, $\bar{g}_{0i} = 0$, and $\bar{g}_{ij} = a^2(t)\delta_{ij}$; $\bar{\varphi}(t)$ is the unperturbed scalar field; and $h_{\mu\nu}$ and $\delta\varphi$ are first-order perturbations. In this paper we will mostly be concerned with the terms in the Lagrangian that are quadratic in perturbations, which are needed for the calculation of Gaussian correlations. Terms of higher order in perturbations that are needed for the calculation of non-Gaussian effects will be considered only briefly.

Because the spatially flat Robertson–Walker metric is also conformally flat it has a vanishing Weyl tensor, and so the Weyl tensor starts with a term of first order in perturbations. This saves us from having to calculate the Weyl tensor to second order in perturbations; we have simply

$$[\Delta\mathcal{L}]^{(2)} = [\sqrt{g}f_1(\varphi)(g^{\mu\nu}\varphi_{,\mu}\varphi_{,\nu})^2]^{(2)} + a^3 f_9(\bar{\varphi})\bar{g}^{\mu\kappa}\bar{g}^{\nu\lambda}\bar{g}^{\rho\eta}\bar{g}^{\sigma\zeta}C^{(1)}_{\kappa\lambda\eta\zeta}C^{(1)}_{\mu\nu\rho\sigma} + f_{10}(\bar{\varphi})\epsilon^{\mu\nu\rho\sigma}\bar{g}^{\kappa\eta}\bar{g}^{\lambda\zeta}C^{(1)}_{\mu\nu\eta\zeta}C^{(1)}_{\rho\sigma\kappa\lambda}, \tag{8}$$

where the superscripts (1) and (2) denote terms of first and second order in perturbations, respectively. Furthermore, the Weyl tensor is traceless, which to first order in perturbations gives

$$C^{(1)}_{i0k0} = a^{-2}C^{(1)}_{ijkj}, \qquad C^{(1)}_{ijj0} = C^{(1)}_{i0i0} = C^{(1)}_{ijij} = 0. \tag{9}$$

Since scalar and tensor fluctuations do not interfere in Gaussian correlations, they will be considered separately.

II. SCALAR FLUCTUATIONS

Here we are interested in terms in the Lagrangian that, after eliminating auxiliary fields, are quadratic in $\mathcal{R}$, the familiar gauge-invariant quantity that is conserved outside the horizon [1]

$$\mathcal{R} \equiv \frac{A}{2} - \frac{H\delta\varphi}{\dot{\bar{\varphi}}}, \tag{10}$$

with A defined by writing the spatial part of the metric perturbation for scalar perturbations in a general gauge as

$$h_{ij}(\mathbf{x},t) = a^2(t)\left[\delta_{ij}A(\mathbf{x},t) + \frac{\partial^2 B(\mathbf{x},t)}{\partial x^i \partial x^j}\right]. \tag{11}$$

Let us consider in turn the contribution of the three terms in Eq. (8) to the quadratic part of the Lagrangian for $\mathcal{R}$.

First, terms in the effective Lagrangian like the first term in Eq. (8) that depend only on φ and $\partial_\mu\varphi\partial^\mu\varphi$ are known to enter into the part of the Lagrangian quadratic in scalar fluctuations only through their effect on the sound speed $c_s(t)$ [12]. That is, after eliminating auxiliary fields,

$$\left[-\frac{\sqrt{g}M_P^2}{2}R + \sqrt{g}P(-\partial_\mu\varphi\partial^\mu\varphi/2,\varphi)\right]^{(2)} = -\frac{M_P^2\dot{H}}{H^2}a^3\left[\frac{1}{c_s^2}\dot{\mathcal{R}}^2 - \frac{1}{a^2}(\vec{\nabla}\mathcal{R})^2\right], \tag{12}$$

where

$$c_s^{-2} = 1 + 2\left[X\frac{\partial^2 P(X,\bar\varphi)}{\partial X^2}\Big/\frac{\partial P(X,\bar\varphi)}{\partial X}\right]_{X=\dot{\bar\varphi}^2/2}. \tag{13}$$

In particular, the first term in Eq. (8) shifts the squared speed of sound by

$$\Delta c_s^2 = \frac{16\dot{H}M_P^2 f_1(\bar\varphi)}{M^4}, \tag{14}$$

corresponding to a second-order perturbation

$$[\sqrt{g}f_1(\varphi)(g^{\mu\nu}\varphi_{,\mu}\varphi_{,\nu})^2]^{(2)} = \frac{16M_P^4 a^3\dot{H}^2 f_1(\bar\varphi)}{M^4H^2}\dot{\mathcal{R}}^2. \tag{15}$$

The other terms in Eq. (8) are greatly simplified by noting that, for scalar modes, C_{ijk0} must take the form of $\delta_{ik}\partial_j - \delta_{jk}\partial_i$ acting on some scalar, so the above trace condition $C^{(1)}_{ijj0} = 0$ implies that $C^{(1)}_{ijk0} = 0$. Hence all we need to evaluate the second term in (8) are the purely spatial components of the Weyl tensor. Using the field equations (5) to eliminate auxiliary fields and second time derivatives, we find after a straightforward though tedious calculation that

$$\begin{aligned} a^3 f_9(\bar\varphi)\bar g^{\mu\kappa}\bar g^{\nu\lambda}\bar g^{\rho\eta}\bar g^{\sigma\zeta}C^{(1)}_{\kappa\lambda\eta\zeta}C^{(1)}_{\mu\nu\rho\sigma} &= a^{-5}f_9(\bar\varphi)[C^{(1)}_{ijkl}C^{(1)}_{ijkl} + 4C^{(1)}_{ijkj}C^{(1)}_{ilkl}] \\ &= \frac{16\dot{H}^2}{3H^2}a^3 f_9(\bar\varphi)\dot{\mathcal{R}}^2. \end{aligned} \tag{16}$$

Comparing this with Eq. (15), we see that the effect of the second term in Eq. (8) is the same as a change in the coefficient f_1 of the first term by an amount

$$\Delta f_1(\varphi) = \frac{M^4}{3M_P^4}f_9(\varphi). \tag{17}$$

Finally, because $C^{(1)}_{ijk0}$ vanishes for scalar modes, the last term in Eq. (8) is

$$\begin{aligned} &f_{10}(\bar\varphi)\epsilon^{\mu\nu\rho\sigma}\bar g^{\kappa\eta}\bar g^{\lambda\zeta}C^{(1)}_{\mu\nu\eta\zeta}C^{(1)}_{\rho\sigma\kappa\lambda} \\ &\quad = f_{10}(\bar\varphi)\epsilon^{ijk0}[4a^{-4}C^{(1)}_{ijlm}C^{(1)}_{k0lm} - 8C^{(1)}_{ijl0}C^{(1)}_{k0l0}] = 0. \end{aligned} \tag{18}$$

We conclude that the leading corrections to the Gaussian correlations of $\mathcal{R}$ are solely of the k -inflation type. This justifies the calculation of the effective Lagrangian for Gaussian scalar correlations in slow-roll inflation in Sec. 3 of Ref. [3] even for generic theories of inflation. In such theories the terms in Eq. (3) that are left out in Ref. [3] can indeed be omitted in calculating the part of the effective Lagrangian quadratic in scalar fluctuations, *not because they are small*, but because as we have seen for scalar Gaussian fluctuations they yield nothing new. But this is not the case for Gaussian tensor fluctuations, and does not seem to be the case when non-Gaussian correlations are considered.

We have so far only considered the terms in the effective action of second order in $\mathcal{R}$, which are needed to calculate Gaussian correlations, but for actions of the k-inflation type, which only involve first derivatives of fields, it is not difficult also to calculate terms in the action of higher order in $\mathcal{R}$, which generate non-Gaussian correlations. For this purpose it is convenient to adopt a gauge in which there are no scalar perturbations to g_{ij}; that is, in which $g_{ij} = a^2(t)[\exp(D(\mathbf{x},t))]_{ij}$, where D_{ij} is a gravitational wave amplitude with $D_{ii} = 0$ and $\partial_i D_{ij} = 0$. In this gauge, Eq. (10) gives $\mathcal{R} = -H\delta\varphi/\dot{\bar\varphi}$. If we tentatively ignore the interaction of the inflaton with gravitational perturbations, and assume that H, f_1, and $\dot{\bar\varphi}$ are varying slowly, then it is trivial, by simply setting φ equal to $\bar\varphi + \delta\varphi$ in $\mathcal{L}$, and using $\dot{H} = -\dot{\bar\varphi}^2(M^2 + 4f_1\dot{\bar\varphi}^2)/2M_P^2$, to write a Lagrangian for $\pi \equiv -\mathcal{R}/H = \delta\varphi/\dot{\bar\varphi}$:

$$\begin{aligned} &\sqrt{g}\left[-\frac{M^2}{2}g^{\mu\nu}\partial_\mu\varphi\partial_\nu\varphi - M_P^2U(\varphi) + f_1(\varphi)(g^{\mu\nu}\partial_\mu\varphi\partial_\nu\varphi)^2\right] \\ &\quad = \bar{\mathcal{L}} + a^3M_P^2\dot{H}(-\dot\pi^2 + a^{-2}(\vec\nabla\pi)^2) + \frac{16a^3M_P^4\dot{H}^2f_1(\bar\varphi)}{M^4}\left(\dot\pi^2 + \dot\pi^3 - \frac{\dot\pi(\vec\nabla\pi)^2}{a^2} + \frac{\dot\pi^4}{4} - \frac{\dot\pi^2(\vec\nabla\pi)^2}{2a^2} + \frac{(\vec\nabla\pi)^4}{4a^2}\right). \end{aligned} \tag{19}$$

This agrees with the result obtained in Eq. (28) of [3], except that here we include terms quartic in π. In [3] the neglect of interactions of the inflaton with gravitational perturbations is justified on the basis of a "high energy" approximation, which amounts to the usual slow-roll approximation that $\epsilon \ll 1$, plus the assumption that, in our terms, $M^2 \gg \epsilon HM_P$, which is much weaker than the assumption $M \gg \sqrt{2\epsilon}M_P$ that we found necessary to treat generic theories of inflation by the methods of effective field theory. We see that the nonquadratic terms in $\mathcal{L}_0$ that can generate non-Gaussian correlations are suppressed in the slow-roll approximation, as found by Maldacena [13], but in ΔL the coefficients of the quadratic and higher order terms are of the same order of magnitude.

III. TENSOR FLUCTUATIONS

Tensor fluctuations appear solely in the perturbation to the purely spatial metric:

$$h_{ij}(\mathbf{x}, t) = a^2(t)[\exp D]_{ij}(\mathbf{x}, t), \qquad D_{ii} = 0, \qquad \partial_i D_{ij} = 0, \tag{20}$$

with $\delta\varphi = 0$. The first term in Eq. (8) involves only the metric components g^{00} and $\mathrm{Det}g_{\mu\nu}$, so it gets no contribution from tensor fluctuations. On the other hand, here the second and third terms in (8) make a nontrivial contribution to the Lagrangian for D_{ij}. Another straightforward calculation (dropping total derivatives) gives these terms as

$$\begin{aligned} &a^3 f_9(\bar\varphi)\bar g^{\mu\kappa}\bar g^{\nu\lambda}\bar g^{\rho\eta}\bar g^{\sigma\zeta}C^{(1)}_{\kappa\lambda\eta\zeta}C^{(1)}_{\mu\nu\rho\sigma} \\ &= f_9(\bar\varphi)[a^{-5}C^{(1)}_{ijkl}C^{(1)}_{ijkl} - 4a^{-3}C^{(1)}_{ijk0}C^{(1)}_{ijk0} \\ &\quad + 4a^{-1}C^{(1)}_{i0k0}C^{(1)}_{i0k0}] \\ &= a^3 f_9(\bar\varphi)\{\dot D_{ik}[2H^2 + 2\nabla^2/a^2]\dot D_{ik} \\ &\quad - 4H\dot D_{ik}(\nabla^2/a^2)D_{ik} + 2D_{ik}(\nabla^4/a^4)D_{ik}\}, \end{aligned} \tag{21}$$

and

$$\begin{aligned} &f_{10}(\bar\varphi)\epsilon^{\mu\nu\rho\sigma}\bar g^{\kappa\eta}\bar g^{\lambda\zeta}C^{(1)}_{\mu\nu\eta\zeta}C^{(1)}_{\rho\sigma\kappa\lambda} \\ &= f_{10}(\bar\varphi)\epsilon^{ijk0}[4a^{-4}C^{(1)}_{lmij}C^{(1)}_{lmk0} - 8a^{-2}C^{(1)}_{l0ij}C^{(1)}_{l0k0}] \\ &= 4f_{10}(\bar\varphi)\epsilon^{ijk0}\frac{\partial}{\partial t}[D_{il}\partial_j\nabla^2 D_{kl}]. \end{aligned} \tag{22}$$

The field equation for the tensor mode (with the term proportional to f_9 dropped for simplicity) is then

$$\ddot D_{il} + 3H\dot D_{il} - (\nabla^2/a^2)D_{il} = -64\pi G\dot f_{10}a^{-3}(\epsilon^{ijk0}\partial_j\nabla^2 D_{kl} + \epsilon^{ljk0}\partial_j\nabla^2 D_{ki}). \tag{23}$$

For a plane wave with comoving wave number $\vec k$ in the 3-direction, the only nonvanishing tensor amplitudes are $D_{11} = -D_{22}$ and $D_{12} = D_{21}$. They satisfy the field equations

$$\ddot D_\pm + 3H\dot D_\pm + (k^2/a^2)D_\pm = \mp 128\pi G(k/a)^3\dot f_{10}D_\pm \tag{24}$$

where $D_\pm \equiv D_{11} \mp iD_{12}$ are the amplitudes with helicity ± 2. As found in Ref. [11], the wave equation depends on helicity because parity is violated.

IV. A NONGENERIC EXAMPLE: GHOST INFLATION

Up to now, we have been concerned with generic theories of inflation, in which the dependence of the action on the inflaton field is unconstrained, and in consequence the characteristic mass M cannot be taken to be much less than $\sqrt{\epsilon}M_P$. For an example of a different sort, we might impose on the action a shift symmetry, under a transformation $\varphi \to \varphi$ + constant, which requires that the Lagrangian density involve only spacetime derivatives of φ rather than φ itself. This possibility was discussed briefly in [9], and in more detail under the name "ghost inflation" in [14]. We will take φ to be normalized so that $\partial_\mu\varphi$ is dimensionless, and has an unperurturbed value at horizon exit of order unity. The term in the Lagrangian density that depends only on $\partial_\mu\varphi$ is then

$$\mathcal{L}_0 = M^4\sqrt{g}P(-\partial_\mu\varphi\partial^\mu\varphi), \tag{25}$$

where $P(X)$ is a power series in X, with coefficients assumed to be of order unity, and M is the characteristic mass of the theory. Since powers of φ are excluded by the shift symmetry, M here can be much smaller than in generic theories of inflation, and, in particular, we will assume that M is much less than the Planck mass M_P. Any additional derivatives acting on $\partial_\mu\varphi$ or on the metric yield factors of order $H \sim M^2/M_P \ll M$, so Eq. (25) along with the Einstein term can be taken as the leading term in $\mathcal{L}$, with any correction terms suppressed by factors of H/M.

Let us first consider a theory in which (25) is the whole Lagrangian density for the scalar field, with no higher-derivative corrections. The field equation for the unperturbed scalar field $\bar\varphi(t)$ in this theory is

$$\frac{d}{dt}(a^3P'(\dot{\bar\varphi}^2)\dot{\bar\varphi}) = 0. \tag{26}$$

As noted in [9], in the limit of late time when $a \to \infty$, either $\dot{\bar\varphi} \to 0$, or $\dot{\bar\varphi} \to v$, where v is a quantity of order unity satisfying $P'(v^2) = 0$. We will consider only the latter case. In Ref. [14] the limit $\bar\varphi = vt$ is supposed to be already reached, in which case interesting fluctuations occur only when higher-derivative correction terms are added to (25). But if we take $\bar\varphi(t)$ to be only close to vt, but not yet there, then we find a nontrivial spectrum of propagating fluctuations even when no correction terms are added to (25). In this case the solution of Eq. (26) [with an appropriate normalization of $a(t)$] has $\dot{\bar\varphi} \to v + a^{-3}$; the speed of sound is $c_s \to \sqrt{1/va^3}$; the expansion rate approaches a limit $H_\infty = (M^2/M_P)\sqrt{P(v^2)/3}$; and the Fourier transform of $\mathcal{R}$ is

$$\mathcal{R}_k \propto a^{-3/2}H^{(1)}_{3/5}\left(\frac{2k}{5H_\infty v^{1/2}a^{5/2}}\right), \tag{27}$$

with a k-independent constant of proportionality. At late times, when the perturbation wave length is outside the acoustic horizon, this approaches a time-independent quantity $\mathcal{R}^o_k$, with

$$|\mathcal{R}^o_k|^2 \propto k^{-6/5}, \tag{28}$$

corresponding to a conventional scalar slope index $n_S = 14/5$, which of course is empirically ruled out. Thus to have a realistic theory of this sort, we must consider corrections to the leading term (25).

The first correction to this Lagrangian density contains just one factor of a second derivative of φ, and in general is of the form

$$\Delta\mathcal{L} = M^3\sqrt{g}Q(-\partial_\mu\varphi\partial^\mu\varphi)\Box\varphi, \tag{29}$$

and is therefore suppressed relative to (25) by factors of order $H/M \approx M/M_P$. (A term proportional to $g^{\rho\kappa}g^{\sigma\lambda}\varphi_{,\kappa}\varphi_{,\lambda}\varphi_{,\rho;\sigma}$ can be put in the form (29) by adding suitable total derivatives. In Ref. [14] these terms were excluded by imposing an additional symmetry under the reflection $\varphi \to -\varphi$, in which case the first correction is quadratic rather than linear in second derivatives of φ.) Once again, we must eliminate the second time derivatives in $\Delta\mathcal{L}$ by setting $\ddot{\varphi}$ equal to the same quantity as given by the field equation derived from the leading part of the Lagrangian

$$P'(-\partial_\nu\varphi\partial^\nu\varphi)\Box\varphi - 2P''(-\partial_\nu\varphi\partial^\nu\varphi)(\partial_\nu\varphi)_{;\mu}\partial^\mu\varphi\partial^\nu\varphi = 0. \tag{30}$$

This is pretty complicated, so for simplicity let us consider the case of a metric fixed in the flat-space Robertson–Walker form. Then after using (30) to eliminate second time derivatives in (29), the correction term is

$$\Delta\mathcal{L} = M^3a^3\left(\frac{2QP''}{P' + 2P''\dot{\varphi}^2}\right)(-2a^{-2}\dot{\varphi}\partial_i\varphi\partial_i\dot{\varphi} + a^{-2}H\dot{\varphi}\partial_i\varphi\partial_i\varphi + a^{-4}\partial_i\varphi\partial_j\varphi\partial_i\partial_j\varphi + 3H\dot{\varphi}^3 - a^{-2}\dot{\varphi}^2\nabla^2\varphi), \tag{31}$$

where Q, P' and P'' all have arguments $-\partial_\mu\varphi\partial^\mu\varphi = \dot{\varphi}^2 - a^{-2}\partial_i\varphi\partial_i\varphi$. This is the correction that has to be added to $\mathcal{L}_0$ in order to find the commutation relations of the field as well as the field equations by canonical quantization.

ACKNOWLEDGMENTS

I am grateful for discussions with N. Arkani-Hamed, J. Distler, J. Meyers, S. Odintsov, S. Paban, D. Robbins, and L. Senatore. This material is based upon work supported by the National Science Foundation under Grant No. PHY-0455649.

---

[1] For reviews with references to the original literature, see V. Mukhanov, *Physical Foundations of Cosmology* (Cambridge University Press, Cambridge, England, 2005); S. Weinberg, *Cosmology* (Oxford University Press, New York, 2008).

[2] This is based on third year WMAP results; D. Spergel *et al.*, Astrophys. J. Suppl. Ser. **170**, 377 (2007).

[3] This is different from the approach followed in an interesting recent paper on the effective field theory of inflation by C. Cheung, P. Creminelli, A. L. Fitzpatrick, J. Kaplan, and L. Senatore, J. High Energy Phys. 03 (2008) 014, whose calculations do not include any of the terms in Eq. (3) following the first term. Their paper does not spell out the rules governing which terms are to be included in the corrections of leading order, but on the basis of a private communication with Senatore, I gather that in judging how much various correction terms are suppressed, Cheung *et al.* do not count spacetime derivatives acting on the background scalar or metric fields in a comoving (or unitary) gauge in which φ equals its unperturbed value $\bar{\varphi}$, but only count derivatives acting on fluctuations in this gauge. In comoving gauge the first term in (3) is $\sqrt{g}f_1(\bar{\varphi})(g^{00}\dot{\bar{\varphi}}^2)^2$, and the factors of $\dot{\bar{\varphi}}$ are not counted as suppressing this term, because $|\dot{\bar{\varphi}}_c|$ is much larger than H^2. But as we have seen, in generic theories of inflation the characteristic mass scale M of the effective field theory must be at least as large as $\sqrt{\epsilon}M_P$, and the quantity $|\dot{\bar{\varphi}}_c|/M$ is then no larger than H, so that the first term in (3) is no less suppressed than the other terms. (Cheung *et al.* do at first include terms involving the extrinsic curvature of the spacelike surface with φ constant, but later drop these terms. The extrinsic curvature is not included here in Eq. (3), because in a general gauge it does not give a local term in the action. But Cheung *et al.* stick to comoving gauge, in which the extrinsic curvature can be expanded in a series of local functions, and these do yield some though not all of the terms in Eq. (3).) The approach of Cheung *et al.* is justified in theories with a much smaller value of M than considered here, which is possible if the dependence of the action on the inflaton field is limited by symmetry principles or other consequences of an underlying theory. This case is discussed briefly at the end of the present paper.

[4] The same list of terms with just four spacetime derivatives (aside from the term that violates parity conservation) has been given by E. Elizalde, A. Jacksenaev, S. D. Odintsov, and I. L. Shapiro, Phys. Lett. B **328**, 297 (1994); Classical Quantum Gravity **12**, 1385 (1995), but not in the context of effective field theory. (In the original preprint of the present paper, written before I had seen the work of Elizalde *et al.*, this list contained an additional term proportional to $\sqrt{g}R^{\mu\nu}\varphi_{,\mu;\nu}$. This term is redundant, because by using the Bianchi identity and discarding total derivatives it may be expressed as a linear combination of the fourth, fifth, and sixth terms listed here in Eq. (3).)

[5] M. Ostrogradski, Mem. Act. St. Petersbourg **VI 4**, 385 (1850). For a modern account, see F. J. de Urries and J. Julve, J. Phys. A **31**, 6949 (1998).

[6] R. Kallosh, J. U. Kang, A. Linde, and V. Mukhanov, J. Cosmol. Astropart. Phys. 04 (2008) 018.

[7] R. S. Arnowitt, S. Deser, and C. W. Misner, in *Gravitation: An Introduction to Current Research*, edited by L. Witten (Wiley, New York, 1962).
[8] See, e. g., J. Z. Simon, Phys. Rev. D **41**, 3720 (1990).
[9] C. Armendáriz-Picón, T. Damour, and V. F. Mukhanov, Phys. Lett. B **458**, 209 (1999).
[10] S. Kawai, M. Sakagami, and J. Soda, Phys. Lett. B **437**, 284 (1998); S. Kawai and J. Soda, Phys. Lett. B **460**, 41 (1999); S. Nojiri, S. D. Odintsov, and M. Sasaki, Phys. Rev. D **71**, 123509 (2005); G. Calcagni, S. Tsujikawa, and M. Sami, Classical Quantum Gravity **22**, 3977 (2005); G. Calcagne, B. de Carlos , and A. De Felice, Nucl. Phys. **B752**, 404 (2006); I. P. Neupane and B. M. N. Carter, J. Cosmol. Astropart. Phys. 06 (2006) 004; G. Cognola, E. Elizalde, S. Nojiri, S. D. Odintsov, and S. Zerbini, Phys. Rev. D **73**, 084007 (2006); B. Leith and I. P. Neupane, J. Cosmol. Astropart. Phys. 05 (2007) 019; S. Tsujikawa and M. Sami, J. Cosmol. Astropart. Phys. 07 (2007) 006; K. Bamba, Z-K Guo, and N. Ohta, Prog. Theor. Phys. **118**, 879 (2007). Some of these articles deal with instabilities produced by a Gauss–Bonnet term, but no instability arises if this term is treated as a correction term in an effective field theory. The first term in Eq. (6) along with a Gauss–Bonnet term equivalent to the second term in Eq. (6) were encountered in a low-energy limit of string theory by Z. K. Guo, N. Ohta, and S. Tsujikawa, Phys. Rev. D **75**, 023520 (2007).
[11] There is a large literature on a parity-violating term of this form. See, for instance, A. Lue, L. Wang, and M. Kamionkowski, Phys. Rev. Lett. **83**, 1506 (1999); S-Y. Pi and R. Jackiw, Phys. Rev. D **68**, 104012 (2003); M. Satoh, S. Kanno, and J. Soda, Phys. Rev. D **77**, 023526 (2008). Leptogenesis due to this term was considered by S. H. S. Alexander, M. E. Peskin, and M. M. Sheikh-Jabbari, Phys. Rev. Lett. **96**, 081301 (2006).
[12] J. Garriga and V. F. Mukhanov, Phys. Lett. B **458**, 219 (1999).
[13] J. Maldacena, J. High Energy Phys. 05 (2003) 013.
[14] N. Arkani-Hamed, P. Creminelli, S. Mukohyama, and M. Zaldarriaga, J. Cosmol. Astropart. Phys. 04 (2001) 001; N. Arkani-Hamed, H-C Cheng, M. A. Luty, and S. Mukohyama, J. High Energy Phys. 05 (2004) 074; S. Mukohyama, J. Cosmol. Astropart. Phys. 10 (2006) 011.

PHYSICAL REVIEW D **81,** 083535 (2010)

Asymptotically safe inflation

Steven Weinberg*

Theory Group, Department of Physics, University of Texas, Austin, Texas 78712, USA

(Received 8 December 2009; published 29 April 2010)

Inflation is studied in the context of asymptotically safe theories of gravitation. Conditions are explored under which it is possible to have a long period of nearly exponential expansion that eventually comes to an end.

DOI: 10.1103/PhysRevD.81.083535 PACS numbers: 04.60.−m, 04.50.Kd, 11.10.Hi, 98.80.Cq

I. INTRODUCTION

Decades ago it was suggested that the effective quantum field theory of gravitation and matter might be asymptotically safe [1], and hence ultraviolet complete. That is, the renormalization group flows might have a fixed point, with a finite-dimensional ultraviolet critical surface of trajectories attracted to the fixed point at short distances. Evidence for a fixed point in the quantum theory of gravitation with or without matter has gradually accumulated through the use of dimensional continuation [2], the large N approximation [3] (where N is the number of matter fields), lattice methods [4], the truncated exact renormalization group [5], and a version of perturbation theory [6]. Recently there has also been evidence that the ultraviolet critical surface is finite dimensional; it has been found that even in truncations of the exact renormalization group equations with more than three (and up to nine) independent coupling parameters, the ultraviolet critical surface is just three dimensional [7]. The condition that physical parameters lie on the ultraviolet critical surface is analogous to the condition of renormalizability in the standard model, and like that condition yields a theory with a finite number of free parameters.

The natural arena for applications of the idea of asymptotic safety is the physics of very short distances, and, in particular, the early universe [8]. In Sec. II we show how to formulate the differential equations for the scale factor in a Robertson-Walker solution of the classical field equations in a completely general generally covariant theory of gravitation. In Sec. III we apply this result to calculate the expansion rate $\bar{H}$ for a de Sitter solution of the classical field equations. We are interested here in solutions for which $\bar{H}$ is of the same order as the scale at which the couplings are beginning to approach their fixed point, or larger. In this case, $\bar{H}$ turns out in the tree approximation to depend strongly on the ultraviolet cutoff, indicating a breakdown of the classical approximation. We deal with this by choosing an optimal cutoff, which minimizes the quantum corrections to the classical field equations. Section IV considers more general time-dependent Robertson-Walker solutions of the classical field equations with an optimal cutoff, and explores the circumstances under which it is possible to have an exponential expansion that persists for a long time but eventually comes to an end. An illustrative example is worked out in Sec. V.

We will work with a completely general generally covariant theory of gravitation. (For simplicity matter will be ignored here.) The effective action with an ultraviolet cutoff Λ takes the form [9]

$$I_\Lambda[g] = -\int d^4x\sqrt{-\mathrm{Det}g}[\Lambda^4 g_0(\Lambda) + \Lambda^2 g_1(\Lambda)R + g_{2a}(\Lambda)R^2 + g_{2b}(\Lambda)R^{\mu\nu}R_{\mu\nu} + \Lambda^{-2}g_{3a}(\Lambda)R^3 + \Lambda^{-2}g_{3b}(\Lambda)RR^{\mu\nu}R_{\mu\nu} + \ldots]. \quad (1)$$

Here we have extracted powers of Λ from the conventional coupling constants, to make the coupling parameters $g_n(\Lambda)$ dimensionless. Because they are dimensionless, these running couplings satisfy renormalization group equations of the form

$$\Lambda\frac{d}{d\Lambda}g_n(\Lambda) = \beta_n(g(\Lambda)). \quad (2)$$

The condition for a fixed point at $g_n = g_{n*}$ is that $\beta_n(g_*) = 0$ for all n. As is well known, the condition for the couplings to be attracted to a fixed point g_{n*} as $\Lambda \to \infty$ can be seen by considering the behavior of $g_n(\Lambda)$ when it is near g_{n*}. In the case where $\beta_n(g)$ is analytic in a neighborhood of g_{n*}, near this fixed point we have

$$\beta_n(g) \to \sum_m B_{nm}(g_m - g_{*m}), \qquad B_{nm} \equiv \left(\frac{\partial\beta_n(g)}{\partial g_m}\right)_*. \quad (3)$$

The solution of Eq. (2) in this neighborhood is

$$g_n(\Lambda) \to g_{*n} + \sum_N u_n^N\left(\frac{\Lambda}{M}\right)^{\lambda_N}, \quad (4)$$

where u^N and λ_N are eigenvectors and corresponding eigenvalues of the matrix B_{nm}:

$$\sum_m B_{nm}u_m^N = \lambda_N u_n^N. \quad (5)$$

It is a physical requirement that the only eigenvectors that are allowed to appear in the sum in Eq. (4) are those for which the real part of the corresponding eigenvalues are

*weinberg@physics.utexas.edu

negative, so that the couplings actually do approach the fixed point. The normalizations of the eigenvectors that do appear in Eq. (4) are free physical parameters, the only free parameters of the theory, except that we can adjust the overall normalization of all the eigenvectors as we like by a suitable choice of the arbitrary mass scale M. If we choose M to make the largest of the u_n^N of order unity, then M is the cutoff scale at which couplings are just beginning to approach their fixed point.

Aside from the illustrative example considered in Sec. V, we will not carry our discussion in this paper to the point of performing numerical calculations, which of course would require some truncation of the series of terms in the action (1). Our purpose here is to lay out the general outlines of such a calculation, for which purpose we do not need to adopt any specific truncation. Our results are worked out in detail for the terms explicitly shown in Eq. (1), but this is only for the purposes of illustration; nothing in this paper assumes the neglect of higher terms. For our purposes here, it makes no difference whether Λ is regarded as a sharp ultraviolet cutoff on loop diagrams to be calculated using the action (1), or as a momentum parameter (usually called k) in a regulator term added to the action, or a sliding renormalization scale.

II. ROBERTSON-WALKER SOLUTIONS

In this section we consider how to find a solution of the classical gravitational field equations for the general action (1), of the flat-space Robertson-Walker form

$$d\tau^2 = dt^2 - a^2(t)d\vec{x}^2. \tag{6}$$

It would be very complicated to derive the ten classical field equations for a general metric that follow from an action like (1), and then specialize to the case of a Robertson-Walker metric. Instead, we can much more easily exploit the symmetries of this metric to derive a *single* differential equation for the Hubble rate $H(t) \equiv \dot{a}(t)/a(t)$. In showing how to derive this differential equation, we will be quite general, not making any use in this section of the assumption of asymptotic safety.

We can use the rotational and translational symmetries of the line element (6) to write the components of the variational derivatives $\delta I_\Lambda/\delta g_{\mu\nu}$ in the form

$$\left[\frac{\delta I_\Lambda[g]}{\delta g_{ij}(x)}\right]_{\rm RW} = \frac{\Lambda^4}{6}\delta_{ij}a^{-2}(t)\mathcal{M}_\Lambda(t), \tag{7}$$

$$\left[\frac{\delta I_\Lambda[g]}{\delta g_{i0}(x)}\right]_{\rm RW} = 0, \tag{8}$$

$$\left[\frac{\delta I_\Lambda[g]}{\delta g_{00}(x)}\right]_{\rm RW} = -\frac{\Lambda^4}{2}\mathcal{N}_\Lambda(t), \tag{9}$$

with the subscript RW indicating that, after taking the variational derivative, the metric is to be set equal to the Robertson-Walker metric defined by (6). (The factors $\Lambda^4/6a^2$ and $\Lambda^4/2$ are inserted in the definitions of $\mathcal{M}_\Lambda$ and $\mathcal{N}_\Lambda$ for future convenience.) Also, the general covariance of the action yields the generalized Bianchi identity

$$0 = \left[\frac{\delta I_\Lambda[g]}{\delta g_{\mu\nu}(x)}\right]_{;\nu}. \tag{10}$$

By using Eqs. (7)–(9) for the Robertson-Walker metric, Eq. (10) is reduced to the condition

$$a^2\dot{a}\mathcal{M}_\Lambda = \frac{d}{dt}(a^3\mathcal{N}_\Lambda). \tag{11}$$

Therefore the gravitational field equations reduce here to a single differential equation,

$$\mathcal{N}_\Lambda(t) = 0, \tag{12}$$

which we see ensures the vanishing of all variational derivatives $\delta I_\Lambda[g]/\delta g_{\mu\nu}$. This result (which holds also in the presence of spatial curvature and matter) is the generalization of the familiar Friedmann equation, which would apply if only the Einstein-Hilbert term $-\sqrt{g}R/16\pi G$ and a vacuum energy term were included in the gravitational action.

We can express $\mathcal{M}_\Lambda$ and then $\mathcal{N}_\Lambda$ in terms of variational derivatives of the action for the Robertson-Walker metric with respect to the scale factor $a(t)$. Because $a(t)$ appears in the Robertson-Walker metric only as a factor $a^2(t)$ in $g_{ij}(\mathbf{x}, t)$, we have

$$\frac{\delta I_\Lambda[g_{\rm RW}]}{\delta a(t)} = \int d^3x 2a(t)\delta_{ij} \times a^3(t)\left[\frac{\delta I[g]}{\delta g_{ij}(\mathbf{x}, t)}\right]_{\rm RW} = V\Lambda^4\mathcal{M}_\Lambda(t)a^2(t), \tag{13}$$

where V is the coordinate space volume (which can be made finite by imposing periodic boundary conditions.) For the flat-space Robertson-Walker metric $(g_{\rm RW})_{\mu\nu}$, the action takes the general form

$$I_\Lambda[g_{\rm RW}] = V\Lambda^4\int dta^3(t)\mathcal{I}_\Lambda(H(t), \dot{H}(t), \ldots), \tag{14}$$

where as usual $H(t) \equiv \dot{a}(t)/a(t)$. Here and in Eqs. (15)–(17) below, the ellipsis indicates a possible dependence of $\mathcal{I}_\Lambda$ on second and higher derivatives of $H(t)$. (Second and higher time derivatives do not occur in $\mathcal{I}_\Lambda$ if the integrand of the action is $\sqrt{-{\rm Det}g}$ times an arbitrary scalar function of the Riemann-Christoffel curvature tensor $R_{\mu\nu\rho\sigma}$, including of course an arbitrary dependence on the curvature scalar and the Ricci tensor, but we do not assume that this is the case.) Comparing Eq. (13) with the result of a straightforward calculation of the variational derivative of the action (14) with respect to $a(t)$ gives

$$\mathcal{M}_\Lambda = 3I_\Lambda - 3H\frac{\partial I_\Lambda}{\partial H} + (3\dot{H} + 9H^2)\frac{\partial I_\Lambda}{\partial \dot{H}} - \frac{d}{dt}\left(\frac{\partial I_\Lambda}{\partial H}\right) + 6H\frac{d}{dt}\left(\frac{\partial I_\Lambda}{\partial \dot{H}}\right) + \frac{d^2}{dt^2}\left(\frac{\partial I_\Lambda}{\partial \dot{H}}\right) + \ldots \tag{15}$$

We note that $a^2\dot{a}\mathcal{M}_\Lambda$ is a time-derivative

$$a^2\dot{a}\mathcal{M}_\Lambda = \frac{d}{dt}\left\{a^3\left[I_\Lambda - H\frac{\partial I_\Lambda}{\partial H} + (-\dot{H} + 3H^2)\frac{\partial I_\Lambda}{\partial \dot{H}} + H\frac{d}{dt}\left(\frac{\partial I_\Lambda}{\partial \dot{H}}\right) + \ldots\right]\right\}. \tag{16}$$

Comparing with Eq. (11), we see that $\mathcal{N}_\Lambda$ equals the term in square brackets in (16), up to a possible term equal to a constant divided by $a^3(t)$. But the term in square brackets is independent of the scale of $a(t)$, as is $\mathcal{N}_\Lambda(t)$, so there can be no term in their difference proportional to $1/a^3(t)$, and thus

$$\mathcal{N}_\Lambda = I_\Lambda - H\frac{\partial I_\Lambda}{\partial H} + (-\dot{H} + 3H^2)\frac{\partial I_\Lambda}{\partial \dot{H}} + H\frac{d}{dt}\left(\frac{\partial I_\Lambda}{\partial \dot{H}}\right) + \ldots \tag{17}$$

The ten classical field equations reduce for the flat-space Robertson-Walker metric to the single requirement that this vanishes.

To evaluate the terms in the action for the Robertson-Walker metric with no spatial curvature that are explicitly shown in Eq. (1), we note that for this metric $R = -12H^2 - 6\dot{H}$ and $R_{\mu\nu}R^{\mu\nu} = 36H^4 + 36H^2\dot{H} + 12\dot{H}^2$. Using these in Eq. (1) and comparing with Eq. (14) gives

$$I_\Lambda = -g_0(\Lambda) + \Lambda^{-2}g_1(\Lambda)(12H^2 + 6\dot{H}) - \Lambda^{-4}g_{2a}(\Lambda)(12H^2 + 6\dot{H})^2 - \Lambda^{-4}g_{2b}(\Lambda)(36H^4 + 36H^2\dot{H} + 12\dot{H}^2) + \Lambda^{-6}g_{3a}(\Lambda)(12H^2 + 6\dot{H})^3 + \Lambda^{-6}g_{3b}(\Lambda)(12H^2 + 6\dot{H})(36H^4 + 36H^2\dot{H} + 12\dot{H}^2) + \ldots, \tag{18}$$

where now the dots denote contributions from terms not shown in (1), some of which involve second and higher derivatives of H. From Eq. (17), we then have

$$\begin{aligned}\mathcal{N}_\Lambda(H, \dot{H}, \ddot{H}, \ldots) = {} & -g_0(\Lambda) + 6\Lambda^{-2}g_1(\Lambda)H^2 - \Lambda^{-4}g_{2a}(\Lambda)(216H^2\dot{H} - 36\dot{H}^2 + 72H\ddot{H}) \\ & - \Lambda^{-4}g_{2b}(\Lambda)(72H^2\dot{H} - 12\dot{H}^2 + 24H\ddot{H}) + \Lambda^{-6}g_{3a}(\Lambda)(-864H^6 + 7776H^4\dot{H} + 3240H^2\dot{H}^2 \\ & - 432\dot{H}^3 + 216H\ddot{H}(12H^2 + 6\dot{H})) + \Lambda^{-6}g_{3b}(\Lambda)(-216H^6 + 2160H^4\dot{H} + 1008H^2\dot{H}^2 - 144\dot{H}^3 \\ & + H\ddot{H}(720H^2 + 432\dot{H})) + \ldots\end{aligned} \tag{19}$$

This is the quantity that must be set equal to zero in finding a flat-space Robertson-Walker solution of the classical gravitational field equations.

III. DE SITTER SOLUTIONS AND OPTIMAL CUTOFF

We can now easily find the condition for a de Sitter solution of the classical field equations, with

$$a(t) \propto e^{\bar{H}t}, \tag{20}$$

where $\bar{H}$ is constant. Setting the quantity (19) equal to zero for $H(t) = \bar{H}$ gives our condition on $\bar{H}$[1]:

$$\begin{aligned}0 = N_\Lambda(\bar{H}) \equiv {} & \mathcal{N}_\Lambda(\bar{H}, 0, 0, \ldots) \\ = {} & -g_0(\Lambda) + 6g_1(\Lambda)(\bar{H}/\Lambda)^2 - 864g_{3a}(\Lambda)(\bar{H}/\Lambda)^6 \\ & - 216g_{3b}(\Lambda)(\bar{H}/\Lambda)^6 + \ldots\end{aligned} \tag{21}$$

It is easy to find solutions of Eq. (21) that have small values of $\bar{H}$, very much smaller than the scale M at which the couplings begin to approach their fixed points. For sufficiently small $\bar{H}$, we can take Λ to be much larger than $\bar{H}$, and yet small enough so that the couplings appearing as coefficients in (1) become independent of Λ, and in particular

$$\Lambda^4 g_0(\Lambda) \to \rho_V, \qquad \Lambda^2 g_1(\Lambda) \to 1/16\pi G_N,$$

where ρ_V and G_N are the conventional, Λ-independent, vacuum energy and Newton constant. Then (21) has the familiar Λ-independent solution

$$\bar{H}^2 = \frac{8\pi G_N \rho_V}{3}.$$

Because of the still mysterious fact that ρ_V is observed to be much less than G^{-2}, this value of $\bar{H}$ is much less than

[1]Note that this is not the result that would be obtained by setting the derivative of $I_\Lambda(\bar{H}, 0, 0, \ldots)$ with respect to $\bar{H}$ equal to zero. For a de Sitter metric with $a(t) = \exp(\bar{H}t)$, the integral over t in the action $I_\Lambda[g]$ diverges at $t = \infty$. If we integrate only from $t = -\infty$ to $t = 0$, the integral $\int dt a^3(t)$ gives a factor $1/3\bar{H}$, but the derivative of $I_\Lambda(\bar{H}, 0, 0, \ldots)/3\bar{H}$ with respect to $\bar{H}$ is not zero; it equals a surface term $(\partial I_\Lambda/\partial \dot{H})_{\bar{H}}$, which again gives Eq. (21).

$G^{-1/2}$, and so radiative corrections and higher terms in (21) can be neglected.

We will instead be interested here in looking for solutions for which $\bar{H}$ is roughly of the order of the scale M at which the couplings begin to approach their fixed points, or larger. In this case, we face a difficult choice: How should we choose Λ? On one hand, if we choose $\Lambda \ll \bar{H}$, then we can expect radiative corrections to the classical result (21) to be unimportant, because $\bar{H}$ provides a natural infrared cutoff in loop diagrams constructed using the action (1). But for $\Lambda \ll \bar{H}$, the sum (21) receives increasing contributions as we include higher and higher terms, and whether or not the series actually converges, it is not useful. On the other hand, if we choose $\Lambda \gg \bar{H}$, then it is reasonable to suppose that the series (21) is dominated by its lowest terms, but for $\Lambda \gg \bar{H}$ there is no reason to suppose that we can neglect radiative corrections to the field equations. Indeed, we can see that radiative corrections to the field equations *are* important here, because where Eq. (21) is dominated by its lowest terms, it gives $\bar{H}$ a strong dependence on Λ. [This is clearest in the case where Λ is so large that the couplings are near their fixed points, in which case (21) gives $\bar{H}$ proportional to Λ.] The whole point of the renormalization group equations (2) is that physical quantities like $\bar{H}$ should be independent of the cutoff, but in general this is true only when radiative corrections are included, and since Eq. (21) gives $\bar{H}$ a strong dependence on Λ when $\Lambda \gg \bar{H}$, radiative corrections evidently can not be neglected.

Ideally, we should leave Λ undetermined, and calculate enough of the radiative corrections to the field equations so that $\bar{H}$ comes out at least approximately independent of Λ. This would not be easy. Instead, we can try to make a judicious choice of Λ to minimize the radiative corrections. We can guess that the optimal Λ is roughly of the order of $\bar{H}$, where radiative corrections are just beginning to be important, and the higher terms in (21) are just beginning to be less important. This sort of guess works quite well in quantum chromodynamics. The radiative corrections to a process like e^+-e^- annihilation into jets of hadrons at an energy E are accompanied with powers of $\ln(E/\Lambda)$, and to avoid large radiative corrections it is only necessary to take $\Lambda \approx E$. In this way, we can use the tree approximation to calculate the annihilation into, say, three jets, with the renormalization scale of the QCD coupling taken of order E. But in our case, radiative corrections are more sensitive to Λ, and we have to make a more careful choice of Λ.

To find an optimal cutoff, we note that in principle we should find $\bar{H}$ by solving the full quantum corrected field equations, which give a result that can be schematically written as

$$\bar{H}_{\text{true}} = \bar{H}(\Lambda) + \Delta\bar{H}(\Lambda), \tag{22}$$

where $\bar{H}(\Lambda)$ is defined as the solution of Eq. (21), and $\Delta\bar{H}(\Lambda)$ represents the effect of radiative corrections. Instead of calculating loop graphs, we can get some idea of the results of such a calculation by using the tree-approximation field equations (21), but with Λ chosen at a local minimum of the radiative corrections to $\bar{H}$. For such an optimal Λ, we have[2]

$$\frac{\partial}{\partial\Lambda}\Delta\bar{H}(\Lambda) = 0. \tag{23}$$

As already mentioned, physical quantities, including the true expansion rate $\bar{H}_{\text{true}}$, must be independent of Λ, so Eq. (23) tells us also that the expansion rate calculated from the classical field equations is stationary at the optimal cutoff

$$0 = \Lambda\frac{\partial}{\partial\Lambda}\bar{H}(\Lambda). \tag{24}$$

By definition, for any Λ we have $N_\Lambda(\bar{H}(\Lambda)) = 0$, and by differentiating this with respect to Λ and using Eq. (24) we find that the condition for an optimal cutoff may be put in the form

$$0 = \Lambda\frac{\partial}{\partial\Lambda}N_\Lambda(\bar{H})|_{\bar{H}=\bar{H}(\Lambda)} = A_\Lambda(\bar{H}(\Lambda)) + B_\Lambda((\bar{H}(\Lambda)), \tag{25}$$

where A_Λ arises from the explicit dependence of $N_\Lambda(\bar{H})$ on $\bar{H}/\Lambda$:

$$\begin{aligned} A_\Lambda(\bar{H}) &\equiv -\bar{H}\frac{\partial}{\partial\bar{H}}N_\Lambda(\bar{H}) \\ &= -12\left(\frac{\bar{H}}{\Lambda}\right)^2 g_1(\Lambda) + 5184\left(\frac{\bar{H}}{\Lambda}\right)^6 g_{3a}(\Lambda) \\ &\quad + 1296\left(\frac{\bar{H}}{\Lambda}\right)^6 g_{3b}(\Lambda) + \dots, \end{aligned} \tag{26}$$

and B_Λ comes from the running of the couplings in N_Λ:

$$\begin{aligned} B_\Lambda(\bar{H}) \equiv &-\beta_0(g(\Lambda)) + 6\beta_1(g(\Lambda))(\bar{H}/\Lambda)^2 \\ &- 864\beta_{3a}(g(\Lambda))(\bar{H}/\Lambda)^6 - 216\beta_{3b}(g(\Lambda)) \\ &\times (\bar{H}/\Lambda)^6 + \dots \end{aligned} \tag{27}$$

We now have two equations, (21) and (25), for the two quantities $\bar{H}$ and Λ, so it is not unreasonable to expect there to be one or more solutions, with both Λ and $\bar{H}$ roughly of order M, the only mass parameter in the theory.

IV. TIME DEPENDENCE

The de Sitter solution found in Sec. II describes a universe that inflates eternally. For a more realistic picture of

[2]This is the weakest point in our discussion. For one thing, we do not know whether the condition (23) gives a local minimum or maximum of the radiative corrections. Worse, even if the radiative corrections are minimized, we do not know that they are small.

inflation, we need a solution that remains close to the de Sitter solution with expansion rate near $\bar{H}$ for a time much longer than $1/\bar{H}$, but that gradually evolves away from the de Sitter solution, so that inflation can come to an end. (We have nothing to say here about the metric before the universe enters into its de Sitter phase.) To find such a solution, we will consider first-order perturbations of the de Sitter solution, of the Robertson-Walker form (6). The expansion rate will take the form

$$H(t) = \bar{H} + \delta H(t), \tag{28}$$

with $|\delta H(t)| \ll \bar{H}$. Keeping only terms in (19) of first order in $\delta H(t)$, the field equation $\mathcal{N}_\Lambda = 0$ becomes

$$c_0(\bar{H}, \Lambda)\frac{\delta H}{\bar{H}} + c_1(\bar{H}, \Lambda)\frac{\delta \dot{H}}{\bar{H}^2} + c_2(\bar{H}, \Lambda)\frac{\delta \ddot{H}}{\bar{H}^3} + \ldots = 0, \tag{29}$$

where

$$c_0(\bar{H}, \Lambda) \equiv \bar{H}\left(\frac{\partial \mathcal{N}_\Lambda}{\partial H}\right)_{\bar{H}} = -A_\Lambda(\bar{H}), \tag{30}$$

with A_Λ given by Eq. (26), and

$$\begin{aligned} c_1(\bar{H}, \Lambda) &\equiv \bar{H}^2\left(\frac{\partial \mathcal{N}_\Lambda}{\partial \dot{H}}\right)_{\bar{H}} \\ &= -216 g_{2a}(\Lambda)\left(\frac{\bar{H}}{\Lambda}\right)^4 - 72 g_{2b}(\Lambda)\left(\frac{\bar{H}}{\Lambda}\right)^4 \\ &\quad + 7776 g_{3a}(\Lambda)\left(\frac{\bar{H}}{\Lambda}\right)^6 + 2160 g_{3b}(\Lambda)\left(\frac{\bar{H}}{\Lambda}\right)^6 \\ &\quad + \ldots, \end{aligned} \tag{31}$$

$$\begin{aligned} c_2(\bar{H}, \Lambda) &\equiv \bar{H}^3\left(\frac{\partial \mathcal{N}_\Lambda}{\partial \ddot{H}}\right)_{\bar{H}} \\ &= -72 g_{2a}(\Lambda)\left(\frac{\bar{H}}{\Lambda}\right)^4 - 24 g_{2b}(\Lambda)\left(\frac{\bar{H}}{\Lambda}\right)^4 \\ &\quad + 2592 g_{3a}(\Lambda)\left(\frac{\bar{H}}{\Lambda}\right)^6 + 720 g_{3b}(\Lambda)\left(\frac{\bar{H}}{\Lambda}\right)^6 \\ &\quad + \ldots, \end{aligned} \tag{32}$$

and so on, with the subscript $\bar{H}$ on partial derivatives meaning that after taking the derivatives we set $H(t) = \bar{H}$. Equation (29) has an obvious solution of the form

$$\delta H \propto \exp(\xi \bar{H} t), \tag{33}$$

where ξ is any root of the equation

$$c_0(\bar{H}, \Lambda) + c_1(\bar{H}, \Lambda)\xi + c_2(\bar{H}, \Lambda)\xi^2 + \ldots = 0. \tag{34}$$

(This is a quadratic equation in the special case in which the integrand of the action is $\sqrt{-\mathrm{Det}g}$ times an arbitrary function of the curvature tensor.) For positive $\mathrm{Re}\xi$, Eq. (33) represents an instability, and the number of e foldings before this instability ends the exponential expansion is $\approx 1/\mathrm{Re}\xi$.

We would generally expect the coefficients in Eq. (34) to be of the same order, in which case typical solutions for ξ would be of order unity, and inflation would either end almost immediately (if $\mathrm{Re}\xi > 0$) or go on forever (if $\mathrm{Re}\xi \leq 0$). But there are various circumstances under which we expect ξ to be much smaller, giving a large number of e foldings before the end of inflation.[3]

(1) If $|c_0|$ is much less than all the other $|c_n|$, then Eq. (34) will have a solution $\xi \simeq -c_0/c_1$, and so much less than unity. In particular, if we now choose Λ to be the optimal cutoff described in the previous section, then we can use the condition (25) and Eq. (30) to write

$$c_0(\bar{H}, \Lambda) = B_\Lambda(\bar{H}). \tag{35}$$

According to Eq. (27), $B_\Lambda(H)$ vanishes if the couplings are at their fixed point, so we can conclude that it is possible to have a long but not eternal period of inflation if the optimal Λ is large enough so that the couplings $g_n(\Lambda)$ are not far from their fixed point. But there is a limit to how close the couplings at the optimum cutoff can be to their fixed point. At the fixed point, the quantities (21) and (25) are both functions of the single parameter $\bar{H}/\Lambda$, and it is not likely that these two functions would vanish at the same value of this parameter.

(2) If the couplings are not very near their fixed point, they are sensitive to the free parameters of the theory that characterize the particular trajectory in coupling-constant space on which the couplings lie, and it is easy to choose these couplings to make $|c_0|$ as small as we like. For instance, where (4) applies, all the couplings are linear in the normalization of the eigenvectors u_n^N, the only free parameters of the theory. In a theory of chaotic inflation, the value of these parameters in any big bang containing observers may be conditioned by the requirement that c_0 should be small enough (and have the right sign) to allow the bang to become big. To be specific, in order for spatial curvature not to interfere with the formation of galaxies it is necessary that the universe should expand enough during inflation so that whatever curvature was present at the beginning of inflation would be decreased enough so that the curvature term in the Friedmann equation should not dominate over the matter term when galaxies form [10]. As is well known, the fact that spatial curvature does not dominate at present requires about 60 to 70 e foldings of inflation [11], and the anthropic requirement that curvature does not interfere with galaxy formation is almost as restrictive.

[3]We are concentrating here on only one mode. In all cases Eq. (34) will have more than one solution, and we are assuming that all modes other than the one (or several) with $\mathrm{Re}\xi$ small and positive either have $\mathrm{Re}\xi \leq 0$ or for some reason are not excited.

But the combination of data from the microwave background, baryon acoustic oscillations, and type Ia supernovae distance-redshift relations has shown [12] that (within 2 standard deviations) the fractional curvature contribution Ω_K to H_0^2 is in the range of -0.0178 to $+0.0066$. It is hard to see any anthropic reason for a number of e foldings large enough to reduce the curvature this much.

(3) Instead of c_0 being anomalously small, it is possible for some or all of the other c_n to be anomalously large, in which case again ξ will be small and the number of e foldings will be large. For instance, we note that c_0 unlike the other c_n does not involve the couplings g_{2a} and g_{2b}, so if these couplings are anomalously large, as in Ref. [6], then c_1, c_2, etc., will be much larger than c_0, and again we will have $|\xi| \simeq |c_0/c_1| \ll 1$.

V. AN EXAMPLE

We will now apply the above results to a classic example of higher derivative theories of gravitation, with action limited to terms with no more than four spacetime derivatives:

$$I_\Lambda[g] = -\int d^4x\sqrt{-\mathrm{Det}g}[\Lambda^4 g_0(\Lambda) + \Lambda^2 g_1(\Lambda)R + g_{2a}(\Lambda)R^2 + g_{2b}(\Lambda)R^{\mu\nu}R_{\mu\nu}]. \tag{36}$$

This theory was studied by Stelle [9] as a possible renormalizable quantum theory of gravitation, and has been considered recently by Niedermaier [6] and by Benedetti *et al.* [13] in connection with asymptotic safety. As is well known, it is possible by using the Gauss-Bonnet identity to put this action in the form used in Refs. [6,13]:

$$I_\Lambda[g] = -\int d^4x\sqrt{-\mathrm{Det}g}[\Lambda^4 g_0(\Lambda) + \Lambda^2 g_1(\Lambda)R + f_a(\Lambda)R^2 + f_b(\Lambda)C^{\mu\nu\rho\sigma}R_{\mu\nu\rho\sigma}], \tag{37}$$

where $C_{\mu\nu\rho\sigma}$ is the Weyl tensor, and

$$f_a = g_{2a} + \frac{g_{2b}}{3}, \qquad f_b = \frac{g_{2b}}{2}. \tag{38}$$

For this action, Eq. (21) gives the expansion rate for a de Sitter solution of the field equations as

$$\bar{H} = \Lambda\sqrt{g_0(\Lambda)/6g_1(\Lambda)}. \tag{39}$$

Instead of trying to find an optimal value of Λ, which minimizes radiative corrections to Eq. (39), here we will simply assume that Λ is large enough so that the couplings $g_n(\Lambda)$ are near their fixed point g_{n*}, and use Eq. (39) to express Λ in terms of $\bar{H}$:

$$\Lambda = \bar{H}\sqrt{6g_{1*}/g_{0*}}, \tag{40}$$

with $\bar{H}$ left undetermined.

The critical question for this sort of theory is whether the de Sitter solution has an instability that ends the exponential expansion after a finite but large number of e foldings. As we have seen, for any small perturbation of the de Sitter solution, $\dot{a}/a$ is a sum of terms with the time dependence $\exp(\xi\bar{H}t)$, with ξ running over the roots of Eq. (34). We are now considering an action whose integrand is $\sqrt{-\mathrm{Det}g}$ times a scalar function of the metric and the Riemann-Christoffel curvature tensor, so as remarked in the previous section, this equation is quadratic:

$$c_0 + c_1\xi + c_2\xi^2 = 0. \tag{41}$$

For the particular action (36), the coefficients are given by

$$c_0 = 12g_{1*}(\bar{H}/\Lambda)^2 = 2g_{0*}, \tag{42}$$

$$c_1 = 3c_2 = (-216g_{2a*} - 72g_{2b*})(\bar{H}/\Lambda)^4 = (-6g_{2a*} - 2g_{2b*})g_{0*}^2/g_{1*}^2, \tag{43}$$

so Eq. (41) reads

$$\xi^2 + 3\xi = A, \tag{44}$$

where

$$A = -\frac{c_0}{c_2} = \frac{3g_{1*}^2}{g_{0*}(3g_{2a*} + g_{2b*})}. \tag{45}$$

We get a realistic picture of inflation if it turns out that A is small and positive. In this case Eq. (44) has a root with $\xi \simeq -3$, corresponding to a perturbation to $\dot{a}/a$ that decays as $\exp(-3\bar{H}t)$, and a root with $\xi \simeq A/3$, corresponding to a slowly growing perturbation, that ends the exponential phase after about $3/A$ e foldings.

Unfortunately, the numerical results obtained in Refs. [6,13] are not encouraging. The calculations of Ref. [6] are expressed in terms of coupling constants λ, g_N, ω, and s, related to the couplings in Eq. (36) by

$$g_0 = 2\lambda/g_N, \qquad g_1 = 1/g_N, \qquad g_{2a} = -(1+\omega)/3s, \qquad g_{2b} = 1/s. \tag{46}$$

Using a version of perturbation theory, Ref. [6] found that for $\Lambda \to \infty$ the parameters ω, λ, and g_N approach the fixed point values

$$\omega_* = -0.0228, \qquad \lambda_* = 12.69, \qquad g_{N*}/(4\pi)^2 = 0.4227, \tag{47}$$

while $s(\Lambda)$ vanishes as

$$s(\Lambda) \to 11.88/\ln(\Lambda/M), \tag{48}$$

where M is some unknown large mass. Then Eq. (45) gives

$$A = -\frac{3s}{2\omega\lambda g_N} \to \frac{0.92}{\ln(\Lambda/M)}, \tag{49}$$

so A is positive, but Λ/M would have to be about 10^8 to give 60 e foldings before inflation ends.

In Ref. [13], by using the truncated exact renormalization group equations, a fixed point is found with (in our notation)

$$g_{0*} = -0.0042, \qquad g_{1*} = -0.0101, \qquad g_{2a*} = -0.0109, \qquad g_{2b*} = 0.01. \tag{50}$$

Using these results in Eq. (45) gives $A = 3.05$. This is positive, but unfortunately not at all small. The two roots of Eq. (44) are $\xi = -3.80$, corresponding to a rapidly decaying mode, and $\xi = 0.80$, corresponding to an instability that ends inflation after only a few e foldings.

ACKNOWLEDGMENTS

I am grateful for discussions with D. Benedetti, W. Fischler, E. Komatsu, M. Niedermaier, and M. Reuter. This material is based in part on work supported by the National Science Foundation under Grant No. PHY-0455649 and with support from The Robert A. Welch Foundation, Grant No. F-0014.

[1] S. Weinberg, in *Understanding the Fundamental Constituents of Matter*, edited by A. Zichichi (Plenum Press, New York, 1977).

[2] S. Weinberg, in *General Relativity*, edited by S. W. Hawking and W. Israel (Cambridge University Press, Cambridge, England, 1979), p. 790; H. Kawai, Y. Kitazawa, and M. Ninomiya, Nucl. Phys. **B404**, 684 (1993); **B467**, 313 (1996); T. Aida and Y. Kitazawa, Nucl. Phys. **B491**, 427 (1997); M. Niedermaier, Nucl. Phys. **B673**, 131 (2003).

[3] L. Smolin, Nucl. Phys. **B208**, 439 (1982); R. Percacci, Phys. Rev. D **73**, 041501 (2006).

[4] J. Ambjørn, J. Jurkewicz, and R. Loll, Phys. Rev. Lett. **93**, 131301 (2004); **95**, 171301 (2005); Phys. Rev. D **72**, 064014 (2005); **78**, 063544 (2008); in *Approaches to Quantum Gravity*, edited by D. Oríti (Cambridge University Press, Cambridge, England, 2009).

[5] M. Reuter, Phys. Rev. D **57**, 971 (1998); M. Reuter, arXiv: hep-th/9605030; D. Dou and R. Percacci, Classical Quantum Gravity **15**, 3449 (1998); W. Souma, Prog. Theor. Phys. **102**, 181 (1999); O. Lauscher and M. Reuter, Phys. Rev. D **65**, 025013 (2001); Classical Quantum Gravity **19**, 483 (2002); M. Reuter and F. Saueressig, Phys. Rev. D **65**, 065016 (2002); O. Lauscher and M. Reuter, Int. J. Mod. Phys. A **17**, 993 (2002); Phys. Rev. D **66**, 025026 (2002); M. Reuter and F. Saueressig, Phys. Rev. D **66**, 125001 (2002); R. Percacci and D. Perini, Phys. Rev. D **67**, 081503 (2003); **68**, 044018 (2003); D. Perini, Nucl. Phys. B, Proc. Suppl. **127**, 185 (2004); D. F. Litim, Phys. Rev. Lett. **92**, 201301 (2004); A. Codello and R. Percacci, Phys. Rev. Lett. **97**, 221301 (2006); A. Codello, R. Percacci, and C. Rahmede, Int. J. Mod. Phys. A **23**, 143 (2008); M. Reuter and F. Saueressig, arXiv:0708.1317; P. F. Machado and F. Saueressig, Phys. Rev. D **77**, 124045 (2008); A. Codello, R. Percacci, and C. Rahmede, Ann. Phys. (N.Y.) **324**, 414 (2009); A. Codello and R. Percacci, Phys. Lett. B **672**, 280 (2009); D. F. Litim, Proc. Sci., QG-Ph (**2008**) 024 [arXiv:0810.3675]; H. Gies and M. M. Scherer, arXiv:0901.2459; D. Benedetti, P. F. Machado, and F. Saueressig, arXiv:0901.2984; arXiv:0902.4630; M. Reuter and H. Weyer, Gen. Relativ. Gravit. **41**, 983 (2009); for a review, see M. Reuter and P. Saueressig, arXiv:0708.1317.

[6] M. R. Niedermaier, Phys. Rev. Lett. **103**, 101303 (2009).

[7] A. Codello, R. Percacci, and C. Rahmede, Int. J. Mod. Phys. A **23**, 143 (2008); Ann. Phys. (N.Y.) **324**, 414 (2009); D. Benedetti, P. F. Machado, and F. Saueressig, Mod. Phys. Lett. A **24**, 2233 (2009); Nucl. Phys. **B824**, 168 (2010).

[8] The implications of asymptotic safety for cosmology have been considered by A. Bonanno and M. Reuter, Phys. Rev. D **65**, 043508 (2002); Phys. Lett. B **527**, 9 (2002); M. Reuter and F. Saueressig, J. Cosmol. Astropart. Phys. 09 (2005) 012; this work differs from that presented here, in that they consider a severe truncation of the gravitational action, including only the cosmological constant and Einstein-Hilbert terms; they include matter as a perfect fluid with a constant equation of state parameter w; and they employ a time-dependent cutoff Λ. For more recent similar work that is somewhat closer in spirit to the present paper, see A. Bonanno and M. Reuter, J. Cosmol. Astropart. Phys. 08 (2007) 024; J. Phys. Conf. Ser. **140**, 012008 (2008).

[9] Higher derivative theories of this sort if used in the tree approximation have long been known to be plagued by ghosts; that is, poles in propagators with residues of the wrong sign for unitarity. This is only if the series of operators in (1) is truncated; otherwise propagator denominators are not polynomials in the squared momentum, and there may be just one pole, or any number of poles. Even with a truncated action, because of the running of the couplings, there is no one Lagrangian that can be used to find the propagator in the tree approximation over the whole range of momenta where the various poles occur, and it is not ruled out that all the poles have the residues of the right sign. For instance, Ref. [6] shows that, in a theory with only the couplings g_1, g_{2a}, and g_{2b}, the residue of the pole in the spin 2 propagator at high mass, which had usually been supposed to have the wrong sign [as for instance in the work of K. S. Stelle, Phys. Rev. D **16**, 953 (1977)], in fact has a sign consistent with unitarity. More generally, Benedetti *et al.* in Ref. [7] point out that for any truncation or no truncation, when we look for a pole at a four-momentum p, we must take the cutoff Λ to be proportional to $\sqrt{-p^2}$, so the denominator of any propagator takes the form $p^2 + m^2(-p^2)$. The function $m^2(-p^2)$ is a constant at sufficiently low $|p^2|$, and of the

form cp^2 for momenta so large that the couplings are near their fixed point, where c is a constant, so the equation $p^2 + m^2(-p^2) = 0$ for the pole position has no solution if $-c > 1$.

[10] B. Freivogel, M. Kleban, M. R. Martinez, and L. Susskind, J. High Energy Phys. 03 (2006) 039.

[11] A. Guth, Phys. Rev. D **23**, 347 (1981).

[12] E. Komatsu *et al.*, Astrophys. J. Suppl. Ser. **180**, 330 (2009).

[13] D. Benedetti, P. F. Machado, and F. Saueressig, Mod. Phys. Lett. A **24**, 2233 (2009); Nucl. Phys. **B824**, 168 (2010).

3.6. Popular Articles

The effort to understand the universe is one of the very few things that lifts human life a little above the level of farce, and gives it some of the grace of tragedy.
Steven Weinberg

Popular books by Weinberg such as *The First Three Minutes*, *Dreams of a Final Theory* and *To Explain the World* ensured that the general public was also kept informed and entertained, as did his frequent columns in the New York Review of Books.

Space limitations mean that separate sections for other important work, such as the foundations of quantum mechanics, had to be omitted, but here we include an article from the New York Review of Books that summarizes his thoughts [3.6.1].

He continued to champion reductionism when wishy-washy notions of emergence gained popularity. While conceding that the fundamental laws that govern elementary particles are of little use in explaining the weather or the stock market, he insists that none of these complex systems are operating outside those fundamental laws. We have included his views on this topic [3.6.2] taken from *To Explain the World.*

Finally, it is a source of great pride that Steve's last publication [3.6.3] should be his contribution to a Special Feature in Proceedings of the Royal Society A: *Quantum gravity, branes and M-theory* dedicated to me on the occasion of my 70th birthday, organized by L. Borsten, A. Marrani, C. N. Pope and K. Stelle.

Reproduced from The New York Review of Books, 19 January 2017 issue. Available online.

The Trouble with Quantum Mechanics

Steven Weinberg

Despite its great successes, arguments continue about its meaning and its future.

The development of quantum mechanics in the first decades of the twentieth century came as a shock to many physicists. Today, despite the great successes of quantum mechanics, arguments continue about its meaning, and its future.

1.

The first shock came as a challenge to the clear categories to which physicists by 1900 had become accustomed. There were particles — atoms, and then electrons and atomic nuclei — and there were fields — conditions of space that pervade regions in which electric, magnetic, and gravitational forces are exerted. Light waves were clearly recognized as self-sustaining oscillations of electric and magnetic fields. But in order to understand the light emitted by heated bodies, Albert Einstein in 1905 found it necessary to describe light waves as streams of massless particles, later called photons.

Then in the 1920s, according to theories of Louis de Broglie and Erwin Schrödinger, it appeared that electrons, which had always been recognized as particles, under some circumstances behaved as waves. In order to account for the energies of the stable states of atoms, physicists had to give up the notion that electrons in atoms are little Newtonian planets in orbit around the atomic nucleus. Electrons in atoms are better described as waves, fitting around the nucleus like sound waves fitting into an organ pipe.[1] The world's categories had become all muddled.

Worse yet, the electron waves are not waves of electronic matter, in the way that ocean waves are waves of water. Rather, as Max Born came to realize, the electron waves are waves of probability. That is, when a free electron collides with an atom, we cannot in principle say in what direction it will bounce off. The electron wave, after encountering the atom,

[1]Conditions on sound waves at the closed or open ends of an organ pipe require that either an odd number of quarter wave lengths or an even or an odd number of half wave lengths must just fit into the pipe, which limits the possible notes that can be produced by the pipe. In an atom the wave function must satisfy conditions of continuity and finiteness close to and far from the nucleus, which similarly limit the possible energies of atomic states.

spreads out in all directions, like an ocean wave after striking a reef. As Born recognized, this does not mean that the electron itself spreads out. Instead, the undivided electron goes in some one direction, but not a precisely predictable direction. It is more likely to go in a direction where the wave is more intense, but any direction is possible.

Probability was not unfamiliar to the physicists of the 1920s, but it had generally been thought to reflect an imperfect knowledge of whatever was under study, not an indeterminism in the underlying physical laws. Newton's theories of motion and gravitation had set the standard of deterministic laws. When we have reasonably precise knowledge of the location and velocity of each body in the solar system at a given moment, Newton's laws tell us with good accuracy where they will all be for a long time in the future. Probability enters Newtonian physics only when our knowledge is imperfect, as for example when we do not have precise knowledge of how a pair of dice is thrown. But with the new quantum mechanics, the moment-to-moment determinism of the laws of physics themselves seemed to be lost.

All very strange. In a 1926 letter to Born, Einstein complained:

> *Quantum mechanics is very impressive. But an inner voice tells me that it is not yet the real thing. The theory produces a good deal but hardly brings us closer to the secret of the Old One. I am at all events convinced that He does not play dice.*[2]

As late as 1964, in his Messenger lectures at Cornell, Richard Feynman lamented, "I think I can safely say that no one understands quantum mechanics."[3] With quantum mechanics, the break with the past was so sharp that all earlier physical theories became known as "classical."

The weirdness of quantum mechanics did not matter for most purposes. Physicists learned how to use it to do increasingly precise calculations of the energy levels of atoms, and of the probabilities that particles will scatter in one direction or another when they collide. Lawrence Krauss has labeled the quantum mechanical calculation of one effect in the spectrum of hydrogen "the best, most accurate prediction in all of science."[4] Beyond atomic physics, early applications of quantum mechanics listed by the physicist Gino Segrè included the binding of atoms in molecules, the radioactive decay of atomic nuclei, electrical conduction, magnetism, and electromagnetic radiation.[5] Later applications spanned theories of semiconductivity and superconductivity, white dwarf stars and neutron stars, nuclear forces, and elementary particles. Even the most adventurous modern speculations, such as string theory, are based on the principles of quantum mechanics.

Many physicists came to think that the reaction of Einstein and Feynman and others to the unfamiliar aspects of quantum mechanics had been overblown. This used to be my view. After all, Newton's theories too had been unpalatable to many of his contemporaries. Newton had introduced what his critics saw as an occult force, gravity, which was unrelated to any sort of tangible pushing and pulling, and which could not be explained on the basis of philosophy or pure mathematics. Also, his theories had renounced a chief aim of Ptolemy

[2]Quoted by Abraham Pais in '*Subtle Is the Lord': The Science and the Life of Albert Einstein* (Oxford University Press, 1982), p. 443.

[3]Richard Feynman, *The Character of Physical Law* (MIT Press, 1967), p. 129.

[4]Lawrence M. Krauss, *A Universe from Nothing* (Free Press, 2012), p. 138.

[5]Gino Segrè, *Ordinary Geniuses* (Viking, 2011).

and Kepler, to calculate the sizes of planetary orbits from first principles. But in the end the opposition to Newtonianism faded away. Newton and his followers succeeded in accounting not only for the motions of planets and falling apples, but also for the movements of comets and moons and the shape of the earth and the change in direction of its axis of rotation. By the end of the eighteenth century this success had established Newton's theories of motion and gravitation as correct, or at least as a marvelously accurate approximation. Evidently it is a mistake to demand too strictly that new physical theories should fit some preconceived philosophical standard.

In quantum mechanics the state of a system is not described by giving the position and velocity of every particle and the values and rates of change of various fields, as in classical physics. Instead, the state of any system at any moment is described by a wave function, essentially a list of numbers, one number for every possible configuration of the system.[6] If the system is a single particle, then there is a number for every possible position in space that the particle may occupy. This is something like the description of a sound wave in classical physics, except that for a sound wave a number for each position in space gives the pressure of the air at that point, while for a particle in quantum mechanics the wave function's number for a given position reflects the probability that the particle is at that position. What is so terrible about that? Certainly, it was a tragic mistake for Einstein and Schrödinger to step away from *using* quantum mechanics, isolating themselves in their later lives from the exciting progress made by others.

2.

Even so, I'm not as sure as I once was about the future of quantum mechanics. It is a bad sign that those physicists today who are most comfortable with quantum mechanics do not agree with one another about what it all means. The dispute arises chiefly regarding the nature of measurement in quantum mechanics. This issue can be illustrated by considering a simple example, measurement of the spin of an electron. (A particle's spin in any direction is a measure of the amount of rotation of matter around a line pointing in that direction.)

All theories agree, and experiment confirms, that when one measures the amount of spin of an electron in any arbitrarily chosen direction there are only two possible results. One possible result will be equal to a positive number, a universal constant of nature. (This is the constant that Max Planck originally introduced in his 1900 theory of heat radiation, denoted h, divided by 4π.) The other possible result is its opposite, the negative of the first. These positive or negative values of the spin correspond to an electron that is spinning either clockwise or counter-clockwise in the chosen direction.

But it is only when a measurement is made that these are the sole two possibilities. An electron spin that has not been measured is like a musical chord, formed from a superposition of two notes that correspond to positive or negative spins, each note with its own amplitude. Just as a chord creates a sound distinct from each of its constituent notes, the state of an electron spin that has not yet been measured is a superposition of the two possible states of definite spin, the superposition differing qualitatively from either state. In this musical

[6]These are complex numbers, that is, quantities of the general form $a + ib$, where a and b are ordinary real numbers and i is the square root of minus one.

analogy, the act of measuring the spin somehow shifts all the intensity of the chord to one of the notes, which we then hear on its own.

This can be put in terms of the wave function. If we disregard everything about an electron but its spin, there is not much that is wavelike about its wave function. It is just a pair of numbers, one number for each sign of the spin in some chosen direction, analogous to the amplitudes of each of the two notes in a chord.[7] The wave function of an electron whose spin has not been measured generally has nonzero values for spins of both signs.

There is a rule of quantum mechanics, known as the Born rule, that tells us how to use the wave function to calculate the probabilities of getting various possible results in experiments. For example, the Born rule tells us that the probabilities of finding either a positive or a negative result when the spin in some chosen direction is measured are proportional to the squares of the numbers in the wave function for those two states of the spin.[8]

The introduction of probability into the principles of physics was disturbing to past physicists, but the trouble with quantum mechanics is not that it involves probabilities. We can live with that. The trouble is that in quantum mechanics the way that wave functions change with time is governed by an equation, the Schrödinger equation, *that does not involve probabilities*. It is just as deterministic as Newton's equations of motion and gravitation. That is, given the wave function at any moment, the Schrödinger equation will tell you precisely what the wave function will be at any future time. There is not even the possibility of chaos, the extreme sensitivity to initial conditions that is possible in Newtonian mechanics. So if we regard the whole process of measurement as being governed by the equations of quantum mechanics, and these equations are perfectly deterministic, *how do probabilities get into quantum mechanics?*

One common answer is that, in a measurement, the spin (or whatever else is measured) is put in an interaction with a macroscopic environment that jitters in an unpredictable way. For example, the environment might be the shower of photons in a beam of light that is used to observe the system, as unpredictable in practice as a shower of raindrops. Such an environment causes the superposition of different states in the wave function to break down, leading to an unpredictable result of the measurement. (This is called decoherence.) It is as if a noisy background somehow unpredictably left only one of the notes of a chord audible. But this begs the question. If the deterministic Schrödinger equation governs the changes through time not only of the spin but also of the measuring apparatus and the physicist using it, then the results of measurement should not in principle be unpredictable. So we still have to ask, how do probabilities get into quantum mechanics?

One response to this puzzle was given in the 1920s by Niels Bohr, in what came to be called the Copenhagen interpretation of quantum mechanics. According to Bohr, in a measurement the state of a system such as a spin collapses to one result or another in a way that cannot itself be described by quantum mechanics, and is truly unpredictable. This

---

[7]Simple as it is, such a wave function incorporates much more information than just a choice between positive and negative spin. It is this extra information that makes quantum computers, which store information in this sort of wave function, so much more powerful than ordinary digital computers.

[8]To be precise, these "squares" are squares of the absolute values of the complex numbers in the wave function. For a complex number of the form $a + ib$, the square of the absolute value is the square of a plus the square of b.

answer is now widely felt to be unacceptable. There seems no way to locate the boundary between the realms in which, according to Bohr, quantum mechanics does or does not apply. As it happens, I was a graduate student at Bohr's institute in Copenhagen, but he was very great and I was very young, and I never had a chance to ask him about this.

Today there are two widely followed approaches to quantum mechanics, the "realist" and "instrumentalist" approaches, which view the origin of probability in measurement in two very different ways.[9] For reasons I will explain, neither approach seems to me quite satisfactory.[10]

3.

The instrumentalist approach is a descendant of the Copenhagen interpretation, but instead of imagining a boundary beyond which reality is not described by quantum mechanics, it rejects quantum mechanics altogether as a description of reality. There is still a wave function, but it is not real like a particle or a field. Instead it is merely an instrument that provides predictions of the probabilities of various outcomes when measurements are made.

It seems to me that the trouble with this approach is not only that it gives up on an ancient aim of science: to say what is really going on out there. It is a surrender of a particularly unfortunate kind. In the instrumentalist approach, we have to assume, as fundamental laws of nature, the rules (such as the Born rule I mentioned earlier) for using the wave function to calculate the probabilities of various results when humans make measurements. Thus humans are brought into the laws of nature at the most fundamental level. According to Eugene Wigner, a pioneer of quantum mechanics, "it was not possible to formulate the laws of quantum mechanics in a fully consistent way without reference to the consciousness."[11]

Thus the instrumentalist approach turns its back on a vision that became possible after Darwin, of a world governed by impersonal physical laws that control human behavior along with everything else. It is not that we object to thinking about humans. Rather, we want to understand the relation of humans to nature, not just assuming the character of this relation by incorporating it in what we suppose are nature's fundamental laws, but rather by deduction from laws that make no explicit reference to humans. We may in the end have to give up this goal, but I think not yet.

Some physicists who adopt an instrumentalist approach argue that the probabilities we infer from the wave function are objective probabilities, independent of whether humans are making a measurement. I don't find this tenable. In quantum mechanics these probabilities do not exist until people choose what to measure, such as the spin in one or another direction. Unlike the case of classical physics, a choice must be made, because in quantum mechanics not everything can be simultaneously measured. As Werner Heisenberg realized, a particle cannot have, at the same time, both a definite position and a definite velocity. The measuring

[9]The opposition between these two approaches is nicely described by Sean Carroll in *The Big Picture* (Dutton, 2016).

[10]I go into this in mathematical detail in Section 3.7 of *Lectures on Quantum Mechanics*, second edition (Cambridge University Press, 2015).

[11]Quoted by Marcelo Gleiser, *The Island of Knowledge* (Basic Books, 2014), p. 222.

of one precludes the measuring of the other. Likewise, if we know the wave function that describes the spin of an electron we can calculate the probability that the electron would have a positive spin in the north direction if that were measured, or the probability that the electron would have a positive spin in the east direction if that were measured, but we cannot ask about the probability of the spins being found positive in both directions because there is no state in which an electron has a definite spin in two different directions.

4.

These problems are partly avoided in the realist — as opposed to the instrumentalist — approach to quantum mechanics. Here one takes the wave function and its deterministic evolution seriously as a description of reality. But this raises other problems.

The realist approach has a very strange implication, first worked out in the 1957 Princeton Ph.D. thesis of the late Hugh Everett. When a physicist measures the spin of an electron, say in the north direction, the wave function of the electron and the measuring apparatus and the physicist are supposed, in the realist approach, to evolve deterministically, as dictated by the Schrödinger equation; but in consequence of their interaction during the measurement, the wave function becomes a superposition of two terms, in one of which the electron spin is positive and everyone in the world who looks into it thinks it is positive, and in the other the spin is negative and everyone thinks it is negative. Since in each term of the wave function everyone shares a belief that the spin has one definite sign, the existence of the superposition is undetectable. In effect the history of the world has split into two streams, uncorrelated with each other.

This is strange enough, but the fission of history would not only occur when someone measures a spin. In the realist approach the history of the world is endlessly splitting; it does so every time a macroscopic body becomes tied in with a choice of quantum states. This inconceivably huge variety of histories has provided material for science fiction,[12] and it offers a rationale for a multiverse, in which the particular cosmic history in which we find ourselves is constrained by the requirement that it must be one of the histories in which conditions are sufficiently benign to allow conscious beings to exist. But the vista of all these parallel histories is deeply unsettling, and like many other physicists I would prefer a single history.

There is another thing that is unsatisfactory about the realist approach, beyond our parochial preferences. In this approach the wave function of the multiverse evolves deterministically. We can still talk of probabilities as the fractions of the time that various possible results are found when measurements are performed many times in any one history; but the rules that govern what probabilities are observed would have to follow from the deterministic evolution of the whole multiverse. If this were not the case, to predict probabilities we would need to make some additional assumption about what happens when humans make measurements, and we would be back with the shortcomings of the instrumentalist approach. Several attempts following the realist approach have come close to deducing

[12]For instance, *Northern Lights* by Philip Pullman (Scholastic, 1995), and the early "Mirror, Mirror" episode of *Star Trek*.

rules like the Born rule that we know work well experimentally, but I think without final success.

The realist approach to quantum mechanics had already run into a different sort of trouble long before Everett wrote about multiple histories. It was emphasized in a 1935 paper by Einstein with his coworkers Boris Podolsky and Nathan Rosen, and arises in connection with the phenomenon of "entanglement."[13]

We naturally tend to think that reality can be described locally. I can say what is happening in my laboratory, and you can say what is happening in yours, but we don't have to talk about both at the same time. But in quantum mechanics it is possible for a system to be in an entangled state that involves correlations between parts of the system that are arbitrarily far apart, like the two ends of a very long rigid stick.

For instance, suppose we have a pair of electrons whose total spin in any direction is zero. In such a state, the wave function (ignoring everything but spin) is a sum of two terms: in one term, electron A has positive spin and electron B has negative spin in, say, the north direction, while in the other term in the wave function the positive and negative signs are reversed. The electron spins are said to be entangled. If nothing is done to interfere with these spins, this entangled state will persist even if the electrons fly apart to a great distance. However far apart they are, we can only talk about the wave function of the two electrons, not of each separately. Entanglement contributed to Einstein's distrust of quantum mechanics as much or more than the appearance of probabilities.

Strange as it is, the entanglement entailed by quantum mechanics is actually observed experimentally. But how can something so nonlocal represent reality?

5.

What then must be done about the shortcomings of quantum mechanics? One reasonable response is contained in the legendary advice to inquiring students: "Shut up and calculate!" There is no argument about how to use quantum mechanics, only how to describe what it means, so perhaps the problem is merely one of words.

On the other hand, the problems of understanding measurement in the present form of quantum mechanics may be warning us that the theory needs modification. Quantum mechanics works so well for atoms that any new theory would have to be nearly indistinguishable from quantum mechanics when applied to such small things. But a new theory might be designed so that the superpositions of states of large things like physicists and their apparatus even in isolation suffer an actual rapid spontaneous collapse, in which probabilities evolve to give the results expected in quantum mechanics. The many histories of Everett would naturally collapse to a single history. The goal in inventing a new theory is to make this happen not by giving measurement any special status in the laws of physics, but as part of what in the post-quantum theory would be the ordinary processes of physics.

One difficulty in developing such a new theory is that we get no direction from experiment — all data so far agree with ordinary quantum mechanics. We do get some help,

[13]Entanglement was recently discussed by Jim Holt in these pages. November 10, 2016. [Jim Holt's article in the New York Review of Books: "Something Faster Than Light? What Is it?" — Ed.]

however, from some general principles, which turn out to provide surprisingly strict constraints on any new theory.

Obviously, probabilities must all be positive numbers, and add up to 100 percent. There is another requirement, satisfied in ordinary quantum mechanics, that in entangled states the evolution of probabilities during measurements cannot be used to send instantaneous signals, which would violate the theory of relativity. Special relativity requires that no signal can travel faster than the speed of light. When these requirements are put together, it turns out that the most general evolution of probabilities satisfies an equation of a class known as Lindblad equations.[14] The class of Lindblad equations contains the Schrödinger equation of ordinary quantum mechanics as a special case, but in general these equations involve a variety of new quantities that represent a departure from quantum mechanics. These are quantities whose details of course we now don't know. Though it has been scarcely noticed outside the theoretical community, there already is a line of interesting papers, going back to an influential 1986 article by Gian Carlo Ghirardi, Alberto Rimini, and Tullio Weber at Trieste, that use the Lindblad equations to generalize quantum mechanics in various ways.

Lately I have been thinking about a possible experimental search for signs of departure from ordinary quantum mechanics in atomic clocks. At the heart of any atomic clock is a device invented by the late Norman Ramsey for tuning the frequency of microwave or visible radiation to the known natural frequency at which the wave function of an atom oscillates when it is in a superposition of two states of different energy. This natural frequency equals the difference in the energies of the two atomic states used in the clock, divided by Planck's constant. It is the same under all external conditions, and therefore serves as a fixed reference for frequency, in the way that a platinum-iridium cylinder at Sèvres serves as a fixed reference for mass.

Tuning the frequency of an electromagnetic wave to this reference frequency works a little like tuning the frequency of a metronome to match another metronome. If you start the two metronomes together and the beats still match after a thousand beats, you know that their frequencies are equal at least to about one part in a thousand. Quantum mechanical calculations show that in some atomic clocks the tuning should be precise to one part in a hundred million billion, and this precision is indeed realized. But if the corrections to quantum mechanics represented by the new terms in the Lindblad equations (expressed as energies) were as large as one part in a hundred million billion of the energy difference of the atomic states used in the clock, this precision would have been quite lost. The new terms must therefore be even smaller than this.

How significant is this limit? Unfortunately, these ideas about modifications of quantum mechanics are not only speculative but also vague, and we have no idea how big we should expect the corrections to quantum mechanics to be. Regarding not only this issue, but more generally the future of quantum mechanics, I have to echo Viola in *Twelfth Night*: "O time, thou must untangle this, not I."

[14]This equation is named for Göran Lindblad, but it was also independently discovered by Vittorio Gorini, Andrzej Kossakowski, and George Sudarshan.

Reproduced from Chapter 15 of *To Explain the World: The Discovery of Modern Science*, Steven Weinberg, Harper/HarperCollins, 2015 (432 pp.). ISBN 9780062346650.

Epilogue: The Grand Reduction

Newton's great achievement left plenty yet to be explained. The nature of matter, the properties of forces other than gravitation that act on matter, and the remarkable capabilities of life were all still mysterious. Enormous progress was made in the years after Newton,[1] far too much to cover in one book, let alone a single chapter. This epilogue aims at making just one point, that as science progressed after Newton a remarkable picture began to take shape: it turned out that the world is governed by natural laws far simpler and more unified than had been imagined in Newton's time.

Newton himself in Book III of his Opticks sketched the outline of a theory of matter that would at least encompass optics and chemistry:

> Now the smallest particles of matter may cohere by the strongest attractions, and compose bigger particles of weaker virtue; and many of these may cohere and compose bigger particles whose virtue is still weaker, and so on for diverse successions, until the progression ends in the biggest particles on which the operations in chemistry, and the colors of natural bodies depend, and which by cohering compose bodies of a sensible magnitude.[2]

He also focused attention on the forces acting on these particles:

> For we must learn from the phenomena of nature what bodies attract one another, and what are the laws and properties of the attraction, before we inquire the cause by which the attraction is perform'd. The attractions of gravity, magnetism, and electricity, reach to very sensible distances, and so have been observed by vulgar eyes, and there may be others which reach to so small distances as to escape observation.[3]

As this shows, Newton was well aware that there are other forces in nature besides gravitation. Static electricity was an old story. Plato had mentioned in the Timaeus that when a piece of amber (in Greek, electron) is rubbed it can pick up light bits of matter. Magnetism was known from the properties of naturally magnetic lodestones, used by the Chinese for geomancy and studied in detail by Queen Elizabeth's physician, William Gilbert. Newton here also hints at the existence of forces not yet known because of their short range, a premonition of the weak and strong nuclear forces discovered in the twentieth century.

In the early nineteenth century the invention of the electric battery by Alessandro Volta made it possible to carry out detailed quantitative experiments in electricity and magnetism, and it soon became known that these are not entirely separate phenomena. First, in 1820 Hans Christian Ørsted in Copenhagen found that a magnet and a wire carrying an electric

[1]I have given a more detailed account of some of this progress in The Discovery of Subatomic Particles, rev. ed. (Cambridge University Press, Cambridge, 2003).

[2]Isaac Newton, Opticks, or A Treatise of the Reflections, Refractions, Inflections, and Colours of Light (Dover, New York, 1952, based on 4th ed., London, 1730), p. 394.

[3]Ibid., p. 376.

current exert forces on each other. After hearing of this result, André-Marie Ampère in Paris discovered that wires carrying electric currents also exert forces on one another. Ampère conjectured that these various phenomena are all much the same: the forces exerted by and on pieces of magnetized iron are due to electric currents circulating within the iron.

Just as happened with gravitation, the notion of currents and magnets exerting forces on each other was replaced with the idea of a field, in this case a magnetic field. Each magnet and current-carrying wire contributes to the total magnetic field at any point in its vicinity, and this magnetic field exerts a force on any magnet or electric current at that point. Michael Faraday attributed the magnetic forces produced by an electric current to lines of magnetic field encircling the wire. He also described the electric forces produced by a piece of rubbed amber as due to an electric field, pictured as lines emanating radially from the electric charges on the amber. Most important, Faraday in the 1830s showed a connection between electric and magnetic fields: a changing magnetic field, like that produced by the electric current in a rotating coil of wire, produces an electric field, which can drive electric currents in another wire. It is this phenomenon that is used to generate electricity in modern power plants.

The final unification of electricity and magnetism was achieved a few decades later, by James Clerk Maxwell. Maxwell thought of electric and magnetic fields as tensions in a pervasive medium, the ether, and expressed what was known about electricity and magnetism in equations relating the fields and their rates of change to each other. The new thing added by Maxwell was that, just as a changing magnetic field generates an electric field, so also a changing electric field generates a magnetic field. As often happens in physics, the conceptual basis for Maxwell's equations in terms of an ether has been abandoned, but the equations survive, even on T-shirts worn by physics students.[i]

Maxwell's theory had a spectacular consequence. Since oscillating electric fields produce oscillating magnetic fields, and oscillating magnetic fields produce oscillating electric fields, it is possible to have a self-sustaining oscillation of both electric and magnetic fields in the ether, or as we would say today, in empty space. Maxwell found around 1862 that this electromagnetic oscillation would propagate at a speed that, according to his equations, had just about the same numerical value as the measured speed of light. It was natural for Maxwell to jump to the conclusion that light is nothing but a mutually self-sustaining oscillation of electric and magnetic fields. Visible light has a frequency far too high for it to be produced by currents in ordinary electric circuits, but in the 1880s Heinrich Hertz was able to generate waves in accordance with Maxwell's equations: radio waves that differed from visible light only in having much lower frequency. Electricity and magnetism had thus been unified not only with each other, but also with optics.

As with electricity and magnetism, progress in understanding the nature of matter began with quantitative measurement, here measurement of the weights of substances participating in chemical reactions. The key figure in this chemical revolution was a wealthy Frenchman, Antoine Lavoisier. In the late eighteenth century he identified hydrogen and oxygen as elements and showed that water is a compound of hydrogen and oxygen, that air is a mixture of elements, and that fire is due to the combination of other elements with oxygen. Also on the basis of such measurements, it was found a little later by John Dalton that the weights with which elements combine in chemical reactions can be understood on the hypothesis that pure chemical compounds like water or salt consist of large numbers of particles (later called molecules) that themselves consist of definite numbers of atoms of pure elements. The water molecule, for instance, consists of two hydrogen atoms and one oxygen atom. In the following decades chemists identified many elements: some familiar, like carbon, sulfur, and the common metals; and others newly isolated, such as chlorine, calcium, and sodium. Earth, air, fire, and water did not make the list. The correct chemical formulas for molecules like water and salt were worked out, in the first half of the nineteenth century, allowing the calculation of the ratios of the masses of the atoms of the different elements from measurements of the weights of substances participating in chemical reactions.

The atomic theory of matter scored a great success when Maxwell and Ludwig Boltzmann showed how heat could be understood as energy distributed among vast numbers of atoms or molecules. This step toward unification was resisted by some physicists, including Pierre Duhem, who doubted the existence of atoms and held that the theory of heat, thermodynamics, was at least as fundamental as Newton's mechanics and Maxwell's electrodynamics. But soon after the beginning of the twentieth century several new experiments convinced almost everyone that atoms are real. One series of experiments, by J. J. Thomson, Robert Millikan, and others, showed that electric charges are gained and lost only as multiples of a fundamental charge: the charge of the electron, a particle that had been discovered by Thomson in 1897. The random "Brownian" motion of small particles on the surface of liquids was interpreted by Albert Einstein in 1905 as due to collisions of these particles with individual molecules of the liquid, an interpretation confirmed by experiments of Jean Perrin. Responding to the experiments of Thomson and Perrin, the chemist Wilhelm Ostwald, who earlier had been skeptical about atoms, expressed his change of mind in 1908, in a statement that implicitly looked all the way back to Democritus and Leucippus: "I am now convinced that we have recently become possessed of experimental evidence of the discrete or grained nature of matter, which the atomic hypothesis sought in vain for hundreds and thousands of years."[4]

But what are atoms? A great step toward the answer was taken in 1911, when experiments in the Manchester laboratory of Ernest Rutherford showed that the mass of gold atoms is concentrated in a small heavy positively charged nucleus, around which revolve lighter negatively charged electrons. The electrons are responsible for the phenomena of ordinary chemistry, while changes in the nucleus release the large energies encountered in radioactivity.

This raised a new question: what keeps the orbiting atomic electrons from losing energy through the emission of radiation, and spiraling down into the nucleus? Not only would this rule out the existence of stable atoms; the frequencies of the radiation emitted in these little atomic catastrophes would form a continuum, in contradiction with the observation that atoms can emit and absorb radiation only at certain discrete frequencies, seen as bright or dark lines in the spectra of gases. What determines these special frequencies?

The answers were worked out in the first three decades of the twentieth century with the development of quantum mechanics, the most radical innovation in physical theory since the work of Newton. As its name suggests, quantum mechanics requires a quantization (that is, a discreteness) of the energies of various physical systems. Niels Bohr in 1913 proposed that

[4]This is from Ostwald's Outlines of General Chemistry, and is quoted by G. Holton, in Historical Studies in the Physical Sciences 9, 161 (1979), and I. B. Cohen, in Critical Problems in the History of Science, ed. M. Clagett (University of Wisconsin Press, Madison, 1959).

an atom can exist only in states of certain definite energies, and gave rules for calculating these energies in the simplest atoms. Following earlier work of Max Planck, Einstein had already in 1905 suggested that the energy in light comes in quanta, particles later called photons, each photon with an energy proportional to the frequency of the light. As Bohr explained, when an atom loses energy by emitting a single photon, the energy of that photon must equal the difference in the energies of the initial and final atomic states, a requirement that fixes its frequency. There is always an atomic state of lowest energy, which cannot emit radiation and is therefore stable.

These early steps were followed in the 1920s with the development of general rules of quantum mechanics, rules that can be applied to any physical system. This was chiefly the work of Louis de Broglie, Werner Heisenberg, Wolfgang Pauli, Pascual Jordan, Erwin Schrödinger, Paul Dirac, and Max Born. The energies of allowed atomic states are calculated by solving an equation, the Schrödinger equation, of a general mathematical type that was already familiar from the study of sound and light waves. A string on a musical instrument can produce just those tones for which a whole number of half wavelengths fit on the string; analogously, Schrödinger found that the allowed energy levels of an atom are those for which the wave governed by the Schrödinger equation just fits around the atom without discontinuities. But as first recognized by Born, these waves are not waves of pressure or of electromagnetic fields, but waves of probability — a particle is most likely to be near where the wave function is largest.

Quantum mechanics not only solved the problem of the stability of atoms and the nature of spectral lines; it also brought chemistry into the framework of physics. With the electrical forces among electrons and atomic nuclei already known, the Schrödinger equation could be applied to molecules as well as to atoms, and allowed the calculation of the energies of their various states. In this way it became possible in principle to decide which molecules are stable and which chemical reactions are energetically allowed. In 1929 Dirac announced triumphantly that "the underlying physical laws necessary for the mathematical theory of a larger part of physics and the whole of chemistry are thus completely known."[5]

This did not mean that chemists would hand over their problems to physicists, and retire. As Dirac well understood, for all but the smallest molecules the Schrödinger equation is too complicated to be solved, so the special tools and insights of chemistry remain indispensable. But from the 1920s on, it would be understood that any general principle of chemistry, such as the rule that metals form stable compounds with halogen elements like chlorine, is what it is because of the quantum mechanics of nuclei and electrons acted on by electromagnetic forces.

Despite its great explanatory power, this foundation was itself far from being satisfactorily unified. There were particles: electrons and the protons and neutrons that make up atomic nuclei. And there were fields: the electromagnetic field, and whatever then-unknown short-range fields are presumably responsible for the strong forces that hold atomic nuclei together and for the weak forces that turn neutrons into protons or protons into neutrons in radioactivity. This distinction between particles and fields began to be swept away in the 1930s, with the advent of quantum field theory. Just as there is an electromagnetic field, whose energy and momentum are bundled in particles known as photons, so there is also an

[5]P. A. M. Dirac, "Quantum Mechanics of Many-Electron Systems," Proceedings of the Royal Society A123, 713 (1929).

electron field, whose energy and momentum are bundled in electrons, and likewise for other types of elementary particles.

This was far from obvious. We can directly feel the effects of gravitational and electromagnetic fields because the quanta of these fields have zero mass, and they are particles of a type (known as bosons) that in large numbers can occupy the same state. These properties allow large numbers of photons to build up to form states that we observe as electric and magnetic fields that seem to obey the rules of classical (that is, non-quantum) physics. Electrons, in contrast, have mass and are particles of a type (known as fermions) no two of which can occupy the same state, so that electron fields are never apparent in macroscopic observations.

In the late 1940s quantum electrodynamics, the quantum field theory of photons, electrons, and antielectrons, scored stunning successes, with the calculation of quantities like the strength of the electron's magnetic field that agreed with experiment to many decimal places.[ii] Following this achievement, it was natural to try to develop a quantum field theory that would encompass not only photons, electrons, and antielectrons but also the other particles being discovered in cosmic rays and accelerators and the weak and strong forces that act on them.

We now have such a quantum field theory, known as the Standard Model. The Standard Model is an expanded version of quantum electrodynamics. Along with the electron field there is a neutrino field, whose quanta are fermions like electrons but with zero electric charge and nearly zero mass. There is a pair of quark fields, whose quanta are the constituents of the protons and neutrons that make up atomic nuclei. For reasons that no one understands, this menu is repeated twice, with much heavier quarks and much heavier electron-like particles and their neutrino partners. The electromagnetic field appears in a unified "electroweak" picture along with other fields responsible for the weak nuclear interactions, which allow protons and neutrons to convert into one another in radioactive decays. The quanta of these fields are heavy bosons: the electrically charged W+ and W−, and the electrically neutral Z 0. There are also eight mathematically similar "gluon" fields responsible for the strong nuclear interactions, which hold quarks together inside protons and neutrons. In 2012 the last missing piece of the Standard Model was discovered: a heavy electrically neutral boson that had been predicted by the electroweak part of the Standard Model.

The Standard Model is not the end of the story. It leaves out gravitation; it does not account for the "dark matter" that astronomers tell us makes up five-sixths of the mass of the universe; and it involves far too many unexplained numerical quantities, like the ratios of the masses of the various quarks and electron-like particles. But even so, the Standard Model provides a remarkably unified view of all types of matter and force (except for gravitation) that we encounter in our laboratories, in a set of equations that can fit on a single sheet of paper. We can be certain that the Standard Model will appear as at least an approximate feature of any better future theory.

The Standard Model would have seemed unsatisfying to many natural philosophers from Thales to Newton. It is impersonal; there is no hint in it of human concerns like love or justice. No one who studies the Standard Model will be helped to be a better person, as Plato expected would follow from the study of astronomy. Also, contrary to what Aristotle expected of a physical theory, there is no element of purpose in the Standard Model. Of course, we live in a universe governed by the Standard Model and can imagine that electrons and the two light quarks are what they are to make us possible, but then what do we make of their heavier counterparts, which are irrelevant to our lives?

The Standard Model is expressed in equations governing the various fields, but it cannot be deduced from mathematics alone. Nor does it follow straightforwardly from observation of nature. Indeed, quarks and gluons are attracted to each other by forces that increase with distance, so these particles can never be observed in isolation. Nor can the Standard Model be deduced from philosophical preconceptions. Rather, the Standard Model is a product of guesswork, guided by aesthetic judgment, and validated by the success of many of its predictions. Though the Standard Model has many unexplained aspects, we expect that at least some of these features will be explained by whatever deeper theory succeeds it.

The old intimacy between physics and astronomy has continued. We now understand nuclear reactions well enough not only to calculate how the Sun and stars shine and evolve, but also to understand how the lightest elements were produced in the first few minutes of the present expansion of the universe. And as in the past, astronomy now presents physics with a formidable challenge: the expansion of the universe is speeding up, presumably owing to dark energy that is contained not in particle masses and motions, but in space itself.

There is one aspect of experience that at first sight seems to defy understanding on the basis of any unpurposeful physical theory like the Standard Model. We cannot avoid teleology in talking of living things. We describe hearts and lungs and roots and flowers in terms of the purpose they serve, a tendency that was only increased with the great expansion after Newton of information about plants and animals due to naturalists like Carl Linnaeus and Georges Cuvier. Not only theologians but also scientists including Robert Boyle and Isaac Newton have seen the marvelous capabilities of plants and animals as evidence for a benevolent Creator. Even if we can avoid a supernatural explanation of the capabilities of plants and animals, it long seemed inevitable that an understanding of life would rest on teleological principles very different from those of physical theories like Newton's.

The unification of biology with the rest of science first began to be possible in the mid-nineteenth century, with the independent proposals by Charles Darwin and Alfred Russel Wallace of the theory of evolution through natural selection. Evolution was already a familiar idea, suggested by the fossil record. Many of those who accepted the reality of evolution explained it as a result of a fundamental principle of biology, an inherent tendency of living things to improve, a principle that would have ruled out any unification of biology with physical science. Darwin and Wallace instead proposed that evolution acts through the appearance of inheritable variations, with favorable variations no more likely than unfavorable ones, but with the variations that improve the chances of survival and reproduction being the ones that are likely to spread.[iii]

It took a long time for natural selection to be accepted as the mechanism for evolution. No one in Darwin's time knew the mechanism for inheritance, or for the appearance of inheritable variations, so there was room for biologists to hope for a more purposeful theory. It was particularly distasteful to imagine that humans are the result of millions of years of natural selection acting on random inheritable variations. Eventually the discovery of the rules of genetics and of the occurrence of mutations led in the twentieth century to a "neo-Darwinian synthesis" that put the theory of evolution through natural selection on a firmer basis. Finally this theory was grounded on chemistry, and thereby on physics, through the realization that genetic information is carried by the double helix molecules of DNA.

So biology joined chemistry in a unified view of nature based on physics. But it is important to acknowledge the limitations of this unification. No one is going to replace the language and methods of biology with a description of living things in terms of individual molecules, let alone quarks and electrons. For one thing, even more than the large molecules of organic chemistry, living things are too complicated for such a description. More important, even if we could follow the motion of every atom in a plant or animal, in that immense mass of data we would lose the things that interest us — a lion hunting antelope or a flower attracting bees.

For biology, like geology but unlike chemistry, there is another problem. Living things are what they are not only because of the principles of physics, but also because of a vast number of historical accidents, including the accident that a comet or meteor hit the Earth 65 million years ago with enough impact to kill off the dinosaurs, and going back to the fact that the Earth formed at a certain distance from the Sun and with a certain initial chemical composition. We can understand some of these accidents statistically, but not individually. Kepler was wrong; no one will ever be able to calculate the distance of the Earth from the Sun solely from the principles of physics. What we mean by the unification of biology with the rest of science is only that there can be no freestanding principles of biology, any more than of geology. Any general principle of biology is what it is because of the fundamental principles of physics together with historical accidents, which by definition can never be explained.

The point of view described here is called (often disapprovingly) "reductionism." There is opposition to reductionism even within physics. Physicists who study fluids or solids often cite examples of "emergence," the appearance in the description of macroscopic phenomena of concepts like heat or phase transition that have no counterpart in elementary particle physics, and that do not depend on the details of elementary particles. For instance, thermodynamics, the science of heat, applies in a wide variety of systems: not just to those considered by Maxwell and Boltzmann, containing large numbers of molecules, but also to the surfaces of large black holes. But it does not apply to everything, and when we ask whether it applies to a given system and if so why, we must have reference to deeper, more truly fundamental, principles of physics. Reductionism in this sense is not a program for the reform of scientific practice; it is a view of why the world is the way it is.

We do not know how long science will continue on this reductive path. We may come to a point where further progress is impossible within the resources of our species. Right now, it seems that there is a scale of mass about a million trillion times larger than the mass of the hydrogen atom, at which gravity and other as yet undetected forces are unified with the forces of the Standard Model. (This is known as the "Planck mass"; it is the mass that particles would have to possess for their gravitational attraction to be as strong as the electrical repulsion between two electrons at the same separation.) Even if the economic resources of the human race were entirely at the disposal of physicists, we cannot now conceive of any way of creating particles with such huge masses in our laboratories.

We may instead run out of intellectual resources — humans may not be smart enough to understand the really fundamental laws of physics. Or we may encounter phenomena that in principle cannot be brought into a unified framework for all science. For instance, although we may well come to understand the processes in the brain responsible for consciousness, it is hard to see how we will ever describe conscious feelings themselves in physical terms.

Still, we have come a long way on this path, and are not yet at its end.[6] This is a grand story — how celestial and terrestrial physics were unified by Newton, how a unified theory of electricity and magnetism was developed that turned out to explain light, how the quantum theory of electromagnetism was expanded to include the weak and strong nuclear forces, and how chemistry and even biology were brought into a unified though incomplete view of nature based on physics. It is toward a more fundamental physical theory that the wide-ranging scientific principles we discover have been, and are being, reduced.

ENDNOTES

[i] Maxwell himself did not write equations governing electric and magnetic fields in the form known today as "Maxwell's equations." His equations instead involved other fields known as potentials, whose rates of change with time and position are the electric and magnetic fields. The more familiar modern form of Maxwell's equations was given around 1881 by Oliver Heaviside.

[ii] Here and in what follows I will not cite individual physicists. So many are involved that it would take too much space, and many are still alive, so that I would risk giving offense by citing some physicists and not others.

[iii] I am here lumping sexual selection together with natural selection, and punctuated equilibrium along with steady evolution; and I am not distinguishing between mutations and genetic drift as a source of inheritable variations. These distinctions are very important to biologists, but they do not affect the point that concerns me here: there is no independent law of biology that makes inheritable variations more likely to be improvements.

[6] To forestall accusations of plagiarism, I will acknowledge here that this last paragraph is a riff on the last paragraph of Darwin's On the Origin of Species.

PROCEEDINGS A

royalsocietypublishing.org/journal/rspa

Editorial

Check for updates

Cite this article: Weinberg S. 2022 Michael J. Duff: a personal reminiscence. *Proc. R. Soc. A* **478**: 20220167.
https://doi.org/10.1098/rspa.2022.0167

Author for correspondence:
Steven Weinberg
e-mail: weinberg@physics.utexas.edu

One contribution to a special feature 'Quantum gravity, branes and M-theory' dedicated to Michael J. Duff on the occasion of his 70th Birthday organized by Leron Borsten, Alessio Marrani, Christopher Pope and Kellogg Stelle.

THE ROYAL SOCIETY PUBLISHING

Michael J. Duff: a personal reminiscence

Steven Weinberg

Department of Physics, The University of Texas at Austin, Austin, TX, USA

I first became acquainted with Mike Duff in the late 1970s. I was developing an idea known as 'asymptotic safety'. According to this idea, all forces including gravitation are described by a quantum field theory with an infinite number of energy-dependent couplings, all couplings that are allowed by symmetry principles. With increasing energy, these couplings tend to run to infinity, except that there may be a finite-dimensional surface of coupling trajectories that are safely attracted to a fixed point as the energy goes to infinity. From the work of Wilson and Fisher, I knew that in scalar field theories in $2 + \varepsilon$ dimensions there is such a fixed point. My problem was to find the conditions under which this would still be true in a theory with gravitons and scalar, vector and spinor fields. Eventually, I found the solution: ask Mike Duff!

As I acknowledged in my article on asymptotic safety, Mike's expertise was tremendously helpful to me. In the following years, I became increasingly aware of his work on supergravity and 'Kaluza–Klein' theories in higher dimensional space–times. So I was delighted when in the early 1980s he agreed to visit our Theory Group here in Austin.

His visit was a smashing success. He gave a superb series of lectures on Kaluza-Klein theories. It was also fun to have him in town. I recall that once my wife and I went with Mike and Chris Pope to wander about Sixth Street, Austin's Barbary Coast. Sitting at a table in a honky-tonk we were visited by a young lady, who offered her professional services. Mike was wonderfully suave and polite in declining the offer.

Since then Mike and I have been mostly on other sides of the water, and I have had little chance to spend time with him. I have of course been aware of his brilliant work on superstrings, supermembranes and all that, which I assume will be described by other contributors to this volume. It was a great pleasure to see him again on my rare visits to London, UK, most recently for a meeting at Imperial College. For personal and professional reasons, I count myself fortunate to know him.

Data accessibility. This article has no additional data.
Competing interests. I declare I have no competing interests.
Funding. No funding has been received for this article.

royalsocietypublishing.org/journal/rspa *Proc. R. Soc. A* **478**: 20220167

4

Complete List of Publications

[1] S. Weinberg, "Michael J. Duff: a personal reminiscence," One contribution to a Special Feature *Quantum gravity, branes and M-theory* dedicated to Michael J. Duff on the occasion of his 70th birthday, organized by L. Borsten, A. Marrani, C. N. Pope and K. Stelle, Proceedings of the Royal Society A: Volume 478, Issue 2259 doi.org/10.1098/rspa.2022.0167

[2] S. Weinberg, "Foundations of Modern Physics," CUP 2021. doi:10.1017/9781108894845

[3] S. Weinberg, "On the Development of Effective Field Theory," Eur. Phys. J. H **46**, no. 1, 6 (2021) doi:10.1140/epjh/s13129-021-00004-x [arXiv:2101.04241 [hep-th]].

[4] S. Weinberg, "Massless particles in higher dimensions," Phys. Rev. D **102**, no. 9, 095022 (2020) doi:10.1103/PhysRevD.102.095022 [arXiv:2010.05823 [hep-th]].

[5] R. Bousso, F. Quevedo and S. Weinberg, "Joseph Polchinski: A Biographical Memoir," [arXiv:2002.02371 [physics.hist-ph]].

[6] S. Weinberg, "Models of Lepton and Quark Masses," Phys. Rev. D **101**, no. 3, 035020 (2020) doi:10.1103/PhysRevD.101.035020 [arXiv:2001.06582 [hep-th]].

[7] R. Flauger and S. Weinberg, "Absorption of Gravitational Waves from Distant Sources," Phys. Rev. D **99**, no. 12, 123030 (2019) doi:10.1103/PhysRevD.99.123030 [arXiv:1906.04853 [hep-th]].

[8] S. Weinberg, "Soft Bremsstrahlung," Phys. Rev. D **99**, no. 7, 076018 (2019) doi:10.1103/PhysRevD.99.076018 [arXiv:1903.11168 [astro-ph.GA]].

[9] S. Weinberg, "Essay: Half a Century of the Standard Model," Phys. Rev. Lett. **121**, no. 22, 220001 (2018) doi:10.1103/PhysRevLett.121.220001

[10] R. Flauger and S. Weinberg, "Gravitational Waves in Cold Dark Matter," Phys. Rev. D **97**, no. 12, 123506 (2018) doi:10.1103/PhysRevD.97.123506 [arXiv:1801.00386 [astro-ph.CO]].

[11] S. Weinberg, "Lindblad Decoherence in Atomic Clocks," Phys. Rev. A **94**, no. 4, 042117 (2016) doi:10.1103/PhysRevA.94.042117 [arXiv:1610.02537 [quant-ph]].

[12] S. Weinberg, "Nambu, at the beginning," PTEP **2016**, no. 7, 07B002 (2016) doi:10.1093/ptep/ptv186

[13] S. Weinberg, "What Happens in a Measurement?," Phys. Rev. A **93**, 032124 (2016) doi:10.1103/PhysRevA.93.032124 [arXiv:1603.06008 [quant-ph]].

[14] S. Weinberg, "Effective field theory, past and future," Int. J. Mod. Phys. A **31**, no. 06, 1630007 (2016) doi:10.1142/S0217751X16300076

[15] S. Weinberg, "Quantum Mechanics Without State Vectors," Phys. Rev. A **90**, no. 4, 042102 (2014) doi:10.1103/PhysRevA.90.042102 [arXiv:1405.3483 [quant-ph]].

[16] S. Weinberg, "Tom Kibble: Breaking ground and breaking symmetries," Int. J. Mod. Phys. A **29**, 1430004 (2014) doi:10.1142/S0217751X1430004X

[17] S. Weinberg, "Goldstone Bosons as Fractional Cosmic Neutrinos," Phys. Rev. Lett. **110**, no. 24, 241301 (2013) doi:10.1103/PhysRevLett.110.241301 [arXiv:1305.1971 [astro-ph.CO]].

[18] S. Weinberg, "Tetraquark Mesons in Large N Quantum Chromodynamics," Phys. Rev. Lett. **110**, 261601 (2013) doi:10.1103/PhysRevLett.110.261601 [arXiv:1303.0342 [hep-ph]].

[19] S. Weinberg, "Particle physics, from Rutherford to the LHC," Int. J. Mod. Phys. A **28**, 1330055 (2013) doi:10.1142/S0217751X1330055X

[20] S. Weinberg, "Particle Physics, from Rutherford to the LHC," doi:10.1142/9789814425810_0001, in "100 Years of Sub-Atomic Physics" at the April 2011 meeting of the American Physical Society, World Scientific.
[21] S. Weinberg, "Minimal fields of canonical dimensionality are free," Phys. Rev. D **86**, 105015 (2012) doi:10.1103/PhysRevD.86.105015 [arXiv:1210.3864 [hep-th]].
[22] S. Weinberg, "Six-dimensional Methods for Four-dimensional Conformal Field Theories II: Irreducible Fields," Phys. Rev. D **86**, 085013 (2012) doi:10.1103/PhysRevD.86.085013 [arXiv:1209.4659 [hep-th]].
[23] S. Weinberg, "Beautiful theories," in Albert Einstein Memorial Lectures https://doi.org/10.1142/7992 — March 2012 Pages: 216. Edited By: Jacob Bekenstein (Hebrew University of Jerusalem, Israel) and Raphael Mechoulam (Hebrew University of Jerusalem, Israel).
[24] S. Weinberg, "Collapse of the State Vector," Phys. Rev. A **85**, 062116 (2012) doi:10.1103/PhysRevA.85.062116 [arXiv:1109.6462 [quant-ph]].
[25] P. Lucas, S. Childress, J. Doe, R. Aymar, S. Chu, E. McMillan, G. Bush, S. Weinberg and A. Jackson, "NOvA Near Detector On the Surface (NDOS)," doi:10.2172/1296761
[26] S. Weinberg, "Particle physics, from Rutherford to the LHC," Phys. Today **64N8**, 29–33 (2011) doi:10.1063/PT.3.1216
[27] S. Weinberg, "Ultraviolet Divergences in Cosmological Correlations," Phys. Rev. D **83**, 063508 (2011) doi:10.1103/PhysRevD.83.063508 [arXiv:1011.1630 [hep-th]].
[28] S. Weinberg, "Pions in Large-N Quantum Chromodynamics," Phys. Rev. Lett. **105**, 261601 (2010) doi:10.1103/PhysRevLett.105.261601 [arXiv:1009.1537 [hep-ph]].
[29] S. Weinberg, "Six-dimensional Methods for Four-dimensional Conformal Field Theories," Phys. Rev. D **82**, 045031 (2010) doi:10.1103/PhysRevD.82.045031 [arXiv:1006.3480 [hep-th]].
[30] S. Weinberg, "Changing views of symmetry," doi:10.1007/978-3-642-30844-4_18 From the PS to the LHC - 50 Years of Nobel Memories in High-Energy Physics pp 233-241 Springer.
[31] S. Weinberg, "Asymptotically Safe Inflation," Phys. Rev. D **81**, 083535 (2010) doi:10.1103/PhysRevD.81.083535 [arXiv:0911.3165 [hep-th]].
[32] S. Weinberg, "Effective Field Theory, Past and Future," PoS **CD09**, 001 (2009) doi:10.22323/1.086.0001 [arXiv:0908.1964 [hep-th]].
[33] S. Weinberg, "Living with Infinities", Published in a collection of the papers of Gunnar Kallen, edited by C. Jarlskog and A. C. T. Wu. [arXiv:0903.0568 [hep-th]].
[34] E. Komatsu, N. Afshordi, N. Bartolo, D. Baumann, J. R. Bond, E. I. Buchbinder, C. T. Byrnes, X. Chen, D. J. H. Chung and A. Cooray, *et al.* "Non-Gaussianity as a Probe of the Physics of the Primordial Universe and the Astrophysics of the Low Redshift Universe," [arXiv:0902.4759 [astro-ph.CO]].
[35] S. Dodelson, R. Easther, S. Hanany, L. McAllister, S. Meyer, L. Page, P. Ade, A. Amblard, A. Ashoorioon and C. Baccigalupi, *et al.* "The Origin of the Universe as Revealed Through the Polarization of the Cosmic Microwave Background," [arXiv:0902.3796 [astro-ph.CO]].
[36] S. Weinberg, "V-A was the key," J. Phys. Conf. Ser. **196**, 012002 (2009) doi:10.1088/1742-6596/196/1/012002
[37] S. Weinberg, "Non-Gaussian Correlations Outside the Horizon II: The General Case," Phys. Rev. D **79**, 043504 (2009) doi:10.1103/PhysRevD.79.043504 [arXiv:0810.2831 [hep-ph]].
[38] S. Weinberg, "Non-Gaussian Correlations Outside the Horizon," Phys. Rev. D **78**, 123521 (2008) doi:10.1103/PhysRevD.78.123521 [arXiv:0808.2909 [hep-th]].
[39] S. Weinberg, "A Tree Theorem for Inflation," Phys. Rev. D **78**, 063534 (2008) doi:10.1103/PhysRevD.78.063534 [arXiv:0805.3781 [hep-th]].
[40] S. Weinberg, "Effective Field Theory for Inflation," Phys. Rev. D **77**, 123541 (2008) doi:10.1103/PhysRevD.77.123541 [arXiv:0804.4291 [hep-th]].
[41] S. Weinberg, "From BCS to the LHC," International Journal of Modern Physics A DOI:10.1142/S0217751X0804038X
[42] S. Weinberg, "From BCS to the LHC," CERN Cour. **48N1**, 17-21 (2008)
[43] S. Weinberg, "Cosmology," OUP 2008.

[44] S. Weinberg, "From BCS to the LHC," Int. J. Mod. Phys. A **23**, 1627–1635 (2008) doi:10.1142/S0217751X0804038X
[45] R. Flauger and S. Weinberg, "Tensor Microwave Background Fluctuations for Large Multipole Order," Phys. Rev. D **75**, 123505 (2007) doi:10.1103/PhysRevD.75.123505 [arXiv:astro-ph/0703179 [astro-ph]].
[46] S. Weinberg, "A No-Truncation Approach to Cosmic Microwave Background Anisotropies," Phys. Rev. D **74**, 063517 (2006) doi:10.1103/PhysRevD.74.063517 [arXiv:astro-ph/0607076 [astro-ph]].
[47] S. Weinberg, "Quantum contributions to cosmological correlations. II. Can these corrections become large?," Phys. Rev. D **74**, 023508 (2006) doi:10.1103/PhysRevD.74.023508 [arXiv:hep-th/0605244 [hep-th]].
[48] S. Weinberg, "Living in the multiverse," in Universe or Multiverse, ed. B. Carr, CUP 2014. [arXiv:hep-th/ [hep-th]].
[49] S. Weinberg, "Quantum contributions to cosmological correlations," Phys. Rev. D **72**, 043514 (2005) doi:10.1103/PhysRevD.72.043514 [arXiv:hep-th/0506236 [hep-th]].
[50] G. 't Hooft, L. Susskind, E. Witten, M. Fukugita, L. Randall, L. Smolin, J. Stachel, C. Rovelli, G. Ellis and S. Weinberg, *et al.* "A theory of everything?," Nature **433**, 257–259 (2005) doi:10.1038/433257a
[51] S. Weinberg, "Einstein's mistakes," Phys. Today **58N11**, 31–35 (2005) doi:10.1063/1.2155755
[52] S. Weinberg, "Must cosmological perturbations remain non-adiabatic after multi-field inflation?," Phys. Rev. D **70**, 083522 (2004) doi:10.1103/PhysRevD.70.083522 [arXiv:astro-ph/0405397 [astro-ph]].
[53] S. Weinberg, "The Making of the standard model," Eur. Phys. J. C **34**, 5–13 (2004) doi:10.1140/epjc/s2004-01761-1 [arXiv:hep-ph/0401010 [hep-ph]].
[54] S. Weinberg, "Can non-adiabatic perturbations arise after single-field inflation?," Phys. Rev. D **70**, 043541 (2004) doi:10.1103/PhysRevD.70.043541 [arXiv:astro-ph/0401313 [astro-ph]].
[55] S. Weinberg, "Damping of tensor modes in cosmology," Phys. Rev. D **69**, 023503 (2004) doi:10.1103/PhysRevD.69.023503 [arXiv:astro-ph/0306304 [astro-ph]].
[56] S. Weinberg, "Adiabatic modes in cosmology," Phys. Rev. D **67**, 123504 (2003) doi:10.1103/PhysRevD.67.123504 [arXiv:astro-ph/0302326 [astro-ph]].
[57] S. Weinberg, "Cosmological fluctuations of short wavelength," Astrophys. J. **581**, 810–816 (2002) doi:10.1086/344441 [arXiv:astro-ph/0207375 [astro-ph]].
[58] S. Weinberg, "Conference summary: 20th Texas symposium on relativistic astrophysics," AIP Conf. Proc. **586**, no. 1, 893 (2001) doi:10.1063/1.1419680 [arXiv:astro-ph/0104482 [astro-ph]].
[59] S. Weinberg, "Fluctuations in the cosmic microwave background. 2. C(l) at large and small l," Phys. Rev. D **64**, 123512 (2001) doi:10.1103/PhysRevD.64.123512 [arXiv:astro-ph/0103281 [astro-ph]].
[60] S. Weinberg, "Fluctuations in the cosmic microwave background. 1. Form-factors and their calculation in synchronous gauge," Phys. Rev. D **64**, 123511 (2001) doi:10.1103/PhysRevD.64.123511 [arXiv:astro-ph/0103279 [astro-ph]].
[61] S. Weinberg, "The Cosmological constant problems," in Dark Matter 2000. [arXiv:astro-ph/0005265 [astro-ph]].
[62] S. Weinberg, "A Priori probability distribution of the cosmological constant," Phys. Rev. D **61**, 103505 (2000) doi:10.1103/PhysRevD.61.103505 [arXiv:astro-ph/0002387 [astro-ph]].
[63] S. Weinberg, "Curvature dependence of peaks in the cosmic microwave background distribution," Phys. Rev. D **62**, 127302 (2000) doi:10.1103/PhysRevD.62.127302 [arXiv:astro-ph/0006276 [astro-ph]].
[64] S. Weinberg, "The quantum theory of fields. Vol. 3: Supersymmetry".
[65] S. Weinberg, "One theory for everything?," Spektrum Wiss. Dossier **2003N1**, 40–47 (2003).
[66] S. Weinberg, "A unified physics by 2050?," Sci. Am. **281N6**, 68–75 (1999) doi:10.1038/scientificamerican1299-68
[67] S. Weinberg, "Nonrenormalization theorems in nonrenormalizable theories," Phys. Rev. Lett. **80**, 3702–3705 (1998) doi:10.1103/PhysRevLett.80.3702 [arXiv:hep-th/9803099 [hep-th]].

[68] S. Weinberg, "Effective field theories in the large N limit," Phys. Rev. D **56**, 2303–2316 (1997) doi:10.1103/PhysRevD.56.2303 [arXiv:hep-th/9706042 [hep-th]].
[69] S. Weinberg, "What is an elementary particle?," SLAC Beam Line **27N1**, 17–21 (1999).
[70] H. Martel, P. R. Shapiro and S. Weinberg, "Likely values of the cosmological constant," Astrophys. J. **492**, 29 (1998) doi:10.1086/305016 [arXiv:astro-ph/9701099 [astro-ph]].
[71] S. Weinberg, "Unified theory of the electroweak interaction. (In German)".
[72] S. Weinberg, "The first elementary particle," Nature **386**, 213–215 (1997) doi:10.1038/386213a0
[73] S. Weinberg, "Theories of the cosmological constant," [arXiv:astro-ph/9610044 [astro-ph]].
[74] S. Weinberg, "What is quantum field theory, and what did we think it is?," [arXiv:hep-th/9702027 [hep-th]].
[75] S. Weinberg, "Electroweak Reminiscences," NATO Sci. Ser. B **352**, 27–36 (1996) doi:10.1007/978-1-4613-1147-8_3
[76] S. Weinberg, "The quantum theory of fields. Vol. 2: Modern applications".
[77] S. Weinberg, "Nature itself".
[78] J. Gomis and S. Weinberg, "Are nonrenormalizable gauge theories renormalizable?," Nucl. Phys. B **469**, 473–487 (1996) doi:10.1016/0550-3213(96)00132-0 [arXiv:hep-th/9510087 [hep-th]].
[79] S. Weinberg, "The Quantum theory of fields. Vol. 1: Foundations".
[80] S. Weinberg, "Strong interactions at low-energies," Lect. Notes Phys. **452**, 1 (1995) doi:10.1007/3-540-59279-2_61 [arXiv:hep-ph/9412326 [hep-ph]].
[81] E. D'Hoker and S. Weinberg, "General effective actions," Phys. Rev. D **50**, R6050–R6053 (1994) doi:10.1103/PhysRevD.50.R6050 [arXiv:hep-ph/9409402 [hep-ph]].
[82] S. Weinberg, "Effective action and renormalization group flow of anisotropic superconductors," Nucl. Phys. B **413**, 567–578 (1994) doi:10.1016/0550-3213(94)90001-9 [arXiv:cond-mat/9306055 [cond-mat]].
[83] L. J. Hall and S. Weinberg, "Flavor changing scalar interactions," Phys. Rev. D **48**, R979–R983 (1993) doi:10.1103/PhysRevD.48.R979 [arXiv:hep-ph/9303241 [hep-ph]].
[84] S. Weinberg, "Dreams of a final theory. (In German)".
[85] S. Weinberg, "Unbreaking symmetries," Conf. Proc. C **930308**, 3–11 (1993).
[86] S. Weinberg, "Three body interactions among nucleons and pions," Phys. Lett. B **295**, 114–121 (1992) doi:10.1016/0370-2693(92)90099-P [arXiv:hep-ph/9209257 [hep-ph]].
[87] S. Weinberg, "Conference summary," AIP Conf. Proc. **272**, 346–366 (2008) doi:10.1063/1.43503 [arXiv:hep-ph/9211298 [hep-ph]].
[88] S. Weinberg, "Changing attitudes and the standard model".
[89] S. Weinberg, "Dreams of a final theory: The Search for the fundamental laws of nature".
[90] S. Weinberg, "The Axial vector coupling of the quark," Phys. Rev. Lett. **67**, 3473–3474 (1991) doi:10.1103/PhysRevLett.67.3473
[91] S. Weinberg, "Using quark and lepton masses to guess the scalar spectrum," UTTG-05-91.
[92] S. Weinberg, "Effective chiral Lagrangians for nucleon - pion interactions and nuclear forces," Nucl. Phys. B **363**, 3–18 (1991) doi:10.1016/0550-3213(91)90231-L
[93] S. Weinberg, "Beyond the standard models".
[94] S. R. Coleman, J. B. Hartle, T. Piran and S. Weinberg, "Quantum cosmology and baby universes. Proceedings, 7th Winter School for Theoretical Physics, Jerusalem, Israel, December 27, 1989 - January 4, 1990".
[95] D. J. Gross, T. Piran and S. Weinberg, "Two-Dimensional Quantum Gravity and Random Surfaces. Proceedings, 8th Winter School for Theoretical Physics, Jerusalem, Israel, December 27, 1990 - January 4, 1991".
[96] S. Weinberg, "Nuclear forces from chiral Lagrangians," Phys. Lett. B **251**, 288–292 (1990) doi:10.1016/0370-2693(90)90938-3
[97] S. Weinberg, "Mended symmetries," Phys. Rev. Lett. **65**, 1177–1180 (1990) doi:10.1103/PhysRevLett.65.1177
[98] S. Weinberg, "Why do quarks behave like bare Dirac particles?," Phys. Rev. Lett. **65**, 1181–1183 (1990) doi:10.1103/PhysRevLett.65.1181

[99] S. Weinberg, "Unitarity Constraints on CP Nonconservation in Higgs Exchange," Phys. Rev. D **42**, 860–866 (1990) doi:10.1103/PhysRevD.42.860

[100] J. C. Wheeler, T. Piran and S. Weinberg, "Supernovae. Proceedings, 6th Winter School for Theoretical Physics, Jerusalem, Israel, December 28, 1988 - January 5, 1989".

[101] S. Weinberg, "Tracing the mechanism of electroweak symmetry breaking," Mod. Phys. Lett. A **5**, 1457–1465 (1990) doi:10.1142/S0217732390001657

[102] S. Weinberg, "Larger Higgs Exchange Terms in the Neutron Electric Dipole Moment," Phys. Rev. Lett. **63**, 2333 (1989) doi:10.1103/PhysRevLett.63.2333

[103] S. Weinberg, "Testing Quantum Mechanics," Annals Phys. **194**, 336 (1989) doi:10.1016/0003-4916(89)90276-5

[104] S. Weinberg, "Precision Tests of Quantum Mechanics," Phys. Rev. Lett. **62**, 485 (1989) doi:10.1103/PhysRevLett.62.485

[105] D. Nelson, T. Piran and S. Weinberg, "Statistical Mechanics of Membranes and Surfaces".

[106] S. Weinberg, "Particle States as Realizations (Linear or Nonlinear) of Space-time Symmetries," Nucl. Phys. B Proc. Suppl. **6**, 67 (1989) doi:10.1016/0920-5632(89)90403-9

[107] S. Weinberg, "The Cosmological Constant Problem," Rev. Mod. Phys. **61**, 1–23 (1989) doi:10.1103/RevModPhys.61.1

[108] T. Piran and S. Weinberg, "Strings and Superstrings. Vol. 3. Porceedings, Winter School for Theoretical Physics, Jerusalem, Israel, December 20, 1985 - January 9, 1986".

[109] S. Weinberg, "Towards The Final Laws of Physics".

[110] S. Weinberg, "Anthropic Bound on the Cosmological Constant," Phys. Rev. Lett. **59**, 2607 (1987) doi:10.1103/PhysRevLett.59.2607

[111] S. Weinberg, "Cancellation of One Loop Divergences in SO(8192) String Theory," Phys. Lett. B **187**, 278–282 (1987) doi:10.1016/0370-2693(87)91096-3

[112] R. P. Feynman and S. Weinberg, "Elementary Particles and The Laws of Physics. The 1986 Dirac Memorial Lectures".

[113] J. N. Bahcall, T. Piran and S. Weinberg, "Dark Matter in The Universe. Proceedings, 4th Jerusalem Winter School for Theoretical Physics, Jerusalem, Israel, December 30, 1986 - January 8, 1987".

[114] S. Weinberg, "Newtonism and Today's Physics".

[115] S. Weinberg, "An Alternative to General Relativity".

[116] S. Weinberg, "Covariant Path Integral Approach to String Theory," UTTG-17-87.

[117] S. Weinberg, "Opportunities for Particle Physics: Solar Neutrinos and Superstrings," Int. J. Mod. Phys. A **2**, 301–317 (1987) doi:10.1142/S0217751X87000132

[118] S. Weinberg, "Superconductivity for Particular Theorists," Prog. Theor. Phys. Suppl. **86**, 43 (1986) doi:10.1143/PTPS.86.43

[119] S. Weinberg, "Particle Physics: Past and Future," Int. J. Mod. Phys. A **1**, 135–145 (1986) doi:10.1142/S0217751X8600006X

[120] T. Piran and S. Weinberg, "Physics in Higher Dimensions. Proceedings, 2nd Jerusalem Winter School for Theoretical Physics, Jerusalem, Israel, December 27, 1984 - January 4, 1985".

[121] S. Weinberg, "Radiative Corrections in String Theories," UTTG-22-85.

[122] S. Weinberg, "Coupling Constants and Vertex Functions in String Theories," Phys. Lett. B **156**, 309–314 (1985) doi:10.1016/0370-2693(85)91615-6

[123] S. Weinberg, "Strings with No Strings," UTTG-28-85.

[124] R. Jackiw, N. N. Khuri, S. Weinberg and E. Witten, "Quantum Field Theory and The Fundamental Problems of Physics. Proceedings, Conference, Shelter Island, USA, June 1–3, 1983".

[125] S. Weinberg, "Particles, Fields, and Now Strings".

[126] S. Weinberg, "Elementary Differential Geometry From A Generalized Standpoint," Conf. Proc. C **841227**, 160–203 (1984).

[127] S. Weinberg, "Causality, Anti-particles and the Spin Statistics Connection in Higher Dimensions," Phys. Lett. B **143**, 97–102 (1984) doi:10.1016/0370-2693(84)90812-8

[128] S. Weinberg, "Genralized Theories of Gravity and Supergravity in Higher Dimensions," UTTG-12-84.

[129] S. Weinberg, "Quasiriemannian Theories of Gravitation in More Than Four-dimensions," Phys. Lett. B **138**, 47–51 (1984) doi:10.1016/0370-2693(84)91870-7
[130] S. Weinberg, "Physics in Higher Dimensions".
[131] S. Weinberg, "The Discovery of Subatomic Particles. (In German)".
[132] S. Weinberg, "The Discovery of Subatomic Particles."
[133] S. Weinberg, "Energy Scales," Phys. Rept. **104**, 107–111 (1984) doi:10.1016/0370-1573(84)90203-5
[134] S. Weinberg, "Unification Through Higher Dimensions," Conf. Proc. C **841031**, 110–114 (1984).
[135] S. Weinberg, "The Ultimate Structure of Matter".
[136] T. Piran and S. Weinberg, "Intersecion Between Elementary Particle Physics and Cosmology: Proceedings of The 1st Jerusalme Winter School for Theoretical Physics, Jerusalem, Israel, December 28 - January 6, 1984".
[137] P. Candelas and S. Weinberg, "Calculation of Gauge Couplings and Compact Circumferences from Selfconsistent Dimensional Reduction," Nucl. Phys. B **237**, 397 (1984) doi:10.1016/0550-3213(84)90001-4
[138] S. Weinberg, "Charges from Extra Dimensions," Phys. Lett. B **125**, 265–269 (1983) doi:10.1016/0370-2693(83)91281-9
[139] S. Weinberg, "Calculation of Fine Structure Constants," Prog. Math. Phys. **9**, 380–394 (1983).
[140] L. J. Hall, J. D. Lykken and S. Weinberg, "Supergravity as the Messenger of Supersymmetry Breaking," Phys. Rev. D **27**, 2359–2378 (1983) doi:10.1103/PhysRevD.27.2359
[141] G. R. Farrar and S. Weinberg, "Supersymmetry at Ordinary Energies. 2. R Invariance, Goldstone Bosons, and Gauge Fermion Masses," Phys. Rev. D **27**, 2732 (1983) doi:10.1103/PhysRevD.27.2732
[142] S. Weinberg, "Upper Bound on Gauge Fermion Masses," Phys. Rev. Lett. **50**, 387 (1983) doi:10.1103/PhysRevLett.50.387
[143] S. Weinberg, "Cosmological Constraints on the Scale of Supersymmetry Breaking," Phys. Rev. Lett. **48**, 1303 (1982) doi:10.1103/PhysRevLett.48.1303
[144] S. Weinberg, "Does Gravitation Resolve the Ambiguity Among Supersymmetry Vacua?," Phys. Rev. Lett. **48**, 1776–1779 (1982) doi:10.1103/PhysRevLett.48.1776
[145] S. Weinberg, "Where is Supersymmetry Broken?," Prog. Phys. **6**, 359–368 (1982) doi:10.1007/978-1-4612-5800-1_27
[146] S. Weinberg, "Supersymmetry at Ordinary Energies. 1. Masses and Conservation Laws," Phys. Rev. D **26**, 287 (1982) doi:10.1103/PhysRevD.26.287
[147] S. Weinberg, "The Decay of The Proton," Sci. Am. **244N6**, 52–63 (1981).
[148] S. Weinberg, "Color and Electroweak Forces as a Source of Quark and Lepton Masses," Phys. Lett. B **102**, 401–407 (1981) doi:10.1016/0370-2693(81)91241-7
[149] S. Weinberg, "Conference Summary".
[150] S. Weinberg, "Why The Renormalization Group is a Good Thing".
[151] S. Weinberg, "The Ultimate Structure of Matter".
[152] J. Preskill and S. Weinberg, "'Decoupling' Constraints on Massless Composite Particles," Phys. Rev. D **24**, 1059 (1981) doi:10.1103/PhysRevD.24.1059
[153] S. Weinberg, "The P Decay. (In German)," Spektrum Wiss. **8**, 30–45 (1981).
[154] S. Weinberg and E. Witten, "Limits on Massless Particles," Phys. Lett. B **96**, 59–62 (1980) doi:10.1016/0370-2693(80)90212-9
[155] S. Weinberg, "Expectations for Baryon and Lepton Nonconservation," HUTP-80-A038.
[156] S. Weinberg, "Varieties of Baryon and Lepton Nonconservation," Phys. Rev. D **22**, 1694 (1980) doi:10.1103/PhysRevD.22.1694
[157] S. Weinberg, "Ultraviolet Divergences in Quantum Theories of Gravitation," in An Einstein Centenary Survey edited by S. W. Hawking and W. Israel, Cambridge University Press 1979.
[158] S. Weinberg, "Effective Gauge Theories," Phys. Lett. B **91**, 51–55 (1980) doi:10.1016/0370-2693(80)90660-7
[159] D. V. Nanopoulos and S. Weinberg, "Mechanisms for Cosmological Baryon Production," Phys. Rev. D **20**, 2484 (1979) doi:10.1103/PhysRevD.20.2484

[160] S. Weinberg, "Cosmological Production of Baryons," Phys. Rev. Lett. **42**, 850–853 (1979) doi:10.1103/PhysRevLett.42.850
[161] S. Weinberg, "Conceptual Foundations of the Unified Theory of Weak and Electromagnetic Interactions," Rev. Mod. Phys. **52**, 515–523 (1980) doi:10.1103/RevModPhys.52.515
[162] S. Weinberg, "Baryon and Lepton Nonconserving Processes," Phys. Rev. Lett. **43**, 1566–1570 (1979) doi:10.1103/PhysRevLett.43.1566
[163] S. Weinberg, "Grand Unification".
[164] S. Weinberg, "Gauge Hierarchies," Phys. Lett. B **82**, 387–391 (1979) doi:10.1016/0370-2693(79)90248-X
[165] S. Weinberg, "Phenomenological Lagrangians," Physica A **96**, no. 1–2, 327–340 (1979) doi:10.1016/0378-4371(79)90223-1
[166] S. Weinberg, "Weak Interactions," Conf. Proc. C **780823**, 907–918 (1978).
[167] S. Weinberg, "Opening Address-Neutrinos '78".
[168] S. Weinberg, "A New Light Boson?," Phys. Rev. Lett. **40**, 223–226 (1978) doi:10.1103/PhysRevLett.40.223
[169] S. Weinberg, "Gauge Theories of Weak, Electromagnetic and Strong Interactions," HUTP-77-A064.
[170] S. Weinberg, "The Problem of Mass," Trans. New York Acad. Sci. **38**, 185–201 (1977) doi:10.1111/j.2164-0947.1977.tb02958.x
[171] S. Weinberg, "Unified Gauge Theories," Experientia Suppl. **31**, 339–352 (1977) HUTP-77-A061.
[172] H. Georgi and S. Weinberg, "Neutral Currents in Expanded Gauge Theories," Phys. Rev. D **17**, 275 (1978) doi:10.1103/PhysRevD.17.275
[173] G. F. Sterman and S. Weinberg, "Jets from Quantum Chromodynamics," Phys. Rev. Lett. **39**, 1436 (1977) doi:10.1103/PhysRevLett.39.1436
[174] B. W. Lee and S. Weinberg, "Cosmological Lower Bound on Heavy Neutrino Masses," Phys. Rev. Lett. **39**, 165–168 (1977) doi:10.1103/PhysRevLett.39.165
[175] J. D. Bjorken, K. D. Lane and S. Weinberg, "The Decay mu -> e + gamma in Models with Neutral Heavy Leptons," Phys. Rev. D **16**, 1474 (1977) doi:10.1103/PhysRevD.16.1474
[176] B. W. Lee and S. Weinberg, "SU(3) x U(1) Gauge Theory of the Weak and Electromagnetic Interactions," Phys. Rev. Lett. **38**, 1237 (1977) doi:10.1103/PhysRevLett.38.1237
[177] S. Weinberg, "Conference Summary".
[178] J. D. Bjorken and S. Weinberg, "A Mechanism for Nonconservation of Muon Number," Phys. Rev. Lett. **38**, 622 (1977) doi:10.1103/PhysRevLett.38.622
[179] S. Weinberg, "The First Three Minutes. A Modern View of the Origin of the Universe".
[180] S. Weinberg, "Critical Phenomena for Field Theorists," doi:10.1007/978-1-4684-0931-4_1
[181] S. L. Glashow and S. Weinberg, "Natural Conservation Laws for Neutral Currents," Phys. Rev. D **15**, 1958 (1977) doi:10.1103/PhysRevD.15.1958
[182] K. D. Lane and S. Weinberg, "Mass Differences of Charmed Hadrons," Phys. Rev. Lett. **37**, 717 (1976) doi:10.1103/PhysRevLett.37.717
[183] S. Weinberg, "Gauge Theory of CP Violation," Phys. Rev. Lett. **37**, 657 (1976) doi:10.1103/PhysRevLett.37.657
[184] S. Weinberg, "Apparent Luminosities in a Locally Inhomogeneous Universe," Print-76-0421 (HARVARD).
[185] S. Weinberg, "Ambiguous Solutions of Supersymmetric Theories (f1)," Phys. Lett. B **62**, 111–113 (1976) doi:10.1016/0370-2693(76)90062-9
[186] S. Weinberg, "Is Nature Simple?"
[187] E. Gildener and S. Weinberg, "Symmetry Breaking and Scalar Bosons," Phys. Rev. D **13**, 3333 (1976) doi:10.1103/PhysRevD.13.3333
[188] S. Weinberg, "Mass of the Higgs Boson," Phys. Rev. Lett. **36**, 294–296 (1976) doi:10.1103/PhysRevLett.36.294
[189] E. C. Poggio, H. R. Quinn and S. Weinberg, "Smearing the Quark Model," Phys. Rev. D **13**, 1958 (1976) doi:10.1103/PhysRevD.13.1958

[190] S. Weinberg, "Implications of Dynamical Symmetry Breaking," Phys. Rev. D **13**, 974–996 (1976) doi:10.1103/PhysRevD.19.1277
[191] S. Weinberg, "The U(1) Problem," Phys. Rev. D **11**, 3583–3593 (1975) doi:10.1103/PhysRevD.11.3583
[192] C. W. Bernard, A. Duncan, J. LoSecco and S. Weinberg, "Exact Spectral Function Sum Rules," Phys. Rev. D **12**, 792 (1975) doi:10.1103/PhysRevD.12.792
[193] S. Weinberg, "Light as a Fundamental Particle," Phys. Today **28N6**, 32–37 (1975) doi:10.1063/1.3069003
[194] S. Weinberg, "Gauge Symmetry Breaking," Conf. Proc. C **750926**, 1–26 (1975) PRINT-76-0484 (HARVARD).
[195] S. Weinberg, "Astrophysical Implications of the New Theories of Weak Interactions," Annals N. Y. Acad. Sci. **262**, 409–421 (1975) doi:10.1111/j.1749-6632.1975.tb31458.x
[196] H. Georgi, H. R. Quinn and S. Weinberg, "Hierarchy of Interactions in Unified Gauge Theories," Phys. Rev. Lett. **33**, 451–454 (1974) doi:10.1103/PhysRevLett.33.451
[197] S. Weinberg, "Problems in Gauge Field Theories," PRINT-74-1313 (HARVARD).
[198] S. Weinberg, "Gauge and Global Symmetries at High Temperature," Phys. Rev. D **9**, 3357–3378 (1974) doi:10.1103/PhysRevD.9.3357
[199] S. Weinberg, "Unified theories of elementary particle interaction," Sci. Am. **231N1**, 50–59 (1974) doi:10.1038/scientificamerican0774-50
[200] S. Weinberg, "Recent progress in gauge theories of the weak, electromagnetic and strong interactions," J. Phys. Colloq. **34**, no. C1, 45–67 (1973) doi:10.1103/RevModPhys.46.255
[201] S. Weinberg, "Current algebra and gauge theories. 2. NonAbelian gluons," Phys. Rev. D **8**, 4482–4498 (1973) doi:10.1103/PhysRevD.8.4482
[202] S. Weinberg, "Current algebra and gauge theories. 1.," Phys. Rev. D **8**, 605–625 (1973) doi:10.1103/PhysRevD.8.605
[203] S. Weinberg, "Where we are now," Science **180**, 276–278 (1973) doi:10.1126/science.180.4083.276
[204] S. Weinberg, "Theory of Weak and Electromagnetic Interactions," Stud. Nat. Sci. **3**, 157–186 (1973) doi:10.1007/978-1-4613-4586-2_7
[205] S. Weinberg, "Perturbative Calculations of Symmetry Breaking," Phys. Rev. D **7**, 2887–2910 (1973) doi:10.1103/PhysRevD.7.2887
[206] S. Weinberg, "Views on Broken Symmetry. (Talk)".
[207] S. Weinberg, "General Theory of Broken Local Symmetries," Phys. Rev. D **7**, 1068–1082 (1973) doi:10.1103/PhysRevD.7.1068
[208] S. Weinberg, "Nonabelian Gauge Theories of the Strong Interactions," Phys. Rev. Lett. **31**, 494–497 (1973) doi:10.1103/PhysRevLett.31.494
[209] S. Weinberg, "New approach to the renormalization group," Phys. Rev. D **8**, 3497–3509 (1973) doi:10.1103/PhysRevD.8.3497
[210] S. Weinberg, "Approximate symmetries and pseudoGoldstone bosons," Phys. Rev. Lett. **29**, 1698–1701 (1972) doi:10.1103/PhysRevLett.29.1698
[211] S. Weinberg, "Electromagnetic and weak masses," Phys. Rev. Lett. **29**, 388–392 (1972) doi:10.1103/PhysRevLett.29.388
[212] S. Weinberg, "Effects of a neutral intermediate boson in semileptonic processes," Phys. Rev. D **5**, 1412–1417 (1972) doi:10.1103/PhysRevD.5.1412
[213] R. Jackiw and S. Weinberg, "Weak interaction corrections to the muon magnetic moment and to muonic atom energy levels," Phys. Rev. D **5**, 2396–2398 (1972) doi:10.1103/PhysRevD.5.2396
[214] S. Weinberg, "Gravitation and Cosmology: Principles and Applications of the General Theory of Relativity".
[215] S. Weinberg, "Multiparticle sum rules," Phys. Rev. D **5**, 900–912 (1972) doi:10.1103/PhysRevD.5.900
[216] S. Weinberg, "Physical Processes in a Convergent Theory of the Weak and Electromagnetic Interactions," Phys. Rev. Lett. **27**, 1688–1691 (1971) doi:10.1103/PhysRevLett.27.1688

[217] S. Weinberg, "Mixing angle in renormalizable theories of weak and electromagnetic interactions," Phys. Rev. D **5**, 1962–1967 (1972) doi:10.1103/PhysRevD.5.1962

[218] S. Weinberg, "Entropy generation and the survival of protogalaxies in an expanding universe," Astrophys. J. **168**, 175 (1971) doi:10.1086/151073

[219] S. Weinberg, "Summing soft pions. ii. statistical approach to unitarity damping," Phys. Rev. D **4**, 2993–2997 (1971) doi:10.1103/PhysRevD.4.2993

[220] S. Weinberg, "Exponentiation and sum rules," Phys. Lett. B **37**, 494–496 (1971) doi:10.1016/0370-2693(71)90354-6

[221] S. Weinberg, "Current Problems in Cosmology," AIP Conf. Proc. **2**, 247–255 (1971) doi:10.1063/1.2948539

[222] I. S. Gerstein, R. Jackiw, S. Weinberg and B. W. Lee, "Chiral loops," Phys. Rev. D **3**, 2486–2492 (1971) doi:10.1103/PhysRevD.3.2486

[223] K. Huang and S. Weinberg, "Ultimate temperature and the early universe," Phys. Rev. Lett. **25**, 895–897 (1970) doi:10.1103/PhysRevLett.25.895

[224] S. Weinberg, "Direct Determination of The Metric From Observed Red Shifts and Distances," MIT-CTP-134.

[225] S. Weinberg, "Summing soft pions," Phys. Rev. D **2**, 674–684 (1970) doi:10.1103/PhysRevD.2.674

[226] S. Weinberg, "Reply to On summing soft pions," Phys. Rev. D **2**, 3085 (1970) doi:10.1103/PhysRevD.2.3085

[227] S. Weinberg, "Dynamic and Algebraic Symmetries".

[228] S. Weinberg, "New links between regge-pole theory and current algebra," Comments Nucl. Part. Phys. **3**, no. 1, 28–34 (1969)

[229] M. Ademollo, G. Veneziano and S. Weinberg, "Quantization conditions for regge intercepts and hadron masses," Phys. Rev. Lett. **22**, 83–85 (1969) doi:10.1103/PhysRevLett.22.83

[230] S. Weinberg, "Feynman rules for any spin. iii," Phys. Rev. **181**, 1893–1899 (1969) doi:10.1103/PhysRev.181.1893

[231] S. Weinberg, "Algebraic structure of superconvergence relations," Phys. Rev. Lett. **22**, 1023–1025 (1969) doi:10.1103/PhysRevLett.22.1023

[232] S. Weinberg, "Algebraic realizations of chiral symmetry," Phys. Rev. **177**, 2604–2620 (1969) doi:10.1103/PhysRev.177.2604

[233] S. Weinberg, "Cosmic Black Body Gravitational Radiation," MIT-CTP-18.

[234] S. Weinberg, "Nonlinear realizations of chiral symmetry," Phys. Rev. **166**, 1568–1577 (1968) doi:10.1103/PhysRev.166.1568

[235] I. S. Gerstein, S. Weinberg and H. J. Schnitzer, "Structure of three-point functions from su(3) x su(3) algebra," Phys. Rev. **175**, 1873–1876 (1968) doi:10.1103/PhysRev.175.1873

[236] S. Weinberg, "Current Algebra 1966–1968," Comments Nucl. Part. Phys. **2**, no. supp, 28–32 (1968).

[237] S. Weinberg, "Current algebra".

[238] H. J. Schnitzer and S. Weinberg, "Current-Algebra Calculation of Hard-Pion Processes: A-1 -> rho+pi and rho -> pi+pi," Phys. Rev. **164**, 1828–1833 (1967) doi:10.1103/PhysRev.164.1828

[239] S. L. Glashow and S. Weinberg, "Breaking chiral symmetry," Phys. Rev. Lett. **20**, 224–227 (1968) doi:10.1103/PhysRevLett.20.224

[240] S. Weinberg, "A Model of Leptons," Phys. Rev. Lett. **19**, 1264–1266 (1967) doi:10.1103/PhysRevLett.19.1264

[241] S. L. Glashow, H. J. Schnitzer and S. Weinberg, "Convergent Calculation of Nonleptonic K Decay in the Intermediate-Boson Model," Phys. Rev. Lett. **19**, 205–208 (1967) doi:10.1103/PhysRevLett.19.205

[242] T. D. Lee, S. Weinberg and B. Zumino, "Algebra of Fields," Phys. Rev. Lett. **18**, 1029–1032 (1967) doi:10.1103/PhysRevLett.18.1029

[243] S. L. Glashow, H. J. Schnitzer and S. Weinberg, "Sum Rules for the Spectral Functions of SU(3) X SU(3)," Phys. Rev. Lett. **19**, 139 (1967) doi:10.1103/PhysRevLett.19.139

[244] S. Weinberg, "Precise relations between the spectra of vector and axial vector mesons," Phys. Rev. Lett. **18**, 507–509 (1967) doi:10.1103/PhysRevLett.18.507
[245] S. Weinberg, "Dynamical approach to current algebra," Phys. Rev. Lett. **18**, 188–191 (1967) doi:10.1103/PhysRevLett.18.188
[246] S. Weinberg, "Current Commutator Calculation of the $K_{\ell 4}$ form Factors," Phys. Rev. Lett. **17**, 336–340 (1966) [erratum: Phys. Rev. Lett. **18**, no. 25, 1178 (1967)] doi:10.1103/PhysRevLett.17.336
[247] S. Weinberg, "Dynamics at infinite momentum," Phys. Rev. **150**, 1313–1318 (1966) doi:10.1103/PhysRev.150.1313
[248] S. Weinberg, "Pion scattering lengths," Phys. Rev. Lett. **17**, 616–621 (1966) doi:10.1103/PhysRevLett.17.616
[249] S. Weinberg, "Current-Commutator Theory of Multiple Pion Production," Phys. Rev. Lett. **16**, no. 19, 879–883 (1966) doi:10.1103/PhysRevLett.16.879
[250] S. Weinberg, "Infrared photons and gravitons," Phys. Rev. **140**, B516–B524 (1965) doi:10.1103/PhysRev.140.B516
[251] S. Weinberg, "Evidence That the Deuteron Is Not an Elementary Particle," Phys. Rev. **137**, B672–B678 (1965) doi:10.1103/PhysRev.137.B672
[252] S. Weinberg, "Photons and gravitons in perturbation theory: Derivation of Maxwell's and Einstein's equations," Phys. Rev. **138**, B988–B1002 (1965) doi:10.1103/PhysRev.138.B988
[253] S. Weinberg, "Do Hyperphotons Exist?," Phys. Rev. Lett. **13**, 495–497 (1964) doi:10.1103/PhysRevLett.13.495
[254] S. Weinberg, "Photons and Gravitons in S-Matrix Theory: Derivation of Charge Conservation and Equality of Gravitational and Inertial Mass," Phys. Rev. **135**, B1049–B1056 (1964) doi:10.1103/PhysRev.135.B1049
[255] S. Weinberg, "Derivation of gauge invariance and the equivalence principle from Lorentz invariance of the S-matrix," Phys. Lett. **9**, no. 4, 357–359 (1964) doi:10.1016/0031-9163(64)90396-8
[256] S. Weinberg, "Systematic Solution of Multiparticle Scattering Problems," Phys. Rev. **133**, B232–B256 (1964) doi:10.1103/PhysRev.133.B232
[257] S. Weinberg, "Feynman Rules for Any Spin," Phys. Rev. **133**, B1318–B1332 (1964) doi:10.1103/PhysRev.133.B1318
[258] S. Weinberg, "Feynman Rules for Any Spin. 2. Massless Particles," Phys. Rev. **134**, B882–B896 (1964) doi:10.1103/PhysRev.134.B882
[259] S. Weinberg, "On The Derivation of Intrinsic Symmetries," PRINT-63-427.
[260] S. Weinberg, "Quasiparticles and the Born Series," Phys. Rev. **131**, 440–460 (1963) doi:10.1103/PhysRev.131.440
[261] S. Weinberg, "Universal Neutrino Degeneracy," Phys. Rev. **128**, 1457–1473 (1962) doi:10.1103/PhysRev.128.1457
[262] S. Weinberg, "Elementary particle theory of composite particles," Phys. Rev. **130**, 776–783 (1963) doi:10.1103/PhysRev.130.776
[263] J. Goldstone, A. Salam and S. Weinberg, "Broken Symmetries," Phys. Rev. **127**, 965–970 (1962) doi:10.1103/PhysRev.127.965
[264] S. Weinberg, "The non-field theory of non-elementary particles".
[265] S. Weinberg, "Cross Sections at High Energies," Phys. Rev. **124**, 2049–2050 (1961) doi:10.1103/PhysRev.124.2049
[266] G. Feinberg and S. Weinberg, "Conversion of Muonium into Antimuonium," Phys. Rev. **123**, 1439–1443 (1961) doi:10.1103/PhysRev.123.1439
[267] G. Feinberg and S. Weinberg, "Law of Conservation of Muons," Phys. Rev. Lett. **6**, 381–383 (1961) doi:10.1103/PhysRevLett.6.381
[268] S. Weinberg, "Muon Absorption in Liquid Hydrogen," Phys. Rev. Lett. **4**, 575–578 (1960) doi:10.1103/PhysRevLett.4.575
[269] S. Weinberg, "New Test for DeltaI=12 in K+ Decay," Phys. Rev. Lett. **4**, 87–89 (1960) [erratum: Phys. Rev. Lett. **4**, 585 (1960)] doi:10.1103/PhysRevLett.4.87

[270] S. Weinberg and S. B. Treiman, "Electromagnetic Corrections to Isotopic Spin Conservation," Phys. Rev. **116**, 465–468 (1959) doi:10.1103/PhysRev.116.465

[271] S. B. Treiman and S. Weinberg, "Interference Effects in Neutral K-Particle Decay," Phys. Rev. **116**, 239–240 (1959) doi:10.1103/PhysRev.116.239

[272] G. Feinberg, P. Kabir and S. Weinberg, "Transformation of muons into electrons," Phys. Rev. Lett. **3**, 527–530 (1959) doi:10.1103/PhysRevLett.3.527

[273] S. Weinberg, "Interference Effects in Leptonic Decays," Phys. Rev. **115**, 481–484 (1959) doi:10.1103/PhysRev.115.481

[274] S. Weinberg and G. Feinberg, "Electromagnetic Transitions Between mu Meson and Electron," Phys. Rev. Lett. **3**, 111–114 (1959) doi:10.1103/PhysRevLett.3.111

[275] S. Weinberg, "High-energy behavior in quantum field theory," Phys. Rev. **118**, 838–849 (1960) doi:10.1103/PhysRev.118.838

[276] S. Okubo, R. E. Marshak, E. C. G. Sudarshan, W. B. Teutsch and S. Weinberg, "Interaction Current in Strangeness-Violating Decays," Phys. Rev. **112**, 665–668 (1958) doi:10.1103/PhysRev.112.665

[277] S. Weinberg, "Charge symmetry of weak interactions," Phys. Rev. **112**, 1375–1379 (1958) doi:10.1103/PhysRev.112.1375

[278] S. Weinberg, "Time-Reversal Invariance and theta20 Decay," Phys. Rev. **110**, 782–784 (1958) doi:10.1103/PhysRev.110.782

[279] R. E. Marshak, S. Okubo, G. Sudarshan, W. B. Teutsch and S. Weinberg, "Divergenceless currents and K-meson decay," Phys. Rev. Lett. **1**, 25 (1958) doi:10.1103/PhysRevLett.1.25

[280] S. Weinberg, "Role of Strong Interactions in Decay Processes," Phys. Rev. **106**, 1301–1306 (1957) doi:10.1103/PhysRev.106.1301

[281] S. Weinberg, "N-V Potential in the Lee Model," Phys. Rev. **102**, 285–289 (1956) doi:10.1103/PhysRev.102.285

[282] S. Weinberg, "Epilogue: The Grand Reduction," in To Explain the World, pp. 256–268. Harper/HarperCollins, 2015 (432 pp.). ISBN 9780062346650.

[283] S. Weinberg, "The Trouble with Quantum Mechanics," in The New York Review of Books. January 19, 2017

About the Editor

Michael Duff is Emeritus Professor of Theoretical Physics and Senior Research Investigator at Imperial College London. He was formerly Abdus Salam Professor of Theoretical Physics and Principal of the Faculty of Physical Sciences. He is recipient of the 2004 Meeting Gold Medal, El Colegio Nacional, Mexico, the 2017 Paul Dirac Gold Medal and Prize, Institute of Physics, UK and the 2018 Trotter Prize, USA. He is a visitor to The Institute of Quantum Science and Engineering and was inducted as 2019 Faculty Fellow of the Hagler Institute for Advanced Study, both at Texas A&M University.

A photo of editor Michael Duff. Photo courtesy of Imperial College, London.

Acknowledgments

I am grateful to KK Phua, Joseph Sebastian Ang and Liang Ning at World Scientific for their patience and efficiency; to Leron Borsten, Lars Brink and Stanley Deser for their suggestions and advice; to Tsvi Piran for tracking down the Jerusalem Winter school photo; to Kelly Stelle for help with Section 3.2; and to Louise and Elizabeth Weinberg for their permission and encouragement.

Printed in Great Britain
by Amazon

41699864R00262